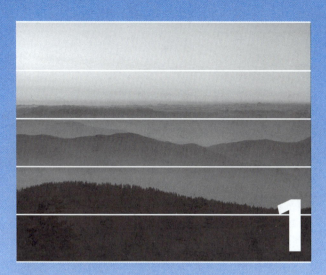

Organizing and Summarizing Univariate Data

Because of the widespread use of computers and the Internet, statistical data appear in almost all fields of study. As a student of statistics you must be able to read, interpret, and apply the results of an analysis of research data. This chapter will introduce you to the basic definitions and techniques used to organize and summarize data. Repetition and practice are essential to familiarize yourself with the statistical terminology and concepts. By working through the examples and exercises, you will better understand how statistical information is produced, used, and evaluated.

Contents

- **STATISTICAL INSIGHT**
1.1 Essential Elements of Statistics
1.2 Displaying Data with Tables and Graphs
1.3 Displaying Numerical Data

1.4 Summarizing Data with Statistics
1.5 Describing a Distribution
1.6 Summary and Review Exercises

The 1872 Hidalgo Stamp

Stamp collectors will tell you that the price of any stamp is based on its relative scarcity, and its scarcity depends on a number of factors. One unusual factor is the thickness of the paper on which the stamp is printed. Paper is sold by weight per ream; a ream contains about 500 pages. Over 100 years ago, when paper was very expensive, manufacturers would keep on hand some extra thick sheets to replace thin sheets in an underweight ream. The goal was to keep the weight up, but at the same time, supply as few sheets as possible. Consequently, a particular stamp could have been printed on a variety of thicknesses of paper (several different reams, each having its own padded sheets).

Mexico began issuing stamps in 1856. Through 1879, almost all stamps carried the image of national hero Miguel Hidalgo y Costilla. One such stamp is the *1872 Hidalgo issue*, which was in circulation until 1874. Today it is estimated that there are less than 100,000 of the original 4,772,000 printed stamps. If you buy one, how can you determine whether it is printed on thick, medium, or thin paper? The 1872 Hidalgo carries a rich history. Between 1979 and 1981, five major collections of the 1872 issue were purchased by the late Walton van Winkler, who measured the thickness (in millimeters) of each of his 485 stamps. If his collection is a representative sample of all of the 1872 Hidalgos, then by using his measurements, we can investigate the distribution of the thicknesses of the stamps. The worksheet is a frequency table of van Winkler's measurements.

x	freq	x	freq	x	freq
0.075	20	0.094	3	0.113	0
0.076	18	0.095	2	0.114	3
0.077	11	0.096	3	0.115	3
0.078	23	0.097	7	0.116	0
0.079	42	0.098	5	0.117	1
0.080	37	0.099	5	0.118	0
0.081	15	0.100	15	0.119	4
0.082	18	0.101	9	0.120	3
0.083	7	0.102	8	0.121	1
0.084	3	0.103	7	0.122	2
0.085	2	0.104	2	0.123	2
0.086	2	0.105	5	0.124	0
0.087	1	0.106	4	0.125	2
0.088	2	0.107	3	0.126	0
0.089	10	0.108	7	0.127	0
0.090	9	0.109	7	0.128	1
0.091	3	0.110	11	0.129	3
0.092	5	0.111	4	0.130	1
0.093	6	0.112	5	0.131	1

Source: Izenman, A., Sommer, C. (1988), Philatelic Mixtures and Multimodal Densities, *Journal of the American Statistical Association*, 83, 941–953.

- What is the median thickness of the stamps? What is the mean thickness of the stamps?
- Are there any unusual thicknesses?
- What is the general shape of the distribution of thickness of the stamps?
- Is there more than one peak?
- How many stamps have thickness less than .075 millimeters?
- You have a Hidalgo stamp with a thickness of .121 millimeter. Is it a rare stamp?

These are a few questions that can be asked about the thickness of the stamps in this collection. See Section 1.6 for further discussion of this Statistical Insight.

WORKSHEET: Stamp.mtw

x	freq	x	freq	x	freq
0.060	1	0.065	1	0.070	26
0.061	0	0.066	1	0.071	20
0.062	0	0.067	0	0.072	32
0.063	0	0.068	1	0.073	11
0.064	2	0.069	7	0.074	10

1.1 Essential Elements of Statistics

Learning Objectives for this section:

❑ Know and understand the basic concepts of: individual unit, population, sample, variable, observation, and data set. Be able to use them to communicate your statistical thoughts.

❑ Know the difference between categorical and numerical variables and be able to classify any variable as one or the other.

❑ Know the properties of discrete and continuous variables and be able to give examples of each.

❑ Understand what is meant by the distribution of a variable and be able to exhibit it with a dotplot.

Statistics is the science of collecting and analyzing numerical information. The goal is to organize and interpret the information in order to reach valid conclusions.

Fundamental Concepts

- A statistics problem involves studying one or more characteristics associated with a group of objects, called **individual units** or **subjects**.
- The collection of all individual units of interest is called the **population of units** and any finite subset is called a **sample**.
- Generally the information obtained from the sample is used to make inferences about the population; thus, it is important that the sample be **representative** of the population.
- If it is necessary that information be obtained from every unit in the population of units, a **census** is conducted.

Consider, for example, the information provided in Table 1.1. The gender, major, class, grade point average (GPA), Scholastic Assessment Test (SAT) score, and the number of courses dropped are given for a group of undergraduate students.

Table 1.1
Data set for a group of undergraduate students: Undergrad.mtw

Gender	Major	Class	GPA	SAT	Number of drops
Female	Physics	Sophomore	3.1	1230	2
Female	Biology	Sophomore	1.9	870	2
Male	Psychology	Senior	3.2	1180	4
Female	Psychology	Sophomore	2.6	1190	2
Female	Music	Senior	2.8	770	4
Male	English	Freshman	2.1	850	1
Male	Physics	Sophomore	3.4	1440	2
.	.	.	.	.	.
.	.	.	.	.	.
.	.	.	.	.	.

The students in the study are the individual units (subjects). Clearly this is not a census (not all students were surveyed). The group of students surveyed is a sample from a larger population of students. Without knowing how the sample was taken, it is impossible to know whether the sample is representative of the population of units.

Variables and Data Sets

- A **variable** is any characteristic that can be measured on each individual unit in a statistical study.

continued

- The complete collection of values (called **observations**) associated with the variables in the study is called a **data set**.
- Data sets are called **univariate, bivariate,** or **multivariate** depending on whether one, two, or more than two variables are listed.

The quantities measured on the students (Gender, Major, Class, etc.) in Table 1.1 are variables; because several variables are listed, it is called a multivariate data set. Notice that some variables take on numerical values, and some, like gender, do not. You must be able to recognize this difference.

Variable Classification

- A **categorical variable** (also called a *qualitative variable*) is a variable whose values are classifications or categories.
- A **numerical variable** (also called a *quantitative variable*) is a variable whose values are numbers obtained from a count or a measurement.

Numerical variables are either discrete or continuous.

- A **discrete variable** (also called a *count variable*) is a numerical variable whose values are obtained from a count. Usually this means that it assumes a finite number of values. If, however, it assumes an infinite number of values, you must be able to associate its values with the counting numbers 1, 2, 3, . . .
- A **continuous variable** (also called a *measurement variable*) is a numerical variable whose values cannot be counted. Usually it is the result of a measurement and assumes all values associated with the numbers in an interval of the real number line.

The values of the first three variables (Gender, Major, Class) are nonnumerical. They simply place the students into different categories; hence, the variables are *categorical variables*. The last three variables (GPA, SAT, Number of Drops) are *numerical variables*. The variables GPA and SAT are the result of measurements and hence are *continuous variables*. The number of courses each student has dropped is a count and therefore a *discrete variable*.

Figure 1.1 illustrates the different types of variables and how they are measured.

Figure 1.1

Variable Type	Example Question	Response
Categorical ———————→ Did you graduate from high school?		Yes No
Numerical ⟨ Discrete ——→ How many math courses did you take in high school?		Count
Continuous —→ What is your grade point average?		Measurement

Example 1.1 Classify the following as either categorical or numerical. If numerical, further classify as discrete or continuous.

a. Age of students entering college
b. Opinion of voters on a political issue
c. Number of children in the families of entering college students

Solution

a. Age is a measurement of time and hence is numerical. Because we report it as a whole number (18, 19, 20, etc.) we are tempted to say that age is a discrete variable. We must remember, however, that the classification is based on the potential values of the variable, not on how we report its value. Because age is time, it is a continuous variable.

b. Opinion is not a measurement but rather a classification, such as for or against; hence, it is categorical. Don't confuse this with the quantity that records the *number* of voters who favor an issue. The number of voters who favor an issue is not a variable at all, but rather a summary of the variable, opinion.

c. The number of children is a *count* and thus a discrete numerical variable.   ■

The amount of information contained in a data set depends on how the variables are measured. For example, recording income in dollars is more informative than recording it simply as low, medium, or high. Income recorded in dollars is numerical, and income recorded as low, medium, and high is categorical. Knowing the amount of information contained in the data may indicate possible procedures to use for analyzing the data.

Example 1.2   Identify the following variables as either categorical or numerical. Comment on the amount of information conveyed in each.

a. Length (in hours) of baseball games.
b. Colors of paint in a paint company's inventory.
c. Ranks of personnel in the military.

Solution

a. Time is a quantity that is measured and thus is a numerical variable. A 4-hour game is twice as long as a 2-hour game; thus, ratios and differences of time are relevant. Any sort of arithmetic operation may be performed on this variable.

b. Color is simply an attribute and thus a categorical variable. Although a numerical code can be assigned to different colors, the values are not numerical, and thus numerical calculations are not meaningful.

c. Rank in the military is not numerical, although it does exhibit a definite ordering. It is a categorical variable, and numerical calculations are not meaningful.   ■

Population of Measurements

In a statistical study the term *population* can have different meanings. We often think of the population as being the entire group of subjects under study, previously defined as the population of units. Another meaning of population, however, relates to the variables measured on the subjects. For each variable in the study we will refer to *all* possibilities of the variable as the **population of measurements** associated with the variable. Thus, each variable measured on the subjects can generate its own population of values. For example, the variables "height" and "weight" of college students (the population of units) have associated with them distinct measurement populations—all possible heights and all possible weights, both of which are infinite sets. Often, continuous variables are called *measurement variables*, leading to the expression *population of measurements*. Unless it is necessary to make a distinction, the expression *population* will be used to refer to either the population of units or the population of measurements. In a similar fashion, the expression *sample* will be used to refer to either the subset of subjects under study or the set of observations (data) collected from the subjects.

Exploratory Data Analysis

The main objective of data analysis is to organize and summarize the information in a data set to make it more comprehensible. **Exploratory data analysis** (EDA) gives us the tools and strategies to examine the data for structure, patterns, or unusual behavior.

Distribution of a Variable

The **distribution** of a variable specifies the values that the variable assumes and how often these values occur.

- For a categorical variable (or a discrete numerical variable) the distribution is displayed in a graph or table that gives the distinct categories (values) that are possible and how often they occur.
- For a continuous variable the distribution is displayed with a graph that illustrates the pattern of the variation in the data.

Example 1.3   The general manager of the Baltimore Orioles baseball team wishes to investigate the possibility of signing a new rookie pitcher. Aware of salary caps and other restrictions imposed by league rules, he wants to study the current salaries of all team members. His accountant enters the 1999 salaries in a Minitab worksheet on a computer, names it **Orioles.mtw**, and creates a simple graph of the data called a **dotplot**. The graph, shown in Figure 1.2, consists of a scale for the salaries with a dot for the salary of each player. Now the accountant sees the distribution of salaries and is able to study the graph for structure, patterns, or unusual behavior.

Figure 1.2

Dotplot for Orioles

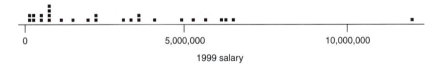

Notice that the salaries are fairly evenly distributed between 0 and $6,000,000; however, there is one salary well above $10,000,000 on the right end of the graph. ∎

When we speak of the *distribution* of a variable we are referring to the *variation* exhibited by the data. In the preceding example, the salaries varied from player to player. The dotplot in Figure 1.2 is a picture of the variation and is an illustration of the distribution of the variable. If all players received the same salary, there would be no distribution and no need for the study. Because of the variation in the salaries, the owner is interested in studying the current salaries of all team members so that he can make intelligent decisions. Keep in mind that measurements will vary from one subject to the next, and one of our main goals is to picture the distribution of the variable with a dotplot or some other graphical display.

We have introduced some basic definitions as well as the concept of exploratory data analysis. Whether or not we have a formal inference in mind, exploratory data analysis should be the first step in processing data. Using numerical quantities and graphs, we inspect the data for patterns, trends, unexpected features, and even errors that may alter our plan of attack. In the next few chapters, you will learn about the methods and techniques necessary for the exploration of data.

COMPUTER TIP

Dotplots

First select **Graph**, and then select **Dotplot**. When the Dotplot window appears, highlight the column you wish to graph, click on **Select**, and then click **OK**. To create side-by-side dotplots with respect to a categorical variable, click on **By variable** and then **Select** the variable.

EXERCISES 1.1   The Basics

1.1   To estimate the percent of students that experience an emotional tragedy during the school year, a psychology professor polled all of her classes. Is the population of units students in general or just students who experience an emotional tragedy? Describe the sample. Is this a reasonable sample if she wishes to generalize the results?

1.2   A sample of 500 university students was selected to obtain their opinions on abortion. What is an individual unit? Describe the sample. Name a categorical variable and a numerical variable that would be relevant in this study.

1.3   Out of 1000 accidents investigated by the Highway Patrol, 526 were found to be alcohol related. What is an individual unit? Is the population of units all accidents or only those that were alcohol related? Name two categorical variables and two numerical variables that would be relevant in this study. Is the sample size 1000 or 526?

1.4   Classify the following variables as categorical or numerical. If a variable is numerical, further classify it as discrete or continuous.
   a. Age of freshmen U.S. senators
   b. Faculty rank
   c. Weight of newborn babies
   d. Per capita income of state residents
   e. Murder rate in a major city
   f. Number of students in a classroom
   g. Brand of television set

1.5   Explain why a sample is preferred to a census in the following situations:
   a. Testing the effectiveness of an experimental drug that is supposed to reduce pain
   b. Determining the potential market for a new product by distributing free samples
   c. Estimating the amount of timber in a forest
   d. Estimating the lifetime of a flashlight battery

1.6   A doctor is keeping a record of patients who need surgery. For each patient, she records gender, age, marital status, occupation, type of surgery needed, and urgency of the surgery. Classify each of these variables as categorical or numerical. Is the data set univariate, bivariate, or multivariate?

1.7   A university placement office keeps track of past graduates of the university. For each former student, workers record gender, major, grade point average at graduation, number of uses of the placement service, and the type of employment. Classify each of these variables as categorical or numerical. Is the data set univariate, bivariate, or multivariate?

1.8   Consider the following year 2000 robbery rates (per 100,000 people) for selected cities with populations exceeding 100,000.

City	Rate
Arlington, Texas	595
Boston, Mass.	2451
Cleveland, Ohio	3084
Dallas, Texas	7046
Fresno, Calif.	1304
Louisville, Ky.	1079
San Jose, Calif.	677

Source: U.S. Department of Justice,
Uniform Crime Report, 2000.

a. Is robbery rate categorical or numerical?
b. Is city a variable or is it a description of the individual units?
c. Is robbery rate a continuous or discrete variable?
d. Reclassify the robbery rate as low (less than 1000), medium (between 1000 and 4000), or high (greater than 4000). How does this affect the amount of information in the data set?
e. What city had the largest robbery rate? The smallest?
f. From the provided information, is it possible to determine how many robberies there were in Boston in 2000?

1.9   In each of the following, identify the variables as either categorical or numerical. If a variable is numerical, classify it as continuous or discrete. Also, describe the population of interest.
a. A psychologist wishes to investigate the threshold reaction time (in seconds) for persons subjected to emotional stress.
b. A sociologist wishes to study the occupations and salaries of Vietnam veterans discharged prior to 1974.

1.10   Give a continuous and a discrete variable that should be measured in a survey of city residents on the issue of gun control. Describe the population of interest.

Interpreting Computer Output

1.11   The worksheet **BBSalaries.mtw** contains 1999 salaries for five major league teams: Baltimore Orioles, Boston Red Sox, California Angels, Chicago White Sox, and Cleveland Indians. So that we have a common scale for comparison purposes, we have constructed *side-by-side dotplots* of the salaries for the four teams. Are there unusual observations in any dotplot? How do the salaries of the Chicago White Sox compare to the salaries of the other teams? Which team pays its players the least?

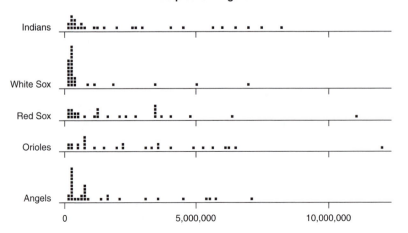

1.12   According to the 2000 *World Almanac and Book of Facts*, there were 50 major aircraft disasters in the 1990s, resulting in 5943 deaths. This compares to only 25 accidents in the 1980s, with 4831 deaths.

   a. Does this mean that it is becoming more dangerous to travel by air?

   b. Is the population of air travelers during the 1980s comparable to the population of air travelers during the 1990s?

   c. Is the total number of fatalities per year (or per 10-year period) a reasonable variable with which to measure the danger of air travel? Can you think of a more valid measure?

   d. The number of deaths per 100 million passenger miles in the 1980s was .094, and in the 1990s it was .0611. What does this say about the increased danger of air travel?

   e. The side-by-side dotplots give the number of fatalities for each aircraft disaster during each 10-year period. What observations can you make about how the number of disasters and the number of fatalities has changed over the years? Data from worksheet: **Airdisasters.mtw**.

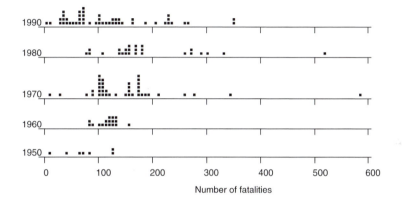

Aircraft Disasters 1950–1990

Computer Exercises

1.13   Based on census data, a study by the Federal Highway Administration shows that commuting times in most urban areas are becoming longer. The list compares the average travel times to work (in minutes) in several major cities.

WORKSHEET: Commute.mtw

City	1980	1990	City	1980	1990
New York	33.7	31.1	San Diego	19.5	22.2
Washington	27.2	29.5	Cincinnati	21.8	22.1
Houston	25.9	29.5	Cleveland	21.6	22.0
Chicago	26.3	28.1	San Antonio	20.2	21.9
Los Angeles	23.6	26.4	Indianapolis	20.8	21.9
Atlanta	24.9	26.0	Sacramento	19.5	21.8
Baltimore	25.3	26.0	Tampa	20.2	21.8
San Francisco	23.9	25.6	Portland, OR	21.4	21.7
New Orleans	24.5	24.4	Charlotte, NC	19.9	21.6
Seattle	22.8	24.3	Norfolk, VA	21.0	21.6
Boston	23.4	24.2	Kansas City, MO	20.7	21.4
Dallas	22.4	24.1	Minneapolis	20.1	21.1
Miami	22.6	24.1	Columbus, OH	20.1	21.2
Philadelphia	24.0	24.1	Hartford, CT	20.1	20.6
Detroit	22.5	23.4	Milwaukee	18.8	20.0
St. Louis	22.6	23.1	Salt Lake City	20.2	19.8
Phoenix	21.6	23.0	Rochester, NY	19.3	19.7
Orlando	20.3	22.9	Providence, RI	18.3	19.6
Pittsburgh	22.8	22.6	Buffalo	19.3	19.4
Denver	22.0	22.4			

Source: Federal Highway Administration.

a. What is an individual unit?

b. Identify the two variables that are measured. What are the values of the variables for Charlotte, NC?

c. How do you suppose the Federal Highway Administration came up with the values?

d. What city had the greatest change in travel time from 1980 to 1990?

e. Create a dotplot of the 1980 data. Describe any unusual features of the dotplot.

f. Create a dotplot of the 1990 data. Describe any unusual features of the dotplot.

g. What general statements can you make about how the commuting times changed from 1980 to 1990?

1.14 Diplomatic immunity covers parking tickets, so foreign diplomats do not bother to pay them. The major offenders in New York are listed in the table. Are Russians the worst offenders? Create dotplots of the number of tickets and the tickets/vehicle/month. Which dotplot gives the best representation of the data to determine who are the worst offenders?

WORKSHEET: Diplomat.mtw

Country	Number of Tickets	Tickets/vehicle/month
Russia	8138	8.9
Nigeria	2556	3.3
Israel	2363	1.6
Indonesia	1582	1.6
Egypt	1421	2.7
Bulgaria	1263	6.6
Brazil	1260	2.1
S. Korea	1239	0.8
Ukraine	956	10.6
Venezuela	919	2.2

Source: Time, November 8, 1993. Figures are from January to June 1993.

1.15 The worksheet: **Undergrad.mtw** contains variables related to a sample of undergraduate students (see Table 1.1).

a. Create side-by-side dotplots of the variable GPA with respect to Class. Describe any differences among the grade point averages of the students in the different classes. In particular, describe differences between the distributions of the freshman and sophomore classes. Does any one class have generally higher grade point averages than the other classes?

b. Compare the distributions of grade point averages for males and females with side-by-side dotplots. Do you think one group has higher grade point averages?

c. Investigate the relationship between SAT scores and the number of courses dropped by constructing side-by-side dotplots of SAT with respect to Number of Drops. What can be said about SAT scores for those students who dropped zero courses? What about those who dropped four courses?

Internet Tasks

1.16 Find the latest Uniform Crime Reports from the Bureau of Justice Statistics web site. Obtain robbery rates for the ten largest cities in your home state. Create a dotplot of the robbery rates and comment on any unusual features in the data.

1.17 From the Census Bureau web site, locate the latest census data. Determine the unemployment rates and personal per capita income rates for all 50 states and Washington, D.C. Create dotplots of the two variables and comment on their distribution.

1.18 Find the *Sporting News* web site. Go to the Baseball section and find the current salaries of the players on your favorite baseball team. Enter the salaries into your statistical computer program and create a dotplot of the salaries. Describe the distribution of the salaries and identify any unusual observations.

1.2   Displaying Data with Tables and Graphs

Learning Objectives for this section:

❏ Learn how to display categorical, discrete, and time series data.

❏ Know how to display data in a frequency table.

❏ Know how to convert the information in a frequency table to a bar graph and/or a pie chart.

❏ Learn how discrete data can be treated like categorical data and be displayed in a frequency table and a bar graph.

❏ Know the characteristics of time series data and how they are displayed.

Categorical Data

The frequency table, bar graph, and pie chart are used to organize and display the distribution of categorical data.

FREQUENCY TABLE   A nationwide telephone poll of 912 adult Americans aged 18+ conducted by the Gallup Organization presented the following question:

> How much progress have we made in dealing with environmental problems in the past few decades?
>
> A great deal____   Only some____   Hardly any____   No opinion____
>
> (Interviews conducted April 3–9, 2000.)

The results of the survey are presented in Table 1.2. The *frequency* is a simple count of the number of subjects falling into the different categories. *Relative frequencies* (in decimal or fraction form) are obtained by dividing the category count by the total count for all categories. *Percents* are then obtained from the relative frequencies. For example, $237/912 = .2599 = 26\%$.

Table 1.2
Gallup telephone poll of 912 adult Americans

How much progress have we made in dealing with environmental problems in the past few decades?

Response	Frequency	Percent
A great deal of progress	237	26%
Only some progress	584	64
Hardly any progress	82	9
No opinion	9	1
Total	912	100%

BAR GRAPH   Frequency table information can be illustrated in a bar graph. The heights of the bars are proportional to the frequencies (or percents) of the categories of the variable. Figure 1.3 is a Minitab bar graph of the Gallup survey.

Figure 1.3
Minitab bar graph of the
Gallup survey

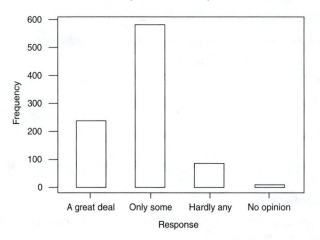

How much progress have we made in dealing with environmental problems in the past few decades?

Example 1.4   According to an FBI report, the number of hate crimes rose to 7725 in 1998. Of those reported, 4321 were motivated by race, 1390 were motivated by religion, 1260 were sexually oriented, and 754 were motivated by ethnicity. Summarize these data in a bar graph.

Solution   The vertical axis must be scaled above 4321, the highest frequency. Bars are constructed at the four motivation categories. The completed graph in Figure 1.4 was initially constructed using Microsoft Excel, but any statistical software package could be employed. Notice how the bar graph illustrates that race is a much greater motivating factor in hate crimes than any of the other categories.

Figure 1.4
Bar graph of hate
crimes reported to the
FBI in 1998

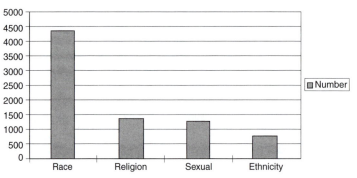

Source: Federal Bureau of Investigation

The bars, either horizontal or vertical, in a bar graph should always have the same width so that the difference in frequencies (or percents) is reflected only in the height of the bars.   ∎

PIE CHART   The pie chart is another useful way to exhibit categorical data. A circle, or pie, is divided into pieces corresponding to the categories of the variable so that the size (angle) of the slice is proportional to the percent of the category. By its nature, all categories of the variable must be included in the pie chart. Normally, the percent is shown in each slice.

Example 1.5   Figure 1.5 is a pie chart of the motivational bias in reported hate crimes in 1998. The pie chart shows that more than half (56%) of all hate crimes are motivated by race.   ■

Figure 1.5
Pie chart of hate crimes reported to the FBI in 1998

Motivation of Reported Hate Crimes in 1998

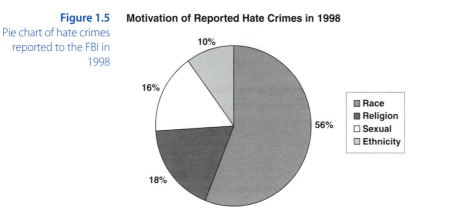

Source: Federal Bureau of Investigation

Bar Graphs and Pie Charts

Bar graphs and pie charts are incomplete unless they have

1. A title to clarify the presented information.
2. A legend to identify the different categories.
3. The source of the information.

Discrete Data

The frequency table and bar graph can also be used to illustrate the distribution of a discrete variable. Because the variable is discrete, we can simply treat the discrete values as different categories and proceed as if it were categorical.

Example 1.6   In 1969 the National Hurricane Center reported a total of 12 hurricanes: in 1982, only 2 were reported. Table 1.3 and worksheet **Hurrican.mtw** give the number of hurricane seasons (from 1950 through 1995) in which the discrete variable, number of reported hurricanes, ranged from 2 to 12. The bar graph in Figure 1.6 shows the distribution of the variable. The possible values of the variable are listed on the horizontal scale and the frequency with which they occur is listed on the vertical scale. Notice, for example, that in only 5 of the 46 seasons were there 9 or more hurricanes reported.   ■

Table 1.3
From worksheet: Hurrican.mtw

Number of hurricanes	2	3	4	5	6	7	8	9	10	11	12
Number of seasons	1	7	9	6	7	7	4	2	1	1	1

Figure 1.6

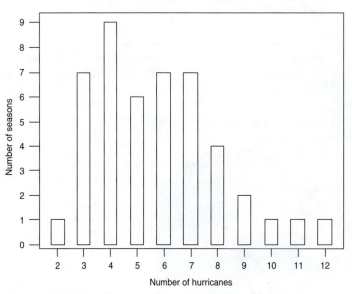

Reported Hurricanes from 1950 through 1995

Source: National Hurricane Center.

Figure 1.7

Time series bar graph

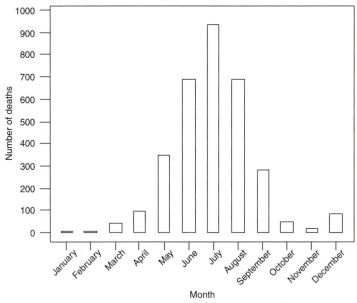

Deaths from Lightning Reported in the United States from 1959 through 1994

Source: Lightning Fatalities, Injuries and Damage Reports in the United States, 1959–1994, NOAA Technical Memorandum NWS SR-193, Dept of Commerce.

Time Series Data

Business and economic data, such as the daily Dow-Jones Average or the quarterly earnings of a company, are often observed at regular time intervals. Any set of data recorded at given time intervals is called **time series data**. The data can be presented numerically and graphically. The bar graph of *monthly* lightning fatalities from 1959 through 1994 in Figure 1.7 is a typical graphical presentation of time series data.

When the time increments of time series data are close together, as in Figure 1.8, a curve is preferred to a bar graph. Often graphs will start the vertical scale at some point above 0 to conserve space. If, however, the graph involves a comparison, as in Figure 1.8, we should begin the vertical scale at 0 in order to make a fair comparison.

Figure 1.8

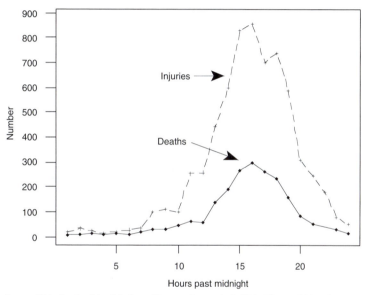

Time of Day of Number of Deaths and Injuries Due to Lightning in the United States from 1959 through 1994

Source: Lightning Fatalities, Injuries and Damage Reports in the United States, 1959–1994, NOAA Technical Memorandum NWS SR-193, Dept of Commerce.

Statistically analyzing time series data involves checking for *trends* and *cycles* in the data.

Example 1.7   As far back as 28 B.C. humans observed dark spots on the sun. In 1848 Rudolf Wolf of Zurich developed the "Wolfer sunspot number" based on the number of groups of sunspots, the total number of sunspots, and a constant associated with the observatory that made the original observation. The number, which represents the average number of sunspots during each year, has been reconstructed as far back as 1700. The Wolfer sunspot numbers from 1700 through 2000 are stored in worksheet: **Sunspot.mtw**. This time series data set has become one of the most widely studied time series ever.

Figure 1.9 is a plot of the sunspot data. If we look closely, we see a dip at the beginning of each century. We also count nine peaks each 100 years, and the pattern of the peaks in the first 100 years repeats each 100 years. This is consistent with the conjecture made by time series analysts, who have suggested that the sunspot data are cyclical (tend to repeat) and have a period of 11 years. ∎

Figure 1.9
Sunspot data

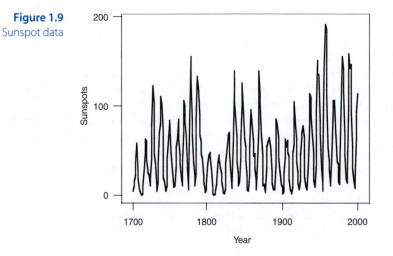

It is beyond the scope of this book to investigate time series data further. For the remainder of the book we will analyze *cross-sectional* data where the observations are taken at one point in time and are assumed to be independent of one another.

COMPUTER TIP

Bar Graphs and Pie Charts

Minitab—First select **Graph**, and then select **Chart**. When the Chart window appears, **Select** the column containing the frequencies and enter in the *Y* variable. Next **Select** the column containing the categories and enter in the *X* variable. If you wish to enter a title, click on **Annotation**, select **Title**, and then type in the title of the graph. If you wish to label the axes, click on **Frame**, select **Axis**, and then label one or both axes. To construct the graph, click **OK**. A Minitab pie chart is constructed in much the same way. First select **Graph**, and then select **Pie Chart**. Proceed as in the construction of the bar graph.

   Excel—To construct bar graphs and pie charts using Microsoft Excel, the data are entered into the worksheet and highlighted. Click on the **Chart Wizard**, and the program will walk you through the steps. Once a bar graph or pie chart is constructed, it is a simple matter to change the format of the graph. For instance, to convert from vertical bars to horizontal bars, click on **Chart Type** and select the horizontal bars.

EXERCISES 1.2   The Basics

1.19   In a random sample of college students, it was found that 124 had blue eyes, 150 had brown eyes, 15 had green eyes, and 103 had hazel eyes. Display these data in a frequency table including percents. From the frequency table, draw a bar graph and a pie chart for the data.

1.20 The Justice Department says that 1 of every 185 women, a total of 572,032, are victims of domestic abuse every year. The rates of domestic violence per 1000 women by age groups are given here:

WORKSHEET: Domestic.mtw

Age group	Rate
12–19	5.8
20–24	15.5
25–34	8.8
35–49	4.0
50–64	0.9

Source: U.S. Department of Justice.

a. Illustrate these data in a bar graph.
b. Is it possible to determine from these rates the actual number of women who are victims of domestic abuse in each of the age groups? What additional information is needed?

1.21 Illustrate the data on domestic violence in Exercise 1.20 in a pie chart.

1.22 According to the Energy Information Administration, the United States produces 30% of the world's nuclear power. France produces 17%, Japan produces 11%, Germany produces 7%, and Russia produces 6%. Illustrate these data in a bar graph. Do these percents sum to 100%? How will you deal with this in the bar graph?

1.23 In 1999 there were 4997 crimes reported in Abilene, Texas. Of those crimes, 12 were murder, 36 were forcible rape, 123 were robbery, 324 were aggravated assault, 1023 were burglary, 3207 were larceny theft, 255 were motor vehicle theft, and 17 were arson. Arrange these data (worksheet: **Abilene.mtw**) in a frequency table that includes percents. (*Source: Uniform Crime Reports*, U.S. Department of Justice.)
a. What crime was most prevalent?
b. What percent of all crimes involved some type of theft?

1.24 The SAT scores for 900 female and 800 male college freshmen are classified as low, medium, or high:

Range	Men	Women
High	190	250
Medium	430	520
Low	180	130

a. Draw a bar graph illustrating the frequencies of the SAT scores for men.
b. Draw a bar graph illustrating the frequencies of the SAT scores for women.
c. Can these two bar graphs be compared as drawn?
d. If the above bar graphs are not comparable, how should they be drawn so that they can be compared?

1.25 Divorce rate is recorded as the number of divorces for every 1000 people. In 1960 the rate was 2.2, in 1970 it was 3.5, in 1980 it was 5.2, in 1990 it was 4.7, and in 2000 it was 4.2. Illustrate these time series data in a bar graph. *Source:* National Center for Health Statistics, *Monthly Vital Statistics Report*, Vol 47, No. 21.

1.26 Fifty families were interviewed and the following numbers of dependent children recorded:

WORKSHEET: Depend.mtw

3 2 2 4 1 1 2 3 4 1 2 0 1 2 1 0 4 2 1 0 0 1 3 0 3 2 2 3 0 3 2 5
0 1 2 1 4 3 0 5 2 0 1 1 2 6 1 2 1 5

Arrange the data in a frequency table that shows the number of families with a given number of dependent children. Draw a bar graph from the frequency table. Does the bar graph appear to be symmetric or does it tail off in one direction?

1.27 The number of defective items produced by the 20 employees in a small business firm are given here:

WORKSHEET: Defectiv.mtw

```
 7   6  10   9   8   7    7   6   8   8
10   7   6   8   8   9   10   9   9   8
```

Arrange the data in a frequency table that shows the number of employees with a given number of defective items. Draw a bar graph from the frequency table. Does the bar graph appear to be symmetric?

1.28 The number of days that the 20 employees in Exercise 1.27 were absent during the year are listed here:

WORKSHEET: Absent.mtw

```
 1   0   4   3   2   0   0   2   0   2
10   0   0   2   3   4   5   0   2   1
```

Arrange the data in a frequency table that shows the number of employees with a given number of absent days. Draw a bar graph from the frequency table. Are there any unusual observations in the data? How would you describe the shape of the distribution?

Interpreting Computer Output

1.29 Housing of Native Americans on the 314 national reservations is often less than desirable. For instances, 20% of American Indian households on reservations do not have indoor plumbing, 18% do not have complete kitchen facilities, and more than half do not have a telephone. Heating the homes is also a serious issue. The table gives the primary heating source for all U.S. homes, homes on reservations, and homes of American Indians who are not living on reservations.

WORKSHEET: Heat.mtw

Heating fuel	American Indians on reservation	All U.S. households	American Indians not on reservations
Utility gas	16%	48%	51%
LP bottled gas	22	9	6
Electricity	19	25	26
Fuel oil	6	9	12
Wood	34	8	4
Other fuel or none	4	1	1

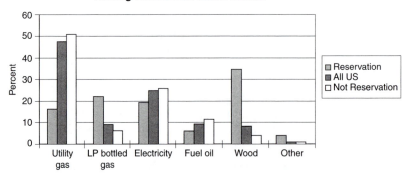

Heating of American Indian Homes

Source: Bureau of the Census, *Housing of American Indians on Reservations*, Statistical Brief 95–11, April 1995.

a. Do the columns sum to 100%? Should they? If they do not, explain why not.
b. From the comparison bar graph, how do the American Indian households, both on and off reservations, compare with all U.S. households? Describe what you see in the bar graph. Are the differences easier to observe in the bar graph or in the table?
c. American Indians on reservations heat mostly with what fuel?
d. All U.S. households heat mostly with what fuel?

1.30 From the accompanying bar graph of weather casualties in 1994 (created from worksheet: **Weather94.xls**), complete the following.

a. What type of weather caused the most casualties?

b. What type of weather caused the least casualties?

c. Approximately how many casualties in 1994 were caused by lightning?

d. Flash flood and river flood together caused approximately how many casualties in 1994?

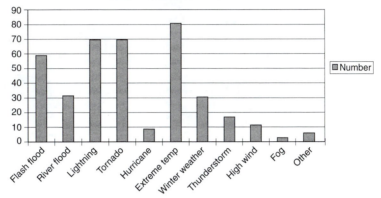

1.31 The following table gives the life expectancy of men and women in the United States. Based on the graph, discuss the difference between the two time series.

WORKSHEET: Life.mtw

Year	Men	Women
1920	53.6	54.6
1930	58.1	61.6
1940	60.8	65.2
1950	65.6	71.1
1960	66.6	73.1
1970	67.1	74.7
1980	70.0	77.5
1990	71.8	78.8

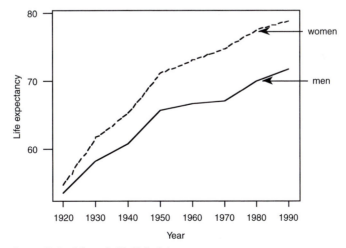

Source: National Center for Health Statistics.

Computer Exercises

1.32   The accompanying pie charts compare the educational backgrounds of parents of entering freshmen at a medium-sized state university. Notice that it is difficult to compare the mothers' and fathers' educational backgrounds. Convert the data (worksheet: **ParentEd.xls**) to a comparison bar graph that better illustrates the comparison of their educational backgrounds.

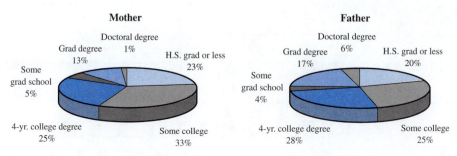

Source: Student Life and Learning, Vol. 4, No. 2, Student Development, Appalachian State University, Boone, NC.

1.33   In the second quarter of 2000, personal income rose 1.7% from the first quarter of 2000. Following is the percent change from the first to second quarter for each state and the District of Columbia:

WORKSHEET: Income.mtw

Ala	1.6%	Mont	1.8
Alaska	1.7	Neb	1.7
Ariz	0.7	Nev	2.5
Ark	1.4	N.H.	0.8
Calif	1.4	N.J.	1.8
Colo	2.1	N.M	1.9
Conn	0.5	N.Y	1.8
Del	2.4	N.C.	1.9
Fla	2.1	N.D.	0.8
Ga	1.7	Ohio	1.5
Hawaii	1.6	Okla	2.0
Idaho	1.3	Ore	1.7
Ill	1.9	Pa	1.4
Ind	1.9	R.I.	1.5
Iowa	2.4	S.C.	2.0
Kan	1.9	S.D.	1.8
Ky	1.6	Tenn	1.8
La	1.5	Texas	2.1
Maine	1.5	Utah	2.2
Md	1.2	Vt	1.5
Mass	1.0	Va	1.2
Mich	2.3	Wash	1.6
Minn	1.8	W.Va	1.5
Miss	1.4	Wis	2.2
Mo	1.9	Wyo	1.0
D.C.	0.3		

Source: U.S Department of Commerce.

Categorize the percent changes into five classes as follows:

Class 1   % change ≤ 0.5
Class 2   0.5 < % change ≤ 1.0
Class 3   1.0 <% change ≤ 1.5
Class 4   1.5 < % change ≤ 2.0
Class 5   % change > 2.0

Make a frequency count of the states that fall into the classes and give the percents. From the frequency table, construct a bar graph.

1.34   The Consumer Price Index (CPI) is a measure of the cost of goods in the United States. The table gives the changes in the CPI from the previous year from 1979 to 1998. Illustrate these time series data with a bar graph. Do you detect a trend in the data? Describe what you see in the graph.

WORKSHEET: CPI.mtw

Year	1979	80	81	82	83	84	85	86	87	88
CPI	13.3	12.5	8.9	3.8	3.8	3.9	3.8	1.1	4.4	4.4

Year	1989	90	91	92	93	94	95	96	97	98
CPI	4.6	6.1	3.1	2.9	2.7	2.3	2.8	3.0	2.3	1.6

Source: Bureau of Labor Statistics.

1.35   The worksheet **DowJones.mtw** contains the closing price and the percent change from the previous year of the Dow-Jones Industrial Average for years 1896 through 2000. Construct a time series graph of the closing prices. Describe what you see in the graph.

1.36   The worksheet lists the winning times (in seconds) for the men's 1500-meter run in the Olympics from 1896 to 1996. (No Olympics were held in 1916, 1940, and 1944 because of world wars; in 1980 the United States and some other countries boycotted the Olympics.)

WORKSHEET: track15.mtw

Year	Time	Year	Time	Year	Time
1896	273.2	1932	231.2	1972	216.3
1900	246.2	1936	227.8	1976	219.2
1904	245.4	1948	229.8	1980	218.4
1908	243.4	1952	225.2	1984	212.5
1912	236.8	1956	221.2	1988	216.0
1920	241.8	1960	215.6	1992	220.1
1924	233.6	1964	218.1	1996	215.8
1928	233.2	1968	214.9		

Source: The World Almanac and Book of Facts, 2000.

a. Draw a time series plot of the data. How did you deal with the 3 years that had no data?
b. Do you find any unusual behavior in the data?
c. Are the data cyclical? Is there a trend?

1.37   Retrieve the data on parking tickets given diplomats in New York that is given in worksheet **Diplomat.mtw** (see Exercise 1.14 in Section 1.1). Use the code variable for the countries and the Chart command in Minitab to create a bar graph of the rate of tickets per vehicle per month. Does the chart tell you who are the greatest offenders? Who are the least offenders?

Internet Tasks

1.38   Worksheet **Hurrican.mtw** lists the number of storms and hurricanes reported from 1950 to 1995. Go to the National Hurricane Center web site and update worksheet **Hurrican.mtw** for 1996 to the present. Construct a bar graph of reported hurricane activity similar to the graph in Figure 1.6.

1.39 From the Federal Justice Statistics Resource Center web site, obtain the latest data on motivations of reported hate crimes. Create a pie chart similar to the one in Figure 1.5.

1.40 Use your favorite search engine and enter the keyword Sunspot. Examine at least three different sources of information on sunspots and write a one-page report. Make sure that you describe in detail what a sunspot is and give information on how sunspots are counted. See if you can duplicate the graph in Figure 1.9.

1.3 Displaying Numerical Data

Learning Objectives for this section:

❏ Learn the mechanics of displaying data with a stem-and-leaf plot and a histogram.

❏ Know why displaying data is important.

❏ Know how to identify an outlier and learn why outliers are important.

❏ Know the different shapes that are used to classify distributions.

❏ Learn the difference between symmetry and skewness.

❏ Learn the characteristics of short-tailed and long-tailed distributions

Stem-and-Leaf Plots

A **stem-and-leaf plot**, much like the *dotplot* introduced in Section 1.1, is a graph that displays the distribution of a numerical variable. Unlike the dotplot, it leaves the data intact for future calculations. It is extremely useful for arranging the observations from smallest to largest so that specific positions within the data set can be found. Example 1.8 describes the construction of a stem-and-leaf plot.

Example 1.8 A random sample of 24 high school seniors was given a college entrance exam that has a maximum score of 100. Construct a stem-and-leaf plot of their scores:

W O R K S H E E T : Entrance.mtw

64 75 81 43 69 75 86 58 63 66 82 62 79 91 83 55 68 74 48 66
84 77 73 59

By observation we see that the scores are two-digit numbers that range from the 40s to the 90s. To construct the display we divide each observation into a *stem* and a *leaf*. In this example, the digits in the tens place of the numbers become the stems, and the digits in the units place will become the leaves of the stem and leaf plot. For example, the first observation, 64, has a stem value of 6 and a leaf value of 4—see Figure 1.10(a). Only the 4 of 64 is plotted on the 6 stem. Figure 1.10(b) shows the completed stem-and-leaf plot after all 24 leaves (corresponding to the 24 observations) are properly plotted.

Figure 1.10

```
4 |                   4 | 3 8              4 | 3 8
5 |                   5 | 8 5 9            5 | 5 8 9
6 | 4                 6 | 4 9 3 6 2 8 6    6 | 2 3 4 6 6 8 9
7 |                   7 | 5 5 9 4 7 3      7 | 3 4 5 5 7 9
8 |                   8 | 1 6 2 3 4        8 | 1 2 3 4 6
9 |                   9 | 1                9 | 1
```

(a) one observation   (b) completed   (c) ordered
plotted   stem-and-leaf plot   stem-and-leaf plot

The stem-and-leaf plot retains the numerical values of the data for future calculations. Some of those calculations involve finding specific positions in the data; this would be relevant only if the data were ordered from smallest to largest. This is accomplished by reordering the leaves of each stem from smallest to largest. Figure 1.10(c) is the resulting *ordered* stem-and-leaf plot. Rotating the stem-and-leaf plot 90 degrees counterclockwise, as in Figure 1.11, we see that the leaves form a graphical picture that shows how the data are *distributed* across a number line. ∎

Figure 1.11

```
         9
         8    9
         6    7    6
         6    5    4
    9    4    5    3
8   8    3    4    2
3   5    2    3    1    1
_____
4   5    6    7    8    9
```

Minitab and most other statistical packages can produce stem-and-leaf plots.

Example 1.9   An ecologist wishes to investigate the level of mercury pollution in a major lake. She catches 25 lake trout and measures the concentration of mercury (measured in parts per million) in each fish. From the following data, construct an ordered stem-and-leaf plot.

WORKSHEET: Mercury.mtw

2.2, 3.4, 3.0, 2.6, 3.8, 1.8, 2.8, 3.2, 3.7, 1.4, 2.7, 3.6, 1.9, 2.2, 3.0, 3.3, 2.3, 1.7, 2.6, 3.5, 3.0, 2.9, 3.4, 3.1, 2.4

Solution   Figure 1.12 gives a Minitab stem-and-leaf plot of the data. Instead of printing all the decimals in the display, Minitab prints Leaf Unit = 0.10. In effect, each leaf value is multiplied by 0.10 in order to reinsert the decimal place; for example, the first entry 1 4 is actually 1.4. The first column in the display gives the depth within the data set of the last entry in each row. For example, the 8 in row three indicates that 2.4 is the eighth observation down in the ordered data set. A count also begins at the bottom of the display. The two counts meet in the middle with a frequency count for the middle category in parentheses. As we shall see, this column is not needed unless we are finding specific location values by hand.

Notice in Figure 1.12 that Minitab has created a *double-stem* stem-and-leaf plot where there are two rows for each stem value 1, 2, and 3. The first category of each is for leaf values 0 through 4 and the second is for leaf values 5 through 9. Likewise, a *five-stem* stem-and-leaf plot (top of page 24) uses five rows for each stem value. The first category is for leaf values 0 and 1, the next category is for leaf values 2 and 3, followed by leaf values 4 and 5, 6 and 7, and finally 8 and 9. Splitting the stems helps spread out the data so that we can better interpret the distribution. In this case, the double-stem display does not spread out the data well enough, so we prefer the five-stem display. ∎

Figure 1.12
Double-stem
stem-and-leaf plot

```
Stem-and-leaf of mercury   N = 25
Leaf Unit = 0.10

   1        1 4
   4        1 789
   8        2 2234
  (5)       2 66789
  12        3 00012344
   4        3 5678
```

Five-stem
stem-and-leaf plot

```
Stem-and-leaf of mercury   N = 25
Leaf Unit = 0.10

  1      1 4
  2      1 7
  4      1 89
  4      2
  7      2 223
  8      2 4
 11      2 667
 (2)     2 89
 12      3 0001
  8      3 23
  6      3 445
  3      3 67
  1      3 8
```

Outliers

Occasionally, one or two scores may be far removed from the rest of the data. These extreme values, called **outliers**, are very influential and should be singled out for further study. Consider the state birth rates given in the next example.

Example 1.10   Following are the 1998 live birthrates per 1000 population for all states and the District of Columbia:

WORKSHEET: Birth.mtw

Ala	14.3	Mont	12.3
Alaska	16.2	Neb	14.2
Ariz	16.8	Nev	16.4
Ark	14.5	N.H.	12.2
Calif	16.0	N.J.	14.1
Colo	15.0	N.M.	15.7
Conn	13.4	N.Y.	14.2
Del	14.2	N.C.	14.8
Fla	13.1	N.D.	12.4
Ga	16.0	Ohio	13.6
Hawaii	14.7	Okla	14.8
Idaho	15.8	Ore	13.8
Ill	15.2	Pa	12.2
Ind	14.4	R.I.	12.7
Iowa	13.0	S.C	14.0
Kan	14.6	S.D.	13.9
Ky	13.8	Tenn	14.3
La	15.3	Texas	17.3
Maine	11.0	Utah	21.5
Md	14.0	Vt	11.1
Mass	13.2	Va	13.9
Mich	13.6	Wash	14.0
Minn	13.8	W.Va	11.5
Miss	15.6	Wis	12.9
Mo	13.9	Wyo	13.0
D.C.	14.7		

Source: National Vital Statistics Report, 48, March 28, 2000, National Center for Health Statistics.

Figure 1.13 shows a Minitab stem-and-leaf plot of the birthrates.

Figure 1.13 Stem-and-leaf plot of 1998 birthrates

```
Stem-and-leaf of 1998rate  N = 51
Leaf Unit = 0.10

   3   11  015
   9   12  223479
  22   13  0012466888999
(16)   14  0001222334567788
  13   15  023678
   7   16  00248
   2   17  3
   1   18
   1   19
   1   20
   1   21  5
```

Notice that the Leaf Unit = 0.10. This means that the observation, 215, is actually 21.5, which clearly is an outlier. After inspecting the data, we see that the outlier is the birthrate for Utah. Without the outlier, the data are very symmetric. If we wish to find the average birthrate across the states, it might be wise to report the average both with and without the Utah birthrate because its value will significantly increase the average. ■

Histograms

The **histogram** is another graph that is used to illustrate the distribution of numerical data. A histogram can be constructed from the stem-and-leaf plot; each stem defines an interval of values called a *class*. The *class limits* are the smallest and largest possible values for the leaves of that stem. For the stem-and-leaf plot in Figure 1.13, we see that the class limits for stem value 11 are 110 and 119, for stem value 12 they are 120 and 129, and so on. Once the class limits are determined, the data can be formulated in a *grouped frequency table* (grouped in the sense that the numerical data are grouped into various classes) in Table 1.4. Remember Leaf Unit = 0.10, so 110 actually means 11.0.

Table 1.4 Grouped Frequency Table 1998 live birthrates per 1000 population

Class Limits	Frequency	Percent
110–119	3	6%
120–129	6	12
130–139	13	25
140–149	16	31
150–159	6	12
160–169	5	10
170–179	1	2
180–189	0	0
190–199	0	0
200–209	0	0
210–219	1	2

The histogram can now be constructed from the grouped frequency table. The classes are scaled off on the horizontal axis. Bars are constructed over each class so that the height of each bar is the frequency of the class, which is marked off on the vertical axis. Figure 1.14 is the completed histogram. Notice that the shape is the same as the shape of the stem-and-leaf plot and the outlier shows up clearly in the histogram.

Figure 1.14
1998 live birthrates per
1000 population

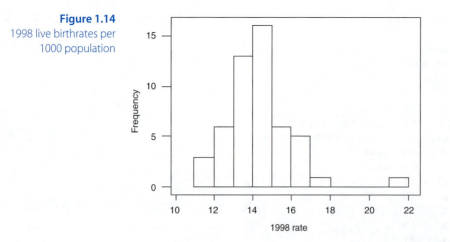

The histogram, like the bar graph for categorical and discrete data, pictures the distribution of the data. The basic difference between the two is that in the histogram (for continuous numerical data) the bars are joined, whereas the bars of a bar graph are separated.

Figure 1.14 is a Minitab histogram obtained by specifying the endpoints of the class limits. Figure 1.15 shows a Minitab histogram of the same data with the midpoints of the classes labeled instead of the endpoints; the vertical axis is percent instead of frequency.

Figure 1.15
1998 live birthrates per
1000 population

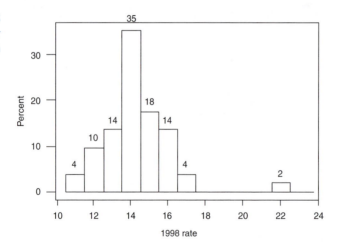

Notice that the histogram's shape is slightly different. The outlier is still prominent, but the rest of the distribution seems more symmetric and peaked. The general appearance is the same, but be aware that the shape of a histogram can be altered by changing the number of classes and/or their width.

A *percent* histogram is important for two reasons. The first is that it removes the histogram's dependence on the size of the data set. This is important when comparing two or more distributions. For example, do women score higher than men on a rote memorization task? If considerably more women are tested than men are, it is not correct to compare their frequency histograms. Percents, however, are independent of group sizes and thus fair comparisons can be made using percent histograms.

The second important fact about a percent histogram is that it can be used in a *probability* sense. From Figure 1.15 we see that 10% of the states have birthrates between 11.5

and 12.5 (interval with midpoint = 12). And by addition, the percent of states having birthrates between 11.5 and 14.5 is 59%.

Density Curve

If the data accurately describe the population, we can picture the overall pattern of the population distribution by drawing a smooth *density curve* over the histogram. Figure 1.16 shows a density curve superimposed over the histogram in Figure 1.15. Like the percent histogram, density curves are drawn so that the total area under the curve represents 100%; hence, the shaded region under the curve between 11.5 and 14.5 is 59%.

Figure 1.16
1998 live birthrates per 1000 population

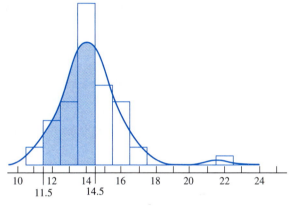

Uses of Density Curves

- Density curves describe the overall pattern of the population distribution.
- The total area under the curve is 100% so population percentages can easily be obtained by finding the area under the curve between a range of values.
- It is easy to compare two or more distributions by graphing their density curves on the same graph.

To compare rote memorization scores for men and women, density curves for both men and women may be graphed together, as illustrated in Figure 1.17.

Figure 1.17
Correct word phrases for men and women in a rote memorization study

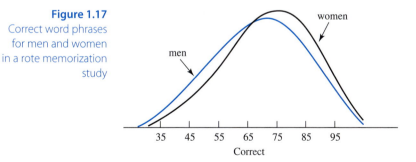

Knowing the overall pattern of a distribution, or its *shape*, is important in determining what statistical procedures are best suited for the collected data. There are several important shapes we must consider.

Important Distribution Shapes

- A distribution is called **unimodal** if its density curve has a single peak. It is called **multimodal** if it has two or more peaks. A **mode** is the numerical value associated with a peak.
- A distribution is said to be **symmetric** if the scores below the center of the distribution are a mirror image of the scores above the center.
- A unimodal distribution that is not symmetric is **skewed**. If it has a long left tail then it is **skewed left**. If it has a long right tail it is **skewed right**.
- A symmetric distribution is called **short tailed** if both tails of its density curve drop off rapidly. It is called **long tailed** if both tails of its density curve are significantly long.

The modes of a distribution are important because they identify the main concentration of data. It is also important that we distinguish between symmetry and skewness.

The best known symmetric distribution is the *normal distribution*, illustrated by the density curve in Figure 1.18. It is commonly referred to as the *bell curve* and provides a good model for many distributions of real data. Properties of the normal distribution will be studied in Chapter 3.

Figure 1.18
Normal density curve

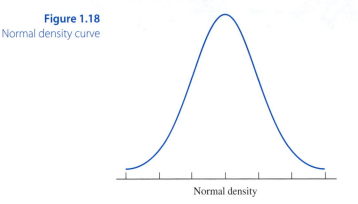

Normal density

Figure 1.19 shows (a) a symmetric-bimodal distribution, (b) a bimodal-with-gap distribution, (c) a skewed-right distribution, and (d) a skewed-left distribution.

Figure 1.19
Distribution Shapes

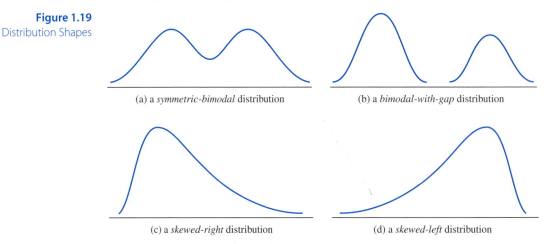

(a) a *symmetric-bimodal* distribution   (b) a *bimodal-with-gap* distribution

(c) a *skewed-right* distribution   (d) a *skewed-left* distribution

The normal distribution is often used to judge other distributions. For example, a distribution is classified short tailed (or long tailed) if its tails are significantly shorter (or longer) than those of a normal distribution. Figure 1.20 shows a short-tailed and a long-tailed distribution in comparison to a normal density curve.

Figure 1.20
A short-tailed, long-tailed, and normal distribution

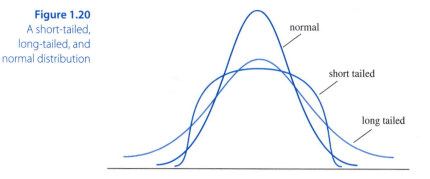

Next, we study several quantities that *numerically* summarize distributions. We will see that the shape of the distribution affects the values we get for these summary measures. Among the most important numerical summaries are measures of center and measures of variability.

C O M P U T E R   T I P

Stem-and-Leaf Plots and Histograms

To create a stem-and-leaf plot, first select **Graph** and then click on **Stem-and-Leaf**. When the Stem-and-Leaf window appears, **Select** the variable and then click **OK**.

A histogram is created in the same manner. First select **Graph** and then click on **Histogram**. When the histogram window appears, **Select** the variable to be graphed and then click **OK**.

Note: The default histogram produced by Minitab may not look exactly like the one in Figure 1.15. To create this graph, at the Histogram window select **Options. . .**, at Type of Histogram choose **Percent**, at Type of Intervals choose **MidPoint**, and at Midpoint/cutpoint positions: type **11 12 13 14 15 16 17 18 19 20 21 22**.

EXERCISES 1.3   The Basics

1.41   Following are Miller Personality Test scores for 20 candidates interviewing for a sales position.

WORKSHEET: Miller1.mtw

21 21 16 26 22 23 31 25 20 25
33 17 27 29 25 22 30 18 23 25

Arrange the data in a stem-and-leaf plot. Should it be a double-stem or five-stem stem-and-leaf plot? Are the data symmetrically distributed?

1.42   A group of 23 students participated in a psychology experiment. Their scores are the numbers of correct responses:

WORKSHEET: Psych.mtw

12 14 18   7   1 15   8 15 10 14 19
14   6 13 14 12 16 10   9 12 15   8 17

Arrange the data in a stem-and-leaf plot. Would you classify the data as being symmetric?

1.43 Following are the daily profits of 20 newsstands. Arrange the data in a five-stem stem-and-leaf plot. Does there appear to be more than one mode in the data?

WORKSHEET: Newsstand.mtw

$81.32 61.47 64.90 70.88 76.02 75.06 76.73 64.21 74.92 77.56 58.01 68.05 73.37 75.41
59.41 65.43 74.76 76.51 65.10 76.02

1.44 An exercise program had 30 members who did exercises 5 days a week for 1 month. Following are the amounts of weight loss of the 30 members (a negative indicates that the person gained weight):

WORKSHEET: Exercise.mtw

5	15	3	−4	8	7	5	10	−3	2	−5	9	5	−4	10
6	−2	7	4	3	−12	−6	11	4	8	−10	9	5	5	2

Construct an ordered stem-and-leaf plot of the amounts of weight loss. Are the data skewed? Does there appear to be more than one mode?

1.45 The prices of regular unleaded gasoline were obtained from 25 service stations around a major city. Construct a stem-and-leaf plot of the data. Are there any unusual features in the data? Describe the distribution.

WORKSHEET: Gasoline.mtw

$1.46 1.52 1.50 1.52 1.48 1.60 1.50 1.49 1.48 1.50 1.49 1.50 1.60 1.59 1.49 1.50 1.52 1.61
1.51 1.50 1.49 1.50 1.48 1.51 1.49

1.46 Following are the weights of 25 soccer players:

WORKSHEET: Soccer.mtw

144 162 197 173 183 129 209 190 117 160 179 177 154 132 151 159 175 154
148 166 184 157 162 150 136

Arrange the data in a stem-and-leaf plot. Construct a histogram from the stem-and-leaf plot with interval widths of 10 and the first lower class limit at 110. Are the data symmetrically distributed?

1.47 Construct a stem-and-leaf plot of the following data, which consist of the maximum number of sit-ups completed by the participants in an exercise class after 1 month in the program. Is there a gap in the data? Would you classify the shape as bimodal?

WORKSHEET: Situp.mtw

24 31 54 62 36 28 37 55 18 27 58 32 37 41 55 39 56 42 29 35

1.48 Create a histogram from the stem-and-leaf plot you constructed in Exercise 1.47. Draw a density curve over the histogram that you feel best describes the population distribution.

1.49 Construct a histogram of the following data, which are the test scores on the first exam in a beginning biology class:

WORKSHEET: Biology.mtw

87 79 94 60 75 94 77 83 68 74 82 73 63 75 77 83 92 57 64 53 53 82 73
90 55 68 72 88 65 78

Draw a density curve over the histogram that you feel best describes the population distribution.

1.50   To estimate the number of trees on a tree farm, a farmer divided the farm into 1000 small grids. He then randomly selected 20 grids and counted the number of trees, with the following results:

WORKSHEET: Trees.mtw

81   96   87   83   99   64   77   63   93   84   102   68   94   81   70   84   92   109   74   86

Construct a stem-and-leaf plot and a histogram of the data. Classify the shape of the distribution.

1.51   A designer of web pages recorded the number of hits a particular page received each day for a 1-month period. Construct a histogram from the stem-and-leaf plot with interval widths of 10 and the first lower class limit at 20. Are the data symmetrically distributed? Do you think it would have been helpful for the designer to include the day of the week the hits were recorded?

```
2 | 3 4 8
3 | 2 4 3 8 4 7
4 | 1 8 9 2 6 7 3 0
5 | 4 5 2 7 5
6 | 7 3 0 8
7 | 5 2 7
8 | 3
```

Interpreting Computer Output

1.52   The following data are the survival times in weeks for 20 male rats that were exposed to a high level of radiation. From the accompanying histogram of the data, would you classify the distribution as bimodal or just skewed left?

WORKSHEET: Rat.mtw

152   152   115   109   137   88   94   77   160   165   125   40   128   123   136   101   62   153   83   69

Source: J. Lawless, *Statistical Models and Methods for Lifetime Data* (New York: Wiley, 1982).

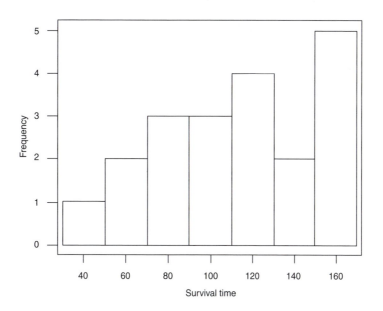

1.53   A high-volume drug screen was designed to find compounds that reduce low-density lipoproteins (LDL) cholesterol in quail. The treatment group of quail was fed a special diet mixed with a drug compound over a specified period of time. The placebo group of quail was fed the same special diet for the same period of time but without the drug compound. Following are the plasma LDL levels for the two groups:

WORKSHEET: Quail.mtw

placebo
64 49 54 64 97 66 76 44 71 89 70 72 71 55 60 62 46 77 86 71
treatment
40   31   50   48   152   44   74   38   81   64

Source: J. McKean, and T. Vidmar (1994), "A Comparison of Two Rank-Based Methods for the Analysis of Linear Models," *The American Statistician, 48,* 220–229.

Placebo		Treatment
	3	18
964	4	048
54	5	0
64420	6	4
7621110	7	4
96	8	1
7	9	
	10	
	11	
	12	
	13	
	14	
	15	2

Shown is a *back-to-back stem-and-leaf plot* in which two stem-and-leaf plots (for the placebo and treatment) are drawn back to back with a common stem. It is useful when comparing distributions. Notice that there is one very large value in the treatment group. The researchers pointed out that this was typical with most of the treatment data in the study. Would you consider it an outlier? Should it be discarded or retained when comparing the characteristics of the two distributions?

Do you think that the treatment drug was effective in reducing the LDL cholesterol in quail?

1.54   The margin of victory in Superbowl I through Superbowl XXXV is given in the following stem-and-leaf plot.

WORKSHEET: Superbowl.mtw

Stem-and-leaf of margin   N = 35
Leaf Unit = 1.0

```
  5      0   13444
 10      0   57779
 16      1   000234
 (9)     1   567777899
 10      2   123
  7      2   579
  4      3   2
  3      3   56
  1      4
  1      4   5
```

a. What was the smallest margin of victory in all of the Superbowls?
b. What was the largest margin of victory in all of the Superbowls?
c. In how many games did the margin of victory exceed 20 points?
d. How would you classify the general shape of the distribution?
e. Would you classify any of the observations as outliers?

Computer Exercises

1.55 A study of toxic waste sites gives the percentage of minorities living in communities with commercial hazardous waste sites (worksheet: **Toxic.mtw**). To better understand the data, construct histograms of all sites and the number of sites near minority communities. How would you classify the shapes of the two distributions? Are they similar in shape?

1.56 Following are cholesterol levels from a sample of 62 subjects from the Framingham Heart Study:

WORKSHEET: Framingh.mtw

```
393  353  334  336  327  300  300  308  283  285  270  270  272
278  278  263  264  267  267  267  268  254  254  254  256  256
258  240  243  246  247  248  230  230  230  230  231  232  232
232  234  234  236  236  238  220  225  225  226  210  211  212
215  216  217  218  200  202  192  198  184  167
```

Source: R. D'Agostino, et. al., (1990) "A Suggestion for Using Powerful and Informative Tests of Normality," *The American Statistician, 44,* 316–321.

a. Construct a stem-and-leaf plot of the data.
b. From the stem-and-leaf plot, define class limits and formulate the data in a grouped frequency table.
c. What percent of the data fall between 200 and 240?
d. Construct a histogram and, by hand, draw a density curve over the histogram that you feel best describes the population distribution.
e. Classify the shape of the distribution.

Internet Tasks

1.57 Mark McGwire made history when he hit 70 home runs in one season in 1998, breaking Roger Maris's record of 61 in 1961. Go to the baseball section of the *Sporting News* web site and find Mark McGwire's career statistics. Make a stem-and-leaf plot of the number of home runs he has hit in each season. Also create a histogram, and comment on the shape of the resulting distribution of home runs. Are there any outliers? In 2001 Barry Bonds broke McGwire's record when he hit 73 home runs. Find Bonds's career home runs and make a back-to-back stem-and-leaf plot with Mark McGwire's career home runs. Compare the two distributions. Who do you think is the better home-run hitter?

1.58 Go to the Census Bureau web site and find the number of doctors per 100,000 population for each of the 50 states and the District of Columbia. Make a histogram of the data and describe its shape. Are there any unusual features, such as outliers, in the data?

1.59 Find the web site for the *Statistical Abstracts of the United States*. Search for infant mortality rate. Construct a histogram of the infant mortality rates for the 50 states and the District of Columbia. Describe its shape and comment on any unusual features in the data.

1.4 Summarizing Data with Statistics

Learning Objectives for this section:

☐ Learn to describe the numerical characteristics of data. In particular, learn about the different measures of location and spread.

☐ Learn to calculate and interpret measures of the center of a distribution—the mean, the median, and the trimmed mean.

☐ Learn to formulate summary measures in a five-number summary diagram.

☐ Learn to calculate and interpret the range, the interquartile range, and the standard deviation.

☐ Know what is meant by degrees of freedom.

☐ Learn how to apply the empirical rule.

☐ Learn about z-scores and how they measure the relative standing of an observation within a data set.

One of the main objectives of statistics is to draw generalizations about the characteristics of a population based on data collected from a random sample. In most cases, it is difficult to work with the complete distribution of values; thus, summary measures are introduced to help answer statistical questions. For example, the comparison of the effects of a certain drug on the reaction time of women to that of men may simplify to a comparison of only two numerical values that correspond to the centers of the two distributions of reaction times. These summary measures can apply either to a sample or to the entire population.

Summary Measures

- A **parameter** is a numerical summary measure that describes a certain characteristic of a population distribution (usually its numerical value is unknown).
- A **statistic** is a numerical summary measure calculated from sample data that describes a certain characteristic of the sample.

Typically, the values of the *parameters* associated with a population distribution are unknown. We must select a random sample from the population and use its summary *statistics* to estimate the unknown population parameters.

Measures of Center

Among the summary measures are *measures of center, measures of specific locations*, and *measures of variability*. We first consider measures of the center of a distribution.

Parameters That Measure the Center of a Population Distribution

- The **population mean**, denoted by the Greek letter μ, is the numerical value that locates the *balance point* of the population distribution.
- The **population median**, denoted by θ, is the numerical value that divides the population distribution in half.

Figure 1.21 illustrates that the *population mean*, μ, is located in the center of the distribution and is the precise point about which the distribution will balance.

Figure 1.21
Location of the population mean

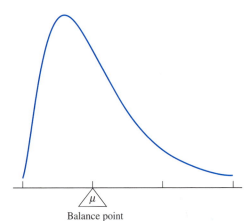

Balance point

Figure 1.22(a) illustrates that the *population median*, θ, is the value that separates the lower 50% of the distribution from the upper 50% of the distribution. In this case, where the distribution is skewed right, the mean is larger than the median. For skewed-left distributions the mean is smaller than the median. Figure 1.22(b) shows that the mean and median are the same when the distribution is symmetric. In this case, either will balance the distribution.

Figure 1.22

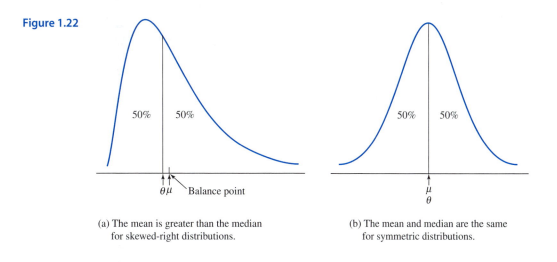

(a) The mean is greater than the median for skewed-right distributions.

(b) The mean and median are the same for symmetric distributions.

Rarely is the entire population of measurements available for calculating the numerical values of its parameters; thus, the values of μ and θ remain unknown. To estimate these unknown quantities, we calculate statistics from the sample data.

Statistics That Measure the Center of the Data

- **Mean**: If a sample consists of observations $x_1, x_2, x_3, \cdots x_n$, the **sample mean** is
  $$\bar{x} = (x_1 + x_2 + \cdots + x_n)/n = \frac{\Sigma x_i}{n}.$$ We will also use $\bar{y}$ to represent the sample mean.
- **Median**: If the sample observations $x_1, x_2, \cdots x_n$ are arranged in order from smallest to largest, the **sample median**, denoted by M, is the middle observation if n is odd, or the average of the two middle observations if n is even. In either case, the median is located at the *position* $(n + 1)/2$ in the ordered data set of n observations.
- **Trimmed Mean**: A $p\%$ **trimmed sample mean**, denoted by $\bar{x}_{T.p}$, is found by calculating $p\%$ of n, trimming that many observations off both ends of the ordered data, and then calculating the mean of the remaining observations. If $p\%$ of n is not a whole number, *round* to the nearest whole number.

Example 1.11   On January 28, 1986, the U.S. space shuttle *Challenger* exploded shortly after liftoff. The Presidential Commission that investigated the accident concluded that the disaster was caused by the failure of an O-ring, which resulted in a combustion gas leak through a field joint on the rocket booster. It was determined that the O-rings did not seal properly at low temperatures. Following are the recorded temperatures on the 24 launches of the shuttle *previous* to the accident:

WORKSHEET: Challeng.mtw

66, 70, 69, 80, 68, 67, 72, 73, 70, 57, 63, 70, 78, 67, 53, 67, 75, 70, 81, 76, 79, 75, 76, 58

Source: Dalal, S. R., Fowlkes, E. B., Hoadley, B. (1989), "Risk Analysis of the Space Shuttle: Pre-Challenger Prediction of Failure," *Journal of the American Statistical Association,* 84, No. 408, 945–957.

Figure 1.23 gives a stem-and-leaf plot and several descriptive statistics calculated for these data.

Figure 1.23
Challenger data
previous to accident

```
Stem-and-leaf of temp   N = 24
Leaf Unit = 1.0

 1        5 3
 3        5 78
 4        6 3
10        6 677789
(6)       7 000023
 8        7 556689
 2        8 01
```

Descriptive Statistics: temp

Variable	N	Mean	Median	TrMean	StDev	SE Mean
temp	24	70.00	70.00	70.27	7.22	1.47

Variable	Minimum	Maximum	Q1	Q3
temp	53.00	81.00	67.00	75.75

The first entry in the Descriptive Statistics display after the variable name is the number of observations, N. Then three statistics that measure the center of the data are given.

$$\text{The sample mean: } \bar{x} = 70$$

$$\text{The sample median: } M = 70$$

$$\text{A 5\% trimmed mean: } \bar{x}_{T.05} = 70.27$$

Just as the population mean balances the population distribution, the sample mean balances the observations in the sample. Thus, placing a fulcrum at the mean, 70, the *Challenger* data will balance.

Because $N = 24$, the median is located in position $(N + 1)/2 = 12.5$, and thus is the average of the 12th and 13th observations. Counting down the leaves of the ordered stem-and-leaf plot in Figure 1.23, we find that the 12th observation is 70 and the 13th observation is also 70, so the median is $M = (70 + 70)/2 = 70$.

The trimmed mean (TrMean) is like the mean, but it excludes the most extreme values in the data set. The highest and lowest 5% of the values (rounded to the nearest integer) are dropped, and the mean is calculated for the remaining values. Minitab automatically calculates a 5% trimmed mean; however, other trimming percents, such as 10% and 20%, are common. Five percent of the 24 observations is 1.2 (which rounds to 1), so the smallest and largest observations are deleted and the remaining 22 observations are averaged to get $\bar{x}_T = 70.27$.

The mean, median, and trimmed mean are all about the same because the data are reasonably symmetric. ∎

Example 1.12   The temperature at liftoff on the day of the *Challenger* accident was 31 degrees. It and the temperatures of previous launches are given in the stem-and-leaf plot in Figure 1.24. The descriptive statistics also include the 31-degree temperature.

Figure 1.24
Challenger data
including 31-degree
temperature

```
Stem-and-leaf of temp   N = 25
Leaf Unit = 1.0

    1     3  1
    1     3
    1     4
    1     4
    2     5  3
    4     5  78
    5     6  3
   11     6  677789
   (6)    7  000023
    8     7  556689
    2     8  01
```

Descriptive Statistics: temp

Variable	N	Mean	Median	TrMean	StDev	SE Mean
temp	25	68.44	70.00	69.52	10.53	2.11

Variable	Minimum	Maximum	Q1	Q3
temp	31.00	81.00	66.50	75.50

Including 31 degrees in the data set reduces the value of $\bar{x}$ from 70 to 68.44 degrees. Because 31 degrees is an outlier on the extreme left tail of the distribution of temperatures, the mean must shift left in order to maintain the balance.

Because $N = 25$, the median is located in position $(N + 1)/2 = 13$. Counting down to the 13th observation, we find that the median is still 70. Including the outlier has not affected the value of the median. Because extreme scores do not affect the value of the median, we say that it is ***resistant*** to the influence of outliers. ∎

A statistic is said to be **resistant** if extreme observations do not affect its numerical value.

The 5% trimmed mean reduces slightly to 69.52 when the 31 degrees is added to the data set. The reduction, however, is not due to the fact that 31 degrees is so small; it is trimmed off and will have no effect on the trimmed mean. The value of TrMean is smaller because the 53-degree observation that was previously trimmed is now included in the computation; we are averaging 23 (5% of 25 is 1.25, which rounds to 1) scores instead of 22. Had the outlier been 15 degrees instead of 31 degrees the trimmed mean, like the median, would be unchanged. It is also a resistant measure. A 10% trimmed mean, however, is more resistant because more data are trimmed; its value is 70.

Which is the best measure of the center of a distribution—the mean, the median, or a trimmed mean?

Unfortunately there is no clear-cut answer to this question; it all depends on the shape of the population distribution. Statistical theory has shown that if the population distribution shape is normal, $\bar{x}$ is the best estimate of the center of the distribution. In most cases, however, we can only speculate as to the shape of the population distribution. We must make a judgment based on sample data. Thus, if the shape of the data does not deviate substantially from the bell-shaped curve, we should use $\bar{x}$ to measure the center of the distribution.

But what about other shapes? Figure 1.25 shows a histogram of a sample from a symmetric long-tailed distribution with mean $\mu = 150$.

Figure 1.25
A long-tailed distribution

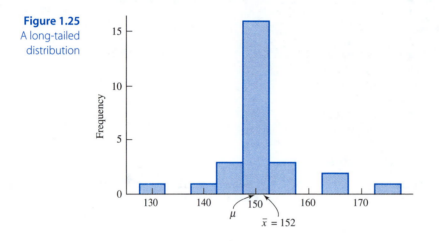

A distribution with long tails tends to produce extreme observations, as is indicated by the observations in the left and right tails of the histogram. Because of the greater number of outliers in the right tail, the value of $\bar{x}$ will be pulled toward them and thus it overestimates the center of the distribution. We see in Figure 1.25 that $\mu = 150$, but because of the extreme scores in the right tail, $\bar{x} = 152$ is not a good approximation of μ. Because a symmetric long-tailed distribution will produce outliers on *either* tail of the distribution, a second sample from this population may produce more outliers on the left tail; then $\bar{x}$ will underestimate μ. Because of this **instability** of $\bar{x}$, we prefer the median or a trimmed mean to the ordinary mean when sampling from a long-tailed distribution. In this example, the sample median is $M = 150$ and the 20% trimmed mean is $\bar{x}_{T.2} = 150.4$. Both provide a good estimate of the population mean, with the median being a perfect estimate.

What this indicates is that there are circumstances under which each statistic will provide a good estimate of the population center. In reality, however, we can narrow the selection down to two choices. We will use the sample mean $\bar{x}$ to estimate the population mean except when there are outliers or strong skewness. In those cases, we will use the sample median to estimate the population median.

Other Measures of Locations

In addition to measures of the center, there are other important measures of location that help describe the characteristics of a distribution. For example, the smallest and largest observations are important in describing the *spread* or *variability* of the data. Equally important are the *quartiles* that divide the data into quarters.

Quartiles Q_1 and Q_3

- The **first sample quartile**, Q_1, is the median of all observations less than or equal to the median. If you order the observations from smallest to largest, then Q_1 is located at position $(n + 1)/4$. If $(n + 1)/4$ is not an integer, the value of Q_1 is found by linear interpolation.
- The **third sample quartile**, Q_3, is the median of all observations greater than or equal to the median. If you order the observations, Q_3 is located at position $3(n + 1)/4$. If $3(n + 1)/4$ is not an integer, the value of Q_3 is found by linear interpolation.
- The **second quartile**, Q_2, is the median.

Note: Minitab's Descriptive Statistics command computes quartiles as we have defined them above. Other statistical software packages may use slightly different methods that will lead to slightly different results. In most cases, however, you need not worry about the difference.

The descriptive statistics display for the *Challenger* data, found in Example 1.11, is reproduced here:

Challenger data previous to accident

Descriptive Statistics: temp

Variable	N	Mean	Median	TrMean	StDev	SE Mean
temp	24	70.00	70.00	70.27	7.22	1.47

Variable	Minimum	Maximum	Q1	Q3
temp	53.00	81.00	67.00	75.75

The second line of the display identifies the smallest and largest observations as **Minimum** and **Maximum**, and then the values of Q_1 and Q_3 are given.

With $N = 24$ the position of Q_1 is $(N + 1)/4 = 6.25$. Thus, Q_1 is 1/4 of the way between the 6th and 7th observations, both of which are equal to 67; this gives

$$Q_1 = (3/4)67 + (1/4)\,67 = 67$$

(*Note:* Because the position 6.25 is closer to 6 than 7, the 6th observation is given three-fourth's weight and the 7th observation is given one-fourth's weight.)

The position of Q_3 is $3(N + 1)/4 = 18.75$. Thus, Q_3 is three-fourth's of the way between the 18th and 19th observations, which are 75 and 76. This gives

$$Q_3 = (1/4)75 + (3/4)76 = 75.75$$

Five-Number Summary Diagram

The quartiles (including the median), along with the smallest and largest scores, are often listed in a *five-number summary diagram*. The following diagram is for the *Challenger* data given in Example 1.11:

Min	Q_1	M	Q_3	Max
53	67	70	75.75	81

Measures of Variability

The information in a five-number summary diagram can be used to compute two important measures of variability.

Measures of Variability from a Five-Number Summary Diagram

- The **range** is the distance between the smallest and largest observations, Max–Min.
- The **interquartile range (IQR)** is the distance between the first and third sample quartiles, Q_3–Q_1.

Because it is determined from the extreme observations, the range is of little use in summarizing the variability of a long-tailed or skewed distribution. The interquartile range, on the other hand, is resistant to the influence of outliers and is the preferred measure of variability for skewed distributions.

Standard Deviation

The most widely accepted measure of variability, however, is the **standard deviation**. It is a sensitive measure of variability that looks at how much the observations deviate from their mean. To calculate the standard deviation, first find the *variance*.

Variance and Standard Deviation

- The **sample variance**, s^2, is the average squared distance of all sample values from the sample mean. It is calculated with the formula

$$s^2 = \frac{\Sigma(x_i - \bar{x})^2}{n - 1}$$

- The **standard deviation**, s, is the square root of variance. It is given by

$$s = \sqrt{\frac{\Sigma(x_i - \bar{x})^2}{n - 1}}$$

- The **population variance**, σ^2, is the average squared distance of all measurements from the population mean. The **population standard deviation** is σ.

SUM OF SQUARES The expression in the numerator of the variance is called a **sum of squares** and measures the *total deviation* of all the data. It is denoted as

$$SS = \Sigma(x_i - \bar{x})^2$$

The difference $(x_i - \bar{x})$ is called a *deviation from the mean*. It represents how far the observation x_i is from $\bar{x}$. For illustration purposes, consider the data set consisting of the following five observations:

$$31, 55, 75, 78, 81$$

The mean of the data is $\bar{x} = 64$. Figure 1.26 shows the deviation from the mean for each observation in the data set. Because $\bar{x}$ is the average of all the measurements and in the middle of the data, there are both positive and negative deviations from the mean. The two deviations below the mean are -33 and -9, and the three deviations above the mean are $+11$, $+14$, and $+17$. Because $\bar{x}$ is the balance point of the data, the negative deviations cancel the positive deviations, and consequently the sum of the deviations is zero ($-33 -9 +11 +14 +17 = 0$). The *squared* deviations (used in finding the sum of squares), however, do not sum to zero and thus give a meaningful measure of how variable the data are about the mean. In this example, the sum of squares is

$$SS = (-33)^2 + (-9)^2 + (+11)^2 + (+14)^2 + (+17)^2 = 1776$$

Figure 1.26

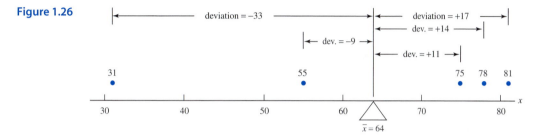

DEGREES OF FREEDOM We find the *average* of the sum of squares, the sample variance, by dividing by $n - 1$ rather than n for two reasons. First, the deviations always sum to zero, so any one of the deviations can be found from the other $n - 1$ deviations. Thus, the value of the sum of squares depends on only the $n - 1$ deviations that are *free* to change from one sample to the next. We say that the sum of squares has $n - 1$ d*egrees of freedom.* Second, we divide by $n - 1$ because a denominator of n would give a sample variance that tends to underestimate the population variance. Dividing by $n - 1$ makes the sample variance a little larger and a better estimate of the population variance.

Because the sample variance comes from a sum of squares, it is in units that are the square of the original units of measurement. That is, if we originally measured our data in miles, then s^2 is in miles squared. For this reason, we almost always report the standard deviation because it is in the original unit of measurement.

The calculations by hand are rather complicated; therefore, in practice we use either a calculator or computer to find the standard deviation. Repeated here are the descriptive statistics displays for the *Challenger* data. The standard deviation of the temperatures previous to the accident, given under the heading StDev, is $s = 7.22$. Including 31 degrees, the temperature on the day of the accident, increases the standard deviation to $s = 10.53$. The larger value of s in the second data set reflects greater variation of the observations from the mean. Notice, however, that the interquartile range is hardly affected by the inclusion of the 31 degrees. Before inclusion of 31 degrees, it is 8.75 and after it is 9.0.

Challenger data previous to accident

Descriptive Statistics: temp

Variable	N	Mean	Median	TrMean	StDev	SE Mean
temp	24	70.00	70.00	70.27	7.22	1.47

Variable	Minimum	Maximum	Q1	Q3
temp	53.00	81.00	67.00	75.75

Challenger data
previous to accident

Descriptive Statistics: temp

Variable	N	Mean	Median	TrMean	StDev	SE Mean
temp	25	68.44	70.00	69.52	10.53	2.11

Variable	Minimum	Maximum	Q1	Q3
temp	31.00	81.00	66.50	75.50

Interpreting the Standard Deviation

Comparing standard deviations from different data sets is important, but how do we interpret variability for a single set of data?

One way is to indicate the percent of the data that is within a specified number of standard deviations of the mean. For example, what percent of the distribution is within one standard deviation of the mean? What percent is within two standard deviations of the mean? The answer, of course, depends on the shape of the distribution. If the histogram is bell shaped, as in Figure 1.27, the *empirical rule* gives the percentages of the distribution that lie within one, two, and three standard deviations of the mean. Theoretically, this rule applies to the normal distribution, but it gives a reasonable approximation for any bell-shaped data.

Figure 1.27
Bell-shaped histogram

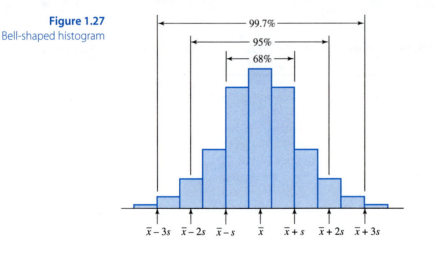

Empirical Rule Applied to Sample Data

If a stem-and-leaf plot, histogram, or similar descriptive tool has a bell-shaped appearance, then:

1. Approximately 68% of the measurements will fall within one standard deviation of the mean. The boundaries are $\bar{x} \pm s$.
2. Approximately 95% of the measurements will fall within two standard deviations of the mean. The boundaries are $\bar{x} \pm 2s$.
3. Essentially, all the measurements (99.7%) will fall within three standard deviations of the mean. The boundaries are $\bar{x} \pm 3s$.

Example 1.13 The recorded temperatures on the 24 launches previous to the *Challenger* accident are given here in a stem-and-leaf plot. Recall that $\bar{x} = 70$ and $s = 7.2$. Use the empirical rule to give an interpretation of the amount of variability in the data.

```
5 | 3
5 | 7 8
6 | 3
6 | 6 7 7 7 8 9
7 | 0 0 0 0 2 3
7 | 5 5 6 6 8 9
8 | 0 1
```

Solution   From the stem-and-leaf plot, it appears that the data are somewhat bell shaped, so it is reasonable to apply the empirical rule, which states that approximately 68% of the measurements should be between

$$\bar{x} - s \quad \text{and} \quad \bar{x} + s$$

which is between

$$70 - 7.2 = 62.8 \quad \text{and} \quad 70 + 7.2 = 77.2$$

We find that 17 of the 24 observations are between 62.8 and 77.2. The actual percent, $17/24 = 70.8\%$, is very close to the specified 68%.

The empirical rule further states that approximately 95% of the measurements should be between

$$\bar{x} - 2s \quad \text{and} \quad \bar{x} + 2s$$

which is between

$$70 - 2(7.2) = 55.6 \quad \text{and} \quad 70 + 2(7.2) = 84.4$$

All but one of the scores falls between 55.6 and 84.4. The actual percent, $23/24 = 95.8\%$, is only slightly greater than the 95% stated by the empirical rule.

Finally, all measurements are between

$$\bar{x} - 3s \quad \text{and} \quad \bar{x} + 3s$$

or between

$$70 - 3(7.2) = 48.4 \quad \text{and} \quad 70 + 3(7.2) = 91.6$$

Thus, the data suggest that the launch temperature is almost always somewhere between 48.4 degrees and 91.6 degrees. What does this say about the 31 degrees on the day of the *Challenger* launch?   ■

Relative Standing

In Example 1.13 we observe that 31 degrees on the day of the *Challenger* launch was an unusually low temperature. Just how low was it? When 31 degrees is included in the data set, the mean becomes $\bar{x} = 68.44$ and the standard deviation becomes $s = 10.53$. (Notice how much larger s is when the outlier is included.) To evaluate a single score, such as the 31 degrees, we calculate its *z-score* (sometimes called a *standard score*) which gives its *relative standing* with respect to the other data.

The **z-score** corresponding to a particular observation x gives the number of standard deviations it is from the mean

The **sample z-score** for x is

$$z = \frac{x - \bar{x}}{s}$$

The **population z-score** for x is

$$z = \frac{x - \mu}{\sigma}$$

The sample z-score for x gives its relative standing within the sample and the population z-score gives its relative standing within the population. Often the population mean μ and standard deviation σ are unknown in which case we can only calculate the sample z-score. We find the sample z-score for 31 degrees to be

$$z = \frac{31 - 68.44}{10.53} = -3.56$$

A negative z-score indicates that the observation is below the mean. Here we see that 31 is 3.56 standard deviations below the mean. Any observation with a z-score larger than 2 (in absolute value) is considered an unusual score. Thus, 31 degrees is an outlier representing an unusually low temperature.

Z-scores can also be used to compare the relative standings of observations from different data sets that have different means and standard deviations. For example, if you wish to compare your test score on a statistics exam with a friend's test score in another class, you should compare the two corresponding z-scores.

C O M P U T E R   T I P

Descriptive Statistics

Under **Stat**, select **Basic Statistics** followed by **Descriptive Statistics**. When the Descriptive Statistics window appears, **Select** the variable(s) and click on **OK**.

EXERCISES 1.4   The Basics

1.60   Based on these data, complete the following:

16   13   7   17   12   10   20   9   22

a.   Find the mean, median, and 5% trimmed mean.
b.   Add 10 to every observation and recompute the three statistics. Explain what happened.
c.   Is it unusual that the mean and 5% trimmed mean are exactly the same? Explain.

1.61 What will be the effect on the standard deviation in Exercise 1.60 if 10 is added to every observation? How will it affect the IQR (interquartile range)?

1.62 Calculate the mean and median of the following data. Next, multiply each observation by 10 and recompute both statistics. What was the effect on the two statistics? In general, scaling the observations up or down by multiplying by a constant has what effect on the mean and median?

```
1.3  6.8  7.8  5.7  9.2  8.4  7.4
```

1.63 What will be the effect on the standard deviation in Exercise 1.62 if each observation is multiplied by 10? How will it affect the IQR?

1.64 The average temperature of the 24 space shuttle launches previous to the *Challenger* accident was found to be 70 degrees in Example 1.11. The temperature at liftoff on the day of the *Challenger* accident was 31 degrees.
a. What effect does including 31 degrees in the data set have on $\bar{x}$?
b. What effect does including 31 degrees have on the median?
c. What effect does including 31 degrees have on the standard deviation?
d. What effect does including 31 degrees have on the IQR?

1.65 Following are the miles per gallon in city driving for minicompact cars in 2001:

WORKSHEET: Epaminicompact.mtw

Model	MPG	Model	MPG
Audi TTCoupe	22	MercedesClk430Cbrit	18
Audi TTCoupeQuattro	20	Mtsbishi EclpseSpdr	22
Audi TTCoupeQuattro	20	Mtsbishi EclpseSpdr	20
BMW 325CI Conv	19	Mtsbishi EclpseSpdr	20
BMW 325CI Conv	19	Mtsbishi EclpseSpdr	19
BMW 330CI Conv	20	Porsche 911 Carrera	17
BMW 330CI Conv	18	Porsche 911 Carrera	17
BMW M3 Conv	16	Porsche 911 Carrera	16
Jaguar XK8 Conv	17	Porsche 911 Carrera	16
Jaguar XKR Conv	16	Porsche 911 Turbo	15
MercedesClk320Cbrlt	20	Porsche 911 Turbo	15

Source: EPA data.

Find the mean and median miles per gallon. Is there much difference between the two? Would a trimmed mean be much different from either the mean or the median? Why, or why not?

Interpreting Computer Output

1.66 A test to measure aggressive tendencies was given to a group of teenage boys who were members of a street gang. The test is scored from 10 to 50, with a high score indicating more aggression. Shown are descriptive statistics and a histogram.

WORKSHEET: Aggress.mtw

```
38  27  44  39  41  26  35  45  39  28  16  37  11  36  33  46  42  37  40  19  24  29  32
34  31  30  32  43
```

Descriptive Statistics: aggres

Variable	N	Mean	Median	TrMean	StDev	SE Mean
aggres	28	33.36	34.50	33.73	8.73	1.65

Variable	Minimum	Maximum	Q1	Q3
aggres	11.00	46.00	28.25	39.75

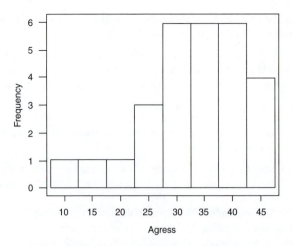

a. Explain why the median is larger than the mean.
b. What is the range? What is the IQR?
c. If 10 is subtracted from every observation, what will be the new mean, median, and trimmed mean?
d. If 10 is subtracted from every observation, what will be the new range, IQR, and standard deviation?
e. In general, what effect does adding to or subtracting a constant from every observation have on measures of center? What effect does it have on measures of variability?

1.67 Construct a five-number summary diagram for the data given in Exercise 1.66. Contrast it with a five-number diagram for the data after subtracting 10 from every observation. Explain the effect of subtracting a constant from each observation on the five-number diagram.

1.68 The Tennessee Self-Concept Scale (a test to measure one's self-confidence) was given to the group of teenage boys in Exercise 1.66. Shown are descriptive statistics and a histogram.

WORKSHEET: Concept.mtw

26 19 23 27 24 33 25 29 14 30 20 25 5 18
7 28 31 37 28 3 20 25 45 29 22 41 34 22

Descriptive Statistics: self

Variable	N	Mean	Median	TrMean	StDev	SE Mean
self	28	24.64	25.00	24.69	9.71	1.83

Variable	Minimum	Maximum	Q1	Q3
self	3.00	45.00	20.00	29.75

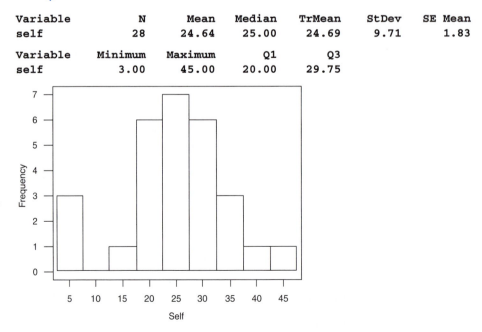

a. Explain why the mean, median, and trimmed mean are all about the same.
b. What is the range? What is the IQR?

c. If every observation is divided by 10 (multiply by 0.1), what will be the new mean, median, and trimmed mean?

d. If every observation is divided by 10, what will be the new range, IQR, and standard deviation?

e. In general, what effect does multiplying or dividing every observation by a constant have on measures of center? What effect does it have on measures of variability?

1.69 A standardized math test is given to 30 students in the tenth grade. Their scores with descriptive statistics and a stem-and-leaf plot are listed here:

WORKSHEET: Math.mtw

44 49 62 45 51 59 57 55 70 64 54 58 65 75 43 42 67 63 71 54 60 53 40
49 52 50 54 61 42 38

Descriptive Statistics: math

Variable	N	Mean	Median	TrMean	StDev	SE Mean
math	30	54.90	54.00	54.73	9.75	1.78

Variable	Minimum	Maximum	Q1	Q3		
math	38.00	75.00	48.00	62.25		

```
Stem-and-leaf of math  N = 30
Leaf Unit = 1.0

   1      3 8
   6      4 02234
   9      4 599
  (7)     5 0123444
  14      5 5789
  10      6 01234
   5      6 57
   3      7 01
   1      7 5
```

a. Do you think that the data are bell shaped? If so, interpret the amount of variability in the data using the empirical rule.

b. Charlie scored 62 on the test. What is his relative standing in the class?

1.70 Shown are descriptive statistics for the following test scores for two beginning statistics classes.

WORKSHEET: Statisti.mtw

Class 1	81	73	86	90	75	80	75	81	85	87	83	75	70
	65	80	76	64	74	86	80	83	67	82	78	76	83
	71	90	77	81	82								
Class 2	87	77	66	75	78	82	82	71	79	73	91	97	89
	92	75	89	75	95	84	75	82	74	77	87	69	96
	65												

Descriptive Statistics: Class1, Class2

Variable	N	Mean	Median	TrMean	StDev	SE Mean
Class1	31	78.58	80.00	78.78	6.73	1.21
Class2	27	80.81	79.00	80.80	9.06	1.74

Variable	Minimum	Maximum	Q1	Q3		
Class1	64.00	90.00	75.00	83.00		
Class2	65.00	97.00	75.00	89.00		

a. Based on the means, which class did better?

b. Based on the medians, which class did better?

c. Based on the standard deviations, which class had greater variability in its test scores?

d. Based on the IQRs, which class had greater variability in its test scores?

e. Do you think there is much difference in the performance of the students in the two classes? Explain.

1.71   In the two statistics classes described in Exercise 1.70, Mary is in class 1 and Alice is in class 2. Both scored 82 on their respective tests. Determine which student scored higher relative to her class.

Computer Exercises

1.72   A class of 30 fifth-graders completed a standardized reading test and received the following scores:

WORKSHEET: Reading.mtw

| 86 | 103 | 92 | 115 | 94 | 102 | 123 | 81 | 108 | 93 | 97 | 105 | 73 | 94 | 117 |
| 83 | 99 | 101 | 98 | 94 | 48 | 106 | 100 | 134 | 98 | 149 | 95 | 67 | 102 | 107 |

a. Construct a histogram and obtain descriptive statistics.

b. Compare the mean, median, and trimmed mean. Based on the histogram, can you explain the relationship among the three measures of the center of the distribution?

c. What is the standard deviation? Does it seem large? Explain.

1.73   The Insurance Institute of Highway Safety claims that car bumpers are getting worse. It crashed 23 midsize four-door sedans into barriers four times each at 5 miles per hour and totaled the damage. Construct a stem-and-leaf plot of the data. Calculate the average repair cost and the standard deviation and give an interpretation based on the empirical rule. What percent of the actual data fall within one standard deviation of the mean? Is this to be expected, according to the empirical rule?

WORKSHEET: Bumpers.mtw

Car	Repair cost	Car cost	Repair
Honda Accord	$ 618	Chevrolet Cavalier	$ 795
Toyota Camry	1304	Saturn SL2	1308
Mitsubishi Galant	1340	Dodge Monaco	1456
Plymouth Acclaim	1500	Chevrolet Corsica	1600
Pontiac Sunbird	1969	Oldsmobile Calais	1999
Dodge Dynasty	2008	Chevrolet Lumina	2129
Ford Tempo	2247	Nissan Stanza	2284
Pontiac Grand Am	2357	Buick Century	2381
Buick Skylark	2546	Ford Taurus	3002
Mazda 626	3096	Oldsmobile Ciera	3113
Pontiac 6000	3201	Subaru Legacy	3266
Hyundai Sonata	3298		

1.74   For the repair cost of the bumpers in Exercise 1.73, there is very little difference among the mean, median, and trimmed mean. Can you explain why? The Buick Skylark required $2546 to repair. How large is this relative to the rest of the data?

 1.75   The cholesterol levels of 62 subjects in the Framingham Heart Study mentioned in Exercise 1.56 of Section 1.3 are stored in worksheet: **Framingh.mtw**.

a. Construct a stem-and-leaf plot of the cholesterol levels.

b. Obtain descriptive statistics for the cholesterol levels.

c. What percent of the actual data falls within one standard deviation of the mean? What percent falls within two standard deviations of the mean? Are these percents consistent with the empirical rule? Why, or why not?

d. The largest cholesterol level is 393. How large is it relative to the rest of the data? Would you consider it an outlier?

1.76   The survival times in weeks for the 20 rats exposed to high levels of radiation discussed in Exercise 1.52 of Section 1.3 are shown here:

WORKSHEET: Rat.mtw

152   152   115   109   137   88   94   77   160   165   125   40   128   123   136   101   62   153   83   69

Source: J. Lawless, *Statistical Models and Methods for Lifetime Data* (New York: Wiley, 1982).

a. Obtain descriptive statistics for the survival times.
b. What percent of the actual data falls within one standard deviation of the mean? What percent falls within two standard deviations of the mean? Are these percents consistent with the empirical rule? Why, or why not?
c. The longest survival time was 165 weeks. Just how large is this score relative to the rest of the data? Would you consider it an outlier?

1.77   A treatment group and a control group of quail were fed a special diet. The treatment group received a drug compound that should reduce levels of low-density lipoproteins (see Exercise 1.53 in Section 1.3). From the following plasma LDL levels, compute summary statistics for each group.

WORKSHEET: Quail.mtw

placebo
64   49   54   64   97   66   76   44   71   89   70   72   71   55   60   62   46   77   86   71

treatment
40   31   50   48   152   44   74   38   81   64

Source: McKean, J., Vidmar, T. (1994), "A Comparison of Two Rank-Based Methods for the Analysis of Linear Models," *The American Statistician, 48,* 220–229.

a. How much difference do you find between the means of the two groups?
b. How much difference do you find between the medians of the two groups?
c. Explain why the means are close and the medians are not.
d. Compare the standard deviations of the two groups. Explain why the standard deviation of the treatment group is so much larger.
e. How large is 152 relative to its group?
f. Do you think we would be justified in removing 152 from its data set?
g. Without 152 is there a large difference between the two means?
h. Do you think the drug compound reduced the levels of low-density lipoproteins?

Internet Tasks

1.78   Go to the web site of the Environmental Protection Agency or another suitable web site and find the current EPA gas mileage ratings for compact cars. Store the data in a statistical worksheet and find the descriptive statistics. Is there much difference between the mean and the median? Which one do you think best represents the center of the distribution?

1.79   Go to the baseball section of the *Sporting News* web site and find career statistics for Mark McGwire and Barry Bonds (see Exercise 1.57). Find descriptive statistics for the number of home runs each player has hit in each season of his career. Which player has averaged more home runs in his career? Also compare their medians. Which player has greater variability in the number of home runs over his career?

1.80   From the web site of the National Association of Realtors or another suitable web site find the median cost of homes in the major cities in your state. Find the descriptive statistics and compare the mean and median cost. Which one do you think best represents the center of the distribution?

1.5 Describing a Distribution

Learning Objectives for this section:

❏ Be able to recognize symmetry and skewness in a distribution with a stem-and-leaf plot, histogram, or similar display.

❏ Be able to examine the length of the tails of the distribution and identify possible outliers.

❏ For symmetric distributions, determine whether the tails are longer or shorter than those of the normal distribution.

❏ Look for unusual characteristics that would adversely affect the summary statistics covered in the last section.

When describing a distribution, we should not be overly influenced by small deviations from established patterns, but rather look at the big picture. We should not insist on a perfect model but one that is closely approximated by the distribution. Describing the important characteristics of a distribution involves answering three basic questions.

Describing Important Characteristics of a Distribution

1. What is the general **shape** of the distribution?
2. Where is the **center** of the distribution?
3. How **spread** out is the distribution?

Shape

Describing the shape of a distribution is one of our most important tasks. To see the shape, look at a stem-and-leaf plot, a histogram, or a dotplot.

- Does the distribution have more than one mode?
- If it is unimodal, is it approximately symmetric or is it skewed?
- If the distribution is symmetric, are the tails longer or shorter than those of a normal distribution?
- Are there outliers in the data set?

Boxplot

Identifying outliers and checking symmetry can be accomplished by constructing a **boxplot**. The boxplot provides an alternate view of the information in a stem-and-leaf plot or histogram and describes the tails of the distribution by identifying outliers in the data. It is constructed based on information in the five-number summary diagram.

Construction of a Boxplot

1. Draw a number line that includes the range of observations.
2. Above the line, draw a box extending from Q_1 to Q_3.
3. Inside the box, draw a line at the median.
4. Outliers are identified as being observations that are more than 1.5(IQR) below Q_1 or above Q_3. That is, observations that are smaller than $Q_1 - 1.5$(IQR) or

---

greater than $Q_3 + 1.5(IQR)$ are classified as outliers and are identified with an asterisk (*).

5. If there are no outliers, extend horizontal line segments (whiskers) from the ends of the box to the smallest and largest observations. If there are outliers, extend the whiskers to the smallest and largest nonoutliers.

Example 1.14   The reading scores for a group of fifth-graders is given in the stem-and-leaf plot in Figure 1.28. From the provided descriptive statistics, complete a five-number summary diagram and identify any outliers. Using the five-number summary diagram and identified outliers, construct the boxplot.

Figure 1.28
```
Stem-and-leaf of reading   N = 30
Leaf Unit = 1.0
    1        4 8
    1        5
    2        6 7
    3        7 3
    6        8 136
   (10)      9 2344457889
   14       10 012235678
    5       11 57
    3       12 3
    2       13 4
    1       14 9
```

Descriptive Statistics: reading

Variable	N	Mean	Median	TrMean	StDev	SE Mean
reading	30	98.80	98.50	98.69	18.92	3.45

Variable	Minimum	Maximum	Q1	Q3
reading	48.00	149.00	92.75	106.25

Solution   From the descriptive statistics we get

Five-number summary diagram

Min	Q_1	M	Q_3	Max
48	92.75	98.5	106.25	149

From the five-number diagram, $IQR = 106.25 - 92.75 = 13.5$.
   To identify outliers, calculate

$$Q_1 - 1.5(IQR) = 92.75 - 1.5(13.5) = 72.50$$

and

$$Q_3 + 1.5(IQR) = 106.25 + 1.5(13.5) = 126.50$$

From the stem-and-leaf plot, two observations, 48 and 67, are smaller than 72.5; hence, they are outliers on the lower tail. Two observations, 134 and 149, are larger than 126.5; therefore, they are outliers on the upper tail. The completed boxplot is given in Figure 1.29. The outliers and the symmetric appearance of the boxplot are indicative of a long-tailed distribution. ∎

Figure 1.29

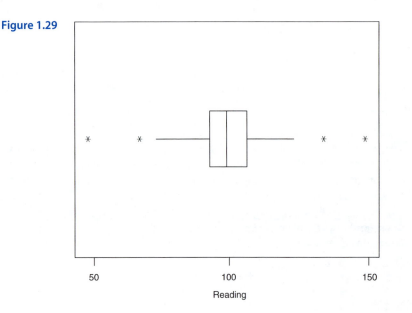

Interpretation of the Boxplot

To examine the length of the tails and classify the shape of a frequency distribution, you must fully understand the information in a boxplot. The box portion of the boxplot contains the middle 50% of the data, with the median dividing the scores in half. The position of the median line gives some indication about the shape of the middle of the distribution. If the median line is close to the center of the box, the middle 50% of the distribution is symmetric. If the median line is toward one end of the box, the middle 50% of the distribution is skewed in the opposite direction.

The lengths of the whiskers (tails) of the boxplot describe the spread and give an indication about symmetry or skewness of the data. If both whiskers are about the same length, the distribution without outliers is symmetric. If one whisker is longer than the other, the distribution is skewed in that direction. Outliers beyond a long whisker are even stronger evidence that the distribution is skewed in that direction.

Center and Spread

After assessing shape, we next give measures of the center and spread of a distribution. As pointed out in Section 1.4, how we measure the center depends on the shape of the distribution. Outliers in one tail of a distribution tend to have an adverse effect on $\bar{x}$ (its value is pulled toward the long tail) and the standard deviation is extremely sensitive to outliers. The median, on the other hand, is resistant to outliers and provides a good estimate of the center of a skewed distribution. If, however, the distribution is reasonably symmetric without outliers, we should use $\bar{x}$ to measure the center and s to measure the spread of the distribution. If the distribution is symmetric with outliers (a long-tailed distribution), both the median and the trimmed means provide good estimates of the center of the distribution. For simplicity, however, we will use the sample median to measure the center when there is severe skewness or an excessive number of outliers. The interquartile range is also resistant to outliers; thus, it should be used in conjunction with the median.

> ## Measuring Center and Spread
>
> - For reasonably symmetric distributions without excessively long tails use the mean, $\bar{x}$, to measure the center and the standard deviation, s, to measure the spread.
> - For skewed distributions or distributions with long tails, use the median, M, to measure the center and the interquartile range, IQR, to measure the spread.

With a boxplot, histogram, and descriptive statistics you should be able to describe the important characteristics of a distribution.

Example 1.15   **R. A. Fisher Data on *Iris Setosa***

A historic data set studied by R. A. Fisher (one of the founding fathers of modern-day statistics) is the measurements of four flower parts on 50 specimens of each of three species of irises. Here are the sepal lengths (in centimeters) of the species *Iris Setosa*:

WORKSHEET: Irises.mtw Column 1

5.1, 4.9, 4.7, 4.6, 5.0, 5.4, 4.6, 5.0, 4.4, 4.9, 5.4, 4.8, 4.8, 4.3, 5.8, 5.7, 5.4, 5.1, 5.7, 5.1, 5.4, 5.1, 4.6, 5.1, 4.8, 5.0, 5.0, 5.2, 5.2, 4.7, 4.8, 5.4, 5.2, 5.5, 4.9, 5.0, 5.5, 4.9, 4.4, 5.1, 5.0, 4.5, 4.4, 5.0, 5.1, 4.8, 5.1, 4.6, 5.3, 5.0

Source: R. A. Fisher (1936), "The Use of Multiple Measurements in Taxonomic Problems," *Annals of Eugenics*, 7, 179–184.

Analyze these data and comment on the characteristics of the distribution.

Solution   The analysis is presented in Figure 1.30. The mean, median, and trimmed mean are very close in value, which suggests symmetry centered at 5 centimeters. The histogram appears symmetric and resembles the shape of a normal distribution. The boxplot also appears symmetric, with one outlier. A sample of size 50 from a normal distribution may very well have one outlier. The exploratory evidence leads us to classify the distribution of sepal lengths of *Iris Setosa* as being close to normally distributed, centered at 5 centimeters with a standard deviation of .3525 centimeters. In Chapter 3, after a more thorough study of the normal distribution, we will have a specific method for checking normality with a normal probability plot.   ■

Figure 1.30   **Descriptive Statistics**

Variable	N	Mean	Median	TrMean	StDev	SEMean
sepalL1	50	5.0060	5.0000	5.0000	0.3525	0.0498

Variable	Min	Max	Q1	Q3
sepalL1	4.3000	5.8000	4.8000	5.2000

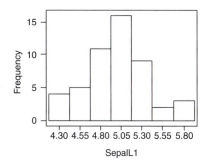

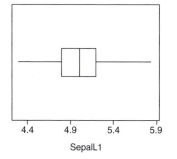

Side-by-Side Boxplots

The boxplot gives a general view of the shape of a distribution without the details of a stem-and-leaf plot or a histogram. It is very useful for comparing two or more distributions. Example 1.16 shows how *side-by-side boxplots* can be used to compare distributions.

Example 1.16   Radiocarbon dating is a method of determining the age of archaeological sites. The data in worksheet: **Archaeo.mtw**, are the "ages" recorded as years before 1983 (the data were obtained in 1983)—that is, B.C. + 1983 for samples taken from one site at the archaeological excavation of the Danebury Iron Age hill fort. Each of the 60 observations is associated with a pottery shard or fragment that has been classified into one of four phases, referred to as Ceramic Phases 1–4. Compare the four phases with side-by-side boxplots. Do the data suggest that the phases correspond to four nonoverlapping periods of time?

Source: Cunliffe, B. (1984), and Naylor and Smith (1988).

Solution   To generate side-by-side boxplots, it is best to use the stacked form of the data; that is, one column of the worksheet contains the data and a second column contains a code indicating the ceramic phase. In the Boxplot window of Minitab, select the column containing the data (C5) in the Y (measurement) blank and select the column containing the phase (C6) in the X (category) blank. Click OK to produce the accompanying graph.

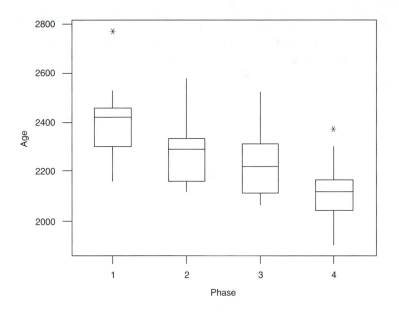

The middle 50% of distributions 1 and 2 are skewed left, but the lengths of the right tails tend to balance that out so that all four distributions are reasonably symmetric. The ages associated with the four phases generally decrease as we go from phase 1 to phase 4. There is a great deal of overlap between phases 2 and 3. This is contrary to the conjecture that the phases correspond to four nonoverlapping periods of time. ∎

Drawing conclusions about multimodal distributions can be misleading because the modes are often the result of some extraneous variable being confounded with the data. If the extraneous variable can be identified, the distribution should be separated into two or more distributions and then analyzed separately.

Example 1.17   Researchers have investigated lead absorption in children of parents who worked in a factory where lead was used to make batteries. Following are the levels of lead in the children's blood (in μg/dl of whole blood):

WORKSHEET: Lead.mtw column 2

38 23 41 18 37 36 23 62 31 34 24 14 21 17 16 20 15 10 45 39 22 35 49
48 44 35 43 39 34 13 73 25 27

Source: Morton, D. et al. (1982), "Lead Absorption in Children of Employees in a Lead-Related Industry," *American Journal of Epidemiology, 155,* 549–555.

Analyze the boxplot and stem-and-leaf plot.

```
Stem-and-leaf of exposed   N = 33
Leaf Unit = 1.0

    3       1 034
    7       1 5678
   13       2 012334
   15       2 57
  (3)       3 144
   15       3 5567899
    8       4 134
    5       4 589
    2       5
    2       5
    2       6 2
    1       6
    1       7 3
```

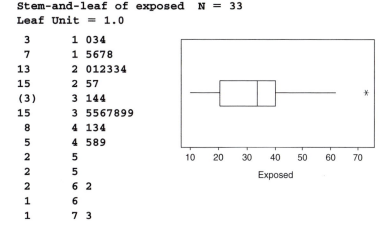

The boxplot indicates that the middle 50% of the distribution is skewed left; but, the lengths of the tails indicate that the overall distribution is skewed right. This suggests that we should take a closer look at the stem-and-leaf plot.

The stem-and-leaf plot appears to have two prominent peaks: one centered at about 20 and the other centered at about 35. The two extreme values, 62 and 73, also suggest an additional mode at 67 or 68. It is interesting that 62 was *not* classified as an outlier. If we feel rather confident that there are indeed two or even three modes in these data, our next job is to determine what caused the different modes. Can we find any characteristic in common for the children with "low" levels of lead and for those children with "high" levels of lead? Insufficient information is provided here to answer this question, but the experimenter conducting the study should explore further by asking questions:

Did the parents of the children with low levels (or high levels) work in the same area of the factory?

What were the parents' specific jobs? Were some parents exposed for greater lengths of time?

What kind of living conditions did the children have outside the factory?

By further analyzing data we hope to identify the extraneous factors that contribute to the different modes.   ■

Important Point: A boxplot will *not* identify multimodality.

As we examine data, one of the main features of our strategy should be flexibility. No two data sets are exactly alike; therefore, there is no standard procedure that works best all the time. We must be able to adapt the analysis to the problem. As we examine the data for structure, we must be able to respond to different patterns that are suggested by our analysis. We must approach the analysis with the goal of answering our questions, but at the same time remain open to the possibility of unexpected features that will prompt new questions.

Example 1.18 Simpson's Paradox

The registrar at their university told the athletic director that in sports where both men and women compete, men maintain a higher overall grade point average than do women. Having some doubts about this conclusion, the athletic director randomly selected 50 male and 50 female student athletes who participate in basketball, soccer, or track. The following data represents the grade point average, the sport they participate in (1 = basketball, 2 = track, and 3 = soccer), and the gender (1 = male, 2 = female) of the students.

WORKSHEET: Simpson.mtw

male

2.78	3	2.80	3	2.39	2	2.71	3	2.46	2
2.64	3	2.74	3	2.29	2	3.10	3	1.95	1
2.77	3	2.75	3	1.96	1	2.57	2	3.02	3
2.86	3	2.92	3	2.00	1	1.99	1	2.32	2
2.84	3	1.83	1	2.90	3	2.41	2	2.85	3
2.43	2	1.93	1	2.29	2	2.76	3	2.93	3
2.32	2	2.76	3	2.39	2	1.99	1	2.24	2
2.03	1	1.88	1	2.10	1	2.70	3	2.69	3
2.74	3	2.67	3	3.02	3	2.30	2	2.34	2
2.67	3	2.82	3	2.42	2	2.63	3	2.43	2

female

2.08	1	2.56	2	2.37	2	2.46	2	3.12	3
2.30	2	2.47	2	3.05	3	2.50	2	2.63	2
3.01	3	2.10	1	2.57	2	2.47	2	2.38	2
2.12	1	2.46	2	2.37	1	2.21	1	3.06	3
2.97	3	2.87	3	2.11	1	2.16	1	2.35	1
2.53	2	3.14	3	2.98	3	2.55	2	2.19	1
2.49	2	2.29	1	2.47	2	2.98	3	2.50	1
2.61	2	2.25	1	2.38	2	2.22	1	2.52	2
2.42	2	2.25	1	3.07	3	2.32	1	2.19	1
2.29	1	3.05	3	3.00	3	3.10	3	2.43	2

Analyze these data by comparing the grade point averages of the male and female student athletes.

Solution Figure 1.31 shows stem-and-leaf plots of the two genders along with side-by-side boxplots.

Figure 1.31

```
Stem-and-leaf of gpa   gender = 1   N = 50
Leaf Unit = 0.010

   2      18 38
   7      19 35699
   9      20 03
  10      21 0
  13      22 499
  19      23 022499
  24      24 12336
  25      25 7
  25      26 34779
  20      27 014456678
  11      28 02456
   6      29 023
   3      30 22
   1      31 0
```

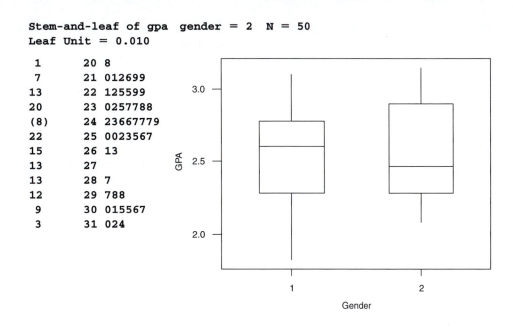

```
Stem-and-leaf of gpa   gender = 2   N = 50
Leaf Unit = 0.010

  1      20  8
  7      21  012699
 13      22  125599
 20      23  0257788
 (8)     24  23667779
 22      25  0023567
 15      26  13
 13      27
 13      28  7
 12      29  788
  9      30  015567
  3      31  024
```

The stem-and-leaf plots of the grade point averages of the two groups of athletes do not reveal any big differences between the men and women. But the boxplots do show that the median grade point average for men is higher than the median grade point average for women. On the surface it appears that the registrar is correct.

Taking a closer look, however, we see that the stem-and-leaf plot for men appears to be trimodal, and the stem-and-leaf plot for the women appears to be bimodal, which indicates an extraneous variable confounded with the data. Investigating further, we decide to look at the grade point averages of the athletes in the different sports.

Figure 1.32 gives side-by-side boxplots of the grade point averages for men and women for the three different sports.

Figure 1.32

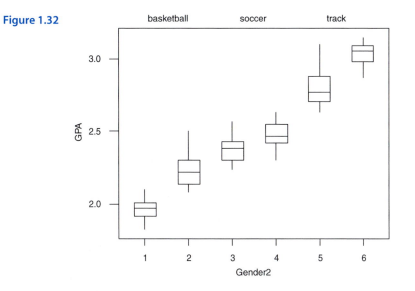

Gender 2 code 1, 3, and 5 are male athletes, and 2, 4, and 6 are female athletes.

We observe from the boxplots that in each sport the female athletes have higher grade point averages than the male athletes. In fact, in basketball women's grade point averages completely surpass those of the men. When all the athletes from the different sports are combined, as shown in Figure 1.31, the median male grade point average exceeds the median female grade point average. When we look at each individual sport, however, just the opposite happens. This phenomenon is know as *Simpson's Paradox*. Can you explain how this can happen? ■

You have just seen that the boxplot and other descriptive tools provide some guidelines for determining the general characteristics of a distribution. Once the shape is identified, you locate the center of the distribution with the mean or median and get a numerical measure of the spread of the distribution with the standard deviation or the IQR. The numerical quantities, or *summary statistics* as they are frequently called, characterize the distribution by distinguishing it from other distributions of the same shape.

C O M P U T E R   T I P

Boxplots

First select **Graph** and then click on **Boxplot**. When the Boxplot window appears, **Select** the variable to be graphed and then click **OK**.

For side-by-side boxplots, store the data in one column of the worksheet and store a code for the different boxplots in another column. **Select** the column containing the data for the *Y* variable and the column containing the code for the *X* variable and then click on **OK**.

EXERCISES 1.5   The Basics

1.81   The descriptive statistics for the aggressive tendency scores for the group of teenage boys in Exercise 1.66 in Section 1.4 are repeated here.

W O R K S H E E T: Aggress.mtw

Descriptive Statistics: aggres

Variable	N	Mean	Median	TrMean	StDev	SE Mean
aggres	28	33.36	34.50	33.73	8.73	1.65

Variable	Minimum	Maximum	Q1	Q3		
aggres	11.00	46.00	28.25	39.75		

a. From the descriptive statistics, construct a five-number summary diagram.
b. Calculate 1.5(IQR) and subtract it from Q_1 and add it to Q_3 to determine whether there are any outliers on either end of the distribution.
c. Sketch a boxplot of the scores and comment on the shape of the distribution.
d. Based on the shape, give measures of the center and spread of the distribution.

1.82   The descriptive statistics for the Tennessee Self-Concept Scale scores for the group of teenagers in Exercise 1.81 that were previously analyzed in Exercise 1.68 in Section 1.4 are repeated here.

WORKSHEET: Concept.mtw

26 19 23 27 24 33 25 29 14 30 20 25   5 18
 7 28 31 37 28   3 20 25 45 29 22 41 34 22

Descriptive Statistics: self

Variable	N	Mean	Median	TrMean	StDev	SE Mean
self	28	24.64	25.00	4.69	9.71	1.83

Variable	Minimum	Maximum	Q1	Q3
self	3.00	45.00	20.00	29.75

a. Construct a stem-and-leaf plot of the data.
b. From the descriptive statistics, construct a five-number summary diagram.
c. Calculate 1.5(IQR) and subtract it from Q_1 and add it to Q_3 to determine whether there are any outliers on either end of the distribution.
d. Sketch a boxplot of the self-concept scores and comment on the shape of the distribution.
e. Based on the shape, give measures of the center and spread of the distribution.

1.83   The reaction times of 31 senior citizens (over 65 years of age) applying for driver's license renewals were measured, with the following results:

WORKSHEET: Senior.mtw

93 105 66 94 64 98 109 71 86 31 101 128 97 85 96 60 94 42 64 99
84 107 77 55 98 80 90 79 90 96 110

a. Construct a stem and leaf plot of the reaction times.
b. Construct a five-number summary diagram.
c. Calculate 1.5(IQR) and subtract it from Q_1 and add it to Q_3 to determine whether there are any outliers on either end of the distribution.
d. Sketch a boxplot of the scores and comment on the shape of the distribution.

1.84   Following is a list of murder rates per 100,000 for 31 cities selected from the South:

WORKSHEET: South.mtw

12, 10, 10, 13, 12, 12, 14, 7, 16, 18, 8, 29, 12, 14, 33,
10, 6, 18, 11, 25, 8, 16, 14, 11, 10, 20, 14, 11, 12, 13, 13

a. Construct a stem-and-leaf plot of the rates.
b. Construct a five-number summary diagram.
c. Calculate 1.5(IQR) and subtract it from Q_1 and add it to Q_3 to determine whether there are any outliers on either end of the distribution.
d. Sketch a boxplot of the scores and comment on the shape of the distribution.

1.85   Exercise 1.73 in Section 1.4 investigated the repair cost for midsize four-door sedans that were crashed into barriers at 5 miles per hour. Following is a five-number summary diagram for repair costs.

Min	Q_1	M	Q_3	Max
618	1456	2129	3002	3298

a. Calculate 1.5(IQR) and determine whether there are any outliers on either end of the distribution.
b. Sketch a boxplot of the repair costs and comment on the shape of the distribution.

Interpreting Computer Output

1.86 Following is a dotplot of the repair costs for the sedans in Exercise 1.85. Based on the dotplot, would you classify the shape of the distribution as multimodal? Did the boxplot you constructed in Exercise 1.85 reveal the different modes? Will a boxplot ever detect multimodality?

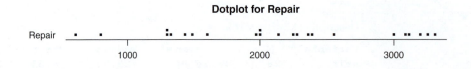

1.87 Shown here are side-by-side boxplots of the test scores for students in the two separate statistics classes described in Exercise 1.70 in Section 1.4 (Worksheet: **Statisti.mtw**). Compare the two distributions. Is one more variable than the other? Which class tends to have the higher scores? Which class has the higher median?

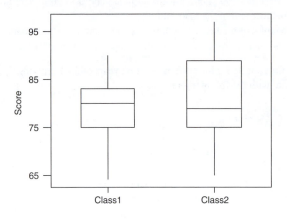

1.88 Shown here are side-by-side boxplots of the plasma LDL levels for a treatment group and a control group of quail that were fed a special diet. The data were given in Exercise 1.53 in Section 1.3 and are in worksheet: **Quail.mtw** (also see Exercise 1.77 in Section 1.4). The treatment group received a drug compound that should reduce low-density lipoproteins. Based on the boxplots, do you think there is evidence that the drug compound reduced LDL levels? What is the effect of the outlier in the treatment group on the mean? In these data, do you think we should compare the means or the medians?

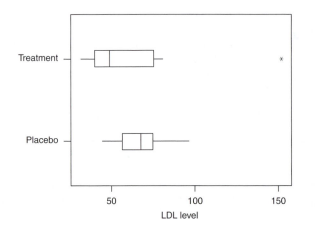

1.89 A consumer group wishes to compare four brands of bathroom scales. A standard weight of 100 pounds is placed on each of the scales five different times. The resulting readings are found in worksheet **Scales.mtw**. Compare the four brands based on the accompanying side-by-side boxplots. Which would you choose if you were purchasing a bathroom scale?

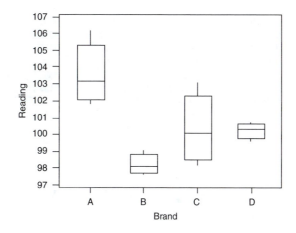

Computer Exercises

1.90 Violent crimes are offenses of murder, forcible rape, robbery, and aggravated assault. Following are the 1993 rates per 100,000 inhabitants for each of the 50 states and the District of Columbia:

WORKSHEET: Crime.mtw

Ala	871.7	Ky	535.5	ND	83.3
Alaska	660.5	La	984.6	Ohio	525.9
Ariz	670.8	Maine	130.9	Okla	622.8
Ark	576.5	Md	1000.1	Ore	510.2
Calif	1119.7	Mass	779.0	Pa	427.0
Colo	578.8	Mich	770.1	RI	394.5
Conn	495.3	Minn	338.0	SC	944.5
Del	621.2	Miss	411.7	SD	194.5
D.C.	2832.8	Mo	740.4	Tenn	746.2
Fla	1207.2	Mont	169.9	Texas	806.3
Ga	733.2	Neb	348.6	Utah	290.5
Hawaii	258.4	Nev	696.8	Vt	109.5
Idaho	281.4	NH	125.7	Va	374.9
Ill	977.3	NJ	625.8	Wash	534.5
Ind	508.3	NM	934.9	W.Va	211.5
Iowa	278.0	NY	1122.1	Wis	275.7
Kan	510.8	NC	681.0	Wyo	319.5

Source: U.S. Department of Justice, Bureau of Justice Statistics, *Sourcebook of Criminal Justice Statistics*, 1993.

a. Construct a stem-and-leaf plot. Is there anything unusual about it?

b. Construct a boxplot. Are there any outliers?

c. Obtain the descriptive statistics. Compare the mean and the median. Are they similar, or is one considerably larger than the other? Can you explain why?

d. The range is much larger than the IQR. What specifically caused this to happen?

e. Based on the above, summarize the shape of the distribution.

f. Which do you think best measures the center of the distribution, the mean or the median? How should we measure the spread of the distribution?

1.91   AFDC (Aid to Families with Dependent Children) is a program funded by federal and state governments to provide assistance to needy families. Following are the average monthly payments per person for families in each of the 50 states and the District of Columbia:

WORKSHEET: Aid.mtw

Ala	$ 57.16	Ky	78.04	N.D.	126.34
Alaska	253.54	La	56.29	Ohio	113.72
Ariz	114.23	Maine	144.74	Okla	104.17
Ark	68.22	Md	119.26	Ore	143.45
Calif	199.57	Mass	192.13	Pa	125.61
Colo	110.86	Mich	144.37	R.I.	182.19
Conn	199.30	Minn	167.91	S.C.	67.05
Del	119.63	Miss	42.29	S.D.	103.79
D.C.	138.33	Mo	91.20	Tenn	59.12
Fla	99.63	Mont	113.24	Texas	56.91
Ga	90.78	Neb	113.75	Utah	123.48
Hawaii	213.96	Nev	104.13	Vt	192.31
Idaho	111.68	N.H.	158.41	Va	99.37
Ill	108.82	N.J.	127.27	Wash	174.26
Ind	86.83	N.M.	103.54	W.Va	85.56
Iowa	134.72	N.Y.	197.61	Wis	155.37
Kan	119.10	N.C.	88.86	Wyo	120.93

Source: U.S. Department of Health and Human Services, 1993.

a. Construct a histogram of the data. Do you observe anything unusual about the histogram?
b. Obtain descriptive statistics. What percent of the actual data falls within one standard deviation of the mean? What percent falls within two standard deviations of the mean? Are these percents consistent with the empirical rule?
c. Construct a boxplot. Are there any outliers?
d. Based on the results obtained above, comment on the shape of the distribution and give measures of the center and spread.

1.92   Following are the percents of the population of each state over the age of 65 in 1998:

WORKSHEET: Elderly.mtw

Ala	13.1	Ky	12.5	N.D.	14.4
Alaska	5.5	La	11.5	Ohio	13.4
Ariz	13.2	Maine	14.1	Okla	13.4
Ark	14.3	Md	11.5	Ore	13.2
Calif	11.1	Mass	14.0	Pa	15.9
Colo	10.1	Mich	12.5	R.I.	15.6
Conn	14.3	Minn	12.3	S.C.	12.2
Del	13.0	Miss	12.2	S.D.	14.3
D.C.	13.9	Mo	13.7	Tenn	12.5
Fla	18.3	Mont	13.3	Texas	10.1
Ga	9.9	Neb	13.8	Utah	8.8
Hawaii	13.3	Nev	11.5	Vt	12.3
Idaho	11.3	N.H.	12.0	Va	11.3
Ill	12.4	N.J.	13.6	Wash	11.5
Ind	12.5	N.M.	11.4	W.Va	15.2
Iowa	15.1	N.Y.	13.3	Wis	13.2
Kan	13.5	N.C.	12.5	Wyo	11.5

Source: U.S. Census Bureau Internet site, February 2000.

a. Construct a stem-and-leaf plot of the percents of residents in the states over 65 years of age. Are there any unusually large or small observations?

    b. Construct a dotplot of the data. Without the largest and smallest observations, what does the shape look like? Can you explain the largest and smallest observations?

    c. Calculate the descriptive statistics. Is there much difference among the mean, median, and trimmed mean? Which do you think we should use to represent the center of the data?

Internet Tasks

1.93 Find the crime index for each of the 50 states and the District of Columbia at the web site of the Bureau of Justice Statistics. Enter the indexes in a statistical worksheet and construct a stem-and-leaf plot and a boxplot. Are there outliers or other unusual behavior in the data? Obtain the descriptive statistics and compare the mean and median. Are they close in value, or is one considerably larger than the other? Is the range much larger than the IQR? Based on this information, summarize the shape of the distribution. Which do you think best measures the center of the distribution, the mean or the median? How should we measure the spread of the distribution?

1.94 Go to the Census Bureau web site and determine the current percents of the population of each state over the age of 65. Enter the percents in a statistical worksheet and construct a histogram and a boxplot. Are there outliers or other unusual behavior in the data? Obtain the descriptive statistics and compare the mean and median. Are they close in value, or is one considerably larger than the other? Is the range much larger than the IQR? Based on this information, summarize the shape of the distribution. Which do you think best measures the center of the distribution, the mean or the median? How should we measure the spread of the distribution?

1.95 Using your favorite search engine, enter the keywords "Space Shuttle *Challenger*." Retrieve information and write a one-page report on the *Challenger* accident. Include as much statistical information about the accident as you can find.

1.6   Summary and Review Exercises

Key Concepts

■ *Statistics* describes methods of organizing, interpreting, and presenting numerical information. In this chapter we have looked at *descriptive methods* of organizing and summarizing the collected data. In later chapters we will look at *inferential statistics*, where conclusions are drawn about the population based on the summarized data.

■ A *population* is the collection of all objects of interest to the statistician. A *sample* is a finite subset of the population. A *census* is a sample consisting of the entire population.

■ A *variable* is a characteristic associated with each unit in a population. An *observation* is a single value assumed by a variable.

■ A data set is a listing of the observed values of one or more variables. A data set with one variable listed is called *univariate*. When two variables are measured on each subject, the data set is called *bivariate*. When several variables are measured on each subject, the data set is called *multivariate*.

■ Variables are classified as either *numerical* or *categorical*. Numerical variables are classified as *discrete* or *continuous*.

■ Data recorded over time are called *time series* data.

■ Categorical data are summarized in a *frequency table, bar graph*, or *pie chart*.

■ Numerical data can be displayed in a *dotplot, stem-and-leaf plot, histogram*, or *boxplot*.

■ Populations and samples are characterized by summary measures. A *parameter* is a summary measure associated with a population. A *statistic* is a summary measure associated with a sample.

■ Summary measures are often classified as *measures of location* and *measures of variability*.

- Among the measures of location are those that measure the center of the distribution, including the *mean*, the *trimmed mean*, and the *median*. Other measures of location are *quartiles* and the *minimum* and *maximum*.

- Measures of variability (spread) include the *range*, the *interquartile range (IQR)*, and the *standard deviation*. Standard deviation can be interpreted with the *empirical rule*.

- The *boxplot* is a graph used to display the data and detect *outliers*.

- Shapes that a distribution can assume include *symmetric*, *skewed*, *multimodal*, *short-tailed*, and *long-tailed*.

To study the distribution of the thickness of the 1872 Hidalgo stamps in the van Winkler collection, we first construct a dotplot of the thickness.

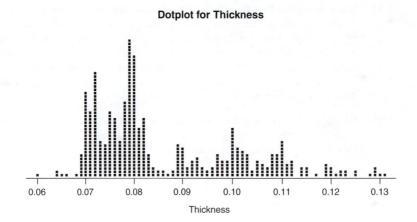

Dotplot for Thickness

From the dotplot we notice seven or more different modes, so we classify the distribution as multimodal. This suggests that the stamps were printed using a variety of thicknesses of paper. The majority of the stamps have a thickness between .07 and .08 millimeters, but we have clusters of thickness at .09, .10, .11, .12, and .13.

From the descriptive statistics we find that the median thickness is .08 millimeters. The mean thickness (.08602 millimeters) is somewhat larger because of the right skewness. The first quartile is $Q_1 = .075$, so 25% of the stamps have thickness less than or equal to .075 millimeters. The third quartile is $Q_3 = .098$, so 25% of the stamps have thickness greater than .098 millimeters.

Descriptive Statistics: thickness

Variable	N	Mean	Median	TrMean	StDev	SE Mean
thicknes	485	0.08602	0.08000	0.08500	0.01496	0.00068

Variable	Minimum	Maximum	Q1	Q3
thicknes	0.06000	0.13100	0.07500	0.09800

How rare is a stamp whose thickness is, say .121 millimeters? By finding the *z*-score we have that it is

$$z = \frac{.121 - .08602}{.01496} = 2.34$$

standard deviations above the average, which is rare. In fact, only 12 of the 485 stamps, or 2.5%, have thicknesses greater than this one.

Questions For Review

Use the following problems to test your skills.

The Basics

1.96 Earlybird Airline is interested in the possibility of opening a new route between Charlotte and Dallas. A survey concerning the issue was sent to 2000 past customers.
a. Describe the population of interest.
b. What is the sample?
c. Give two variables of interest.

1.97 A stock market investor is interested in oil stocks. She collects last year's price/earnings ratio on ten selected oil stocks.
a. Describe the population of interest.
b. What is the sample?
c. Give two variables of interest.

1.98 A university is interested in the success its placement office has in placing graduates in their chosen fields. A survey concerning the issue was sent to 500 past graduates.
a. Describe the population of interest.
b. What is the sample?
c. Give three variables of interest.

1.99 Identify the following variables as either categorical or numerical. If numerical, further identify them as continuous or discrete.
a. Amount of tar in a cigarette
b. SAT score for entering freshmen
c. Brand of breakfast cereal
d. Number of gold medals won by the United States
e. Ratings of one's favorite TV shows
f. Number of defective computer parts

1.100 A survey by the American Council of Life Insurance of 516 adults on the issue of birth control showed the following results:

Birth Control	Percent
Completely for it	50
Somewhat for it	28
Somewhat against it	10
Completely against it	11
Undecided	1

Illustrate these data in a bar graph.

1.101 The money that Americans spend on alcoholic beverages is distributed as follows:

Beverage	Percent
Beer	52
Wine	23
Liquor	25

Illustrate these data in a pie chart.

1.102 Classify the following variables as categorical or numerical. If numerical, further classify them as discrete or continuous.
a. Number of vehicles owned by a family
b. Homicide rate in a major city
c. Socioeconomic status
d. Concentration of PCB in a chemical spill
e. Duration of a kidney transplant operation

1.103 Calculate the mean for the following sample of nine measurements: 8, 12, 18, 16, 9, 10, 2, 8, 7. Verify that the deviations from the mean sum to zero. Calculate the sum of squares and then the standard deviation.

1.104 Following are the ages of 25 executives in the banking industry:

WORKSHEET: Executiv.mtw

35 45 63 42 59 45 50 62 36 64 50 26 51 54 45 59 57 64 28 38 48 42 61 54 60

a. Construct a stem-and-leaf plot of the ages. Are there any unusual observations?
b. From the stem-and-leaf plot, construct a frequency table and draw a histogram.
c. Construct a five-number summary diagram and find the IQR.
d. Calculate 1.5(IQR) and subtract it from Q_1 and add it to Q_3 to determine whether there are any outliers on either end of the distribution.
e. Sketch a graph of a boxplot of the data.

1.105 A survey of 1000 adults (conducted by R. H. Bruskin Associates) found that 720 thought the legal drinking age should be 21 years old, 50 said 20 years old, 60 said 19 years old, 110 said 18 years old, and 60 were undecided.
a. Organize these data in a frequency table.
b. Display the data in a bar graph.
c. Display the data in a pie chart.

1.106 On December 2, 1982, Barney Clark became the first human to receive an artificial heart. He was in surgery for 7 hours. William Schroeder, the second recipient on November 25, 1984, was in surgery for 6.5 hours. On February 17, 1985, Murray Haydon, the third artificial heart recipient, was in surgery for 3.5 hours. Suppose you were to keep records on the artificial heart recipients. List three numerical and two categorical variables relevant to the study.

1.107 In reference to Exercise 1.106, 15 artificial heart transplants were performed, and the duration (in hours) was recorded for each transplant.

WORKSHEET: Artifici.mtw

7.0 6.5 3.5 3.8 3.1 2.8 2.5 2.6 2.4 2.1 1.8 2.3 3.1 3.0 2.5

a. Organize these data in a stem-and-leaf plot. Excluding the duration associated with the first two operations (7.0 and 6.5), what do the remaining data look like?
b. Calculate the mean and median of all 15 observations. Next, calculate the mean and median without the two extreme observations. What impact does removing the two outliers have on the two statistics?
c. How long should a new patient undergoing a transplant expect to be in surgery?

1.108 For the heart transplant data in Exercise 1.107, calculate the standard deviation both with and without the two extreme observations. Describe the effect the outliers have on the standard deviation. What should a new patient be told about the variability in duration for the surgery?

1.109 Name three statistics that measure the center of a data set. Name three statistics that measure the variability of a data set.

1.110 Following are the education levels (years of formal education) of nine workers at a plant:

8   8   1   12   14   9   12   14   12

a. Find the mean and the median.
b. Find the 10% trimmed mean.
c. Find the standard deviation.

1.111 Following are the numbers of reported serious reactions per million doses of vaccines in 11 southern states:

WORKSHEET: Vaccine.mtw

Alabama	3.9
Arkansas	27.8
Florida	5.0
Georgia	73.5
Louisiana	24.8
Mississippi	12.1
North Carolina	54.9
Oklahoma	35.3
South Carolina	11.2
Tennessee	138.8
Texas	9.2

Source: Centers for Disease Control, Atlanta, Georgia.

a. Organize the data in a stem-and-leaf plot.
b. Construct a five-number summary diagram and find the IQR.
c. Calculate 1.5(IQR) and subtract it from Q_1 and add it to Q_3 to determine whether there are any outliers on either end of the distribution.

1.112 A market analyst is studying the percents of the computer market held by the major manufacturers. She records each corporation name, the area of the market in which it is active (personal computers, software, etc.), the assets of the company, and the percent of the market it controls. Classify corporate name, market area, assets, and market percent as categorical or numerical.

1.113 The 226,361 poisons reported to 16 poison control centers that serve 11% of the U.S. population were distributed as follows:

WORKSHEET: Poison.mtw

Type poison	No. reported
Drugs	150,857
Cleaning agent	22,347
Plants	22,326
Cosmetics	13,192
Insecticides	8438
Alcohol	9201

Source: Centers for Disease Control, Atlanta, Georgia.

Organize these data in a bar graph.

1.114 To evaluate the side effects of a drug, 70 people were asked to do a certain task after receiving a standard dose of the drug. Following are the number of trials necessary to complete the task:

Trials	Frequency
18	2
17	4
16	9
15	7
14	5
13	7
12	8
11	10
10	12
9	6
Total	70

    a. Is the variable number of trials discrete or continuous?

    b. Illustrate the data in a bar graph.

    c. Calculate the mean and median for the number of trials to complete the task. Which do you think best represents the center of the data?

    d. Calculate the IQR for the number of trials to complete the task.

1.115 Classify the following as either categorical or numerical:

    a. Parent's occupation

    b. Family size

    c. ISBN catalog number used by libraries for books

    d. Type of weapon carried to school

    e. Discount rate on airline tickets

1.116 Following are reading scores on the California Achievement Test for a group of third-graders:

WORKSHEET: CAT.mtw

48 54 52 73 66 45 70 73 62 58 59 54 54 68 50 49 61

    a. Organize the data in an ordered stem-and-leaf plot.

    b. Complete a five-number summary diagram and find the IQR.

    c. Calculate 1.5(IQR) and subtract it from Q_1 and add it to Q_3 to determine whether there are any outliers on either end of the distribution.

    a. Sketch a graph of a boxplot of the data. Based on the length of the tails, would you classify this distribution as short-tailed?

1.117 In the finishing department of a furniture company, the quality of the finish is classified as excellent, good, fair, needs improvement, and redo. Sixty-five items were classified as follows: 17 excellent, 22 good, 13 fair, 5 needs improvement, and the rest redo. Organize these data in a frequency table and draw a bar graph.

Interpreting Computer Output

1.118 Graduation rates for student athletes at schools in the Southeastern Conference are given in worksheet **Graduate.mtw**. From the accompanying bar graph:

    a. What school had the highest graduation rate among student athletes?

    b. What school had the lowest graduation rate for student athletes?

    c. The disparity between the graduation rates of Louisiana State and Kentucky is dramatic. Do you feel that this bar graph gives a fair comparison of the graduation rates at the different schools?

    d. What would you do to the bar graph to show a more realistic comparison?

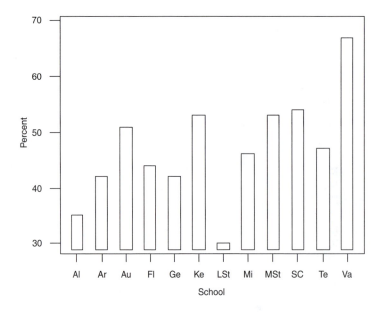

1.119   Modern technology, regulations, malpractice premiums, and physician fees all contribute to the spiraling cost of health care. Following are the fees for an appendectomy for a random sample of 20 hospitals in North Carolina:

WORKSHEET: Append.mtw

$3821 3981 3931 5498 5582 6046 4257 4591 4775 6163 3840 4053 4844 6266
4026   2478 4347 5104 4673 6389

Source: North Carolina Medical Database Commission, August 1994.

Following are descriptive statistics for the physician fees.

Descriptive Statistics: fee

Variable	N	Mean	Median	TrMean	StDev	SE Mean
fee	20	4733	4632	4767	1016	227

Variable	Minimum	Maximum	Q1	Q3
fee	2478	6389	3992	5561

a. Is there much difference between the mean and the median? What was the estimated cost of an appendectomy in North Carolina in 1994?

b. What are the limits that are within two standard deviations of the mean? Are there any observations outside these limits? What are they?

1.120   Following are test grades in a beginning statistics class:

WORKSHEET: Grades.mtw

76 73 81 65 83 90 77 60 67 76 84 72 57 64 71 78 87 76 92 75 58 65 79 86 95 81 71 68 74

Based on the histogram, what is the general shape of the distribution of test scores? Would you use the mean and standard deviation or the median and *IQR* to measure center and spread? Explain your reasoning.

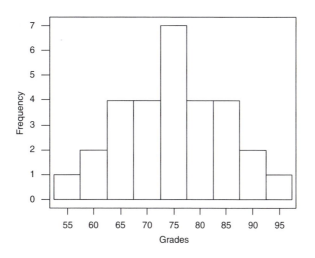

1.121   Following are side-by-side boxplots of the 1980 and 1990 commuting times in 39 major cities found in worksheet **Commute.mtw** (see Exercise 1.13 in Section 1.1). Identify the outliers in the boxplots. The median commuting time is greater in which year? The variability in commuting times is greater in which year? Caution: Your answer may depend on how you measure variability. Briefly describe how commuting times have changed from 1980 to 1990.

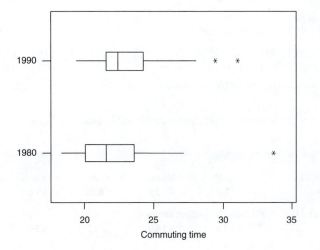

1.122 Following are the weekly rentals for 45 apartments in a large metropolitan area:

WORKSHEET: Rentals.mtw

100	130	130	305	175	155	150	95	295	210	80	270	135	130	335	230	235	75
90	285	65	345	110	135	185	300	70	250	125	180	150	305	170	95	90	145
90	160	130	80	490	235	75	60	425									

Descriptive Statistics: rent

Variable	N	Mean	Median	TrMean	StDev	SE Mean
rent	45	179.4	150.0	171.6	101.4	15.1

Variable	Minimum	Maximum	Q1	Q3
rent	60.0	490.0	95.0	242.5

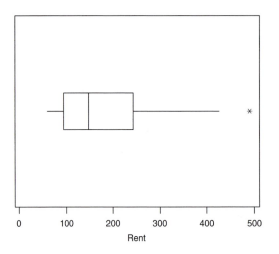

a. Based on the boxplot, what is the general shape of the distribution?
b. From the descriptive statistics, is there much difference between the mean and the median? Which do you think best represents the center of the distribution? Explain.

c. Are any scores more than three standard deviations from the mean? If so, will they affect the value of either the mean or the median? Explain.

d. Give your estimates of the center and spread of the distribution of weekly rentals.

1.123 Data collected from the states by the Bureau of Justice Statistics indicate that violent offenders released from state prisons in 1992 served 48% of their sentences. Following are the sentences (in months) of a sample of 41 prisoners convicted of a homicide offense:

WORKSHEET: Sentence.mtw

months

117 188 172 145 173 159 123 136 115 190 158 147 117 196 169 160 135 163 176 155 126 133 150
146 157 137 180 134 135 128 209 144 164 185 168 121 195 149 148 216 139

Source: U.S. Department of Justice, Bureau of Justice Statistics, *Prison Sentences and Time Served for Violence*, NCJ-153858, April 1995.

```
Stem-and-leaf of months   N = 41
Leaf Unit = 1.0

  3      11 577
  7      12 1368
 14      13 3455679
 20      14 456789
 (5)     15 05789
 16      16 03489
 11      17 236
  8      18 058
  5      19 056
  2      20 9
  1      21 6
```

Descriptive Statistics: months

Variable	N	Mean	Median	TrMean	StDev	SE Mean
months	41	155.07	150.00	154.08	25.86	4.04

Variable	Minimum	Maximum	Q1	Q3
months	115.00	216.00	135.00	172.50

a. From the accompanying stem-and-leaf plot, how would you classify the shape of the distribution?

b. From the descriptive statistics, do you detect a large difference between the measures of the center of the data?

c. Use the empirical rule to find limits for the length of 95% of the prison sentences. Does it seem that there is a wide disparity in sentence lengths for these prisoners?

1.124 Given here are the graduation rates for student athletes and all students who entered schools in the Big Ten Conference in 1993–94:

WORKSHEET: BigTen.mtw

School	Student	Athlete
Illinois	75	65
Indiana	68	66
Iowa	62	71
Michigan	82	68
Michigan State	64	56
Minnesota	51	65

continued

WORKSHEET: BigTen.mtw
continued

School	Student	Athlete
Northwestern	92	93
Ohio State	56	50
Penn State	80	78
Purdue	64	72
Wisconsin	74	55

Source: NCAA Graduation Rates Report, 2000.

Descriptive Statistics: 1993–94students, 1993–94athletes

Variable	N	Mean	Median	TrMean	StDev	SE Mean
1993–94s	11	69.82	68.00	69.44	12.11	3.65
1993–94a	11	67.18	66.00	66.22	11.84	3.57

Variable	Minimum	Maximum	Q1	Q3
1993–94s	51.00	92.00	62.00	80.00
1993–94a	50.00	93.00	56.00	72.00

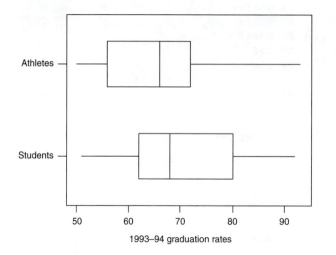

1993–94 graduation rates

a. Based on the side-by-side boxplots, what can you say about the differences between the graduation rates of athletes and all students in Big Ten schools?
b. How do you classify the shapes of the two distributions?
c. Would you use means or medians to compare the centers of the two distributions?
d. From the descriptive statistics, do you detect a difference between the centers of the two distributions? Explain.

Computer Exercises

1.125 Fertility rate is the number of births a woman can expect in her childbearing years. An average rate of 2.12 is needed to keep the population constant. Following are the fertility rates in all states, including the District of Columbia:

WORKSHEET: Fertility.mtw

Ala	1.9	Calif	1.9	D. C.	1.5
Alaska	2.3	Colo	1.8	Fla	1.7
Ariz	2.1	Conn	1.5	Ga	1.9
Ark	2.0	Del	1.8	Hawaii	2.1

continued

WORKSHEET: Fertility.mtw
continued

Idaho	2.5	Mo	1.9	Pa	1.6
Ill	1.9	Mont	2.1	R.I.	1.5
Ind	1.8	Neb	2.0	S.C.	1.8
Iowa	2.0	Nev	1.8	S.D.	2.4
Kan	2.0	N.H.	1.7	Tenn	1.7
Ky	1.9	N.J.	1.6	Texas	2.1
La	2.2	N.M.	2.2	Utah	3.2
Maine	1.7	N.Y.	1.6	Vt	1.7
Md	1.6	N.C.	1.6	Va	1.6
Mass	1.5	N.D.	2.1	Wash	1.8
Mich	1.8	Ohio	1.8	W.Va	1.8
Minn	1.9	Okla	2.0	Wis	1.9
Miss	2.2	Ore	1.8	Wyo	2.4

Source: Population Reference Bureau.

a. Construct a stem-and-leaf plot of the fertility rates. Are there any unusual observations? For what state(s)?
b. Obtain descriptive statistics for the fertility rates. Is there much difference between the mean and the median? Explain.
c. Construct a five-number summary diagram. What is the IQR? Use it to check for outliers.
d. Construct a boxplot.
e. What is the general shape of the distribution? Without outliers, what is the general shape of the distribution?
f. What is your best estimate of the average fertility rate? Is it large enough to keep the population constant?

1.126 Following are the birthplaces, ages at inauguration, and ages at death of the first 43 presidents of the United States:

WORKSHEET: Presiden.mtw

President	Birth	Inaug Age	Age at Death
G. Washington	VA	57	67
J. Adams	MASS	61	90
T. Jefferson	VA	57	83
J. Madison	VA	57	85
J. Monroe	VA	58	73
J. Q. Adams	MASS	57	80
A. Jackson	S.C.	61	78
M. Van Buren	N.Y.	54	79
W. Harrison	VA	68	68
J. Tyler	VA	51	71
J. Polk	N.C.	49	53
Z. Taylor	VA	64	65
M. Fillmore	N.Y.	50	74
F. Pierce	N.H.	48	64
J. Buchanan	PA	65	77
A. Lincoln	KY	52	56
A. Johnson	N.C.	56	66
U. Grant	OH	46	63
R. Hayes	OH	54	70

continued

WORKSHEET: Presiden.mtw
continued

President	Birth	Inaug Age	Age at Death
J. Garfield	OH	49	49
C. Arthur	VT	50	57
G. Cleveland	N.J.	47	71
B. Harrison	OH	55	67
G. Cleveland	N.J.	55	71
W. McKinley	OH	54	58
T. Roosevelt	N.Y.	42	60
W. Taft	OH	51	72
W. Wilson	VA	56	67
W. Harding	OH	55	57
C. Coolidge	VT	51	60
H. Hoover	IA	54	90
F.D. Roosevelt	N.Y.	51	63
H. Truman	MO	60	88
D. Eisenhower	TEX	62	78
J. Kennedy	MASS	43	46
L. Johnson	TEX	55	64
R. Nixon	CAL	56	81
G. Ford	NEB	61	
J. Carter	GA	52	
R. Reagan	ILL	69	
G.H. Bush	MASS	64	
B. Clinton	ARK	46	
G.W. Bush	CONN	54	

a. Construct a pie chart of the birth states of the presidents. What state has had the most presidents? Of the 50 states, how many have never produced a president?
b. Construct stem-and-leaf plots of the inaugural ages and the ages at death. Are they similar in shape?
c. Construct side-by-side dotplots of the inaugural ages and the ages at death with a common number line. Does this provide additional information about how the two distributions are related?

1.127   Disposal of toxic waste is a major concern for U.S. manufacturing plants. One measure of the environmental performance of a plant is its "toxic intensity," which is the pounds of toxic waste released (into the environment or captured and transferred to another site) per $1000 of shipments of its product. Successful plants release fewer pounds of toxins per $1000 of shipments than the average of other plants that manufacture the same product. Following are the toxic intensities measured for plants whose product line is herbicidal preparations:

WORKSHEET: Disposal.mtw

1.45   1.38   4.37   2.97   1.06   0.44   2.20   1.25   2.23   3.96   9.12   3.49   5.64   1.74   1.69   4.33
0.28   2.43   5.46   3.57   2.03   1.39   3.35   4.33   2.68   2.48   0.82   3.47   0.33

Source: Bureau of the Census, *Reducing Toxins*, Statistical Brief SB/95–3, February 1995.

a. Display the data in a stem-and-leaf plot. Do any plants have an unusually high toxic intensity?
b. Display a boxplot and a histogram of the data. Are there any outliers? How do you classify the shape of the distribution?
c. Find descriptive statistics for the data. Is there much difference between the mean and median? Which do you think best represents the toxic intensity for these plants?
d. Based on your answers to b and c, give measures of the center and spread for the distribution of toxic intensity.

1.128 The average speed (in miles per hour) for the winners of the Indianapolis 500 for the years 1961–1999 are as follows:

WORKSHEET: Indiapol.mtw

Year	Speed	Year	Speed
1961	139.130	1981	139.085
1962	140.293	1982	162.026
1963	143.137	1983	162.117
1964	147.350	1984	163.621
1965	151.388	1985	152.982
1966	144.317	1986	170.722
1967	151.207	1987	162.175
1968	152.882	1988	144.809
1969	156.867	1989	167.581
1970	155.749	1990	185.984
1971	157.735	1991	176.457
1972	162.962	1992	134.477
1973	159.036	1993	157.207
1974	158.589	1994	160.872
1975	149.213	1995	153.616
1976	148.725	1996	147.956
1977	161.331	1997	145.827
1978	161.363	1998	145.155
1979	158.899	1999	153.176
1980	142.862		

Source: The World Almanac and Book of Facts, 2000, p. 1004.

Retrieve the data and construct a time series plot. Compare the trend in speeds for the first 14 years with the trend in the last 25 years. Can any generalizations be made about the speeds over the entire 39-year period?

2

Summarizing Relationships between Variables

In Chapter 1, we studied characteristics of univariate data. Now we investigate questions regarding relationships between two variables. For example, is there a relationship between the birth weights of children and their adult heights? Is there a relationship between students' grades and their study habits? Furthermore, if there is a relationship, what form does it take? How strong is the relationship? To answer these questions, our first task is to organize and summarize the bivariate data. Graphical tools such as the *scatterplot* are used to examine the data and visually determine whether there is an association between the numerical variables. The *correlation coefficient*, a summary statistic, is used to measure numerically the degree of linear association between the numerical variables. Having established that there is a linear relationship between the variables, we use the *least squares method* to determine the equation of the line. Finally, in Section 2.5, we illustrate relationships between two categorical variables with a *contingency table*.

Contents

■ **STATISTICAL INSIGHT**
2.1 Scatterplots
2.2 Correlation
2.3 Least Squares Regression

2.4 Assessing the Fit of a Line
2.5 Relationships among Categorical Variables
2.6 Summary and Review Exercises

Does This Graph Explain Why George W. Bush Is the 43rd President of the United States?

The graph gives the vote count for candidate Pat Buchanan versus the total votes cast in each county in Florida. In the closest presidential race in history, the final vote count in Florida gave Bush a 537-vote margin over Al Gore, Bush's Democratic opponent. By capturing Florida's 25 electoral votes, Bush won the presidency with 271 electoral votes to 266 for Gore. Had Gore won 538 more votes in Florida, he would have become the 43rd president of the United States.

WORKSHEET: Florida2000.mtw

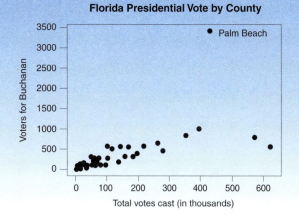

Florida Presidential Vote by County

Here are some important questions we should ask about these data:

■ Do you recognize a pattern in the data?
■ Is there a positive or negative trend?
■ Are there any outliers?
■ Will a straight line fit the data?
■ How strong is the relationship?
■ Does this graph reveal any discrepancies in the voting pattern of counties in Florida?
■ Can you explain how Gore would have won the presidency if Palm Beach County had voted like other counties in Florida?

After completing this chapter, you should be able to answer these and other interesting questions. Refer to Section 2.6 for a complete analysis of these data.

2.1 Scatterplots

Learning Objectives for this section:

❑ Know how to construct a scatterplot for bivariate data.

❑ Know what it means for data in a scatterplot to cluster close to a line.

❑ Be able to recognize trends in the data graphed in a scatterplot.

❑ Know the difference between univariate and bivariate outliers.

❑ Know the characteristics of an influential observation.

❑ Understand how transformations of data can sometimes help describe relationships.

❑ Know how to graph and interpret numerical versus categorical data.

When we study the relationship between two (or more) variables, we must determine whether they are numerical or categorical variables. We will study numerical versus numerical, numerical versus categorical, and in Section 2.5, categorical versus categorical variables.

Numerical versus Numerical Variables

When both variables are numerical, their bivariate data are displayed in a *scatterplot*, as illustrated in the following example.

Example 2.1 Does watching too much television affect math scores? Educators have suggested that the social conditions in American homes contribute heavily to the decline in children's academic performance. The Educational Testing Service completed a report, *America's Smallest School: The Family*, that suggests watching television has a serious effect on school performance. In the report, state averages on a national math test were compared to the percent of students in the state that watch 6 or more hours of television per day. The following data were collected for eighth-graders in 37 states, two territories, and Washington, D.C.:

WORKSHEET: TV.mtw

State	Percent	test	State	Percent	test
Ala	18	253	Mont	6	279
Alaska	na	na	Neb	9	276
Ariz	12	258	Nev	na	na
Ark	20	256	N.H.	7	273
Calif	11	256	N.J.	13	270
Colo	9	267	N.M.	11	256
Conn	12	270	N.Y.	17	261
Del	18	262	N.C.	21	251
Fla	19	255	N.D.	6	281
Ga	17	258	Ohio	11	264
Hawaii	23	252	Okla	14	263
Idaho	7	272	Ore	9	272
Ill	14	261	Pa	10	262
Ind	11	267	R.I.	12	261
Iowa	8	278	S.C.	na	na
Kan	na	na	S.D.	na	na
Ky	14	256	Tenn	na	na
La	19	246	Texas	15	258
Maine	na	na	Utah	na	na
Md	19	261	Vt	na	na
Mass	na	na	Va	16	264
Mich	14	264	Wash	na	na
Minn	7	276	W.Va	16	256
Miss	na	na	Wis	8	274
Mont	na	na	Wyo	7	272
D.C.	33	233	Vir Is	27	218
			Guam	20	233

Source: Educational Testing Services.

To construct a scatterplot of these data, draw a graph with the percent who watch more than 6 hours of television per day recorded on the *x*-axis (horizontal scale) and the test scores recorded on the *y*-axis (vertical scale). Each pair (*x,y*) of observations is plotted as a point on the graph. For example, Alabama's data (18,253) is plotted first. Figure 2.1 is the completed scatterplot of the data.

Figure 2.1
Scatterplot of percent
versus test scores

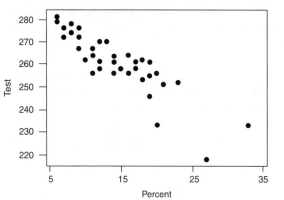

The scatterplot shows a relationship between test scores and the percent that watch more than 6 hours of television per day. We see a downward linear (straight-line) trend between the two numerical variables. That is, the data tend to cluster about a straight line with negative slope. This means that the greater the percent that watch more than 6 hours of television per day, the lower the test scores. ∎

A scatterplot is useful for displaying trends in the data and exploring relationships that might exist between the two variables. Following are some important points to consider when exploring a scatterplot.

Exploring a Scatterplot

- If the values of one variable (test scores) *depend* on the values of the other (hours watching television), graph the dependent values on the vertical axis. Otherwise it doesn't matter which variable goes on the vertical axis.
- Examine the scatter for a pattern. Do the points tend to *cluster* about a line? Are there any unusual observations?
- Look for specific trends. If the variables are related, what is the form of the relationship? How strong is the relationship? Can we predict the value of one variable from the other?

Positive and Negative Relationships

A relationship like the one in Figure 2.1, where one variable decreases as the other increases, is called a *negative* relationship. In a *positive* relationship, higher values of one variable tend to be associated with higher values of the other variable.

Bivariate Outliers

Univariate outliers were defined as observations that are far removed from the remainder of the data. In a scatterplot, if an observation pair is far removed from the overall pattern of the remaining data, it is called a *bivariate outlier*. Care must be exercised when interpreting bivariate data that have outliers.

Example 2.2   The number of convictions reported by U. S. attorneys' offices varies widely from district to district. Some claim to be understaffed and unable to prosecute all crimes. Worksheet: **Attorney.mtw** gives a comparison of 88 districts by staff per 1 million population and convictions per 1 million population.

If there are outliers in the data it is a good idea to always graph the data both with and without the outliers.

Figure 2.2 is a scatterplot of the staff per million by convictions per million. Because the point (212,838), Washington D.C., is an outlier (located in the upper right-hand corner), the scale for both the x- and y-axes must be extended to graph the point. Consequently, the rest of the data are so compact that they are difficult to interpret. Figure 2.3 is a scatterplot of the same data without the outlier, and you can see that there is a moderate upward linear trend. Note also that there is a large degree of scatter in the graph, indicating that the association between the two variables is weak. ∎

Figure 2.2
U.S. Attorney data with Washington, D.C.

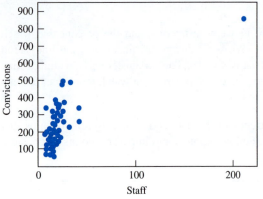

Figure 2.3
U.S. Attorney data without Washington, D.C.

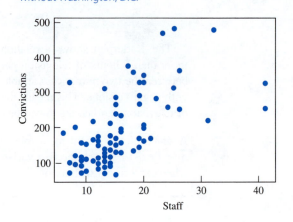

In the next section, we measure the strength of the association between two variables with the *correlation coefficient*. There we will see that outliers such as the Washington, D.C., observation are called *influential observations* and will have a dramatic effect on the value of the correlation.

Figure 2.4 shows that an observation can be a *bivariate outlier* but not an outlier in either of the other variables when considered separately. The identified observation is clearly removed from the main scatter of the data, and yet it is still within the range of the x data and within the range of the y data.

Figure 2.4
Bivariate outlier

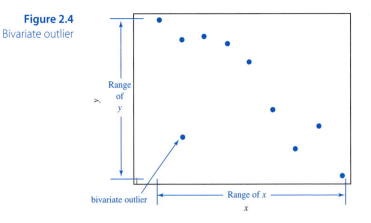

Transforming Data

Occasionally a simple transformation of the data, such as taking a logarithm or square root of each observation, can reveal relationships among variables that might otherwise go undetected.

Example 2.3   Are the brain weights of animals related to their body weights? In other words, does it require a larger brain to govern a heavier body? The brain weights (in grams) and the body weights (in kilograms) of 28 animals are given here.

WORKSHEET: Brain.mtw

Species	Body wt(kg)	Brain wt(g)
Mountain beaver	1.35	8.1
Cow	465.0	423.0
Gray wolf	36.33	119.5
Goat	27.66	115.0
Guinea pig	1.04	5.5
Diplodocus	11,700.0	50.0
Asian elephant	2547.0	4603.0
Donkey	187.1	419.0
Horse	521.0	655.0
Potar monkey	10.0	115.0
Cat	3.3	25.6
Giraffe	529.0	680.0
Gorilla	207.0	406.0
Human	62.0	1320.0
African elephant	6654.0	5712.0
Triceratops	9400.0	70.0
Rhesus monkey	6.8	179.0
Kangaroo	35.0	56.0
Hamster	0.12	1.0
Mouse	0.023	0.4
Rabbit	2.5	12.1
Sheep	55.5	175.0
Jaguar	100.0	157.0
Chimpanzee	52.16	440.0
Brachiosaurus	87,000.0	154.5
Rat	0.28	1.9
Mole	0.122	3.0
Pig	192.0	180.0

Source: P. Rousseeuw and A. Leroy, *Robust Regression and Outlier Detection* (New York: Wiley, 1987).

Because of the large variation in body weights (from .023 kg to 87,000 kg) and brain weights (from .4 to 5712 g), a standard scatterplot would be so distorted that no reasonable pattern could be detected. After applying logarithms (base 10) to the data, however, we see a definite pattern as shown in Figure 2.5.

Figure 2.5

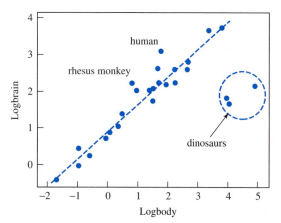

Notice the main scatter about the dotted line in the graph. Also notice the three outliers associated with the dinosaurs. These are bivariate outliers, but they are not univariate outliers with respect to either of the individual variables. ∎

Other Relationships

Often there are relationships among variables other than linear relationships. The scatterplot helps identify those relationships.

Example 2.4 The *Ronald H. Brown* is a state-of-the-art oceanographic and atmospheric research ship operated by NOAA (National Oceanic and Atmospheric Administration). Scientists aboard the *Ron Brown* conducted two CTD (Conductivity, Temperature, and Depth) drops on October 23 and November 3, 1998. The data in worksheet: **Ronbrown2.mtw** gives temperature, salinity, and density every 10 meters to a depth of 500 meters and every 20 meters to a depth of 1000 meters. To view a video that shows the CTD being lowered into the water, visit this web site: *http://classroomatsea.noaa.gov/ronbrown/ctd*

Figure 2.6

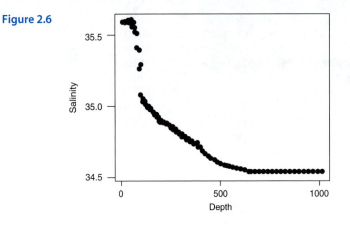

The relationship we see between salinity level and ocean depth in the scatterplot in Figure 2.6 is clearly not linear, yet it does exhibit a definite pattern between the variables. From the scatterplot we see that salinity is constant for a depth of a few meters, then begins a dramatic drop when the depth increases from approximately 50 meters to 100 meters. It again begins a downward trend until a depth of 500 meters where it levels off to about 34.5 and stays at that level from 500 meters to 1000 meters. We may not be able to find a functional relationship between these two variables but with the scatterplot we have a very good idea of the salinity level at different depths of the ocean. ∎

Categorical versus Numerical Variables

A variation of the scatterplot can be used to display bivariate data in which one variable is numerical and the other is categorical. This is illustrated in the following example.

Example 2.5 At the end of a semester, a university professor asked her students to grade her teaching on a scale from A to F. She also asked them to record their present grade point averages. Graph the data that were obtained:

Teacher's rating	B	C	C	A	B	D	C	C	C	F	A	C	D	B	B	F
Grade pt average	3.2	2.4	3.6	4.0	3.6	1.2	3.2	4.0	2.0	0.4	3.6	3.2	2.8	2.4	4.0	1.6

Solution   These data are graphed with the teacher's grades on the horizontal scale and the students' GPAs on the vertical scale (see Figure 2.7). The data are plotted by placing a dot at the intersection of the grade (rating) and the GPA for each of the 16 students.

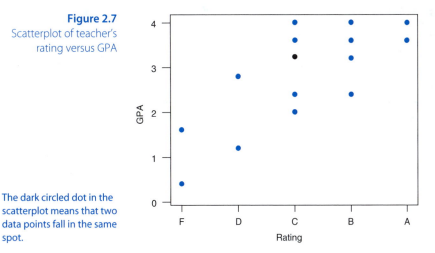

Figure 2.7
Scatterplot of teacher's rating versus GPA

The dark circled dot in the scatterplot means that two data points fall in the same spot.

Note that, as might be expected, there is an upward trend; students that rate the teacher higher tend to have higher GPAs. To observe a trend like this, it is necessary that we order the categorical variable (teacher's rating) from smallest to largest; otherwise, we could list the categories in the opposite direction and observe a downward trend.   ■

If there is a large amount of data, boxplots of the numerical data can be constructed at each category of the categorical variable. The resulting graph of side-by-side boxplots provides a comparison of the different distributions of data at each category of the categorical variable. The boxplots may give a more informative view of the data without the scatter in a scatterplot.

Example 2.6   The university professor in Example 2.5 had a large lecture class of 250 students. A scatterplot of individual grade point averages versus the teacher's ratings, as in Example 2.5, would be unusually cluttered. With the aid of a computer, however, it is a simple matter to construct side-by-side boxplots of the students' GPAs for each of the ratings F, D, C, B, and A. Figure 2.8 is such a graph, created with Minitab from the worksheet: **Ratings.mtw**.   ■

Figure 2.8
GPA distributions versus teacher's rating

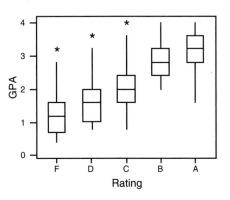

> ### C O M P U T E R   T I P
>
> ### Scatterplots
>
> After the data have been stored in the Minitab worksheet, select **Graph** and click on **Plot**. At the Plot window, **Select** the X and Y variables and click **OK**.

EXERCISES 2.1 The Basics

2.1 A study was made to relate aptitude test scores to productivity in a factory after employees had 3 months of personnel training. The following results were obtained by testing eight randomly selected applicants and later measuring their productivity. Draw a scatterplot of aptitude scores versus productivity and comment on the relationship.

WORKSHEET: Aptitude.mtw

Applicant	1	2	3	4	5	6	7	8
Aptitude score	9	17	13	19	20	23	12	15
Productivity	23	35	29	33	40	38	25	31

2.2 The robbery rate varies from precinct to precinct. From the following data, construct a scatterplot and comment on the relationship between the robbery rate and the percent of low-income residents in the precinct.

WORKSHEET: Precinct.mtw

Precinct	1	2	3	4	5	6	7	8
Robbery rate	20	41	165	88	60	120	65	81
Percent low income	4.9	7.1	10.1	11.8	13.5	14.8	16.2	11.2

2.3 A record of maintenance cost is kept on nine cash registers in a major department store chain. Construct a scatterplot and comment on the relationship between the variables.

WORKSHEET: Register.mtw

Age (years)	6	7	1	3	6	2	5	4	3
Cost (dollars)	92	181	23	40	126	35	86	72	51

2.4 It is suspected that the life span of a particular electronic component used in a spacecraft is dependent on the heat it experiences. Six such components were tested at various levels of heat. Construct a scatterplot and comment on the relationship between the variables.

WORKSHEET: Lifespan.mtw

Heat (C°)	50	100	150	200	250	300
Life span(hours)	875	884	762	424	365	128

Interpret Computer Output

2.5   The following data give the amount of energy consumed in a home versus the size of the home.

WORKSHEET: Energy.mtw

Size of home(sq ft)	2820	2500	2350	2000	1950	1875	1740	1650	1490	1350	1270	1200
Kilowatt-hours/mo	1975	1952	1894	1841	1769	1674	1590	1505	1386	1220	1089	1042

From the scatterplot, comment on the relationship that you see between the variables. Does it matter which variable is located on the *x*-axis of the scatterplot? In describing the relationship, pay particular attention to when the size of the home is less than 2000 square feet. What happens when the size is greater than 2000 square feet? Would a straight line describe the relationship between the two variables? Explain.

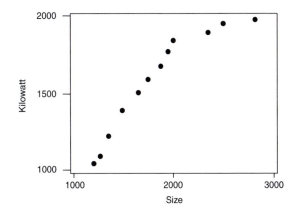

2.6   The following scatterplot gives the number of deaths due to lightning strikes in relation to damage reports. Is there a positive or negative trend in the data? Would a straight line fit the data? Is this relationship strong or weak?

WORKSHEET: Ligntmonth.mtw

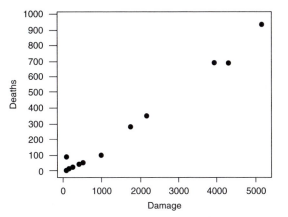

Source: Lightning Fatalities, Injuries and Damage Reports in the United States, 1959–1994, NOAA Technical Memorandum NWS SR-193, Dept of Commerce.

2.7   Monthly sunspot activity for the years 1979 to 1999 is graphically illustrated in the following side-by-side boxplots. Is there a positive or negative trend in the data? Are the data cyclical?

WORKSHEET: Yearsunspot.mtw

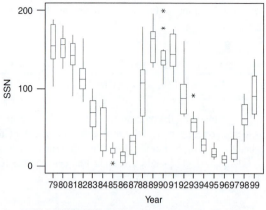

Source: NASA/Marshall Space Flight Center, Huntsville, AL 35812.

2.8 The following side-by-side boxplots give the price per gallon for diesel fuel in 1999 and early 2000 for four regions of the country. In comparison to the national average, what can be said about how the price varies from one region to the next?

WORKSHEET: Diesel.mtw

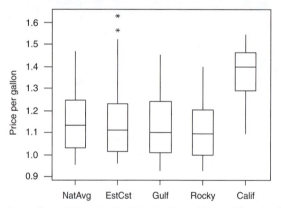

Source: Energy Information Administration, National Energy Information Center: 1000 Independence Ave., SW, Washington, D.C., 20585.

2.9 Describe the relationship that you see in the following scatterplot of temperature versus depth of the ocean. The data, in worksheet: **Ronbrown1.mtw**, were obtained by scientists aboard the ship *Ron Brown* on October 23 and November 3, 1998. (See Example 2.4.)

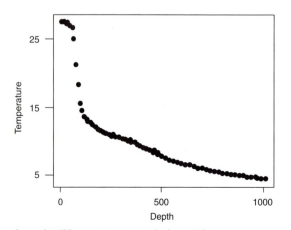

Source: http://classroomatsea.noaa.gov/ronbrown/ctdactvy.

Computer Exercises

2.10 A study of the association between reading ability and IQ scores was conducted by a reading coordinator in a large public school system. A random sample of 14 eighth-grade students was given a reading achievement test and an IQ test. The scores are recorded in the table:

WORKSHEET: Readiq.mtw

Reading score	42	35	61	28	48	46	59	21	47	29	65	37	35	53
IQ score	105	110	122	92	112	100	120	85	125	96	130	90	107	120

a. Organize these data in a scatterplot.
b. Are there any bivariate outliers in the data?
c. Does it appear that a straight line would fit the data reasonably well?
d. On your scatterplot, sketch in a straight line that fits the data.

2.11 To understand some of the problems facing the American farmer, construct scatterplots of the following data that relate the price of a bushel of wheat to the national weekly earnings of production workers by first plotting weekly earnings versus year and then plotting wheat price versus year.

WORKSHEET: Wheat.mtw

Year	1980	1981	1982	1983	1984
Wheat price	$3.91	3.65	3.55	3.54	3.39
Weekly earnings	$235.10	255.20	267.26	280.70	292.86

Year	1985	1986	1987	1988	1989
Wheat price	$3.08	2.42	2.57	3.72	3.72
Weekly earnings	$299.09	304.85	312.50	322.02	334.24

Year	1990	1991	1992	1993	1994
Wheat price	$2.61	3.00	3.24	3.26	3.45
Weekly earnings	$345.35	353.98	363.61	373.64	385.86

Year	1995	1996	1997	1998	
Wheat price	$4.55	4.30	3.38	2.65	
Weekly earnings	$394.34	406.61	424.89	441.84	

Source: The World Almanac and Book of Facts, 2000.

a. Describe how weekly earnings have changed since 1980.
b. Prior to 1987 there was a definite downward trend in the price of wheat. Does that trend appear linear?
c. What happened to the price of wheat in 1988, 1989, and 1990?
d. What happened to the price of wheat between 1990 and 1995?
e. What has happened to the price of wheat since 1995?
f. Name two variables that you think affects the price of wheat.

2.12 The following data relates pay raises from the previous year for salaried employees to the inflation rate:

WORKSHEET: Inflatio.mtw

Year	1976	1977	1978	1979	1980	1981	1982	1983	1984	1985	1986	1987
Raise	7.3%	8.0	8.4	8.3	8.1	8.9	5.9	4.4	3.7	3.0	2.2	2.5
Inflation	4.9%	6.7	9.0	13.3	12.5	8.9	3.8	3.8	3.9	3.8	1.1	4.4

Year	1988	1989	1990	1991	1992	1993	1994	1995	1996	1997	1998	
Raise	3.3	4.1	3.6	3.1	2.4	2.5	2.7	2.8	3.4	3.9	4.0	
Inflation	4.4	4.6	6.1	3.1	2.9	2.7	2.7	2.5	3.3	1.7	1.6	

Source: Bureau of Labor Statistics.

a. Construct a scatterplot of the pay raises versus the inflation rates.
b. Describe the pattern that you see in the scatterplot.
c. Describe the pattern when the inflation rate is below 6%.
d. What happens when the inflation rate goes above 6%? In particular, what can you say about the pay raises when the inflation rate is between 6.5% and 14%?

2.13 Using the data in Exercise 2.12, code the inflation rate as low if it is less than 4%, as middle if it is greater than or equal to 4% and less than 6.5%, and as high if it is greater than or equal to 6.5%.
a. Create side-by-side boxplots of the pay increase for the three categories of inflation rate.
b. What do you observe about the variability in pay raises for the three categories of inflation rate?

2.14 J. D. Power and Associates ranks cars and trucks every year according to the number of problems reported during the first 3 months of ownership. The following data, obtained from Power's rating, represent the number of problems reported per 100 cars:

WORKSHEET: JDPower.mtw

Car	1994	1995	Car	1994	1995
Infiniti	75	55	Lexus	54	60
Acura	101	64	Honda	92	71
Toyota	69	74	Mercedes	91	79
BMW	114	82	Geo	134	82
Nissan	99	83	Volvo	108	86
Subaru	136	88	Buick	91	90
Cadillac	104	91	Saturn	78	91
Lincoln	76	100	Oldsmobile	109	101
Chevrolet	138	105	Mazda	115	106
Pontiac	138	110	Mercury	86	115
Ford	112	116	Jaguar	144	132
Plymouth	122	137	Dodge	144	144
Saab	180	155	Mitsubishi	150	164
Eagle	155	168	Volkswagen	158	183
Hyundai	193	195			

Source: USA Today, May 25, 1995.

a. Construct a scatterplot of the 1995 rating versus the 1994 rating.
b. Are there any bivariate outliers in the data?
c. Is there a positive or negative trend? What does it mean in terms of the number of problems reported for the vehicles in 1995 versus the number reported in 1994?
d. Does it appear that a straight line would fit the data reasonably well?
e. If yes, sketch in a straight line that fits the data.

2.15 The brain weights and body weights of the 28 animals listed in Example 2.3 are given in worksheet: **Brain.mtw**.
a. Construct a scatterplot of brain weight versus body weight. Describe the problem of graphing the original data.
b. Construct a scatterplot of log(brain wt) versus log(body wt). By transforming each variable, have you produced a scatterplot that shows a definite relationship between the two variables?

2.16 In World War II, the actual number of German submarines sunk each month by the U.S. Navy did not exactly match the number reported by the Navy. Following are results for 16 months:

WORKSHEET: Submarin.mtw

Month	1	2	3	4	5	6	7	8	9	10	11	12	13	14	15	16
Reported by Navy	3	2	4	2	5	5	9	12	8	13	14	3	4	13	10	16
Actual count	3	2	6	3	4	3	11	9	10	16	13	5	6	19	15	15

Source: F. Mosteller, S. Fienberg, and R. Rourke, *Beginning Statistics with Data Analysis* (Reading, MA: Addison-Wesley, 1983).

a. Construct a scatterplot with the actual count on the vertical (*y*) axis and the reported count on the horizontal (*x*) axis.
b. Is there a reasonable linear pattern to the data? Is there a positive or negative trend?
c. Are there any bivariate outliers in the data?
d. Do you see two separate clusters in the scatterplot?
e. Describe the variability of the actual count when the reported count was 5 or less. Describe the variability of the actual count when the reported count was greater than 5.

Internet Tasks

2.17   Go to the J. D. Power and Associates web site. Obtain the current ratings of cars listed in worksheet: **JDPower.mtw**. Add the data to the worksheet and construct a scatterplot of the current ratings with those in 1995. Describe the relationship between the two variables. Are there any outliers? Is there a linear trend?

2.18   The Environmental Protection Agency (EPA) conducts yearly gas mileage tests for all new cars. Data obtained on each make of automobile includes the number of cylinders, the displacement (size) of the engine, weight, average city gas mileage, highway gas mileage, and a combined figure. Go to the EPA web site and find the latest figures for a class of automobiles such as compacts or subcompacts. Construct a scatterplot of the combined gas mileage versus the displacement of the engine. Investigate the relationship between the two variables. Is there a linear trend? Are there any outliers?

2.19   Search the Internet for more information on the 2000 presidential election. Find a graph similar to the one in the Statistical Insight that gives the vote count for Buchanan versus the vote count for Gore. Write a paragraph describing how close the election was and how the outcome might have been different.

2.2   Correlation

Learning Objectives for this section:

- ❏ Know how to calculate the correlation coefficient.
- ❏ Learn the basic properties of the correlation coefficient.
- ❏ Know the effects of outliers on correlation.
- ❏ Know when outliers are classified as influential observations.
- ❏ Understand that correlation does not necessarily mean causation.
- ❏ Know that a lurking variable may explain the correlation between two variables.

Numerical versus Numerical Variables

If two numerical variables are related in such a way that the points in a scatterplot tend to fall in a straight line, we say they are *linearly related*. In this section, we discuss a numerical measure of the degree of linearity between the two variables.

Chapter 1 introduced the concept of summarizing a univariate data set with summary statistics such as the mean and standard deviation. In summarizing bivariate data, the means and standard deviations of the two samples are computed independently of each other, and thus tell us nothing about how the two variables are related. What we need to numerically evaluate the linear relationship between two variables is a statistic that is calculated on the bivariate data. This is accomplished with the *correlation coefficient*.

The **correlation coefficient**, denoted by r, measures the direction and strength of the linear relationship between two numerical variables. It is given by the equation

$$r = \frac{1}{(n-1)} \sum \left(\frac{x_i - \bar{x}}{s_x} \right) \left(\frac{y_i - \bar{y}}{s_y} \right)$$

where $\bar{x}$ and s_x are the mean and standard deviation of the x-values and $\bar{y}$ and s_y are the mean and standard deviation of the y-values.

The formula for r is a bit complicated and normally we will use a calculator or computer to compute its value. Notice that the means and standard deviations of the x and y data are needed in the formula. Using $\bar{x}$ and s_x each x value is standardized, and using $\bar{y}$ and s_y each y value is standardized. The correlation is the average (divided by $n - 1$) of the products of these standardized x and y values.

Example 2.7   Using the data on watching television and math scores in Example 2.1, indicate how the correlation coefficient is calculated and then find the correlation.

Solution   Here are the descriptive statistics on the two variables:

Descriptive Statistics: percent, test

Variable	N	N*	Mean	Median	TrMean	StDev
percent	40	13	14.000	13.500	13.556	5.996
test	40	13	261.13	261.50	262.06	12.89

Variable	SE Mean	Minimum	Maximum	Q1	Q3
percent	0.948	6.000	33.000	9.000	18.000
test	2.04	218.00	281.00	256.00	271.50

Notice that data are missing on 13 states. The descriptive statistics are calculated from the remaining 40 observations. Using the means and standard deviations of the variables percent and test we calculate standardized values for each pair of observations. The correlation is obtained by averaging the products of the standardized values. Here are the first two products:

Alabama   Arizona

$$\left(\frac{18 - 14}{5.996} \right) \left(\frac{253 - 261.13}{12.89} \right) + \left(\frac{12 - 14}{5.996} \right) \left(\frac{258 - 261.13}{12.89} \right) + \cdots$$

The resulting correlation is given in this printout:

Correlations: percent, test

```
Pearson correlation of percent and test = -0.862
```

Notice that the correlation is negative and close to -1. This suggests that a straight line with negative slope will describe the relationship between these two variables. ∎

Interpretation of r

To properly interpret a correlation coefficient, you must understand the basic properties of r:

Properties of the Correlation Coefficient

- The value of r measures the strength of the *linear* relationship between x and y and will always be between -1 and $+1$.
- The closer r is to -1 or $+1$, the stronger the linear relationship between x and y. In fact, when all the data fall exactly on a straight line there is a correlation of $+1$ if the line has positive slope and -1 if the line has negative slope.
- If r is zero, then x and y are not linearly related. They may be related by something other than a straight line.
- The value of r does not change when the units of measurement are changed.
- A positive value for r indicates a positive relationship and a negative r indicates a negative relationship.
- Outliers will adversely affect the value of r.

Figure 2.9 illustrates scatterplots of data with four different values of r. In the first scatterplot, where $r = +.05$, there is very little relationship between x and y. The second scatterplot, where $r = +.99$, shows an almost perfect straight line relationship between x and y. The third and fourth scatterplots show the same degree of association with $r = +.7$ and $r = -.7$. The only difference between the two is that when $r = +.7$ there is a positive linear relationship between x and y (as one variable increases the other one also increases) and when $r = -.7$ there is a negative linear relationship (as one variable increases, the other decreases).

Caution: There could be a nonlinear relationship between x and y and r could be near zero. Figure 2.10 illustrates a perfect quadratic relationship between x and y and yet the correlation is 0. This is a reminder that the correlation coefficient measures *linear* relationships.

Figure 2.9
Scatterplot for various values of r

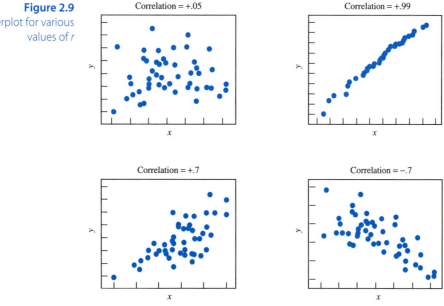

Figure 2.10

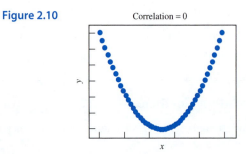

Correlation = 0

The following example shows two variables that are highly correlated.

Example 2.8   What's in a brand name? *Financial World* magazine, using its own complex formula, estimated how much the following brand names would be worth in cash. How are these estimated values related to each company's revenue? The table gives the brand name, its value in billions of dollars, and the company's revenue in billions:

WORKSHEET: Name.mtw

Brand name	Value	Revenue	Brand name	Value	Revenue
Marlboro	31.2	15.4	Wrigley's	1.5	1.0
Coca-Cola	24.4	8.4	Schweppes	1.4	1.3
Budweiser	10.2	6.2	Tampax	1.4	0.6
Pepsi-Cola	9.6	5.5	Heinz	1.3	0.8
Nescafe	8.5	4.3	Quaker	1.2	1.1
Kellogg	8.4	4.7	Colgate	1.2	1.1
Winston	6.1	3.6	Gordon's	1.1	0.6
Pampers	6.1	4.0	Hermes	1.0	0.5
Camel	4.4	2.3	Kleenex	0.8	0.7
Campbell	3.9	2.4	Carlsberg	0.7	0.8
Nestle	3.7	6.0	Haagen-Dazs	0.6	0.5
Hennessy	3.0	0.9	Fisher-Price	0.6	0.6
Heineken	2.7	3.5	Nivea	0.6	0.9
Johnnie Wal	2.6	1.5	Sara Lee	0.5	0.8
Louis Vuitt	2.6	0.9	Oil of Olay	0.5	0.6
Hershey	2.3	2.6	Planters	0.5	0.7
Guinness	2.3	1.8	Green Giant	0.4	1.0
Barbie	2.2	0.8	Jell-o	0.4	0.3
Kraft	2.2	2.8	Band-Aid	0.2	0.2
Smirnoff	2.2	1.0	Ivory	0.2	0.4
Del Monte	1.6	2.3	Birds Eye	0.2	0.3

Source: Financial World.

Construct a scatterplot and find the correlation between value and revenue.

Solution   Figure 2.11 shows a scatterplot of the data. It appears that the data fall close to a straight line. From Minitab, the correlation is found to be +.94, which is a very high positive correlation.   ∎

Figure 2.11
Brand data

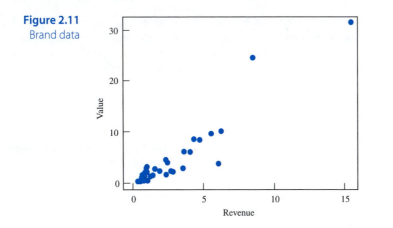

There are occasions when we expect that two variables are correlated but in fact the data do not support our intuition.

Example 2.9 Are complaints with airlines correlated with their ability to arrive on time? Listed here are the percentage of on-time arrivals and the number of complaints per 1000 passengers for 11 major airlines.

WORKSHEET: Airline.mtw

Airline	Percent on time	Complaints (per 1000)
Alaska	91.1	5.4
American	85.8	3.6
American West	90.8	4.0
Continental	87.2	4.6
Delta	85.7	4.6
Northwest	91.1	4.3
Pan Am	88.3	4.6
Southwest	93.5	3.6
TWA	88.4	5.4
United	87.2	4.9
USAir	87.3	4.4

Source: Transportation Department.

Figure 2.12
Airline data

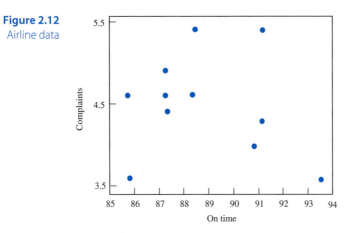

Figure 2.12 is the scatterplot of the data; the correlation is found to be $-.15$. The correlation coefficient is negative, which agrees with our intuition that as the percentage of on-time arrivals increases, the number of complaints should go down. However, the correlation of $-.15$ is not nearly as strong as one would suspect. We should not attempt to fit a straight line to these data.   ∎

Effect of Outliers

It was pointed out in Section 2.1 that outliers make it difficult to construct a scatterplot. Similarly, outliers sometimes have a dramatic effect on the value of the correlation coefficient.

Example 2.10   Figure 2.13 is a plot of the airline data given in Example 2.9 with one additional observation. Suppose there is one more airline that is on schedule 100% of the time and has only one complaint per 1000 passengers. With the addition of that one point, the correlation jumps from the previous value of $-.15$ to $-.74$. The increased correlation indicates that now a straight line will fit the data. In fact, the added observation, a clear outlier, has great influence on the direction of the line. It is an *influential observation* and should be singled out for individual study. We should exercise caution when interpreting bivariate data with outliers.   ∎

Influential observation

Figure 2.13
Airline data with
addition of (100, 1)

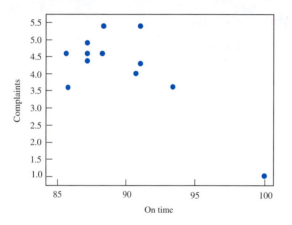

Correlation and Cause

A positive correlation between two variables means that large values of one variable tend to be associated with large values of the other variable. This does not necessarily mean that the large values of the first variable *caused* the large values of the other variable. Correlation measures the association between the variables, not the causal effect. For example, it has been shown that there is a positive correlation between class attendance and course grades. Does this mean that if you consistently attend class your grades will improve?

Not necessarily; it may be that the association we see between class attendance and grades is through a third variable, called a *lurking variable*. A lurking variable is a variable that has an effect on the response but is not included in the study. Motivation, though difficult to measure, may explain the association between class attendance and course grade. A motivated student attends class and a motivated student gets good grades. Just attending class is not enough to get good grades.

Lurking variable

Because two variables are related through a lurking variable does not mean that we should ignore the association between them. Universities all over the country use SAT scores for admission purposes because of the association between SAT scores and college grade point averages. Several lurking variables explain the association between SAT scores and grade point averages—motivation, study habits, home environment and, of course, intelligence. The highly motivated student with good study habits will achieve a higher SAT score and, most likely, a higher grade point average. It is much easier for the university to measure SAT scores than to measure motivation, study habits, home environment, and intelligence. By measuring the SAT score, it is, in essence, measuring all the other variables at the same time.

As we examine data, one of the main features of our strategy should be flexibility. No two data sets are exactly alike, and therefore no standard procedure works best all the time. We must be able to adapt the analysis to the problem. As we examine the data for structure, we need to respond to different patterns that are suggested by our analysis. We must approach the analysis with the goal of answering our questions, but at the same time, be open to the possibility of unexpected features that will prompt new questions.

Because of the ease with which we can analyze data with a computer, we must be willing to look at our data in any number of ways. For example, if the managers in charge of the launch of the *Challenger* had not ignored relevant data, it is generally agreed that the fatal launch would not have taken place.

Example 2.11   *Challenger* **Accident**

In seven previous flights of the space shuttle *Challenger*, there was damage to at least one O-ring, as documented in these data:

WORKSHEET: Challeng.mtw

flight	date	temp	failures
1	4/12/81	66	0
2	11/12/81	70	1
3	3/22/82	69	0
4	6/27/82	80	(hardware lost at sea)
5	11/11/82	68	0
6	4/4/83	67	0
7	6/18/83	72	0
8	8/30/83	73	0
9	11/28/83	70	0
41-b	2/3/84	57	1
41-c	4/6/84	63	1
41-d	8/30/84	70	1
41-g	10/5/84	78	0
51-a	11/8/84	67	0
51-c	1/24/85	53	3
51-d	4/12/85	67	0
51-b	4/29/85	75	0
51-g	6/17/85	70	0
51-f	7/29/85	81	0
51-i	8/27/85	76	0
51-j	10/3/85	79	0
61-a	10/30/85	75	2
61-b	11/26/85	76	0
61-c	1/12/86	58	1
61-i	1/28/86	31	(*Challenger* accident)

Source: Dalal, S. R., Fowlkes, E. B., Hoadley, B., (1989) Risk Analysis of the Space Shuttle: Pre-*Challenger* Prediction of Failure, *Journal of the American Statistical Association*, 84, (408), 945–957.

The scatterplot in Figure 2.14 of the number of failures versus the temperature at launch for the seven flights with at least one O-ring failure reveals no clear linear trend. The correlation between temperature and number of failures is only $r = -0.263$. Based on this graph and the small correlation coefficient, the managers recommended to proceed with the launch, although the forecasted temperature for the day of the launch was 31° F. The coldest temperature for any previous launch was 53° F, and with that there were three failures.

Figure 2.14
Plot where there was damage to at least one O-ring; $r = -0.263$

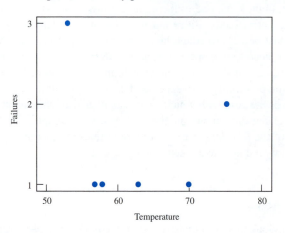

The fatal mistake was the failure to consider the 16 flights where there was *no* O-ring damage. With the exception of one data point (75,2), the scatterplot of all the data in Figure 2.15 shows a definite downward linear trend between the number of damaged O-rings and the temperature. The correlation between temperature and the number of failures using all the data is $r = -.561$. If we exclude the one outlier, (75,2), the correlation is $r = -0.714$. Had the managers considered this scatterplot and observed these correlations, surely they would have canceled the launch. ∎

Figure 2.15
Plot with all data points; $r = -0.561$

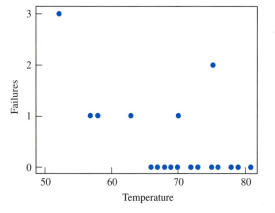

COMPUTER TIP

Correlation

To correlate data in two columns of the Minitab worksheet, first choose the **Stat** menu, select **Basic Statistics**, and click on **Correlation**. When the Correlation window appears, **Select** the columns and click on **OK**. If more than two columns are selected, a matrix of correlations appears.

EXERCISES 2.2 The Basics

2.20   Karl Pearson, who developed the correlation coefficient, gave the following historic data set on the heights (in inches) of brothers and sisters.

WORKSHEET: Pearson.mtw

Family	Height of brother	Height of sister
1	71	69
2	68	64
3	66	65
4	67	63
5	70	65
6	71	62
7	70	65
8	73	64
9	72	66
10	65	59
11	66	62

Source: Pearson, K. and Lee, A. (1902–3), On the Laws of Inheritance in Man, *Biometrika, 2,* 357.

a. Should one of the variables be considered the response variable?
b. Construct a scatterplot of the data. Does it matter which variable is placed on the vertical axis? Is there a linear trend? Describe the scatter in the data.
c. Calculate the correlation between the two variables. Is its value consistent with the appearance of the scatterplot?

2.21   The following data compare a child's age with the number of gymnastic activities successfully completed:

WORKSHEET: Gym.mtw

Age	2	3	4	4	5	6	7	7
Number of activities	5	5	6	3	10	9	11	13

a. Construct a scatterplot and sketch the straight line that fits the data.
b. Calculate the correlation coefficient. Is its value consistent with the appearance of the scatterplot?
c. Is there a positive or negative trend? Is that what you would normally expect?
d. Can you identify any lurking variables that would explain the correlation between age and number of activities completed?

2.22   If one were to measure the relationship between shoe size and vocabulary for elementary school children there would probably be a high positive correlation. Does this mean that bigger feet cause an increase in one's vocabulary? Is there a lurking variable that would explain the association that we observe between these two variables?

Interpreting Computer Output

2.23   These data produced the following scatterplot:

WORKSHEET: Correlat.mtw

X	42	61	12	71	52	48	74	65	53	63	55	94	19
Y	75	49	95	64	83	84	38	58	81	47	78	51	93

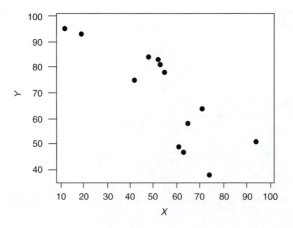

a. Do you see a linear trend? Is it positive or negative?
b. Sketch a straight line through the data. Is there more variability in the Y data when X is large or when X is small? What does this say about predicting Y when X is large?
c. Are there any data points that you would classify as outliers?
d. Is the correlation positive or negative?
e. If every X value is increased by 10 and every Y value is increased by 15, how will the appearance of the scatterplot change?
f. If X and Y are increased as in part e, how will the correlation change?

2.24   The data in worksheet: **Energy.mtw** give the amount of energy consumed in a home versus the size of the home. A listing of the data and a scatterplot is given in Exercise 2.5 in Section 2.1. The correlation between the size of the home and the kilowatts consumed per month is $+.94$. What does this suggest about the relationship between the two variables?

2.25   Exercise 2.11 in Section 2.1 presented the prices of wheat and the national weekly earnings of production workers for the years 1980–1998. Based on the following scatterplot from that exercise, do you think there is a high linear correlation between the two variables? Explain your answer. Should we be looking for a linear trend in these data?

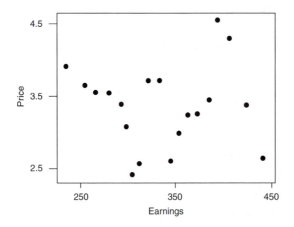

2.26   In Exercise 2.1 in Section 2.1, you constructed the following scatterplot relating aptitude test scores and productivity in a factory. Based on the scatterplot, is the correlation between the two variables closer to .5 or .9? Explain your reasoning. Does a straight line fit the data reasonably well?

WORKSHEET: Aptitude.mtw

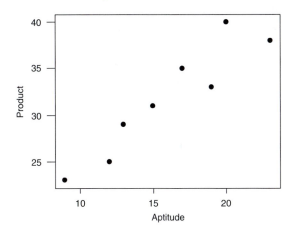

Computer Exercises

2.27   Recent studies suggest that women who smoke during pregnancy affect the birth weights of their newborn infants. To study this issue further, a sample of 16 women smokers were asked to estimate the average number of cigarettes they smoke per day. Following are their estimates and the birth weights of their children:

WORKSHEET: Cigarett.mtw

Cigarettes	22	16	4	19	42	8	12	30	14	16	5	20	32	2	15	48
Birth weight	6.4	7.2	8.1	6.9	6.1	8.4	7.6	6.5	8.4	8.1	8.5	6.6	6.0	7.9	7.1	5.5

a. Construct a scatterplot of the data.
b. Does there appear to be a linear trend in the data? Is it positive or negative?
c. Calculate the correlation coefficient. Is it positive or negative? What does this mean in terms of the linear trend?

2.28   There is an extreme outlier corresponding to Marlboro in the data in Example 2.8 (worksheet: **Name.mtw**). Remove the outlier and recalculate the correlation. Is there a significant change in the value of the correlation? Explain what effect this outlier has on the correlation coefficient. Should it be classified as an influential observation? Does the removal of Coca-Cola also affect the correlation? Is it an influential observation?

2.29   In Exercise 2.12 in Section 2.1, the private pay raise from the previous year was compared to the inflation rate for the years 1976–1998. From the scatterplot in that exercise, does there appear to be a linear trend? What does the correlation coefficient tell you about the linear trend? From the data, stored in worksheet: **Inflatio.mtw**, calculate the correlation coefficient and evaluate the linear trend.

2.30   Violent crimes are the offenses of murder, forcible rape, robbery, and aggravated assault. Following are the rates per 100,000 inhabitants for each of the 50 states and the District of Columbia for 1983 and 1993.

WORKSHEET: Crime.mtw

State	1983	1993	State	1983	1993	State	1983	1993
Ala	416.0	871.7	Calif	772.6	1119.7	D.C.	1985.4	2832.8
Alaska	613.8	660.5	Colo	476.4	578.8	Fla	826.7	1207.2
Ariz	494.2	670.8	Conn	375.0	495.3	Ga	456.7	733.2
Ark	297.7	576.5	Del	453.1	621.2	Hawaii	252.1	258.4

continued

WORKSHEET: Crime.mtw
continued

State	1983	1993	State	1983	1993	State	1983	1993
Idaho	238.7	281.4	Mo	477.2	740.4	Pa	342.8	427.0
Ill	553.0	977.3	Mont	212.6	169.9	RI	355.2	394.5
Ind	283.8	508.3	Neb	217.7	348.6	SC	616.8	944.5
Iowa	181.1	278.0	Nev	655.2	696.8	SD	120.0	194.5
Kan	326.6	510.8	NH	125.1	125.7	Tenn	402.0	746.2
Ky	322.2	535.5	NJ	553.1	625.8	Texas	512.2	806.3
La	640.9	984.6	NM	686.8	934.9	Utah	256.0	290.5
Maine	159.6	130.9	NY	914.1	1122.1	Vt	132.6	109.5
Md	807.1	1000.1	NC	409.6	681.0	Va	292.5	374.9
Mass	576.8	779.0	ND	53.7	83.3	Wash	371.8	534.5
Mich	716.7	770.1	Ohio	397.9	525.9	W.Va	171.8	211.5
Minn	190.9	338.0	Okla	423.4	622.8	Wis	190.9	275.7
Miss	280.4	411.7	Ore	487.8	510.2	Wyo	237.2	319.5

Source: Department of Justice, *Uniform Crime Reports for the U.S. in 1983 and 1993*, pp. 52–63 and 366.

a. Construct a scatterplot of the 1993 crime rate versus the 1983 crime rate.
b. Are there any outliers in the data?
c. Would the removal of outliers change the relationship between the 1983 crime rate and the 1993 crime rate?
d. Would the removal of outliers have an effect on the correlation coefficient?
e. Calculate the correlation both with and without the outlier and compare the results. Would you classify the outlier as an influential observation?

2.31 In Exercise 2.14 in Section 2.1, a scatterplot was constructed of the 1994 and 1995 ratings of cars by J. D. Power and Associates. The scatterplot showed a modest amount of scatter, and it appeared that a straight line would fit the data reasonably well. From the data, stored in worksheet: **JDPower.mtw**, calculate the correlation between the 1994 and 1995 ratings. Based on the scatterplot and the correlation, do you still think that a straight line fits the data?

Internet Tasks

2.32 Find the most recent issue of the Uniform Crime Reports from the Department of Justice web site. Find the latest violent crime rates for each of the 50 states and the District of Columbia. Enter the data into worksheet: **Crime.mtw** and construct scatterplots of these data with the 1983 and 1993 data. Describe the patterns you see in the two scatterplots. Calculate all pairwise correlations between the variables. Which year, 1983 or 1993, is most closely related to the current data?

2.33 Revisit the J. D. Power and Associates web site. In Exercise 2.17 you were asked to obtain the current ratings of cars listed in worksheet: **JDPower.mtw**. Calculate the correlation between the current car ratings and the 1995 ratings. Based on the scatterplot and the correlation, do you think that a straight line fits the data?

2.3 Least Squares Regression

Learning Objectives for this section:

❑ Know the difference between a response variable and a predictor variable.

❑ Understand that the primary goal of linear regression is to fit a straight line to the data.

❑ Know that the method of least squares is used to find the equation of the straight line.

❑ Know how to find the least squares regression line.

❏ Be able to find the residuals and calculate the sum of squares due to error.

❏ Be able to use the regression equation to predict future observations.

❏ Understand extrapolation and why it should be avoided if possible.

Numerical versus Numerical Variables

A scatterplot is used to visually detect relationships in bivariate data. If it appears that a straight line will fit data in a scatterplot, we measure the strength of the linear relationship with the correlation coefficient. Next, using regression analysis, we find the equation of the straight line that numerically describes the relationship so that predictions can be made. Unlike correlation, in regression we must distinguish between the *response (dependent) variable* and the *predictor (independent) variable*.

Regression Variables

- A **response variable**, also called a dependent variable, is the variable we wish to predict or describe based on the values of another variable.
- The **predictor variable**, also called the independent variable, is the variable that is used to predict the response variable.
- In the context of regression, the response variable is labeled y and the predictor variable is labeled x.

Example 2.12   Grades in high school and college are measured with the cumulative grade point average (GPA). College admissions officers attempt to predict the success of applicants based on an assumed relationship between the high school and college GPAs. Here, college GPA is the response variable y, and high school GPA is the predictor variable x.   ∎

As noted in the two previous sections, the most common relationship that exists between two variables is a straight line. To investigate the relationship, the first step is to plot the data in a scatterplot. If a linear relationship seems plausible, the value of the correlation coefficient, r, will shed light on the predictive power of the straight line.

Example 2.13   Following are the high school GPAs and the college GPAs at the end of the freshman year for ten different students:

WORKSHEET: Gpa.mtw

Student	1	2	3	4	5	6	7	8	9	10
High School GPA	2.7	3.1	2.1	3.2	2.4	3.4	2.6	2.0	3.1	2.5
College GPA	2.2	2.8	2.4	3.8	1.9	3.5	3.1	1.4	3.4	2.5

Graph the data in a scatterplot and then comment on the relationship between the two variables.

Solution   The college GPA is the response variable and is labeled on the vertical axis. The scatterplot in Figure 2.16 shows that the college GPA increases as the high school GPA increases. In fact, the dots appear to cluster along a straight line. The correlation coefficient is $r = +0.844$, which indicates that a straight line is a reasonable relationship between the two variables.   ∎

Figure 2.16
Scatterplot of high
school GPA versus
college GPA

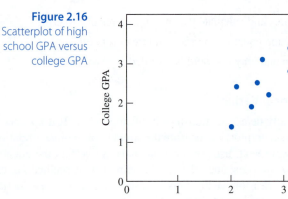

Fitting a Straight Line to Data: The Method of Least Squares

The equation of a straight line is

$$y = b_0 + b_1 x$$

where b_0 is the y-intercept and b_1 is the slope of the line. The y-intercept is the value of y when $x = 0$, and the slope gives the change in y relative to the change in x. To find the equation, we must find b_0 and b_1.

If the data points in Example 2.13 had fallen exactly in a straight line, there would be an exact straight-line equation relating college GPA (y) to high school GPA (x) and we would say there is no *error* in obtaining y from x. However, there is no reason to expect the points of the scatterplot to fall perfectly on a straight line. There could still be a linear relationship between the variables when the points fall off the straight line because of unknown error. For example, two people with the same high school GPA most likely will end up with different college GPAs because of other (lurking) variables, such as study habits and motivation. When these other variables cannot be measured (or we choose not to measure them) their effects are lumped together into what we call *residual error*.

From the equation of the line that *best* fits the data,

$$\hat{y} = b_0 + b_1 x$$

we can compute a *predicted y* for each value of x and then measure the error of the prediction. We use $\hat{y}$ (read "y hat") as the predicted y to distinguish it from the actual y.

Predicted Values and Residuals

- From the data point (x_i, y_i) the observed value of y is y_i and the **predicted value of y**, $\hat{y}_i$, is obtained by using the equation of the best-fit line:

$$\hat{y}_i = b_0 + b_1 x_i$$

continued

- The error of the prediction, e_i (also called the *residual*) is the difference in the actual y_i and the predicted $\hat{y}_i$. That is, the **residual** associated with the data point (x_i, y_i) is

$$e_i = y_i - \hat{y}_i$$

The residuals associated with the data points are depicted as vertical line segments in Figure 2.17. When we draw the best-fit line that comes as close as possible to all the observations, some of the residuals are positive (those above the line) and some are negative (those below the line). In fact, the sum of the residuals is zero because the line is positioned so that the negative residuals cancel the positive residuals. To measure the total error, the residuals are summarized into a single value by squaring (to avoid the problem of canceling residuals) each and summing them all. The resulting value is called the *sum of squares due to error*.

Figure 2.17
Fitted line and residuals

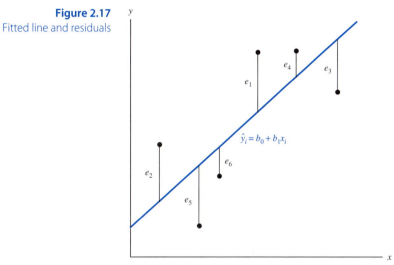

The **sum of squares due to error** (also called the residual sum of squares) is given by

$$\text{SSE} = \Sigma(y_i - \hat{y}_i)^2$$

Finding b_0 and b_1

To find the *regression line* that best fits the data, we must find values for b_0 and b_1. Substituting for $\hat{y}_i$, we have

$$\text{SSE} = \Sigma \, (y_i - b_0 - b_1 x_i)^2$$

which clearly depends on b_0 and b_1. The *method of least squares* chooses the values of b_0 and b_1 so that the value of SSE is as small as possible. (This is desirable because SSE is a combined measure of the discrepancy between the actual y and the predicted $\hat{y}$.) The formulas for the values of b_0 and b_1, obtained via the method of least squares, follow.

> The **least squares regression line** is
>
> $$\hat{y} = b_0 + b_1 x$$
>
> where
>
> $$b_1 = \frac{\sum(x_i - \bar{x})(y_i - \bar{y})}{\sum(\bar{x}_i - \bar{x})^2} = r\left(\frac{s_y}{s_x}\right)$$
>
> and
> $$b_0 = \bar{y} - b_1 \bar{x}$$
>
> The second expression for b_1 is obtained from the standard deviations, s_x and s_y, and the correlation r.

Example 2.14   Find the least squares regression line $\hat{y} = b_0 + b_1 x$ for the GPA data in Example 2.13.

Solution   From the data on high school GPA (x), we find

$$\bar{x} = 2.71 \text{ and } s_x = .4771$$

From the data on college GPA (y), we find

$$\bar{y} = 2.70 \text{ and } s_y = .7616$$

Previously we found $r = .844$. Substituting in these values, we find

$$b_1 = r\left(\frac{s_y}{s_x}\right) = .844\,\frac{.7616}{.4771} = 1.347$$

$$b_0 = \bar{y} - b_1 \bar{x} = 2.70 - (1.347)(2.71) = -0.950$$

The regression line that relates college GPA to high school GPA is therefore

$$\hat{y} = -0.950 + 1.347x$$

Figure 2.18 shows the scatterplot with a graph of the straight line. ■

Figure 2.18

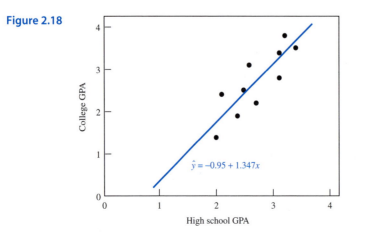

$\hat{y} = -0.95 + 1.347x$

College GPA (vertical axis)

High school GPA (horizontal axis)

Example 2.15   Find the residuals for the ten data points in Example 2.13 and then calculate SSE.

Solution   First we substitute the x_i values into the predicting equation found in Example 2.14: $\hat{y} = -0.950 + 1.347x$. The result is the predicted $\hat{y}_i$'s. They, along with the residuals, are listed in Table 2.1.

Table 2.1

x_i	y_i	$\hat{y}_i$	e_i
2.7	2.2	2.687	−0.487
3.1	2.8	3.225	−0.425
2.1	2.4	1.878	0.522
3.2	3.8	3.360	0.440
2.4	1.9	2.282	−0.382
3.4	3.5	3.629	−0.129
2.6	3.1	2.552	0.548
2.0	1.4	1.744	−0.344
3.1	3.4	3.225	0.175
2.5	2.5	2.417	0.083

Squaring the residuals and summing, we find that

$$\text{SSE} = \Sigma(y_i - \hat{y}_i)^2$$

$$= (-0.487)^2 + (-0.425)^2 + \cdots + (0.083)^2 = 1.5023$$

Because the values of b_0 and b_1, and hence the predicting equation, were obtained by the method of least squares, we can be assured that this value of SSE is as small as it possibly can be for the original data, (x_i, y_i). ∎

Prediction and Extrapolation

A very important reason for finding a regression line is to predict new values of the response variable from different given values of the predictor variable.

Example 2.16   In Example 2.8 the correlation between the value of a brand name and a company's revenue was found to be $+.94$. Find the least squares regression line to predict the value of a brand name from the company's revenue. Use the equation to predict values of brand names when the revenues are $5 billion, $10 billion, and $20 billion.

Solution   Shown here is computer output of the regression analysis of value using revenue as the predictor variable:

Regression Analysis

```
The regression equation is
value = -0.889 + 2.02 revenue

Predictor        Coef       Stdev     t-ratio          p
Constant      -0.8889      0.4174       -2.13      0.039
revenue        2.0244      0.1158       17.49      0.000

s = 2.096     R-sq = 88.4%      R-sq(adj) = 88.1%
```

```
Analysis of Variance

SOURCE          DF          SS          MS          F          p
Regression       1       1344.1      1344.1      305.82      0.000
Error           40        175.8         4.4
Total           41       1519.9

Unusual Observations
Obs.    revenue     value        Fit    Stdev.Fit    Residual    St.Resid
  1        15.4     31.200     30.287       1.553       0.913       0.65 X
  2         8.4     24.400     16.116       0.779       8.284       4.26R
 11         6.0      3.700     11.257       0.539      -7.557      -3.73R

R denotes an obs. with a large st. resid.
X denotes an obs. whose X value gives it large influence.
```

The printout is much more involved than we need at this time. Later, in Chapter 9, we will learn more about this output. Notice, however, that the regression equation is

$$\text{value} = -0.889 + 2.02\,\text{revenue}$$

Substituting the values 5, 10, and 20 for revenue and computing values yield the following predicted values of brand names when the revenues are $5, $10, and $20 billion:

Revenue	Predicted Value
$5 billion	$-0.889 + 2.02(5) = 9.211$ billion
$10 billion	$-0.889 + 2.02(10) = 19.311$ billion
$20 billion	$-0.889 + 2.02(20) = 39.511$ billion

∎

Remember that the regression line is obtained from the observed sample. Using the regression line to predict outside the range of the observed data is called *extrapolation*. Caution should be exercised when you attempt to extrapolate the line beyond the range of available data. We have no information about how the variables are related beyond the observed data. In Example 2.8, predicting the value of a brand name for a revenue of $20 billion is probably not wise. The maximum observed revenue is $15.4 billion for Marlboro, which in itself is way beyond the rest of the data and should be viewed as an outlier. There is no information to suggest that the straight line will continue in the same pattern beyond $15 billion.

COMPUTER TIP

Least Squares Regression Line

Store the bivariate data, (x_i, y_i), in two columns of the worksheet. For example, store the x values in C1 and the y values in C2. From the **Stat** menu, select **Regression** and click on **Regression** again. From the Regression window, **Select** the **Response** variable (y) and **Select** the **Predictor** variable (x). If you wish to compute the residuals, select **Storage** and click on **Residuals**. Finally click on **OK**.

EXERCISES 2.3   The Basics

2.34. Identify the response and predictor variables in the following:
   a. A study of the relationship between air pollution and high blood pressure.
   b. A study of the relationship between mental retardation and lead poisoning.
   c. A study of the relationship between the percent of drivers who wear seat belts and the death rate for automobile accidents.
   d. A study of the relationship between the suicide rate and alcohol consumption among teenagers.
   e. A study of the relationship between per capita income and public education expenditures of the states.

2.35 Previously (Exercises 2.1 and 2.26) we studied the relationship between aptitude test scores and productivity in a factory. Here are summary statistics from worksheet: **Aptitude.mtw**.

Aptitude:	Mean = 16	Standard Deviation = 4.63
Productivity:	Mean = 31.75	Standard Deviation = 5.97
Correlation = .936		

   a. Find the least squares regression equation.
   b. Based on the correlation, do you think the line fits the data?
   c. Find the residuals associated with two observations: (9,23) and (20,40).
   d. Find the predicted productivity for a worker whose aptitude test score is 15.
   e. Should we use the equation to predict productivity for a worker whose aptitude test score is 40?

2.36 Here are descriptive statistics associated with the study of the relationship between watching television and math scores described in Example 2.1.

Descriptive Statistics: percent, test

Variable	N	N*	Mean	Median	TrMean	StDev
percent	40	13	14.000	13.500	13.556	5.996
test	40	13	261.13	261.50	262.06	12.89

Variable	SE Mean	Minimum	Maximum	Q1	Q3
percent	0.948	6.000	33.000	9.000	18.000
test	2.04	218.00	281.00	256.00	271.50

In Example 2.7 we found the correlation between the two variables to be $r = -0.862$.
   a. Find the least squares regression equation.
   b. Based on the correlation, do you think the line fits the data?
   c. Find the residuals associated with two observations: Iowa (8,278) and Guam (20,233).
   d. Find the predicted math test score when the percent that watch 6 or more hours of television per day is 15.
   e. Should we use the equation to predict math test scores when the percent is 40?

2.37 Data corresponding to the life span of a particular electronic component and the heat to which it is exposed are given here and in Exercise 2.4 of Section 2.1.

WORKSHEET: Lifespan.mtw

Heat (C°)	50	100	150	200	250	300
Life span(hours)	875	884	762	424	365	128

   a. Find the least squares regression equation.
   b. Find the residuals.
   c. Calculate SSE.
   d. Find the predicted life span for a component exposed to 180°.

Interpreting Computer Output

2.38   Data relating robbery rate and the percent of low-income residents in eight precincts are given in Exercise 2.2 of Section 2.1. Using the data in worksheet: **Precinct.mtw** we obtain this fitted line plot.

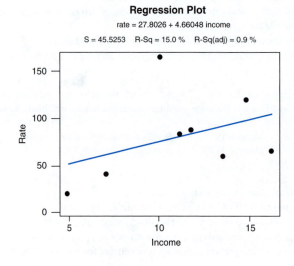

a. What is the least squares regression equation?
b. Do you think the line fits the data?
c. Would you use the equation in part **a** to predict robbery rate in a precinct with 10% low income?
d. If your answer in part **c** is yes, give your prediction. If your answer is no, what suggestions can you make to help predict robbery rate?

2.39   Data corresponding to maintenance costs on nine cash registers and their ages are recorded in Exercise 2.3 of Section 2.1. Using the data in worksheet: **Register.mtw** we obtained this fitted line plot.

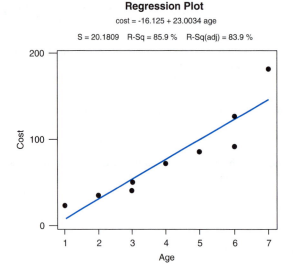

a. What is the least squares regression equation?
b. Does the line fit the data?
c. Would you use the equation to predict the maintenance cost of a cash register whose age is 5 years? What about 10 years old?

   d. If your answer in part **c** is yes, give your predictions. If your answer is no, what suggestions can
      you make to help predict maintenance cost?
   e. Find the residuals associated with two observations: (1,23) and (5,86).

2.40   The accompanying computer printout is associated with the J. D. Power and Associates ratings of
           1994 and 1995 cars described in Exercise 2.14 in Section 2.1 and Exercise 2.31 in Section 2.2.

WORKSHEET: JDPower.mtw

```
Correlation of 1994 and 1995 = 0.822

Regression Analysis

The regression equation is
1995 = 2.2 + 0.910 1994

Predictor          Coef        Stdev      t-ratio          p
Constant           2.22        14.64         0.15      0.880
1994             0.9098       0.1213         7.50      0.000

s = 21.72     R-sq = 67.6%     R-sq(adj) = 66.4%

Analysis of Variance

SOURCE            DF          SS          MS          F          p
Regression         1       26537       26537      56.26      0.000
Error             27       12735         472
Total             28       39272

Unusual Observations

Obs.     1994        1995         Fit    Stdev.Fit    Residual    St.Resid
29        193      195.00      177.82        10.17       17.18        0.90 X

X denotes an obs. whose X value gives it large influence.
```

   a. What is the correlation between the ratings of 1994 and 1995 cars?
   b. What is the least squares regression equation?
   c. Does it make sense to use the equation to predict 1995 ratings based on 1994 ratings?

2.41   Study the accompanying scatterplot that compares the crime rates (violent crimes per 100,000 popu-
           lation) for the different states, including the District of Columbia, to the percent of the population in
           the states without a high school degree.

WORKSHEET: Educat.mtw

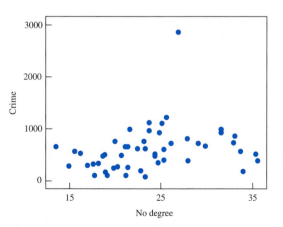

   a. Is there a trend in the data?
   b. Is the one bivariate outlier a univariate outlier with respect to the percent without a degree, with
      respect to the crime rate, or with respect to both variables?

    c. Do you think the outlier is from one of the states or the District of Columbia?

    d. Is it appropriate to fit the least squares regression line to these data? Explain.

Computer Exercises

2.42   Listed in the table are the grade point averages, SAT math scores, and final exam grades in college algebra for a group of 20 sophomores.

WORKSHEET: Sophomor.mtw

Student	GPA	SAT	Final Exam
1	2.6	510	84
2	2.1	460	77
3	3.5	680	94
4	1.7	390	45
5	2.2	420	71
6	2.9	510	89
7	2.3	370	65
8	3.2	550	90
9	2.3	420	82
10	2.8	470	87
11	3.9	700	99
12	2.2	450	83
13	1.4	380	63
14	2.7	440	75
15	2.8	460	79
16	3.0	520	85
17	1.8	430	70
18	2.0	410	72
19	3.6	650	98
20	2.5	500	75

    a. Determine the correlation between all pairs of variables.

    b. Do you think that SAT or GPA is a better predictor of the Final Exam score?

    c. Obtain the least squares regression equation relating Final Exam to the predictor you chose in part **b**.

    d. Given the correlation between SAT and GPA, does it really matter which variable you use as the predictor? Explain.

2.43   The number of convictions reported by U.S. attorneys' offices and the size of their staffs were considered in Example 2.2 of Section 2.1. Worksheet: **Attorney.mtw** gives data from 88 districts with one notable outlier corresponding to the Washington, D.C., attorney's office.

    a. Determine the correlation between convictions and staff both with and without the outlier. What effect did the outlier have on the correlation? Would you consider it an influential observation?

    b. To describe the relationship between convictions and staff, do you think that the outlier should be removed from the data set?

    c. Obtain the least squares regression equation relating convictions to staff size based on your answer to part **b**.

2.44   The brain weights and body weights of the 28 animals listed in Example 2.3 are given in worksheet: **Brain.mtw**. In Example 2.3, after taking logarithms of both variables, we noticed a definite linear pattern. We also noted three distinct outliers corresponding to the dinosaurs listed in the data set.

    a. Obtain the least squares regression equation relating log(brain wt) to log(body wt), first with the dinosaur data and then without.

    b. Plot both lines on their respective scatterplots. Describe the effects of the outliers on the regression equation.

2.45   In Exercise 2.16 in Section 2.1 we compared the actual number of German submarines sunk during World War II to the number reported by the Navy. Using the data in worksheet: **Submarin.mtw**, find the linear regression equation relating the actual count to the reported count. Construct a fitted line plot to see how well the line fits the data. Do you feel that the line adequately describes the relationship between the actual and reported number of submarines sunk during World War II?

2.46 The accompanying data give the number of icebergs sighted each month south of Newfoundland and south of the Grand Banks in 1920. Assess the relationship between the two. Do you think that a linear regression equation can be used to predict the number of icebergs in Newfoundland based on the number in the Grand Banks? If so, obtain the equation and use it to predict the number in Newfoundland when there are 15 in the Grand Banks.

WORKSHEET: Iceberg.mtw

Month	Jan	Feb	Mar	Apr	May	Jun	Jul	Aug	Sep	Oct	Nov	Dec
Newfoundland	3	10	36	83	130	68	25	13	9	4	3	2
Grand Banks	0	1	4	9	18	13	3	2	1	0	0	0

Source: N. Shaw, *Manual of Meteorology*, Vol. 2 (London: Cambridge University Press 1942), 7; and F. Mosteller and J. Tukey, *Data Analysis and Regression* (Reading, MA: Addison-Wesley, 1977).

Internet Tasks

2.47 Find the most recent issue of the Uniform Crime Reports from the Department of Justice web site. Find the crime rates (violent crimes per 100,000 population) for the different states, including the District of Columbia. From the Census Bureau web site, find the percent of the population in the states without a high school degree. Calculate the least squares regression equation relating the two variables. Based on the correlation and a scatterplot, do you think the line adequately fits the data?

2.48 From the *Statistical Abstract of the United States* web site, locate the most frequently requested data. Identify two variables that you think should be related. Calculate the least squares regression equation relating the two variables. Based on the correlation and a scatterplot, do you think the line adequately fits the data?

2.4 Assessing the Fit of a Line

Learning Objectives for this section:

❑ Be able to construct and evaluate residual plots.

❑ Understand how outliers affect the regression line.

❑ Know how influential observations affect the regression line.

❑ Know and understand how to interpret the coefficient of determination, r^2.

Having found the least squares regression line, next we ask: How well does it fit the data? That is, does the line effectively describe the relationship between x and y, or would some nonlinear relationship be better? Because the residuals measure the discrepancy between the line and the observed data, a plot of the residuals versus the predictor variable (x) graphically illustrates the deficiencies of the line as a model of the x–y relationship.

Residual Plots

A **residual plot** is a graph of the residuals versus the predictor variable x. When you examine a residual plot, look for the following:

• If the residuals oscillate randomly in a narrow band about the horizontal line at residual = 0.0, the linear equation has satisfactorily explained the variability in the response variable.

• If the residuals show a distinct pattern, the linear equation has not satisfactorily explained the variability in the response variable.

continued

- If the residuals show a curved pattern, the relationship is not linear and a higher order equation needs to be considered.
- If the residuals become more (or less) spread out as x increases, predictions of y for large x will not be accurate.
- Observations with large residuals are outliers but influential observations may not have large residuals.

Example 2.17   In Example 2.15 in Section 2.3 we found the residuals that resulted from fitting a straight line to predict college GPA based on high school GPA. Figure 2.19 is a plot of the residuals versus the predictor variable. Notice that the residuals oscillate in a somewhat random pattern about the horizontal line at residual = 0.0. This pattern is typical of data that do not deviate substantially from the model under study. It appears, in this case, that the straight line has explained most of the trend in these data.   ■

Figure 2.19

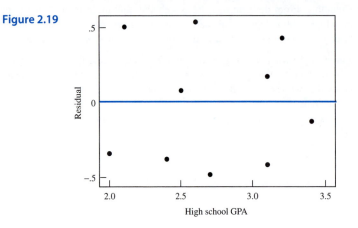

Often the straight-line model adequately describes the relationship between x and y and the residual plot bears this out, as in the previous example. However, there can be relationships that are not linear, as in the following example.

Example 2.18   The tensile strength of Kraft paper (in pounds per square inch) was measured for different percentages of hardwood in the batch of pulp that was used to produce the paper.

WORKSHEET: Hardwood.mtw

tensile	6.3	11.1	20.0	24.0	26.1	30.0	33.8	34.0	38.1	39.9
hardwood	1.0	1.5	2.0	3.0	4.0	4.5	5.0	5.5	6.0	6.5
tensile	42.0	46.1	53.1	52.0	52.5	48.0	42.8	27.8	21.9	
hardwood	7.0	8.0	9.0	10.0	11.0	12.0	13.0	14.0	15.0	

Source: G. Joglekar, et al., "Lack-of-Fit Testing When Replicates Are Not Available," *The American Statistician*, 43(3), (1989), 135–143.

Does a straight line fit the data well enough that the tensile strength of the paper can be predicted from the percentage of hardwood?

Solution   Figure 2.20 shows the residual plot after a straight line is fit to the data. It is clear that the pattern of the residuals is not random. It exhibits a *curvilinear* pattern. Often the scatterplot will reveal such nonlinear relationships, but if one overlooks the scatterplot and continues to fit a straight line to the data, the resulting residual plot will magnify the pattern and, as in this case, suggest that the straight line is an inadequate model. It is evident here that a higher order equation is needed to explain the relationship between x and y.   ■

Figure 2.20
Residual plot

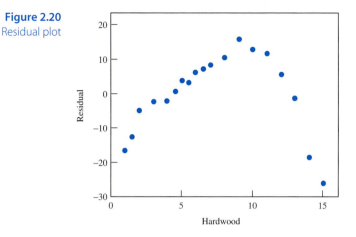

Example 2.19   The correlation between the percent of eighth-graders that watch 6 or more hours of television per day and scores on a national math test was found to be -0.862 in Example 2.7. This high correlation suggests that a straight line will adequately fit the data. Use the residuals, however, to detect possible problems in using a linear equation to predict math scores from the percent that watch 6 or more hours of television per day.

Solution   Figure 2.21 gives a residual plot after fitting a straight line to the data. Notice that the residuals "fan out" as the percent values increase. In other words, the variability of the residuals increases as the percent increases. This makes it more difficult to get accurate predictions with the linear equation when the percent that watch more than 6 hours of television per day is large.   ■

Figure 2.21

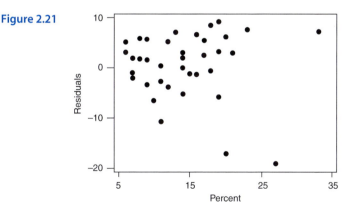

Effects of Outliers

Outliers, if present in the data, can have an effect on the regression line.

Example 2.20   In Example 2.3, we observed a linear relationship (after taking logarithms) between the brain weights and body weights of animals. Figure 2.22 is a scatterplot of the log data (worksheet: **Brain.mtw**) with the regression line drawn in as a solid line. Notice how the line is pulled down toward the three outliers associated with the dinosaur observations.

This is a major concern with least squares regression analysis. It is clearly not *resistant* to outliers. In this example, the dotted line is perhaps a more accurate description of the relationship between the two variables. ∎

Figure 2.22

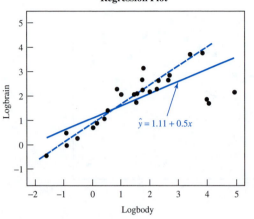

Regression Plot

$\hat{y} = 1.11 + 0.5x$

Influential Observations

In the regression setting, an outlier may or may not have an influence on the equation of the regression line. If an outlier still lies in the same plane as the main scatter of the data, it will not be an *influential observation*. If, however, it is an outlier in only the *x* variable, it probably is an influential observation and will have a significant effect on the regression equation. As suggested for all outliers, you should investigate influential observations for validity. Most computer packages have formal diagnostic procedures for identifying and evaluating outliers and influential observations. The computer printout in Example 2.16 identifies three unusual observations. All three are bivariate outliers, two of which have large residuals (see Figure 2.23). The third, Marlboro, has a significant effect on the regression line and is classified as an influential observation. It, as is the case with most influential observations, has a small residual because the line is pulled toward it.

Figure 2.23
Brand name data

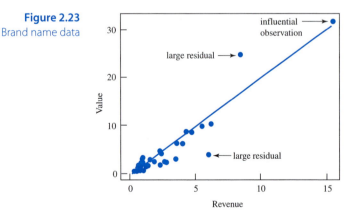

Coefficient of Determination

Another useful way to evaluate how well the straight line fits the data involves the correlation coefficient. Recall that it measures the strength of the linear association between *x* and *y*. Interpreting its value, however, can be difficult. A valid interpretation can be obtained from its square, r^2. Treating *y* as *depending* on *x*, we have that r^2 is the proportion of the variability in *y* that is explained by *x* through a linear relationship. The quantity r^2 is referred to as the *coefficient of determination*.

> The **coefficient of determination,** r^2, is the percent of the variability in the response variable that is explained by the predictor variable.

Example 2.21   Calculate the coefficient of determination for the data in Example 2.13 and give an interpretation.

Solution   Example 2.13 states that the correlation between high school GPA and college GPA is $r = .844$, so, the coefficient of determination is $r^2 = .712$. Thus, 71.2% of the variability in college GPAs is explained by the linear relationship that they have with the high school GPAs.   ∎

EXERCISES 2.4 The Basics

2.49   In Exercise 2.37 in Section 2.3 we found the following residuals resulting from fitting a straight line to the data corresponding to the life span of a particular electronic component as it is exposed to heat.

WORKSHEET: Lifespan.mtw

Heat (C°)	50	100	150	200	250	300
Life span (hours)	875	884	762	424	365	128
Residuals	−100.143	69.714	108.571	−68.571	33.286	−42.857

Plot the residuals in a residual plot. Is there a pattern to the residuals, or do they appear random? Do you think that the straight line adequately describes the relationship between the variables?

2.50   The data are PSAT and SAT scores for a group of high school seniors. The FITS are the predicted SAT scores from a linear regression equation. Find the residuals, plot them, and comment on the relationship between the variables.

WORKSHEET: Psat.mtw

Student	1	2	3	4	5	6	7
PSAT	760	1150	820	1060	950	1320	750
SAT	920	1100	1050	1340	1060	1500	920
FITS	938.44	1284.18	991.63	1204.40	1106.88	1434.89	929.58

2.51   From data relating aptitude test scores to productivity in a factory, we found the following least squares regression equation in Exercise 2.35 of Section 2.3.

WORKSHEET: Aptitude.mtw

Applicant	1	2	3	4	5	6	7	8
Aptitude score	9	17	13	19	20	23	12	15
Productivity	23	35	29	33	40	38	25	31

The regression equation is
product = 12.4 + 1.21 aptitude

a. Use the equation to find the FITS (predicted productivity) and the resulting residuals.
b. Construct a residual plot. Do the residuals appear random, or is there a pattern?
c. The correlation coefficient was previously found to be +.936. What is the coefficient of determination?
d. With this information, is it reasonable to assume that the straight line is the correct equation relating aptitude and productivity?

2.52   Data relating robbery rate and the percent of low-income residents considered in Exercise 2.2 of Section 2.1 have a correlation coefficient of only .388. What is the value of the coefficient of determination? What does this say about the fit of the line to the data?

Interpreting Computer Output

2.53   A study of the association between reading ability and IQ scores was conducted by a reading coordinator in a large public school system. The data from 14 eighth-grade students who were given a reading achievement test and an IQ test were listed in Exercise 2.10 in Section 2.1. The following fitted line plot was obtained from that data:

WORKSHEET: readiq.mtw

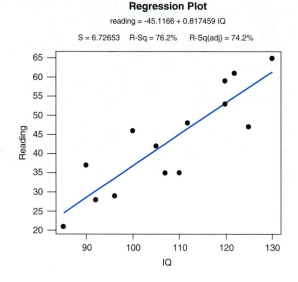

Regression Plot

reading = -45.1166 + 0.817459 IQ

S = 6.72653   R-Sq = 76.2%   R-Sq(adj) = 74.2%

a.   What is the least squares regression equation?
b.   Does the line fit the data?
c.   What is the value of the coefficient of determination? Give an interpretation.
d.   Would you use the equation in part **a** to predict reading ability based on IQ?
e.   If your answer in part **d** is yes, give your predictions for IQs of 100 and 120. What are the residuals for two students with observations (100,46) and (120,59)?

2.54   Shown here is a residual plot associated with the least squares regression equation found in Exercise 2.39 of Section 2.3. Have we properly described the relationship between the age and maintenance cost of a cash register?

WORKSHEET: Register.mtw

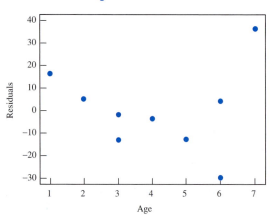

2.55 The scatterplot constructed in Exercise 2.5 in Section 2.1 showed how the amount of energy in a home is related to the size of the home. The scatterplot showed that a straight line is a reasonable fit for the data. Furthermore, the correlation between the size of the home and the kilowatts consumed per month is +.94, which again suggests that the straight line is a reasonable fit. Does the following residual plot support this discussion?

WORKSHEET: Energy.mtw

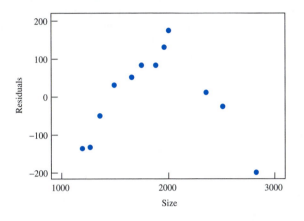

2.56 A straight line can be fit to any data set involving two variables. Following is a residual plot that resulted from fitting a straight line to the data corresponding to salinity level and depth of the ocean obtained aboard the *Ron Brown* in Example 2.4 in Section 2.1. There we stated that the relationship is clearly not linear. Does the residual plot support this statement?

WORKSHEET: RonBrown2.mtw

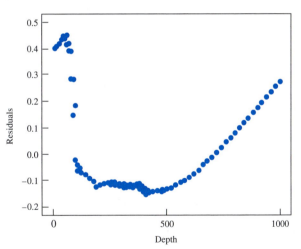

Computer Exercises

2.57 Worksheet: **Attorney.mtw** gives data from 88 U.S. attorneys' offices on the number of convictions and their staff size. There is one notable outlier corresponding to the Washington, D.C., attorney's office. In Exercise 2.43 of Section 2.3 you were to obtain the least squares regression equation of convictions based on staff size.

    a. Construct a residual plot based on the regression equation containing the Washington, D.C., data. Is there an unusual pattern?

    b. Construct a residual plot based on the regression equation without the outlier. Describe the pattern.

2.58   The brain weights and body weights of the 28 animals listed in Example 2.3 of Section 2.1 are given in worksheet: **Brain.mtw**. In Exercise 2.44 of Section 2.3 we obtained the linear regression equation of log(brain wt) versus log(body wt) both with and without the outliers corresponding to the dinosaurs listed in the data set.
    a. Compare the coefficient of determinations for the two linear regression equations.
    b. Construct residual plots for both equations. Describe the patterns shown in the two residual plots.

2.59   In Exercise 2.45 of Section 2.3 we found the linear regression equation relating the actual number of German submarines sunk during World War II to the number reported by the Navy.
    a. Using the obtained equation and the data in worksheet: **Submarin.mtw**, find the residuals and plot them in a residual plot. Describe the pattern.
    b. What is the value of the coefficient of determination? What can be said about how well the line fits the data?
    c. Does the line adequately describe the relationship between the actual and reported number of submarines sunk during World War II?

2.60   In Exercise 2.46 of Section 2.3 we found a linear regression equation relating the number of icebergs sighted each month south of Newfoundland and the number sighted south of the Grand Banks in 1920. Using the data in worksheet: **Iceberg.mtw**, construct a residual plot. Do you observe any unusual behavior in the residual plot? What is the value of the coefficient of determination? Describe what this means in terms of the linear relationship between the variables.

2.61   Worksheet: **Elderly.mtw** contains the percents of the population of each state over the age of 65 for the years 1985 and 1998.
    a. Construct a scatterplot of the 1998 percent versus the 1985 percent. Describe the pattern.
    b. Are there any outliers? If so, would you classify them as influential observations?
    c. Calculate the coefficient of determination between the two variables. What percent of the variability in the 1998 data is explained by the variability in the 1985 data? Do you think that a straight line adequately describes the relationship between the two?
    d. Would the removal of outliers increase, decrease, or not affect the coefficient of determination?
    e. Obtain the regression equation relating the two percents. Based on the slope of the regression equation, has the percent of the population over the age of 65 increased or decreased over this 13-year period?

Internet Tasks

2.62   Revisit the Environmental Protection Agency (EPA) web site. In Exercise 2.18 you were asked to find the latest figures on combined gas mileage and the displacement of the engine for a class of automobiles such as compacts or subcompacts. Using the least squares regression equation between the two variables, find the residuals and plot them in a residual plot. Describe the pattern. What is the value of the coefficient of determination? What can be said about how well the line fits the data?

2.63   Are public school teachers' average salaries related to state per capita revenue? Go to the National Education Association web site and find the current average teachers' salaries for the 50 states and the District of Columbia. Next, go to the Census Bureau web site and find the current state per capita revenues. Using the least squares regression equation between the two variables, find the residuals and plot them in a residual plot. Describe the pattern. What is the value of the coefficient of determination? What can be said about how well the line fits the data?

2.5   Relationships among Categorical Variables

Learning Objectives for this section:

❏ Know how to organize bivariate categorical data into a contingency table.

❏ Understand marginal totals and how they are used to find marginal percents.

❏ Understand the concept of independent categorical variables.

❏ Know how to construct and interpret a segmented bar graph.

❏ Know how to illustrate bivariate data in a comparison bar graph.

Categorical versus Categorical Variables

We summarized a single categorical variable in a frequency table that consisted of the various categories with their corresponding counts. When two variables are categorical, their bivariate data are displayed in a two-way frequency table called a *contingency table*. The entries in the table are counts of the number of observations falling into the *cross-classification* of the categories of the two variables. The word *contingency* is used because if there is an association between the two variables, we can say that the values assumed by one variable are *contingent* (dependent) on the values of the other variable.

Example 2.22   When customers take their cars back to the dealership for the 15,000-mile servicing, do they get more than they bargained for? A survey by *U.S. News and World Report* found that many dealers routinely sell far more maintenance services than manufacturers recommend. Customers pay for unnecessary inspections, cleaning, lubrication, and adjustments. Researchers randomly selected 122 dealerships in seven metropolitan areas: Philadelphia, Miami, Chicago, Dallas-Fort Worth, Denver, Los Angeles, and Seattle. Their findings are summarized in Table 2.2.

WORKSHEET: Dealers.xls

Table 2.2
Service rendered by 122 auto dealers

		Service Rendered		
		Number that replace parts before they are needed	Number that perform only services recommended by manufacturer	Total
	Honda	19	2	21
	Toyota	3	16	19
	Mazda	12	9	21
Type of	Ford	8	13	21
Dealership	Dodge	11	10	21
	Saturn	4	15	19
	Total	57	65	122

Analyzing data in a contingency table involves comparing the percents in the various categories. For example, it is easy to see that 21/122 = 17.2% of the dealerships in the study are Honda dealerships and 19/122 = 15.6% are Saturn dealerships. Furthermore, 57/122 = 46.7% of the dealerships unnecessarily replace parts. These *marginal percents* are obtained from the *marginal totals*, which are the row and column totals.

More interesting, however, is whether the type of service rendered is contingent (dependent) on the type of dealership. If not, we would say that the service and the type of dealership are *independent* and hence the service should be the same for all dealerships. That is, we would expect 57/122 or 46.7% of each type of dealership to perform unnecessary replacements. In the case of Honda, however, 19 of 21, or 90.5% perform unnecessary replacements. On the other hand, only 3 of 19, or 15.8% of the Toyota dealers perform unnecessary replacements. These percentages being so different from 46.7% suggest that service is contingent on the type of dealership. ∎

In Example 2.22, there are six types of dealerships and two types of service rendered; hence, we call the table a 6 × 2 contingency table. In a 6 × 2 contingency table, there are 12 *cells* for the 12 possible combinations of the two categorical variables. To evaluate completely, we should compare the percentages in *all* cells in the contingency table. Table 2.3 (on page 120) shows all percentages based on the row totals.

Table 2.3
Service rendered
by 122 auto
dealers

		Service Rendered		
		Number that replace parts before they are needed	Number that perform only services recommended by manufacturer	Total
	Honda	19 (90.5%)	2 (9.5%)	21
	Toyota	3 (15.8%)	16 (84.2%)	19
	Mazda	12 (57.1%)	9 (42.9%)	21
Type of	Ford	8 (38.1%)	13 (61.9%)	21
Dealership	Dodge	11 (52.4%)	10 (47.6%)	21
	Saturn	4 (21.1%)	15 (78.9%)	19
	Total	57 (46.7%)	65 (53.3%)	122

The percentages for Mazda, Ford, and Dodge do not differ significantly from the overall percentage of 46.7%; however, the percentages for Honda, Toyota, and Saturn are probably large enough to suggest an association between the type of dealership and the service rendered. A formal evaluation of the possible association between categorical variables will be presented in Chapter 8.

With service rendered having only two categories, a *segmented bar graph* can be used to illustrate the association we have observed between the two variables. In Figure 2.24 the segments show the numbers in the two service categories for each of the dealerships. Notice the difference between the Honda and Toyota bars. If the variables were independent, all the bars would have a rather uniform appearance like Mazda, Ford, and Dodge.

Figure 2.24
Segmented bar graph of
the type service rendered
by 122 auto dealers

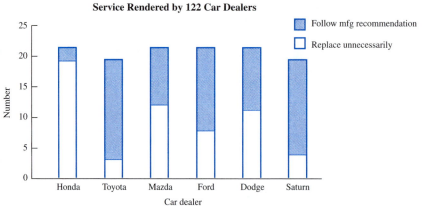

Example 2.23 illustrates that when we compare two or more categories in a contingency table, it is important to compare the percentages and not the frequencies. Percents are used because their values are independent of the sizes of the data sets.

Example 2.23 A sample of 1000 university students who live in dorms was surveyed about housing on campus. The students were asked to rate their current housing on the following scale:

1—very desirable
2—desirable
3—sufficient
4—livable
5—undesirable

The results showed that 120 students chose category 1, 180 chose category 2, 360 chose category 3, 240 chose category 4, and the remaining chose category 5. A second sample of 400 students who live off campus was given the same survey. Of that group, 32 chose category 1, 60 chose category 2, 160 chose category 3, 120 chose category 4, and the remaining chose category 5. Organize these data in a table listing frequencies and percents for the two groups.

Solution

Table 2.4
Student opinion of living conditions

Opinion	Dorm		Off Campus	
	Frequency	Percent	Frequency	Percent
very desirable	120	12%	32	8%
desirable	180	18	60	15
sufficient	360	36	160	40
livable	240	24	120	30
undesirable	100	10	28	7
Total	1,000	100%	400	100%

Note in Table 2.4 that twice as many students (240 versus 120) who live in dorms classified their living conditions as "livable." When expressed as a percent, however, more off-campus students (30% versus 24%) classified their conditions as "livable." ■

Bivariate data can also be displayed in a comparison bar graph.

Example 2.24 Using the percents for the two groups of students in Example 2.23, we can draw the comparison bar graph in Figure 2.25. Notice that the title at the top identifies the data, the legend on the right side identifies the two groups, and the vertical axis is labeled percent. The percents printed at the top of the bars are optional but often help clarify the data. ■

Figure 2.25
Comparison bar graph

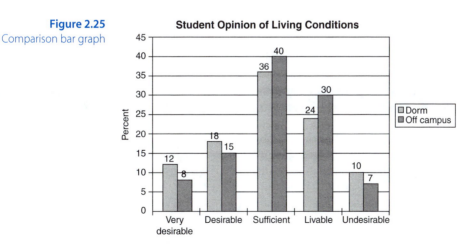

We can analyze three categorical (qualitative) variables in a three-way contingency table by using a two-way table of two of the variables at each level of the third variable, as demonstrated in Example 2.25.

Example 2.25 On April 22, 1987, the U.S. Supreme Court upheld Georgia's death penalty system by a 5-to-4 margin. The case involved the conviction of a black man for the murder of a white policeman in a 1978 robbery. The issue in question is whether race is a determining factor in the death penalty. David Baldus, a professor of law, and George Woodworth, a professor of statistics, presented the following three-way contingency table as part of their evidence.

| | | Race of Defendant | | | | |
| | | White | | Black | | |
		White victim	Black victim	White victim	Black victim	Total
Death	Yes	58	2	50	18	128
Sentence	No	687	62	178	1420	2347
	Total	745	64	228	1438	2475

Source: "Supreme Court Ruling on Death Penalty," *Chance*, 1 (1988) 7–8.

Newspaper accounts of the decision noted that the court upheld the death penalty *despite* the statistical evidence of racial disparity in its application. With just a few simple calculations, we can understand the statistical evidence presented by Baldus and Woodworth. First, collapse the table down to the following two-way table involving the variables race of the victim and death sentence:

		White victim	Black victim	Total
Death	Yes	108	20	128
Sentence	No	865	1482	2347
	Total	973	1502	2475

We now see that the percent of time that the death penalty is given when the victim is white is $108/973 = 11.1\%$, and the percent of time the death penalty is given when the victim is black is only $20/1502 = 1.3\%$. This suggest that the death penalty is almost 10 times more likely to be given when the victim is white than when the victim is black. Based on these data, it appears that Baldus and Woodworth had a strong case. ■

COMPUTER TIP

Chart Wizard

The Chart Wizard in Microsoft *Excel* is used to create graphs like the ones in this section. To create a chart, enter the data in the *Excel* worksheet as a table and follow the four steps of the *Chart Wizard*.

EXERCISES 2.5   The Basics

2.64   A sample of 120 employed people are classified according to gender and occupation. There were 48 women: 9 are blue-collar workers, 23 are white-collar workers, 4 are farmworkers, and the rest were classified as other. Of the 72 men, 31 are blue-collar workers, 22 are white-collar workers, 15 are farmworkers, and the rest are classified as other. Arrange these data in a bivariate contingency table.

a. What percent are women?

b. What percent of the women are blue-collar workers?

    c. What percent of the men are blue-collar workers?

    d. What percent are female blue-collar workers?

    e. Why are the answers to parts **b** and **d** different?

2.65   Draw a comparison bar graph depicting the frequency of men and women in each occupation group for the data in Exercise 2.64.

2.66   One hundred psychology majors were classified according to gender and class level. Ten were lower division women, 20 were upper division women, 40 were lower division men, and 30 were upper division men.

    a. Arrange the data in a bivariate table.

    b. Calculate the marginal totals.

    c. What percent of the women are lower division?

    d. What percent of the men are lower division?

    e. What percent of the lower division are women?

    f. Are the lower division students more likely to be women or men?

    g. Are the men more likely to be lower or upper division students?

2.67   Draw a segmented bar graph depicting the frequency of men and women in each class level for the data in Exercise 2.66.

2.68   Nobel prize–winner Linus Pauling (1901–1994) gave these incidences of colds among 279 French skiers who were given either vitamin C or a placebo.

	Cold	No Cold
Vitamin C	17	122
Placebo	31	109

    a. What percent of the vitamin C group had colds?

    b. What percent of the placebo group had colds?

    c. Based on these percentages do you think that vitamin C was beneficial in reducing the incidence of colds among these skiers?

2.69   Draw a comparison bar graph showing the incidence of cold and no cold among the vitamin C and placebo groups for the data in Exercise 2.68.

2.70   In Example 2.22, the service rendered by car dealerships was classified as replacing parts unnecessarily or performing services recommended by manufacturers. Collapse the dealerships into two categories, foreign cars and American cars. From the new table determine the following:

    a. Of the dealerships that replace parts unnecessarily, what percent are foreign dealerships?

    b. Of the dealerships that replace parts unnecessarily, what percent are American? Which are more likely to replace parts unnecessarily, foreign or American car dealerships?

    c. Of the foreign dealerships, what percent replace parts unnecessarily?

    d. Should the answers to parts **a** and **c** be the same? Explain.

2.71   From the three-way contingency table concerning Georgia's death penalty presented in Example 2.25, determine the following:

    a. What percent of the time is the death penalty given when the defendant is black and the victim is white?

    b. What percent of the time is the death penalty given when the defendant is white and the victim is black?

    c. What do the answers to **a** and **b** suggest about racial disparity in the application of the death penalty?

Interpreting Computer Output

2.72 A local election is to be held on the legislation of sales of alcoholic beverages. City residents were asked: Should the sale of alcoholic beverages within the city limits be legalized? The survey yielded these results:

	Yes	No	No Opinion	Total
Male	55	16	4	75
Female	53	26	9	88
Total	108	42	13	163

a. What percent of the respondents are male?
b. What percent of the respondents are female?
c. What percent answered yes to the survey?
d. What percent of the males answered yes to the survey?
e. What percent of the females answered yes to the survey?
f. Of the respondents who answered yes, what percent were male?
g. Is the following a comparison or segmented bar graph?

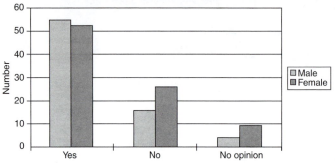

2.73 A survey of 500 students, 100 faculty members, and 30 administrators revealed that 360 students oppose the present parking policy, 60 faculty oppose it, and only 5 administrators oppose it. Arrange the data in a bivariate table by attitude (favor, oppose) and status (student, faculty, administrator).

a. What percent of those sampled oppose the present parking policy?
b. What percent of the students oppose the policy?
c. What percent are faculty and oppose the policy?
d. Why does the following graph not give a realistic comparison of the opinions of the students, faculty, and administration on the parking policy?
e. What should be done to the bar graph so that it will give a realistic comparison of the opinions of students, faculty, and administrators on the parking policy?

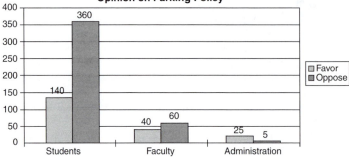

2.74 A sample of 1200 university students consisting of 700 men and 500 women was surveyed on the issue of the possession of marijuana.

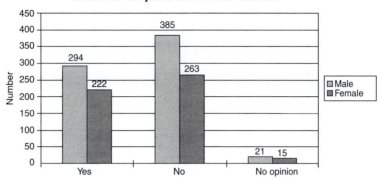

Do you believe that the possession of a small amount of marijuana is a criminal offense?

a. How many responded yes to the question?
b. How many responded no to the question?
c. How many women, compared to men, responded yes?
d. How does the percent of women answering yes compare to the percent of men?
e. Which is the correct interpretation of the results, **c** or **d**?

2.75 Women received 53% of all the bachelor's degrees awarded in 1990 compared with only 43% in 1970. Moreover, women are entering male-dominated fields. In 1990, for example, 13.8% of all engineering degrees were awarded to women compared with less than 1% in 1970. The table lists the percentage of bachelor's degrees awarded women in different fields in 1970 and 1990.
a. What areas show a dramatic change from 1970 to 1990?
b. What areas remained about the same from 1970 to 1990?
c. What percent of the bachelor's degrees in health were awarded to men in 1990?
d. Would the differences show up better in comparison pie charts or comparison bar graphs?
e. Would a graphical display illustrate the differences better than the table?

WORKSHEET: Degree.mtw

Field	Percent awarded, 1970	Percent awarded, 1990
Health	78.0	84.3
Education	75.0	78.1
Foreign languages	73.4	73.4
Psychology	43.3	71.5
Fine arts	57.3	67.5
Life sciences	27.8	50.7
Business	8.7	46.7
Social science	37.1	44.2
Physical sciences	13.6	31.2
Engineering	0.7	13.8
All fields	43.1	53.2

Source: U.S. Department of Health and Human Services, National Center for Education Statistics.

Computer Exercises

2.76 The following data are from a poll by George Gallup on people's attitude toward the treatment of the possession of small amounts of marijuana as a criminal offense in the United States in 1980. The question was: Do you think the possession of small amounts of marijuana should or should not be treated as a criminal offense?

WORKSHEET: Gallup.mtw

	Should be treated as a criminal offense	Should not be treated as a criminal offense	No opinion
National	43%	52%	5%
Gender			
Male	42%	53%	5%
Female	44%	51%	5%
Education			
College	30%	67%	3%
High School	45%	50%	5%
Grade School	58%	33%	9%
Age			
18–24	27%	67%	6%
25–29	26%	70%	4%
30–49	45%	52%	3%
50-older	54%	39%	7%
Religion			
Protestant	49%	47%	4%
Catholic	39%	55%	6%

Source: George H. Gallup, *The Gallup Opinion Index Report No. 179* (Princeton, NJ: The Gallup Poll, July 1980), p. 15.

a. Using only the levels of the gender variable, draw a bar graph illustrating the opinions on the issue of the possession of marijuana.
b. Using only the education level, draw a bar graph showing the opinions on the issue of the possession of marijuana.
c. Using only the age level, draw a bar graph illustrating the opinions on the issue of the possession of marijuana.

2.77 Hodgkin's disease is a cancer of the lymph nodes. In one study, 538 patients with the disease were classified by histological type and by their response to treatment after 3 months. The histological types are LP = lymphocyte predominance, NS = nodular sclerosis, MC = mixed cellularity, and LD = lymphocyte depletion. The relationship between histological type and response to treatment may be examined from the table.

WORKSHEET: Hodgkin.mtw

		Response			
		Positive	Partial	None	Total
	LP	74	18	12	104
Histological	NS	68	16	12	96
type	MC	154	54	58	266
	LD	18	10	44	72
Total		314	98	126	538

Source: I. Dunsmore, F. Daly, *Statistical Methods, Unit 9, Categorical data,* Milton Keynes: The Open University, 18.

a. Of the patients with histological type LP, what percent had a positive response?
b. Of the patients with positive responses, what percent have histological type LP?
c. Why are the percentages in parts **a** and **b** so different?
d. What percent of all patients had no response to treatment?

2.78 The data in the table were collected on 1398 children to determine whether carriers of the bacterium *Streptococcus pyogenes* have larger tonsils than noncarriers.

WORKSHEET: Tonsils.mtw

		Carrier Status		Total
		Carrier	Noncarrier	
Tonsil size	Normal	19	497	516
	Large	29	560	589
	Very large	24	269	293
Total		72	1326	1398

Source: W. J. Krzanowski, *Principles of Multivariate Analysis* (Oxford: Oxford University Press, 1988), p. 269.

a. Of the children with very large tonsils, what percent are carriers?
b. Of the children with normal tonsils, what percent are carriers?
c. Of all carriers, what percent have very large tonsils and what percent have normal tonsils?
d. Does the evidence present here show that there is a relationship between tonsil size and the carrier status of these children?

2.79 Worksheet: **Abilene.mtw** contains crime index information for 1992 and 1999. Create a comparison bar graph to compare the types of crimes in 1992 to 1999.
a. What crimes increased from 1992 to those in 1999?
b. What crimes decreased from 1992 to 1999?
c. Did any crimes stay about the same?
d. What crime is most prevalent? Least prevalent?

Internet Tasks

2.80 Go to the web site of the Gallup Organization. Find the results of a current issue survey where the data are organized by two variables in a contingency table. Use a segmented bar graph and comparison pie charts to illustrate the data.

2.81 Find the web site for the *Statistical Abstract of the United States* of the Bureau of the Census. Find a categorical variable that has at least two categories and is classified according to gender. Arrange the data in a contingency table and construct a comparison bar graph. Discuss whether men and women respond differently to the categories of the variable.

2.6   Summary and Review Exercises

Key Concepts

■ Bivariate data in which both variables are numerical are displayed in a *scatterplot*. Categorical bivariate data are displayed in a *contingency table*.

■ *Pearson's correlation coefficient* is a summary measure of the degree of association between two variables.

■ In regression analysis, the independent variable is called the *predictor variable* and the dependent variable is called the *response variable*.

■ The equation of the least squares regression line is

$$\hat{y} = b_0 + b_1 x$$

where

$$b_1 = \frac{\Sigma(x_i - \bar{x})(y_i - \bar{y})}{\Sigma(x_i - \bar{x})^2} = r\left(\frac{s_y}{s_x}\right)$$

and

$$b_0 = \bar{y} - b_1 \bar{x}$$

- The predicted value of y is

$$\hat{y}_i = b_0 + b_1 x_i$$

- The residual associated with the data point (x_i, y_i) is

$$e_i = y_i - \hat{y}_i$$

- The sum of squares due to error is

$$\text{SSE} = \Sigma(y_i - \hat{y}_i)^2$$

- Examining residuals is a way of assessing the fit of a line.

- The coefficient of determination is r^2, the square of the correlation coefficient.

Statistical Insight Revisited

The scatterplot of the vote count for Pat Buchanan versus the total votes cast in each county in Florida is reproduced here. The data are in worksheet: **Florida2000.mtw.**

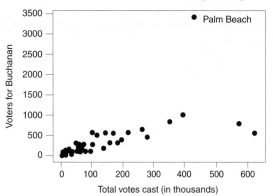

Florida Presidential Vote by County

We see a distinct positive linear trend in the data. There are, however, three outliers. One outlier, Palm Beach County, is clearly classified as an influential observation. A straight line will fit the data, but should we include the outliers? The correlation between Buchanan and Total is 0.678, but without the outliers it jumps to a staggering 0.922. It is clear that three counties did not follow the voting trend of the other counties in Florida. Palm Beach County, in particular, cast over 3400 votes for Buchanan when no other county cast more than 1100 votes for him. In fact, the median count for Buchanan was only 114 votes.

The regression equation that fits the data is

```
BUCHANAN = 52.7 + 0.00231 Total
```

Without outliers it is not much different

```
BUCHANAN = 45.1 + 0.00223 Total
```

Using either equation the predicted vote count for Buchanan in Palm Beach County is less than 1053, some 2300 votes less than what he actually got. Because Al Gore lost Florida by 537 votes, it is clear that he would have won Florida and the presidency had Palm Beach County voted like other counties in Florida.

Questions for Review

Use the following questions to test your skills:

The Basics

2.82 Graph the equation $y = 15.3 - 4.7x$. Find the residual for (2,3.8).

2.83 Identify the response variable and predictor variable in the following studies:
a. A study involving the yield of a wheat crop and the amount of rainfall.
b. A study of health care costs and the number of new AIDS cases.
c. A study of the crime rate and poverty rate.

2.84 Two hundred people were interviewed to determine if they think children of working mothers are adequately cared for. Their opinions are given here:

	Men	Working Mothers	Nonworking Mothers
Yes	28	44	18
No	32	14	64

a. What percent of the sample are women?
b. What percent believe the children are adequately cared for?
c. What percent of the working mothers feel that the children are adequately cared for?
d. Of those who think children are not adequately cared for, what percent are women?

2.85 Draw a bar graph depicting the frequency of those who responded yes and no among the men, working mothers, and nonworking mothers from the data in Exercise 2.84.

2.86 Members of a group of students selected randomly from the student body were asked to identify their major. There were 58 women: 9 were science majors, 15 were business majors, 22 were education majors, and the rest were liberal arts majors. There were 44 men: 17 were science majors, 15 were business majors, 5 were education majors, and the rest were liberal arts majors.
a. Arrange the data in a bivariate frequency table.
b. What percent are science majors?
c. What percent of the women are education majors?
d. What percent of the education majors are women?

2.87 Draw a segmented bar graph illustrating the number of men and women in each of the four majors using the bivariate frequency table found in Exercise 2.86.

2.88 Listed here are the 1992, 1993, 1997, and 1998 rankings of our favorite breeds of dogs.

WORKSHEET: Dogs.mtw

Dog	1992	1993	1997	1998
Labrador	1	1	1	1
Rottweiler	2	2	2	4
Shepherd	4	3	3	3
Spaniel	3	4	8	13
Retriever	6	5	4	2
Poodle	5	6	5	7
Beagle	7	7	6	6
Dachshund	8	8	7	5
Dalmatian	15	9	12	30
Shetland	9	10	15	15
Pomeranian	12	11	10	10
Yorkshire	14	12	9	9

continued

WORKSHEET: Dogs.mtw
continued

Dog	1992	1993	1997	1998
Shih Tzu	11	13	11	11
Schnauzer	13	14	14	14
Chow	10	15	35	39
Chihuahua	16	16	12	8
Boxer	17	17	13	12
Husky	18	18	16	18
Doberman	20	19	22	22
Springer	19	20	25	27

Source: The World Almanac and Book of Facts, 2000.

Find the correlation between the 1992 and 1993 rankings. Find the correlation between the 1997 and 1998 rankings. Was there a greater change in rankings from 1992 to 1993 or from 1997 to 1998?

2.89   It is suspected that the more people there are in a family, the less the cost per person per week for groceries. To evaluate this, a marketing institute randomly sampled 20 families:

WORKSHEET: Family.mtw

Number in family	2	2	1	3	4	3	2	4	1	3	5	2	2	3	4	1	2	6	3	2
Cost per person	78	85	88	76	72	74	79	69	79	75	68	82	78	72	76	90	84	67	77	79

a. Draw a scatterplot of the number in the family and the cost per person for food.
b. Find the correlation between the number in the family and the cost per person for food.
c. Is there a linear trend to the data? Is it positive or negative?
d. Find the least squares regression line that fits the data.
e. Predict the cost for groceries for a family of four.

2.90   A group of college students diagnosed as having dyslexia were asked to attend a reading workshop. From a reading test, the number of words read per minute was recorded for each student:

WORKSHEET: Dyslexia.mtw

Words/min	Age	Gender	L/R (handed)	Weight (lb)	Height (in)	Children in family
165	21	M	L	165	70	2
201	18	F	R	115	66	1
75	19	F	R	138	65	4
124	19	M	R	187	72	3
105	20	F	L	100	61	2
143	18	M	R	210	71	1
126	19	M	L	178	69	1
92	20	M	R	155	68	3

a. Identify each variable as either numerical or categorical.
b. Identify the numerical variables as either continuous or discrete.
c. Draw a scatterplot of words/minute by handedness.

2.91   Suppose the following display represents 500 automobile accidents that occurred in a large city:

	No fatalities	At least one fatality
Involved alcohol	68	142
No alcohol	194	96

a. Fill in the marginal totals.
b. What percent of the accidents involved alcohol?
c. What percent of the accidents with at least one fatality involved alcohol?
d. What percent of the accidents not related to alcohol had at least one fatality?
e. What percent of the accidents involved alcohol and had at least one fatality?
f. Based on the above percents, do you think alcohol and fatalities are related?

Interpreting Computer Output

2.92 Shown here is the scatterplot given in Figure 2.11 that describes the relationship between value of a brand name product and the company's revenue. There are three outliers in the data.

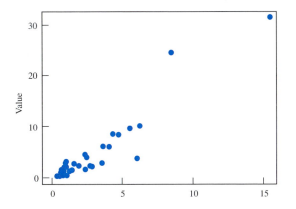

Refer to Example 2.8 and label each outlier by its brand name. Describe in detail what effect the outliers will have on the least squares regression equation. Which one will have the most effect on the equation of the regression line? Should one, two, or all three outliers be removed before calculating the least squares regression equation?

2.93 Four hundred employees of a medium-sized company were asked to rate their satisfaction with their job as happy, OK, or unhappy and indicate their marital status. The accompanying contingency table shows the numbers in the various categories:

		Marital Status				
		Single	Married	Divorced	Widowed	Total
	Happy	46	46	28	5	125
Satisfaction	OK	30	64	42	18	154
with Job	Unhappy	12	42	62	5	121
	Total	88	152	132	28	400

a. What percent of the married employees are happy with their jobs?
b. What percent of those who are happy with their jobs are married?
c. What percent of the employees are married and happy with their jobs?
d. Which of the four groups is most happy with their jobs?

2.94 Here are graduation rates for students and student athletes who entered schools in the Big Ten Conference in 1993–94:

WORKSHEET: BigTen.mtw

School	Student	Athlete
Illinois	75	65
Indiana	68	66
Iowa	62	71

continued

WORKSHEET: BigTen.mtw
continued

School	Student	Athlete
Michigan	82	68
Michigan St	64	56
Minnesota	51	65
Northwestern	92	93
Ohio State	56	50
Penn State	80	78
Purdue	64	72
Wisconsin	74	55

Source: NCAA Graduation Rates Report, 2000.

a. Explain the pattern you see in the scatterplot of the graduation rates of all students versus student athletes.
b. Based on the correlation coefficient, do you think there is a relationship between these two variables?

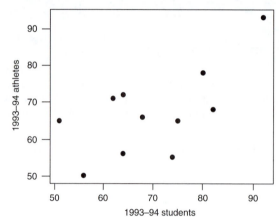

Correlations: 1993–94students, 1993–94athletes
Pearson correlation of students and athletes = 0.641

2.95 In 1990, 53.8% of American Indian adults living on reservations were high school graduates or had even more education. This compares to 75% in the general population. Is the educational attainment of American Indians related to their income or to their poverty rate? Following are the percents of persons 25 years or older with a high school diploma or higher, the per capita income, and the percent in poverty on ten reservations:

WORKSHEET: Indian.mtw

Reservation	Percent high school	Per capita income	Poverty rate
Blackfeet, MT	66.3	$4718	50.1
Hopi, AZ	62.6	4566	49.4
Rosebud, SD	59.3	3739	60.4
Zuni Pueblo, AZ-NM	55.4	3904	52.5
Pine Ridge, NE-SD	55.2	3115	66.6
San Carlos, AZ	49.4	3173	62.5
Fort Apache, AZ	48.3	3805	52.7
Papago, AZ	47.3	3113	65.7
Navajo, AZ-NM-UT	41.1	3735	57.8
Gila River, AZ	37.3	3176	64.4

a. Describe the pattern shown in the scatterplot of per capita income versus the percent with a high school diploma.
b. Describe the pattern shown in the scatterplot of poverty rate versus percent with a high school diploma.
c. Does either scatterplot in part **a** or **b** show a distinct linear relationship?
d. Based on the following correlation matrix, what does the correlation between the percent with a high school diploma and the per capita income say about the relationship between the two? What about the relationship between percent with a high school diploma and the poverty rate? Is poverty rate and per capita income related? Explain.

Correlations: highsch, income, poverty

```
              highsch    income
income         0.696
poverty       -0.542    -0.928
```

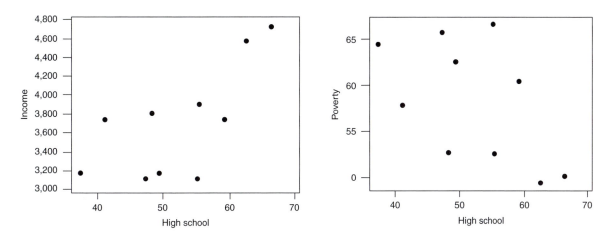

Computer Exercises

2.96 Following are anxiety scores before a major math test and the math test scores for a group of tenth-grade students:

WORKSHEET: Anxiety.mtw

Anxiety	12	16	6	22	17	14	8	27	18	19	11	9	21	3	15	17	22	19	13	5
Math test	75	56	91	48	69	73	88	65	72	65	88	94	48	99	78	65	58	52	70	90

a. Compute the correlation between the anxiety scores and the test scores.
b. Find the least squares regression line that relates the anxiety scores and the test scores.
c. Store the residuals in a column and plot these against the anxiety scores. Do the residuals exhibit a random pattern?

 2.97 The *IAAF/ATFS Track and Field Statistics Handbook* for the 1984 Los Angeles Olympics listed the national records for women in the 100-meter, 200-meter, and 400-meter races.
Source: Dawkins (1989).

a. From worksheet: **Track.mtw**, construct a scatterplot of the 100-meter and 200-meter records.
b. Construct a scatterplot of the 100-meter and 400-meter records.
c. Which plot has the greater scatter? Why?
d. Which set of variables, the one in **a** or the one in **b**, will have the higher correlation?
e. Calculate the two correlations to verify your answer to part **d**.
f. Does a straight line describe the relationship between the 100- and 200-meter records? How about between the 100- and 400-meter records?

2.98 The following data are the mean annual levels of Lake Victoria Nyanza for the years 1902–1921 and the number of sunspots in each year:

WORKSHEET: Victoria.mtw

Year	Level	Sunspot
1902	−10	5
1903	13	24
1904	18	42
1905	15	63
1906	29	54
1907	21	62
1908	10	49
1909	8	44
1910	1	19
1911	−7	6
1912	−11	4
1913	−3	1
1914	−2	10
1915	4	47
1916	15	57
1917	35	104
1918	27	81
1919	8	64
1920	3	38
1921	−5	25

Sources: N. Shaw, *Manual of Meteorology*, Vol. 1 (London: Cambridge University Press, 1942), p. 284; and F. Mosteller, and J. W. Tukey, *Data Analysis and Regression* (Reading, MA: Addison-Wesley, 1977).

a. Identify the response and predictor variables.
b. Construct a scatterplot of the data with the predictor variable on the horizontal axis.
c. Do you detect a linear trend in the data? Is it positive or negative?
d. Are there any outliers in the data?
e. Calculate the correlation coefficient.
f. Find the least squares regression equation that relates the level of the lake to the number of sunspots.
g. Predict the level of the lake when the number of sunspots is 50. Would you use this equation to predict the level when the number of sunspots is 200? Explain.

Probability and Probability Distributions

Recall that it is usually impractical to investigate a population in its entirety. Therefore, we resort to investigating a sample and inferring from it to the population. Because the decisions or predictions made about a population are based on sample information, a degree of uncertainty is involved. That uncertainty is measured with probability.

In this chapter, we introduce the vocabulary and laws of probability and present problems that illustrate how probabilities are determined. In later chapters, you will see how probability relates to statistical inference.

Contents

■ STATISTICAL INSIGHT
3.1 Basic Probability Concepts
3.2 The Binomial Distribution

3.3 The Normal Distribution
3.4 Summary and Review Exercises

Let's Make a Deal: The Monty Hall Three-Door Problem

For years, fans of the popular *Let's Make a Deal* television show have watched Monty Hall give contestants the chance to choose, among three doors, the one that concealed the prize of the day. Behind two of the doors were gag gifts, but the other door concealed a valuable prize. After the contestant chose a door, Monty opened one of the other doors, revealing a gag gift. The contestant was then asked whether he or she wished to stay with the original choice or switch to the other closed door. What should the contestant do? Is it better to stay with the original choice or switch to the other closed door? Or does it really matter? The answer, of course, depends on whether contestants would improve their chances of winning by switching doors. In particular, what is the probability of winning by switching and what is the probability of winning by staying?

Columnist Marilyn vos Savant posed this question to her readers in the popular *Parade* magazine column "Ask Marilyn" (1990a). Thousands responded, including many Ph.D.'s in mathematics and statistics (1990b, 1991a, 1991b). Most respondents, laypersons and experts alike, concluded that the probabilities of winning were the same for switching and for staying. Are they right?

In this chapter, we will develop the tools necessary to understand this "three-door problem," as well as other interesting probability problems. See Section 3.4 for a discussion of this Statistical Insight.

3.1   Basic Probability Concepts

Learning Objectives for this section:

- ❏ Learn the basic terms associated with the study of probability.
- ❏ Be able to list the outcomes in a sample space.
- ❏ Know what an event is and be able to determine whether it has occurred.
- ❏ Be able to assign probabilities to the individual outcomes in a sample space.
- ❏ Be able to calculate the probability of an event.
- ❏ Know the difference between mutually exclusive events and independent events.
- ❏ Know how to use the additive, complement, and multiplication laws of probability.
- ❏ Know how to simulate data with the Random Data command.

One of the most important goals of statistics is to make inferences about a population from the information contained in a sample. Once an inference is made, however, we must measure the reliability of that inference. This is accomplished with probability. First, we introduce some basic terminology.

- • An **experiment** is the process of making an observation or taking a measurement.
- • The collection of all possible outcomes of an experiment is called the **sample space, S**.

If the experiment is rolling a die, the sample space is

$$S = \{1,2,3,4,5,6\}$$

If the experiment is tossing a coin, the sample space is

$$S = \{\text{head, tail}\}$$

If the experiment is weighing a person, the sample space (assuming no one weighs more than 600 pounds) is

$$S = \{x \mid x \text{ is a real number between 0 and 600}\}$$

- Any subset of the sample space is called an **event**.
- An event is said to **have occurred** if any one of its elements is the outcome when the experiment is conducted.

In the die-rolling experiment, the event A that an odd number occurs is $A = \{1,3,5\}$. The event A occurs if, when the die is rolled, any one of the three possibilities 1, 3, or 5 comes up.

Example 3.1   According to the Bureau of the Census, 50% of all homes in Dallas are heated with gas (*Statistical Brief SB/95-7*, May, 1995). For three randomly selected homes in Dallas, list the outcomes of the sample space corresponding to whether they are heated with gas. Also list the outcomes that make up the following events:

Event A: Exactly two of the three heat with gas

Event B: Only one heats with gas

Event C: All three heat with gas

Event D: At least one of the three heats with gas

Solution   Each home will either heat with gas (g) or not (n). If the first selected home heats with gas, the second can either heat with gas or not; or, if the first does not heat with gas, the second can again heat with gas or not. So, for two homes, there are four possibilities:

$$\{(g,g), (g,n), (n,g), (n,n)\}$$

The third home heats with gas or not, so for each of these four possibilities there are two possibilities, resulting in eight different possible outcomes. The sample space is

$$S = \{(g,g,g),(g,g,n),(g,n,g),(n,g,g),(g,n,n),(n,g,n),(n,n,g),(n,n,n)\}$$

where, for example, (g,n,g) represents the outcome that the first and third homes heat with gas and the second does not.

A *tree diagram*, shown in Figure 3.1, helps list the possible outcomes.

Figure 3.1

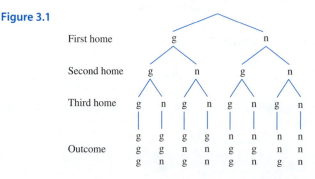

From the outcomes in the sample space or the tree diagram, we can list the outcomes of the listed events:

$$A = \{(g,g,n),(g,n,g),(n,g,g)\}$$

$$B = \{(g,n,n),(n,g,n),(n,n,g)\}$$

$$C = \{(g,g,g)\}$$

$$D = \{(g,n,n),(n,g,n),(n,n,g),(g,g,n),(g,n,g),(n,g,g),(g,g,g)\}$$

Notice that the 50% figure is not used to list the outcomes in the sample space or the events. Later we will use the 50% figure to assign probabilities to the outcomes.   ■

Example 3.2   In Example 3.1 suppose that three randomly selected homes are owned by the Jones, Brown, and Smith families, and it is determined that the Jones home is heated with gas, the Brown home is heated with gas, and the Smith home is not heated with gas. List which of the four events, A, B, C, and D, occurred and which did not occur.

Solution   If we assume that the homes are selected in the order Jones, Brown, and Smith, the outcome we observed is (g,g,n). This outcome is listed in events A and D; so, events A and D occurred but events B and C did not.   ■

Probability of an Individual Outcome

The likelihood of the occurrence of an event depends on its probability. First, we will look at the probability of the individual outcomes in the sample space.

The probability of an outcome of an experiment can be described as the *relative frequency* with which the outcome occurs if we repeat the experiment a *large* number of times. For example, if we toss a fair coin, we say that the probability of getting a head on the coin is 50% or 1/2 because half of the time the coin should land on a head and half of the time it should land on a tail. This does not mean that out of 10 tosses, 5 should be heads and 5 should be tails; rather, if we tossed the coin 10,000 times, the number of heads should be near 5000 (i.e., the relative frequency of heads should be near .5).

We will write the probability of an outcome as a number between 0 and 1 or as a percent (1/2, .5, 50% all mean the same thing). The closer it is to 0, the *less likely* the outcome is to occur, and the closer it is to 1, the *more likely* the outcome is to occur. If its probability is 0, the outcome cannot happen, and if its probability is 1, the outcome is certain to occur.

Example 3.3 In a particular card game, all 52 cards are dealt to four players. Suppose you are dealt four aces. What is the probability that one of the other three players has an ace?

Solution Because there are only four aces in the deck and you have all four, no one else can have an ace. The outcome is impossible; therefore, the probability is 0.   ■

Example 3.4 Ten people, four men and six women, are in a meeting. At random, they select a person to preside. What is the probability that the person selected is a woman?

Solution There are two possible outcomes to the experiment. The selected person is either a man or a woman. If we repeatedly selected a large number of times, we would expect to select a woman about 60% of the time because six of the ten people are women. Thus, the probability that a woman is selected is .6.

> ### Assigning Probabilities to Individual Outcomes
>
> In assigning probabilities to the individual outcomes in a sample space, two conditions must be satisfied:
>
> 1. The probability of each outcome must be between 0 and 1, inclusive.
> 2. The probabilities of all outcomes in the sample space must sum to 1.

Often outcomes of an experiment are *equally likely*. For example, if we roll a fair die, each of the six sides are equally likely to be the outcome of the experiment. Because they are equal and must sum to 1, each is assigned a probability of 1/6. In general, if the sample space has n outcomes and they are equally likely, each outcome is assigned a probability of $1/n$.   ■

Example 3.5 Suppose students who are registered for the fall semester are randomly assigned to one of six sections of freshman English. Because the students are randomly assigned to the sections, the six outcomes—{1}, {2}, {3}, {4}, {5}, {6}—are equally likely to occur. Therefore, we assign a probability of 1/6 to each because there is one chance out of six of getting any one of the sections. If we sum the probabilities of all outcomes, we get 6/6 = 1, which says that one of the outcomes will occur when the experiment is conducted. Notice the similarity between this example and the die-rolling experiment.   ■

Probability of an Event

Once probabilities are assigned to all outcomes of the sample space, we can find the probabilities of other events.

> The **probability of an event A** is the sum of the probabilities of the outcomes in A. We write it as P(A).

We saw in Example 3.5 that if students are randomly assigned to one of six sections of freshman English, we can assign probability 1/6 to each of the six outcomes. Then the probability of event A, that a student is assigned to an odd-numbered section, is

$$P(A) = P(\{1\}) + P(\{3\}) + P(\{5\}) = \frac{1}{6} + \frac{1}{6} + \frac{1}{6} = \frac{1}{2}$$

Example 3.6   In the game of craps the "shooter" wins on the first roll of a pair of fair dice if the sum of the two dice is 7 or 11. Calculate the probability that the shooter wins on the first roll.

Solution   The sample space for the roll of a pair of dice is

$$S = \{(1,1),(1,2),(1,3),(1,4),(1,5),(1,6), (2,1),(2,2),(2,3),(2,4),(2,5),(2,6),$$
$$(3,1),(3,2),(3,3),(3,4),(3,5),(3,6), (4,1),(4,2),(4,3),(4,4),(4,5),(4,6),$$
$$(5,1),(5,2),(5,3),(5,4),(5,5),(5,6), (6,1),(6,2),(6,3),(6,4),(6,5),(6,6)\}$$

Assuming both dice are fair, we have 36 outcomes that are equally likely—that is, all have probability 1/36. The event of interest is that a total of 7 or 11 turns up. In set notation, the event of interest is

$$A = \{(1,6),(2,5),(3,4),(4,3),(5,2),(6,1),(5,6),(6,5)\}$$

Adding the probabilities of the outcomes in A, we have

$$P(A) = \frac{1}{36} + \frac{1}{36} + \cdots + \frac{1}{36} = 8\left(\frac{1}{36}\right) = \frac{8}{36}$$

So the probability is 8/36 or 2/9 that the shooter wins on the first roll.   ■

Example 3.6 can be used to illustrate an alternate way of calculating probabilities when the outcomes in the sample space are equally likely. There are a total of 36 outcomes in the sample space, 8 of which make up event A. By counting the number of outcomes in the event and dividing by the number of total outcomes in the sample space we have P(A) = 8/36.

Probability of an Event When Outcomes Are Equally Likely

Let A be an event in a finite sample space S where each outcome is equally likely to occur. If $n(A)$ is the number of outcomes in A and $n(S)$ is the number of outcomes in S, then

$$P(A) = \frac{n(A)}{n(S)}$$

Example 3.7   Consider the problem of whether homes in Dallas are heated with gas. For three randomly selected homes, we found in Example 3.1 that the sample space of possible outcomes is

$$S = \{(g,g,g),(g,g,n),(g,n,g),(n,g,g),(g,n,n),(n,g,n),(n,n,g),(n,n,n)\}$$

Assign probabilities to the eight outcomes in the sample space and calculate the probabilities of the events listed in Example 3.1.

Solution   Because 50% of the homes heat with gas, the outcome (g,g,g), for example, is just as likely as, say (n,n,n). In other words, the eight outcomes are equally likely. From Example 3.1 we have

Event A: Exactly two of the three heat with gas = {(g,g,n),(g,n,g),(n,g,g)}

Event B: Only one heats with gas = {(g,n,n),(n,g,n),(n,n,g)}

Event C: All three heat with gas = {(g,g,g)}

Event D: At least one of the three heats with gas = {(g,n,n),(n,g,n),(n,n,g),(g,g,n),(g,n,g),(n,g,g),(g,g,g)}

Because there are three outcomes in event A we have P(A) = 3/8. Similarly, we have P(B) = 3/8, P(C) = 1/8, and P(D) = 7/8.   ■

Caution: Before using the above method of finding probabilities, verify that the sample space is finite and the outcomes are equally likely.

Regardless of whether the outcomes are equally likely, the following steps are useful when calculating the probability of any event.

Calculating the Probability of an Event

1. Define the experiment and list the outcomes in the sample space.
2. Assign probabilities to the outcomes such that each is between 0 and 1 and they sum to 1.
3. List the outcomes of the event of concern.
4. Sum the probabilities of the outcomes that are in the event of concern.

Mutually Exclusive Events

Combining the outcomes from other events forms most events of interest. This was true in Example 3.6 when we were interested in the probability of a total of 7 *or* 11 when rolling a pair of fair dice. We could have approached the problem as the combination of two events. Event A is rolling a 7 and event B is rolling an 11. The event of interest then is event (A or B). We could also speak of the event (A and B). Notice, however, that there would be no outcomes in the event (A and B) because it is impossible to roll both a 7 and an 11 on a single toss. In this case we say that event A and event B are disjoint or *mutually exclusive*.

Events A and B are said to be **mutually exclusive** if they have no outcomes in common.

Mutually exclusive means the occurrence of one event precludes the possibility of the other event occurring. When events are mutually exclusive, this additive law applies.

The Additive Law for Mutually Exclusive Events

Let A and B be two mutually exclusive events. Then

$$P(A \text{ or } B) = P(A) + P(B)$$

Applying the additive law to the above example, we have

$$P(7 \text{ or } 11) = P(7) + P(11) = \frac{6}{36} + \frac{2}{36} = \frac{8}{36}$$

The additive rule for mutually exclusive events also extends to multiple events. For example, the probability of rolling a total that exceeds 8 with a pair of dice is

$$P(\text{total} > 8) = P(9) + P(10) + P(11) + P(12)$$

(Notice that it is impossible to roll a total greater than 12.) Another useful probability law involves the *complement* of an event.

> The **complement of event A** is the collection of all outcomes in the sample space that are not in A. It is denoted as A^c.

Because A^c is all outcomes not in A, the two events are mutually exclusive and make up the entire sample space. Accordingly, we have the complement law.

> ### The Complement Law
>
> Let A be an event with probability denoted by $P(A)$. Then
>
> $$P(A^c) = 1 - P(A).$$

Thus, if the probability of an event is known, the probability of its complement is also known. Simply stated,

$$P(\text{not rolling a } 7) = 1 - P(\text{rolling a } 7) = 1 - \frac{6}{36} = \frac{30}{36}$$

Example 3.8   The Bureau of the Census states that 50% of all homes in Dallas are heated with gas. Out of six randomly selected homes in Dallas, what is the probability that at least one home is heated with gas?

Solution   With three randomly selected homes, we saw in Example 3.1 that there are eight different outcomes. Recall that the sample space is

$$S = \{(g,g,g),(g,g,n),(g,n,g),(n,g,g),(g,n,n),(n,g,n),(n,n,g),(n,n,n)\}$$

If there are four homes, each outcome is of the form (g,g,g,g), (g,g,g,n), and so on, and there are 16 of them. The tree diagram in Figure 3.2 helps list all 16.

Figure 3.2

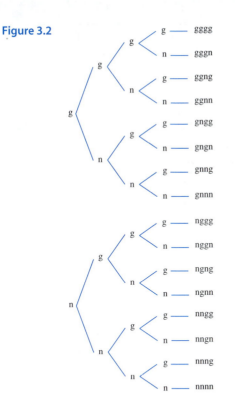

If there are five homes, there are 32 different outcomes of the form (g,g,g,g,g), (g,g,g,g,n), and so on. We see that the number of outcomes goes up by a power of 2 because there are two possibilities for each home; namely, they heat with gas (g) or they do not (n). Consequently, for six homes there are 64 different outcomes of the form (g,g,g,g,g,g), (g,g,g,g,g,n), and so on. Because 50% heat with gas, each sequence of g's and n's has the same chance of occurring, and therefore all the outcomes are equally likely. Because there are 64 of them, each has probability 1/64.

The event of interest (call it A) is that at least one of the six homes heats with gas. Event A contains 63 of the 64 outcomes in the sample space. Certainly we do not want to list all 63. However, the complement of A, A^c, contains only one outcome:

$$A^c = \{nnnnnn\} \text{ and } P(A^c) = \frac{1}{64}$$

Using the complement law, we have

$$P(A) = 1 - P(A^c) = 1 - \frac{1}{64} = \frac{63}{64}$$

We see that there is a very high probability that at least one home heats with gas.   ■

If two events are not mutually exclusive, they must have outcomes in common which make up the event (A and B). In this case, we have the more general additive law.

The General Additive Law

Let A and B be any two events. Then

$$P(A \text{ or } B) = P(A) + P(B) - P(A \text{ and } B)$$

Figure 3.3

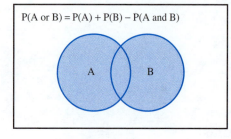

$P(A \text{ or } B) = P(A) + P(B) - P(A \text{ and } B)$

The overlapping section in Figure 3.3 between the two events represents the outcomes in common, which is event (A and B). When we add P(B) to P(A), the outcomes in (A and B) are added in twice, once with P(A) and once with P(B). Hence, to compute P(A or B), we must subtract P(A and B) one time from the sum of P(A) and P(B).

Example 3.9   In Example 3.5, students were randomly assigned to one of six different sections of freshman English—Sections 1, 2, 3, 4, 5, or 6. Let events A and B be defined as follows:

A: Susan is assigned to an even-numbered section.

B: Susan is assigned to a section numbered lower than 3.

Find the probability of the event (A or B).

Solution   Because students are randomly assigned to the sections, each assignment has probability 1/6. Consequently,

$$P(A) = P(\{2\}) + P(\{4\}) + P(\{6\}) = \frac{1}{6} + \frac{1}{6} + \frac{1}{6} = \frac{3}{6}$$

and

$$P(B) = P(\{1\}) + P(\{2\}) = \frac{1}{6} + \frac{1}{6} = \frac{2}{6}$$

Recall that the event (A and B) contains outcomes that are common to both A and B; thus,

$$P(A \text{ and } B) = P(\{2\}) = \frac{1}{6}$$

Using the additive law, we have

$$P(A \text{ or } B) = P(A) + P(B) - P(A \text{ and } B) = \frac{3}{6} + \frac{2}{6} - \frac{1}{6} = \frac{2}{3}$$

Note that in adding P(A) and P(B) we are adding in P({2}) twice. However, we have subtracted it out one time when we subtract P(A and B) = P({2}). ∎

Independent Events

Mutually exclusive events A and B cannot both occur simultaneously because they have no outcomes in common. Let us now consider events that may occur simultaneously, but if one occurs, it does not affect the probability of the other occurring. That is, the likelihood of event A occurring is unaffected by the occurrence of event B. In those cases, we say that events A and B are *independent*.

Events A and B are said to be **independent** if the occurrence of one does not affect the probability of the occurrence of the other. Otherwise, A and B are **dependent**.

If we toss a quarter and a dime in the air, it is possible that both will land with heads facing up. The fact that the quarter landed with heads up, however, has no effect on what happens to the dime. Their outcomes are independent. That is to say, both have a 1/2 chance of landing with heads up, and the probability remains 1/2 for each coin regardless of what happens to the other coin. Because there are four equally likely outcomes in the sample space, S = {hh, ht, th, tt}, there is a 1/4 chance that both land with heads up {hh}. Notice that we get 1/4 if we multiply 1/2 by 1/2. This is a result of the *multiplication law* for independent events.

The Multiplication Law for Independent Events

Let A and B be two independent events. Then

$$P(A \text{ and } B) = P(A) * P(B)$$

Example 3.10 Let us reconsider the problem of whether homes in Dallas are heated with gas. For three randomly selected homes, we found in Example 3.1 that the sample space of possible outcomes is

$$S = \{(g,g,g),(g,g,n),(g,n,g),(n,g,g),(g,n,n),(n,g,n),(n,n,g),(n,n,n)\}$$

Instead of assuming that 50% of all homes are heated with gas, suppose only 30% are heated with gas. Assign probabilities to the eight outcomes in the sample space.

Solution The outcome {(g,g,g)} means all three homes heat with gas, and each heats with gas with a 30% probability. One home heating with gas is independent of any other home heating with gas, so using an extension of the multiplication law, we have

$$P(\{g,g,g\}) = P(\text{first heats with gas})P(\text{second heats with gas})P(\text{third heats with gas})$$

$$= (.3)(.3)(.3)$$

$$= .027$$

In a similar manner, outcome {(g,g,n)} should be assigned probability $(.3)(.3)(.7) = .063$, as should outcomes {(g,n,g)} and {(n,g,g)}, because each consists of two homes heating with gas and the other one not. Outcomes {(g,n,n)}, {(n,g,n)}, and {(n,n,g)} should each be assigned probabilities $(.3)(.7)(.7) = .147$. Finally, outcome $\{(n,n,n)\}$ should be assigned probability $(.7)(.7)(.7) = .343$. Table 3.1 gives the probabilities associated with each outcome.

Table 3.1

Outcome	Probability
$\{(g,g,g)\}$	$(.3)(.3)(.3) = .027$
$\{(g,g,n)\}$	$(.3)(.3)(.7) = .063$
$\{(g,n,g)\}$	$(.3)(.7)(.3) = .063$
$\{(n,g,g)\}$	$(.7)(.3)(.3) = .063$
$\{(g,n,n)\}$	$(.3)(.7)(.7) = .147$
$\{(n,g,n)\}$	$(.7)(.3)(.7) = .147$
$\{(n,n,g)\}$	$(.7)(.7)(.3) = .147$
$\{(n,n,n)\}$	$(.7)(.7)(.7) = .343$
Total	1.000

If we had used 50% instead of 30%, notice that $P(\{g,g,g)\}) = (1/2)(1/2)(1/2) = 1/8$. In fact, all outcomes would have probability $(1/2)(1/2)(1/2) = 1/8$ because g and n both have probability 1/2. In other words, the outcomes are equally likely and each has probability 1/8. ■

Example 3.11   In Example 3.1 events A, B, C, and D were defined as follows:

Event A: Exactly two of the three heat with gas

Event B: Only one heats with gas

Event C: All three heat with gas

Event D: At least one of the three heats with gas

If 30% of the homes heat with gas, determine the probabilities of events A, B, C, and D.

Solution   Event A consists of outcomes

$$A = \{(g,g,n), (g,n,g), (n,g,g)\}$$

From Table 3.1 each outcome has probability .063. Using the additive law for mutually exclusive events, we have

$$P(A) = .063 + .063 + .063 = .189$$

Event B consists of

$$B = \{(g,n,n), (n,g,n), (n,n,g)\}$$

Each outcome has probability .147, so

$$P(B) = .147 + .147 + .147 = .441$$

Event C consists of only $C = \{(g,g,g)\}$.   Thus,

$$P(C) = .027$$

Finally, event D consists of all outcomes *except* {(n,n,n)} which has probability .343. Applying the complement law, we have

$$P(D) = 1.000 - .343 = .657$$ ■

The concepts of mutually exclusive events and independent events are often confused. Note in the box that if events A and B are independent, we have a formula for finding the probability that both occur. If, on the other hand, the events are mutually exclusive, it is impossible that both events occur at the same time.

- If events A and B are mutually exclusive, then P(A and B) = 0.
- If events A and B are independent, then P(A and B) = P(A)P(B).

When two events are not independent, the probability of one event occurring depends on whether the other event has occurred. Calculating probabilities in this case involves the *conditional probability* of event A given that B has occurred. The applications of probability to statistics that we need in the remainder of this book, however, involve independent events; thus, we will not discuss conditional probability.

COMPUTER TIP

Random Data

Random Data under the Calc menu in Minitab simulates data from a number of probability distributions, among which is the discrete uniform distribution. The discrete uniform distribution generates integer values (1, 2, 3, 4, etc.) such that all values have an equal chance of occurring. To simulate observations from the discrete uniform distribution, first select the **Calc** menu, choose **Random Data**, and then click on **Integer**. When the Integer Distribution window appears, choose the number of observations to **Generate**, the column to **Store the data**, and the **Minimum and Maximum values** for the data. For example, to simulate the roll of a die, the minimum value is 1 and the maximum value is 6; that is, roll a number between 1 and 6.

The **Tally** command can be used to give a frequency table that summarizes how many times each value comes up in the simulation. To execute a tally, first select the **Stat** menu, choose **Tables**, and then click on **Tally**. When the Tally window appears **Select columns** and click **OK**. If you wish to see percents click the **Percents** button.

Example 3.12   Simulate 18 rolls of a fair die. Ideally, we would expect each of the six possible integers to turn up three times in 18 rolls. Determine the frequency of occurrence of each possible outcome. Next, simulate 18,000 rolls of a fair die and determine the frequency of occurrence of each possible outcome. What do these results illustrate?

Solution   In the Integer Distribution window, generate **18** observations, store in **C1**, choose minimum value **1** and maximum value **6**, and then click **OK**. Repeat the process, but this time generate **18,000** observations and store in **C2**. To summarize how many times each value came up in the two simulations, at the Tally window, **select both columns** and click **OK**. This is a summary of the results of the simulations.

Tally for Discrete Variables: C1, C2

C1	Count	Percent	C2	Count	Percent
1	1	5.56	1	2992	16.62
2	4	22.22	2	3004	16.69
3	2	11.11	3	3033	16.85
4	2	11.11	4	2905	16.14
5	8	44.44	5	3032	16.84
6	1	5.56	6	3034	16.86
N=	18		N=	18000	

In the first simulation we expected each of the six possible values to come up three times each; however, numbers 1 and 6 appeared only once and number 5 appeared eight times. On the other hand, the counts in the second simulation are close to what we expect (about 3000 each). This illustrates that probability reflects the long-term likelihood of the occurrence of an outcome. Anything can happen in the short-run but as we repeat a process over and over a pattern emerges that represents the true probability of occurrence of the outcome. For practice, take a real die and toss it 18 times and see what happens. If you do not have plans for the weekend, toss it 18,000 times. ∎

EXERCISES 3.1   The Basics

3.1   Based on these experiments, compute the following probabilities.
   a. Roll a fair die. What is the probability that you will observe a number greater than 4?
   b. There are 200 names in a bowl and one is drawn as the winner. Assuming that your name is in the bowl, what is the probability that you win?
   c. Suppose there are three doctors in a group of ten people. If we select a person at random, what is the probability that he or she is a doctor?
   d. A card is drawn from an ordinary deck of 52 cards. What is the probability that it is red?

3.2   In a particular card game, all 52 cards are dealt to four players who are matched up as partners. Suppose you are dealt three aces. What is the probability that your partner has the other ace? Remember that there are three other players, one of whom is your partner.

3.3   There are three states on the West Coast: Washington, Oregon, and California.
   a. If we randomly choose one of the states, what is the probability that we select California?
   b. If the residents of the three states are grouped together and we randomly select a resident, is the probability that the person is a California resident the same as your answer in part a? Explain.

3.4   A basketball player has a 50% chance of making a free throw. Assuming attempts are independent, what is the probability that she makes two shots in a row?

3.5   Roulette is played by spinning a ball on a round table that is divided into 38 slots of equal size. The slots are numbered 00, 0, 1, 2, 3, . . . , 35, 36. The slots are also colored as follows:

   Red: 1,3,5,7,9,12,14,16,18,19,21,23,25,27,30,32,34,36
   Black: 2,4,6,8,10,11,13,15,17,20,22,24,26,28,29,31,33,35
   Green: 00,0

   a. Find the probability of the ball falling into a black numbered slot on a single spin.
   b. Find the probability that the ball falls into a slot that is neither red nor black on a single spin.
   c. Suppose you bet on black on three consecutive plays. What is the probability you win on all three plays?

3.6 In a physical fitness study, subjects are randomly assigned to five different exercise groups. List the outcomes of possible exercise groups in the sample space and the following events:

A: Assigned to group 3
B: Assigned to one of the first three groups
C: Assigned to group 4 or 5
D: Assigned to a group between 2 and 5 exclusively

Assign probabilities to the outcomes in the sample space and then compute the probabilities of the different events.

3.7 A person conducting a poll randomly chooses one of two houses. He then randomly chooses either a man or woman from the house. List the outcomes in the sample space of possibilities.

3.8 In the roll of a pair of dice, we are concerned with the event W that a total of 7 is rolled. One of the dice turns up on an even number. What must the other die be for event W to occur?

3.9 Suppose $S = \{2,3,4,5,6,7,8\}$, $A = \{3,5,7\}$, $B = \{4,5,6\}$, and $C = \{3,4,5,6,7\}$. List the outcomes in the following events:

a. A or B
b. A and B
c. A and C
d. C^c
e. A^c and C
f. A or B or C
g. A and B and C
h. What must the outcome be in order for event (A and B) to occur?
i. Are A and B mutually exclusive? Why?

3.10 For events A and B, $P(A) = .5$, $P(B) = .4$, and $P(A \text{ and } B) = .3$.

a. Find $P(A^c)$.
b. Find P(A or B).
c. Are A and B mutually exclusive? Why?
d. Are A and B independent? Why?

3.11 For events A and B, $P(A) = .4$ and $P(B) = .3$.

a. If A and B are mutually exclusive, find P(A or B).
b. If A and B are independent, find P(A or B).
c. If events A and B are mutually exclusive, explain why $P(A \text{ and } B) = 0$.

3.12 A manufacturer of golf clubs has plants in Phoenix, Denver, and Memphis. The Memphis plant produces 60% of the company's clubs, and the remaining 40% are evenly divided between the Phoenix and Denver plants. A set of clubs is randomly selected from a Kmart in Amarillo, Texas.

a. Are the events produced in Memphis, produced in Denver, and produced in Phoenix mutually exclusive?
b. What is the probability that either the Memphis or Denver plants produced the clubs?
c. What is the probability that the Memphis plant did not produce the clubs?

3.13 Assume that 60% of the students at a university are women. Three students are selected at random. List the outcomes in the sample space of possible genders of the three students. Are the outcomes equally likely? Assign probabilities to the outcomes if 50% of the students are women.

3.14 The big wheel on *The Price Is Right* game show is marked off in 20 equal increments from 5 cents to 1 dollar. The contestant gets two tries to spin a total as close as possible to 1 dollar without going over. Suppose a contestant gets 60 cents on the first spin.

a. If she spins again, what is the probability she gets a total of 1 dollar?
b. What is the probability she goes over 1 dollar?

3.15 Student loan applications are either approved or disapproved. If three students apply, list the sample space of possible outcomes. If 60% of the applications are approved, are the outcomes equally likely? If 50% of the applications are approved, are the outcomes equally likely? Assuming 50% are approved, assign probabilities to the outcomes. What is the probability that at least two of three loans are approved?

3.16   Suppose two fair dice are rolled; we are concerned with these events:

A: The sum on the two up faces is even.
B: The sum on the two up faces is less than 6.

Are events A and B independent? Are they mutually exclusive?

3.17   Two rescue teams set out to find a lost camper in the Grand Canyon. Team A has a 30% chance of finding the hiker, and team B has a 40% chance of finding the hiker. Assume the two rescue teams act independently. What is the probability that the hiker will be rescued?

3.18   Suppose a man has a certain trait that will be passed on to his offspring with 30% probability. His wife has the same trait, and she will pass it to her offspring with 20% probability. What is the probability that their child will have the trait?

3.19   A trial lawyer claims that she wins 80% of her cases. Out of two independent cases, what is the probability that she wins both, assuming that her claim is true? What is the probability that she loses both? What is the probability that she wins only one?

Interpreting Computer Output

3.20   The accompanying table on local jail inmates and facilities, by size of facility, was produced by the Bureau of Justice Statistics. It appeared in the bulletin *Jails and Jail Inmates 1993–94*, April 1995.

Local jail inmates and facilities, by size of facility

	Inmates		Facilities	
Size of facility	Number	Percent	Number	Percent
Fewer than 50	34,332	7.5	1874	56.7
50–99	37,135	8.1	545	16.5
100–149	31,293	6.8	253	7.7
150–249	41,472	9.0	218	6.6
250–499	73,938	16.1	209	6.3
500–999	90,481	19.7	129	3.9
1000–1499	44,000	9.6	35	1.1
1500–1999	30,764	6.7	18	.5
2000 or more	76,389	16.6	23	.7

a. Do the percents sum to 100 percent?
b. What percent of the inmates are in facilities with fewer than 50 inmates?
c. What percent of the facilities have fewer than 50 inmates?
d. What percent of the inmates are housed in facilities that have fewer than 250 inmates?
e. What percent of the facilities have fewer than 250 inmates?
f.   If we randomly select an inmate, what is the probability he or she is in a facility with fewer than 250 inmates?
g. If we randomly select a facility, what is the probability that it has fewer than 250 inmates?

3.21   A woman has a blue suit that she wears 30% of the time. Let B correspond to her wearing the blue suit and N mean that she does not. For two days, list the outcomes of the sample space. Are the outcomes equally likely? Following are the results of a simulation of the experiment 1000 times. Column C1 gives the number of days out of two that she wears the blue suit. Based on the simulation, what is the probability that she wears the blue suit two days in a row? What is the probability that she wears it only one of the two days?

Tally for Discrete Variables: C1

C1	Count
0	522
1	401
2	77
N=	1000

Now, assign probabilities to the outcomes listed in the sample space and compute the above probabilities. How do these theoretical probabilities compare with the simulation?

3.22 There are three traffic lights on your way home. Assume that each light is either red (R) or green (G) and that it is green with probability .7. List the outcomes of the sample space. Are the outcomes equally likely? Following are the results of a simulation of the experiment 1000 times. Column C2 gives the number of green lights you make out of the three lights. Based on the simulation, what is the probability that you make all three green lights? What is the probability that you stop at all three lights? What is the probability you make exactly one green light?

Tally for Discrete Variables: C2

C2	Count
0	23
1	194
2	433
3	350
N=	1000

Now, assign probabilities to the outcomes listed in the sample and compute the above probabilities. How do the theoretical probabilities compare with the simulation?

3.23 In 1990, 25% of all American Indians lived on reservations, mostly in the Rocky Mountain states. The accompanying bar graph was obtained from data from the Bureau of the Census.

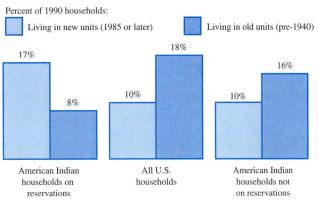

New Homes Versus Old Homes

Percent of 1990 households:

☐ Living in new units (1985 or later)   ☐ Living in old units (pre-1940)

Source: Bureau of the Census, U. S. Department of Commerce, *Housing of American Indians on Reservations, Statistical Brief SB/95–10,* April 1995.

a. From the information provided, what percent of American Indians on reservations live in new units?
b. From the information provided, what percent of American Indians not on reservations live in new units?
c. Can the two percents in **a** and **b** be added to find the percent of American Indians (reservation or not) who live in new units?
d. Of all U.S. households, what percent do not live in either new or old units?

3.24   Here are the results of Gallup polls concerning the environment conducted in April 1990 and in April 2000.

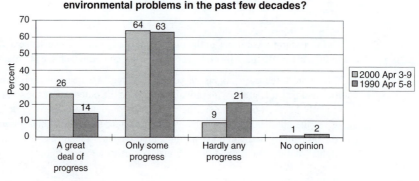

How much progress have we made in dealing with environmental problems in the past few decades?

Source: The Gallup Organization
http://www.gallup.com/poll/surveys/2000.

a. Are the four categories listed mutually exclusive?
b. In what categories have the percents changed significantly?
c. What percent in 2000 felt that we made only some or hardly any progress in dealing with the environment?
d. Generally speaking, did Americans feel that we dealt better with environmental problems in 2000 or in 1990? Explain your reasoning.

3.25   Here are results of Gallup polls concerning crime conducted in February 1990 and in October 2000.

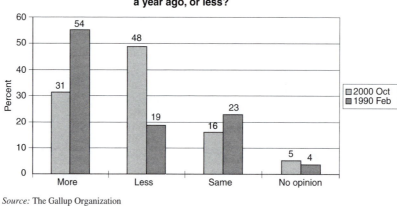

Is there more crime in your area than there was a year ago, or less?

Source: The Gallup Organization
http://www.gallup.com/poll/surveys/2000.

a. Do the percents add to 100% as they should?
b. In 1992 over half of Americans felt that there was more crime in their area than the year before. What was that percent in 1998?
c. What percent felt that crime was less or the same in 1992? What about 1998?
d. Do you think that the months the polls were taken had any effect on the percents?
e. Do you feel that the overall attitude about crime was better in 1998 or in 1992? Explain your reasoning.

Computer Exercises

3.26   For the game of roulette (see Exercise 3.5) assign the number 37 to 00 and the number 38 to 0 so that an integer from 1 to 38 is assigned to each of the 38 slots. Use the Random Data command in Minitab to simulate 38 rolls of the roulette wheel. Out of the 38 rolls, how many times did the ball fall in the black slot? How many times did it fall in the red slot? The green slot? Next, simulate 3800 rolls and use the Tally command to summarize the results. Out of the 3800 rolls, how often should we expect each value to turn up? How close are the actual results to what you would expect?

3.27   Simulate 2000 rolls of a fair die and determine the frequency of occurrence of each possible outcome. Based on the simulation, what is the probability of rolling a 3? Would you say that the possibilities are equally likely? What is the probability that it turns up odd and greater than 4?

3.28   Simulate 2000 rolls of a pair of fair dice by simulating one die in column 1 and the other die in column 2; then add the two columns together. Determine the frequency of occurrence of each possible outcome. Based on the simulation, what is the probability of rolling a total of 7? What is the probability that the total is 7 or 11?

3.2   The Binomial Distribution

Learning Objectives for this section:

❏ Understand the concept of a random variable.

❏ Know what is meant by the probability distribution of a random variable.

❏ Know the characteristics of a Bernoulli population.

❏ Know the characteristics of a binomial experiment and be able to identify a binomial random variable.

❏ Know how to find probabilities associated with a binomial random variable.

❏ Know how to find and use the mean and standard deviation of a binomial random variable.

❏ Know how to simulate Bernoulli and binomial data.

According to the Census Bureau, 50% of all homes in Dallas are heated with gas. In Example 3.8, we learned that if we randomly select six homes, the sample space of possibilities consists of 64 different outcomes of the form

$$(g,g,g,g,g,g), (g,g,g,g,g,n), (g,g,g,g,n,g), \ldots$$

It is not an easy matter to list all 64 possibilities; moreover, we may not even want to list them all. The outcome (g,n,n,g,n,g), for example, means that the first, fourth, and sixth homes heat with gas and the second, third, and fifth do not. Instead of concerning ourselves with distinct outcomes such as this, it is much simpler to consider just the number of homes out of six that are heated with gas. That is, instead of considering the sample space of outcomes of the form (g,g,g,g,g,g), (g,g,g,g,g,n), (g,g,g,g,n,g), and so on, we need concern ourselves only with the numerical values 0, 1, 2, 3, 4, 5, and 6, which is much simpler. This is accomplished with a *random variable*.

> • A **random variable**, which is denoted by a letter such as X or Y, is a rule that represents the possible numerical values associated with the outcomes of an experiment.
> • A table or function that lists all of the possible values of a random variable and their associated probabilities is called the **probability distribution** of the random variable.

Example 3.13   Let the variable X be the number of homes out of three that heat with gas in Dallas. Find the probability distribution for X and graph the distribution.

Solution   Recall from Example 3.1 that the sample space is

$$S = \{(g,g,g),(g,g,n),(g,n,g),(n,g,g),(g,n,n),(n,g,n),(n,n,g),(n,n,n)\}$$

We see there are eight possibilities, and because 50% of the homes heat with gas, all of the eight possibilities are equally likely. Of the eight, only one (n,n,n), corresponds to zero homes heating with gas, in which case $X = 0$ and the probability is 1/8. Three of the eight possibilities correspond to $X = 1$ (exactly one heats with gas), and therefore the probability is 3/8. Continuing in this fashion, we have the following completed probability distribution of X:

X	0	1	2	3
$P(X)$	1/8	3/8	3/8	1/8

Observe that the probabilities sum to 1, which is always the case with a probability distribution table. The graph of this distribution is shown in Figure 3.4.   ■

Figure 3.4   P(X)

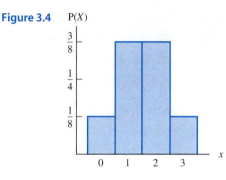

Bernoulli Population

The distribution described in Example 3.13 and graphed in Figure 3.4 is an example of the *binomial distribution*. At each randomly selected home, the outcome is one of two possibilities—the home heats with gas or it does not. As we have seen, many experiments share this same characteristic; namely, the outcome is only one of two possibilities. This is what we commonly call a *Bernoulli population*, after Jacob Bernoulli (1654–1705), the first of a series of gifted mathematicians of the Bernoulli family.

- A **Bernoulli population** is a population in which each element is one of two possibilities. The two possibilities are usually designated as *success* and *failure*.
- A **Bernoulli trial** is observing one element in a Bernoulli population.

Example 3.14   a. Tossing a coin is an example of a Bernoulli population because each toss results in a head (success) or a tail (failure).

b. The homes in Dallas either heat with gas (success) or do not (failure). The choice to heat with gas or not is a Bernoulli population.

c. In a local election, the voters either favor a candidate for mayor or do not. We can view the population of voter choices as a Bernoulli population, where a success is one who favors the candidate and a failure is one who opposes the candidate.   ■

COMPUTER TIP

Bernoulli Trials

The **Bernoulli** command in Minitab will simulate the outcomes of a series of Bernoulli trials. Select the **Calc** menu, choose **Random Data**, and then click on **Bernoulli**. When the Bernoulli Distribution window appears, specify the number of observations to **Generate**, give the **column(s) to store** the data, give the **Probability of success**, and then click **OK**. The outcomes are coded as a 0 or a 1, where the probability of a 1 on any given trial is the specified probability of success.

Example 3.15   Suppose a field-goal kicker has an 80% success rate inside the 35-yard line. Simulate eight field-goal kicks inside the 35 during a game. Determine the number of successes. Simulate eight kicks inside the 35 for ten consecutive games. Determine his success rate in each game and his success rate for the ten-game season.

Solution   The results of a simulation of eight kicks with a success rate of 80% are

ROW	C1	C2	C3	C4	C5	C6	C7	C8
1	1	0	0	1	1	1	0	1

Out of eight kicks, he missed three: the second, third, and seventh kicks.

Following are the results of the ten-game simulation:

ROW	C1	C2	C3	C4	C5	C6	C7	C8
1	0	0	1	1	1	1	1	1
2	1	1	1	1	1	1	1	1
3	1	0	1	1	0	1	1	1
4	1	1	1	1	0	1	1	1
5	1	0	1	0	1	1	1	1
6	1	0	1	1	1	1	1	1
7	1	1	1	1	1	1	1	1
8	1	1	1	0	1	0	1	1
9	0	1	1	1	1	1	1	1
10	1	1	1	0	1	1	1	1

In game 1, the kicker missed the first two tries and made the rest, he made all eight kicks in game 2, and so on. Following is a summary with the kicker's success rate in each game.

Game	#hits	#misses	success rate
1	6	2	75%
2	8	0	100
3	6	2	75
4	7	1	87.5
5	6	2	75
6	7	1	87.5
7	8	0	100.0
8	6	2	75
9	7	1	87.5
10	7	1	87.5
Total	68	12	85%

In no simulation did the kicker hit exactly 80% of his kicks (of course, this is impossible with only eight kicks); rather, the success rate ranged from 75% to 100%. It is important to realize that, although the population rate is 80%, there is no guarantee that any sample will duplicate the population rate exactly. If we combine all 80 kicks, the overall success rate is 85%. What do you think the simulation would yield if we simulated 800 kicks? An important lesson is that the sample results will vary from sample to sample even though the population rate is constant. This will become more apparent as we study sampling distributions in the next chapter. ■

Binomial Experiments and Random Variables

A random sample from a Bernoulli population consists of n objects, where each object is a success or a failure. The result, which is a sequence of Bernoulli trials, is called a *binomial experiment*.

- A **binomial experiment** is an experiment that consists of n repeated independent Bernoulli trials in which the probability of success on each trial is π and the probability of failure on each trial is $1 - \pi$.
- The random variable X, which gives the number of successes in the n trials of a binomial experiment, is called a **binomial random variable**. The sample space of values of X is

$$S_X = \{0, 1, 2, \ldots, n\}$$

(Note: We use π as a symbol for the probability of success in a binomial experiment. It should not be confused with the number 3.14159 that we usually associate with the symbol π.)

Thus, an experiment that consists of repeatedly drawing independently from a Bernoulli population is called a binomial experiment. The outcomes are labeled as success or failure and the number of possible successes out of the n trials is a binomial random variable.

Example 3.16   Are the following random variables binomial random variables? Explain why, or why not.

a. Forty percent of all airline pilots are over 40 years of age. Out of a random sample of 15 pilots, let the random variable X be the number of pilots who are over 40 years of age.

b. Suppose a salesperson makes sales to 20% of her customers. One day she counted her customers until she made a sale. Let X be the number of customers until her first sale.

c. A room contains six women and four men. Three people are selected to form a committee. Let the random variable X be the number of women on the committee.

Solution

a. Each of the 15 pilots constitutes a trial, and each trial results in one of two outcomes: Success is being over 40 years of age and failure is not being over 40 years of age. The trials are independent because one pilot's age has no relationship to another pilot's age. The probability of a success is $\pi = .4$ and of a failure is $(1 - \pi) = .6$, and these probabilities remain the same from trial to trial because 40% of all pilots are over 40 years of age. Thus, the conditions of a binomial experiment are met. Because X is the number of successes, it is a binomial random variable.

b. The salesperson may never make a sale; hence, the sample space for X is

$$S_X = \{1, 2, 3, \ldots\}$$

which clearly is not the sample space of a binomial random variable. If X had been defined as the number of sales out of, say, 50 customers, then X would have been a binomial random variable because it would have been the number of successes out of 50 trials.

c. The probability of a woman being selected on the first pick is 6/10. On the second pick, however, it is 5/9 if a woman was selected the first time or 6/9 if a man was selected the first time. Clearly, the trials are dependent, and hence the experiment does not satisfy the conditions of a binomial experiment. The random variable X is not a binomial random variable. ■

COMPUTER TIP

Random Data: Binomial

The outcomes of a binomial random variable can be simulated by simulating Bernoulli trials, as before, and then counting the number of successes. If, however, we wish to repeat the process several times, we should use the **Binomial distribution**.

Under the **Calc** menu, select **Random Data**, and click on **Binomial**. At the Binomial Distribution window, specify the number of observations to **Generate**, what **Column(s) to store the data**, give the **Number of trials** and the **Probability of success**, and click on **OK**.

Example 3.17   Suppose we randomly select six homes in Dallas and determine whether they heat with gas. The number of successes (heat with gas) out of the six homes is a binomial random variable with $n = 6$ and $\pi = .5$. Simulate the experiment 60 times and summarize the results.

Solution   A home heating with gas is called a success. Each home is treated as a trial that results in a success or failure. Out of six homes, we wish to find the number of successes, where the probability of success is .5 (assuming that 50% of Dallas homes heat with gas). We repeat the binomial experiment 60 times and tally the results. From the tally on page 158 we see that on one occasion none of the six homes heats with gas. On two occasions only one of the six homes heats with gas. On 12 occasions two homes heat with gas. On 19 occasions three homes heat with gas, and so on. In simulating the experiment 60 times, only once did all six homes heat with gas.

Data Display

```
C1
3    2    3    5    4    3    3    4    5    4
3    4    4    4    0    3    4    4    5    3
5    3    2    3    5    4    2    2    4    4
3    4    1    4    3    2    3    2    3    2
1    5    5    3    4    2    4    4    2    5
3    3    2    4    6    3    2    3    2    3
```

```
C1      Count
 0         1
 1         2
 2        12
 3        19
 4        17
 5         8
 6         1
N=        60
```
■

Determining Binomial Probabilities

In the simulation in Example 3.17 we can calculate the relative frequency of occurrence of each of the values of the random variable by dividing the count by 60. The resulting percents are approximations to the actual probabilities of observing the different values of the random variable X = number of homes out of 6 that are heated with gas. Simulating the experiment 6000 times would give more accurate approximations. Using the laws of probability, however, we can determine the theoretical probabilities.

In order to determine the theoretical probabilities and the resulting probability distribution of X, we must be able to count the number of outcomes in the sample space that are associated with each value of X. As pointed out in Example 3.8, there are 64 equally likely outcomes of the form (gggggg), (gggggn), (ggggng), and so on in the sample space. Only one (gggggg), corresponds to $X = 6$; hence $P(X = 6) = 1/64$. Realizing that there are six outcomes with five g's, (ggggggn), (ggggng), (gggngg), (ggnggg), (gngggg), and (ngggggg), we have $P(X = 5) = 6/64$. But now, how many of the 64 outcomes have exactly four g's? The answer is found with the *combinations formula*.

The number of **combinations** of k objects taken from n objects is given by

$$\binom{n}{k} = \frac{n!}{k!(n-k)!}$$

where $n!$ is read "n factorial" and is given by

$$n! = n(n-1)(n-2)(n-3)\cdots(2)(1) \quad \text{and} \quad 0! = 1.$$

Thus, there are $\binom{6}{4} = 6!/4!(6-4)! = 15$ combinations that have four g's. Continuing with the combinations formula, we come up with the following table, which is the probability distribution table for X:

X	0	1	2	3	4	5	6
$P(X)$	1/64	6/64	15/64	20/64	15/64	6/64	1/64

The preceding example is a case where the outcomes in the sample space are equally likely. The outcomes are equally likely because the probability of a home being heated with gas is 50%; that is, $\pi = 0.5$. What adjustments must be made when $\pi \neq 0.5$? Recall that in Example 3.10, we changed the percent from 50% to 30% and reduced the number of homes to three to see what would happen. Table 3.1, repeated here, gives the probabilities for the eight possible outcomes.

Outcome	Probability
{(g,g,g)}	(.3)(.3)(.3) = .027
{(g,g,n)}	(.3)(.3)(.7) = .063
{(g,n,g)}	(.3)(.7)(.3) = .063
{(n,g,g)}	(.7)(.3)(.3) = .063
{(g,n,n)}	(.3)(.7)(.7) = .147
{(n,g,n)}	(.7)(.3)(.7) = .147
{(n,n,g)}	(.7)(.7)(.3) = .147
{(n,n,n)}	(.7)(.7)(.7) = .343
Total	1.000

Let Y = number of homes out of three that are heated with gas when $\pi = 0.3$.

The random variable Y will be 0 only if we observe the outcome (n,n,n) in the sample space. Therefore,

$$P(Y = 0) = .343$$

Three outcomes—(g,n,n), (n,g,n), and (n,n,g)—are associated with the $Y = 1$. Each of the three has probability (.3)(.7)(.7) = 0.147. Therefore,

$$P(Y = 1) = 3(0.147)$$

The completed probability distribution table for Y is

Y	$P(Y)$
0	1(.343) = .343
1	3(.147) = .441
2	3(.063) = .189
3	1(.027) = .027

To generalize, suppose there are n independent Bernoulli trials with π being the probability of success on each trial. For there to be exactly k successes, there must be $n - k$ failures. Moreover, there are

$$\binom{n}{k}$$

different possibilities for the k successes. Because the probability of each success is π, the probability of any given arrangement of k successes and $n - k$ failures is

$$\pi^k (1 - \pi)^{n-k}$$

Thus,

Binomial Probability Distribution

The probability of k successes in n independent Bernoulli trials is

$$P(k) = \binom{n}{k} \pi^k (1 - \pi)^{n-k}$$

Computing these probabilities can become rather tedious if n is very large. In fact, in most cases we use a computer to compute the probabilities; or, for certain values of n and π, the probabilities are given in the binomial table (Table B.1 in Appendix B).

To aid our discussion of the binomial table, an excerpt is reproduced in Table 3.2. This is the section of the table corresponding to $n = 10$ independent trials. Down the left column we see values for k of 0, 1, 2, 3, 4, 5, 6, 7, 8, 9, and 10. These numbers correspond to k successes out of the ten trials. Along the top we see values for π of .10, .20, .30, and so on to .90. (Table B.1 also includes values for .01, .05, .95, and .99.) Remember that π stands for the probability of a single success in the Bernoulli population. The body of the table gives the probabilities corresponding to the particular values of k and π for $n = 10$. For example, the probability of four successes out of ten trials when $\pi = .7$ is .037.

Table 3.2

$n = 10$

$k \mid \pi$	.10	.20	.30	.40	.50	.60	.70	.80	.90
0	.349	.107	.028	.006	.001	.000+	.000+	000+	.000+
1	.387	.268	.121	.040	.010	.002	.000+	.000+	.000+
2	.194	.302	.233	.121	.044	.011	.001	.000+	.000+
3	.057	.201	.267	.215	.117	.042	.009	.001	.000+
4	.011	.088	.200	.251	.205	.111	.037	.006	.000+
5	.001	.026	.103	.201	.246	.201	.103	.026	.001
6	.000+	.006	.037	.111	.205	.251	.200	.088	.011
7	.000+	.001	.009	.042	.117	.215	.267	.201	.057
8	.000+	.000+	.001	.011	.044	.121	.233	.302	.194
9	.000+	.000+	.000+	.002	.010	.040	.121	.268	.387
10	.000+	.000+	.000+	.000+	.001	.006	.028	.107	.349

(.000+ means that the probability rounded to three decimal points is 0. It is not exactly 0, however, because that would mean that the outcome is impossible. The "+" means it is slightly greater than 0.)

Example 3.18   Suppose that the probability of successfully rehabilitating a convicted criminal in a penal institution is .4. If we let R represent the number successfully rehabilitated out of a random sample of ten convicted criminals, then R is a binomial random variable with $n = 10$ independent trials and $\pi = .4$. The probability distribution for R is given in the binomial probability distribution table.

a. Find the probability that six of the ten prisoners are successfully rehabilitated.
b. Find the probability that no more than two of the ten are successfully rehabilitated.

Solution
a. Focusing on the $\pi = .4$ column in Table 3.2, we see the probabilities associated with all the values of k when $n = 10$. Thus, the probability that six of the ten prisoners are rehabilitated is .111.
b. No more than two implies that k is 0, 1, or 2. Thus, to find the probability, we simply add those probabilities in the .4 column that correspond to the values 0, 1, and 2 in the k column:

$$.006 + .040 + .121 = .167$$

There is a 16.7% chance that two or fewer criminals are successfully rehabilitated. ∎

If we plot all the probabilities for $n = 10$ and $\pi = .4$, we get a graph of this binomial distribution, shown in Figure 3.5.

Figure 3.5

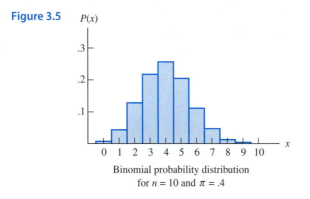

Binomial probability distribution
for $n = 10$ and $\pi = .4$

If π were only .1, we could go down the .10 column and get all the probabilities associated with the possible values of k. Figure 3.6(a) is a graph of this distribution. Note that for small values of π, the distribution is heavily concentrated on the lower values of k, which means that a smaller number of successes is more likely. Figure 3.6(b) shows that for large values of π (in this case $\pi = .9$), the binomial distribution is heavily concentrated on the larger values of k, which means that a greater number of successes is more likely. Note also that the distributions for $\pi = .1$ and $\pi = .9$ are mirror images of each other, as is the case for $\pi = .2$ and $\pi = .8$, and so on. If $\pi = .5$, the binomial distribution is symmetric, as shown in Figure 3.7.

Figure 3.6

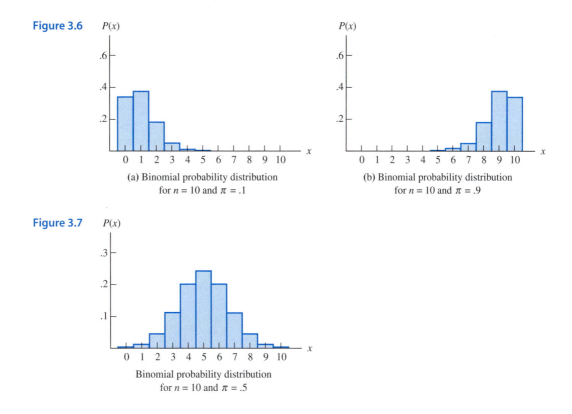

(a) Binomial probability distribution
for $n = 10$ and $\pi = .1$

(b) Binomial probability distribution
for $n = 10$ and $\pi = .9$

Figure 3.7

Binomial probability distribution
for $n = 10$ and $\pi = .5$

Mean and Standard Deviation of the Binomial Distribution

It is apparent that the values of n and π are very important because they determine the various binomial distributions. They are the *parameters* for the binomial distribution. From the parameters we get other characteristics of the distribution, such as the mean and the standard deviation.

If X has a **binomial distribution** with parameters n and π, the mean and standard deviation of X are

$$\text{Mean} = n\pi$$

$$\text{Standard deviation} = \sqrt{n\pi(1 - \pi)}$$

Example 3.19   In Example 3.18, find the mean and the standard deviation of the variable R; that is, find the mean number of convicted criminals successfully rehabilitated out of the ten prisoners. Also, find the standard deviation.

Solution   The parameters are $n = 10$ and $\pi = .4$. Therefore, the mean is

$$n\pi = 10(.4) = 4$$

and the standard deviation is

$$\sqrt{n\pi(1 - \pi)} = \sqrt{10(.4)(.6)} = 1.55$$

So we expect four of the ten prisoners to be successfully rehabilitated, and the standard deviation is 1.55 prisoners.   ■

An interesting feature of the binomial distribution is what happens when n increases and π remains constant. Figure 3.8 gives the binomial distributions for $\pi = .2$ and $n = 5$, 10, 20, and 30. Note that as n increases from 5 to 30, the distribution becomes more bell shaped. Recall that the empirical rule can be applied to bell-shaped distributions.

Figure 3.8

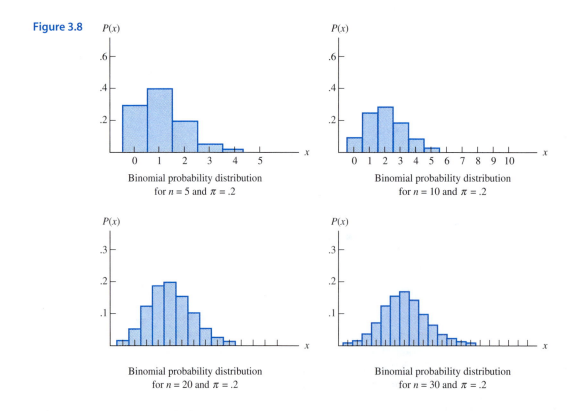

Binomial probability distribution
for $n = 5$ and $\pi = .2$

Binomial probability distribution
for $n = 10$ and $\pi = .2$

Binomial probability distribution
for $n = 20$ and $\pi = .2$

Binomial probability distribution
for $n = 30$ and $\pi = .2$

Example 3.20 A certain drug manufacturer claims that its vaccine is 80% effective; that is, each person who is vaccinated stands an 80% chance of developing immunity. Suppose 100 people are vaccinated. If we let V be the number that develops immunity, then V is a binomial random variable with parameters $n = 100$ and $\pi = .8$, and thus the mean is

$$n\pi = 100(.8) = 80$$

so we expect 80 to develop immunity. The standard deviation is

$$\sqrt{n\pi(1 - \pi)} = \sqrt{100(.8)(.2)} = \sqrt{16} = 4$$

Because n is large, the distribution of V should be close to bell shaped, so using the empirical rule, we can say that approximately 95% of the distribution is between 72 and 88. That is, two standard deviations below the mean is

$$80 - 2(4) = 80 - 8 = 72$$

and two standard deviations above the mean is

$$80 + 2(4) = 80 + 8 = 88$$

Thus, we know that if $\pi = .8$, then out of 100 people, we expect between 72 and 88 to develop immunity. Anything outside those limits is considered unusual. ∎

COMPUTER TIP

Binomial Distribution

Suppose 68% of all requests for financial aid are approved by a university. Determine the probabilities associated with the number of approvals being 7, 8, 9, 10, 11, or 12 out of 20 requests.

First enter 7, 8, 9, 10, 11, and 12 into a column, say C1, in the Data worksheet. From the **Calc** menu, select **Probability Distributions**, and then select **Binomial**. When the Binomial Distribution window appears, check **Probability**, key in the **Number of trials** = 20 and the **Probability of success** = .68, and specify C1 as the **Input Column**. If you wish to store the probabilities in, say C2, specify it as **Optional Storage**. When you check **OK**, the probabilities will appear in the Session window or the optional storage column that you have specified.

The contents of the Session window will be:

Probability Density Function

Binomial with n = 20 and p = 0.680000

x	P(X = x)
7.00	0.0019
8.00	0.0066
9.00	0.0188
10.00	0.0440
11.00	0.0849
12.00	0.1354

EXERCISES 3.2   The Basics

3.29   The Red Cross has 15 blood donors and is searching for type B blood. Is B, the number out of the 15 who have type B blood, a binomial random variable? Explain your reasoning.

3.30   Let X be the SAT math score of a student selected at random from the student body of your school. Is X a binomial random variable? Explain your reasoning.

3.31   Suppose X is a random variable with the following probability distribution table:

X	1	3	5	7	9
$P(X)$	1/15	2/15	3/15	4/15	5/15

a. What value of X is most likely to occur?
b. What is the probability that X is even?
c. What is the probability that X is less than 7?
d. Is X a binomial random variable?

3.32   An experiment consists of tossing three fair coins. Let X be the random variable that is the number of heads in the three tosses. Is X a binomial random variable? Find the probability distribution of X.

3.33   Suppose your skill at playing a particular video game is such that you stand a 40% chance of winning each time you play. Suppose you play three games and either win (W) or lose (L).
a. List the sample space of outcomes.
b. Assign probabilities to the outcomes.
c. Let X = number of games you win. What is the sample space for X?
d. Give the probability distribution table for X.
e. What is the probability that you win at least two of the three games?

3.34   For each of the following, find the mean and standard deviation and then graph the binomial distribution.
a. $n = 10$ $\pi = .8$
b. $n = 15$ $\pi = .3$
c. $n = 6$ $\pi = .5$
d. $n = 4$ $\pi = .9$

3.35   If Y is a binomial random variable with n and π given as follows, calculate the unknown probability.
a. $n = 10$ $\pi = .3$ $P(Y \leq 4)$
b. $n = 6$ $\pi = .8$ $P(Y > 3)$
c. $n = 15$ $\pi = .9$ $P(Y > 13)$
d. $n = 5$ $\pi = .2$ $P(Y < 3)$

3.36   Assume that 60% of a college's student loan applications are approved. Ten applications are chosen at random.
a. What is the probability that eight or more are approved?
b. How many applications are expected to be approved?
c. What is the standard deviation of the number approved out of ten applications?

3.37   A new television show has a 20% chance of being successful. NBC will introduce eight new shows this season. What is the probability that fewer than three will be successful? How many are expected to succeed?

3.38   A presidential aide believes that half of the members of Congress favor a particular action taken by the president. If this is true, what is the probability that out of a sample of 15 members of Congress, more than eight favor the action? Of the 15, how many are expected to favor the action?

3.39   Which of the following are binomial random variables?
a. The number of accidents per week involving alcohol in your state.
b. The number of violent crimes per month committed in your city.
c. The number of successful heart transplants out of five patients.

    d. The length of a prison term for possession of marijuana.

    e. The number of approved food stamp recipients out of 50 applications.

3.40 According to the Census Bureau, 40% of the 1.6 million men with children whose mothers were absent were awarded child support in 1991 (*Statistical Brief SB/95–16*, June 1995). Out of 18 randomly selected men with children whose mothers were absent, what is the probability that less than half are awarded child support?

3.41 The probability distribution for the number of daily requests for assistance from a poison control center is

No. of requests	0	1	2	3	4	5	6
Probability	.1	.1	.2	.3	.1	.1	.1

The mean number of requests and the standard deviation are 2.9 and 1.7, respectively. Sketch a graph of the probability distribution and locate the interval $\mu \pm 2\sigma$ on the graph. Does approximately 95% of the distribution lie inside the interval? According to the empirical rule, should 95% of the distribution lie inside the interval?

3.42 An oil-well drilling company generally drills four wells per month. The probability distribution for the number of successful attempts out of the four wells is given by

No. of successes	0	1	2	3	4
Probability	.1	.4	.3	.1	.1

The mean number of successful wells and the standard deviation are 1.7 and 1.1, respectively. Sketch a graph of the probability distribution and locate the interval $\mu \pm 2\sigma$ on the graph. Does approximately 95% of the distribution lie inside the interval? According to the empirical rule, should 95% of the distribution lie inside the interval?

3.43 One-fourth of all pregnancies end in abortion (*Statistical Abstracts of the United States*, 1994, p. 83). Suppose that three pregnant women are randomly selected. Let Y be the number, out of the three, who get an abortion. Find the probability distribution of Y. Find the mean and standard deviation of Y. Sketch a graph of the probability distribution and locate the interval $\mu \pm 2\sigma$ on the graph. Does approximately 95% of the distribution lie inside the interval? According to the empirical rule, should 95% of the distribution lie inside the interval?

3.44 The mortality rate for a certain disease is 30%. Of ten patients who have the disease, what is the probability that more than half will die from the disease? Of the ten patients, how many are expected to die from the disease?

3.45 A new drug, Nimodipine, holds considerable promise of providing relief for those people suffering from migraine headaches who have not responded to other drugs. Clinical trials have shown that 90% of patients with severe migraines experience relief of their pain without suffering allergic reactions or side effects. Suppose 16 migraine patients try Nimodipine.

    a. What is the probability all 16 experience relief?

    b. What is the probability at least 14 experience relief?

    c. How many are expected to experience relief?

Interpreting Computer Output

3.46 According to the Census Bureau, 70% of all Americans have health coverage with a private insurance plan (*Statistical Brief SB/94–28*, October 1994). Out of a random sample of 15 Americans, what is the probability that all 15 have a private insurance plan? What is the probability that 10 or more have health coverage with a private insurance plan? Use the results of the following simulation of 300 random samples of 15 Americans regarding health insurance coverage to estimate these probabilities. That is, out of the 300 samples, how many times did all 15 Americans have a private insurance plan? How many times did 10 or more have a private insurance plan? Calculate relative frequencies, and compare to the theoretical probabilities.

Tally for Discrete Variables: C1

C1	Count
6	2
7	14
8	30
9	35
10	66
11	62
12	59
13	23
14	5
15	4
N=	300

3.47 Of all persons who live inside poverty areas, the Census Bureau has reported that 56% are white (*Statistical Brief SB/95–13*, June 1995). From a random sample of 50 persons who live in poverty areas, what is the probability that 35 or more are white? What is the probability that fewer than 20 are white? Use the results of the following simulation of 200 random samples of 50 persons who live in poverty areas to estimate these probabilities, and then compare to the theoretical probabilities. Out of a sample of 50, how many do you expect to be white? What number came up most frequently in the simulation?

Tally for Discrete Variables: C1

C1	Count
18	1
19	1
20	2
21	2
22	4
23	10
24	13
25	13
26	15
27	29
28	18
29	21
30	19
31	16
32	11
33	8
34	3
35	4
36	9
37	1
N=	200

3.48 An experiment consists of rolling five dice and observing how many turn up on 6. The number of 6's that turn up is a binomial random variable with $n = 5$ and $\pi = 1/6 = .16666. \ldots$. From this simulation of the experiment 200 times determine the following:
a. How many times did all five dice turn up on 6?
b. How many times did a 6 fail to turn up?
c. Explain why the counts for 3, 4, and 5 are zero.

Tally for Discrete Variables: C1

C1	Count
0	85
1	90
2	25
3	0
4	0
5	0
N=	200

Computer Exercises

3.49 Use the **Random Data > Bernoulli** command to simulate the toss of five coins **(store in C1-C5)** and count how many times a head comes up **by summing C1-C5 and storing in C6**. Repeat the simulation 3200 times (3200 different **rows** in the worksheet) and determine the number of heads in each of the 3200 simulations. Out of the 3200, how many times did you get no heads? One head? Two heads? Three heads? Four heads? Five heads? Construct a frequency table (include percents) of the number of heads out of the 3200 simulations.

3.50 In 1994 the Bureau of Justice Statistics reported that 10% of all jail inmates were women (*Jails and Jail Inmates*, *Bulletin NCJ-1511651*, April 1995). With the **Random Data > Bernoulli** command, simulate the outcomes from five randomly selected inmates and count how many times a woman appears (remember only 10% are women). Repeat the simulation 3200 times (3200 different columns in the worksheet) and determine the number of women in each of the 3200 simulations. Out of the 3200, how many times did you get no women? One woman? Two women? Three women? Four women? Five women? Construct a frequency table (include percents) of the number of women out of the 3200 simulations.

3.51 Distributions such as the ones given in Exercises 3.41 and 3.42 can be simulated with the **Random Data > Discrete** command in Minitab. First, store the values of the random variable in one column of the worksheet (C1) and the corresponding probabilities in an adjacent column (C2). Under the **Calc** menu, select **Random Data** and click on **Discrete**. When the Discrete Distribution window appears, fill in the **Generate** box with the desired number of simulated values. Specify the column to **Store** the values. Give the column containing the **Values** and the column containing the **Probabilities** and click on **OK**. Simulate 20 days of the number of daily requests for assistance from the poison control center given in Exercise 3.41. Make a frequency table (including percents) of the values. How closely does it agree with the theoretical distribution?

3.52 Use the simulation procedure outlined in Exercise 3.51 to simulate 24 months of drilling four oil wells each month by the oil-well drilling company given in Exercise 3.42. Make a frequency table (including percents) of the values. What percent of the time were all four wells successful? What percent of the time were no wells successful? Do the simulated values agree with the theoretical values?

3.3 The Normal Distribution

Learning Objectives for this section:

- ❑ Know the basic properties of a probability density function.
- ❑ Know the characteristics of the normal distribution.
- ❑ Know how to apply the empirical rule to normal or close to normal distributions.
- ❑ Know how to find probabilities associated with the normal distribution.
- ❑ Know how to find percentiles associated with the normal distribution.
- ❑ Understand normal probability plots and how they help diagnose the shape of a distribution.

A binomial random variable assumes a discrete set of values, $(0, 1, 2, 3, \ldots, n)$, and thus is a *discrete random variable*. Its probability function assigns probabilities to the discrete values in the range of the random variable. *Continuous random variables*, on the other hand, assume all values in an interval of the number line. For this reason, we are unable to assign probabilities as we do in the discrete case. For continuous random variables, a *probability density function* or *density curve* is defined so that probabilities are represented by areas under the curve.

A **probability density function**, $f(y)$, that describes the probability distribution for a continuous random variable Y has the following properties:

1. $f(y) \geq 0$
2. The total area under the probability density curve is 1.00, which corresponds to 100%.
3. $P(a \leq Y \leq b) =$ area under the probability density curve between a and b.

Example 3.21   Let the variable Y represent the distance of a home run in major league baseball. According to collected data, the mean is approximately 420 feet. Suppose that the probability density function given by the density curve in Figure 3.9 represents the distribution of home runs in the majors.

Figure 3.9
Distribution of home runs in the majors (in feet)

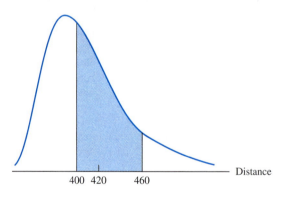

The shaded area represents the percent of home runs between 400 feet and 460 feet. The shaded area covers approximately 40% of the total area, so we write

$$P(400 \leq Y \leq 460) \cong .40 \qquad \blacksquare$$

Because probabilities are represented by areas under the curve, it does not matter whether the endpoints of the interval are included. In other words, we have

$$P(a < Y < b) = P(a \leq Y \leq b)$$

Observe that the probability density curve in Example 3.21 is centered at about 420, is somewhat spread out, and tails off to the right. The shape of a probability density curve indicates what values of the random variable are most likely to occur in a sample. Two important characteristics of a random variable are its mean and its standard deviation. The mean of a continuous random variable, like that of a discrete random variable, is the value of the random variable that will balance the frequency curve. Both the mean and the standard deviation of a continuous random variable can be calculated from the probability density function, but this is beyond the scope of this book.

The Normal Distribution

One of the most commonly observed random variables is the *normal random variable*. Its probability density curve, introduced in Section 1.3, is shown in Figure 3.10. The distribution is bell shaped; symmetric about its mean, μ; and the amount of spread is determined by the standard deviation, σ.

Figure 3.10
A normal probability density curve

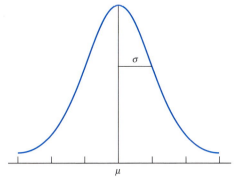

Notice in Figure 3.10 that the curve is concave downward in the middle and concave upward on the tails. One standard deviation is the distance out from the mean where the curve changes from concave down to concave up. Three standard deviations on either side of the mean covers almost all of the distribution. If the standard deviation is increased, the density curve becomes more spread out. Also, if the mean were changed, the density curve would shift so that it would be centered at the mean. Thus, these two *parameters* change the appearance of the density curve, which is why we say that they characterize the normal distribution. Figure 3.11 shows three normal curves that have different means and standard deviations. We see that by changing the two parameters, μ and σ, we can describe any number of different normal distributions.

Figure 3.11
Probability density curves for three different normal distributions

In addition to being symmetric about its mean, the normal density curve continues infinitely in both directions. However, most of the distribution lies within three standard deviations on either side of μ. In fact, the empirical rule, which we have already studied, was derived from the normal distribution. Figure 3.12 illustrates the empirical rule as it applies to the normal distribution.

The Empirical Rule

For a normal distribution with mean μ and standard deviation σ:

- 68.26% of the distribution is between $\mu - \sigma$ and $\mu + \sigma$
- 95.44% of the distribution is between $\mu - 2\sigma$ and $\mu + 2\sigma$
- 99.74% of the distribution is between $\mu - 3\sigma$ and $\mu + 3\sigma$

If the probability distribution is bell shaped (close to normal), the percents are approximate.

Figure 3.12
Normal Density Curve

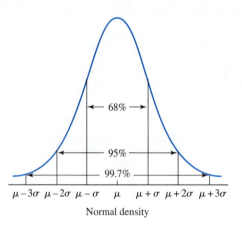

Normal density

The normal distribution is important because it provides a model for many real-world distributions. Such measurements as aptitude test scores, physical measurements such as heights and weights, and random error in production processes may follow the normal distribution.

Example 3.22   Scores on the Stanford-Binet IQ test are assumed to be normally distributed with a standardized mean of 100 and a standard deviation of 16. What percent of the population have IQs between 100 and 116?

Solution   First, we suggest that a picture be drawn of the distribution and that the desired area be shaded. Figure 3.13 shows a normal curve centered at 100, with the desired area from 100 to 116 shaded. Observe that 116 is 16 points or exactly one standard deviation above the mean. From the empirical rule, we know that 68.26% of the distribution is between $\mu - \sigma$ and $\mu + \sigma$, so 34.13% is between μ and $\mu + \sigma$; hence, 34.13% of the population have IQs between 100 and 116. ∎

Figure 3.13

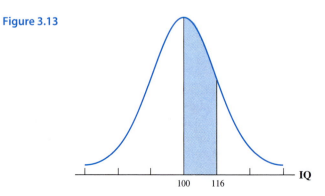

Example 3.23   Assume that the time it takes a bank teller to serve a customer is normally distributed with a mean of 30 seconds and a standard deviation of 10 seconds. What percent of the customers are served in less than 20 seconds?

Solution   Figure 3.14 is a graph of the distribution in which the desired area (below 20) is shaded.

Figure 3.14

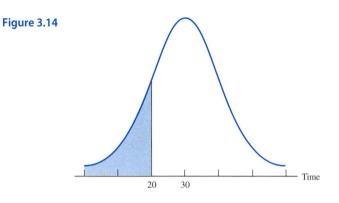

Because of the symmetry of the curve, 50% of the area is below the mean of 30. Since 20 is 10 points, or one standard deviation, below 30, we know that 34.13% of the area lies between 20 and 30. Thus,

$$P(\text{Time} < 20) = .5000 - .3413 = .1587$$

Just under 16% of the customers are served in less than 20 seconds.   ■

In the preceding two examples, the probability was determined by using the empirical rule, which gives probabilities associated with one, two, or three standard deviations above or below the mean. But suppose we are not dealing with an integer number of standard deviations above or below the mean? For example, what percent of the population has an IQ (on the Stanford-Binet test) below 124? The probability is illustrated in Figure 3.15.

Figure 3.15

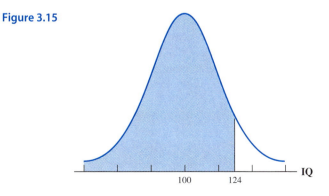

Clearly, 124 is 24 points above the mean, which is not an exact multiple of the standard deviation of 16; thus, we cannot apply the empirical rule.

We know, however, that 24 points above μ represents $24/16 = 1.5$ standard deviations above μ. So the problem becomes: What percent of a normal distribution lies below a point that is 1.5 standard deviations above the mean?

To solve problems of this type, we first determine the number of standard deviations that a given score is from the mean. This is accomplished by calculating the *z-score*.

z-Score

Recall from Chapter 1 that the **z-score** for a score *x* is

$$z = \frac{x - \mu}{\sigma}$$

It gives the number of standard deviations that a score is from the mean. A negative *z*-score indicates that the score is below the mean, and a positive *z*-score indicates that the score is above the mean. The distribution of *z* has a mean of 0 and a standard deviation of 1 and is known as the *standard normal distribution*.

Finding Probabilities Associated with z-Scores

The *standard* normal probability distribution table in Appendix B (Table B.2) is constructed to find probabilities associated with *z*-scores. Each entry in Table B.2 corresponds to the area under the curve below a point that is *z* standard deviations from the mean. A negative *z* value corresponds to a score that is below the mean and a positive *z* value corresponds to a score above the mean. A portion of Table B.2 corresponding to positive values is reproduced in Table 3.3 for our discussion:

Table 3.3

z	.00	.01	.02	.03	.04	.05			
.0	.5000	.5040	.5080	.5120	.5160	.5199	—	—	—
+.1	.5398	.5438	.5478	.5517	.5557	.5596	—	—	—
+.2	.5793	.5832	.5871	.5910	.5948	.5987	—	—	—
+1.4	.9192	.9207	.9222	.9236	.9251	.9265	—	—	—
+1.5	.9332	.9345	.9357	.9370	.9382	.9394	—	—	—

To complete the problem, we simply find +1.5 in the *z* column and locate the associated probability of .9332. Thus, 93.32% of the area lies below a score that is 1.5 standard deviations above the mean. Because an IQ score of 124 is 1.5 standard deviations above the mean, we can say that 93.32% of the population have IQs below 124.

We can also say that 43.32% of the population have IQs between 100 and 124 because 50% of the total area is below 100. Subtracting 50% from 93.32% gives 43.32%. Furthermore, because of the symmetry of the normal distribution about the mean, the area from 76 to 100 is the same as the area from 100 to 124. Therefore, 2(.4332) = .8664, or 86.64% have IQs between 76 and 124. Alternatively, we could find the area below 76, subtract it from the area below 124, and get the same answer.

The probabilities associated with any normal distribution, standard or otherwise, can be found from the probabilities associated with the *z*-scores in the standard normal distribution table. The additional columns in Table B.2 are for *z*-scores carried out to two decimal places.

Example 3.24   An important measurement in the textile industry is the tensile strength of a produced material. Suppose that the tensile strength (in pounds per square inch) of a roll of woven polypropylene is normally distributed with a mean of 89 and a standard deviation of 4. What is the probability that the tensile strength of a randomly selected piece is somewhere between 80 and 100 pounds per square inch?

Solution   The probability we are seeking is represented by the shaded area in Figure 3.16.

Figure 3.16

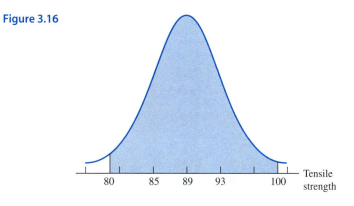

We first find A_1, the area to the left of 100, and then A_2, the area to the left of 80. We then subtract A_2 from A_1 to find the desired probability. To find A_1 we compute the z-score for 100 as follows:

$$z = \frac{100 - 89}{4} = +2.75$$

From Table B.2 we find that $A_1 = .9970$. The z-score for 80 is

$$z = \frac{80 - 89}{4} = -2.25$$

From Table B.2 we find that $A_2 = .0122$. So the probability that the tensile strength is between 80 and 100 is

$$A_1 - A_2 = .9970 - .0122 = .9848$$ ■

By using the complement rule for probabilities, we can find areas in the upper tail of the normal distribution.

Example 3.25   Use Example 3.24 to find the probability that the tensile strength of the sample of woven polypropylene is greater than 95 pounds per square inch.

Solution   The probability is the area above 95 in Figure 3.17. To find that area, we first find its complement, the area below 95. This area can be determined by finding the z-score for 95 and looking up the associated probability. Thus,

$$z = \frac{95 - 89}{4} = +1.5$$

which has a corresponding probability of .9332. Therefore, the desired area is

$$A = 1.0000 - .9332 = .0668$$

The probability is .0668 that the tensile strength of the material is greater than 95 pounds per square inch.   ■

Figure 3.17

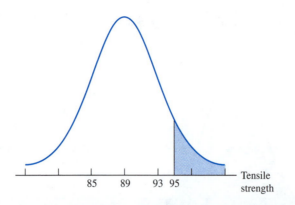

85 89 93 95   Tensile
strength

COMPUTER TIP

Normal Distribution

It is extremely easy to find probabilities associated with a given z-score with Minitab. First store the z-score (or several z-scores) in a column of the worksheet. From the **Calc** menu, choose **Probability Distribution** followed by **Normal**. When the Normal Distribution window appears, click on **Cumulative probability** and then enter the **mean**, **standard deviation**, and the **input column** where you stored the z-scores. Click on **OK**, and the probability will appear in the Session window.

The following procedure will help when you work probability problems involving normally distributed populations.

To Work Probability Problems for Normally Distributed Populations

1. Draw a graph of a normal curve, label the mean, and shade the desired area.
2. Find the number of standard deviations the given score is from the mean by finding the z-score.
3. Find the associated probability for the z-score in the standard normal probability table.
4. Relate the result to the problem at hand.

Finding Percentiles

Often, instead of finding the probability associated with a certain score, we wish to find the score corresponding to a probability.

Example 3.26   Consider again the tensile strength of a roll of woven polypropylene. In Example 3.24, we assumed that it is normally distributed with a mean of 89 and a standard deviation of 4. Find the value of b so that 80% of the distribution lies below it. In other words, if X is a normal random variable with mean 89 and standard deviation 4, find b so that $P(X \leq b) = .80$.

Solution   This is the reverse of the previous examples. Here we are given the probability and asked to find the associated score. We know that b is above the mean 89 because 50% of the distribution lies below the mean. Figure 3.18 illustrates that b is the value with .8000 area under the curve below it.

Figure 3.18

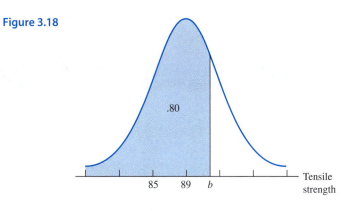

Looking in the z-table, we find a z-score of .84 corresponding to .7995 (the closest value to .8000). Hence, b is .84 standard deviations above 89. We have that

$$b = 89 + (.84)(4) = 89 + 3.36 = 92.36$$

We can say that 80% of the woven polypropylene will have a tensile strength less than 92.36 pounds per square inch. This value of b is called the 80th percentile.   ■

> The ***p*th percentile** is the value in the population such that $p\%$ of the distribution lies at or below that value.

Example 3.27   What IQ score on the Stanford-Binet corresponds to the 95th percentile?

Solution   The 95th percentile is the IQ score such that 95% of all IQ scores are below it. From Figure 3.19 we see that we are looking for the score above the mean such that .95 of the area is below that score.

Figure 3.19

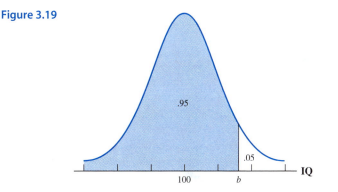

The z-score corresponding to an area of .9500 is approximately 1.64. Thus, we can say that the 95th percentile is 1.64 standard deviations above the mean. Since the standard deviation is 16, the 95th percentile score is

$$b = 100 + 1.64(16) = 100 + 26.24 = 126.24$$ ■

Just as we use the computer to simulate Bernoulli and binomial distributions, we can simulate data drawn from a normally distributed population.

C O M P U T E R   T I P

Random Data: Normal

To simulate a random sample from a normal population with a specified mean and standard deviation, under the **Calc** menu, select **Random Data** and click on **Normal**. At the Normal Distribution window, specify the **Number of observations** to generate, what **Columns to store the data**, give the **Mean** and the **Standard deviation**, and click on **OK**.

Example 3.28   Assuming the mean IQ is 100 and the standard deviation is 16, simulate 80 IQs and construct a histogram of the results. Does the histogram resemble a normal distribution?

Solution   Figure 3.20 shows the histogram of the randomly generated data. It appears to approximate a normal distribution.   ■

Figure 3.20

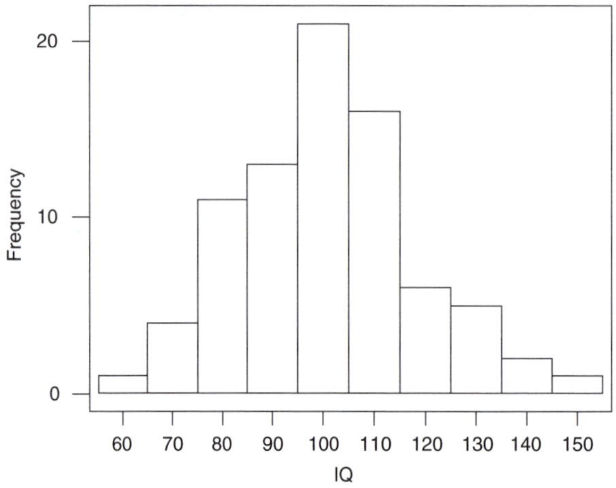

Checking Normality

In Section 1.3 we looked at several of the most common distribution shapes that appear in the daily practice of statistics. Being able to detect the shape of the distribution from sample information is vital. As you will see later, many inferential procedures are based on the fact that the sample is from a normally distributed population. Therefore, detecting normality is of utmost importance. We have seen how a histogram gives the general appearance of the data and how the boxplot describes the length of the tails of a distribution. Now, we introduce the *normal probability plot*, a very sensitive graphical procedure for checking normality.

Normal Probability Plot

The normal probability plot is a graph of the sample data against the values we would expect if the sample had come from a normal population. If the sample is indeed from a normal population, the graph should be close to a straight line. Deviations from a straight line indicate nonnormality. Determining whether or not the plot exhibits a linear pattern is somewhat subjective, and therefore one should not rule out normality unless the evidence is obvious. Figure 3.21 illustrates normal probability plots together with boxplots for data generated from several different theoretical distributions. The boxplots are given to show symmetry versus skewness and the length of the tails of the distributions. To help judge whether or not the points in a probability plot exhibit a linear pattern, a confidence band is graphed about the straight line. We will assume normality as long as most plotted points fall within the confidence band.

Figure 3.21

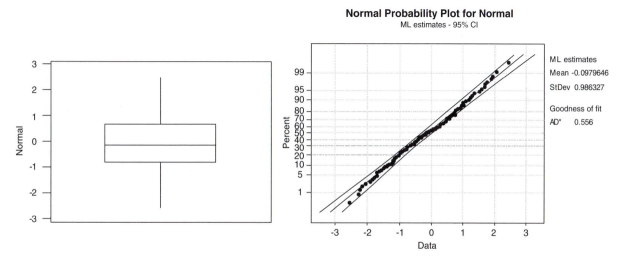

(a) Sample of size 200 from a normally distributed population with mean 0 and standard deviation 1.

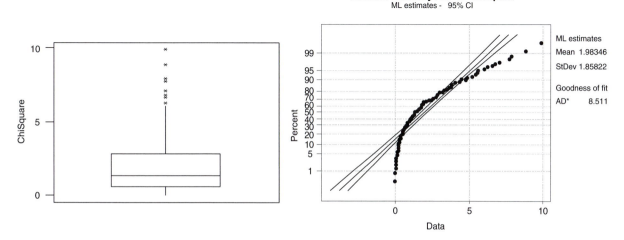

(b) Sample of size 200 from a skewed right distribution—Chi-square with 2 degrees of freedom

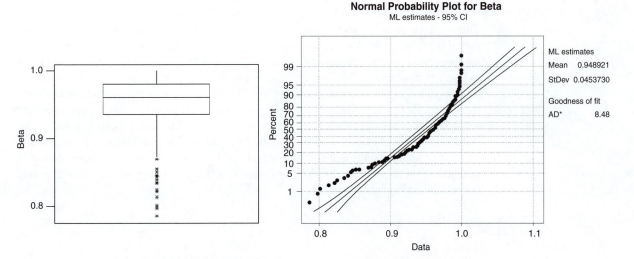

(c) Sample of size 200 from a skewed left distribution—Beta (20,1)

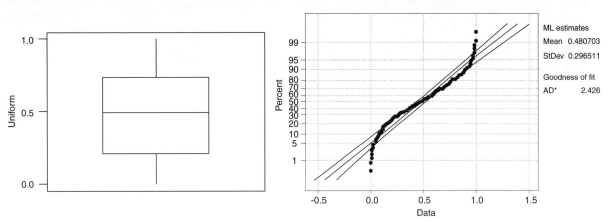

(d) Sample of size 200 from a short-tailed distribution—Uniform (0,1)

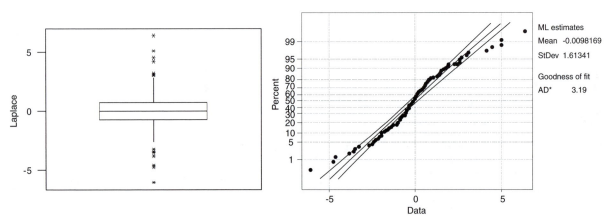

(e) Sample of size 200 from a long-tailed distribution—Laplace

Notice first in Figure 3.21(a) that the probability plot of data from a normal population does exhibit a straight-line behavior with most points falling within the confidence band. Figure 3.21(b) and (c) illustrate that the left tail of the probability plot curves downward for a skewed-right distribution and the right tail curves upward for a skewed-left distribution. Figure 3.21(d) shows the left tail curving down and the right tail curving up for a short-tailed distribution. Figure 3.21(e) exhibits an S-shape behavior for a long-tailed distribution.

By hand, the computations necessary to construct a probability plot are very tedious. We, however, will use Minitab (or almost any other statistical package) to construct the graph for us.

C O M P U T E R   T I P

Normal Probability Plot

To construct a normal probability plot, first the data must be stored in a column of the Minitab worksheet. From the **Graph** menu, select **Probability Plot**. When the Probability Plot window appears, **Select** the variable and click **OK**.

The construction of the normal probability plot is illustrated in the next example.

Example 3.29   Check the random data generated in Example 3.28 for normality with a normal probability plot.

Solution   We have the data stored in C1 of the worksheet and named IQ. Figure 3.22 gives the resulting normal probability plot.

Figure 3.22

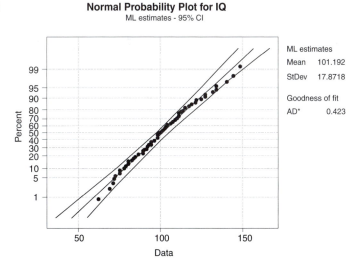

Normal Probability Plot for IQ
ML estimates - 95% CI

The closeness of the plotted points to the straight line (all points fall within the confidence band) indicates that the data are coming from a normally distributed population.   ∎

The normal distribution is very important because many physical measurements and natural phenomena are closely approximated by it. However, a more fundamental reason for studying the normal distribution involves the theoretical properties of the *sample* mean, which allows us to make statistical inferences about the population mean. This will be discussed at length in Chapter 4.

EXERCISES 3.3 The Basics

3.53 Suppose z has a standard normal distribution. Find the percent of the distribution in each case:
a. below $z = 2.0$
b. below $z = 2.6$
c. below $z = 1.36$
d. below $z = -2.0$
e. between $z = -1.42$ and $z = 1.25$
f. between $z = -2.82$ and $z = -0.58$

3.54 Suppose X is a normally distributed random variable with a mean of 8 and a standard deviation of 3. Find
a. $P(5 < X < 10)$
b. $P(X > 9)$
c. Find b such that $P(X \le b) = .9$.

3.55 What is the probability that it takes the bank teller in Example 3.23 longer than 1 minute to serve a customer? Recall that $\mu = 30$ and $\sigma = 10$.

3.56 Based on the Stanford-Binet IQ test (see Example 3.22), what percent of the population has an IQ above 120? Between 80 and 120? What is the probability of an IQ more than two standard deviations from the mean?

3.57 How do you measure the relative magnitude of an arbitrary measurement that is taken from a population?

3.58 Suppose a test of coordination for first-graders is scored so that the mean for all first-graders is 50 and the standard deviation is 15. If we assume further that the distribution is normal, what percent of the first-graders have the following scores?
a. Below 30?
b. Between 40 and 70?
c. Above 75?

3.59. Suppose that the test scores for a college entrance exam are normally distributed with a mean of 450 and a standard deviation of 100.
a. What percent of those who take the exam score between 350 and 550?
b. A student who scores above 400 is automatically admitted. What percent score above 400?
c. The upper 5% receive scholarships. What score must they make on the exam to get a scholarship?

3.60 A job-satisfaction index score for nurses is normally distributed with a mean of 50 and a standard deviation of 10. What is the probability that a nurse selected at random has an index score higher than 55?

3.61 Suppose the mean account in an investment firm is $8000 and the standard deviation is $2000. Assuming that the distribution of accounts is normal, what percent of the accounts are in the following ranges?
a. Below $ 9000?
b. Below $ 5000?
c. Above $13,000?
d. Between $3000 and $6000?

3.62 A public health department closes the beach when the contamination level index is in the 80th percentile. From research, we know that the index level is normally distributed with a mean of 160 and a standard deviation of 20.
a. What is the probability the index level exceeds 190?
b. What is the probability the index level is between 150 and 170?
c. What is the index level when the beach is closed?

3.63   Employees of a company are given a test that is distributed normally with mean 100 and standard deviation 5. The top 5% will be awarded top positions with the company. What score is necessary to get one of the top positions?

Interpreting Computer Output

3.64   Using the data in Exercise 1.90 in Section 1.5 on violent crime rates (worksheet: **Crime.mtw**), we have constructed the following normal probability plot. Use this result to comment on the shape of the distribution. Does your conclusion agree with your assessment in Exercise 1.90 in Section 1.5?

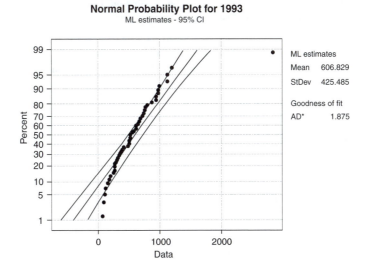

3.65   A normal probability plot of Ceramic Phase 4 of the radiocarbon ages studied in Example 1.16 is given here.

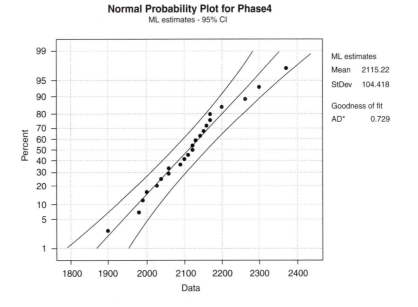

Judging from the appearance of this normal probability plot, is there evidence to suggest that the data are not normally distributed?

Computer Exercises

3.66   Using the data in worksheet: **Salary.mtw** on the starting salaries of Ph.D. psychologists, construct a normal probability plot. Use the results to comment on the shape of the distribution.

3.67   Using the data in Exercise 1.83 in Section 1.5 on reaction times of 31 senior citizens (worksheet: **Senior.mtw**), construct a normal probability plot. Use the results to comment on the shape of the distribution. Does your conclusion agree with your assessment in Exercise 1.83 in Section 1.5?

3.68   Using the data in Exercise 1.91 in Section 1.5 on monthly payments for families in the Aid to Families with Dependent Children (worksheet: **Aid.mtw**), construct a normal probability plot. Use the results to comment on the shape of the distribution. Does your conclusion agree with your assessment in Exercise 1.91 in Section 1.5?

3.69   Using the data in Exercise 1.56 in Section 1.3 on cholesterol values of subjects in the Framingham heart study (worksheet: **Framingh.mtw**), construct a normal probability plot. Use the results to comment on the shape of the distribution. Does your conclusion agree with your assessment in Exercise 1.56 in Section 1.3?

3.4   Summary And Review Exercises

Key Concepts

- An *experiment* is an activity of making an observation or taking a measurement that leads to a collection of outcomes. The set of all outcomes is called the *sample space*. Any subset of the sample space is called an *event*.

- *Probability* is a number between 0 and 1 that describes the likelihood with which an event is to occur. If probabilities can be assigned to the outcomes in a sample space, the probability of any other event A is the sum of the probabilities that have been assigned to the outcomes that are in event A.

- Several laws of probability apply to events. The *additive law* is

$$P(A \text{ or } B) = P(A) + P(B) - P(A \text{ and } B)$$

  For *mutually exclusive* events A and B, $P(A \text{ and } B) = 0$.

- The *complement law* says that

$$P(A^c) = 1 - P(A).$$

- Independent events are such that if one event occurs, it does not affect the probability of the other event occurring.

- The *multiplicative law for independent events* states that

$$P(A \text{ and } B) = P(A)P(B).$$

- The terms *mutually exclusive* and *independent* are sometimes confused. If events A and B are mutually exclusive, then $P(A \text{ and } B) = 0$. If events A and B are independent, then $P(A \text{ and } B) = P(A)P(B)$.

- A *random variable* is a rule that assigns numerical values to the outcomes of an experiment. If the sample space of values of the random variable is a discrete set of values, the random variable is called *discrete*. If the sample space of values is a continuous set of values, it is called a *continuous random variable*.

- The *probability distribution* of a discrete random variable is a table or function that lists the values of the variable and the probability with which it assumes those values. This table can be used to find probabilities of events.

- The *mean* of a random variable is an average value of the random variable. It is the value that balances the distribution. The *standard deviation* is a measure of the variability associated with the random variable. The amount of variability of a random variable can be interpreted with the *empirical rule* if the distribution is close to bell shaped.

- A *binomial experiment* consists of several independent trials in which each trial results in either a success or a failure. The *binomial random variable* is a discrete variable that represents the number of successes in a binomial experiment. Probabilities associated with the binomial random variable can be found in the binomial probability tables (Table B.1) in Appendix B.

- The *normal random variable* is a continuous random variable that is associated with the measurements of some numerical variable. The *normal probability density curve* (that corresponds to probabilities) is a bell-shaped curve that is symmetrical. The mean and standard deviation are the parameters that distinguish between normal random variables. Probabilities associated with the normal distribution can be found by determining how far a score is from the mean. To do this, we calculate the z-score and then find the associated probability in the standard normal tables (z-table, Table B.2) in Appendix B.

Statistical Insight Revisited

To better understand the three-door problem presented in the Statistical Insight, conduct the following experiment.

a. Randomly generate a door for the valuable prize by generating a 1, 2, or 3 with the Random Data > Integer command. Simulate 30 games by generating 30 random values and store the results in column 1.
b. Randomly generate a guess by the contestant by again generating a 1, 2, or 3. Simulate 30 guesses to go along with the 30 winning doors in part **a** and store the results in column 2.
c. Check the 30 games by comparing the corresponding values in columns 1 and 2. If the numbers match, Monty would show one of the other two doors that conceal gag gifts, and the winning strategy is to stay with the original choice. If, on the other hand, the numbers do not match— say a 1 and a 3—then Monty would show door 2, which conceals a gag gift, and the winning strategy is to switch doors.
d. Out of the 30 games, how many times was the winning strategy to switch and how many times was the winning strategy to stay?
e. Based on this simulation, would you advise the contestant to switch or to stay?
f. Simulate 300 games and see what happens.

To develop the theoretical probability of winning by switching and by staying in the three-door problem, consider the following sample space analysis:

Location of prize	Possible guesses of the contestant	Monty opens door	Winning strategy
1	1	2 or 3	remain
1	2	3	switch
1	3	2	switch
2	1	3	switch
2	2	1 or 3	remain
2	3	1	switch
3	1	2	switch
3	2	1	switch
3	3	1 or 2	remain

a. In the table, have all possibilities of where the actual prize is located and all possible guesses been considered? How many are there?
b. Out of the total number of possibilities, how many win with the remain strategy? How many win with the switch strategy?
c. What is the probability of winning by remaining? What is the probability of winning by switching?
d. Do the results of the above simulations follow these theoretical probabilities?

Questions for Review   Use the following problems to test your skills:

The Basics   Multiple choice

3.70   Suppose X is a binomial random variable with $n = 10$ and $\pi = .8$. Then
a. the expected value of X is 8.
b. the standard deviation of X is 1.6.
c. X gives the number of trials until the first success.
d. all of the above.

3.71   Suppose X is a binomial random variable with $n = 10$ and $\pi = .8$. The probability that X is less than 7 is
a. $.322^+$ b. .201   c. .088   d. $.121^+$

3.72   Suppose X is a normal random variable with $\mu = 25$ and $\sigma = 5$. The probability that X exceeds 32 is
a. .4192   b. .9192   c. .0808   d. 1.4

3.73   Suppose there are five locations to drill wells and each location has a .6 chance of having water. What is the probability that all wells have water?
a. .6   b. .30   c. $.5^6$ d. $.6^5$

3.74   Suppose a class contains five history majors, seven English majors, and eight psychology majors. If three students are to be selected, how many elements are there in the sample space of possible majors for the students?
a. 20   b. 27   c. 3   d. 8

3.75   Suppose a class contains five history majors, seven English majors, and eight psychology majors. If two students are randomly selected, what is the probability that both are psychology majors?
a. .15   b. .4   c. .33   d. .16

3.76   A room contains five women and five men. Two are selected at random. What is the probability both are women?

3.77   Suppose 25% of all science majors return for an advanced degree within 5 years after graduation. A random sample of three science majors is selected.
a. List the elements in the sample space of those who return for an advanced degree.
b. Are the outcomes equally likely? Why, or why not?
c. What is the probability that none returns for an advanced degree within 5 years after graduation?
d. What is the probability that at least two return within 5 years?

3.78   Suppose X has the following distribution:

X	1	2	3	4	5
$P(X)$	.2	.3	.3	.1	.1

a. Find the probability that X is odd.
b. Find the probability that X is greater than 3.
c. What is the probability that X is less than or equal to 2?

3.79   A man has three keys, one of which will open a door. If he randomly picks one key after another until he finds the one to open the door, what is the probability that he will open the door on the first try? What is the probability that it will take more than three tries to open the door?

3.80   A tetrahedra (regular four-sided polyhedron) has four sides numbered 1, 2, 3, and 4. A pair of fair tetrahedra are tossed.
a. How many outcomes are possible?
b. Are they equally likely?
c. A variable is defined to be the sum of the two numbers on the two tetrahedra. What are the possible outcomes of the variable?

d. Are they equally likely?

e. We win in a game if we roll a 5 or 7 on the first roll. What is the probability that we win on the first roll?

3.81 A circular game board is divided into five equal slots numbered 1, 2, 3, 4, and 5. The odd numbers are colored red, and the even numbers are colored black. A marble is spun around the board and randomly falls in one of the slots.

a. What is the probability that it lands on red?

b. Suppose the slots are such that an even number is twice as likely as an odd number. What is the probability that the marble lands on red?

3.82 Suppose 30% of the new employees hired by a computer firm are women. Three new employees are selected at random.

a. List the sample space of possible genders of the three employees.

b. Assign a probability to each of the outcomes.

c. Suppose A is the event that exactly two of the three selected are women. List the outcomes that make up A.

d. What is P(A)?

3.83 The student council for the School of Science and Math has one representative from each of the five academic departments: biology (B), chemistry (C), mathematics (M), physics (P), and statistics (S). Two of these students are to be randomly selected for inclusion on a university-wide student committee.

a. What are the ten possible outcomes?

b. All outcomes are equally likely. What is the probability of each?

c. What is the probability that one of the committee members is the statistics student?

d. What is the probability that both members are from laboratory science departments?

3.84 Identify each of the following random variables as either discrete or continuous:

a. The amount of water pumped into a tank overnight.

b. The number of tickets drawn from a barrel until a lottery winner is found.

c. The cost of a week's groceries for a family of four.

d. The number of patients seen by the emergency room over a 24-hour period.

e. The length of life of a computer chip.

3.85 A device to measure one's resistance to pain has a scale that is assumed to be normally distributed with a mean of 30 and a standard deviation of 5. What percent of those using the device score from 22 to 30? Above 34?

3.86 Sketch a graph of the distribution of a binomial variable when $n = 5$ and $\pi = .3$.

3.87 True or false? The following experiments satisfy the conditions of a binomial experiment.

a. Select 20 random voters and record whether they favor reelection of the president.

b. Select eight tickets from a barrel and record the names on the tickets.

c. Draw three balls without replacement from an urn containing five red and seven white balls and record the color.

d. Roll a pair of fair dice ten times and each time observe whether the total is 7.

e. A drug to relieve pain is administered to ten patients. After 2 hours, each patient reports that he is better, worse, or not changed.

f. The IQ of 20 college students is recorded.

3.88 It is believed that about 70% of convicted felons have a history of juvenile delinquency. Ten convicted felons are chosen at random.

a. What is the probability that no more than five have a history of juvenile delinquency?

b. What is the expected number that will have a history of juvenile delinquency?

3.89 Suppose the grade point average of students at a university is normally distributed with a mean of 2.5 and a standard deviation of .5.

a. What percent of the students have grade point averages higher than 3.4?

b. If the top 10% in grade point average make the dean's list, what grade point average do they need?

3.90   Suppose 30% of the student body lives off campus. Five students are selected at random. Let the random variable X = the number of students who live off campus out of the five selected.
   a.   What kind of random variable is X?
   b.   Give the probability distribution table for X and graph it.
   c.   How many students do you expect to live off campus out of the five selected?
   d.   What is the probability that no more than two students live off campus?

3.91   Suppose a reading ability exam for 12-year-olds is normally distributed with a mean of 40 and a standard deviation of 6. What is the probability that a 12-year-old will score in the following ranges?
   a.   Above 60
   b.   Below 45
   c.   Between 50 and 70

3.92   A survey of students showed that 30% are in favor of a ban of alcohol on campus, 52% are in favor of eliminating hazing of pledges, and 20% favor both measures. What is the probability that a student selected at random favors one or the other? Based on these percents, are the events alcohol on campus and hazing of pledges mutually exclusive? Are they independent?

3.93   Suppose a variable, such as IQ or SAT, is distributed normally with a standardized mean of 50 and a standard deviation of 15.
   a.   What percent should score between 30 and 59?
   b.   What percent should score greater than 74?
   c.   What is the 90th percentile?

3.94   a.   A test has two true/false questions on it. If a student guesses, what is the probability he gets both right?
   b.   A test has six multiple-choice problems, and each problem has four choices. If a student guesses, how many problems is she expected to get right?

3.95   In the game of craps, the "shooter" loses on the first roll of a pair of fair dice if the two dice are double 1's or double 6's and wins on the first roll if the two dice total 7 or 11.
   a.   What is the probability that the shooter loses on the first roll?
   b.   What is the probability that the shooter neither loses nor wins on the first roll?

3.96   In the game "three strikes you're out" on *The Price Is Right* game show, seven circular chips are placed in a bag. Four of the chips are white and have numbers on them corresponding to the digits in the price of an automobile. The other three chips are red and have X's on them corresponding to strikes. If the contestant draws the four digits of the price of the car and places them in order before drawing the three strikes, he wins the car. On the first draw, what is the probability that the contestant gets a white chip? If he gets a white chip and guesses, what is the probability that he selects the right position of the number in the price of the car?

3.97   An urn contains a red, a green, and a black marble. Two marbles are drawn from the urn with replacement. List the outcomes in the sample space and the following events:

   A: Both marbles are red.
   B: None are red.
   C: One is red.

3.98   Drivers can select any one of three pumps at a gas station. If two drivers enter the station at the same time (they obviously cannot use the same pump), list the sample space of possible selections for the two drivers.

3.99   A fair die is rolled. If it comes up even, the die is rolled again; if it comes up odd, a fair coin is tossed.
   a.   List the sample space and assign probabilities to each of the outcomes.
   b.   What is the probability of a head on the coin?

3.100   A beer drinker is asked to rank three unmarked glasses of beer according to taste. List the outcomes of the sample space. What must be true in order to assume that the outcomes are equally likely? Assuming that they are, assign probabilities to them.

3.101 A shopper wishes to buy two pairs of shoes but cannot decide among four different pairs. List the outcomes of the sample space of possibilities. If she randomly chooses the two pairs, assign probabilities to the outcomes.

3.102 Eighty students who are candidates for an honor society are classified according to gender and class:

	Men	Women	Total
Freshman	16	14	30
Sophomore	24	26	50
Total	40	40	80

Are the events "being a sophomore" and "being a woman" independent?

3.103 A professional football quarterback has a 60% completion record this season. Assume that pass attempts are independent. Consider his next ten passes.
a. What is the probability that he will complete from seven to nine passes?
b. How many is he expected to complete?
c. What is the standard deviation of the number he is to complete?

3.104 An estimated 2.3 million people are poisoned each year by dangerous chemicals and products found in the home. Sixty-four percent involve children under the age of 6. Out of the next four calls into the poison control center, what is the probability at least two are for children under the age of 6?

3.105 Ninety percent of the trees planted by a landscaping firm survive. What is the probability that 10 or more of the 14 trees just planted will survive?

3.106 Suppose it cost $1 to play a game in which you have a .01 chance of winning $10. Is it to your advantage to play the game? If you play 100 times, how much should you expect to gain?

3.107 Only a third of California homeowners carry earthquake insurance on their homes. A random sample of three homeowners is selected.
a. List the sample space of possibilities of those who have earthquake insurance. Code them S = have and F = do not have.
b. Assign probabilities to the outcomes.
c. What is the probability that at least two of the three will have insurance?
d. What is the probability that none of the three has insurance?
e. What is the expected number out of three that have earthquake insurance?

Computer Exercises

3.108 Using the Random Data > Integer command, simulate 2000 rolls of a single die and store them in C1 of the worksheet. Repeat the process, and store the 2000 results from a second die in C2 of the worksheet. Add C1 to C2 and store the sum in C3. The results in C3 should simulate 2000 rolls of a pair of fair dice. Use the Tally command to summarize the results. From the simulation, what do you estimate the probability of rolling a total of 7 to be? How about a total of 11?

3.109 Open worksheet: **Dice.mtw**. In column 1 you will find the possibilities of the roll of a pair of dice. In column 2 you will find the associated theoretical probabilities. Do these percentages agree with your simulation in Exercise 3.108? Use the Random Data > Discrete command (see Exercise 3.51 in Section 3.2) to simulate the 2000 rolls of a pair of dice. Use the Tally command to summarize the results. Compare this simulation with the one obtained in Exercise 3.108 and with the theoretical probabilities in column 2.

3.110 In Exercise 3.81, a circular game board is divided into five equal slots numbered 1, 2, 3, 4, and 5. The odd numbers are colored red, and the even numbers are colored black. Simulate the game 2000 times and tally the results. Are the outcomes fairly evenly distributed? Should they be? Out of the 2000 times, what percent of the time did red come up? Is your result close to the correct percentage of 60%?

3.111 In Exercise 3.90, five students are randomly selected and the number who live off campus is recorded. The probability of living off campus is 30%. Simulate 1000 realizations of the random variable X that gives the number out of five who live off campus. Tally the results. Does the simulation seem consistent with the probability distribution found in Exercise 3.90 part b?

3.112   In Exercise 3.91, you were given that scores on a reading exam for 12-year-old children are normally distributed with a mean of 40 and a standard deviation of 6. Simulate 2000 reading scores and construct a histogram of the results. Does the histogram have a bell-shaped appearance? Check the normality of the data with a normal probability plot. Is there evidence to show that the data are not normally distributed?

3.113   Figure 3.8 shows four binomial distributions, all with $\pi = .2$ but with $n = 5, 10, 20,$ and 30. To determine the probabilities for $n = 5$ and $\pi = .2$, store the values 0, 1, 2, 3, 4, and 5 in column 1 of the worksheet. Using the Probability Distributions > Binomial command, generate the probabilities in column 2. Use the Plot command to graph the distribution. Is the graph similar to the histogram in Figure 3.8 when $n = 5$? Repeat the process for $n = 10$ (store values 0, 1, 2, . . . , 10 in column 1) and compare with the histogram in Figure 3.8 when $n = 10$. Repeat for $n = 20$ and $n = 30$. Do the graphs become more symmetrical and bell shaped as n increases?

Sampling and Sampling Distributions

The computed value of a statistic depends on the sample that is observed. In other words, for each different sample, there is possibly a different value for the statistic. Thus, a statistic may be viewed as a random variable whose distribution corresponds to the different values that the statistic could assume in repeated sampling.

The distributions, called sampling distributions, of two important statistics are presented in this chapter. You will see that each can be approximated by the normal distribution. Also, through examples, you will see how these sampling distributions are used to make generalizations from a sample to a population.

Contents

■ STATISTICAL INSIGHT
4.1 Principles of Sampling
4.2 The Sampling Distribution of $\bar{x}$

4.3 The Sampling Distribution of the Sample Proportion p
4.4 Summary and Review Exercises

Does Your Vote Count?

We often think that our votes don't really count when it comes to important issues. This was far from the truth in the presidential election of 2000. In the closest race ever, George W. Bush was able to hold on to a narrow margin in the state of Florida to get its 25 electoral votes. Bush won the presidency with 271 electoral votes to 266 for Al Gore. Ironically, Gore won the popular vote 50,996,116 to 50,456,169, or by a margin of 539,947 votes.

The vote in Florida was very close and highly contested. When the final votes were cast on November 7, 2000, Bush was ahead by 1725 votes out of almost 6 million votes cast. On a mandatory recount, his margin dropped to 930 votes. The Gore campaign contested the outcome all the way to the U.S. Supreme Court. The final certified vote count gave Bush a 537-vote margin over Gore. So, 537 votes in the state of Florida determined that George W. Bush would be the 43rd President of the United States instead of Al Gore.

Florida popular vote count

Bush	2,912,790	48.85%	Bush by 537 votes
Gore	2,912,253	48.84%	
Others	138,067	2.32%	

National popular vote count

Bush	50,456,169	47.94%	Gore by 539,947 votes
Gore	50,996,116	48.45%	
Others	3,801,080	3.61%	

Could these results have been predicted prior to the election? Exactly, no; to a certain extent, yes. Prior to the election, most polls showed Bush in the lead, as demonstrated in this table:

Notice that these polls were taken a week before the election and thus do not reflect changes brought about by later events. For example, it was announced on November 3 that Bush had been arrested in 1976 for drunken driving. Public opinion about that event was not reflected in these polls. Most likely, these polls reflected voters' opinions when they were taken 1 week before the election. A poll taken closer to November 7, 2000, would more accurately reflect the outcome of the election. For example, a *Los Angeles Times* poll of 8132 voters, as they exited 140 polling places across the nation, showed Bush with 48% (actual 47.94%) and Gore with 49% (actual 48.45%).

A poll of the public is simply a means by which we can predict individuals' opinions on an issue at that point in time. If events change their opinion, obviously the poll would have to be taken again to reflect those events. Even polls taken concurrently will have different outcomes, as demonstrated in the four polls above. (The percentage for Bush, for example, varies from 45% to 49%.) Many factors affect a poll's outcome. First and foremost, the poll should be based on a random sample of the population; otherwise, we cannot depend on the outcome because it is likely to be biased. But even with well-selected samples, outcomes of polls differ because of the wording of questions, the order of the questions, the method of collecting data, and the size of the sample. But more important, outcomes of polls differ because they are taken from different samples.

So, don't expect two polls on the same issue to produce exactly the same results. Estimates vary from sample to sample, and they do not have to be the same to be valid estimates. Polls have a built-in margin of error. Once an estimate is given, we should factor in the margin of error to get an interval of values that likely cover the true characteristics of the population. In the Statistical Insight Revisited in Section 4.4, we will take a look at how results vary from sample to sample.

Poll	Bush	Gore	Date of poll	Number polled	Margin of error
ABC News	49%	45%	Oct 30–Nov 1	1032	3%
CNN–USA Today–Gallup	47	43	Oct 30–Nov 1	2123	2%
MSNBC–Reuters–Zogby	45	42	Oct 30–Nov 2	1200	3%
Marist	49	44	Nov 1–Nov 2	623	4.5%

4.1 Principles of Sampling

Learning Objectives for this section:

❏ Know the basic principles of sampling and how they are used to obtain a representative sample.

❏ Know the difference between a survey and an experiment.

❏ Understand the concept of bias as it relates to surveys and experiments.

❏ Know that we should avoid using convenience and volunteer samples.

❏ Learn the definition of a simple random sample and how to obtain one.

❏ Know how systematic, stratified, and cluster sampling are used in survey sampling.

Statistical inference is the process of drawing generalizations about a population based on the information in the sample. It is quite clear that those generalizations will be only as good as the sample. Thus, it is imperative that we have sound sampling principles.

Principles of Sampling

Important Steps for Sound Sampling Principles

- Determine the objectives of the study you are undertaking.
- Carefully identify the population.
- Choose the variables that you will measure in the study.
- Decide on an appropriate *design* for producing the data.
- Collect the data.

DETERMINE OBJECTIVES In any study, statistical or otherwise, we must understand the problem and know what results we are seeking. Specifying the objectives is of the utmost importance if we are to draw meaningful conclusions from the study.

IDENTIFY THE POPULATION If we are to make generalizations from a sample, the population must be explicitly described in order to obtain a sample that provides accurate information about the characteristics of the population. For example, if we are interested in persons that are poverty stricken, will we consider only those on welfare or should we attempt to identify those whose net worth is below a certain level? If we sample only welfare recipients, our results will generalize to the population of welfare recipients, and not necessarily to all those living in poverty.

CHOOSE VARIABLES To accomplish the stated objectives of a statistical study, we must determine what will be measured and how it will be measured. For example, suppose we are interested in studying the issue of substance abuse on campus. The main objective is to determine whether there is a substance abuse problem. We are interested in what substances are used and what percent of the student body uses each one, and whether the percentages differ for men and women. A poll of the student body would involve determining the gender of the respondents and their substance use habits. We might also be interested in whether the percentages change for the different academic classes, whether family history of substance use is a factor, and so on. This means that several variables must be measured on each respondent. In order not to overlook an important issue, we should attempt to identify all relevant variables prior to collecting any data. Also, if there is some doubt as to whether to measure a certain quantity, it is best to go ahead and obtain data rather than later regretting that it had not been measured.

STATISTICAL DESIGN Having stated the objectives, identified the population, and determined the variables that will be measured, we turn our attention to the design of the statistical study. Statistical designs generally fall into two categories: **surveys** and **experiments**.

The poll of the student body described earlier is an example of a *survey*. In fact, all opinion polls are surveys. Polling organizations such as Gallup, Harris, Roper, ABC/*Washington Post*, and *TIME*/CNN conduct polls almost daily. Most of their polls are "random digit dialing" telephone surveys in which telephone numbers are randomly selected by a computer. In a survey, the already existing opinions or facts about a group of people are solicited. Unlike an experiment, there is no control over the respondents' behavior; they are simply asked to respond in a way that reflects their opinion at that point in time. Their responses are then tabulated for analysis.

An *experiment*, on the other hand, attempts to determine a cause-and-effect relationship between two or more variables. The experimenter plays an active roll in the study by controlling the environment and by administering a treatment to the subjects. In an experiment, for example, a scientist administers a drug to a group of subjects in an attempt to study its effect. After the treatment is applied to the subjects, measurements are taken and then analyzed.

COLLECT THE DATA The design of a statistical study, whether it is a survey or an experiment, also specifies how the data will be collected. One possibility is to collect data from everyone in the population; that is, to take a census. We desire information about the population; therefore, if we have a listing of all subjects (a *sampling frame*), then why not take a census? Because a sample is much smaller than the population, it is generally more cost effective and less time consuming to take a sample. If we carefully select the sample, the quality of the collected data may be superior to that obtained from a census. It is quite possible that a sample will give more accurate results.

Quality data, however, are difficult to obtain. Even with the best of intentions, errors (measurement errors, sampling errors, recording errors) can creep into data. Measurement and recording errors can be minimized by a careful attention to detail when tabulating results. Exercising sound sampling principles can control sampling errors. We should not use **convenience samples**, which are groups that are conveniently available for study. And we should avoid situations that allow the respondents to be *self-selected*, as is the case in a **volunteer sample**.

The primary goal is to obtain a sample whose characteristics match those of the population under study. If the sample is **representative** of the population in those factors that are relevant to the study, the results will be accurate and useful. If we are conducting a political survey, for example, the weight of the subject probably is irrelevant; however, the gender and race of the subject *are* relevant. If the population consists of 50% women and 30% blacks, the sample should consist of approximately 50% women and 30% blacks so the sample will be representative with respect to gender and race. Otherwise, the sample is **biased**; for example, if more than 50% are women.

> **Bias** is a systematic tendency of the sample to misrepresent the population.

Example 4.1 Suppose we wish to study the television viewing habits of the general population. Our sample should be representative of the population with respect to variables such as gender, race, education level, age, income, IQ, and occupation of subject. ∎

Simple Random Sample

One way to obtain a representative sample and reduce the effects of sampling bias is to *randomly* select the sample in such a way that every sample of that size has the same chance of being chosen. The resulting sample is called a *simple random sample*.

> A **simple random sample** of size n consists of n elements chosen from the population in such a way that all samples of that size have the same chance of being selected.

There are several ways to obtain a simple random sample. A popular method is to use *physical mixing*, also referred to as *lottery sampling*. Each element in the population is given an identification tag. The tags are thoroughly mixed in a barrel, and the sample is selected from the barrel, one observation at a time. If the mixing is thorough, each element that remains in the population has the same chance of being selected at each draw. If n tags are drawn, that group of n is just as likely as any other group of n, resulting in a simple random sample of size n.

If there is a sampling frame of the population, a simple random sample can be selected by enumerating the population and using random digits to choose the members of the sample. Prior to the widespread use of computers, tables of random digits were commonly used to randomly select elements from a population. Today most people use a computer, or in some cases a calculator, to generate random numbers like those found in a table of random digits.

COMPUTER TIP

Random Numbers

A computer or calculator can be used to generate random numbers. Many scientific calculators have a random number key, labeled [RND#], that produces a three-digit random number when pressed. If six-digit numbers are desired, then press the key twice and put the two numbers together.

To generate random numbers with Minitab, from **Calc** on the menu bar select **Random Data** followed by **Integer**. When the Integer Distribution window appears, specify the desired number of random numbers to **Generate**, select the column to **Store** the numbers, give the **Minimum** and **Maximum** possibilities for the random numbers and then click **OK**.

Other Sampling Procedures

It is important to realize that the statistical methods presented later in this text are developed with the assumption that the sample is a simple random sample. For some surveys, however, the simple random sample may not be desirable or economically feasible. We end this section with three sampling procedures that are commonly used in survey sampling.

SYSTEMATIC SAMPLE   Suppose we have a sampling frame of 10,000 students listed by some identification number. If we start at a random place within the first 100 names on the list and then select every 100th name, the result is a *systematic random sample* of size 100.

It is possible to use systematic sampling without having a tangible sampling frame. In conjunction with a census, to obtain a 10% sample of homes, the U.S. Bureau of the Census might start at a random place and select every tenth home for further study. In this manner, it is possible to sample the households in a city without knowing how many households are in the city.

Systematic sampling should be avoided when the data in the population are of a periodic nature. For example, suppose we want to sample the daily receipts of a large department store. A one-in-seven systematic sample would result in a biased sample because the sample would consist of the daily receipts for only one particular day of the week. This problem could be avoided by taking a one-in-ten systematic sample. However, it is best to use some method other than systematic sampling when the population is periodic.

STRATIFIED SAMPLE   A simple random sample of size 1000 could produce 600 women and 400 men. If the population is evenly divided among women and men, this sample is biased toward women. To obtain a representative sample, divide the population into the two strata (women and men) and select random samples of size 500 from each stratum. The resulting sample of 500 women and 500 men is a *stratified random sample*. The strata can be geographical regions, religious preference, race, income bracket, political party, or gender, as mentioned above. The stratification can be based on any variable relevant to the survey. Stratification will guarantee that the sample is similar to the population in those characteristics on which one chooses to stratify.

Example 4.2   Suppose bank officials wish to sample customer savings accounts. Suppose, furthermore, that they know that 5% of the accounts are over $50,000, 20% are between $10,000 and $50,000, 25% are between $5000 and $10,000, and 50% are below $5000. Using stratification, they can assure themselves that a sample of 100 accounts will have accounts of all possible categories.   ■

Often, stratified random sampling is accompanied by *proportional* sample size. This means, for example, that if stratum 1 is twice as large as stratum 2, the random sample from stratum 1 will be twice as large as the sample from stratum 2. In other words, the sample size for a stratum is proportional to the population size in that stratum.

Example 4.3   Using proportional sampling in Example 4.2, the bank officials would sample so as to have 5 of the desired 100 accounts from the over-$50,000 category. Similarly, 20 accounts would be from the $10,000 to $50,000 category, 25 from the $5000 to $10,000 category, and the remaining 50 accounts from the below-$5000 category. Not only does the sample consist of accounts from all possible categories; the sample sizes also are proportional to the stratum sizes.   ■

A requirement of proportional sampling is that the stratum sizes be known. This information can be obtained only from a census. Thus, it is important that up-to-date census information be available.

CLUSTER SAMPLE   The expense of sampling is often a major factor in a statistical survey. *Cluster sampling* is an economical way of selecting a sample by choosing groups of subjects, called *clusters*, and then surveying every individual unit in the cluster. Suppose we wish to sample the residents of a large city. With a city map, divide the entire city into blocks or clusters of households. Obtain a simple random sample of clusters, and then interview each resident of each selected cluster. Clearly, this is a more cost-effective method than traveling around the city looking for individuals in a simple random sample. On the other hand, cluster sampling may be susceptible to sampling bias. A certain ethnic group may not be represented in the sample simply because its cluster is not selected. Also, the individual units in a cluster usually are very homogeneous (similar age, income, educational background, recreational interests, etc.), and thus a cluster sample is less informative than a simple random sample of the same size. With cluster sampling we may be unable to get a true cross section of the population.

EXERCISES 4.1   The Basics

4.1   Advice columnist Ann Landers asked her readers to clip a questionnaire about sexual attitudes out of the newspaper, complete it, and return it to her. To what population would this study generalize? Is this a reasonable sample if she wishes to generalize the results to all American adults?

4.2   To estimate the murder rates in major U.S. cities, a law enforcement agency determined the rates in the 12 largest cities last year. Describe the sample. Do you think the figures obtained from this sample would generalize to the major cities in the U.S.? Why, or why not?

4.3   To assess the possible success of a new shopping center, a questionnaire was mailed to a sample of people obtained from the mailing list of the local chamber of commerce. The questionnaire asked whether the respondent thought the availability of merchandise was adequate in the city. Describe the population of interest. Do you think that taking every tenth name from the chamber of commerce list is a reasonable way to obtain a sample? Explain your answer. Other than the chamber of commerce list, can you think of other lists to use for the sample?

4.4   How would you select a sample of 100 students from your student body so that you would obtain representatives of each class? What sampling technique would you use?

4.5   The university newspaper reported on a survey conducted by the office of student development on the percent of students who use their student government representative to convey their feelings on university issues.
a. What is the population of interest?
b. Explain in some detail how you would take a sample for this survey.
c. Identify the sampling procedure if a personal interview was conducted with randomly selected students as they left the cafeteria.
d. What is the sampling procedure if a survey was sent to every student on the second floor of four randomly selected dorms?
e. What is the sampling procedure if a survey was sent to the first name on each page of the student directory?

4.6   A marketing analyst is asked to study the buying habits of shoppers of a national chain store. Suppose there are 200 stores around the country.
a. Describe the population of interest.
b. Describe a sampling procedure for obtaining a representative sample.

4.7   A popular sampling procedure conducted by the television networks on the day of an election is *exit polling*. Randomly selected individuals are asked for whom they voted as they exit preselected polling booths around the country. Is this systematic, convenience, stratified, or cluster sampling? Explain.

4.8   Identify the sampling technique in the following cases:

   a. To obtain a sample of books in the library, the librarian randomly selects a book from the first rack. He then selects a book from that same position from each rack in the library.

   b. To obtain a sample of voters from the county, ten voters are randomly selected from each precinct.

   c. A local television station asks people to send in their names on a postcard to win a free trip to Disney World. The postcards are placed in a large cage that is rotated several times, and then ten winning cards are selected.

   d. To investigate the attitudes of the general population on the subject of nuclear weapons, *Time* magazine takes a random sample of its readers.

4.9   A large department store has its charge accounts listed alphabetically by the customer's last name. Accountants want to estimate the total amount of unpaid balances. Should systematic, stratified, or simple random sampling be used? Discuss the advantages of your choice.

4.10   Explain how bias might enter into the following statistical study: To evaluate energy consumption for heating residential buildings, a random sample of 50 buildings was observed for *one* heating season by a regional electrical utility company.

4.11   Discussion question—1936 presidential election: A national survey that yielded unreliable results occurred in the 1936 presidential election. A national magazine of the time, the *Literary Digest*, had successfully predicted every presidential election since 1916. For the 1936 election, the magazine mailed out 10 million sample ballots to people whose names were taken from subscription lists of magazines, telephone directories, and automobile registrations. More than 2.3 million ballots were returned, the largest number of people ever to respond to a survey. From the responses on the 2.3 million cards, the *Digest* predicted that Alf Landon would defeat Franklin Roosevelt by a margin of 57% to 43%. It predicted that Roosevelt would receive only 161 of the 531 electoral votes. As it turned out, Roosevelt was given 523 of the 531 electoral votes for a landslide victory with 62% of the popular vote. This marked the end of a 20-year period in which the *Literary Digest* had successfully predicted the presidents. (The *Digest* went bankrupt shortly thereafter.) Discuss reasons why the *Digest*'s prediction was erroneous.

Computer Exercises

4.12   With your calculator, obtain 15 random numbers between 0 and 500. What do you do with the random numbers produced by the calculator that are greater than 500? Explain how you would use the random numbers to select a random sample of size 15.

4.13   With your computer, obtain 500 random numbers between 1 and 10,000. If you had a sampling frame of 10,000 students, explain how you would use the 500 random numbers to select a sample.

4.2   The Sampling Distribution of $\bar{x}$

Learning Objectives for this section:

❏ Know the difference between a parameter and a statistic.

❏ Know what is meant by the sampling distribution of a statistic.

❏ Be able to describe the sampling distribution of $\bar{x}$ when sampling from a normally distributed population.

❏ Know how the Central Limit Theorem describes the sampling distribution of $\bar{x}$ when sampling from a population that is not normally distributed.

❏ Know how to use simulations to illustrate the Central Limit Theorem.

❏ Be able to apply the Central Limit Theorem to work problems involving the sample mean, $\bar{x}$.

Having obtained a random sample from a population, we next decide which **statistic**, or numerical quantity calculated from the observations in the sample, best estimates an unknown population **parameter**. For example, which of the following three statistics would you use to estimate the parameter, μ.

1. $\bar{x}$, the sample mean
2. $\bar{x}_\mathrm{T}$, a trimmed sample mean
3. M, the sample median

To answer this question we must understand the concept of the *sampling distribution* of a statistic.

> The **sampling distribution of a sample statistic** is the probability distribution associated with the various values that the statistic could assume in repeated sampling.

The sampling distribution shows how the values of the statistic vary from sample to sample.

Unlike the value of a population parameter, which is usually unknown and may never be known, the value of a statistic *is* known because it is calculated from the collected data. We must realize, however, that the value assumed by the statistic depends on the particular sample observed. A large number of samples can be chosen from a population, and each will yield its own value of the statistic. The distribution of the potential values that the statistic assumes from all possible samples is its *sampling distribution*. In this section, we will investigate the sampling distribution of the statistic, $\bar{x}$.

Distribution of $\bar{x}$ when Sampling from a Normal Population

Consider, for the moment, a population that is normally distributed with a mean of 50 and a standard deviation of 15. The distribution curve is shown in Figure 4.1.

Figure 4.1
Population distribution with mean 50 and standard deviation 15

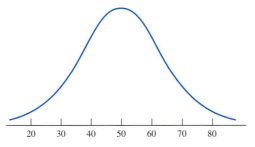

If we take a random sample from this population, most observations will be between 20 and 80 because 20 is two standard deviations below the mean of 50 and 80 is two standard deviations above the mean of 50.

Figure 4.2 is a histogram of a sample of 100 observations selected at random (simple random sample) from the population, with the population frequency curve superimposed over the histogram. Most observations do indeed fall between 20 and 80, with the shape of the histogram similar to the shape of the frequency curve of the population, as expected.

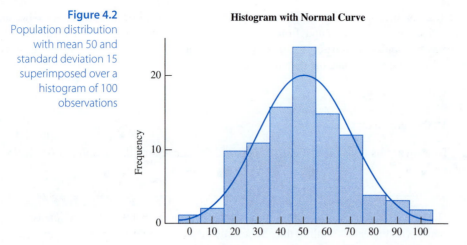

Histogram with Normal Curve

Descriptive Statistics: C1

Variable	N	Mean	Median	TrMean	StDev	SE Mean
C1	100	49.19	48.55	48.83	19.97	2.00

Variable	Minimum	Maximum	Q1	Q3
C1	4.00	98.00	35.55	62.20

From the descriptive statistics for the sample, we see that the mean of this sample is 49.19, which is close to 50, the mean of the population. Certainly another sample of 100 observations will not yield the same histogram or the same mean, but we would expect to see something similar. In fact, from the following descriptive statistics for a second sample, we see that the mean is 50.71, which is also close to 50.

Descriptive Statistics: C1

Variable	N	Mean	Median	TrMean	StDev	SE Mean
C1	100	50.71	49.68	50.35	17.57	1.76

Variable	Minimum	Maximum	Q1	Q3
C1	13.10	95.55	37.29	64.65

Other samples will yield still different means, but again the means should be close to the population mean, μ. Just how close depends on the *sampling variability* of $\bar{x}$. It is through the variability of the sampling distribution of the statistic that we can determine how precise an estimate of a parameter will be. The amount of variability associated with the sampling distribution of a statistic is measured by its *standard deviation*. The smaller the standard deviation, the more *stable* the statistic is as an estimate of the parameter. To evaluate its stability, we consider the distribution of all potential values that $\bar{x}$ could assume in repeated sampling. In particular, we ask:

1. What is the general shape of the distribution of potential values $\bar{x}$ can assume?
2. Are the potential values of $\bar{x}$ centered about a certain quantity?
3. How much variability is associated with the potential values of $\bar{x}$; that is, what is its standard deviation?

These questions are answered with a description of the sampling distribution of $\bar{x}$. Prior to sampling, the statistic can be thought of as a random variable because different samples can lead to different values of the statistic. The probability distribution of that random variable is what we refer to as the sampling distribution. If we know the sampling distribution of $\bar{x}$, we can find its mean and standard deviation, and the probabilities associated with the various values that it can assume. Therefore, we can answer the three questions posed above.

Returning to the problem at hand, recall that the first random sample of size 100 had a mean of 49.75 and a second random sample of 100 had a mean of 50.89. We have two *realizations* of the random variable $\bar{x}$. Figure 4.3 gives a histogram and a description of 2000 realizations of the random variable $\bar{x}$ obtained from 2000 different random samples, each of size 100. The histogram of these 2000 $\bar{x}$'s gives the general appearance of the theoretical sampling distribution of $\bar{x}$.

Descriptive Statistics: xbar

Variable	N	Mean	Median	TrMean	StDev	SE Mean
xbar	2000	50.031	50.030	50.032	1.497	0.033

Variable	Minimum	Maximum	Q1	Q3
xbar	44.430	54.948	48.970	51.058

Figure 4.3
Histogram of 2000 $\bar{x}$'s

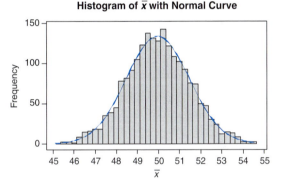

Notice that the values of $\bar{x}$ tend to mound up in normal fashion around 50, the mean of the population. If we think about it, the 100 observations in each sample are nearly all somewhere between 20 and 80, so each mean should be around 50, the mean of the population. Also recall that the standard deviation of the population is 15, but the standard deviation of the sampling distribution of $\bar{x}$ doesn't tend to be so great. In fact, most realizations of $\bar{x}$ lie between 47 and 53. If 95% of the $\bar{x}$'s range from 47 to 53, as they appear to do, the standard deviation is 1.5. (Working backward, we know that if the standard deviation is 1.5, then *two* standard deviations on each side of 50 will create a range from 47 to 53.) We show later that the standard deviation of the sampling distribution of $\bar{x}$ is given by $\sigma/\sqrt{n} = 15/\sqrt{100} = 1.5$.

From the descriptive statistics in Figure 4.3, we see that the 2000 simulated values have a mean of 50.031 (very close to 50) and a standard deviation (which is an estimate of the standard deviation of $\bar{x}$) of 1.497, which is very close to 1.5. Furthermore, the smallest realization of $\bar{x}$ is 44.430 and the largest is 54.948.

Finally, the normal probability plot of the 2000 simulated values in Figure 4.4 indicates that the sampling distribution is normally distributed.

Figure 4.4

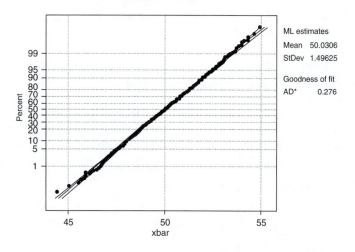

Normal Probability Plot for xbar
ML estimates - 95% CI

We are now in a position to answer the three questions about the sampling distribution of $\bar{x}$ when sampling from a normally distributed population with a mean of 50 and a standard deviation of 15.

1. The general shape is bell shaped; in fact, it is normally distributed.
2. The distribution of potential values of $\bar{x}$ is centered at 50, the mean of the population.
3. The standard deviation is $\sigma/\sqrt{n} = 15/\sqrt{100} = 1.5$, where σ is the standard deviation of the population.

Caution: We are referring to two different standard deviations in this discussion. First, we have the standard deviation of the population, σ, and second, the standard deviation of the sampling distribution of $\bar{x}$, which is given by $\sigma/\sqrt{n}$.

The preceding discussion is based on a specific example. In particular, the population is normally distributed and centered at 50, with a standard deviation of 15 and a sample size of 100. However, the results can be generalized for any sample size and arbitrary means and standard deviations.

Sampling Distribution of $\bar{x}$ When Sampling from a Normally Distributed Population

Let $\bar{x}$ be the mean of a sample of size n from a normally distributed population that has mean μ and standard deviation σ. For all sample sizes n, the sampling distribution of $\bar{x}$:

1. Is exactly normally distributed.
2. Is centered at μ, the mean of the population.
3. Has a standard deviation of $\sigma/\sqrt{n}$, where σ is the standard deviation of the population.

Distribution of $\bar{x}$ When Sampling from a Nonnormal Population

Next we ask: What is the sampling distribution of $\bar{x}$ when we sample from a population that is not normally distributed? Figure 4.5 shows the distribution of a population that is skewed right (not normally distributed) and has a mean $\mu = 5$ and a standard deviation $\sigma = \sqrt{10}$.

Figure 4.5

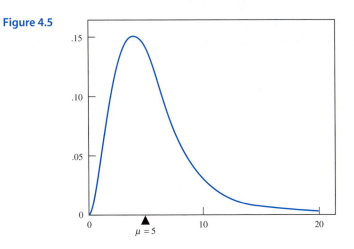

We will examine the sampling distribution of $\bar{x}$ when samples of size $n = 100$ are randomly selected from this population. As before, we randomly generate 2000 realizations of $\bar{x}$ to simulate its sampling distribution. Figure 4.6 is a histogram of the 2000 values of $\bar{x}$.

Descriptive Statistics: xbar

Variable	N	Mean	Median	TrMean	StDev	SE Mean
xbar	2000	4.9945	4.9793	4.9928	0.3184	0.0071

Variable	Minimum	Maximum	Q1	Q3
xbar	3.9953	6.0241	4.7723	5.2102

Figure 4.6

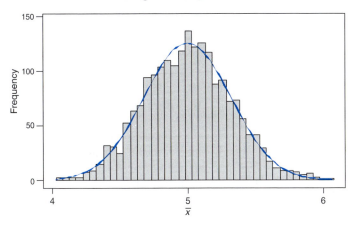

Histogram of $\bar{x}$ with Normal Curve

Although the population is skewed right, the simulated sampling distribution is almost symmetric, is centered at 5, and is not nearly so spread out as the population distribution. The population assumes values from 0 to 20, whereas most of the values of $\bar{x}$ are somewhere between 4 and 6. Recall that the standard deviation of the population is $\sigma = \sqrt{10} = 3.16$. The standard deviation of the simulated values is only 0.3184, which is very close to $\sigma/\sqrt{n} = 3.16/\sqrt{100} = 0.316$. It appears that the sampling distribution of $\bar{x}$ is close to being normally distributed with a mean equal to the population mean and a standard deviation of $\sigma/\sqrt{n}$.

The Central Limit Theorem

What we have observed in this simulation is a result of one of the most important theorems in statistics: the Central Limit Theorem. What was stated previously about the sampling distribution of $\bar{x}$ when sampling from normally distributed populations generalizes to almost any population. The only difference is that the sampling distribution is not exactly normal but is only approximately normally distributed and the sample size must be rather large. The larger the sample size, the more closely the distribution is approximated by a normal distribution.

Central Limit Theorem

Let $\bar{x}$ be the mean of a sample of size n from a population with an unknown distribution. When n is relatively large, the sampling distribution of $\bar{x}$ is approximately normally distributed. The approximation becomes better as the sample size increases.

Generally, the more the population deviates from a normal distribution, the larger the sample size must be to conclude that the sampling distribution of $\bar{x}$ is approximately normally distributed. In most cases, however, a sample size of 30 or more observations is adequate.

Sampling Distribution of $\bar{x}$ When Sampling from a General Population Distribution

Let $\bar{x}$ be the mean of a sample of size n from a population that has mean μ and standard deviation σ. When the sample size, n, is sufficiently large*, the sampling distribution of $\bar{x}$:

1. Is approximately normally distributed.
2. Is centered at μ, the mean of the population.
3. Has a standard deviation of $\sigma/\sqrt{n}$, where σ is the standard deviation of the population.

*In most cases, a sample size of 30 or more is sufficient.

To see how the shape of the sampling distribution is affected by the shape of the population and the sample size, we will simulate the sampling distribution of $\bar{x}$ based on three different sample sizes (5, 10 and 30) from three different distributions. Each simulation is based on 2000 realizations of $\bar{x}$ computed from 2000 different random samples.

Figure 4.7(a) gives the three different population distributions. The first is the normal distribution with mean 50 and standard deviation 15 studied previously. Second is the symmetric uniform distribution in which all values are equally probable. The last is a highly skewed distribution called the lognormal. Figure 4.7(b) shows histograms of the 2000 realizations of $\bar{x}$ when the sample sizes are 5. For the normal and uniform distributions, both of which are symmetric, we see that the sampling distribution is also symmetric but has less variability. In the case of the skewed lognormal distribution, the sampling distribution is less skewed and less variable. Remember that the standard deviation of $\bar{x}$ is $\sigma/\sqrt{n}$, which gets smaller as n increases because n is in the denominator. In Figure 4.7(c), we see the sampling distributions when the sample sizes are 10. Again, when the distribution is symmetric, the sampling distribution is symmetric, and even in the nonsymmetric lognormal case, the sampling distribution is becoming more symmetrical as the sample size increases. In all three cases, the variability of the sampling distribution is decreasing as the sample size is increasing. In Figure 4.7(d), we see that when the sample size is 30, the sampling distribution is more bell shaped with a smaller standard deviation. With the aid of boxplots in Figure 4.7(e), we see how the sampling distributions become more symmetric with smaller standard deviations as the sample size increases from 5 to 10 to 30. The normal probability plots for the sampling distributions when $n = 30$ in Figure 4.7(f) show that the distributions are approximately normally distributed in the normal and uniform cases. Because of the severe skewness of the lognormal distribution, however, it may take a sample size as large as $n = 100$ to remove the skewness and get a reasonable normal approximation of the sampling distribution. As guaranteed by the Central Limit Theorem, however, there is a sample size n for which each of the three sampling distributions can be reasonably approximated with a normal curve centered at μ with a standard deviation of $\sigma/\sqrt{n}$.

Again, we point out that in the first case, when the population is normally distributed, the sampling distribution of $\bar{x}$ is *exactly* normally distributed for *all* sample sizes. It is only approximately normal when sampling from a nonnormal population. The approximation becomes better as the sample size becomes larger. The more the population deviates from normality, the larger the sample size must be in order to get a reasonable approximation. In most cases, a sample size of 30 or more is sufficient.

Figure 4.7

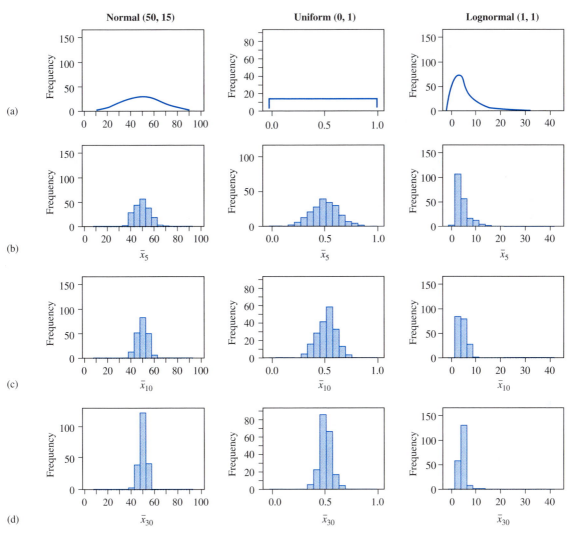

Figure 4.7
continued

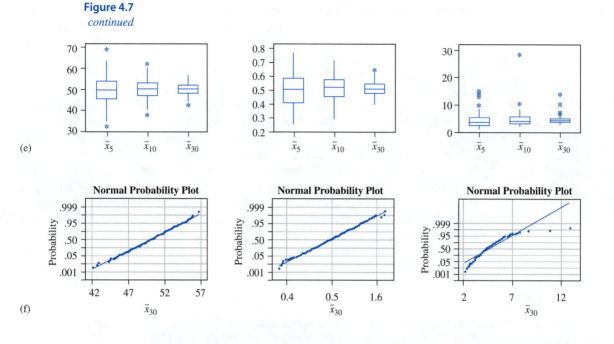

(e)

(f)

COMPUTER TIP

Simulations

The above simulations were obtained by using the Random Data command in Minitab. Under the **Calc** menu, select **Random Data** and then choose the distribution. When the Distribution window appears, fill in the **Generate** box with the desired number of rows of data. It is possible to store random data into several columns at the same time by specifying the different columns in the **Store in column(s)** window. To generate the data, click **OK**.

For these simulations it is best to consider the rows as the different random samples and then use the row means as the different realizations of $\bar{x}$. For example, to generate 2000 random samples each of size 30, we generate 2000 rows of data and store in C1–C30. To compute the 2000 realizations of $\bar{x}$, under the **Calc** menu, select **Row Statistics**. In the Row Statistics window, click on **Mean** and then specify the **Input Variables** as C1–C30. **Store** the **results** in a column—say, C100—and click **OK**. We are then able to construct a histogram or a boxplot of the 2000 different values of $\bar{x}$ in C100 to get a simulation of the sampling distribution of $\bar{x}$ when the sample size is 30.

Applications of the Central Limit Theorem

Example 4.4 The federal program, Aid to Families with Dependent Children (AFDC), made a mean monthly payment of $546 to families in New York in 1993 (U.S. Department of Health and Human Services). A social worker in upstate New York would like to know something about the monthly payments to welfare recipients in her region of the state. The mean monthly payment of all welfare recipients in her region is the unknown population parameter, μ, which she wants to estimate. She decides to take a random sample of 100 welfare recipients from her region, calculate the sample mean $\bar{x}$, and use it as an estimate of μ. The sample mean turns out to be \$528, so she estimates the mean payment to all welfare recipients to be \$528. If the standard deviation of all welfare recipients is \$200—that is, $\sigma = 200$—then how close will her estimate be to the true mean payment to all welfare recipients in her region?

Solution   Because $\sigma = 200$ and $n = 100$, the standard deviation of $\bar{x}$ is

$$\frac{\sigma}{\sqrt{n}} = \frac{200}{\sqrt{100}} = 20$$

Thus, we can be reasonably assured that the observed $\bar{x}$ will be within \$40 (two standard deviations) of the population mean; that is, the mean monthly payment for all welfare recipients in her region is most likely between \$488 and \$568. Notice that the mean monthly payment of \$546 paid to welfare recipients in all of New York State is inside the interval of values.   ■

Example 4.5   Suppose in Example 4.4 that the social worker had a random sample of size 400 instead of 100. How close then would her estimate be to the true mean income level of all welfare recipients in her region?

Solution   With a sample size of 400, the standard deviation of $\bar{x}$ is

$$\frac{\sigma}{\sqrt{n}} = \frac{200}{\sqrt{400}} = 10$$

So now 95% of the $\bar{x}$'s, and in particular her estimate, are within \$20 (instead of \$40) of the population mean. We see that by increasing the sample size fourfold, the standard deviation is halved. The accuracy of the estimate improves as the sample size increases.   ■

Knowing the sampling distribution of the sample mean, $\bar{x}$, we are able to evaluate how close a certain $\bar{x}$ is to μ once it is calculated from the sample.

Example 4.6   Census data from the National Center for Education Statistics (U.S. Department of Education, 1997) indicate that the distribution of annual income for schoolteachers in the United States has a mean of \$39,400 and a standard deviation of \$4000. This means that the majority of schoolteachers earn between \$31,400 and \$47,400 (two standard deviations below and above the mean). Now suppose we randomly select 64 teachers from a certain state, and we find that their mean income is \$37,900. Can anything be said about the income of teachers in that state?

Solution   Suppose the salaries in this state are comparable to the salaries across the United States. Then, from the Central Limit Theorem, we know that the sampling distribution of $\bar{x}$ (obtained from samples of size 64 from this state) is approximately normally distributed with

$$\text{mean} = 39{,}400 \text{ and standard deviation} = 4000/\sqrt{64} = 500$$

Thus, 95% of all potential $\bar{x}$'s should fall between 38,400 and 40,400 (within two standard deviations of 39,400). We observed $\bar{x} = 37{,}900$, a full three standard deviations below the mean. This is unusual if the mean salary in the state is \$39,400. We are led to believe that the mean of the distribution of teacher's salaries in this state is not \$39,400, but is somewhat below the national mean.   ■

Probability problems involving $\bar{x}$ can be worked if we know its sampling distribution, as indicated in the next example.

Example 4.7   The mean length of stay in a certain Alcoholics Anonymous clinic is 17 days, and the standard deviation of the length of stay is 3 days. A random sample of 36 patients is chosen.

a. Find the probability that the mean stay is more than 18.5 days.
b. What is the probability that the mean stay is between 16 and 19 days?

Solution   We have from the Central Limit Theorem that $\bar{x}$ is approximately normally distributed with

$$\text{mean} = 17 \qquad \text{standard deviation} = \frac{\sigma}{\sqrt{n}} = \frac{3}{\sqrt{36}} = .5$$

a. To find the probability that $\bar{x} > 18.5$, we find the z-score associated with 18.5, as follows:

$$z = \frac{18.5 - 17.0}{.5} = 3.0$$

From the standard normal probability table, we find that .9987 of the area lies to the left of a point that is three standard deviations above the mean. Using the complement rule, we have that the probability of observing an $\bar{x} > 18.5$ is

$$1.0000 - .9987 = .0013$$

The probability that the mean stay is longer than 18.5 days is .0013, extremely small.

b. To find the probability that $\bar{x}$ lies somewhere between 16 and 19, we must work two problems: First find the area below 16 and then find the area below 19. Finally, subtract the first from the second. The z-score associated with 16 is

$$z = \frac{16.0 - 17.0}{.5} = -2.0$$

From the normal probability table, we find an area of .0228 associated with a z-score of -2. The z-score associated with 19 is

$$z = \frac{19.0 - 17.0}{.5} = 4.0$$

From the normal probability table, we find an area of .99997 associated with a z-score of 4.0. Consequently, the desired probability is

$$.99997 - .0228 = .97717$$

which is very large. It is very likely that the mean length of stay is between 16 and 19 days. In fact, with approximately 95% probability, the average stay is somewhere between 16 and 18 days (two standard deviations on either side of 17).   ■

The sampling distribution also allows us to determine which of the potential values of the statistic are reasonable and which are unlikely. Thus, when we obtain a value for a particular statistic, knowing the sampling distribution of that statistic allows us to determine immediately whether that value agrees with our ideas about the population.

Example 4.8   Returning to the welfare problem in Example 4.4, suppose a conjecture is made that the mean monthly payment to the welfare recipients in the social worker's region is significantly less than $546, which is the mean amount paid to all New York welfare recipients. What is your opinion of this conjecture?

Solution   The mean amount paid to all New York recipients is $546. Based on the sample mean, $\bar{x} = \$528$, of the 100 recipients randomly selected in her region, we determined in Example 4.4 that the mean payment to all recipients in the social worker's region is between $488 and $568. Because the mean amount paid in her region could be as high as $568 and the mean amount paid to all New York recipients is $546, we cannot support the conjecture that the recipients in her region are paid less.   ■

EXERCISES 4.2   The Basics

4.14   What statistic is identified by the bold value in the following?
a.   Half of a sample of 50 workers in a plant make **$320** or less per week.
b.   In a nationwide survey of 3800 adults, **66%** said that elected officials don't care what people think about political issues.
c.   The mean weight loss for patients on a diet plan was **4.6** pounds.

4.15   Is the bold value a parameter or a statistic?
a.   The difference between the percent of seniors and juniors who have cars on campus is **9%**.
b.   According to a survey of 6000 service stations in 1995, the average price per gallon of gasoline was **$1.22**, which was 3 cents less than in 1994.
c.   Half (**50%**) of all Americans are involved in alcohol-related accidents in their lifetimes.

4.16   A random sample of size n is selected from a population with a mean of 25 and a standard deviation of 8. For each value of n, give the mean and standard deviation of the sampling distribution of $\bar{x}$.
a. $n = 10$ b. $n = 16$
c. $n = 30$ d. $n = 100$

In which of these can we be assured that the sampling distribution will be adequately approximated by a normal distribution?

4.17   The national norm of a science test for tenth-graders has a mean of 75 and a standard deviation of 20. A sample of 100 tenth-grade students from the New York City public school system had an average of 72 on the test. Obtain bounds that would almost certainly include the mean score of New York City tenth-graders.

4.18   The Bayley Scale of Infant Development, which has a mean of 100 and a standard deviation of 16, was given to a group of 36 infants who were exposed to a new educational procedure developed at a university. Their average score was 105.6. Is there evidence that this group of children had an unusually high average on the test?

4.19   An insurance company's records show that the mean payout for all automobile claims is $1800 and the standard deviation is $400. Suppose 90 claims are filed in one week. What is the probability that the average is more than $1900?

4.20   The average length of a felony sentence imposed by state courts for trafficking in drugs in 1992 was 72 months. (Bureau of Justice Statistics, *Felony Sentences in State Courts, 1992*, Bulletin NCJ-151167, January, 1995). If the standard deviation is 10 months, is it unusual for a random sample of 50 subjects accused of trafficking in drugs to have an average sentence longer than 75 months? To answer this question, find the probability that a sample of size 50 will have a sample mean above 75.

4.21   It takes General Motors an average of 31 hours of labor to build a car (*Term*, Vol. 4, No. 1, January, 1995). Suppose that the duration of time is normally distributed with $\sigma = 2$ hours.
   a.   What is the probability that a single car, selected at random, will take between 28 and 34 hours to build?
   b.   What is the probability that the mean of the times for a sample of 25 cars will be between 28 and 34 hours?

4.22   A standardized science test has a national mean of 250 and a standard deviation of 50. The test is to be given to a group of prospective science majors. What is the standard deviation of the estimate, $\bar{x}$, calculated from a sample of size 64? From a sample of size 400?

4.23   During a 3-month period in Asheville, North Carolina, the ambulance response time was recorded on 236 calls. The average response time was 12.63 minutes. If the standard deviation of all response times is 5 minutes, give an interval that you are reasonably sure (95% sure) contains the population ambulance response time for Asheville.

4.24   The incubation temperature to hatch ostrich eggs for a Type SR-50 incubator is set at 99°F and is allowed to vary with a standard deviation of 2°F. Each hour the temperature is recorded at 50 randomly selected times. If the average of the 50 measurements is less than 98.5 or greater than 99.5, an alarm goes off. What is the probability of a false alarm? That is, what is the probability that the average is not between 98.5 and 99.5 when the actual temperature is 99 degrees?

Interpreting Computer Output

4.25   The histogram is the result of a simulation of the sampling distribution of $\bar{x}$ from a random sample of size 100 when $\sigma = 20$. Based on the simulation, does the sampling distribution appear approximately normally distributed? What is the value of μ? What is the value of the standard deviation of $\bar{x}$? Between what two values do 95% of the possibilities for $\bar{x}$ fall?

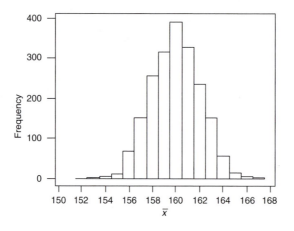

Computer Exercises

4.26   Simulate 2000 random samples of size 16 from a normally distributed population with a mean of 30 and a standard deviation of 8.
   a.   Determine the sample mean of each of the 2000 samples.
   b.   Construct a histogram of the 2000 sample means. Does it appear to be normally distributed?
   c.   Construct a normal probability plot of the 2000 sample means. Is normality plausible?
   d.   Compute descriptive statistics for the 2000 sample means.
   e.   What is the mean of the 2000 sample means? Is it close to the population mean? Should it be?
   f.   What is the standard deviation of the 2000 sample means? Is it close to the population standard deviation? Should it be?

4.27   Repeat Exercise 4.26, but this time each random sample should be of size 64. After answering parts **a–e**, compare these results to those found in Exercise 4.26. What general observations can you make?

4.28   The exponential distribution is a skewed-right distribution that is often used to model lifetimes or waiting times. It has a single parameter, which is the reciprocal of the mean. That is, if the parameter is 1/2, the mean is 2. Simulate 2000 random samples of size 16 from an exponential distribution with a mean of 5.

a. Determine the sample mean of each of the 2000 samples.
b. Construct a histogram of the 2000 sample means. Does it appear to be normally distributed?
c. Construct a normal probability plot of the 2000 sample means. Is normality plausible?
d. Compute descriptive statistics for the 2000 sample means.
e. What is the mean of the 2000 sample means? Is it close to the population mean? Should it be?

4.29   Repeat Exercise 4.28, but this time each random sample should be of size 64. After answering parts **a–e**, compare these results to those found in Exercise 4.28. What general observations can you make?

4.3   The Sampling Distribution of the Sample Proportion p

Learning Objectives for this section:

❑ Know the characteristics of a Bernoulli population.

❑ When sampling from a Bernoulli population, know how the Central Limit Theorem is used to describe the sampling distribution of the sample proportion, p.

❑ Know how to use simulations to illustrate the sampling distribution of p.

❑ Be able to apply the Central Limit Theorem to problems involving the sample proportion p.

❑ Know the relationship between the proportion of success and the binomial random variable.

❑ Know how the sample size affects the sampling distribution of p.

❑ Know the effect of different values of π on the sampling distribution of p.

❑ Be able to use the normal distribution to approximate binomial probabilities.

Recall from Chapter 3 that a Bernoulli population is one in which each element is either a success or a failure. We are generally interested in the proportion of successes, denoted by π. To estimate π from a sample, it seems reasonable to calculate the sample proportion, p. In order to evaluate how well p estimates π, we need to study its sampling distribution.

Consider the problem of predicting the percent of registered voters who favor an incumbent mayor seeking reelection. The percent of *all* registered voters in favor of the mayor is the parameter π, which we wish to predict. The statistic p, which is used to estimate the unknown value of π, is the percent of voters in a sample who are in favor of the mayor. Suppose that from a sample of 1000 registered voters, 380 favor the incumbent mayor for reelection. Then we say that the statistic *realized* the value

$$p = \frac{380}{1000} = .38$$

Using p to estimate π, we are led to believe that approximately 38% of all registered voters favor the mayor for reelection.

If we select another sample of 1000 registered voters, however, it is almost certain that there will *not* be exactly 380 in favor of the mayor as before. Just as $\bar{x}$ varies from sample to sample, so does the sample proportion p. So, is it reasonable to use the value of $p = .38$ to estimate the unknown value of π? The answer is yes, if the *sampling variability* of p is small; that is, the values it assumes from the different samples should be reasonably close to one another. If the statistic assumes widely varying values from sample to sample, the conclusions we draw about the population parameter will be less than reliable because we do not know which of the different values of the statistic are close to the parameter. To evaluate the closeness of the statistic to the parameter we must understand the sampling variability of p. The three questions we asked about the sampling distribution of $\bar{x}$ may also be asked of the sampling distribution of p:

1. What is the general shape of the distribution of potential values that p can assume?
2. Are the potential values of p centered about a certain quantity?
3. How much variability is there associated with the potential values of p?

The following version of the Central Limit Theorem applies to the sample proportion:

Central Limit Theorem Applied to the Sample Proportion p

If the sample size, n, is sufficiently large, the sampling distribution of p:

1. Is approximately normally distributed.
2. Is centered at π, the true proportion of successes in the Bernoulli population.
3. Has a standard deviation of $\sqrt{\pi(1 - \pi)/n}$.

The next example illustrates the Central Limit Theorem as it applies to p.

Example 4.9   Suppose the proportion of voters in favor of the mayor seeking reelection is $\pi = .4$. With a computer, simulate a random sample of 1000 voters and calculate p, the proportion in the 1000 who favor the mayor seeking reelection. Repeat the process 2000 times; that is, simulate 2000 random samples each of size 1000. With a histogram, describe the results of the simulation.

Solution   A histogram of the 2000 realizations of p should resemble the sampling distribution of p when $n = 1000$. Figure 4.8 shows the results of the simulation.

Figure 4.8

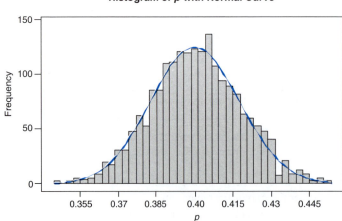

Histogram of p with Normal Curve

Observe that the histogram appears to be approximately normally distributed; is centered around .4, the true proportion in the population; and exhibits very little variability.

In fact, almost all of the potential values of p are somewhere between .35 and .45, which is very close to π. Knowing that $\pi = .4$, we can calculate the standard deviation of the sampling distribution of p to be

$$\sqrt{\frac{\pi(1 - \pi)}{n}} = \sqrt{\frac{(.4)(.6)}{1000}} = .0155$$

Two standard deviations on either side of .4 gives the interval (.37, .43). We see from the simulation that approximately 95% of the potential values of p are indeed between .37 and .43.   ■

To perform the simulation in Example 4.9 we can use the **Random Data > Bernoulli** command in Minitab. This would require, however, a worksheet that has 2000 rows and 1000 columns (each row represents one of the 2000 samples that has a size of 1000). Instead, use the following relationship between the proportion of successes and the number of successes.

Relationship between the Proportion of Successes and a Binomial Random Variable

If p represents the proportion of successes in a random sample of size n from a Bernoulli population with parameter π, the number of successes

$$X = np$$

is a binomial random variable with parameters n and π.

This follows from the fact that if

$$p = \text{number of successes}/n$$

then we can multiply both sides by n to get

$$np = \text{number of successes}$$

Also recall that the *number of successes* in a random sample of size n from a Bernoulli population with parameter π was defined (in Chapter 3) to be a binomial random variable with parameters n and π. So instead of simulating 1000 Bernoulli values and calculating the proportion of successes as described previously, we can simulate one binomial variable with $n = 1000$ and then divide by n to get a realization of p.

COMPUTER TIP

Simulating Binomial Data

To simulate a binomial random variable, under the **Calc** menu select **Random Data** followed by **Binomial**. Once the Binomial Distribution window appears, specify the number of rows of data in the **Generate** box and specify a column for the data in the **Store in column(s)** box. Fill in the **Number of trials** (n) and the **Probability of success** (π) and click on **OK**.

Example 4.10   **The Effect of Sample Size on the Distribution of p**

To illustrate how the sampling distribution of p is affected by the sample size, simulate 2000 realizations of p when $\pi = .2$ and $n = 50, 100, 200, 400,$ and 1000. Construct a histogram for each of the sample sizes and then compare them with boxplots.

Solution   Figure 4.9 shows the simulated results.

Figure 4.9

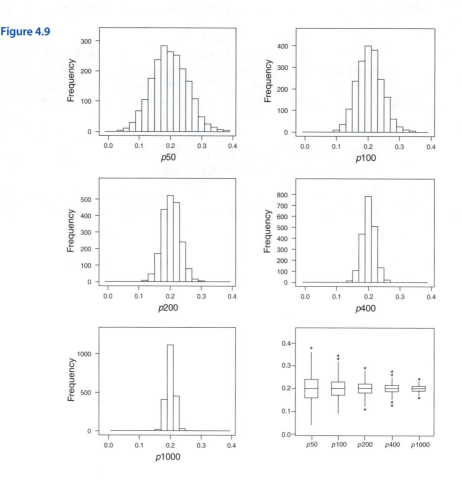

Notice that in all five simulations the sampling distribution of p is centered at $\pi = .2$. As n increases from 50 to 1000 the sampling distribution becomes more bell shaped and the standard deviation decreases. The boxplots clearly illustrate that the sampling distributions become less variable and collapse down on $\pi = .2$ as n increases. ∎

The reason that the standard deviation decreases as the sample size increases, as in the case of the standard deviation of $\bar{x}$, is that n appears in the denominator. That is, the standard deviation of p,

$$\sqrt{\frac{\pi(1 - \pi)}{n}}$$

will, for fixed values of π, decrease as n increases. In fact, if n is increased fourfold, the standard deviation will be halved.

The Effect of π on the Standard Deviation

Next, fix n and see how different values of π affect the standard deviation. Consider the following table of values of the standard deviation for various values of π when $n = 1600$.

π	.2	.3	.4	.5	.6	.7	.8
$\sqrt{\dfrac{\pi(1-\pi)}{1600}}$	.01	.0115	.0122	.0125	.0122	.0115	.01

The standard deviation increases until we get to $\pi = .5$ and then it decreases. The *maximum* standard deviation occurs when $\pi = .5$. The sample size $n = 1{,}600$ was arbitrary; it is easy to see that this scheme is true for any sample size. Also, note the symmetry in the table; the standard deviation is the same for $\pi = .2$ and .8, for $\pi = .3$ and .7, and the same for $\pi = .4$ and .6.

Example 4.11   A *Statistical Brief* (SB/95-7) issued in May 1995 by the Bureau of the Census reported the percent of housing units in major metropolitan areas that are heated with gas. The percents ranged from a low of 6.6% in Fort Lauderdale to a high of 91.9% in Salt Lake City. The percent for Tucson, Arizona, was not given. Suppose a random sample of $n = 400$ housing units in Tucson gives a sample proportion of 37% being heated with gas. If π denotes the Bernoulli proportion of *all* housing units in Tucson that are heated with gas, how close do you think the 37% is to π?

Solution   Because a random sample of size 400 is large, according to the Central Limit Theorem the sampling distribution of p tends to mound up around π. Moreover, the standard deviation is maximum when $\pi = .5$, so we know that the standard deviation is no more than

$$\sqrt{\frac{(.5)(.5)}{400}} = \frac{.5}{20} = .025$$

Thus, we feel confident that 95% of all possible values for p will lie within $2(.025) = .05$, or 5%, of the true value of π. So we believe that the value of $p = .37$ should be within $\pm.05$ of π. That is, we estimate that π is somewhere within the interval $(.32, .42)$.   ■

Example 4.12   Recent studies have shown that 40% of the adult population in the United States believe that abortion is permissible under any circumstances. In a random sample of 1000 adults, 450 said abortion is permissible under any circumstances. Does this cast doubt on the initial claim?

Solution   If the true proportion of adults who believe that abortion is permissible is 40%, the Central Limit Theorem states that the sampling distribution of the sample proportion p is approximately normal with mean .4 and standard deviation $\sqrt{(.4)(.6)/1000} = .0155$. So, we would expect approximately 95% of the potential p's to be between

$$.4 - 2(.0155) \text{ and } .4 + 2(.0155)$$

which reduces to the interval

$$(.369, .431)$$

However, the p we obtained from our sample was

$$p = \frac{450}{1000} = .45$$

which clearly is not in the interval; it is more than two standard deviations above .4. Consequently, we have sufficient evidence to question the validity of the claim that $\pi = .40$. ∎

Because of the Central Limit Theorem and the relationship between the number of successes, X, and the proportion of successes, p, ($X = np$), we have the next rule.

Normal Approximation of the Binomial Distribution

When n becomes large, the binomial distribution can be reasonably approximated with a normal distribution that has a mean of $n\pi$ and a standard deviation of $\sqrt{n\pi(1 - \pi)}$.

For a good approximation, n should be large enough so that both $n\pi > 5$ and $n(1 - \pi) > 5$.

Example 4.13   It has been reported that 46% of all homes in the United States have more than one television set. Out of a random sample of 500 homes, what is the probability that less than half have more than one television set?

Solution   If we let X be the number of homes that have more than one television set, the probability that less than half of the homes in a sample of 500 have more than one television is written as $P(X < 250)$.

Because the distribution of X is approximately normal with a mean of $n\pi$ and a standard deviation of $\sqrt{n\pi(1 - \pi)}$, we will determine the probability by finding the z-score corresponding to 250 and then looking up the associated probability in the standard normal probability table. We have

$$z = \frac{x - n\pi}{\sqrt{n\pi(1 - \pi)}} = \frac{250 - 500(.46)}{\sqrt{500(.46)(.54)}} = 1.79$$

From Table B.2 in Appendix B the associated probability is .9633, so that

$$P(X < 250) = .9633$$

It is highly probable that fewer than half the homes in the sample have more than one television set.

An alternate solution to this problem is to consider the *proportion* of homes with more than one television set instead of the *number* of homes with more than one television set. A proportion, p, calculated from a sample of 500 homes would have a sampling distribution that is approximately normal with a mean of .46 and a standard deviation of $\sqrt{(.46)(.54)/500} = .0223$. The probability that less than half of the homes in a sample of 500 have more than one television can be written as

$$P(p < .5)$$

As before, find the z-score, but this time the form is

$$z = \frac{p - \pi}{\sqrt{\pi(1 - \pi)/n}}$$

Substituting in values, we get the same z-score:

$$z = \frac{.5 - .46}{\sqrt{(.46)(.54)/500}} = 1.79$$

and hence the same answer. In other words, we have

$$P(p < .5) = P(X < 250) = .9633$$ ∎

This section and the previous section described the sampling distributions of two often-used statistics. Several examples illustrated different applications. What we have learned will be useful information as we proceed to statistical inferences about the two population parameters, μ and π.

EXERCISES 4.3   The Basics

4.30   Briefly describe the characteristics of a Bernoulli population and give two examples.

4.31   A random sample of size n is selected from a Bernoulli population with 70% successes. For each value of n, give the mean and standard deviation of the sampling distribution of p.
   a. $n = 100$ b. $n = 400$ c. $n = 1000$ d. $n = 1600$

4.32   A random sample of 1600 is selected from a Bernoulli population. For each value of π, find the mean and standard deviation of the sampling distribution of p.
   a. $\pi = .1$ b. $\pi = .3$ c. $\pi = .5$
   d. $\pi = .7$ e. $\pi = .9$ f. $\pi = .575$

4.33   Determine the maximum standard deviation of p for the following sample sizes.
   a. 25   b. 100   c. 200
   d. 500   e. 1000   f. 2000

4.34   A new drug is proposed as a treatment for lung cancer. A sample of 100 patients is to be tested. What is the maximum standard deviation of the proportion of successfully treated patients? What would be the maximum standard deviation of the estimate if the sample size were 900?

4.35   A random sample of 1000 adults in a regional survey revealed that 23% of the residents in western states dine out on a credit card. How close is the 23% to the true proportion of Westerners who dine out on a credit card? That is, make a general statement about the variability of p as an estimate of π.

4.36   AIDS–related deaths accounted for one-third of all deaths of state prison inmates during 1993 (Bureau of Justice Statistics, Bulletin NCJ-152765, August 1995). There were 121 deaths from all causes in state prisons in New Jersey in 1993. What is the probability that 45 or more of those deaths were AIDS related?

4.37   The Census Bureau reported that 70% of all Americans have hospitalization coverage by a private insurance plan (*Statistical Brief* SB94/28, October 1994). In a large hospital there were 200 patients admitted one week. What is the probability that more than 65% of the 200 patients have a private insurance plan?

4.38   The poverty cutoff line for a family of four was $14,335 in 1992. Fifty-three percent of all families in Tunica, Mississippi, are below that poverty line. Is it unusual that a sample of 100 families in Tunica will have more than 60 families below the poverty line?

4.39   President Reagan received 59% of the popular vote in the November 1984 election, for a landslide victory.
  a. From a random sample of 400 voters, what is the probability that more than 240 voted for Reagan?
  b. In 1992 President Clinton received 43% of the popular vote. From a random sample of 400 voters, what is the probability that more than 240 voted for Clinton?
  c. Explain why there is such a big difference in the answers to parts **a** and **b**.

4.40   Suppose 40% of the adult residents in North Dakota favor the death penalty. In a simple random sample of 100 North Dakota adult residents, what is the probability that more than 50% of them favor the death penalty?

4.41   A report from the president's office stated that more than half of the nation agrees with the current foreign policy. However, a CNN/Gallup poll of 1008 Americans found only 44% support the President's foreign policy. Did the 44% simply occur by chance or is there statistical evidence that the report is in error?

4.42   One-third of all ex-convicts return to jail within 3 years. One state is trying a new rehabilitation system for prisoners. After 200 prisoners who had participated in the new system were released, only 48 returned within 3 years. Is there statistical evidence that the new system is working to reduce the number of repeat offenders?

4.43   According to *Racing Update*, three out of four racehorse owners lose money. A racing partnership had 11 of its 36 horses show a profit. Are the results of the partnership substantially better than what would normally be expected from 36 randomly selected horses?

4.44   A *USA Today*/CNN/Gallup nationwide poll of 1022 adults conducted on September 6–7, 1994, showed that 34% approved of President Clinton's handling of foreign policy. The article stated that the prediction was accurate to within ±3 percentage points.
  a. How many of the 1022 adults approve of the president's handling of foreign policy?
  b. What statistic does the 34% represent?
  c. What was the standard deviation of the statistic?
  d. Is it possible that a second sample of 1022 adults would yield a different percent approving the president's foreign policy?
  e. Most samples of 1022 adults would give a percent somewhere between what two values?

Interpreting Computer Output

4.45   The histogram is the result of a simulation of the sampling distribution of the sample proportion p from a random sample of size 400. Based on the simulation, does the sampling distribution appear approximately normally distributed? What is the value of π? What is the approximate value of the standard deviation of p? Between what two values do 95% of the possibilities for p fall?

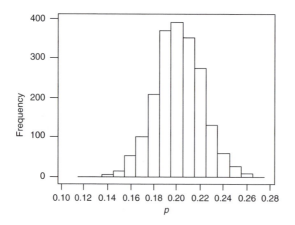

Computer Exercises

4.46 The Census Bureau reported that 70% of all Americans have hospitalization coverage through a private insurance plan (see Exercise 4.37). Simulate 2000 random samples of 200 patients admitted to a large hospital and calculate the proportion that has hospitalization coverage through a private insurance plan. Using $\pi = .7$ and $n = 200$ calculate the standard deviation of p. Based on a histogram of your simulation, does the distribution appear normally distributed? Where is the distribution centered? Does approximately 95% of the 2000 realizations of p lie within two standard deviations of $\pi = .7$?

4.47 Exercise 4.40 stated that 40% of the adult residents in North Dakota favor the death penalty. Simulate a random sample of 100 adult residents of North Dakota and calculate p, the proportion that favors the death penalty. Repeat the process 2000 times; that is, simulate 2000 random samples each of size 100. With a histogram, describe the results of the simulation. From the histogram, try to approximate the probability that more than 50% of the adult residents favor the death penalty. Does your answer agree with the probability found in Exercise 4.40?

4.48 Simulate 2000 realizations of p when $\pi = .8$ and $n = 20$ and again when $n = 1000$. Based on histograms, do the simulated distributions appear to be normally distributed? Construct side-by-side boxplots. What observations can you make about the boxplots?

4.49 Simulate 2000 realizations of p when $n = 1000$ and $\pi = .8$ and again when $\pi = .2$. Based on histograms, do the simulated distributions appear to be normally distributed? Construct side-by-side boxplots. What observations can you make about the boxplots?

4.4 Summary and Review Exercises

Key Concepts

- The *sampling distribution* of a statistic describes the variability associated with the statistic in repeated sampling. If the sampling distribution of a statistic is somewhat stable from sample to sample, it is possible to evaluate the statistic.

- As with any distribution, describing the sampling distribution of a statistic involves three tasks:
  1. Describe the general shape of the distribution.
  2. Give a measure of where the distribution is centered.
  3. Give a measure of the amount of variability in the distribution.

- The variability in the sampling distribution of a statistic is measured by the *standard deviation* of the statistic.

- The *Central Limit Theorem* (extremely important) describes the sampling distribution of the sample mean, $\bar{x}$. It states that regardless of the shape of the population distribution, the sampling distribution of $\bar{x}$ is approximately normal for sufficiently large samples. In addition, the mean of the sampling distribution is the same as the mean of the population distribution, and the standard deviation is $\sigma/\sqrt{n}$.

- The Central Limit Theorem also applies to the sampling distribution of the sample proportion, p. For sufficiently large sample sizes, its sampling distribution is approximately normally distributed with a mean of π and a standard deviation of $\sqrt{\pi(1 - \pi)/n}$.

Statistical Insight Revisited

In the Statistical Insight at the beginning of this chapter, four different polls attempted to predict the percent of the popular vote George W. Bush would get in the 2000 presidential election. A poll of 1069 people will produce a margin of error of 3%. Use the **Random Data > Bernoulli** command to simulate the results of 20 polls in C1–C20 where each poll has a sample size of 1069. Use the actual percent of 48 for the probability of success. Complete a tally of your results.

On average 95% of all random samples should produce results within the margin of error. How many of your 20 polls have a sample proportion that falls within 3% of 48%, that is, between 45% and 51%? What results do you expect if the sample size is increased? Repeat the experiment with samples of size 4000 and see what happens. Describe the results.

In this simulation with $n = 1069$, only one column, C17, contains a percent that is outside the interval from 45% to 51%. As expected, 95% of the intervals fall in the acceptable range.

Tally for Discrete Variables: C1–C20

C1	Count	Percent	C2	Count	Percent	C3	Count	Percent
0	561	52.48	0	567	53.04	0	531	49.67
1	508	47.52	1	502	46.96	1	538	50.33
N=	1069		N=	1069		N=	1069	

C4	Count	Percent	C5	Count	Percent	C6	Count	Percent
0	562	52.57	0	562	52.57	0	578	54.07
1	507	47.43	1	507	47.43	1	491	45.93
N=	1069		N=	1069		N=	1069	

C7	Count	Percent	C8	Count	Percent	C9	Count	Percent
0	541	50.61	0	574	53.70	0	558	52.20
1	528	49.39	1	495	46.30	1	511	47.80
N=	1069		N=	1069		N=	1069	

C10	Count	Percent	C11	Count	Percent	C12	Count	Percent
0	544	50.89	0	533	49.86	0	541	50.61
1	525	49.11	1	536	50.14	1	528	49.39
N=	1069		N=	1069		N=	1069	

C13	Count	Percent	C14	Count	Percent	C15	Count	Percent
0	556	52.01	0	562	52.57	0	579	54.16
1	513	47.99	1	507	47.43	1	490	45.84
N=	1069		N=	1069		N=	1069	

C16	Count	Percent	C17	Count	Percent	C18	Count	Percent
0	576	53.88	0	521	48.74	0	567	53.04
1	493	46.12	1	548	51.26	1	502	46.96
N=	1069		N=	1069		N=	1069	

C19	Count	Percent	C20	Count	Percent
0	554	51.82	0	541	50.61
1	515	48.18	1	528	49.39
N=	1069		N=	1069	

Questions for Review

Use the following problems to test your skills:

The Basics

4.50   In 1968 Nixon won 43.4% of the popular vote and Hubert Humphrey won 42.7% of the popular vote. Are these figures parameters or statistics?

4.51   According to the U.S. Department of Health and Human Services, the death rate due to automobile accidents in 1993 was 15.9 per 100,000. Is 15.9 a parameter or a statistic?

4.52   A Gallup poll in September 1968 found that 43% favored Nixon and 28% favored Humphrey. Are these figures parameters or statistics?

4.53   According to 1995 figures, 17.6% of households in the United States earn more than $50,000. Is it unusual that a random sample of 200 households would have more than 40 with incomes in excess of $50,000? Calculate the probability that more than 40 households would have incomes in excess of $50,000.

4.54   Suppose the mean of a population is 65 and the standard deviation is 15. For samples of size 100, give the mean and standard deviation of the sampling distribution of $\bar{x}$. Will the sampling distribution be adequately approximated by a normal distribution? Why?

4.55   A random sample of size 100 is selected from a population with mean μ and standard deviation σ. For each value of μ and σ, find the mean and standard deviation of the sampling distribution of $\bar{x}$:
  a. $\mu = 10, \sigma = 2$ b. $\mu = 10, \sigma = 4$
  c. $\mu = 50, \sigma = 12$ d. $\mu = 50, \sigma = 24$

4.56   Which of the following does the sampling distribution of a statistic describe?
  a. The shape of the distribution of all potential values of the statistic in repeated sampling
  b. The center of all potential values of the statistic in repeated sampling
  c. The amount of variability associated with all potential values of the statistic in repeated sampling
  d. All of the above

4.57   Suppose, over a period of time, we took a large number of samples, each of size 100, from a population and calculated the average, $\bar{x}$, of each sample. If the $\bar{x}$'s were concentrated around 8000 and varied from 7600 to 8400, what would be an approximate value of the standard deviation of $\bar{x}$?

4.58   Jerry Rice is considered by many to be the best wide receiver in National Football League history. He has averaged 75 receptions each year of his NFL career, and he has averaged 17 yards per reception. If the standard deviation of the yards gained for each reception is $\sigma = 8$ yards, is it reasonable to expect him to average 20 yards per reception for a season in which he catches 75 passes?

4.59   A large automobile garage has determined that the average time to install new brakes on an automobile is 56.7 minutes with a standard deviation of 9.3 minutes. What is the probability that a random sample of 36 installations had an average exceeding one hour?

4.60   Scores on a standardized test of mathematical ability have a mean of 70 and a standard deviation of 15. Is it unusual that a random sample of 40 scores will average more than 75? Calculate the probability that a random sample of 40 will average more than 75.

4.61   A sample of n observations is randomly selected from a Bernoulli population with parameter π. Describe the sampling distribution of the sample proportion p by stating the Central Limit Theorem.

4.62   The cure rate for colorectal cancer is 70% if detected early enough. Out of a random sample of 40 patients, what is the probability that more than 30 are cured?

4.63   What is the maximum standard deviation of the sample proportion when the sample size is 200?

4.64   The *Corporate Travel/RIT* reports that traveling employees spend an average of $167.21 a day on lodging, rental car, and food. A company has 60 employees who spent an average of $148.75 on travel. If the standard deviation of the daily amount spent traveling is $50, what can be said about the average amount spent by the 60 employees?

4.65   In 1992 the average sentence for homicide was 149 months (Bureau of Justice Statistics, *Prison Sentences and Time Served for Violence*, NCJ-153858, April 1995). A random sample of size 100 is selected from the population of homicide convictions. With a standard deviation of 40 months, approximate the probability that the sample mean sentence, $\bar{x}$, falls in each range:
  a. Greater than 145
  b. Less than 155
  c. Between 151.2 and 159.6

4.66   It is not known what percent of the student body favors construction of a new building on campus. However, if 60% are in favor, what is the probability of each of the following when we have a random sample of 30 students?
  a. More than half are in favor.
  b. Twenty-five or more are in favor.
  c. All are in favor.

4.67   Suppose the scores on a management test are normally distributed with a mean of 75 and a standard deviation of 15.
  a. What percent of the scores are above 80?
  b. What percent are between 84 and 95?
  c. What is the 80th percentile?
  d. What is the probability that the average of 25 scores is above 80?

4.68 The average number of days spent in a North Carolina hospital for a coronary bypass in 1992 was 9 days, and the standard deviation was 4 days (North Carolina Medical Database Commission, *Consumer's Guide to Hospitalization Charges in North Carolina Hospitals*, August 1994). What is the probability that a random sample of 30 patients will have an average stay longer than 9.5 days?

4.69 A zoologist is interested in estimating the life span of the white-tailed deer. Assuming the standard deviation in the life span is 2.5 years, how large a sample is needed so that the standard deviation of his estimate is .4 years?

4.70 Suppose a bimodal population has a mean of 20 and a standard deviation of 4. A sample of size 100 is to be randomly selected from the population, and the sample mean is to be computed. Prior to selection of the sample, tell how the potential values of the sample mean will be distributed. Give the mean and standard deviation of the sampling distribution of the sample mean.

4.71 An index score is given to new nurses after they have been on the job for 2 months. From past records, the mean index score is 45 and the standard deviation is 8 points. What is the probability that a group of 35 new nurses will have an average index score less than 43?

4.72 An airline company has historical data that suggest that the mean number of passengers on its flights is 212 and the standard deviation is 42 passengers. What is the probability that the next 50 flights will average less than 200 passengers?

4.73 In the eastern states, 12% of those dining out use a credit card to charge their dinner. Out of a random sample of 160 people from eastern states, what is the probability that fewer than 15 will use a credit card to pay for their dinner?

4.74 We wish to estimate the proportion of students who have smoked marijuana on at least three different occasions. What will be the maximum standard deviation of our estimate if we randomly select 200 students? How large a sample is necessary for the maximum standard deviation to be no more than 5%?

4.75 According to the National Crime Victimization Survey, there were 43.6 million criminal victimizations in 1993. Of the victims of these violent crimes, 29% stated that a firearm was involved in the crime. From a random sample of 200 criminal victimizations, is it conceivable that more than 65 involve firearms? Explain your answer.

4.76 In 1989, 52% of all Ph.D.s awarded by U.S. colleges went to foreign students. Of the 32 Ph.D.s awarded by the university system of a particular state, 14 were awarded to U.S. students. Was this an unusually low percentage being awarded to U.S. students? (Hint: Calculate the probability that 14 or fewer in a sample of 32 Ph.D.s would be awarded to U.S. students.)

4.77 The owner of a local ski resort would like to estimate the mean daily amount of money spent at the resort by the guests. Assuming that $15 is a reasonable estimate of the standard deviation of the amounts spent by guests, how large a sample is needed so that the standard deviation of the estimated mean daily amount of money spent is $2?

4.78 The state highway patrol would like to estimate the average speed of motorists on a certain section of the interstate. How large a sample is necessary so that the standard deviation of its estimate is 1 mile per hour? Assume the standard deviation of the speeds of the motorists is 12 miles per hour.

4.79 A study showed that typical college students study an average of 8 hours per week with a standard deviation of 3 hours. From a random sample of 100 students, it was found that they studied on average 9.2 hours per week. Do these students study significantly more than typical college students, or did the average of 9.2 happen purely by chance? Explain your answer.

4.80 To illustrate the variability associated with a statistic, conduct the following project: From the classified ads section of your local Sunday newspaper, randomly select 20 houses that are for sale. Calculate the average price of the 20 houses. From the same newspaper, randomly select another 20 houses and again calculate the average. Do this several times and then compare the averages you found. Remember that the average cost of a house in your city for a given week is not variable; it is a fixed quantity, μ. However, the $\bar{x}$'s you calculate will vary; each one is a separate estimate of μ. The variability associated with the $\bar{x}$'s is the main idea of the sampling distribution of $\bar{x}$.

Computer Exercises

4.81 The amount of dye dispensed in a gallon of blue paint at the auto paint store is normally distributed with a mean of 2 ounces and a standard deviation of .5 ounces. Simulate 2000 random samples, each of size 5, from the population and calculate the sample mean and sample median of each sample.

a. Construct histograms with a common scale for the sampling distribution of the sample mean and the sample median.
b. Where are the two sampling distributions centered?
c. Describe the general shape of the two distributions.
d. Which distribution is more variable?
e. Can you think of a reason why one distribution is more variable?

4.82 Repeat Exercise 4.81, except this time each random sample should be of size 30. Compare your results with those from Exercise 4.81. What basic difference did the increase in sample size cause?

4.83 The amount of time one has to wait for a bus at a particular bus stop is exponentially distributed with a mean of 5 minutes. Simulate 2000 random samples, each of size 5, from the population and calculate the sample mean and sample median of each sample.

a. Construct histograms with a common scale for the sampling distributions of the sample mean and the sample median.
b. Where are the two sampling distributions centered?
c. Describe the general shape of the two distributions.
d. Which distribution is more variable?
e. Can you think of a reason why the distributions are skewed right?

4.84 Repeat Exercise 4.83, except this time each random sample should be of size 100. Compare your results to those from Exercise 4.83. What basic difference did the increase in sample size cause?

4.85 A survey of 10,000 youths by the National Center for Health Statistics (NCHS) in 1993 showed that 16% of boys and girls aged 12 to 18 smoke. Over half said they expected to quit within a year. Worksheet: **Kidsmoke.mtw** contains the results of a similar survey of 1000 youths. Column 1 indicates the gender of the child (0 = female, 1 = male), and column 2 indicates whether they smoke (0 = no, 1 = yes).

a. Find the percent of young people who smoke.
b. What is the percent of girls who smoke?
c. What is the percent of boys who smoke?
d. Are there an equal number of boys and girls in the study?
e. Which group has the higher smoking rate?
f. Are these results similar to those from the NCHS survey?

5

Introduction to Inference and Confidence Interval Estimation

In the practice of statistics, we are concerned with the task of extracting information from data. By properly analyzing data, we are able to detect patterns and recognize relationships that give us a better understanding of the data. The preceding chapters introduced the tools that are commonly used to explore data. Graphical displays such as histograms, boxplots, and scatterplots are used to look at the overall pattern produced by the data. Numerical summaries such as the mean, median, standard deviation, and correlation are used to identify specific characteristics of the data. Earlier chapters introduced these exploratory tools in a rather disjointed fashion and employed in a purely descriptive sense. These same tools, however, can be employed as a preliminary to formal statistical inference. Thus, as we begin our study of inferential statistics, we hope to use these tools collectively to develop a strategy for analyzing data and conducting formal inferences. Our first task will be that of inferring a general shape for the population distribution.

Contents

- **STATISTICAL INSIGHT**
5.1 Describing the Parent Distribution
5.2 Estimating Population Parameters
5.3 Confidence Interval for a Bernoulli Proportion

5.4 Confidence Interval for a Population Mean
5.5 Confidence Interval for a Population Median
5.6 Summary and Review Exercises

The Cost of a New Home in the United States

A home is the most expensive item most of us will ever buy. In fact, many people may never be able to afford one. The cost of a new home depends on several factors, including labor, material, land, and financing. In 1949, financing accounted for only 5% of the total cost of a new home. Today, it accounts for close to 20% of the total cost. Because of labor, material, and land costs, the price of a new home can vary significantly from one region of the country to another. For example, a 1997 housing cost survey by Better Homes and Gardens Real Estate Service reported the mean cost of a new home in the East was $198,600; in the West, $180,600; in the Midwest, $129,700; and in the South, $116,500. The mean price difference between the East and South was over $82,000.

Instead of reporting the mean cost of a new home, the National Association of Realtors reports the median cost of a new home. It reported in 2000 that the median cost of a new home in the West was $175,700; in the Northeast, $136,400; in the Midwest, $118,600; and in the South, $121,700. The median price difference between the West and Midwest was $57,100.

So we see that not only do costs vary across regions; the choice of statistic used to estimate the cost also varies. Which is the better measure of the price of a new home—the mean or the median? We will see that the shape of the parent distribution is a determining factor in choosing a parameter to represent the center of the distribution.

Furthermore, after deciding on a parameter to represent the center, should we be satisfied with just a single number as an estimate of a new home's cost? It might be more beneficial to prospective purchasers to give a range of values that contain the center of the distribution with a specified degree of confidence. In this chapter, we take a closer look at these issues.

The list below gives the 2000 median prices of single-family homes in 65 metropolitan statistical areas, as defined by the U.S. Office of Management and Budget, across the United States. The regions (in order) 1, 2, 3, and 4 are the Northeast, Midwest, South, and West, respectively. In Section 5.6, we will analyze these data further.

WORKSHEET: Homes.mtw Median Housing Cost in 2000

City	Region	Price	City	Region	Price
Akron, OH	2	$100,900	Des Moines, IA	2	113,500
Albuquerque, NM	4	128,000	Detroit, MI	2	137,000
Anaheim, CA	4	300,800	El Paso, TX	3	78,200
Atlanta, GA	3	125,400	Grand Rapids, MI	2	$112,700
Baltimore, MD	1	145,200	Hartford, CT	1	149,900
Baton Rouge, LA	3	106,400	Honolulu, HI	4	289,000
Birmingham, AL	3	122,200	Houston, TX	3	105,800
Boston, MA	1	255,000	Indianapolis, IN	2	106,600
Bradenton, FL	3	114,300	Jacksonville, FL	3	97,000
Buffalo, NY	1	78,800	Kansas City, MO	2	119,400
Charleston, SC	3	130,700	Knoxville, TN	3	106,800
Chicago, IL	2	166,700	Las Vegas, NV	4	134,300
Cincinnati, OH	2	124,000	Los Angeles, CA	4	204,800
Cleveland, OH	2	121,300	Louisville, KY	2	112,300
Columbia, SC	3	107,900	Madison, WI	2	147,700
Columbus, OH	2	126,900	Memphis, TN	3	104,800
Corpus Christi, TX	3	82,400	Miami, FL	3	138,200
Dallas, TX	3	118,900	Milwaukee, WI	2	137,900
Daytona Beach, FL	3	82,500	Minneapolis, MN	2	141,000
Denver, CO	4	182,700	Mobile, AL	3	91,900

continued

223

City	Region	Price	City	Region	Price
Nashville, TN	3	116,000	St Louis, MO	2	99,900
New Haven, CT	1	141,900	Salt Lake City, UT	4	137,700
New Orleans, LA	3	104,400	San Antonio, TX	3	89,900
New York, NY	1	220,600	San Diego, CA	4	251,400
Oklahoma City, OK	3	80,800	San Francisco, CA	4	418,600
Omaha, NE	2	115,300	Seattle, WA	4	226,100
Orlando, FL	3	107,300	Spokane, WA	4	99,400
Philadelphia, PA	1	113,300	Syracuse, NY	1	74,500
Phoenix, AZ	4	130,900	Tampa, FL	3	103,800
Pittsburgh, PA	1	87,800	Toledo, OH	2	98,900
Portland, OR	4	166,700	Tulsa, OK	3	93,600
Providence, RI	1	129,400	Washington, DC	1	177,500
Sacramento, CA	4	136,600			

Source: National Association of Realtors.

5.1 Describing the Parent Distribution

Learning Objectives for this section:

❏ Know what is meant by the underlying or parent distribution.

❏ Know the three important characteristics of a distribution.

❏ Know that we attempt to classify the shape of a distribution as symmetric or skewed.

❏ Understand that other shapes, such as multimodal, may require special study.

❏ Know how we measure center and variability of symmetric distributions.

❏ Know how we measure center and variability of skewed distributions.

❏ Be able to investigate the shape of a distribution and then choose appropriate parameters to measure center and variability.

Data analysis, referred to as *exploratory data analysis* (EDA) by some, is designed to provide us with a strategy for investigating statistical questions. It is more than just a collection of summary measures and graphic tools that are used to look at data; it is a philosophy by which we attempt to extract knowledge from data. When we begin a statistics problem, we may have specific questions in mind that we wish to answer by investigating the data. Most frequently, however, our questions are not fully developed, and thus we look to the data with the idea that they will suggest our course of action. Even when we have a well-structured path of analysis, a careful exploration of the data can turn up unexpected features that suggest new and better questions to improve our analysis of the data.

Formal statistical inference is based on specific mathematical models and assumptions. When the model is inaccurate or assumptions are violated, the inference procedures can lead to false conclusions. Data analysis gives us diagnostic tools that are used to assess the adequacy of the model and to check assumptions. It is important that we use these tools to carefully analyze the data prior to making any formal inferences.

In an inference problem, we must decide which parameter best describes the population characteristic of interest. For a Bernoulli population the task is straightforward; we are concerned with π, the proportion of success. For a *measurement* population, however, we must decide which parameters best describe the center and the variability. The choice depends on the shape of the *parent distribution*.

> The distribution of the measurements in the original population is called the **underlying** or **parent distribution**.

To describe the parent distribution we are concerned with three characteristics:

- The general *shape* of the distribution
- A measure of the *center* of the distribution
- A measure of *variability* of the distribution

Determining the Shape of the Parent Distribution

Our first task is to examine the shape of the parent distribution. In previous chapters, we introduced several diagnostic tools for this purpose:

- The stem-and-leaf plot and histogram give the general appearance of the collected data, which, assuming we have a representative sample, resembles the shape of the parent distribution.
- The boxplot shows skewness and symmetry and identifies outliers.
- The normal probability plot checks for normality as well as identifies other unusual behavior in the data.

Symmetric or Skewed

For a unimodal (single-peaked) distribution, we first attempt to classify the shape as either symmetric or skewed. Many statistical inference procedures that we consider later require that the parent distribution be close to normally distributed. Thus, having determined that a distribution is symmetric, we must also be able to determine whether the data are coming from a normally distributed population.

Example 5.1 *Challenger* Accident

The U.S. space shuttle *Challenger* accident (see Examples 1.11 and 1.12) occurred because of failure of the O-rings that seal the field joints of the rocket boosters. It was well known at the time that the rubber O-rings, which are designed to prevent the release of hot gases produced during combustion, are quick to recover their shape after compression in warm weather and are slow to recover in cold weather. The stem-and-leaf plot for the launch temperatures is given here. Comment on the distributional shape after you analyze a boxplot and a normal probability plot.

Worksheet: Challeng.mtw

```
Stem-and-leaf of temp N= 25
Leaf Unit = 1.0

    1     3   1
    1     3
    1     4
    1     4
    2     5   3
    4     5   78
    5     6   3
   11     6   677789
   (6)    7   000023
    8     7   556689
    2     8   01
```

Solution   Figure 5.1 shows a boxplot and a normal probability plot of the data. The box-plot identifies the 31° observation as a clear outlier on the left tail, suggesting that the distribution is skewed-left. The normal probability plot has the general appearance of a skewed-left distribution.

Figure 5.1
Challenger data with 31 degrees

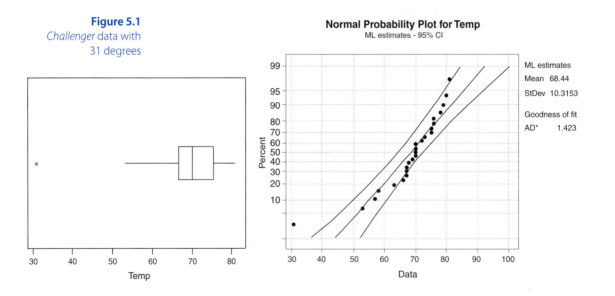

On closer investigation, however, we see in Figure 5.2 that without the 31-degree temperature included in the data set, the normal probability plot suggests that the distribution is possibly normally distributed. (All plotted points are within the confidence bands.) Hence, by carefully examining the data, we see that without the 31-degree temperature the parent distribution is possibly normally distributed and temperatures below about 55 degrees are extreme temperatures. The 31-degree temperature on the day of the *Challenger* launch is most certainly an anomaly.   ∎

Figure 5.2

Challenger data without 31 degrees

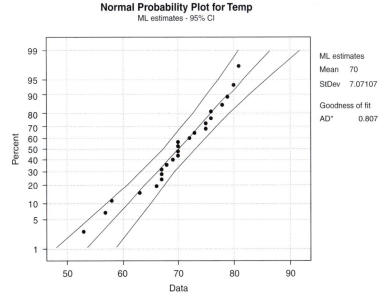

Normal Probability Plot for Temp
ML estimates - 95% CI

ML estimates
Mean   70
StDev   7.07107

Goodness of fit
AD*   0.807

Multimodal

Most often the distributions we will encounter in practice are either symmetric or skewed. There are other possibilities, however, such as multiple modes or gaps in the data, and these shapes can be identified with a proper analysis of the data.

Example 5.2 **Clustering of Galaxies**

A popular theory in astronomy is that after the Big Bang, matter expanded at a tremendous rate, and because of the local attraction of matter, the galaxies were formed. It is believed that gravitational pull would cause some clustering of galaxies; would the gravitational pull be great enough, however, to form superclusters of galaxies surrounded by large voids? To help answer this question, we need to measure the distances among galaxies—not a trivial task. If the universe is indeed expanding, then points more distant from our galaxy are moving at greater velocities; that is, we can estimate distances by measuring velocities. The next data are the velocities (measured in kilometers per second) of 82 galaxies found in the Corona Borealis region. Is there evidence in the data to suggest that there are superclusters of galaxies?

WORKSHEET: Galaxie.mtw

9172, 9558, 10406, 18419, 18972, 19330, 19440, 19541, 19846, 19914, 19989, 20179, 20221, 20795, 20875, 21492, 21921, 22209, 22314, 22746, 22914, 23263, 23542, 23711, 24289, 24990, 26995, 34279, 9350, 9775, 16084, 18552, 19052, 19343, 19473, 19547, 19856, 19918, 20166, 20196, 20415, 20821, 20986, 21701, 21960, 22242, 22374, 22747, 23206, 23484, 23666, 24129, 24366, 25633, 32065, 9483, 10227, 16170, 18600, 19070, 19349, 19529, 19663, 19863, 19973, 20175, 20215, 20629, 20846, 21137, 21814, 22185, 22249, 22495, 22888, 23241, 23538, 23706, 24285, 24717, 26960, 32789

Source: K. Roeder, "Density Estimation with Confidence Sets Exemplified by Superclusters and Voids in the Galaxies," *Journal of the American Statistical Association*, 85 (1990), 617–624.

Solution   From the appearance of the boxplot and the normal probability plot in Figure 5.3, we should classify the shape of the distribution as symmetric with long tails. The boxplot clearly shows numerous outliers on each tail and the normal probability plot gives the appearance of a long-tailed distribution. This would mean that the distances among galaxies tend to mound up around a common value, with some galaxies far removed from the center.

Figure 5.3

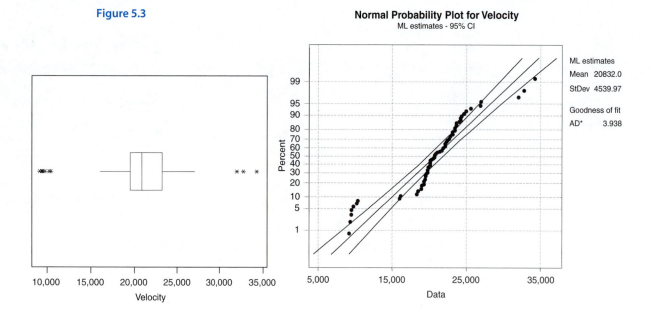

A histogram and a dotplot, however, shown in Figure 5.4, reveal something over-looked by the boxplot and the normal probability plot. The histogram reveals three distinct modes. The dotplot shows as many as seven different modes. If the distribution is multimodal, the galaxies are gathering in different clusters. Each mode represents a cluster of galaxies. The question then is whether the modes are "real" or just random variations in the data.

Figure 5.4

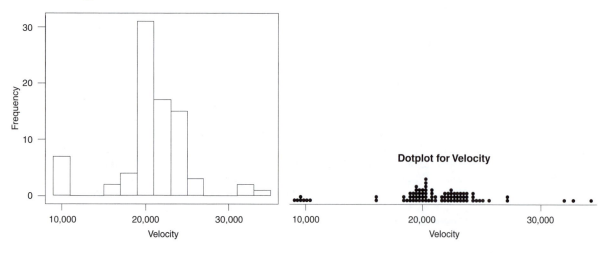

Very sophisticated data analysis procedures, beyond the scope of this book, can be used to assess the significance of the modes. But even with simple exploratory tools we can study what is perhaps an important scientific discovery in astronomy. Incidentally, Roeder (1990) concluded that there are somewhere between three and seven significant modes in these data. ∎

Measuring the Center and Variability of a Distribution

Using sample data we are able to predict the shape of a population distribution. Knowing the general shape, we next investigate the numerical characteristics of the distribution. That is, based on the shape, we identify those population parameters that best describe the center and the amount of variability in the distribution.

SYMMETRIC DISTRIBUTIONS   For symmetric distributions all natural measures of center, including the mean and median, coincide. Either one would serve as a good measure of the center. In most cases, we identify the population mean, μ, as the measure of center of a symmetric distribution. There are occasions, however, where we will identify the population median, θ, as the measure of the center of a symmetric distribution.

The most common parameter used to measure variability is the population standard deviation, σ. Because it is a measure of variation about the population mean, it should be employed when μ is the parameter used to measure the center of a distribution.

SKEWED DISTRIBUTIONS   Because of its nature, a skewed distribution has extreme observations in one tail of its distribution. In order for the mean to balance the distribution, its numerical value is pulled toward those extreme observations, and consequently may give an unrealistic measure of the center of the distribution. On the other hand, the median is unaffected by extreme scores. For this reason, the population median is the preferred measure of the center of a skewed distribution.

To measure variability in skewed distributions, recall that the population median, θ, divides the population frequency distribution in half, and the population quartiles, θ_1 and θ_3, further subdivide it into quarters. The distance between the first and third quartiles, the IQR $= \theta_3 - \theta_1$, is unaffected by extreme observations, and thus should be used to measure variability in a skewed distribution.

OTHER DISTRIBUTIONS   There are distributions that are neither symmetric nor skewed. The galaxy data in Example 5.2 is a prime example. Rather than trying to identify a single measure of the center of a multimodal distribution, we should attempt to identify the extraneous factor that is creating the different modes and then analyze separately the subgroups corresponding to each mode.

Some distributions, such as the *Challenger* data in Example 5.1, appear reasonably symmetric but also have an outlier on one tail of the distribution. What should we do about the outlier? Often an outlier is the result of an error in measurement or recording the data. If that is the case, it should be corrected or removed from the data set. If, however, it is a legitimate observation, the decision to include or exclude is left up to the experimenter.

Example 5.3   **Choosing Parameters**

Following is a sample of starting salaries of 50 chemistry majors selected at random from a large midwestern university.

WORKSHEET: Chemist.mtw

39400,	39720,	37600,	39500,	40520,	40230,	38100,	40400,
33200,	39950,	40200,	42330,	42410,	40530,	44200,	42670,
40300,	36950,	39760,	41460,	37950,	41500,	46250,	39510,
37980,	40350,	39280,	38300,	43100,	36930,	40890,	42840,
37760,	36950,	41250,	39970,	38960,	40680,	35480,	39940,
41760,	38640,	34100,	41320,	38130,	38600,	37720,	43550,
41380,	39900						

Using a histogram, boxplot, and normal probability plot classify the shape of the parent distribution. Next, identify parameters that measure the center and variability of the distribution.

Solution   The histogram and boxplot indicate that the parent distribution is symmetric. All but one of the plotted points in the normal probability plot is within the confidence bands, suggesting that the data are coming from a normally distributed population. As such, we choose the population mean, μ, and the population standard deviation, σ, as measures of the center and variability of the distribution. Because the entire population of measurements is unavailable, we will use the sample data (in Section 5.2) to estimate the numerical values of μ and σ.   ■

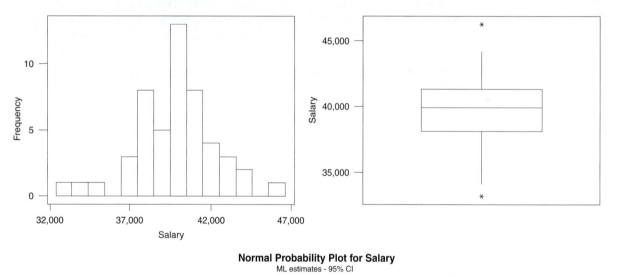

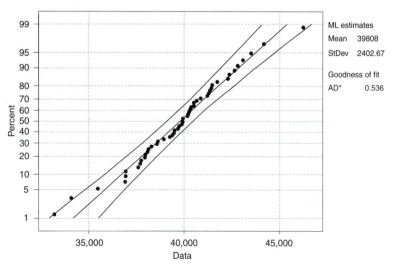

Example 5.4   *Challenger* **Accident**

In the analysis of the *Challenger* temperatures in Example 5.1, we concluded, with the inclusion of the 31-degree temperature the day of the *Challenger* accident, that the distribution is skewed left. Based on this conclusion, the population median, θ, best represents the center and the IQR best represents the variability of the distribution. If we choose to exclude the crucial 31-degree temperature, the distribution is close to symmetric. For symmetric distributions the mean and median coincide. Because the distribution may occasionally produce an outlier, the median—being a resistant measure—is probably a more reasonable measure of a "typical" temperature for a space shuttle launch.   ∎

EXERCISES 5.1   The Basics

5.1   In each of the following identify the bold value as the parameter π, μ, θ, or σ.
   a.   The average IQ is **100** as measured by the Stanford-Binet IQ test.
   b.   The salaries of teachers in our state do not vary much. More than half are within **±$3000** of the mean.
   c.   A score of **20** on a test divides the upper 50% from the lower 50%.
   d.   The leading cause of death among those aged 15 to 19 is car accidents, which account for **45%** of teenage deaths.

5.2   If the parent distribution is symmetric with tails that are not excessively long, which parameter best represents the center of the distribution, μ or θ?

5.3   An animal behaviorist is studying the time it takes an animal to perform a certain behavioral task. The variable measured is the time from the beginning of the experiment until each individual animal has performed. The researcher is interested in the mean time of performance. Some animals are slow to respond and, in fact, some may never respond properly, which makes the calculation of the mean impossible. What parameter is a meaningful measure of the "typical" time of performance?

5.4   When researchers study the response time to a drug or poison, some subjects will react very quickly, whereas others do not react for a long time. What is a meaningful measure of the average response time?

5.5   A survey of 400 randomly selected homes in a large community reveals that a television set is turned on an average of 6.2 hours per day. Does this mean that all sets are on 6.2 hours per day? Does it mean that the average number of hours is 6.2 in all large communities? What about small or rural communities?

5.6   The visual acuity of a group of subjects is tested under different doses of a drug. The visual acuity measurements for drug dose equal to 5 units are listed here:

WORKSHEET: Visual.mtw

.36   .41   .55   .33   .28   .25   .17   .14   .23
.27   .35   .37   .25   .24   .13   .19   .25   .34

   a.   Construct a stem-and-leaf plot of the data.
   b.   Construct a five-number summary diagram.
   c.   Construct a boxplot.
   d.   Based on the stem-and-leaf plot and the boxplot, comment on the shape of the parent distribution.

5.7   A group of 25 college students applying for graduate school take the Miller Personality Test for admission purposes and receive these scores:

WORKSHEET: Miller.mtw

21, 18, 20, 25, 23, 19, 30, 24, 29, 14, 25, 22, 35, 26, 23, 16, 33, 25, 22, 18, 34, 22, 31, 27, 25

   a.   Construct a stem-and-leaf plot of the data.
   b.   Construct a five-number summary diagram.
   c.   Construct a boxplot.
   d.   Based on the stem-and-leaf plot and the boxplot, comment on the shape of the parent distribution.

Interpreting Computer Output

5.8 Worksheet: **EpaTwoSeater.mtw** gives the 2001 EPA fuel-efficiency ratings for all two-seater vehicles. Based on this boxplot of the city miles per gallon, how do you classify the shape of the distribution?

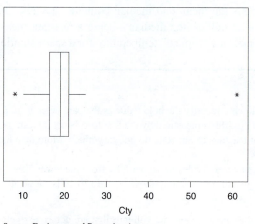

Source: Environmental Protection Agency.

5.9 Here are descriptive statistics and a dotplot for the city EPA fuel-efficiency ratings of two-seater vehicles given in Exercise 5.8.

Descriptive Statistics: cty

Variable	N	Mean	Median	TrMean	StDev	SE Mean
cty	36	18.86	19.00	17.97	8.38	1.40

Variable	Minimum	Maximum	Q1	Q3
cty	8.00	61.00	16.25	21.00

Dotplot for Cty

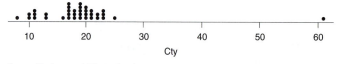

Source: Environmental Protection Agency.

a. What information can you gather from the dotplot that was not available in the boxplot given in Exercise 5.8?
b. Give an explanation for the different modes.
c. Of the mean, median, and trimmed mean, which gives a more representative measure of the fuel efficiency of these vehicles?

5.10 The figure shows a normal probability plot of the Miller Personality test scores given in Exercise 5.7. Does the plot indicate that the data are from a normally distributed population? Does this agree with your conclusion in Exercise 5.7?

Normal Probability Plot for Miller
ML estimates - 95% CI

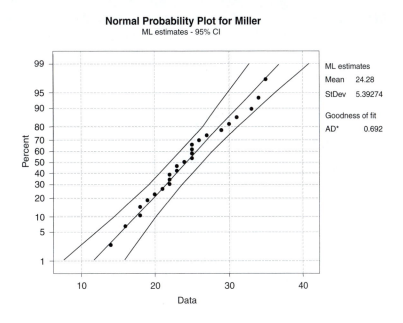

ML estimates
Mean   24.28
StDev   5.39274

Goodness of fit
AD*   0.692

5.11 A social scientist examined the records at several hospitals and recorded the following ages of women at the birth of their first child:

WORKSHEET: Firstchi.mtw

30, 18, 35, 22, 23, 22, 36, 24, 23, 28, 19, 23, 25, 24, 33, 21, 24, 19, 33, 23, 19, 32,
21, 18, 36, 21, 25, 17, 21, 24, 39, 22, 23, 18, 22, 28, 18, 15, 25, 21, 23, 26, 38, 24,
20, 36, 27, 21, 28, 26, 22, 28, 33, 18, 17, 21, 15, 20, 16, 21, 23, 15, 20, 38, 16, 24,
42, 22, 24, 24, 20, 17, 26, 39, 22, 21, 28, 20, 29, 14, 25, 20, 19, 17, 21, 24, 26

Based on the appearance of the boxplot and normal probability plot, how do you classify the shape of the parent distribution? Is it reasonable to suggest that the parent distribution is normally distributed? Why or why not?

Normal Probability Plot for Age
ML estimates - 95% CI

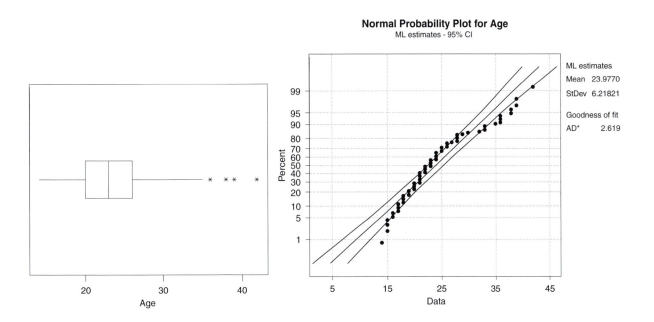

ML estimates
Mean 23.9770
StDev 6.21821

Goodness of fit
AD*   2.619

5.12   Old Faithful is a geyser located in the Yellowstone National Park in Wyoming. It is a hot-water spring that becomes unstable and spews hot water and steam into the air. It is called Old Faithful because the eruptions follow a rather predictable pattern. Data on the wait times between successive eruptions of Old Faithful were obtained between August 1 and August 15, 1985, and are stored in the worksheet: **Faithful.mtw**. The histogram is for 299 wait times (in minutes) associated with the 300 observed eruptions.

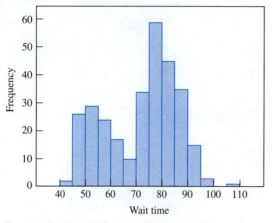

Source: A. Azzalini and A. Bowman, "A Look at Some Data on the Old Faithful Geyser," *Journal of the Royal Statistical Society,* Series C, *39* (1990), 357–366.

What observations can you make from this histogram? Certainly, Old Faithful is not 100% predictable. There is going to be some variability in the wait times between eruptions, but is the histogram what one would expect? Shouldn't the wait times be bell shaped? If visitors to the park came up to you and asked how long they should expect to wait for an eruption, what would you tell them? Here are the descriptive statistics associated with the data.

Descriptive Statistics

Variable	N	Mean	Median	TrMean	StDev	SEMean
time	299	72.314	76.000	72.439	13.890	0.803

Variable	Min	Max	Q1	Q3
time	43.000	108.000	59.000	83.000

Would you tell them that they should expect to wait about 72 minutes (72.314 to be exact) from the last eruption? Is 72 minutes a reasonable measure of the expected wait time? We can see clearly from the histogram that 72 minutes is not very likely to occur.

Geophysicists have studied Old Faithful and have determined that there are two types of eruptions: those of short duration and those of long duration. The short eruptions have a duration just less than 3 minutes, and the long eruptions have a duration in excess of 3 minutes. A long-duration eruption tends to clear out the geyser tube of hot water, which causes a longer wait time (the time it takes to heat the new water to the boiling point) before the next eruption. Thus, a long wait time follows a long eruption and a short wait time follows a short eruption. So, if the last eruption lasted more than 3 minutes, visitors should expect to have a long waiting time before the next eruption. Here are descriptive statistics for the wait times associated with short and long eruptions:

Descriptive Statistics

Variable	Eruption	N	Mean	Median	TrMean	StDev	SEMean
time	short 1	110	57.109	55.000	56.612	8.148	0.777
	long 2	189	81.164	81.000	81.263	7.303	0.531

Variable	Eruption	Min	Max	Q1	Q3
time	short 1	43.000	77.000	50.000	61.250
	long 2	59.000	108.000	77.000	87.000

Following a short eruption, how long should a visitor expect to wait before the next eruption? Following a long eruption, how long should a visitor expect to wait before the next eruption? Do these values seem consistent with the histogram?

Computer Exercises

5.13 A *tort* is a claim arising from personal injury or property damage caused by the negligent or intentional act of another person or business. Over half (60%) of all tort cases are automobile accident cases. Other tort cases include injury caused by dangerous conditions of residential or commercial property, medical malpractice, injury caused by defective products, and injury caused by toxic substances. The data presented here are drawn from court files from a sample of 45 of the nation's 75 largest counties. They include the county, the average number of months to process a tort case, the population of the county, the number of torts, and the rate per 100,000 residents. We will analyze the number of months to process a tort case.

WORKSHEET: Tort.mtw

County	Months	Population	Torts	Rate
maricopa, az	11	2209567	9914	449
pima, az	17	690202	3346	485
alameda, ca	22	1307572	3258	249
contra costa, ca	15	840585	2469	294
fresno, ca	42	705613	2364	335
los angeles, ca	12	3485398	21954	630
orange, ca	21	2484789	18297	736
san bernadino, ca	30	1534343	7583	494
san francisco, ca	21	728921	3467	476
santa clara, ca	20	1528527	5148	337
ventura, ca	22	686560	1917	279
fairfield, ct	23	267099	2303	862
hartford, ct	19	858831	4184	487
dade, fl	13	2007972	7122	355
orange, fl	15	714579	1700	238
palm beach, fl	14	900655	4194	466
fulton, ga	13	665765	1470	221
honolulu, hi	16	863117	1795	208
cook, il	34	5139341	20573	400
dupage, il	16	816116	1884	231
marion, in	14	812835	1725	212
jefferson, ky	14	670837	1543	230
essex, ma	28	669984	1843	275
middlesex, ma	27	1394408	5735	411
norfolk, ma	19	620957	1733	279
suffolk, ma	22	639192	5023	786
worchester, ma	24	708164	1673	236
oakland, mi	12	1118611	5279	472
wayne, mi	12	2096179	16739	799
hennepin, mn	13	1041332	2485	239
st. louis, mo	21	1000690	1684	168
bergen, nj	20	834983	6830	818
essex, nj	21	773420	10544	363
middlesex, nj	19	684456	7080	034
new york, ny	25	1489066	9150	614
cuyahoga, oh	12	1411209	9589	679
franklin, oh	16	992095	3951	398
allegheny, pa	20	1334396	5430	407
philadelphia, pa	24	1552572	18283	178
bexar, tx	18	1233096	3087	250

continued

WORKSHEET: Tort.mtw
continued

County	Months	Population	Torts	Rate
dallas, tx	13	1913395	6411	335
harris, tx	23	2971755	11483	386
fairfax, va	16	877531	3370	384
king, wa	13	1557537	5057	325
milwaukee, wi	10	951884	3234	340

Source: U.S. Department of Justice, *Tort Cases in Large Counties,* Bureau of Justice Statistics Special Report, April 1995.

a. Create a histogram and boxplot of the months variable. What is the general shape of the distribution?
b. Generate descriptive statistics for the months variable. Compare the mean and median. If you classified the distribution as being skewed right in part **a**, the mean should be greater than the median. Is it? Why not?
c. Create a dotplot of the data. Do you see two, or maybe three modes? Should we classify this distribution as skewed right or multimodal? Can you explain the different modes in these data?
d. How would this multimodal distribution cause the mean to be smaller than the median in an otherwise skewed-right distribution?

5.14   It is suspected that schizophrenia causes changes in the activity of a substance called dopamine in the central nervous system. Twenty-five patients with schizophrenia were classified as psychotic or nonpsychotic after being treated with an antipsychotic drug. Cerebrospinal fluid was take from each patient and assayed for dopamine b-hydroxylase (DBH) activity. Compare the DBH activity (units are nmol/(ml)(h)/(mg) of protein) for the two groups. For ease of calculation, the data values in the worksheet may be multiplied by 10,000 so that .0104, for example, would become 104.

WORKSHEET: Dopamine.mtw

nonpsych (group 1)
.0104, .0105, .0112, .0116, .0130, .0145, .0154, .0156, .0170, .0180, .0200, .0200, .0210, .0230, .0252
psychotic (group 2)
.0150, .0204, .0208, .0222, .0226, .0245, .0270, .0275 .0306, .0320

Source: D. E. Sternberg, D. P. Van Kammen, and W. E. Bunney, "Schizophrenia: Dopamine b-Hydroxylase Activity and Treatment Response," *Science, 216* (1982), 1423–1425.

a. Create boxplots of the two samples. Are the distributions symmetric or skewed?
b. Construct normal probability plots of the two groups. Does each group appear to be normally distributed?
c. Calculate descriptive statistics for the two groups. Is there much difference between the means? Between the medians?
d. From the appearance of the boxplots and the descriptive statistics, does one group tend to have higher DBH activity? Which group? Was the antipsychotic drug effective?

5.2   Estimating Population Parameters

Learning Objectives for this section:

- Know the difference between an estimator and an estimate.
- Know what it means for an estimator to be unbiased.
- Understand how estimators are evaluated through their variability.
- Know how to estimate a Bernoulli proportion.
- Know when to estimate a population mean as opposed to a population median.
- Know how to estimate population variability.

Having chosen a parameter to describe a particular characteristic of a population distribution, we must choose a statistic that will be reasonably close to the unknown value of the parameter. Obviously we want to choose the one that generally gives the best estimate of the parameter. The statistic that we decide to use to estimate the parameter is called an *estimator* of the parameter.

An **estimator** of a parameter is a statistic, the values of which should be close to the true value of the parameter. The actual numerical value that the estimator assumes from the collected data (the sample) is called the **estimate**.

Example 5.5 **Choosing an Estimator**

In Example 5.3 we concluded that the distribution of starting salaries of chemistry majors at a large midwestern university is symmetric and very close to being normally distributed. We concluded that the population mean is the best measure of the center of the distribution, and a typical salary. Now we must decide which statistic is the best estimator of the mean salary. Here is a list of estimators along with the resulting estimate:

1. If the estimator is $\bar{x}$, the estimate is 39,808.
2. If the estimator is M, the estimate is 39,945.
3. If the estimator is $\bar{x}_{T.05}$, then the estimate is 39,855. ∎

In the preceding example, we must decide which estimator is the most reliable for estimating the mean salary. That is, which estimator gives consistently closer values to the unknown parameter? By studying the sampling distribution of an estimator, we can investigate the various values it can assume and possibly determine whether its potential values are "close" to the value of the parameter.

Unbiased Estimator

Recall from Chapter 4 that the sampling distribution of $\bar{x}$ is approximately normally distributed (when the sample size is sufficiently large) and centered at the population mean μ. The potential values of $\bar{x}$ cluster symmetrically about μ; thus, there is no systematic tendency for $\bar{x}$ to overestimate or underestimate μ. In other words, $\bar{x}$ is an *unbiased estimator* of μ.

If the sampling distribution of an estimator has a mean equal to the parameter being estimated, it is called an **unbiased estimator** of the parameter. If the mean of the sampling distribution is not equal to the parameter, the estimator is **biased**.

Unbiasedness is a major criteria used for choosing an estimator. It is not the only criteria, however. When the parent distribution is continuous and symmetric, the sample median and the trimmed means are also unbiased estimators of μ. So, deciding which to use to estimate μ in a given situation requires that we look further at the sampling distribution of the possible estimators.

Variability of Estimators

The possible values of an unbiased estimator are centered at the parameter being estimated. But unless the standard deviation of the estimator is small, there is no guarantee that its value from a particular sample will be close to the parameter being estimated. For example, estimator A in Figure 5.5 is an unbiased estimator of μ but has a high probability of considerable deviation from μ. This is because its standard deviation, which measures the variability in its sampling distribution, is large. On the other hand, unbiased estimator B in Figure 5.5 has a sampling distribution with a smaller standard deviation. Consequently, the potential values (from the various samples) of estimator B have a higher probability of being close to μ.

Figure 5.5
Sampling distribution of two unbiased estimators

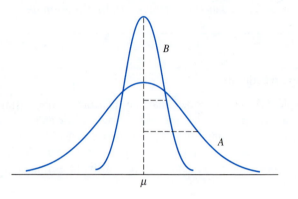

Restricting ourselves to unbiased estimators, we can say that the smaller the standard deviation, the more tightly clustered will be the potential values of the estimator about the unknown parameter. If the standard deviation of an unbiased estimator were zero, every sample would yield the same value for the estimator, which would coincide with the value of the parameter. That is, we could estimate the unknown parameter with 100% accuracy. However, no estimators (except in trivial cases) have zero standard deviation. What we strive for is an unbiased estimator that has a standard deviation smaller than the standard deviation of any other unbiased estimator.

The first question we should ask ourselves in an estimation problem is whether the population of interest is a *Bernoulli* population or a *measurement* population. Recall that a Bernoulli population is one where each element in the population is either a success or a failure and we are concerned with only one parameter, the proportion of successes. A measurement population, on the other hand, is one where the elements are the result of some measurement taken on the experimental units. The problem is more complicated because we must address the question of shape of the distribution and then identify appropriate parameters that measure the center and spread of the distribution. We will now look at these estimation problems in more detail.

Bernoulli Proportion

To estimate π, the proportion of successes in a Bernoulli population, we should use p, the proportion of successes in the sample. As indicated in Example 5.6, it is an unbiased estimator of π and the best estimator.

Example 5.6   *p* Is an Unbiased Estimator of π

Consider a Bernoulli population with unknown parameter π. Let p be the proportion of successes in a sample of size n. The Central Limit Theorem says that if the sample size is large then the sampling distribution of p is approximately normal. Furthermore, it has a mean of π and a standard deviation of $\sqrt{\pi(1 - \pi)/n}$. Figure 5.6 illustrates the sampling distribution of p. Because the sampling distribution is centered at π, it is an unbiased estimator of π. And because its standard deviation, $\sqrt{\pi(1 - \pi)/n}$, has n in the denominator, the variability of p can be made very small by increasing the sample size. In fact, its standard deviation is smaller than the standard deviation of any other unbiased estimator of π. ∎

Figure 5.6
Sampling distribution
of p

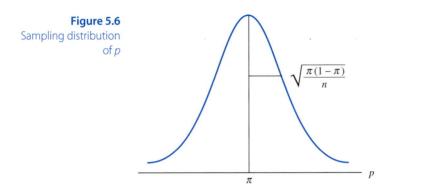

Example 5.7   Estimating a Population Proportion

What percent of the population feel they must wait too long when they visit their family physicians? A random sample of 400 patients revealed that 280 felt they had to sit in the waiting room too long. Give an estimate of π, the proportion of the population that believes they must wait too long. Using the estimate, give an estimate of the standard deviation of p.

Solution   Because 280 of the 400 indicated that they had excessive waiting periods, our estimate of π is

$$p = \frac{280}{400} = .70$$

or 70 percent.

Using .7 we have $\sqrt{.7(1 - .7)/400} = .023$ as an estimate of the standard deviation of p. ∎

Population Mean

Measurement populations require that we investigate the center of the distribution. Previously we pointed out that all natural measures of the center of a symmetric distribution coincide. Thus, the problem of estimating the center of a symmetric distribution becomes one of estimating the point of symmetry. For most symmetric distributions the population mean is the parameter we choose to estimate. And because of the next example, we often use $\bar{x}$ to estimate it.

Example 5.8   For Normal Distributions, $\bar{x}$ is the Best Estimator of μ

As previously observed, $\bar{x}$ is an unbiased estimator of μ. Furthermore, the standard deviation is always

$$\text{Stdev}(\bar{x}) = \frac{\sigma}{\sqrt{n}}$$

If, additionally, the parent distribution is *normal*, no other unbiased estimator has a smaller standard deviation. In other words, when the parent distribution is normal, $\bar{x}$ is the best estimator of μ. In Example 5.5, the best estimate of the mean salary is $\bar{x} = 39,808$. ■

If the normality assumption in Example 5.8 is dropped, it is possible that other estimators of the center of the distribution have a smaller standard deviation than the sample mean. We can't say for sure because this depends on the distribution of the parent population. In most cases, however, the sample mean is the preferred estimator of the center. One notable exception is when the parent distribution is symmetric with *long tails*. In this case, it can be shown that the sample median has a standard deviation smaller than $\text{Stdev}(\bar{x}) = \sigma/\sqrt{n}$, and thus is a better estimator of the center than is $\bar{x}$.

Population Median

In many applications, the population mean is the parameter we wish to estimate. There are situations, as pointed out above, where the median is a better representation of the center of the distribution and we wish to estimate it.

SYMMETRIC LONG-TAILED DISTRIBUTION   If the parent distribution is symmetric with long tails (compared with the normal distribution), $\bar{x}$ is *not* the best estimator of the point of symmetry because it is very sensitive to outliers. And with a long-tailed distribution, it is likely that outliers will appear in any sample. The sample median, M, on the other hand, is resistant to outliers and, in this case, has a smaller standard deviation than $\bar{x}$. Therefore, we should identify the population median, θ, as the point of symmetry and use M to estimate it when there is evidence that the parent distribution is symmetric with long tails.

Example 5.9   Estimating the Center of a Long-Tailed Distribution

A kilowatt-hour of electricity is the amount of energy required to burn ten 100-watt light bulbs for one hour. Following are basic rates per kilowatt-hour for each of the 50 states and the District of Columbia. Construct a stem-and-leaf plot, boxplot, and normal probability plot. Comment on the shape of the distribution. Choose an appropriate measure of the center and use the data to give an estimate.

WORKSHEET: Kilowatt.mtw

Ala	6.34	Ky	5.63	ND	5.61
Alaska	8.38	La	6.25	Ohio	7.34
Ariz	7.35	Maine	6.98	Okla	6.11
Ark	6.68	Md	6.84	Ore	3.88
Calif	6.74	Mass	8.44	Pa	7.36
Colo	6.14	Mich	6.71	RI	9.02
Conn	9.12	Minn	6.21	SC	6.00
Del	9.17	Miss	6.08	SD	6.07
D.C.	6.31	Mo	6.01	Tenn	4.80
Fla	7.26	Mont	4.21	Texas	7.17
Ga	5.94	Neb	5.91	Utah	6.88
Hawaii	11.29	Nev	6.02	Vt	6.33
Idaho	3.58	NH	8.97	Va	6.57
Ill	8.47	NJ	10.01	Wash	3.38
Ind	6.11	NM	7.34	W.Va	5.50
Iowa	6.72	NY	10.40	Wis	6.70
Kan	7.10	NC	6.19	Wyo	5.20

Source: The Energy Information Administration.

Solution   A complete analysis follows:

Character Stem-and-Leaf Display

```
Stem-and-leaf of rate      N = 51
Leaf Unit = 0.10

     1     3  3
     3     3  58
     4     4  2
     5     4  8
     6     5  2
    11     5  56699
    25     6  00000111122333
    (9)    6  567777889
    17     7  1123333
    10     7
    10     8  344
     7     8  9
     6     9  011
     3     9
     3    10  04
     1    10
     1    11  2
```

Descriptive Statistics

Variable	N	Mean	Median	TrMean	StDev	SEMean
rate	51	6.761	6.570	6.717	1.610	0.225

Variable	Min	Max	Q1	Q3
rate	3.380	11.290	6.010	7.340

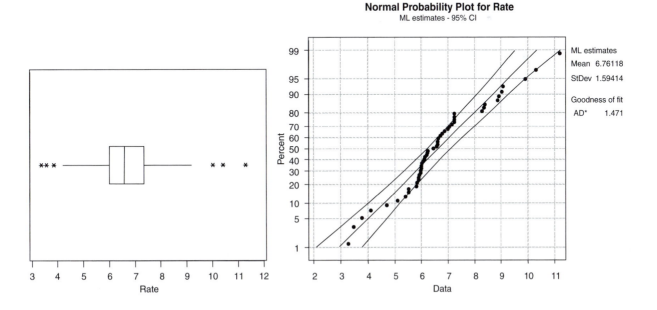

The stem-and-leaf plot and boxplot look symmetrical. The appearance of the normal probability plot rules out normality. The boxplot strongly suggests a long-tailed distribution, which is consistent with the normal probability plot. Therefore, we should classify this distribution as a long-tailed distribution. As such, we choose to estimate the population median to be M = 6.57. ∎

SKEWED DISTRIBUTIONS If the parent distribution is skewed, the median may be a better indicator of the center of the distribution. In such cases, the sample median should be used to estimate the population median.

Example 5.10 **Estimating the Center of a Skewed Distribution**

A psychological test for measuring racial prejudice is given to a random sample of 25 high school students. Analyze the data and comment on the shape of the parent distribution. Identify a parameter that best describes a "typical" score and estimate it from the data.

WORKSHEET: Prejudic.mtw

59, 54, 41, 51, 87, 42, 65, 42, 44, 46, 74, 41, 58, 83, 58, 47, 62, 48, 48, 45, 72, 79, 52, 61, 48

Solution The following stem-and-leaf plot shows skewness to the right. Because the distribution is skewed, the median is a more representative measure of the center of the distribution and a typical score. From the stem-and-leaf plot, we find the estimate of a typical score to be the sample median, M = 52.

Stem-and-Leaf Display: prejud

```
Stem-and-leaf of prejud     N = 25
Leaf Unit = 1.0

    5      4 11224
   11      4 567888
   (3)     5 124
   11      5 889
    8      6 12
    6      6 5
    5      7 24
    3      7 9
    2      8 3
    1      8 7
```
∎

There are numerous other estimators of the center of a distribution. In most cases, however, we can simplify the problem to the following two choices.

Estimating the Center of a Distribution

1. The sample mean, $\bar{x}$, is the recommended estimator of the population center when the parent distribution is normal, near normal, or symmetric with tails that are *not* excessively long.
2. The sample median, M, is the recommended estimator of the population center when the parent distribution is symmetric with long tails or is skewed in either direction.

Population Variance

Example 5.11 points out that s^2 is an unbiased estimator of the population variance, σ^2. Also, it is well known that the standard deviation of the sampling distribution of s^2

becomes arbitrarily small as the sample size increases. For these reasons, it is the most widely used estimator of the population variance.

Example 5.11 s^2 **Is an Unbiased Estimator of** σ^2

Suppose that we wish to estimate the population variance, σ^2. A reasonable estimator of σ^2 is the sample variance, which is given by the formula

$$s^2 = \sum \frac{(x_i - \bar{x})^2}{n - 1}$$

It can be shown with more advanced statistical theory that the sampling distribution of s^2 has mean σ^2; that is, s^2, as defined above, is an unbiased estimator of σ^2. With a divisor of n instead of $n - 1$, it is a biased estimator and tends to underestimate σ^2. It is for this reason that s^2 has a divisor of $n - 1$. ∎

> ## Estimator of the Population Variance
>
> The sample variance, s^2, is an unbiased estimator of the population variance, σ^2.

Standard Deviation

Just because s^2 is an unbiased estimator of σ^2, it does not follow that s is an unbiased estimator of σ. The bias in using s to estimate σ is slight, however, and becomes negligible for large samples. For lack of a better estimator, we will use the standard deviation, s, to estimate the population standard deviation, σ.

Because the computation of s involves squared deviations, it is affected by outliers even more than $\bar{x}$. The interquartile range, on the other hand, is resistant and not affected by outliers. Hence, when there are outliers, as in long-tailed or skewed distributions, the IQR seems to be a reasonable estimator of the population variability.

In the following sections we will expand the topic of estimation by finding confidence interval estimates of three of the above parameters.

EXERCISES 5.2   ## The Basics

5.15   Explain why p is an unbiased estimator of π.

5.16   Suppose the parent population is *not* normal and has mean μ.
a. Does $\bar{x}$ have to be an unbiased estimator of μ?
b. Is there possibly an estimator of μ with a smaller standard deviation than $\bar{x}$?

5.17   Sketch a picture of the sampling distribution of a biased estimator of μ.

5.18   The manufacturer of a computerized airline ticket reservation system has contracted to deliver a system that will check out reservations in a mean time of 40 seconds. Identify the parameter of interest. Give two possible statistics that might be used to estimate this parameter. From the following sample of 20 reservations, calculate the values of your two statistics. Which do you prefer?

WORKSHEET: Ticket.mtw

40, 45, 29, 58, 48, 39, 30, 50, 33, 42, 63, 31, 27, 47, 58, 32, 51, 34, 39, 51

5.19   The time it takes a subway to travel from the airport to downtown was recorded on 30 different trips. Identify a parameter of interest. From the sample, calculate the values of two statistics that might be used to estimate the parameter. Which do you prefer?

WORKSHEET: Subway.mtw

34.4, 41.6, 39.9, 40.1, 40.4, 39.7, 37.6, 39.2, 40.1, 38.8, 41.2, 37.9, 40.3, 42.5, 38.9, 39.5, 39.9, 40.4, 42.5, 41.1, 39.4, 37.7, 40.2, 40.1, 39.3, 39.6, 40.0, 41.3, 40.4, 38.3

Interpreting Computer Output

5.20   The survival time of a skin graft is extremely important to burn victims. The worksheet shows survival times (in days) of closely and poorly matched skin grafts on the same burn patient. Of interest to the statistician is the difference in the survival times.

WORKSHEET: Skin.mtw

Patient	1	2	3	4	5	6	7	8	9	10	11
Close	37	19	57	93	16	22	20	18	63	29	60
Poor	29	13	15	26	11	17	26	21	43	15	40

Source: R. F. Woolson and P. A. Lachenbruch, "Rank Tests for Censored Matched Pairs," *Biometrika*, 67 (1980), 597–606.

a. Examine the stem-and-leaf plot of the difference scores. How do you classify the general shape?

Character Stem-and-Leaf Display

```
Stem-and-leaf of differ    N = 11
Leaf Unit = 1.0

    2       -0 63
   (4)       0 5568
    5        1 4
    4        2 00
    2        3
    2        4 2
    1        5
    1        6 7
```

b. Examine the boxplot and normal probability plot. Based on these graphs, how do you classify the general shape? Does this agree with your answer in part **a**?

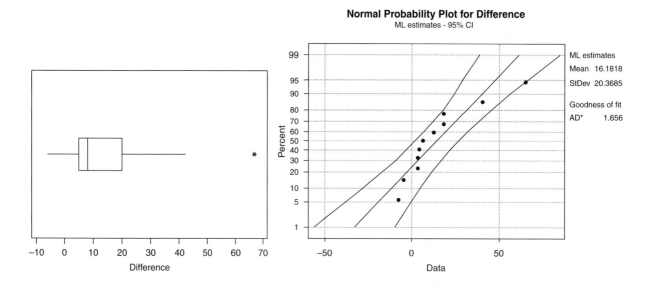

Normal Probability Plot for Difference
ML estimates - 95% CI

ML estimates
Mean 16.1818
StDev 20.3685

Goodness of fit
AD*   1.656

c. Based on the results obtained in parts **a** and **b**, from the following descriptive statistics, what is your estimate of the center of the distribution of difference scores? What is your estimate of the amount of variability in the difference scores?

Descriptive Statistics

Variable	N	Mean	Median	TrMean	StDev	SEMean
differ	11	16.18	8.00	13.00	21.36	6.44

Variable	Min	Max	Q1	Q3
differ	-6.00	67.00	5.00	20.00

5.21 Here are the educational levels (in years) of a sample of 40 auto workers in a plant in Detroit.

WORKSHEET: Detroit.mtw

22, 16, 11, 11, 8, 12, 21, 8, 12, 12, 17, 14, 12, 5, 10, 8, 6, 12, 8, 12,
11, 10, 9, 14, 13, 12, 10, 10, 12, 1, 11, 13, 14, 10, 12, 12, 12, 12, 9, 12

Based on the following analysis, comment on the shape of the underlying parent distribution of the educational levels, and estimate the mean of the distribution.

Descriptive Statistics

Variable	N	Mean	Median	TrMean	StDev	SEMean
educ	40	11.400	12.000	11.306	3.699	0.585

Variable	Min	Max	Q1	Q3
educ	1.000	22.000	10.000	12.000

Character Stem-and-Leaf Display

```
Stem-and-leaf of educ    N = 40
Leaf Unit = 1.0

     1      0 1
     1      0
     2      0 5
     3      0 6
     9      0 888899
    18      1 000001111
   (15)     1 222222222222233
     7      1 444
     4      1 67
     2      1
     2      2 1
     1      2 2
```

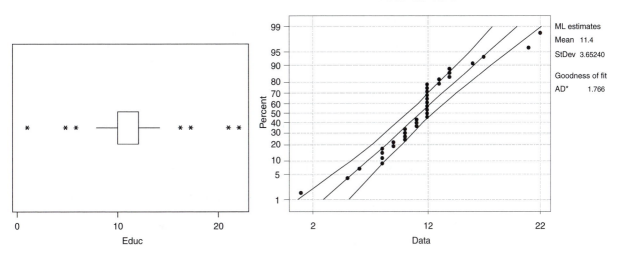

Normal Probability Plot for Educ
ML estimates - 95% CI

ML estimates
Mean 11.4
StDev 3.65240

Goodness of fit
AD* 1.766

5.22   A sociologist who is studying ethnic problems randomly selects 50 Puerto Rican families in Miami. She records these weekly family incomes and performs an analysis that includes a stem-and-leaf plot, boxplot, and descriptive statistics:

WORKSHEET: Puerto.mtw

150, 280, 175, 190, 305, 380, 290, 300, 170, 315, 280, 255, 335, 180, 200, 210, 350, 360, 550, 225, 245, 175, 180, 260, 320, 345, 180, 225, 275, 280, 325, 350, 190, 220, 265, 270, 335, 310, 270, 240, 235, 355, 370, 260, 280, 350, 390, 250, 375, 250

Stem-and-Leaf Display: income

```
Stem-and-leaf of income    N = 50
Leaf Unit = 10

    9      1 577788899
   17      2 01222344
  (14)     2 55566677788889
   19      3 001122334
   10      3 555567789
    1      4
    1      4
    1      5
    1      5 5
```

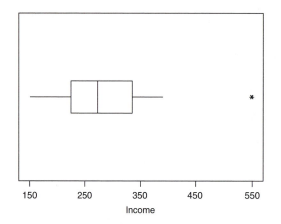

Descriptive Statistics: income

Variable	N	Mean	Median	TrMean	StDev	SE Mean
income	50	277.5	272.5	274.1	75.9	10.7

Variable	Minimum	Maximum	Q1	Q3
income	150.0	550.0	223.8	335.0

a. Investigate the stem-and-leaf plot and boxplot. Are there any outliers in the data? What is the general shape of the distribution?
b. Without outliers, what is the general shape of the distribution?
c. From the descriptive statistics, compare the mean, median, and trimmed mean. Why is the mean so much larger than the median?
d. Of the descriptive statistics in part **c**, which one do you think best represents the center of the distribution?
e. Which descriptive statistic best represents the variability of the distribution?

5.23   The sociologist in Exercise 5.22 has a random sample of 225 names from the list of registered voters in Miami and finds that 45 are Puerto Rican. Estimate the percent of registered voters in Miami who are Puerto Rican.

Computer Exercises

5.24   To study the distance a professional golfer can drive a golf ball, the following distances in yards were recorded by 20 pros.

WORKSHEET: Golf.mtw

259, 270, 248, 262, 271, 255, 261, 242, 251, 238, 273, 271, 265, 268, 251, 273, 265, 241, 239, 254

a. Construct a stem-and-leaf plot and boxplot. What is your general impression about the shape of the underlying parent distribution?
b. Construct a normal probability plot. Does it suggest that the parent distribution might be short tailed? Does this assessment agree with your analysis in part **a**?
c. Comment on the general shape of the underlying parent distribution, and estimate the center of the distribution.

5.25   A survey of freshmen was conducted to find out how many hours per week they studied for their courses. The sample of 50 freshmen revealed these scores:

WORKSHEET: Study.mtw

30 34 32 40 33 25 15 29 37 30 28 42 28 34 60 35 20 35 44 5 32 25 45 39
40 40 41 38 50 30 20 35 28 43 30 35 41 35 38 45 10 39 30 20 40 36 25
37 35 65

a. Construct a stem-and-leaf plot and boxplot. What is your general impression about the shape of the underlying parent distribution?
b. Does a normal probability plot suggest that the parent distribution is long tailed? Does this assessment agree with your analysis in part **a**?
c. Examine the descriptive statistics. Why is the mean and trimmed mean the same?
d. Which statistic do you think best estimates the center of the distribution?
e. Which statistic do you think best estimates the variability of the distribution?

5.26   A standardized exam for math competency has a national mean of 70 and a standard deviation of 12. The exam is given to 31 entering freshmen at a small community college with the results listed here:

WORKSHEET: Mathcomp.mtw

61, 67, 73, 68, 76, 82, 90, 83, 75, 55, 53, 70, 92, 75, 100, 61, 46, 65, 90, 77, 95, 73, 85, 68, 75, 87, 70, 69, 65, 50, 70

a. Construct a stem-and-leaf plot and boxplot. What is your general impression about the shape of the underlying parent distribution?
b. Does a normal probability plot suggest that the parent population is normally distributed? Does this agree with your analysis in part **a**?
c. Comment on the general shape of the underlying parent distribution and estimate the center of the distribution.

5.27   Surface-water salinity measurements were taken in a bottom-sampling project in Whitewater Bay, Florida. Geographic considerations led geologists to believe that the salinity variation should be normally distributed. If this is true, it means there is free mixing and interchange between open marine water and fresh water entering the bay.

WORKSHEET: Salinity.mtw

46 37 62 59 40 53 58 49 60 56 58 46 47 52 51 60 46 36 34 51 60 47 40 40 35 49
48 39 36 47 59 42 61 67 53 48 50 43 44 49 46 63 53 40 50 78 48 42

Source: J. Davis, *Statistics and Data Analysis in Geology,* 2nd ed. (New York: John Wiley, 1986).

a. Construct a stem-and-leaf plot of the data. Does it appear bell shaped?
b. Construct a normal probability plot. Does it rule out normality?
c. If the above analyses suggest that the salinity level is approximately normally distributed, what estimator do you recommend for the mean salinity level? What is the estimate?

5.3   Confidence Interval for a Bernoulli Proportion

Learning Objectives for this section:

❑ Understand the concept of a confidence interval.

❑ Know the basic form of a confidence interval.

❑ Know and understand standard error of a statistic.

❑ Know how to construct a confidence interval for a Bernoulli proportion.

❑ Know how to interpret confidence intervals.

❑ Understand the concepts of validity and precision of confidence intervals.

❑ Know how to find the sample size that will assure a given bound on the margin of error when estimating a Bernoulli proportion.

An estimator of a population parameter produces a single numerical value from the information in a random sample. That value becomes the estimate of the parameter. Because each sample produces its own numerical value, the estimator is subject to sample variability and may assume values that are somewhat removed from the true value of the parameter. Thus, instead of assuming that our sample will produce an exact estimate of an unknown parameter, we give a *confidence interval* of values that should contain the true value of the parameter with a high degree of confidence.

> A **confidence interval** for a population parameter is an interval of possible values for the unknown parameter. The interval is computed from sample data in such a way that we have a high degree of confidence that the interval contains the true value of the parameter. The confidence, stated as a percent, is the **confidence level**.

In practice, confidence interval estimates of unknown parameters are usually given in the form

$$\text{estimate} \pm \text{margin of error}$$

The *margin of error* is the maximum error that one would expect in the estimate for a specified confidence level. For example, the Bureau of Labor Statistics estimates the number of unemployed in a certain area to be 1500 ± 300. The estimate of the number unemployed is 1500, yet it can be off by as much as ± 300. The Bureau is rather confident that the actual number is somewhere between 1200 and 1800. If a degree of confidence can be attached to this interval, we have a confidence interval estimate for the unknown value of the parameter.

Three determinations must be made to develop a confidence interval:

- A good estimator of the parameter
- The sampling distribution (or approximate sampling distribution) of the estimator
- The desired confidence level, usually stated as a percent

Knowledge of the sampling distribution of the estimator allows us to make a probability statement that involves the estimator and the parameter with the desired confidence level.

Confidence Interval for π

A *USA Today*/CNN/Gallup nationwide telephone poll of 1022 adults reported that 42% approved of the president's handling of the economy. Because the respondents either approve or do not, the population can be considered a Bernoulli population consisting of a collection of successes and failures. If we let π be the proportion of successes (approvals), then 42% is an estimate of its numerical value. The actual value of π is the proportion of *all* adults who approve of the president's handling of the economy

In Example 5.6, you learned that the sample proportion, p, is an unbiased estimator of π. As stated above, to develop the confidence interval, we need to know the sampling distribution of the estimator. In most situations when we are estimating a proportion, the sample size is large; for example, the opinion poll consisted of 1022 adults. When n is large, the Central Limit Theorem applies and says that the sample proportion, p, has a sampling distribution that is approximately normal (see Section 4.3). Furthermore, the mean of the sampling distribution is π and the standard deviation is $\sqrt{\pi(1-\pi)/n}$.

The normal probability table and Figure 5.7 show that a normally distributed variable will be within ± 1.96 standard deviations of its mean with 95% probability. To see this, we place 95% of the distribution in the middle, leaving 2.5% on each tail. If we look up the z-score with .025 (2.5%) on the left tail of the normal probability table, we will find -1.96. Because of the symmetry of the normal distribution, we have .025 above $+1.96$; leaving .95 (or 95%) between -1.96 and $+1.96$.

Figure 5.7

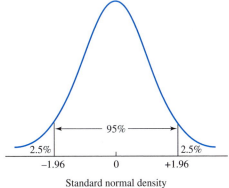

Standard normal density

Because p is approximately normally distributed and its standard deviation is $\sqrt{\pi(1-\pi)/n}$, it will be within

$$\pm 1.96 \sqrt{\frac{\pi(1-\pi)}{n}}$$

of its mean, π, with 95% probability.

Notice, however, that the standard deviation, $\sqrt{\pi(1 - \pi)/n}$, depends on the unknown π. That being the case, we are unable to compute numerical values for the confidence interval. Because the sample proportion p estimates π, however, the standard deviation can be estimated by replacing π with p. The result

$$SE(p) = \sqrt{\frac{p(1 - p)}{n}}$$

is called the *standard error of the estimator p.*

The **standard error of a statistic** is the standard deviation of its sampling distribution when all unknown population parameters have been estimated.

(Note: Some statisticians call $\sqrt{\pi(1 - \pi)/n}$ the standard error and then refer to $\sqrt{p(1 - p)/n}$ as the *estimated* standard error.)

The margin of error in estimating π with 95% probability is then found to be

$$\pm 1.96SE(p) = \pm 1.96\sqrt{\frac{p(1 - p)}{n}}$$

Adding and subtracting the margin of error to the estimator p, we have the 95% confidence interval for π:

$$p \pm 1.96\sqrt{\frac{p(1 - p)}{n}}$$

Substituting in the values from the *USA Today*/CNN/Gallup poll, we have

$$p = .42 \quad \text{Margin of error} = \pm 1.96\sqrt{\frac{(.42)(.58)}{1022}} = \pm 0.0303$$

(The poll reported the margin of error as ± 3 percentage points.)

The result says that we are 95% confident that the true proportion of adults who approve of the president's handling of the economy is somewhere between $.42 - .03$ and $.42 + .03$—that is, between 39% and 45%.

Suppose we desire a confidence level other than 95%. For example, suppose we desire 98% confidence, or maybe only 80% confidence. The estimator is still p, but what about the margin of error?

We see from Figure 5.8 that z^* denotes the value of a standard normal variable chosen so that $\alpha/2$ of the probability lies above it; or $(1 - \alpha)100\%$ of the probability lies between $-z^*$ and $+z^*$. By choosing different values for α, we can specify different confidence levels. For example, if $\alpha = .02$, then $1 - \alpha = .98$, so that $(1 - \alpha)100\% = 98\%$. That is, we are interested in a 98% confidence interval when we choose $\alpha = .02$.

Figure 5.8
Definition of z^*

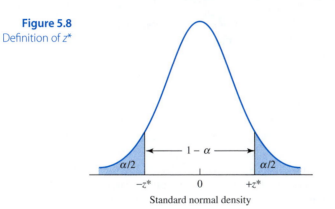

Standard normal density

We call z^* the *upper critical value* for the standard normal distribution. Here are a few values of z^* obtained from the standard normal probability table:

Upper Critical Values for z^*

$1 - \alpha$	.80	.85	.90	.95	.98	.99
z^*	1.28	1.44	1.645	1.96	2.33	2.58

Substituting z^* for 1.96, we have that the margin of error in using p to estimate π with $(1 - \alpha)100\%$ confidence is

$$\text{ME}(p) = \pm z^* \sqrt{\frac{p(1 - p)}{n}}$$

That is, the margin of error is found by multiplying the z^* critical value by the standard error of the estimator.

We have the following general statement:

A large-sample $(1 - \alpha)100\%$ confidence interval for π based on p is given by the limits

$$p \pm z^* \sqrt{\frac{p(1 - p)}{n}}$$

Example 5.12   To investigate the proportion of smokers who believe that smoking should be banned from public buildings, a researcher randomly samples 500 adult American smokers and finds that 155 believe smoking should be banned from public buildings. Obtain a 99% confidence interval for the true proportion of smokers who believe that smoking should be banned from public buildings.

Solution   The form of a 99% confidence interval for π, the true population proportion, is

$$p \pm 2.58 \sqrt{\frac{p(1 - p)}{n}}$$

The sample reveals that $p = 155/500 = .31$ and thus $(1 - p) = .69$, so the interval for π becomes

$$.31 \pm 2.58 \sqrt{\frac{(.31)(.69)}{500}} = .31 \pm .053$$

The interval is from .257 to .363 and we are 99% confident that this interval does contain the proportion of adult American smokers who believe that smoking should be banned from public buildings. ∎

Interpreting Confidence Intervals

Recall from Chapter 3 that the probability that is assigned to an event is relevant only before the experiment is conducted. For example, you may have a 1/10,000 chance of winning a new car in a raffle, but after the winning number is chosen, you either won the car or you did not. The same can be said about confidence intervals. The probability that is attached to an interval is only relevant prior to sampling. After we sample and calculate the endpoints of the interval, the probability statement no longer applies. That is, in Example 5.12, it is meaningless to say that π lies between .257 and .363 with 99% probability. Either π is between the two values or it is not; there is no probability associated with it. What is meant by a 99% confidence interval is that if 100 different intervals are obtained from 100 different samples, it is likely that approximately 99 of those intervals will contain π and 1 will not. To illustrate, we will use the computer to simulate 100 confidence intervals and then count how many actually contain the value of π and how many do not.

Example 5.13 Suppose in Example 5.12 we know that the true value of π is .3. That is, somehow we know that 30% of all adult American smokers believe that smoking should be banned from public buildings. Let us now simulate 100 random samples, each of size 500, and then calculate, for each sample, the proportion of the 500 who believe smoking should be banned from public buildings. From each simulated proportion, we can then construct a 99% confidence interval for π. Knowing that $\pi = .3$, we can then determine how many of the 100 intervals successfully contain π. To simulate the number of successes (banning smoking in public buildings) out of 500 Bernoulli trials, we generate a value of a binomial random variable with $n = 500$ and $\pi = .3$. Dividing the results by 500 gives a realization of p, from which we obtain a confidence interval. For each of the 100 realizations of p, we constructed the 99% confidence intervals for π listed in Table 5.1. From the table, we find that two of the intervals do not contain $\pi = .3$; consequently, 98 of the 100 do contain $\pi = .3$. Notice that we did not get 99 out of 100. Just because an event has a 99% probability, there is no guarantee that the event will occur *exactly* 99 times out of every 100. In a second simulation of 100 intervals, for example, all 100 intervals may contain $\pi = .3$. ∎

Table 5.1

(0.257, 0.363)	(0.230, 0.334)	(0.228, 0.332)	(0.223, 0.325)
(0.268, 0.376)	(0.249, 0.355)	(0.259, 0.365)	(0.260, 0.368)
(0.230, 0.334)	(0.228, 0.332)	(0.232, 0.336)	(0.285, 0.395)
(0.276, 0.384)	(0.255, 0.361)	(0.249, 0.355)	(0.208, 0.308)
(0.240, 0.344)	(0.259, 0.365)	(0.264, 0.372)	(0.226, 0.330)
(0.257, 0.363)	(0.266, 0.374)	(0.249, 0.355)	(0.234, 0.338)
(0.247, 0.353)	(0.211, 0.313)	(0.264, 0.372)	(0.268, 0.376)
(0.245, 0.351)	(0.274, 0.382)	(0.241, 0.347)	(0.228, 0.332)
(0.247, 0.353)	(0.260, 0.368)	(0.240, 0.344)	(0.270, 0.378)
(0.274, 0.382)	(0.274, 0.382)	(0.238, 0.342)	(0.234, 0.338)
(0.234, 0.338)	(0.255, 0.361)	(0.249, 0.355)	(0.259, 0.365)
(0.245, 0.351)	(0.268, 0.376)	(0.251, 0.357)	(0.260, 0.368)
(0.278, 0.386)	(0.262, 0.370)	(0.259, 0.365)	(0.260, 0.368)
(0.234, 0.338)	(0.247, 0.353)	(0.274, 0.382)	(0.247, 0.353)

continued

Table 5.1
continued

(0.253, 0.359)	(0.226, 0.330)	(0.243, 0.349)	(0.215, 0.317)
(0.268, 0.376)	(0.224, 0.328)	(0.255, 0.361)	(0.272, 0.380)
(0.238, 0.342)	(0.238, 0.342)	(0.253, 0.359)	(0.257, 0.363)
(0.240, 0.344)	(0.266, 0.374)	(0.213, 0.315)	(0.230, 0.334)
(0.234, 0.338)	(0.240, 0.344)	(0.266, 0.374)	(0.243, 0.349)
(0.282, 0.390)	(0.219, 0.321)	(0.204, 0.304)	(0.270, 0.378)
(0.234, 0.338)	(0.245, 0.351)	(0.245, 0.351)	(0.262, 0.370)
(0.243, 0.349)	(0.282, 0.390)	(0.283, 0.393)	(0.240, 0.344)
(0.236, 0.340)	(0.238, 0.342)	(0.247, 0.353)	(0.305, 0.415) X
(0.276, 0.384)	(0.257, 0.363)	(0.191, 0.289) X	(0.253, 0.359)
(0.208, 0.308)	(0.249, 0.355)	(0.240, 0.344)	(0.247, 0.353)

To display a confidence interval graphically, we mark off the endpoints on a line graph and put a circle at the location of the estimate of the parameter. The first seven intervals listed in Table 5.1 are graphed in Figure 5.9. Notice that the intervals are approximately the same length but are centered in different locations. All seven intervals do cover the true value $\pi = .3$.

Figure 5.9

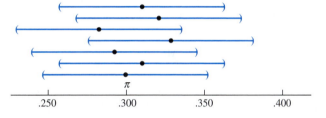

From the simulation experiment in Example 5.13, do not be led to believe that in practice we calculate 100 different intervals and find which ones contain the parameter. To the contrary, we calculate only *one* interval from the *one* random sample that we have chosen. After it is found, however, we know from the above simulation that there is a very good chance that the interval does indeed contain π. Unfortunately, there is always the chance that it does not, but, as we have seen, this is rare.

C O M P U T E R   T I P

Simulating Binomial Data

To execute the simulation in Example 5.13, from the **Calc** menu select **Random Data > Binomial**. Once the Binomial Distribution window appears **Generate** 100 rows of data and **Store in column** C1. In the **Number of trials** box, type 500 and in the **Probability of success** box, type .3 and then click **OK**. Using the Minitab Calculator, divide C1 by 500 and store the resulting values of p in C2. To compute the endpoints of the 100 confidence intervals, perform the following expressions using the Minitab calculator:

```
c3 = c2 - 2.58*sqrt(c2*(1-c2)/500)
c4 = c2 + 2.58*sqrt(c2*(1-c2)/500)
c3 = round(c3,3)
c4 = round(c4,3)
```

(The third and fourth lines are to round the results to three decimal places.)

Another important point about confidence intervals is that once an interval is obtained, no preferential treatment is given to any value in the interval. True, p is the center of the confidence interval for π and is the estimate of π, but for interpretation purposes, it is just one of the values in the interval.

Validity and Precision of Confidence Intervals

Its *validity* and *precision* judge the usefulness of a confidence interval. The confidence level measures validity, which is the probability that the interval will contain the true value of the parameter. The length of the interval measures precision. Because the length is twice the margin of error (the interval is usually estimator±margin of error), the smaller the margin of error, the more precise the interval.

In general, validity and precision compete with each other. Because the confidence level measures validity, it would be desirable to have a 100% confidence level—that is, a guarantee that the parameter is somewhere in the interval. There would be *no* precision, however, because the interval would have to contain all possible values; we could say with 100% confidence that π is somewhere between 0 and 1. Because this is unreasonable, the validity is fixed at a level—say, 95%—and then, given the assumptions, we search for an interval that has a small margin of error.

Recall that the margin of error of p is

$$\text{ME}(p) = \pm z^* \sqrt{\frac{p(1-p)}{n}}$$

Clearly for a fixed sample size n, the only way to reduce the margin of error (increase precision) is to decrease the value of z^*, which in effect reduces validity. On the other hand, to raise validity—say, from 95% confidence to 99% confidence—the z^* critical value increases from 1.96 to 2.58, which increases the margin of error and decreases precision.

It appears that it is not possible to increase precision while maintaining a fixed validity level. Because the sample size, n, appears in the denominator, however, it is possible to decrease the margin of error and hence increase precision by increasing n. In fact, we can have as much precision as we desire if we can afford the increased sample size. Remember, though, that large samples are usually costly.

To determine the sample size for a desired confidence level and a specific bound B on the margin of error, we set the equation for the margin of error equal to B and then solve for n.

Sample Size Determination for Estimating π

The sample size required to estimate π with a $(1 - \alpha)100\%$ confidence interval so that the bound on the margin of error is no more than B is

$$n = \left[\frac{z^*}{B} \sqrt{p(1-p)} \right]^2$$

(If the computation yields a decimal expression, always round the results up to the next whole number in order not to exceed the desired bound on the margin of error.)

Notice that the solution depends on the value of the sample proportion, p. At this point, however, we do not have a sample; we are trying to find an appropriate sample size. This being the case, we must guess a value for p. One approach is to use the value of p that may have been obtained in a similar study. If that is unavailable, we recall that the standard deviation of p, $\sqrt{\pi(1 - \pi)/n}$, is maximized when we choose $\pi = 0.5$. The same applies to the margin of error; that is, the margin of error, $z^*\sqrt{p(1 - p)/n}$, is maximized when we choose $p = 0.5$. By choosing $p = 0.5$ we guarantee that the margin of error will not exceed the specified bound B.

Example 5.14   The estimated margin of error for the 99% confidence interval in Example 5.12 is .053. Find the sample size necessary to reduce that margin to .03.

Solution   To retain the same validity (99%), the margin of error is

$$2.58\sqrt{\frac{p(1 - p)}{n}}$$

which now must also equal .03. That is,

$$.03 = 2.58\sqrt{\frac{p(1 - p)}{n}}$$

which must be solved for n. The solution is

$$n = \left[\frac{2.58}{.03}\sqrt{p(1 - p)}\right]^2$$

To solve this equation we need a value for p. We can take the conservative approach and use $p = .5$ because it maximizes the standard error; or, in this case, we can use $p = .31$, which is the estimate obtained from the sample in Example 5.12. Because we have an estimate of $\pi(p = .31)$ from a previous study, we will use it. Substituting, we have

$$n = \left[\frac{2.58}{.03}\sqrt{(.31)(.69)}\right]^2$$

So $n = (39.77)^2 = 1582.0044$.

Thus, to estimate π with a 99% confidence interval so that the margin of error is no greater than .03, we need a sample size of 1583.

Using the conservative approach with $p = .5$, we get

$$n = \left[\frac{2.58}{.03}\sqrt{(.5)(.5)}\right]^2 = 43^2$$

$$n = 1849$$ ∎

The results of this section are summarized in the box.

Large-Sample Confidence Interval for π

Application: A random sample from a Bernoulli population

Estimator: p, the sample proportion

Standard error of the estimator: $\text{SE}(p) = \sqrt{p(1 - p)/n}$

continued

A $(1 - \alpha)100\%$ confidence interval for π is given by the limits

$$p \pm z^* \sqrt{\frac{p(1 - p)}{n}}$$

For a given bound B on the margin of error, the sample size required for estimating a population proportion is

$$n = \left[\frac{z^*}{B} \sqrt{p(1 - p)} \right]^2$$

EXERCISES 5.3 The Basics

5.28   What will be the effect on a 95% confidence interval for π if the confidence level is changed from 95% to 98%?

5.29   What will be the effect on the margin of error in a 95% confidence interval for π if the sample size is changed from 100 to 400?

5.30   To learn whether financial aid is a determining factor in a student's choice of a college, the College Board conducted a survey of 1183 students. They determined that 722 high school graduates entered their first choice of college regardless of financial aid offers. Find a 99% confidence interval for the proportion who did not use financial aid as a determining factor in their college choice. Is a 95% confidence interval wider or narrower than the interval you found? Explain your answer.

5.31   What is the margin of error in the interval estimate of Exercise 5.30? What sample size will reduce the margin of error to no greater than .03?

5.32   Of 900 people treated with a new drug, 180 experienced an allergic reaction. Estimate with a 90% confidence interval the proportion of the population that would experience an allergic reaction. How large a sample is necessary to ensure that the margin of error is no greater than .03?

5.33   The state of Massachusetts matched bank statements with welfare recipients and found that of 1000 welfare recipients, 200 had accounts ranging from $20,000 to $50,000, 50 had accounts ranging from $50,000 to $100,000, and 10 had accounts in excess of $100,000. The state says that if a person has a bank account in excess of $20,000, he or she should not receive welfare benefits. Estimate with a 99% confidence interval the percent of welfare recipients who should not receive welfare benefits. How large a sample is necessary to ensure that the margin of error is no greater than .03?

5.34   A nationwide poll of 1000 students by the Gordon S. Black Corporation for *USA Today* in 1996 showed that 880 believed that education is very important to their future (www.usatoday.com, May 13, 1996). Find a 90% confidence interval for the proportion of all students who believe that education is very important to their future.

5.35   In 1998, the Census Bureau estimated that 16.3% of all Americans had no health insurance coverage (U.S. Bureau of the Census, Current Population Survey [CPS], March 1998). The CPS is a nationwide monthly survey of about 60,000 households conducted by the Census Bureau. In the report, the estimates and standard errors were given for the percents not covered by health insurance in several states:

State	Percent	Standard Error
Arizona	24.2	0.9
Colorado	15.1	0.8
Florida	17.5	0.5
Montana	19.6	0.9
New Jersey	16.4	0.6
New Mexico	21.1	0.9

Source: U.S. Census Bureau.

Find individual 95% confidence intervals for the percent without health insurance in each of the states. Graph the intervals on a common number line. Describe any differences you observe among the percents without health coverage in the different states.

5.36   A Roper poll of 2000 American adults showed that 1440 thought that chemical dumps are among the most serious environmental problems. Estimate with a 98% confidence interval the proportion of the population who consider chemical dumps among the most serious environmental problems. After obtaining the interval, can we say that there is a 98% chance that the population proportion is inside the interval? Explain your answer.

5.37   A national survey of students in grades 6–12 was conducted in the spring of 1993 as part of the 1993 National Household Educational Survey by the U.S. Department of Education. Based on responses from 6504 students, these percents of students stated that they had witnessed a crime or threat at school:

Student's Race/Ethnicity	Percent	Standard Error
White, non-Hispanic	57	2.6
Black, non-Hispanic	56	3.4
Hispanic	51	2.3
Other races	48	4.8

Source: U. S. Department of Education.

Construct 99% confidence intervals for the percent in each race/ethnicity category that had witnessed a crime or threat. Graph the intervals on a common number line. Which group had the highest percent witnessing a crime or threat? Can you detect any differences among the groups?

5.38   A 1996 poll of 1200 African American adults conducted by Yankelovich Partners, Inc., for the *New Yorker* found that 708 think that the American dream has become impossible to achieve. Construct a 90% confidence interval for the proportion of all African American adults who feel this way. What is the maximum margin of error for the survey? How large a sample will yield a maximum margin of error no greater than .03?

5.39   It is believed that 30% of all adults are overweight. How large a sample is necessary to estimate the true proportion of adults who are overweight with a 95% confidence interval so that the margin of error of the estimate is no greater than 2 percentage points?

5.40   The nationwide poll of 1000 students by the Gordon S. Black Corporation for *USA Today* (reported in Exercise 5.34) found that 60% of the students think that behavior of students on school buses is a serious problem. How many students must be surveyed so that the margin of error of a 98% confidence interval estimate of all students who feel this way is no greater than 3%?

5.41   Suppose one wishes to estimate the percent of American citizens who receive some sort of government aid. How large a sample is needed so that the maximum margin of error of a 95% confidence interval estimate is no greater than 2%?

5.42   On September 22, 1993, President Clinton presented his health care plan to Congress. The next day Yankelovich Partners, Inc., took a poll of 500 adult Americans for *TIME*/CNN. It reported that 36% thought increased federal involvement in the U.S. health care system would make the system worse and 33% foresaw improvement (given the margin of sampling error, that amounts to a statistical tie).

What is meant by the statement in parentheses: "given the margin of sampling error, that amounts to a statistical tie"? The article in *Time* (November 8, 1993) gave the sampling error as ±4.5%. How was that figure reached? Is it the correct value for the sampling error? Assuming that it is the correct value, give your interpretation of it.

Interpreting Computer Output

5.43   The list below gives the results of a simulation of 100 different 95% confidence intervals for π when it is known that the true value of π is .3. Mark the intervals that do not contain .3. How many of the intervals contain .3? Is this a reasonable number, given that these are 95% confidence intervals?

(0.295, 0.377)	(0.268, 0.348)	(0.269, 0.351)	(0.271, 0.353)
(0.277, 0.359)	(0.244, 0.324)	(0.273, 0.355)	(0.250, 0.330)
(0.256, 0.336)	(0.271, 0.353)	(0.266, 0.346)	(0.271, 0.353)
(0.268, 0.348)	(0.241, 0.319)	(0.266, 0.346)	(0.291, 0.373)
(0.248, 0.328)	(0.262, 0.342)	(0.254, 0.334)	(0.248, 0.328)
(0.244, 0.324)	(0.212, 0.288)	(0.250, 0.330)	(0.262, 0.342)
(0.254, 0.334)	(0.287, 0.369)	(0.254, 0.334)	(0.314, 0.398)
(0.225, 0.303)	(0.252, 0.332)	(0.268, 0.348)	(0.208, 0.284)
(0.283, 0.365)	(0.316, 0.400)	(0.273, 0.355)	(0.233, 0.311)
(0.254, 0.334)	(0.239, 0.317)	(0.254, 0.334)	(0.223, 0.301)
(0.269, 0.351)	(0.279, 0.361)	(0.250, 0.330)	(0.264, 0.344)
(0.275, 0.357)	(0.281, 0.363)	(0.273, 0.355)	(0.243, 0.321)
(0.223, 0.301)	(0.250, 0.330)	(0.231, 0.309)	(0.239, 0.317)
(0.258, 0.338)	(0.256, 0.336)	(0.295, 0.377)	(0.266, 0.346)
(0.235, 0.313)	(0.291, 0.373)	(0.306, 0.390)	(0.216, 0.292)
(0.260, 0.340)	(0.281, 0.363)	(0.244, 0.324)	(0.289, 0.371)
(0.298, 0.382)	(0.295, 0.377)	(0.250, 0.330)	(0.235, 0.313)
(0.237, 0.315)	(0.246, 0.326)	(0.266, 0.346)	(0.283, 0.365)
(0.254, 0.334)	(0.250, 0.330)	(0.250, 0.330)	(0.283, 0.365)
(0.269, 0.351)	(0.243, 0.321)	(0.244, 0.324)	(0.262, 0.342)
(0.248, 0.328)	(0.243, 0.321)	(0.285, 0.367)	(0.281, 0.363)
(0.279, 0.361)	(0.277, 0.359)	(0.300, 0.384)	(0.293, 0.375)
(0.262, 0.342)	(0.262, 0.342)	(0.275, 0.357)	(0.285, 0.367)
(0.285, 0.367)	(0.262, 0.342)	(0.254, 0.334)	(0.273, 0.355)
(0.285, 0.367)	(0.237, 0.315)	(0.269, 0.351)	(0.246, 0.326)

Computer Exercises

5.44   Assume that $\pi = .7$. With a computer, simulate 100 different 98% confidence intervals for π, each with a sample size of 500. Determine how many actually contain .7. Repeat the simulation with the sample size being 100. Was the number of intervals containing .7 affected by the change in sample size? What effect does sample size have on a confidence interval?

5.45   Assume that $\pi = .3$. With a computer, simulate 100 different 90% confidence intervals for π, each with a sample size of 200. Determine how many actually contain .7. How does the sample size of 200 affect the confidence intervals?

5.4   Confidence Interval for a Population Mean

Learning Objectives for this section:

❑ Know how to formulate a confidence interval for μ when the standard deviation of the population is known.

❑ Be able to find the sample size that will guarantee a specified margin of error when finding a confidence interval for μ.

❑ Understand the basic properties of Student's t distribution and how it differs from the standard normal distribution.

❑ Know how to formulate a confidence interval for μ when the standard deviation of the population is not known.

❑ Know the rules for applying the t-interval.

Having developed a confidence interval for the Bernoulli proportion π in the previous section, we turn our attention to developing confidence intervals for the center of a measurement distribution. When the parent distribution is symmetric without unusually long tails, we use the population mean μ to measure the center. We now develop a confidence interval estimate of μ based on the sample mean $\bar{x}$.

Confidence Interval for μ Based on $\bar{x}$ When σ Is Known

Suppose, for the moment, that the parent population is normally (or nearly normally) distributed and its standard deviation, σ, is a known quantity. Under these conditions, the sampling distribution of $\bar{x}$ is also normally distributed with a mean μ and a standard deviation given by $\sigma/\sqrt{n}$. Because σ is known, we also have

$$SE(\bar{x}) = \frac{\sigma}{\sqrt{n}}$$

The normal probability table shows that a normal variable will be within ±1.96 standard deviations of its mean with 95% probability. Because $\bar{x}$ is normally distributed, the margin of error in using $\bar{x}$ to estimate μ with 95% probability is

$$\pm1.96SE(\bar{x}) = \pm1.96\frac{\sigma}{\sqrt{n}}$$

Adding and subtracting the margin of error to the estimator, we have that a 95% confidence interval estimate of μ is

$$\bar{x} \pm 1.96\frac{\sigma}{\sqrt{n}}$$

For confidence intervals with confidence levels other than 95%, we replace 1.96 with the appropriate z^* value from the z-table.

The requirement that the parent population be normally distributed can be relaxed somewhat because of the Central Limit Theorem. For a parent distribution of any shape, the sampling distribution of $\bar{x}$ is approximately normally distributed when the sample size is large enough. Thus, we have the z-interval:

The z-Interval

When the population distribution is approximately normal (or the sample size is large) and the standard deviation σ is known, a $(1-\alpha)100\%$ confidence interval for μ based on $\bar{x}$ is given by the limits

$$\bar{x} \pm z^*\frac{\sigma}{\sqrt{n}}$$

Example 5.15   Tensile strength is the ability of a material to resist rupture when pressure is applied under specified conditions to one of its sides by an instrument. In the manufacture of a certain woven polypropylene, previous production runs indicate that the tensile strength is approximately normally distributed. The process is operating properly when the standard deviation of the tensile strength is $\sigma = 4$ pounds per square inch. Measurements of the tensile strength on a random sample of 40 rolls of woven polypropylene produced a mean of 87.3 pounds (information from Dr. Fred Morgan, J. P. Stevens & Co., Inc.). Find a 98% confidence interval for μ, the mean tensile strength of the material.

Solution   Because a 98% confidence interval is desired, we have .01 probability on each tail, which corresponds to z* = 2.33. The desired confidence interval becomes

$$\bar{x} \pm 2.33 \, \frac{\sigma}{\sqrt{n}}$$

Substituting in the numerical values, we have

$$87.3 \pm 2.33 \, \frac{4}{\sqrt{40}} = 87.3 \pm 1.474$$

which is 85.826 to 88.774. We are 98% confident that this interval contains the mean tensile strength of the woven polypropylene.   ■

COMPUTER TIP

z-Interval

After we have stored data in the worksheet, it is a simple procedure to construct a z-interval. From the **Stat** menu, choose **Basic Statistics** and then select **1-Sample Z**. When the 1-Sample Z window appears, **Select** the **Variable** containing the data, give the value for the population standard deviation in the **Sigma** box, and click on **Options**. In the 1-Sample Z Options window, specify the desired **Confidence level** and click on **OK**. To execute the command, click **OK** again.

Because of the Central Limit Theorem, the results of the preceding discussion apply to a distribution of any shape as long as the sample size is large enough. Recall, however, that in a severely skewed distribution, the sample size must be very large for the Central Limit Theorem to overcome the skewness. Furthermore, the population median θ, not μ, is the preferred parameter to represent the center of most skewed distributions. Additionally, $\bar{x}$ performs badly as an estimator of μ when the parent distribution is symmetric with extremely long tails. Here again, we do not recommend the confidence interval for μ based on $\bar{x}$ but rather a confidence interval for θ, which is presented in Section 5.5. Thus, the applications of the z-interval are rather limited. The most restrictive limitation, of course, is that the population standard deviation, σ, must be known. We will overcome this limitation shortly, but first we will look at the problem of sample size determination.

Determining Sample Size

Using the formula for the margin of error, we can determine the sample size necessary to guarantee a given margin of error for a given confidence level.

Example 5.16   The precision of the 98% confidence interval obtained in Example 5.15 is quantified by its margin of error, which is 1.474. Suppose we wish to maintain the same validity—namely, 98%—but increase the precision so that the margin of error is only 1.2 pounds per inch. How large should the sample size be?

Solution   The margin of error for a 98% confidence interval is

$$2.33 \, \frac{\sigma}{\sqrt{n}}$$

which now must also equal 1.2. Knowing that $\sigma = 4$ (which was given in the problem), we can solve this equation for n:

$$1.2 = 2.33 \frac{4}{\sqrt{n}}$$

We have

$$\sqrt{n} = \frac{(2.33)(4)}{1.2} = 7.77$$

So

$$n = (7.77)^2 = 60.37$$

Rounding up to the next whole number (to maintain the specified margin of error), we need a sample of size 61. ■

Sample Size Determination for Estimating μ

The sample size required to estimate μ with a $(1 - \alpha)100\%$ confidence interval so that the bound on the margin of error is no more than B, is

$$n = \left[\frac{z^* \sigma}{B} \right]^2$$

If necessary, round to the next whole number.

In Example 5.16, we were able to determine the sample size because the value of σ was known. In most cases, however, the value of σ is not available and must be estimated. Normally, the sample standard deviation, s, is used to estimate σ. Here, however, we are trying to determine the sample size for the random sample, in which case s is not available to us. We must find some other means of estimating σ. Perhaps a previous study gives a sample standard deviation that can be used; or, the empirical rule can be employed to give a quick estimate of the standard deviation.

Estimating σ with the Range

Recall that for distributions that are somewhat normal in shape, the empirical rule states that most (95%) of the distribution lies within two standard deviations of the mean. Thus, the range should cover roughly four standard deviations. Consequently, dividing the range by 4 gives an approximate estimate of the standard deviation. If we believe that the range covers the entire distribution, we should probably divide by 6 instead of 4 because the empirical rule states that virtually all of the distribution (99.7%) is within 3 standard deviations of the mean.

Example 5.17   Each February in Indianapolis, prospective National Football League players meet to try out for the NFL draft. A determining factor is a player's time to run the 40-yard dash. In 1996, the fastest time recorded was 4.36 seconds, by Leeland McElroy of Texas A&M. The slowest time was 5.6 seconds. A scout for an NFL team would like to know how many measurements for a particular player should be used to estimate the player's mean time in the dash to within .15 seconds with a 90% confidence interval.

Solution   Because a 90% confidence interval is desired, we have

$$z^* = 1.645 \text{ and Margin of error} = 1.645 \frac{\sigma}{\sqrt{n}}$$

From the 1996 data (which are typical for recent years), the range is $5.6 - 4.36 = 1.24$. An estimate of σ is

$$\frac{\text{Range}}{4} = \frac{1.24}{4} = .31$$

Substituting .15 seconds for the bound on the margin of error and .31 for σ, we have

$$n = \left[\frac{(1.645)(.31)}{.15}\right]^2 = 11.56$$

Thus, the scout will need a sample of 12 measurements of the times from each player in the 40-yard dash.   ■

The *t*-Interval

The development of the z-interval relies on the assumption that the population standard deviation σ is a known quantity. In most situations, however, the numerical value of σ is unknown to the experimenter and must be estimated with the sample data.

As before, we begin our discussion by assuming that the parent distribution is normally distributed with a mean μ and a standard deviation σ. The only difference now is that we do not know the numerical value of σ. Again, under these assumptions, the sampling distribution of $\bar{x}$ is exactly normally distributed with mean μ and a standard deviation of $\sigma/\sqrt{n}$.

Computing the standardized version of $\bar{x}$ (the z-score for $\bar{x}$), we have that

$$z = \frac{\bar{x} - \mu}{\sigma/\sqrt{n}}$$

has a *standard normal distribution*. If we replace σ with its estimate s in the above formula, what is the effect on the distribution?

Student's *t* Distribution

This is precisely the question posed by the chemist W. S. Gosset in 1908. In particular, Gosset, who worked for an Irish brewing company, was concerned with the distribution of

$$t = \frac{\bar{x} - \mu}{s/\sqrt{n}}$$

We will use the letter t to distinguish it from z. Notice that the only difference between t and z is that s has replaced σ. Because his employer did not allow its chemists to publish their own research, Gosset published his work under the pseudonym "Student." Thus, the distribution of t became known as *Student's t distribution*.

To better understand Gosset's problem and the effect that replacing σ with s has on the distribution of t, we will use the computer to simulate the sampling distributions of both z and t. We first generate 2000 different random samples of size 10 from a normally distributed population that has a mean $\mu = 50$ and a standard deviation $\sigma = 15$. For each of the 2000 random samples we compute $\bar{x}$ and s and then compute z and t. In computing z, we use $\sigma = 15$ for all 2000 cases. In computing t we use the sample standard deviation, s, which may change from sample to sample.

Figure 5.10

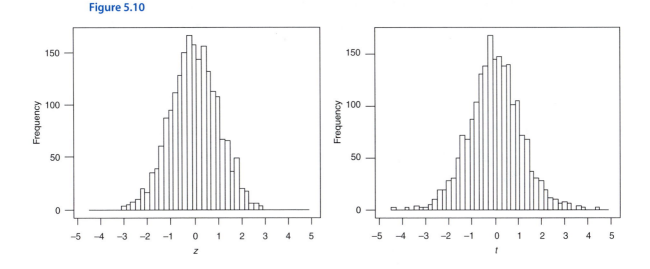

Figure 5.10 gives histograms of the 2000 realizations of z and of t. Notice that the simulated distribution of z looks very much like the standard normal distribution. It is symmetrically distributed about zero and spreads out from about -3 to $+3$. The simulated distribution of t is also symmetrically distributed about zero but is spread out slightly more than the distribution of z. The side-by-side boxplots in Figure 5.11 illustrate the symmetry about zero and the greater variance in the t distribution. We see that z is indeed normally distributed but the distribution of t has tails that are longer than those of a normal distribution.

Figure 5.11

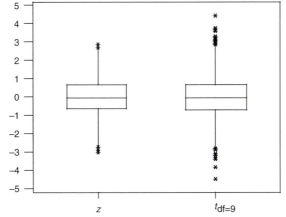

Gosset, of course, did not have a computer to simulate distributions, nor did he have boxplots to illustrate variability. He observed the difference, however, between z and t and noticed that the shape of the t distribution changes when the sample size changes. As n gets larger the standard deviation of t gets closer to 1, which is the standard deviation of z. In fact, as n becomes infinitely large, the t distribution approaches the standard normal distribution. If n is small, however, the t distribution has thicker and longer tails than does the standard normal distribution.

Recall from Section 1.4 that the standard deviation, s, has $n - 1$ degrees of freedom. Because s is in the denominator of t, we say that the t distribution also has $n - 1$ degrees of freedom. In place of saying that the shape of the *t* distribution depends on the sample size, it is more common to say that its shape depends on the degrees of freedom.

Side-by-side boxplots of simulated *t* distributions in Figure 5.12 show how the spread of the *t* distribution decreases as the degrees of freedom increase from 4 to 29. Notice that if the degrees of freedom are 30 or more, there is not much difference in the spreads of the *t* and the *z* distributions.

Figure 5.12

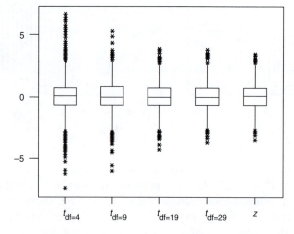

Just as z^* is the critical value of the standard normal variable such that $\alpha/2$ of the area lies on the tail above it, t^* denotes the critical value of the *t* variable with $n - 1$ degrees of freedom such that $\alpha/2$ of the area lies on the tail above it. Because of the symmetry about zero, the limits that include the middle $(1 - \alpha)100\%$ are $-t^*$ to $+t^*$ (see Figure 5.13).

Figure 5.13
A *t* distribution with
$n - 1$ degrees of
freedom

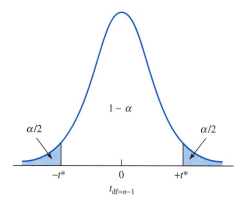

Because the *t* distribution has longer tails than the standard normal distribution, the limits $\pm t^*$ that include the middle $(1 - \alpha)100\%$ will naturally be larger than the limits $\pm z^*$ found in the normal table.

Table B.3 in Appendix B gives the values of t^* for different tail probabilities and different values of the degrees of freedom (abbreviated df). For example, for df $= 10$, $t^* = 2.228$ has .025 probability above it. Because the *t* distribution is symmetric about zero, the middle 95% will be between -2.228 and $+2.228$. Recalling that the middle 95% of a standard normal is between -1.96 and $+1.96$, we see that the *t* distribution is indeed spread out more than the standard normal distribution.

Example 5.18   Using Table B.3, find t^* that has .05 probability above it for a t distribution with 15 degrees of freedom.

Solution   With 15 degrees of freedom, the upper 5% critical value is found to be $t^* = 1.753$. Because the distribution is symmetric about zero, the lower 5% critical value is $-t^* = -1.753$. We then have that the middle 90% of the distribution lies between -1.753 and $+1.753$. ■

Confidence Interval for μ When σ Is Unknown

To develop the confidence interval for μ based on $\bar{x}$ when σ is unknown, we must replace the z^* critical value found in the z-table with a t^* critical value found in Table B.3.

The t-Interval

When the population distribution is close to normal, a $(1 - \alpha)100\%$ confidence interval for μ is given by the limits

$$\bar{x} \pm t^* \frac{s}{\sqrt{n}}$$

When to Use the t-Interval

The derivation of the t distribution assumed that the parent population is normally distributed. In practice, however, perfect normally distributed populations never happen. Fortunately, statistical research has shown that the t-interval is remarkably *robust* (validity and precision are unaffected by moderate departures from normality) except when the parent distribution is severely skewed or has long tails. In other words, moderate departures from normality are permissible when applying the t-interval.

Prior to using the t-interval, therefore, we should check whether there are unusual violations of the normality assumption. This is done quite easily with the data analysis procedures given in Section 5.1. If a normal probability plot (or some other procedure) indicates the presence of a large number of outliers or severe skewness, we should avoid using the t-interval.

For nonnormal populations, recall that when the sample size is reasonably large, the Central Limit Theorem guarantees that the sampling distribution of $\bar{x}$ is approximately normally distributed. This, along with the fact that when the sample size is large, the sample standard deviation s is a close approximation to σ, assures the validity of the t-interval. In other words, for large sample sizes, we need not be unusually cautious about the nonnormality of the population.

Follow these rules when deciding whether to use the t-interval procedure.

Rules for Applying the t-Interval

- If the sample size is less than 15, the population distribution should be close to normal.
- If the sample size is between 15 and 30, the t-interval can be used when the population distribution is reasonably symmetric and has no outliers.
- If the sample size is greater than 30, the t-interval procedure is permissible except when there are extreme outliers or strong skewness.

For those situations when the *t*-interval is not appropriate, you should consider the procedures in Section 5.5.

Confidence Interval for μ Based on $\bar{x}$ When σ Is Unknown— the *t*-Interval

Application: Symmetric population distributions with tails that are not excessively long. If the sample size is small, the population distribution should not deviate substantially from normality.

Estimator: $\bar{x}$, the sample mean

Standard error of the estimator:

$$\mathrm{SE}(\bar{x}) = \frac{s}{\sqrt{n}}$$

A $(1 - \alpha)100\%$ confidence interval for μ is given by the limits

$$\bar{x} \pm t^* \frac{s}{\sqrt{n}}$$

where t^* is the upper $\alpha/2$ critical value found in the *t*-table with degrees of freedom $n - 1$.

Example 5.19   In Example 1.15 in Section 1.5, the distribution of sepal lengths (in centimeters) of the species *Iris Setosa* (worksheet: **Irises.mtw** column 1) was shown to be close to normally distributed. In Figure 5.14, the normal probability plot of the data reaffirms that conjecture by showing no significant departures from normality. Find a 99% confidence interval for the mean sepal length of *Iris Setosa*.

Figure 5.14

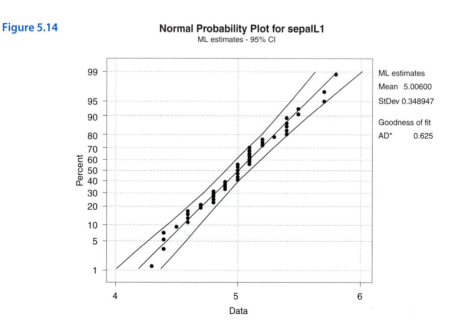

Normal Probability Plot for sepalL1
ML estimates - 95% CI

ML estimates
Mean 5.00600
StDev 0.348947

Goodness of fit
AD*   0.625

Solution   Because the data are reasonably close to being normally distributed, we can find a confidence interval based on the Student's t distribution. It takes the form

$$\bar{x} \pm t^* \frac{s}{\sqrt{n}}$$

Because a 99% interval is requested, we have .005 on the upper tail of the t distribution. With degrees of freedom $= n - 1 = 49$ we find

$$t^* = 2.678$$

From the data, we find

$$\bar{x} = 5.006 \qquad s = 0.3525$$

Thus, the 99% confidence interval for μ becomes

$$5.006 \pm 2.678 \frac{0.3525}{\sqrt{50}} = 5.006 \pm .1335$$

This gives the interval from 4.8725 to 5.1395. Thus, we are 99% confident that the mean sepal length of *Iris Setosa* is somewhere between 4.8725 centimeters and 5.1395 centimeters.   ■

Based on the normal probability plot in Figure 5.14 there is strong evidence that the parent distribution is close to normal. Thus, we have satisfied the basic requirement for the t-interval. Additionally, the sample size is $n = 50$; this alone would have satisfied the requirement unless there is strong skewness or evidence of symmetric long tails. Because we have a reasonably large sample size *and* the parent distribution is close to normal, we are assured of the accuracy of the confidence interval.

C O M P U T E R   T I P

t-Interval

After the data are stored in the worksheet, to execute the t-interval select the **Stat** menu, choose **Basic Statistics**, and then select **1-Sample *t***. In the 1-Sample t window, **Select** the **Variable** where the data are stored, click on **Options**. In the 1-Sample t Options window specify the desired **Confidence level** and click **OK**. To execute the command, click **OK** again.

Figure 5.15 is the Minitab output from Worksheet: **Irises.mtw column 1**. Note that there is a slight difference between the interval we obtained in Example 5.19 and the one found by Minitab. The difference is because we rounded the degrees of freedom to 50 (Table B.3 does not have 49 degrees of freedom). The difference is negligible and should be ignored.

Figure 5.15   **One-Sample T: sepalL1**

```
Variable     N     Mean    StDev    SE Mean        99.0% CI
sepalL1     50    5.0060   0.3525    0.0498    ( 4.8724, 5.1396)
```

Graphically displaying confidence intervals can help evaluate their precision.

Example 5.20   From random samples of 12 measurements of three different players' times in the 40-yard dash, the NFL scout referred to in Example 5.17 found these results:

Player	average time	standard deviation
1	4.56	.36
2	4.48	.32
3	4.62	.33

Find individual 90% confidence intervals for each of the players' mean time in the 40-yard dash. Graph the intervals on a common number line.

Solution   If we assume the times have a distribution that is close to normal, the confidence intervals should be *t*-intervals. The margin of error for a 90% confidence interval when the sample size is 12 is

$$\pm 1.796 \frac{s}{\sqrt{12}}$$

From the collected data, we have the following 90% confidence intervals for the mean times for the three players:

Player 1	4.56±.187	(4.37, 4.75)
Player 2	4.48±.166	(4.31, 4.65)
Player 3	4.62±.171	(4.45, 4.79)

Based on the times collected in 12 trials, we are 90% confident that the mean time for each player is somewhere in their given interval. Figure 5.16 gives a graphical illustration of the confidence intervals. We can clearly see the lengths of the intervals and thus evaluate their precision.   ■

Figure 5.16
Displaying confidence intervals

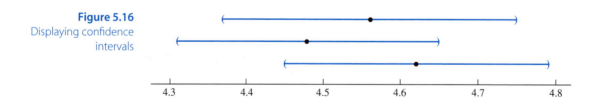

EXERCISES 5.4   The Basics

5.46   A random sample of size n is selected from a population with unknown mean μ and standard deviation $\sigma = 20$. Calculate a 95% confidence interval for μ based on $\bar{x}$ for each of these samples:
a. $n = 30, \bar{x} = 94.3$ b. $n = 45, \bar{x} = 96.4$
c. $n = 100, \bar{x} = 95.6$ d. $n = 200, \bar{x} = 95.8$

5.47   A random sample of 60 observations selected from a population with unknown mean μ and standard deviation $\sigma = 50$ yielded $\bar{x} = 465.8$.
a. Find a 95% confidence interval for μ based on $\bar{x}$.
b. Find a 99% confidence interval for μ based on $\bar{x}$.
c. Which interval has greater validity?
d. Which interval has greater precision?

5.48 Find the following confidence intervals for μ based on $\bar{x}$ from the given information:
  a. $n = 36$, $\bar{x} = 72.4$, $\sigma = 11.2$, confidence level $= .95$
  b. $n = 64$, $\bar{x} = 128.3$, $\sigma = 32.4$, confidence level $= .98$
  c. $n = 100$, $\bar{x} = 465$, $\sigma = 112$, confidence level $= .99$
  Which interval has greater validity? Which interval has greater precision?

5.49 The illustration shows 95% and 99% confidence intervals for μ based on $\bar{x}$ using the same data. Which one (A or B) is the 95% confidence interval? Explain your answer.

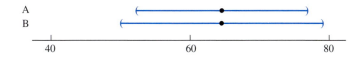

5.50 Two 95% confidence intervals for μ based on $\bar{x}$ are shown here. One has a sample size $n = 80$ and the other has a sample size $n = 120$. If we assume the standard deviations of the two populations are close to each other, which interval (A or B) has $n = 120$? Explain your answer. Why are the intervals centered in different locations?

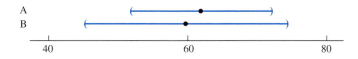

5.51 A city planner randomly samples 100 apartments in the inner city area to estimate the mean living area per apartment. Find a 95% confidence interval based on $\bar{x}$ for the mean living area if the sample yielded an average of 1325 square feet. Assume that the standard deviation σ is 42 square feet. Having found the interval, can you say that there is a 95% chance that the mean living area is within the interval? Explain your answer.

5.52 To determine the diameter of Venus, an astronomer makes 36 measurements of the diameter and finds $\bar{x} = 7848$ miles. Assuming $\sigma = 310$ miles, find a 95% confidence interval estimate based on $\bar{x}$ of the diameter of Venus. What sample size is required so that the margin of error in determining the diameter of Venus is only 50 miles?

5.53 A consumer research group sampled 100 handheld video games, all of the same make and model. The sample mean life (hours of operation before failure) was 560 hours. Assuming the standard deviation σ is 35 hours, construct a 90% confidence interval estimate based on $\bar{x}$ of the true mean life span of the video games. Is a 95% confidence interval wider or narrower than the interval you obtained? Explain your answer.

5.54 To study the birth weights of infants whose mothers smoke, a physician records the weights of 100 newborns whose mothers smoke. Find a 98% confidence interval based on $\bar{x}$ for the mean birth weight of children of smoking mothers if $\bar{x}$ was found to be 6.1 pounds. Use $\sigma = 2.1$ pounds. What sample size is required so that the margin of error in determining the mean birth weight is only .3 pound?

5.55 An ad for the Philips Longer Life Outdoor Floodlight gave an average life of 2500 hours. A random sample of 100 floodlights gave an average life of 2418 hours and a standard deviation of 325 hours. Find a 99% confidence interval for the mean life of the floodlights. Does the interval contain 2500 hours? What does this state about the advertiser's claim?

5.56 A banker would like to estimate the average amount of the loans made to farmers in her community for the past growing season. She randomly selects 40 accounts and finds that the average loan is $78,460 and the standard deviation is $22,000. She proceeds to find a 90% confidence interval for the mean loan amount. Should she have checked the underlying parent distribution for skewness or long tails before finding the confidence interval? Explain. What were the endpoints of the interval that she found?

5.57 On August 20, 1984, a skull and almost complete skeleton of a 12-year-old male *Homo erectus* who died some 1.6 million years ago was found on the west side of Lake Turkana in northern Kenya. It is believed that the 5′6″ youth would have matured to 6′ had he lived (*USA Today*, October 19, 1984). This is contrary to the assumption that early humans were shorter than modern man. *Homo erectus* has been found in Java, China, and various African sites. Suppose we were able to determine the adult height of five *Homo erectus* skeletons to be 64, 70, 73, 69, and 70 inches. Assume that the parent distribution of heights is near normal. Find a 95% confidence interval for the mean adult height of *Homo erectus*. Does the interval contain 6 feet? Is the normality condition required for constructing the *t*-interval? Is validity a problem with this interval? Is precision a problem with this interval?

Interpreting Computer Output

5.58 A new anesthetic was tested on ten patients who were to undergo surgery. These are their recovery times (in hours):

WORKSHEET: Anesthet.mtw

2.6   3.0   2.8   3.1   3.5   2.9   3.1   2.7   2.9   3.3

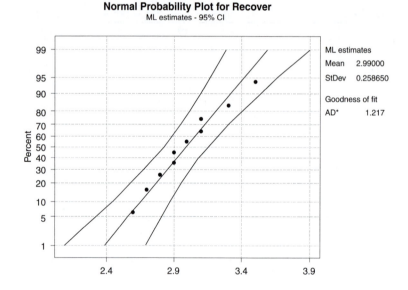

Normal Probability Plot for Recover
ML estimates - 95% CI

ML estimates
Mean   2.99000
StDev   0.258650

Goodness of fit
AD*   1.217

From the accompanying normal probability plot, can we assume that the recovery time is normally distributed? Is this condition required for a *t*-interval? Construct a 90% confidence interval for the mean recovery time for the new drug. Give your interpretation of the interval.

5.59 An air pollution index for 15 randomly selected days during the summer months was recorded for a major western city.

WORKSHEET: Pollutio.mtw

57.6,   61.2,   59.4,   65.6,   58.3,   44.7,   63.2,   48.8,   55.7,   59.0,   64.3,   43.2,   59.7,   71.2,   45.6

Based on the stem-and-leaf display, can we assume that the parent distribution of the index is symmetric? If we desire a confidence interval for the mean air pollution index, should we check the distribution for normality?

```
Stem-and-leaf of index     N = 15
Leaf Unit = 1.0

    2      4 34
    4      4 58
    4      5
   (6)     5 578999
    5      6 134
    2      6 5
    1      7 1
```

Descriptive Statistics

Variable	N	Mean	Median	TrMean	StDev	SEMean
index	15	57.17	59.00	57.16	8.21	2.12

Variable	Min	Max	Q1	Q3
index	43.20	71.20	48.80	63.20

Assuming that the conditions for a *t*-interval are met, use the information in the descriptive statistics to estimate the mean air pollution index with a 98% confidence interval.

5.60 These are the cholesterol values from a sample of 62 subjects from the Framingham Heart Study presented in Exercise 1.56 in Section 1.3:

WORKSHEET: Framingh.mtw

393 353 334 336 327 300 300 308 283 285 270 270 272 278 278 263 264 267 267 267 268 254 254 254 256 256 258 240 243 246 247 248 230 230 230 230 231 232 232 232 234 234 236 236 238 220 225 225 226 210 211 212 215 216 217 218 200 202 192 198 184 167

Source: R. D'Agostino et. al., "A Suggestion for Using Powerful and Informative Tests of Normality," *The American Statistician, 44* (1990), 316–321.

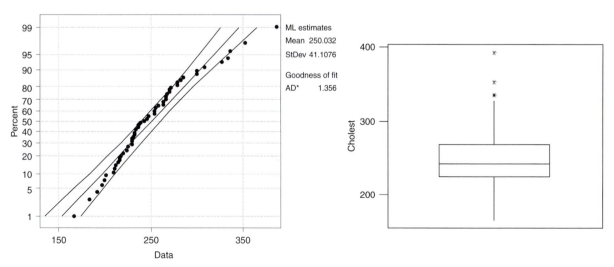

Notice that the accompanying normal probability plot has a slight curvature and thus does not fit the straight line. What does this mean about the shape of the parent distribution? Does the boxplot give the same conclusion? Considering the apparent shape of the parent distribution, should we use the population mean or median to represent a typical cholesterol value?

5.61   These descriptive statistics are for the cholesterol levels given in Exercise 5.60. Suppose we decide to estimate the population mean. From the descriptive statistics, calculate a 95% confidence interval for μ based on $\bar{x}$. Is the sample size large enough to maintain the validity of the t-interval?

Descriptive Statistics

Variable	N	Mean	Median	TrMean	StDev	SEMean
cholest	62	250.03	241.50	247.80	41.44	5.26

Variable	Min	Max	Q1	Q3
cholest	167.00	393.00	225.00	268.50

5.62   In Exercise 1.52 in Section 1.3, the survival times in weeks were given for 20 male rats that were exposed to a high level of radiation:

WORKSHEET: rat.mtw

152 152 115 109 137 88 94 77 160 165 125 40 128 123 136 101 62 153 83 69

Source: J. Lawless, *Statistical Models and Methods for Lifetime Data* (New York: Wiley, 1982).

Based on the accompanying normal probability plot, is it reasonable to assume that the survival time is normally distributed? Should the data be normally distributed in order to construct a t-interval?

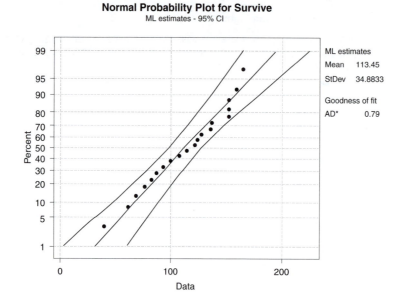

Assuming that the conditions for a t-interval are met, use the information in the descriptive statistics below to estimate the mean survival time with a 95% confidence interval.

Descriptive Statistics

Variable	N	Mean	Median	TrMean	StDev	SEMean
survive	20	113.45	119.00	114.67	35.79	8.00

Variable	Min	Max	Q1	Q3
survive	40.00	165.00	84.25	148.25

5.63   Ozone, a prevalent photochemical oxidant, has been linked to forest decline and severe crop loss. Column 3 of worksheet **Tablrock.mtw** contains ozone concentrations that were obtained on Mt.

Mitchell in Yancey County, North Carolina, in September, 1992 (data provided by Dr. Thomas C. Rhyne). We provide the descriptive statistics, boxplot, and normal probability plot of the ozone concentration here.

Descriptive Statistics

Variable	N	N*	Mean	Median	TrMean	StDev	SEMean
O3	629	90	23.693	20.000	22.183	19.276	0.769

Variable	Min	Max	Q1	Q3
O3	0.000	254.000	11.000	33.000

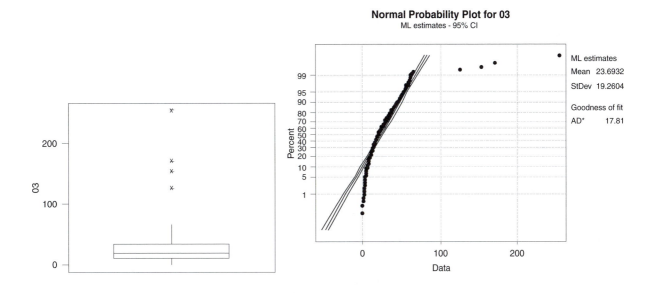

a. What do the boxplot and normal probability plot tell about the distribution of the data?
b. What is the sample size? What effect does the sample size have on the decision to find a confidence interval for the mean ozone concentration based on the t-distribution?
c. Using the descriptive statistics, find a 99% confidence interval for the mean ozone concentration.
d. Do you think the shape of the population distribution affects the validity of the confidence interval?

Computer Exercises

5.64 Surface-water salinity measurements were taken in a bottom-sampling project in Whitewater Bay, Florida, and analyzed in Exercise 5.27 of Section 5.2:

WORKSHEET: Salinity.mtw

46 37 62 59 40 53 58 49 60 56 58 46 47 52 51 60 46 36 34 51 60 47 40 40 35 49
48 39 36 47 59 42 61 67 53 48 50 43 44 49 46 63 53 40 50 78 48 42

Source: J. Davis, *Statistics and Data Analysis in Geology*, 2nd ed. (New York: Wiley, 1986).

In the analysis of the data given previously, we concluded that the salinity variation should be approximately normally distributed. Is this a required condition in order to find a confidence interval for the mean salinity? Proceed to find a 99% confidence interval for the mean salinity level.

5.65   The weekly family incomes of a random sample of 50 Puerto Rican families in Miami were analyzed in Exercise 5.22 of Section 5.2:

WORKSHEET: Puerto.mtw

150 280 175 190 305 380 290 300 170 315 280 255 335 180 200 210 350 360 550 225
245 175 180 260 320 345 180 225 275 280 325 350 190 220 265 270 335 310 270 240
235 355 370 260 280 350 390 250 375 250

In the previous analysis, after one outlier (550) was removed, the remainder of the data appeared to
be short tailed. If we remove the outlier, are there any serious consequences in finding a confidence
interval for μ based on $\bar{x}$? Does the fact that the sample size is 50 have any affect on your decision?
Remove the outlier and construct a 90% confidence interval estimate for the mean weekly income
for the Puerto Rican families of Miami.

5.66   Exercise 1.53 in Section 1.3 described a special diet that was mixed with a drug compound designed
to reduce low-density lipoprotein (LDL) cholesterol and fed to a treatment group of quail. A
placebo group of quail was fed the same special diet for the same period of time but without the
drug compound.

WORKSHEET: Quail.mtw

placebo
64 49 54 64 97 66 76 44 71 89 70 72 71 55 60 62 46 77 86 71
treatment
40 31 50 48 152 44 74 38 81 64

Source: J. McKean and T. Vidmar, "A Comparison of Two Rank-Based Methods for the Analysis of Linear Models,"
The American Statistician, 48 (1994), 220–229.

Examine normal probability plots of the plasma LDL levels for the two groups. Does it appear that
the parent population of the placebo group is normally distributed? What about the treatment group?
It was pointed out that there is one very large value in the treatment group. The researchers stated
that this was typical with data in the study. Remove that one observation and examine a normal
probability plot for the remaining data in the treatment group. By removing the one observation, can
we assume that the parent population of the treatment group is normally distributed? If we wish to
construct confidence intervals for the placebo and treatment groups, is it advisable to remove the
one extreme outlier in the treatment group and find *t*-intervals?

5.67   The accompanying figure shows individual *t*-intervals for the quail data in Exercise 5.66. The first
set of intervals is a comparison of the placebo and treatment groups when no data are removed from
the data sets. The second set is the same comparison with the one extreme outlier in the treatment
group removed. Without removing the outlier, do you detect a difference between the placebo and
treatment groups? After removing the outlier, do you detect a difference between the two groups?
What conclusion should we draw from these data?

```
Individual 95% CIs For Mean

Level     N  Mean   StDev ---+---------+---------+---------+---------+------
placebo  20  67.20  13.97                      (------*------)
treatmen 10  62.20  35.42 (------------------------*------------------------)
                          ---+---------+---------+---------+---------+----------
                             40       50        60        70        80


Individual 95% CIs For Mean

Level     N  Mean   StDev ---+---------+---------+---------+---------+------
placebo  20  67.20  13.97                      (------*------)
treatmen  9  52.22  17.08   (------------*-------------)
                          ---+---------+---------+---------+---------+-------
                             40       50        60        70        80
```

5.68   Investing in coins can be very profitable as well as risky. Had one invested $10,000 in coins in 1980, in five years the investment would have grown to $23,602. Over this five-year period, coins were the best investment available, with more than a 27% per year return. Investing in the New York Stock Exchange would have yielded an average rate of return of 10.7% per year. Suppose that an investor calculates the yearly return on each of 12 possible investments with these results:

WORKSHEET: Coins.mtw

12.6	9.8	13.2	11.6	12.1	10.7
14.6	10.4	18.4	11.2	27.0	10.7

What does a normal probability plot say about the shape of the parent distribution? In choosing a parameter to measure a "typical" return on an investment, what do you recommend: the mean or the median? Why? What condition must be satisfied in order to calculate a t-interval? Do you think that condition is met with these data?

5.5   Confidence Interval for a Population Median

Learning Objectives for this section:

❑ Based on the shape of the parent distribution, be able to decide whether the population mean or the population median best describes the center of the distribution.

❑ Know how to find a small-sample confidence interval for the population median.

❑ Know how to find a large-sample confidence interval for the population median.

❑ Know that the procedures presented in the section are based on the one-sample sign test.

The t-interval, covered in the preceding section, is used to find a confidence interval for the center of a population that has a reasonably symmetric distribution without long tails. In cases where the sample size is rather small, the population distribution should not deviate substantially from the normal distribution. In fact, it has been pointed out that the precision of the confidence interval based on $\bar{x}$ is significantly reduced when the parent distribution has long tails. In Section 5.2, we suggested estimating the population median instead of the population mean when the parent distribution is significantly skewed or symmetric with long tails. This section shows how to construct confidence intervals for the population median given small and large samples.

Small Sample Confidence Interval for the Population Median

The interval we give is based on the binomial distribution and is commonly referred to as the confidence interval obtained from the *sign test* (a nonparametric text will give an explanation of the sign test). To obtain a confidence interval for the population median, θ, we first need to order the sample as in an ordered stem-and-leaf plot. Then the confidence interval is formed from two of the sample observations, C_L and C_H. C_L is located by counting up from the low end of the ordered data, and C_H is located by counting down from the high end of the ordered data. The distance we count in from each end (position of C) is found from the binomial table with $\pi = .5$, or more quickly from Table 5.2. If, for example, we desire a 95% confidence interval for the population median and the sample size is $n = 15$, then Table 5.2 gives:

Position of C = 4

Thus, C_L is the fourth observation from the low end of the ordered data and C_H is the fourth observation from the high end of the ordered data.

Table 5.2
Position of *C*

| | 90% | | 95% | | 99% | |
	Position of C	Achieved confidence	Position of C	Achieved confidence	Position of C	Achieved confidence
n						
4	1	87.50%				
5	1	93.75				
6	1	96.88	1	96.88		
7	2	87.50	1	98.44		
8	2	92.97	2	92.97	1	99.22
9	2	96.09	2	96.09	1	99.61
10	3	89.06	2	97.85	1	99.80
11	3	93.46	3	93.46	2	98.83
12	4	85.40	3	96.14	2	99.37
13	4	90.77	3	97.75	2	99.66
14	4	94.26	4	94.26	3	98.71
15	5	88.15	4	96.48	3	99.26
16	5	92.32	5	92.32	3	99.58
17	6	85.65	5	95.10	4	98.73
18	6	90.37	5	96.91	4	99.25
19	6	93.64	6	93.64	4	99.56
20	7	88.47	6	95.86	5	98.82

As stated, the construction of a confidence interval for the population median is based on the binomial distribution. Because of the discrete nature of the binomial distribution the achieved confidence level of the confidence interval is seldom exactly equal to the desired confidence interval. By using Table 5.2, however, we obtain an achieved confidence level as close as possible to the desired confidence level.

The binomial probabilities with $\pi = .5$ can be accurately approximated with the normal distribution when $n > 20$, so we use a sample size of 20 to distinguish between large and small samples when finding a confidence interval for the population median.

Example 5.21   A study was undertaken to investigate the number of days to recover after being treated with a new drug for patients who suffer from a certain type of viral disease. From the data on recovery time, in days, recorded in the stem-and-leaf plot, obtain a 95% confidence interval for the median number of days to recovery.

```
0 | 5 6 7 8 8 9
1 | 1 2 2 3
1 | 7 8
2 |
2 | 8
3 | 2
3 | 7
```

Solution   From the stem-and-leaf plot, it appears that the data are skewed right. Therefore, we choose to find a confidence interval for the median number of days to recovery. From Table 5.2, with 15 observations and 95% confidence, we find:

$$\text{Position of } C = 4$$

Thus, C_L, the fourth value from the low side, is 8 and C_H, the fourth value from the high side, is 18. Therefore, the 95% confidence interval (actual achieved confidence is 96.48%) for the median number of days is given by the limits (8, 18).   ∎

We should point out that the confidence level given in Table 5.2 is only approximate because of the discrete nature of the binomial distribution. For example, the exact confidence level for the interval obtained in Example 5.21 is found, using the binomial probability table, to be 96.48%, which is reasonably close to 95%. It is as close to 95% as we can get because we must count to the fourth observation; we cannot count to a fraction of an observation.

Large-Sample Confidence Interval for the Population Median

We now construct a large-sample confidence interval for the population median, θ. Assuming that we have a large sample, we can use the Central Limit Theorem and approximate the binomial distribution with the normal distribution.

To construct the interval, we proceed as in the small-sample case by first ordering the sample from smallest to largest values. Then the $(1 - \alpha)100\%$ confidence interval is formed (as in the small-sample case) by two of the sample observations, C_L and C_H. C_L is located by counting up from the low end of the ordered data and C_H is located by counting down from the high end of the ordered data (the same way we found C_L and C_H in the small-sample case). The distance we count in from each end, the position of C, is approximated by the normal distribution.

Locating Endpoints of a Confidence Interval for the Median

To find a $(1 - \alpha)100\%$ confidence interval for the population median when the sample size is greater than 20, we use the following formula to find the location of the endpoints of the interval:

$$\text{Position of } C = \frac{n - z^*\sqrt{n}}{2}$$

where z^* is the upper $\alpha/2$ critical value found in the standard normal distribution table.

Usually the position of C will not be a whole number, so we round up to the next whole number.

Example 5.22   In Example 5.10, it was determined that the parent distribution of the scores on the psychological test used for measuring racial prejudice (worksheet: **Prejudic.mtw**) is skewed right and that the population median best represents the center of the distribution. Find a 90% confidence interval for the population median, θ.

Solution   The sample size in Example 5.10 was given to be $n = 25$. For a 90% confidence interval, we have $z^* = 1.645$. Substituting these values in the formula for position of C, we find:

$$\text{Position of C} = \frac{[n - z^* \sqrt{n}]}{2} = \frac{(25 - 1.645\sqrt{25})}{2}$$
$$= 8.4$$
$$\cong 9$$

Figure 5.17
```
Stem-and-leaf of prejud    N = 25
Leaf Unit = 1.0

     5      4  11224
    11      4  567888
    (3)     5  124
    11      5  889
     8      6  12
     6      6  5
     5      7  24
     3      7  9
     2      8  3
     1      8  7
```

Notice the skewness of the data in Figure 5.17. The ninth observation from the low end of the stem-and-leaf plot of the data is $C_L = 48$ and the ninth observation from the high end is $C_H = 59$. Thus, we are 90% confident that the population median is somewhere inside the interval from 48 to 59.   ■

COMPUTER TIP

Confidence Interval for the Median

After the data are stored in the worksheet, the confidence interval for the median is found by selecting the **Stat** menu, choosing **Nonparametrics**, and then selecting **1-Sample Sign**. In the 1-Sample Sign window, **Select** the **Variable** where the data are stored, click on **Confidence interval**, specify the desired **Level**, and click **OK**.

Figure 5.18 gives the computer results for the data in Example 5.22. Minitab prints three confidence intervals that have a confidence level close to the requested 90%. The first confidence interval has an achieved confidence level below 90% (.8922) and the third confidence interval has an achieved confidence level above 90% (.9567). The middle confidence interval with an achieved confidence level of the requested 90% is obtained by nonlinear interpolation (NLI) between the first and third intervals. Notice that the first interval is the one we got in Example 5.22.

Figure 5.18   **Sign CI: prejud**

```
Sign confidence interval for median

                        ACHIEVED
          N    MEDIAN   CONFIDENCE   CONFIDENCE  INTERVAL    POSITION
prejud   25    52.00      0.8922     (  48.00,   59.00)          9
                          0.9000     (  47.94,   59.12)        NLI
                          0.9567     (  47.00,   61.00)          8
```

EXERCISES 5.5   The Basics

5.69   For the following sample sizes, determine the position of the two observations in an ordered stem-and-leaf plot that form a 95% confidence interval for the population median, θ.
  a. $n = 10$ b. $n = 12$
  c. $n = 15$ d. $n = 19$

5.70   For a sample size of 18, determine the position of the two observations in an ordered stem-and-leaf plot that form a confidence interval for the population median, θ, with each confidence level:
  a. 90%   b. 95%   c. 99%

5.71   From the ordered stem-and-leaf plots, construct 95% confidence intervals for the population median:

a.
```
4 | 0 3 5
5 | 0 2 3 5 5 7 9
6 | 1 2 5
7 | 4
8 | 1
9 | 2
```

b.
```
 4 | 0
 5 | 0 2 5 9
 6 | 1 2 4 5 5 8
 7 | 4 7 9
 8 | 1
 9 | 2 5
10 | 1
```

c.
```
 5 | 0 2 9
 6 | 1 4 5 8
 7 | 4 9
 8 | 1
 9 | 2
10 | 1 4
```

5.72   Shoplifting and employee theft cost retailers more than $2 billion annually. These are the values of merchandise found in the possession of shoplifters apprehended in a department store over a busy weekend:

$38   20   12   100   22   45   5   75   150   20   25   19

Construct a 90% confidence interval for the median value of shoplifted merchandise.

5.73   Medical malpractice awards often exceed $1 million. Because some malpractice awards are extremely large, it may be appropriate to report the median award. From these data (recorded in $1000s) construct a 90% confidence interval for the median malpractice award:

WORKSHEET: Malpract.mtw

760   380   125   250   2800   450   100   150   2000
180   650   275   850   1700   1500   3000   390

Also construct a 90% confidence interval based on $\bar{x}$ and plot the two intervals on the same line graph. Compare the two intervals. Which do you think better represents the typical amount of malpractice awards?

5.74   For the following sample sizes, determine the position in an ordered stem-and-leaf plot of the two observations that form a 95% confidence interval for the population median, θ.
  a. $n = 25$ b. $n = 50$ c. $n = 100$ d. $n = 200$

5.75 From the ordered stem-and-leaf plots, construct a 95% confidence interval for the population median, θ.

a.
0	5
1	2 8
2	1 2 3 5 7 8
3	1 3 4 5 5 6 9
4	2 5 8 8 9
5	1 4
6	3
7	4

b.
0	0 0 1 5
1	2 4 5 5 8 9 9
2	1 2 3 5
3	3 4 6
4	2 5
5	1 4
6	3
7	4
8	1

c.
0	0 0 0 1 5 7 9
1	2 5 8 9
2	1 3 5
3	3 4 6
4	2 5 7
5	1
6	3
7	4
8	1

5.76 An experiment was designed to examine the distribution of the times (in minutes) to breakdown of an insulating fluid under various levels of voltage stress (E. Soofi, N. Ebrahimi, and M. Habibullah, 1995). Here are the fluid breakdown times at 34 kilovolts.

WORKSHEET: Fluid.mtw

0.19 0.78 0.96 1.31 2.78 3.16 4.15 4.67 4.85 6.50 7.35 8.01 8.27 12.06 31.75 32.52 33.91 36.71 72.89

a. Construct a stem-and-leaf plot to verify that the data are skewed.
b. Construct a 95% confidence interval for the median breakdown time.

5.77 The time to repair the body of an automobile involved in a wreck can vary from reasonably short to rather long when special parts must be ordered. It is reasonable to expect that the distribution of times for repairs is skewed positively. From the following recorded times (in hours) for repairing 22 automobiles, verify that the data are skewed and construct a 98% confidence interval for the median time of a repair:

WORKSHEET: Repair.mtw

10.3 4.1 8.6 2.1 3.7 5.6 11.4 5.7 3.8 4.5 2.7 22.6 5.5 9.6 4.8 3.9 10.4 4.8 5.9 6.7 7.6 12.0

Interpreting Computer Output

5.78 The normal probability plot in Exercise 5.68 in Section 5.4 indicates that the distribution of the returns on the investments studied by the investor is skewed right. It is recommended that the "typical" return be measured with the population median. From the stem-and-leaf plot, construct a 90% confidence interval for the median return on investment.

```
Stem-and-leaf of coins    N = 12
Leaf Unit = 1.0

    1      0 9
    6      1 00011
    6      1 223
    3      1 4
    2      1
    2      1 8
    1      2
    1      2
    1      2
    1      2 7
```

5.79 The female alcoholic generally begins drinking at a later age than does the male alcoholic. Here are the ages at which 14 female alcoholics began drinking:

WORKSHEET: Alcohol.mtw

18	16	24	14	19	22	28
16	19	21	22	35	20	15

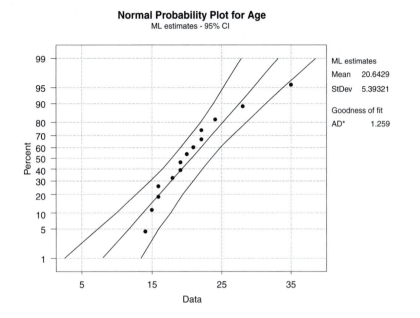

Notice the curvature in the normal probability plot. What does this suggest about the shape of the parent distribution? Under these conditions, it might be best to estimate the population median instead of the population mean. Find a 99% confidence interval for the median age.

5.80 Here is a list of 25 scores on the first test in a statistics class taught at a small midwestern university.

WORKSHEET: Test1.mtw

61, 38, 64, 70, 81, 11, 61, 92, 77, 47, 98, 76, 83,
58, 97, 78, 14, 80, 41, 65, 90, 81, 98, 72, 78

From a stem-and-leaf plot and boxplot, classify the shape of the parent distribution and give an estimate of the typical score.

```
Stem-and-leaf of test1    N = 25
Leaf Unit = 1.0
    2      1 14
    2      2
    3      3 8
    5      4 17
    6      5 8
   10      6 1145
   (6)     7 026788
    9      8 0113
    5      9 02788
```

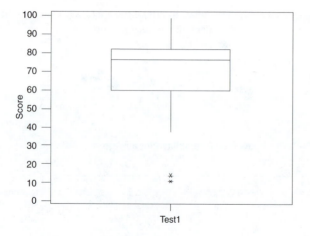

Computer Exercises

5.81 Top executives are in demand by the major corporations. From the random sample of advertised salaries listed here, construct a 95% confidence interval for the median salary offered to general managers in 1999:

WORKSHEET: Manager.mtw

$95,000	85,000	65,500	98,000	150,000	75,000
60,000	78,000	100,000	75,500	90,000	85,000
150,000	120,000	55,000	70,000	85,500	120,000
85,000	99,900	110,000	150,000	95,000	60,000
185,000	75,000				

5.82 According to the National Association of Realtors, the median price of a single-family home in the U.S. was $73,600 in 1984. Ten years later, in 1993, the median price had risen to $104,200. Following are the median prices for 1984 and 1993 in 37 markets across the U.S.:

WORKSHEET: Housing.mtw

	1984	1993		1984	1993
Albany, NY	$ 52,400	109,900	Memphis, Tenn	65,200	84,200
Anaheim, Calif	134,900	222,200	Miami	84,400	98,000
Atlanta	64,600	94,000	Milwaukee	69,600	98,400
Baltimore	65,200	113,200	Minneapolis-St Paul	75,100	96,400
Birmingham, Ala	66,600	89,000	Nashville, Tenn	64,000	87,000
Boston	102,000	165,200	New York metro area	106,900	168,000
Chicago	77,500	131,300	Oklahoma City	63,600	61,100
Cincinnati	59,600	85,600	Philadelphia	59,500	108,900
Cleveland	65,600	89,200	Providence, R.I.	61,400	112,500
Columbus, Ohio	60,400	89,300	Rochester, N.Y.	62,100	83,300
Dallas-Fort Worth	83,400	89,600	St. Louis	64,400	80,900
Denver	85,000	96,300	Salt Lake City	67,600	79,000
Detroit	48,000	92,200	San Antonio, Texas	71,600	72,600
Fort Lauderdale	76,000	99,500	San Diego	102,900	175,500
Houston	79,600	78,000	San Francisco	132,600	249,300
Indianapolis	54,200	83,800	San Jose, Calif	121,600	297,100
Kansas City, Mo	57,600	79,100	Tampa, Fla	60,800	69,900
Los Angeles	115,300	199,700	Washington	92,900	153,500
Louisville, K	49,500	69,900			

Source: National Association of Realtors.

To compare the cost of housing in 1984 and 1993, use your computer to construct summary diagrams such as stem-and-leaf plots, dotplots, and histograms of the two samples. Compute descriptive statistics for each sample and compare them. Construct normal probability plots to shed light on the length of the tails of the two distributions. Finally, construct side-by-side boxplots. From the information, classify the shapes of the distributions and give the parameter that you think would be best to compare the centers of the two distributions. Find 98% confidence intervals for the median costs in 1984 and in 1993. Graph both intervals on a common number line. Discuss the differences between the two intervals.

5.83 Trawl gear for fishing in the ocean consists of a cone-shaped net towed behind a vessel at a speed similar to a walking pace. Fish that are caught in the trawl work themselves to the end of the net to a portion called the codend. The size and shape of the codend determines what size fish will be caught and what size fish will escape the net. The list gives the size and the number of haddock caught with a 35-mm diamond-mesh (small-mesh) codend and a 87-mm diamond-mesh (large-mesh) codend. The data are in frequency table form, with the first column listing the length of the fish and the second and third columns giving the number of fish of the various lengths caught.

WORKSHEET: Fish.mtw

Length (cm)	Small mesh	Large mesh
24	1	0
25	1	0
26	3	0
27	14	1
28	30	5
29	49	19
30	60	29
31	50	51
32	70	71
33	108	120
34	88	118
35	84	107
36	68	78
37	37	52
38	33	40
39	12	17
40	5	17
41	6	14
42	10	10
43	1	4
44	6	6
45	2	2
46	1	5
47	0	1

Source: R. Millar, "Estimating the Size-Selectivity of Fishing Gear by Conditioning on the Total Catch," *Journal of the American Statistical Association*, 87 (1992), 962–968.

a. What is the median-length fish caught by the two sizes of codend?
b. What is the interquartile range for the two distributions?
c. Does it appear that the lengths of fish caught by the two codends are different?
d. Calculate individual 99% confidence intervals for the median-length fish caught by the two different codends. Describe any differences between the two confidence intervals.

5.6   Summary and Review Exercises

Key Concepts

- Describing a population distribution via the collected data involves three things: the shape of the distribution, a measure of location, and a measure of variability. Important *distribution shapes* include near normal, skewed, symmetric, and multimodal.

- The *stem-and-leaf plot, histogram*, and *boxplot* are used to describe the general shape of a distribution. The *normal probability plot* can be used to compare the distribution to the normal distribution.

- Once the shape of a distribution is determined, we then decide which parameters best describe the center and the variability.

- Once a parameter is decided on, we are then in a position to determine what statistic best estimates the parameter. In making this decision, we must be familiar with the desirable characteristics of an *estimator* (the statistic used to estimate the parameter).

- Parameters we wish to estimate include μ, θ, and σ for measurement populations and π for a Bernoulli population.

- A confidence interval for the population proportion, π, is given by

$$p \pm z^* \sqrt{\frac{p(1 - p)}{n}}$$

- If the distribution is normal, near normal, or symmetric with tails that are not excessively long, the confidence interval for the mean (the z-interval) takes the form

$$\bar{x} \pm z^* \frac{\sigma}{\sqrt{n}}$$

  when the population standard deviation, σ, is known.

- If the population standard deviation is unknown, the confidence interval for the mean (the t-interval) takes the form

$$\bar{x} \pm t^* \frac{s}{\sqrt{n}}$$

- If the distribution is skewed, the best measure of the center is the median and therefore the confidence interval should be for the population median. The confidence interval is formed from two sample observations, (C_L, C_H), where C_L is found by counting from the low end of the ordered data and C_H is found by counting from the high end of the ordered data. The distance we count in is given by the position of C. For large samples,

$$\text{Position of C} = \frac{[n - z^*\sqrt{n}]}{2}$$

  For small samples, the position of C is found in Table 5.2.

Statistical Insight Revisited

Housing costs in worksheet: **Homes.mtw** are classified into four regions of the country (Northeast—1, Midwest—2, South—3, and West—4) for the years 1994 and 2000. Examine the costs in the year 2000. Comment on the shape of the distribution and estimate the center of the distribution.

  Based on the boxplot and normal probability plot we classify the distribution of housing costs in the year 2000 as heavily skewed right. As such, the population median is a better measure of the center of the distribution than is the mean. Thus, our estimate of the median housing costs is $M = \$119,400$. Notice that the mean, $\$136,494$, is much larger. It is inflated from the outliers on the upper tail of the distribution.

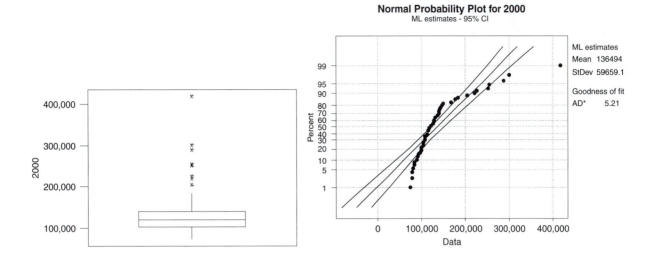

Normal Probability Plot for 2000
ML estimates - 95% CI

ML estimates
Mean 136494
StDev 59659.1

Goodness of fit
AD* 5.21

Descriptive Statistics: 2000

Variable	N	Mean	Median	TrMean	StDev	SE Mean
2000	65	136494	119400	129359	60123	7457

Variable	Minimum	Maximum	Q1	Q3
2000	74500	418600	104100	141450

Descriptive Statistics

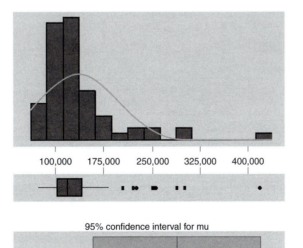

95% confidence interval for mu

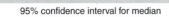

95% confidence interval for median

Variable: 2000

Anderson-Darling Normality Test

A-squared:	5.079
P-value:	0.000

Mean	136494
StDev	60123
Variance	3.61E+09
Skewness	2.47406
Kurtosis	7.65200
N	65

Minimum	74500
1st quartile	104100
Median	119400
3rd quartile	141450
Maximum	418600

95% confidence interval for mu

121596	151392

95% confidence interval for sigma

51272	72697

95% confidence interval for median

112349	130542

Sign CI: 2000

Sign confidence interval for median

	N	Median	Achieved Confidence	Confidence interval	Position
2000	65	119400	0.9175	(112700, 129400)	26
			0.9500	(112349, 130542)	NLI
			0.9528	(112300, 130700)	25

Notice the skewness in the histogram in the Descriptive Statistics graph. Also, we see that the 95% confidence interval for the population median, (112349, 130542), is considerably shorter and centered lower than the confidence interval for the population mean, (121596, 151392). We are reasonably confident that the median cost of housing in 2000 in the United States is somewhere between $112,349 and $130,542.

The side-by-side boxplots show that the cost of housing rose substantially between 1994 and 2000. Both distributions are similarly skewed and we see that that the median cost rose from $91,300 in 1994 to $119,400 in 2000.

Descriptive Statistics: 1994, 2000

Variable	N	Mean	Median	TrMean	StDev	SE Mean
1994	65	107158	91300	100858	47713	5918
2000	65	136494	119400	129359	60123	7457

Variable	Minimum	Maximum	Q1	Q3
1994	66200	355000	79850	113550
2000	74500	418600	104100	141450

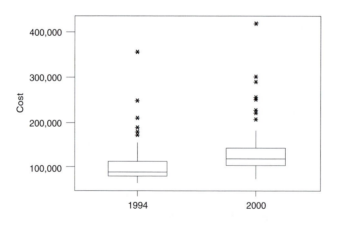

If we examine the difference in housing cost between 1994 and 2000, we see a symmetric long-tailed distribution in the histogram and normal probability plot. Again, for distributions of this type, the median is the preferred measure of the center of the distribution. Therefore, we estimate that the change in median cost from 1994 to 2000 is $25,000.

Descriptive Statistics: difference

Variable	N	Mean	Median	TrMean	StDev	SE Mean
differen	65	29335	25000	27739	28966	3593

Variable	Minimum	Maximum	Q1	Q3
differen	−66000	171700	16550	36100

Normal Probability Plot for Difference
ML estimates - 95% CI

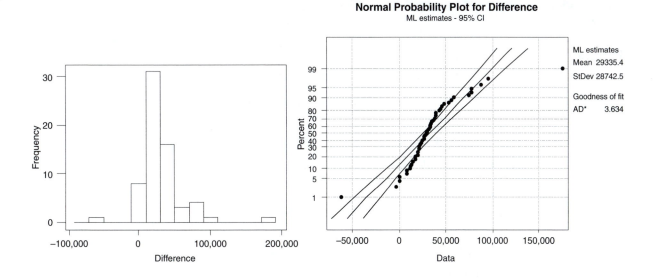

ML estimates	
Mean	29335.4
StDev	28742.5
Goodness of fit	
AD*	3.634

Finally, from the side-by-side boxplots of the housing costs for the four regions, we see how the housing costs vary across the nation. We see that the cost in the South is the least and does not vary much across the region. The median cost is greater in the West, but there is much variability in the cost across this region.

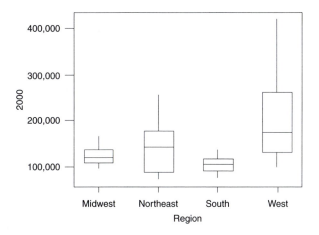

Descriptive Statistics: 2000 by Region

Variable	Region	N	Mean	Median	TrMean	StDev
2000	Midwest	17	122471	119400	121093	18593
	Northeas	11	143082	141900	138267	57217
	South	23	104748	105800	104419	16251
	West	14	200500	174700	190750	88975

Variable	Region	SE Mean	Minimum	Maximum	Q1	Q3
2000	Midwest	4509	98900	166700	109450	137450
	Northeas	17252	74500	255000	87800	177500
	South	3388	78200	138200	91900	116000
	West	23780	99400	418600	133450	260800

To test your skills, answer the following questions.

The Basics

5.84 What statistic would you use to estimate the center of a distribution if a boxplot and a normal probability plot of the sample suggest each of these parent populations?
a. Approximately normal in shape?
b. Symmetric but has tails significantly longer than the normal distribution?
c. Symmetric but has tails significantly shorter than the normal distribution?
d. Highly skewed left?
e. Highly skewed right?

5.85 Identify the bold value as the parameter π, μ, θ, or σ.
a. According to Ford Motor Company, **82%** of the functions in its cars are computer controlled.
b. General auto mechanics earn, on average, **$30,000** a year to start and those with top diagnostic skills sometimes earn as much as $100,000.
c. According to the Census Bureau (*Statistical Brief* SB/95–20, August 1995), half of all people stay in poverty more than **4.3** months.
d. **Eighty** percent of the adult population drives cars.

5.86 Porosity measurements (percent) on twenty samples of Tensleep Sandstone, Pennsylvanian, from Wind River Basin and Bighorn Basin in Wyoming are listed here:

WORKSHEET: Porosity.mtw

15 10 15 23 18 26 24 18 19 21 13 17 15
23 27 29 18 27 20 24

Source: Davis, J. C. (1986), *Statistics and Data Analysis in Geology*, 2nd edition, pages 63–65.

a. Construct a stem-and-leaf plot and a five-number summary diagram for these data. Determine whether there are outliers in the distribution.
b. Use the information in **a** to construct a boxplot. From the boxplot, make a general statement about the shape of the distribution. What parameter best describes a "typical" score? What is your estimate of the parameter?

5.87 We wish to estimate with a 95% confidence interval the proportion of television viewers who would watch a one-hour national news program. How large a sample is necessary to allow our precision of estimation to be ± 3 percentage points?

5.88 You are studying the percent of convicted felons who have a history of juvenile delinquency.
a. How large a sample do you need to estimate the percent to within .03 with a 99% confidence interval?
b. A sample of 1600 convicted felons revealed that 1120 had a history of juvenile delinquency. Find a 99% confidence interval for the true percent. Interpret the results.

5.89 A random sample of 64 homes in a small community found that an average of 160 gallons of heating oil were consumed over a given period of time. If it is known that the amount of heating oil consumed is close to normally distributed with a standard deviation of 32 gallons, find a 95% confidence interval based on $\bar{x}$ for the mean number of gallons of heating oil consumed over this period by all residents of the community.

5.90 From a random sample of 2400 college students, 960 believe that the penalties for the use of marijuana should be reduced.
a. Find a 99% confidence interval for the proportion of all college students who believe the penalties should be reduced.
b. How large a sample is needed so that the margin of error of the estimate is no greater than 2%?

5.91 When the flow characteristics of oil through a valve are determined, the inlet oil temperature is measured in degrees Fahrenheit. Here is a sample of 12 readings:

WORKSHEET: Inletoil.mtw

93 99 97 99 94 91 93 90 89 92 90 93

Construct a 95% confidence interval based on $\bar{x}$ for the mean temperature. What assumption is required for the interval to be 95% valid? Do you think the assumption is violated?

5.92 In a study to determine whether alcoholism has a genetic basis, several genetic markers were observed in a group of 50 Caucasian alcoholics. For 10 of the 50, the antigen B15 was present. Estimate with a 90% confidence interval the proportion who have this antigen in the population of Caucasian alcoholics.

5.93 How large a sample is needed to estimate patient's mean hospital costs per day to within $30 with a 90% confidence interval if the population distribution is near normal with a standard deviation of $120?

5.94 A reading teacher would like to estimate the mean reading speed of students in the fifth grade. A sample of 16 students had an average reading speed of 285 words per minute and a standard deviation of 48. To construct a confidence interval for the mean reading speed based on $\bar{x}$, what assumption must be made about the parent population? Assuming the assumption is satisfied, find a 95% confidence interval for the mean reading speed.

5.95 A pollster believes that approximately 40% of the registered voters will favor a particular issue; however, he wishes to estimate the true proportion with a 98% confidence interval.
   a. How many people will he need to interview so that the error of estimation is no greater than 4%?
   b. The pollster found 260 in favor of the issue. What is the 98% confidence interval for the true proportion?

5.96 A sample of 35 hotels in Miami yielded an average daily rate of $182.40 and a standard deviation of $53.50. Find a 90% confidence interval based on $\bar{x}$ for the mean hotel rate in Miami.

5.97 In the early 1980s, samples of Tylenol were found to be tampered with and contained traces of poison. Tylenol's manufacturer designed a new safety bottle and launched a mass advertising campaign to convince its customers that its product was safe. After the advertising campaign, 320 out of a random sample of 400 Tylenol users said they would continue to use the product. Find a 98% confidence interval for the proportion of all Tylenol users who would continue to use the product.

5.98 An instrument used to measure anxiety has a standardized mean of 50 and a standard deviation of 15. The instrument will be used to estimate the average anxiety level of emotionally disturbed children. How large a sample is necessary to be sure that the standard error of the estimate will be no greater than 2 points?

5.99 A company wishes to estimate the proportion of accounts that are paid on time.
   a. How large a sample is needed to estimate the true proportion to within 3% with a 95% confidence interval?
   b. Construct a 95% confidence interval if 300 out of 400 accounts were paid on time.

5.100 A survey of 17,000 seniors in approximately 140 public and private high schools nationwide (by the University of Michigan's Institute for Social Research) showed that 10,540 had tried an illegal drug. Construct a 99% confidence interval for the proportion of all high school seniors who have tried an illegal drug.

5.101 A Michigan State University survey of 549 employers found that 220 indicated that they sought students who had practical work experience, such as co-op or internship experience. Construct a 90% confidence interval for the percent of all employers who prefer work experience prior to graduation.

5.102 A survey of 49 randomly selected graduates from the class of 1999 with bachelor's degrees in chemical engineering found the average starting salary was $58,394 and the standard deviation was $8482. Based on this random sample, find a 99% confidence interval for the mean starting salary of all chemical engineers.

Interpreting
Computer
Output
5.103 Here is a normal probability plot of the porosity measurements given in Exercise 5.86. Does the plot indicate that the data are from a normal population? If not, how do you classify the shape of the distribution? Are the conditions for the construction of a t-interval satisfied? If so, construct a 90% confidence interval for the mean porosity percent in these two basins.

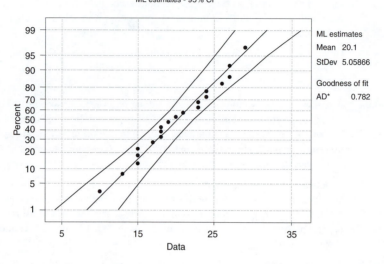

Normal Probability Plot for Porosity
ML estimates - 95% CI

5.104 The table gives the median charges for a coronary bypass at 17 hospitals in the state of North Carolina.

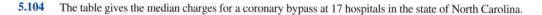

WORKSHEET: Bypass.mtw

hospital	charge	hospital	charge
Carolinas Med Ct	38578	New Hanover Regional	35831
Duke Med Ct	31935	Pitt Co. Memorial	34001
Durham Regional	34465	Presbyterian	27445
Forsyth Memorial	24810	Rex	31419
Frye Regional	35144	Moses Cone Memorial	34695
High Point Region	29245	NC Baptist	29154
Memorial Mission	29473	Univ of North Carolina	36745
Mercy	34376	Wake County	31163
Moore Regional	32428		

Source: Consumer's Guide to Hospitalization Charges in North Carolina Hospitals,
North Carolina Medical Database Commission, Raleigh, NC, August 1994.

a. Here is a stem-and-leaf plot of the charges. Do you detect severe skewness in the data?

```
Stem-and-leaf of charge     N = 17
Leaf Unit = 1000

    1      2 4
    2      2 7
    5      2 999
    8      3 111
   (1)     3 2
    8      3 444455
    2      3 6
    1      3 8
```

b. If we wish to construct a confidence interval for the center of the distribution, would it be worthwhile to check for normality? Explain.

c. Here is a boxplot of the medical charges for a coronary bypass.

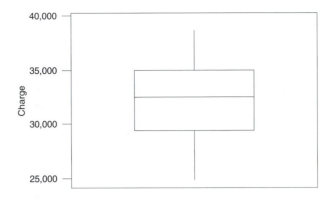

Use the boxplot to make a general statement about the shape of the distribution and identify a parameter that would best measure the center of the distribution. Give an estimate of the parameter.

5.105   Here is a normal probability plot of the coronary bypass data given in Exercise 5.104. Does the plot indicate that the data are from a normal population? If not, how would you classify the shape of the distribution? Are the conditions for the construction of a *t*-interval satisfied? If so, construct a 90% confidence interval for the mean charge for a coronary bypass.

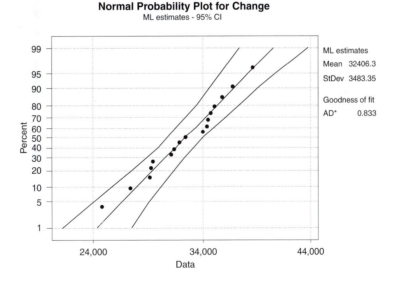

5.106   Following are the annual food expenditures for a random sample of 40 single households in the state of Ohio.

WORKSHEET: Food.mtw

$2845, 3170, 2352, 4978, 3820, 2475, 3160, 5780, 2175, 2648, 2872, 4250, 3970, 2534, 6870, 2734, 2847, 4670, 5176, 3640, 2765, 1180, 3679, 3320, 7580, 2416, 3743, 2830, 3127, 3249, 2648, 1976, 2784, 3869, 2086, 5587, 3420, 2645, 8147, 4367

Source: Bureau of Labor Statistics.

Using this descriptive statistic display, evaluate the skewness of the distribution. Which confidence interval, the one for the mean or for the median, is best suited for these data? Should we check for normality? Explain.

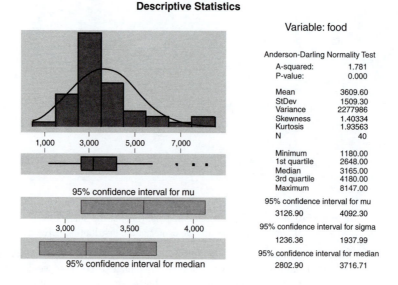

Descriptive Statistics

Variable: food

Anderson-Darling Normality Test

A-squared:	1.781
P-value:	0.000

Mean	3609.60
StDev	1509.30
Variance	2277986
Skewness	1.40334
Kurtosis	1.93563
N	40

Minimum	1180.00
1st quartile	2648.00
Median	3165.00
3rd quartile	4180.00
Maximum	8147.00

95% confidence interval for mu

3126.90   4092.30

95% confidence interval for sigma

1236.36   1937.99

95% confidence interval for median

2802.90   3716.71

95% confidence interval for mu

95% confidence interval for median

5.107 Here is a normal probability plot of the food expenditures given in Exercise 5.106. Does the plot indicate that the data are from a normal population? If not, how would you classify the shape of the distribution? Does this agree with your conclusion in Exercise 5.106?

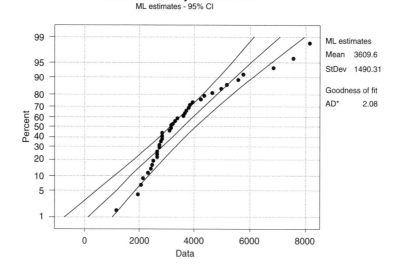

Normal Probability Plot for Food
ML estimates - 95% CI

ML estimates
Mean   3609.6
StDev   1490.31

Goodness of fit
AD*   2.08

5.108 The number of hazardous toxic waste sites in each of the 50 states is recorded in worksheet: **Toxic.mtw**. Shown here is a normal probability plot of the number of hazardous sites. Notice the curvature in the plot and that it does not fit the straight line very well. What does this say about the shape of the parent distribution? Do you recommend investigating the mean number of sites or the median number of sites?

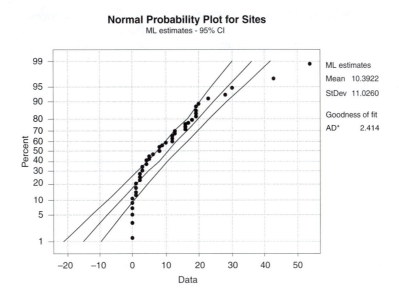

Normal Probability Plot for Sites
ML estimates - 95% CI

ML estimates
Mean 10.3922
StDev 11.0260

Goodness of fit
AD*   2.414

5.109   The stem-and-leaf plot shows the number of hazardous waste sites discussed in Exercise 5.108 and given in worksheet: **Toxic.mtw**. Based on these data, construct a 98% confidence interval for the median number of hazardous waste sites.

Character Stem-and-Leaf Display

```
Stem-and-leaf of sites     N = 51
Leaf Unit = 1.0

  21    0 000000111112222333444
  (8)   0 55668889
  22    1 0222233
  15    1 666789999
   6    2 03
   4    2 8
   3    3 0
   2    3
   2    4 3
   1    4
   1    5 4
```

5.110   In the Statistical Insight at the beginning of Chapter 1, we learned that the value of a rare stamp is affected by the thickness of the paper on which the stamp was printed. Recorded in worksheet: **Stamp.mtw** are the thicknesses of the 1872 Hidalgo stamp that was issued in Mexico. From these descriptive statistics, construct a 99% confidence interval for the mean thickness of the 1872 Hidalgo.

Descriptive Statistics

Variable	N	Mean	Median	TrMean	StDev	SEMean
thicknes	485	0.08602	0.08000	0.08500	0.01496	0.00068

Variable	Min	Max	Q1	Q3
thicknes	0.06000	0.13100	0.07500	0.09800

5.111   The Old Faithful geyser, located in Yellowstone National Park in Wyoming, has two types of eruptions: a short eruption somewhat less than 3 minutes in duration and a long eruption with a duration in excess of 3 minutes (see Exercise 5.12 in Section 5.1). The waiting time between eruptions consequently has a bimodal distribution. In worksheet: **Faithful.mtw**, the wait times are classified into two categories: a short eruption and a long eruption. From these descriptive statistics, construct individual 95% confidence intervals for the mean wait time between eruptions for the two types of eruptions. Graph the two intervals on a common number line. Is the precision of the two intervals about the same? If they are different, explain why.

Descriptive Statistics

Variable	Eruption	N	Mean	Median	TrMean	StDev	SEMean
time	1	110	57.109	55.000	56.612	8.148	0.777
	2	189	81.164	81.000	81.263	7.303	0.531

Variable	Eruption	Min	Max	Q1	Q3
time	1	43.000	77.000	50.000	61.250
	2	59.000	108.000	77.000	87.000

Computer Exercises

5.112   The 1999 annual salary of each state governor is given below:

WORKSHEET: Governor.mtw

State 1999	salary	State 1999	salary
Alabama	94655	Montana	83672
Alaska	81648	Nebraska	65000
Arizona	95000	Nevada	117000
Arkansas	60000	New Hampshire	90547
California	165000	New Jersey	85000
Colorado	90000	New Mexico	90000
Connecticut	78000	New York	179000
Delaware	107000	North Carolina	113656
Florida	117240	North Dakota	76879
Georgia	115939	Ohio	119235
Hawaii	94780	Oklahoma	101140
Idaho	92500	Oregon	88300
Illinois	140132	Pennsylvania	105035
Indiana	77200	Rhode Island	95000
Iowa	104352	South Carolina	106078
Kansas	91742	South Dakota	87276
Kentucky	97067	Tennessee	85000
Louisiana	95000	Texas	115345
Maine	70000	Utah	93000
Maryland	120000	Vermont	115763
Massachusetts	90000	Virginia	124855
Michigan	127300	Washington	132000
Minnesota	120303	West Virginia	90000
Mississippi	83160	Wisconsin	115699
Missouri	112755	Wyoming	95000

Source: The 2000 World Almanac and Book of Facts.

Complete an analysis of the data by constructing a histogram and normal probability plot and obtain descriptive statistics. Comment on the shape of the distribution of governors' salaries across the states, and give a measure of the center of the data.

5.113   Violent crime rates for all 50 states and the District of Columbia are given in worksheet **Crime.mtw**. An analysis of these data in Exercise 1.90 of Section 1.5 determined that there was one outlier, the rate for the District of Columbia. From the data construct 95% confidence intervals for the mean with and without the one outlier. What are the basic differences between the two intervals? Which one is the better representation of a 95% confidence interval for the mean? Explain why.

5.114 These are the average teacher's salaries across the states for school years 1973–74, 1983–84, and 1993–94.

WORKSHEET: Teacher.mtw

	Average salary				Average salary		
State	1973–74	1983–84	1993–94	State	1973–74	1983–84	1993–94
Ala	$ 9226	$18,000	$27,000	Mont	$ 9429	$20,657	$27,600
Alaska	15,667	36,564	44,700	Neb	9174	18,785	27,200
Ariz	10,414	21,605	31,200	Nev	11,549	23,000	33,900
Ark	7820	16,929	26,600	N.H.	9613	17,376	33,200
Calif	13,113	26,403	40,200	N.J.	11,920	23,044	41,000
Colo	10,131	22,895	33,100	N.M.	9100	20,760	26,700
Conn	11,030	22,624	47,000	N.Y.	13,371	26,750	43,300
Del	11,304	20,925	34,500	N.C.	10,223	18,014	29,200
D.C.	N/A	27,659	41,300	N.D.	8493	20,363	24,500
Fla	10,018	19,545	31,100	Ohio	10,107	21,421	33,300
Ga	9392	18,505	29,500	Okla	8238	18,490	25,300
Hawaii	11,112	24,357	34,500	Ore	10,180	22,833	34,100
Idaho	8383	18,640	26,300	Pa	10,921	22,800	38,700
Ill	11,871	23,345	36,500	R.I.	11,407	24,641	36,000
Ind	10,508	21,587	34,800	S.C.	8654	17,500	28,300
Iowa	9863	20,140	29,200	S.D.	8150	16,480	23,300
Kan	8894	19,598	30,700	Tenn	8840	17,900	28,600
Ky	8295	19,780	30,900	Texas	8920	20,100	29,000
La	9166	19,100	27,000	Utah	9146	20,256	26,500
Maine	9238	17,328	30,100	Vt	8932	17,931	33,600
Md	11,741	24,095	39,500	Va	9919	19,867	31,900
Mass	11,121	22,500	37,300	Wash	11,295	24,780	34,800
Mich	12,545	28,877	41,100	W.Va	8467	17,482	27,400
Minn	11,122	24,480	33,700	Wis	10,830	23,000	35,200
Miss	7604	15,895	24,400	Wyo	9668	24,500	30,400
Mo	9530	19,300	28,900	Average	10,778	22,019	34,100

Source: National Education Association.

Clearly the average salary for 1973–74 is lower than 1983–84, which in turn is lower than 1993–94, but other than that, are the distributions similar? How variable are the salaries? What are the shapes of the three distributions? To investigate these questions, use your computer to construct histograms and descriptive statistics. Finally, construct a side-by-side boxplot and normal probability plots to shed light on the length of the tails of the three distributions. From this information, compare the shapes of the three distributions.

5.115 Below are the national track records for women in the 100-meter, 200-meter, and 400-meter race. (The reference also gives the men's records.) Study the distributions of the three races by constructing histograms and boxplots of each sample. Do the distributions appear to be normally distributed? Are all three similar in shape? Construct normal probability plots for each sample. Calculate descriptive statistics for each of the three samples. Are the means, medians, and trimmed means about the same for each of the three samples? Is the amount of variability about the same for each of the three races? Are the standard deviations about what you expected? Write a short paragraph explaining what you found out by analyzing these data.

WORKSHEET: Track.mtw

country	100m	200m	400m	country	100m	200m	400m
Argentina	11.61	22.94	54.50	Italy	11.29	23.00	52.01
Australia	11.20	22.35	51.08	Japan	11.73	24.00	53.73
Austria	11.43	23.09	50.62	Kenya	11.73	23.88	52.70
Belgium	11.41	23.04	52.00	Korea	11.96	24.49	55.70
Bermuda	11.46	23.05	53.30	Dprkorea	12.25	25.78	51.20
Brazil	11.31	23.17	52.80	Luxembourg	12.03	24.96	56.10
Burma	12.14	24.47	55.00	Malaysia	12.23	24.21	55.09
Canada	11.00	22.25	50.06	Mauritius	11.76	25.08	58.10
Chile	12.00	24.52	54.90	Mexico	11.89	23.62	53.76
China	11.95	24.41	54.97	Netherlands	11.25	22.81	52.38
Colombia	11.60	24.00	53.26	Nz	11.55	23.13	51.60
Cookis	12.90	27.10	60.40	Norway	11.58	23.31	53.12
Costa	11.96	24.60	58.25	Png	12.25	25.07	56.96
Czech	11.09	21.97	47.99	Philippines	11.76	23.54	54.60
Denmark	11.42	23.52	53.60	Poland	11.13	22.21	49.29
Domrep	11.79	24.05	56.05	Portugal	11.81	24.22	54.30
Finland	11.13	22.39	50.14	Rumania	11.44	23.46	51.20
France	11.15	22.59	51.73	Singapore	12.30	25.00	55.08
Gdr	10.81	21.71	48.16	Spain	11.80	23.98	53.59
Frg	11.01	22.39	49.75	Sweden	11.16	22.82	51.79
Gbni	11.00	22.13	50.46	Switzerland	11.45	23.31	53.11
Greece	11.79	24.08	54.93	Taipei	11.22	22.62	52.50
Guatemala	11.84	24.54	56.09	Thailand	11.75	24.46	55.80
Hungary	11.45	23.06	51.50	Turkey	11.98	24.44	56.45
India	11.95	24.28	53.60	USA	10.79	21.83	50.62
Indonesia	11.85	24.24	55.34	USSR	11.06	22.19	49.19
Ireland	11.43	23.51	53.24	Wsamoa	12.74	25.85	58.73
Israel	11.45	23.57	54.90				

Source: Dawkins, B. (1989), "Multivariate Analysis of National Track Records," *The American Statistician*, *43* (2), 110–115.

5.116 Blood pressure is a complex quantitative human trait that is affected by environmental as well as genetic factors. Red blood cell sodium-lithium countertransport (SLC) is correlated with blood pressure and is of interest to geneticists because it is easier to study than blood pressure. Notable is the general distribution shape of SLC and a measure of the center of the distribution. Following is the SLC activity measured on 190 individuals from six large English extended families.

WORKSHEET: SLC.mtw

0.467	0.430	0.192	0.192	0.293	0.160	0.164	0.126	0.328	0.202	0.282	0.328
0.247	0.132	0.138	0.224	0.512	0.221	0.252	0.193	0.263	0.186	0.346	0.219
0.177	0.349	0.272	0.245	0.213	0.197	0.229	0.245	0.210	0.281	0.175	0.273
0.439	0.471	0.451	0.237	0.313	0.136	0.245	0.391	0.349	0.158	0.252	0.416
0.232	0.183	0.254	0.195	0.141	0.151	0.073	0.300	0.231	0.075	0.208	0.267
0.187	0.244	0.245	0.231	0.167	0.337	0.251	0.209	0.181	0.411	0.191	0.288
0.280	0.119	0.394	0.443	0.423	0.534	0.393	0.273	0.149	0.225	0.159	0.170
0.329	0.183	0.262	0.250	0.179	0.329	0.253	0.270	0.310	0.321	0.333	0.284
0.380	0.222	0.178	0.265	0.289	0.199	0.309	0.279	0.194	0.203	0.139	0.162
0.251	0.619	0.343	0.155	0.340	0.332	0.412	0.218	0.304	0.261	0.206	0.231
0.182	0.267	0.198	0.191	0.258	0.179	0.197	0.188	0.202	0.150	0.201	0.255
0.293	0.255	0.189	0.414	0.292	0.253	0.168	0.295	0.215	0.213	0.267	0.216

continued

WORKSHEET: SLC.mtw
continued

0.264	0.138	0.239	0.288	0.311	0.414	0.462	0.361	0.623	0.199	0.215	0.321
0.273	0.259	0.206	0.376	0.228	0.155	0.186	0.097	0.179	0.174	0.386	0.393
0.198	0.243	0.326	0.250	0.590	0.461	0.361	0.321	0.236	0.139	0.316	0.313
0.263	0.180	0.184	0.354	0.264	0.269	0.171	0.359	0.338	0.163		

Source: Roeder, K., (1994), "A Graphical Technique for Determining the Number of Components in a Mixture of Normals," *Journal of the American Statistical Association*, 89, 487–495.

a. Construct a stem-and-leaf plot and a boxplot of the data. Based on these two graphs, how would you describe the shape of the distribution?
b. Construct a normal probability plot. Would you reject normality based on this plot? How would you describe the shape? Does it agree with your answer to part **a**?
c. Do you think that the mean or median best describes the center of the distribution? Explain your reasoning.
d. Give your estimate of the center.
e. How would you measure the variability in the distribution? What is your estimate?

5.117 The Census Bureau maintains housing data on 335 metropolitan areas across the United States. Here are the monthly rental costs in those metro areas with 1 million or more population:

WORKSHEET: Metrent.mtw

790 529 490 646 656 426 491 367 406 421 456 431 455 575 428 406 413 425 626 493 680 447 479 778 397 503 583 480 642 524 516 465 366 437 562 466 531 379 380 611 709 773 516 415 448 667

Source: U.S. Bureau of the Census, *Housing in Metropolitan Areas, Statistical Brief* SB/94/19, September 1994.

a. Construct a boxplot of the data. Does the data appear to violate the normality assumption?
b. To construct a confidence interval for the mean rental cost, is the normality assumption required?
c. Construct a 99% confidence interval for the mean rental cost.

5.118 Using the **Random Data** command with the **Normal** distribution, randomly generate 20 samples of 50 observations each from a normal population with mean 70 and standard deviation 12. Construct 20 separate 90% confidence intervals for μ using $\sigma = 12$ and the **Zinterval** command.
a. How many of the 20 intervals contain μ, which is specified to be 70?
b. How many intervals would you expect to contain μ?
c. If the confidence level had been 99% instead of 90%, would the interval be longer or shorter?
d. Would you expect more of the intervals to contain μ if the confidence level is 99%?

5.119 Into the Minitab worksheet, recall worksheet: **Fish.mtw** which contains the length and number of fish caught with a small-mesh and a large-mesh codend (the base of a fishing net). From the **Graph** menu, select the **Plot** command. Using the length of the fish as the X variable and the frequency count as the Y variable, graph the two distributions. Under **Display**, select **Connect** so that the points in the graph are connected. What are the general shapes of the two distributions? To estimate the centers of the two distributions with confidence intervals, on what statistic should we base the interval? Are confidence intervals for the medians (see Exercise 5.83 in Section 5.5) of the two distributions appropriate, or would you recommend confidence intervals for the means of the two distributions? Calculate individual 99% confidence intervals for the two means and compare the results to the results found in Exercise 5.83 in Section 5.5.

5.120 Worksheet **Archaeo.mtw** contains radiocarbon dates of pottery samples taken from an archaeological excavation of the Danebury Iron Age hill fort (see Example 1.16 in Section 1.5). The data are generally accepted by archaeologists as being classified into one of four phases, which are called Ceramic Phases 1–4. Verify that all four samples are from populations that are reasonably close to being normally distributed. Construct individual 95% t-intervals from the four samples and graph them on a common line. Does it appear from the four confidence intervals that the phases correspond to four nonoverlapping periods of time?

6

Hypothesis Testing

Many of our statistical questions can be answered with estimation procedures; however, others may require verification with hypothesis testing procedures.

Hypothesis testing is a means by which statistical decisions are made. For example, a new medication will be marketed if it is more than 70% effective. This statement can be formulated in a hypothesis and evaluated with data from a random sample. If the data support the hypothesis, the decision will be made to market the product. In this chapter, we will investigate hypothesis-testing problems involving a single population. In Chapter 7, we will investigate those situations that involve two populations. We address several different statistical problems that can be solved with a test of hypothesis.

Contents

- **STATISTICAL INSIGHT**
6.1 Introduction to Hypothesis Testing
6.2 Testing a Population Proportion
6.3 Testing a Population Mean
6.4 Testing a Population Median
6.5 Summary and Review Exercises

EPA Mileage Charts

How reliable is the Environmental Protection Agency (EPA) mileage chart? Can someone who buys a Honda Civic expect to get 40 miles per gallon in highway driving, as suggested by the EPA's annual mileage ratings?

One obvious way to answer this question is to drive the car and check the mileage. Just as the EPA cautions, however, a motorist's actual mileage can vary, depending on such factors as the amount of traffic, the weather, and car maintenance. Therefore, one should check the mileage on several different occasions under a variety of conditions.

Suppose that the mileage of a Honda Civic was checked on 35 different occasions with these miles per gallon recorded:

WORKSHEET: **Honda.mtw**

39.3 40.8 30.6 34.5 37.8 40.4 43.6 36.3 40.7 36.3
31.7 36.6 39.2 41.3 39.2 31.9 34.5 45.2 41.0 42.2
41.8 36.5 40.2 31.5 36.3 41.2 35.8 38.3 39.7 40.2
38.4 42.4 40.6 41.3 39.5

Is there evidence in these data to suggest that the average mileage for the Honda Civic is less than 40 miles per gallon? In Section 6.5 we will answer this question.

6.1 Introduction to Hypothesis Testing

Learning Objectives for this section:

❏ Know how to formulate null and alternate hypotheses.

❏ Be able to determine which test statistic to use in a testing procedure.

❏ Understand the concept of and the calculation of the p-value.

❏ Know what the level of significance is and the criteria for rejection of the null hypothesis.

❏ Be able to interpret Type I and Type II errors.

❏ Be able to test a population mean with a z-test.

Hypothesis testing is an area of statistical inference in which we evaluate a conjecture (which we will call a *hypothesis*) about some characteristic of the parent population. Usually the hypothesis concerns an unknown parameter of the population. Consider, for example, the hypothesis that the mean farm income of all farmers in a certain rural county is $15,000 per year. This hypothesis is concerned with the population mean and can be written as $\mu = 15{,}000$, where μ denotes the true mean farm income of all farmers in the rural county.

> The statement being tested in a test of hypothesis is called the **null hypothesis** and is denoted as H_0. Because it usually is stated as an equality ($\mu = 15{,}000$), the null hypothesis is referred to as the hypothesis of "no difference," meaning that the difference between the parameter and the value being tested is zero; that is, $\mu - 15{,}000 = 0$.

In a test of the null hypothesis, data are collected and analyzed to assess the strength of evidence against the null hypothesis. If the evidence casts doubt about the truth of the null hypothesis, it will be rejected in favor of the *alternative* hypothesis.

The **alternative hypothesis**, denoted as H_a, is what is believed to be true if the null hypothesis is false. Usually the person conducting the research wishes to establish that there is a difference between the parameter and the value being tested, and thus the alternative is also called the **research hypothesis**.

Suppose, in the preceding example, that the researcher believes that the mean farm income of the county farmers is greater than $15,000. This is formulated as $\mu > 15{,}000$, and because this is what the researcher wishes to support, it is a statement of the alternative hypothesis. Thus, the null hypothesis is

$$H_0: \mu = 15{,}000 \text{ and the alternative hypothesis is } H_a: \mu > 15{,}000$$

Example 6.1   A conjecture is made that the mean starting salary for computer science graduates is $45,000 per year. You believe it is less than $45,000. Formulate the null and alternate hypotheses to evaluate the claim.

Solution   The parameter about which the conjecture is made is the *mean* starting salary, μ. The statement that "you believe it is less than $45,000" is formulated as $\mu < 45{,}000$. Because this is the statement you wish to support, it is placed in the alternative (research) hypothesis. Consequently, we have

$$H_0: \mu = 45{,}000 \quad \text{versus} \quad H_a: \mu < 45{,}000 \qquad \blacksquare$$

Example 6.2   The standard medication for a certain disease is effective in 60% of all cases. A pharmaceutical company believes that its new drug is more effective than the old treatment. Formulate the null and alternative hypotheses to test whether there is statistical evidence to support the new drug.

Solution   Let π denote the proportion of cases for which the new drug is effective. The pharmaceutical company feels that the new drug is more effective than the standard medication; therefore, the research hypothesis will be that $\pi > .60$. Thus,

$$H_0: \pi = .60 \quad \text{versus} \quad H_a: \pi > .60 \qquad \blacksquare$$

The alternative hypotheses in the preceding examples are *one-sided alternatives* because the researcher is interested in deviations from the null hypothesis only in a specified direction. If deviations from the null hypothesis in either direction are of concern, it is called a *two-sided alternative*.

Example 6.3   Researcher Alfredo Morabia of University Hospital in Geneva, Switzerland, was interested in comparing the proportion of smokers among women with breast cancer and women without breast cancer. Morabia's report, in the May 1996 issue of the *American Journal of Epidemiology*, is based on a study of 244 women with breast cancer and 1032 women without the disease. Set up null and alternative hypotheses to evaluate the claim that there is no difference between the two proportions.

Solution   Let π_1 represent the proportion who smoke among women with breast cancer, and let π_2 represent the proportion who smoke among women without breast cancer. The null hypothesis says that there is no difference between the two proportions, and the alternative hypothesis says that there is a difference (π_1 could be greater than or less than π_2). Thus, we have

$$H_0: \pi_1 = \pi_2 \quad \text{versus} \quad H_a: \pi_1 \neq \pi_2 \text{ (This is a two-sided alternative.)} \qquad \blacksquare$$

The Test Statistic

Having formulated the null and alternate hypotheses, we next develop the test procedure to assess the evidence against H_0. If the evidence is attributable to factors other than chance, then H_0 will be rejected in favor of the alternative, H_a.

An important step in developing the test procedure is to determine an appropriate *test statistic*—that is, a statistic from the sample that seems appropriate for testing the null hypothesis.

> A **test statistic** is a statistic, calculated from the sample data, which is used to test the null hypothesis.

The choice of test statistic depends on the parameter being tested and the underlying parent distribution. Just as in confidence interval estimation, one must use exploratory data analysis techniques to describe the underlying distribution so that an appropriate test statistic can be chosen. In Example 5.8 you learned that if the underlying parent distribution is near normal, the statistic $\bar{x}$ is the best estimator of μ. Similarly, under the normality assumption, using $\bar{x}$ as a test statistic results in the best test procedure for testing the population mean μ. If the parent distribution is not normal, however, $\bar{x}$ may not be the best test statistic for testing μ. For example, if the parent distribution is symmetrical with long tails, a better test statistic is the sample median or one of the trimmed sample means. Generally speaking, the test statistic we will use is the estimator of the parameter that was found in Chapter 5.

Let us return to the example of the mean farm income of the farmers in the rural county. As stated, the parameter of concern is the population mean, and we wish to test

$$H_0: \mu = 15{,}000 \quad \text{versus} \quad H_a: \mu > 15{,}000$$

Assuming that the parent distribution is nearly normal, we will use $\bar{x}$ as the test statistic. Suppose a random sample of 36 farmers yielded an average farm income of $\bar{x} = \$20{,}000$. We recall that sample means vary from the population mean, but if $\mu = 15{,}000$, is it reasonable to observe an $\bar{x}$ this large? Because of the large difference between $\bar{x}$ and μ in this case, we have statistical evidence to reject the null hypothesis. If, on the other hand, the random sample gave an average of $\bar{x} = \$15{,}100$, we could not say for sure that $\mu = 15{,}000$, but it is doubtful that there is sufficient evidence to reject it. Thus, an important element in testing this hypothesis is the difference between the observed value of the test statistic $\bar{x}$ and the hypothesized value of μ.

The Sampling Distribution of the Test Statistic

The strategy we will use to evaluate the difference is to first determine the sampling distribution of the test statistic under the assumption that H_0 is true. With the sampling distribution, it is then possible to determine what values of the test statistic are unlikely if the null hypothesis is true. Figure 6.1 illustrates the sampling distribution of $\bar{x}$ when it is assumed that $\mu = 15{,}000$. Notice that $\bar{x}_{\text{obs}} = \$15{,}100$ is not that unusual and thus would not lead to the rejection of H_0. (We use $\bar{x}_{\text{obs}}$ to denote the value of $\bar{x}$ that was observed from the random sample.)

Figure 6.1
Sampling distribution
of $\bar{x}$ when $\mu = 15{,}000$

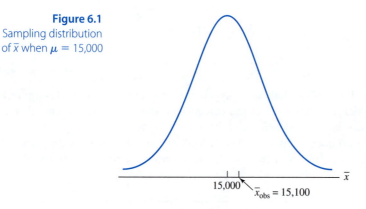

Recall that when the parent distribution is near normal, or the sample size is large (Central Limit Theorem), the sampling distribution of $\bar{x}$ is approximately normally distributed with a mean the same as the mean of the parent population and a standard deviation of $\sigma/\sqrt{n}$. Assuming H_0 is true, we have that the mean of the parent population is 15,000, and thus the sampling distribution is drawn as a normal curve centered at 15,000.

Because the alternative hypothesis is $\mu > 15{,}000$, we should not reject the null hypothesis for any value of $\bar{x}$ that is smaller than 15,000. In fact, we should reject H_0 only for those values of $\bar{x}$ that are *considerably* greater than 15,000. This test with a one-sided alternative is called a *right-tailed test* because the region for rejection of H_0 is on the right tail of the sampling distribution.

The *p*-Value

To continue the example, suppose the researcher selects a random sample of 36 county farmers and finds that their average farm income is $\bar{x}_{\text{obs}} = \$16{,}200$. Is this convincing evidence to reject H_0 and conclude that $\mu > 15{,}000$? To assess the strength of evidence against the null hypothesis, we will calculate the probability of observing a value of the test statistic *at least as extreme as* the value actually observed from the sample. This probability, called the *p-value*, is computed under the assumption that the null hypothesis is true. Thus, a large probability will support the truth of the null hypothesis, and a small probability provides evidence that the null hypothesis is false and should be rejected.

> The **p-value** is the probability (computed assuming H_0 is true) of observing a value of the test statistic at least as extreme as that given by the actual observed data. The smaller the *p*-value, the stronger is the evidence against the null hypothesis.

Figure 6.2 illustrates the *p*-value, which is given by

$$p\text{-value} = P(\bar{x} \geq \bar{x}_{\text{obs}}) = P(\bar{x} \geq 16{,}200)$$

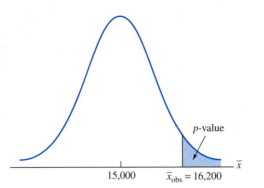

Figure 6.2
Sampling distribution
of $\bar{x}$ and
$p\text{-value} = P(\bar{x} \geq \bar{x}_{obs})$

Knowing that the sampling distribution of $\bar{x}$ is near normal with mean 15,000 and standard deviation $\sigma/\sqrt{n}$, we can find the p-value in the normal probability table by determining the distance between $\bar{x}_{obs} = 16,200$ and 15,000 with a z-score.

The z-Test

The z-score,

$$z = \frac{\bar{x} - 15,000}{\sigma/\sqrt{n}}$$

is called the *standardized test statistic*, and the p-value is given by

$$p\text{-value} = P(\bar{x} \geq \bar{x}_{obs}) = P(z \geq z_{obs})$$

Calculating the observed z-score for this *z-test* requires that we know the population standard deviation, σ. In most applications, σ is not known and must be estimated from the sample data. We will address this issue in Section 6.3, but for the time being, suppose we know that $\sigma = \$3000$. Substituting values, we have

$$z_{obs} = \frac{\bar{x}_{obs} - 15,000}{\sigma/\sqrt{n}} = \frac{16,200 - 15,000}{3000/\sqrt{36}} = 2.4$$

Already the evidence suggests that the null hypothesis should be rejected. A z-score of 2.4 means that 16,200 is 2.4 standard deviations above 15,000, which is highly unlikely. In fact, looking up a z-score of 2.4 in the normal distribution table, we find that .9918 of the area lies *below* 2.4, so

$$p\text{-value} = 1.0 - .9918 = .0082$$

which is a very small probability. If $\mu = 15,000$, it is highly unlikely that one would find $\bar{x}_{obs} = 16,200$. But the researcher did find $\bar{x}_{obs} = 16,200$, so we are left to conclude that $\mu > 15,000$. The farther $\bar{x}_{obs}$ is from 15,000, the larger will be z_{obs} and consequently, the smaller will be the p-value. Thus, the smaller the p-value, the stronger is the evidence that H_0 should be rejected.

Level of Significance

Just how small should a p-value be before we reject H_0? Prior to collecting data, experimenters will often establish a preset criterion that must be achieved in order to reject H_0. A probability, called *the level of significance* and denoted by the Greek letter α, determines how small the *p*-value should be before we reject the null hypothesis. If the *p*-value is smaller than α, as shown in Figure 6.3, $\bar{x}_{obs}$ is far enough removed from the hypothesized value of the parameter ($\mu = 15,000$) that we reject the null hypothesis. In practice, the level of significance is set at a small probability such as $\alpha = .05$. Then if the *p*-value $< .05$ we say, "The results are *statistically significant* (P $< .05$)." If the *p*-value $> .05$ we say, "The results are *insignificant*." In some applications, the criterion for statistical significance may be as lenient as $\alpha = .10$; in others, it may be set at a more stringent $\alpha = .01$.

Figure 6.3
Sampling distribution of
$\bar{x}$ with *p*-value $< \alpha$

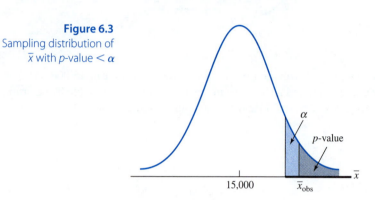

Setting a specified value for α prior to testing makes sense if the situation is such that we must make a clear-cut decision about the hypotheses. For example, a pharmaceutical company may decide to market a new drug if its test results are statistically significant at the $\alpha = .01$ level. Or, in the area of *acceptance sampling*, a manufacturer of log homes accepts logs from a milling company if the mean moisture content of the logs is less than a specified amount. The manufacturer decides to reject the logs if the test results of the moisture content are statistically significant at the $\alpha = .05$ level.

If, on the other hand, we are testing simply to assess the strength of the statistical evidence against the null hypothesis, then reporting the *p*-value will suffice. In this context, the evaluation of a hypothesis is referred to as a *test of significance*.

Interpreting the Results

In most situations, the following criterion may be used for rejecting the null hypothesis.

Criterion for Rejection of H_0

1. If *p*-value $> .10$, fail to reject H_0 and declare the results insignificant.
2. If $.05 <$ *p*-value $\leq .10$, you may reject H_0, but the results are only mildly significant.
3. If $.01 <$ *p*-value $\leq .05$, reject H_0 and declare the results significant.
4. If *p*-value $\leq .01$, reject H_0 and declare the results highly significant.

Because the *p*-value $= .0082$ in the farm income example we conclude by saying, "On the basis of a random sample of 36 county farmers, there is *highly significant* evidence that the mean farm income of all farmers in the county is greater than \$15,000."

Testing hypotheses, then, is a problem of deciding between the null and the alternative hypotheses based on the information contained in a random sample. Because the alternative is the hypothesis that the researcher believes to be true, the goal will be to reject H_0 in favor of H_a. To summarize, there are five basic steps in the hypothesis-testing procedure:

Procedure Steps for Testing Hypotheses

1. Formulate the null and alternate hypotheses.
2. Decide on an appropriate test statistic.
3. Determine the sampling distribution of the test statistic under the assumption that the null hypothesis is true.
4. Calculate the p-value and determine whether it is sufficiently small to reject the null hypothesis.
5. Interpret the results in a way that a nonstatistician could understand.

Example 6.4 The scores on a college placement exam in mathematics are assumed to be normally distributed with a mean of 70 and a standard deviation of 18. The exam is given to a random sample of 50 high school seniors who have been admitted to college. Their average on the exam was 67. If this is a true random sample, is the evidence sufficient to suggest that the population mean score is lower than 70?

Solution

STEP 1

Let μ denote the true population mean of the placement exam. We then wish to see whether there is evidence that $\mu < 70$. This is the research (alternative) hypothesis. The null and alternative hypotheses are

$$H_0: \mu = 70 \quad \text{versus} \quad H_a: \mu < 70$$

STEP 2

Because the parent distribution is assumed to be normal, the best test statistic is the sample mean $\bar{x}$.

STEP 3

Given that we know the numerical value of σ and we assume that $\mu = 70$, the standardized form of the test statistic

$$z = \frac{\bar{x} - 70}{\sigma/\sqrt{n}}$$

has a standard normal distribution.

STEP 4

Because the alternative hypothesis is $\mu < 70$, the p-value is calculated on the left tail of the sampling distribution of $\bar{x}$, as illustrated in Figure 6.4. This test with a one-sided alternative is called a *left-tailed test*.

Figure 6.4
A left-tailed test

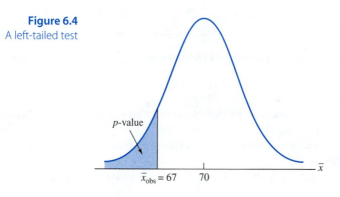

The p-value is the probability of observing a value of $\bar{x}$ less than or equal to $\bar{x}_{obs}$ when the true population mean is assumed to be 70; that is, p-value $= P(\bar{x} \le 67)$. Substituting in values, we have

$$z_{obs} = \frac{67 - 70}{18/\sqrt{50}} = -1.18$$

and thus,

$$p\text{-value} = P(\bar{x} \le 67) = P(z \le -1.18) = .1190$$

A p-value this large exceeds any reasonable level of significance α, and thus we fail to reject H_0.

STEP 5

Conclusion: Based on the results of a random sample of 50 high school seniors, there is insufficient evidence to say that the mean score on the college placement exam is lower than 70. ∎

Decision Making and Testing Errors

(Optional) If we are testing a hypothesis for the purpose of decision making, the level of significance α establishes the criterion for rejection. As shown in Figure 6.5, α identifies a *rejection region* for the test statistic.

> The **rejection region** consists of those values of the test statistic that will lead to the rejection of the null hypothesis. The **size** of the rejection region is the level of significance α.

Figure 6.5
Sampling distribution of $\bar{x}$ with rejection region and level of significance

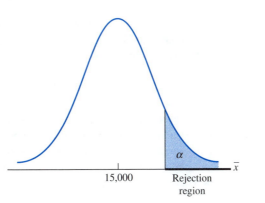

15,000   Rejection
                region

If $\bar{x}_{obs}$ falls in the rejection region, the decision is to reject the null hypothesis at the α level of significance. Because sample results vary from sample to sample, however, the evidence provided by a particular sample can result in an incorrect decision. To understand this, we draw an analogy between hypothesis testing and a trial by jury. The null hypothesis says, "The defendant is not guilty," and the jury must decide whether the presented evidence (observed sample) is convincing enough to warrant conviction (or rejection) of the defendant. Even though we hope that the jury reaches the correct decision in all cases, we are aware that an innocent person can be convicted or a guilty person freed. Similarly, two errors are possible when testing hypotheses. The first type error occurs when a null hypothesis is rejected (convicted) when, in fact, it is true (innocent). The second type error occurs when the null hypothesis is not rejected (not convicted) when, in fact, it is false (guilty).

> A **Type I error** is rejecting a null hypothesis that is true. The probability of committing a Type I error is the level of significance α.
> A **Type II error** is failing to reject a null hypothesis that is false. The probability of committing a Type II error is denoted as β.

Table 6.1 illustrates four possibilities, of which two are correct decisions and two are incorrect decisions. If H_0 is true, either our decision is correct or we make a Type I error. If H_0 is false, either our decision is correct or we make a Type II error. Because it is impossible for H_0 to be both true and false, we risk at most one of the errors.

Table 6.1

		Null Hypothesis	
		True	False
Decision	Reject H_0	Type I error	Correct decision
	Fail to Reject H_0	Correct decision	Type II error

An important concern here is the probabilities of the two types of errors. Recall that the size of the rejection region of a test is the level of significance α. As illustrated in Figure 6.5, α is the probability that the test statistic falls in the rejection region when the null hypothesis is true. In other words, α is the probability of a Type I error. Because we wish not to make this error, we choose a small value, such as $\alpha = .05$, for the level of significance. This means that the test procedure, if used repeatedly with different samples, will wrongly reject the null hypothesis about 5 times out of 100. Why don't we make α extremely small—say, .00001—

Rarely would we set the level of significance at a value greater than .10 because seldom do we want the chance of a Type I error to exceed 10%.

by reducing the size of the rejection region? The result would be a very small rejection region, and consequently we would accept H_0 more frequently, thus increasing our chances of making a Type II error. That is, decreasing α automatically increases β, the probability of a Type II error. Also, increasing α automatically decreases β. Because of this relationship between α and β, it is not possible to make both values arbitrarily small. Thus, in decision making we fix α at a reasonable level of significance and then attempt to minimize β by increasing the sample size. The level set for α depends on the seriousness of the two types of errors. If making a Type I error results in a serious consequence, α should be set at a low level. If making a Type II error results in a more serious consequence, α should be set at a high level. Thus, we can, in effect, control both probabilities by setting α at a specified level.

Suppose the decision is whether we should accept or reject a parachute for a sky dive. The hypotheses are

$$H_0: \text{The parachute will open} \quad \text{versus} \quad H_a: \text{The parachute will not open}$$

A Type I error is committed if we reject H_0 when it is true; that is, we say the parachute will not open when in fact it will. On the other hand, a Type II error is committed if we accept H_0 when it is false; that is, we say the parachute will open when in fact it will not. Clearly, in this case, we do not want to make a Type II error! We want the chance of the Type II error to be extremely small, so we would set α at a high level. In other situations, the opposite might be true, and the Type I error is the more serious error.

Example 6.5   A manufacturer of log homes accepts logs from a milling company if the mean moisture content of the logs does not exceed 10% on her moisture reading equipment. The manufacturer randomly selects a sample of the logs, examines them, and rejects the shipment if the test results do not meet her standard. Formulate the null and alternative hypotheses to determine whether the manufacturer should reject the shipment. Interpret Type I and Type II errors and determine which is more serious.

Solution   Let μ denote the mean moisture content of the shipment. The null hypothesis says the shipment meets the standard established by the manufacturer, and the alternative says the shipment does not meet the standard. Therefore,

$$H_0: \mu = 10 \quad \text{versus} \quad H_a: \mu > 10$$

Rejecting the null hypothesis amounts to rejecting the shipment. Rejecting it when it is true, or committing a Type I error, means rejecting the shipment when the moisture content is acceptable. A Type II error accepts the shipment when the moisture content does not meet the standard. Which is more serious: rejecting a good shipment of logs or accepting a bad shipment of logs? The answer depends on whether you are the manufacturer or the milling company that supplies the logs. To the manufacturer, accepting a bad shipment of logs (a Type II error) is clearly more serious. To the supplier of the logs, however, a Type I error means that a shipment of good logs is rejected. ∎

EXERCISES 6.1   The Basics

6.1   In each case, formulate the null and alternative hypotheses involving the population mean, μ.
a. You wish to show that μ exceeds 50.
b. You wish to show that μ is less than 50.
c. You wish to find evidence against the claim that μ is 50 when you believe it is greater than 50.
d. You wish to find evidence against the claim that μ is 50 when you believe it is less than 50.
e. You wish to test the claim that μ is 50.

6.2   In each case, formulate the null and alternative hypotheses involving the Bernoulli proportion, π.
   a. You wish to show that π exceeds 70%.
   b. You wish to show that π is less than 70%.
   c. You wish to find evidence against the claim that π is 70% when you believe it is greater than 70%.
   d. You wish to find evidence against the claim that π is 70% when you believe it is less than 70%.
   e. You wish to test the claim that π is 70%.

6.3   A claim is made that 60% of the adult population feels that there is too much violence on television. Assuming you feel this figure is too high, set up the null and alternative hypotheses to evaluate the claim.

6.4   It has been asserted that by the year 2010, 20% of all American adults will be drawing retirement benefits. If you believe that the percent is greater than 20%, which is the correct way to formulate the hypotheses?
   a. $H_o: p = .20$ versus $H_a: p > .20$
   b. $H_o: p \geq .20$ versus $H_a: p < .20$
   c. $H_o: \pi = .20$ versus $H_a: \pi > .20$
   d. $H_o: \pi \geq .20$ versus $H_a: \pi < .20$

6.5   Explain each of the following:
   a. If the null hypothesis is rejected, does this mean that the alternative is true?
   b. If the null hypothesis is not rejected, does this mean that it is true?
   c. If the null hypothesis is rejected at the 1% level of significance, will it also be rejected at the 5% level?

6.6   Suppose a testing procedure leads to the rejection of the null hypothesis. Is it possible that a Type I error was committed? Is it possible that a Type II error was committed? Explain.

6.7   Does $\alpha + \beta = 1$? Why, or why not?

6.8   During a group study session, one student says, "I know what a p-value is. It is the probability that the null hypothesis is true." A second student says, "No, it is the probability that the null hypothesis is rejected." Discuss these two interpretations of the p-value.

6.9   In each of the following, interpret Type I and Type II errors and decide which is more serious.
   a. The null hypothesis says, "The patient will recover."
   b. The null hypothesis says, " The fire is out."

6.10   In the oil industry, a wildcat is an experimental oil well that is drilled in an area that is not known to be successful. The wildcat success rate is about 10% in the United States. For this reason, the oil industry considers the consequences of failing to drill in locations where there is oil to be more serious than drilling dry holes. Suppose the null hypothesis states that the wildcat will not produce oil and the alternative states that the wildcat will be productive. Interpret the Type I and Type II errors. Which would oil-industry executives consider more serious?

6.11   Suppose a geologist believes that the mean porosity measurement of Tensleep Sandstone from the Bighorn Basin, Wyoming, exceeds 18%. To find statistical evidence to support his theory, he computes the porosity percents on ten core samples of Tensleep Sandstone. State the null and alternative hypotheses required to evaluate his theory. Explain the situation if the geologist makes a Type II error.

6.12   Draw a normal curve and shade in the p-value area as in Figure 6.2 for these test-of-hypothesis problems. Calculate the p-values.
   a. $H_0: \mu = 15$
       $H_a: \mu < 15$ $n = 48$ $\bar{x} = 14.2$ $\sigma = 4.1$
   b. $H_0: \mu = 120$
       $H_a: \mu > 120$ $n = 100$ $\bar{x} = 124.6$ $\sigma = 16.3$
   c. $H_0: \mu = 650$
       $H_a: \mu > 650$ $n = 250$ $\bar{x} = 694$ $\sigma = 235.7$

6.13   An instructor gives her class an examination that, as she knows from years of experience, yields $\mu = 78$ and $\sigma = 7$. Her present class of 35 students scores an average of 81.4. Is she correct in assuming that this is a superior class? Is it necessary that the parent distribution be approximately normally distributed?

Interpreting Computer Output

6.14   Here is the computer printout of a test of hypothesis about the mean age of food stamp recipients.

z-Test

```
Test of mu = 40.000 vs mu not = 40.000
The assumed sigma = 5.00
```

Variable	N	Mean	StDev	SE Mean	Z	P-Value
C1	37	38.378	4.609	0.822	−1.97	0.049

a. State the null and alternative hypotheses.
b. Was the population standard deviation given or was it estimated?
c. How large was the sample?
d. What was the average of the sample? Was it significantly different from the hypothesized mean age?
e. What was the p-value? Should the null hypothesis be rejected?
f. State your conclusion.

6.15   Suppose the mean entrance exam score for incoming freshmen at a large university is 550 and the standard deviation is 120. A sample of 90 students from this year's freshmen class had an average of 582.6. Is this unusual? Use the computer printout to perform a test of significance.

z-Test

```
Test of mu = 550.0 vs mu > 550.0
The assumed sigma = 120
```

Variable	N	Mean	StDev	SE Mean	Z	P-Value
C1	90	582.6	112.3	12.6	2.57	0.0051

a. State the null and alternative hypotheses.
b. Was the population standard deviation given or was it estimated?
c. What was the difference between the sample mean and the hypothesized value of the population mean? Was the difference significant?
e. What was the p-value? Should the null hypothesis be rejected?
f. State your conclusion.

6.2   Testing a Population Proportion

Learning Objectives for this section:

☐ Be able to test a population proportion with a large sample.

☐ Be able to identify a right-tailed, a left-tailed, and a two-tailed alternative.

☐ Understand the equivalence of confidence interval estimation and the two-tailed test.

Often we wish to evaluate the proportion of successes in a Bernoulli population. Recall that a Bernoulli population is one in which each outcome is classified as either a success or failure, and we are concerned with π, the proportion of successes. In this section, conjectures about π are investigated with a test of hypothesis. For example, is it true that 69% of all American drivers wear seat belts? Is it true that no more than 40% of school-age children receive regular dental care? These conjectures can be evaluated with a test of hypothesis about the Bernoulli proportion π.

Let us evaluate the conjecture made by the pharmaceutical company in Example 6.2. We wish to test

$$H_0: \pi = .60 \quad \text{versus} \quad H_a: \pi > .60$$

As outlined in Section 6.1, after formulating the null and alternative hypotheses, we must determine an appropriate test statistic for testing the hypothesis. As in confidence interval estimation, the best estimator of π is the sample proportion p; hence, p is the test statistic.

Because of the alternative hypothesis, this is a right-tailed test, and consequently we reject H_0 when the observed value of p is considerably greater than .60. Figure 6.6 illustrates the sampling distribution of the test statistic p.

Figure 6.6
Sampling distribution
of p

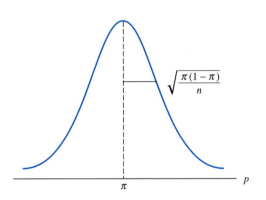

The graph is drawn as a normal curve because, with a large sample, the Central Limit Theorem guarantees that the sampling distribution of p is approximately normal. Furthermore, the mean and standard deviation of the sampling distribution are π and $\sqrt{\pi(1 - \pi)/n}$, respectively.

Suppose that the drug company investigates 200 cases and finds that the new drug is effective in 134 cases ($p_{obs} = 134/200 = .67$). Is this enough evidence to reject H_0 and say that $\pi > .60$?

To assess the evidence against H_0, we calculate the p-value, which is illustrated in Figure 6.7 and given by

$$p\text{-value} = P(p \geq p_{obs}) = P(p \geq .67)$$

Figure 6.7
Sampling distribution
of p and p-value $=$
$P(p \geq .67)$

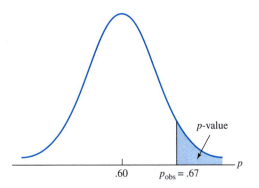

Because the sampling distribution of p is approximately normal with mean π and standard deviation $\sqrt{\pi(1 - \pi)/n}$, the standardized test statistic is

$$z = \frac{p - \pi}{\sqrt{\pi(1 - \pi)/n}}$$

and the p-value is given by

$$p\text{-value} = P(p \geq p_{obs}) = P(z \geq z_{obs})$$

Under the assumption that H_0 is true ($\pi = .60$) we have

$$z_{obs} = \frac{p_{obs} - .60}{\sqrt{(.6)(.4)/200}} = \frac{.67 - .60}{.0346} = 2.02$$

Looking up a z-score of 2.02 in the normal probability table, we find that .9783 of the area lies below $+2.02$, so the p-value $= 1.0 - .9783 = .0217$.

Thus, we reject H_0 at any level of significance greater than .0217. This is strong evidence that the null hypothesis is false. In conclusion, based on a random sample of size 200, there is statistical evidence that the new drug is more effective than the standard medication.

Testing a More General H_0

Suppose that in the previous example the hypotheses had been stated as

$$H_0: \pi \leq .60 \quad \text{versus} \quad H_a: \pi > .60$$

instead of

$$H_0: \pi = .60 \quad \text{versus} \quad H_a: \pi > .60$$

Recall that the p-value is calculated under the assumption that the null hypothesis is true. When H_0 states that $\pi = .60$, the procedure is straightforward because we can use the specified value of $\pi = .60$ in the z-score and proceed to find the p-value. If H_0 states that $\pi \leq .60$, however, which value of π do we use to calculate the z-score? Should we, perhaps, choose .50 in the calculation of the z-score, or just any value of π that is less than .60? In Figure 6.8 we see two sampling distributions of p: one assuming that $\pi = .60$ and the other assuming that π is some value $m < .60$ (the dashed curve). Because the p-value is the tail probability beyond the observed test statistic, p_{obs}, it is clear that the p-value is greatest when we assume that $\pi = .60$.

Figure 6.8
The p-value assuming that $\pi = .60$ or a value $m < .60$

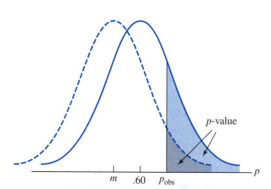

Thus, when testing the more general null hypothesis

$$H_0: \pi \leq \pi_0$$

we will compute the *p*-value under the assumption

$$H_0: \pi = \pi_0$$

so as to obtain its maximum value. In this sense, testing

$$H_0: \pi \le \pi_0 \quad \text{versus} \quad H_a: \pi > \pi_0$$

The important inequality is in the statement of the alternative hypothesis H_a because that determines the direction of the test.

is equivalent to testing

$$H_0: \pi = \pi_0 \quad \text{versus} \quad H_a: \pi > \pi_0$$

Unless stated otherwise, we will give the null hypothesis in the more general form.

Two-Tailed Test

To this point we have only considered tests with one-sided alternatives. We now look at the *two-tailed test*. For the Bernoulli proportion π, the hypotheses are

$$H_0: \pi = \pi_0 \quad \text{versus} \quad H_a: \pi \ne \pi_0$$

In light of the two-sided alternative, we could conceivably reject the null hypothesis in favor of the alternative if p_{obs} is either significantly greater than π_0 or significantly less than π_0. Clearly, p_{obs} cannot fall on both tails of the sampling distribution, so let us suppose that it falls on the upper tail. If this were a one-sided alternative, the *p*-value would be the area under the curve on the upper tail, as in a right-tailed test. But because this is a two-tailed test, we must account for both tails by doubling the probability. If

$$p_{\text{obs}} > \pi_0$$

we calculate the *p*-value as in a right-tailed test and then double it. In a similar fashion, if

$$p_{\text{obs}} < \pi_0$$

we calculate the *p*-value as in a left-tailed test and then double it.

Figure 6.9 illustrates the calculation of the *p*-value for the two-tailed test.

Figure 6.9
The two-tailed test

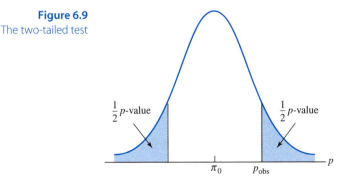

Example 6.6   A recent report claimed that 20% of all college graduates find a job in their chosen field of study. A survey of a random sample of 500 graduates found that 110 obtained work in their field. Is there statistical evidence to refute the claim?

Solution   If we let π denote the percent of college graduates who find a job in their field of study, the null and alternative hypotheses are

$$H_0: \pi = .20 \quad \text{versus} \quad H_a: \pi \neq .20$$

The test statistic is p, the proportion of successes in the sample. The observed value of p is found to be

$$p_{obs} = \frac{110}{500} = .22$$

The observed value of the standardized test statistic is

$$z_{obs} = \frac{p_{obs} - .20}{\sqrt{(.2)(.8)/500}} = \frac{.22 - .20}{.018} = 1.11$$

If this were a one-sided alternative, the p-value would be

$$P(z \geq 1.11) = 1.0 - .8665 = .1335$$

But because this is a two-sided alternative, we have

$$p\text{-value} = 2(.1335) = .267$$

Clearly, this is insignificant evidence to reject the null hypothesis. On the basis of a random sample of 500 college graduates, there is no evidence to refute the claim that 20% find work in their field of study.   ■

Testing with Confidence Intervals

In the notation for confidence intervals and tests of hypotheses, it is no coincidence that we use the same Greek letter α in a $(1 - \alpha)100\%$ confidence interval and for the level of significance in a test of hypothesis. In fact, it is possible to carry out a two-tailed test of hypothesis by constructing a confidence interval for the unknown parameter.

Equivalence of Confidence Intervals and Two-Tailed Tests

The null hypothesis $H_0: \pi = \pi_0$ versus the alternative $H_a: \pi \neq \pi_0$ is rejected at an α level of significance if and only if the hypothesized value π_0 falls outside a $(1 - \alpha)100\%$ confidence interval for π.

Example 6.7   A news report in a major eastern city stated that 80% of all violent crimes in that city involve firearms. A survey of all violent crimes in the city for the past two years revealed that of 283 violent crimes, 240 involved firearms. Determine with a confidence interval whether the news report is correct.

Solution   The claim made by the news report is that the percent is 80; therefore, we set up the hypothesis as a two-tailed test.

$$H_0: \pi = .80 \quad \text{versus} \quad H_a: \pi \neq .80$$

Instead of finding the p-value, however, we will find a 95% confidence interval for π. Recall that the form of a 95% confidence interval for π is

$$p \pm 1.96 \sqrt{\frac{p(1-p)}{n}}$$

The sample reveals that

$$p_{obs} = \frac{240}{283} = 0.848$$

so the interval for π becomes

$$.848 \pm 1.96 \sqrt{\frac{(.848)(.152)}{283}} = .848 \pm .042$$

Thus, the 95% confidence interval for π is (.806, .890). Notice that the hypothesized value $\pi_0 = .80$ is outside the interval. Therefore, the null hypothesis is rejected at the 5% level of significance.

Will it also be rejected at the 1% level of significance? To find out, we construct a 99% confidence interval for π. The interval becomes

$$.848 \pm 2.58 \sqrt{\frac{(.848)(.152)}{283}} = .848 \pm .055$$

Thus, the 99% confidence interval for π is (.793, .903). This time the hypothesized value $\pi_0 = .80$ is inside the confidence interval, which means that the null hypothesis cannot be rejected at the 1% level of significance. Figure 6.10 shows that $\pi_0 = .80$ is covered by the 99% confidence interval but not by the 95% confidence interval.   ■

Figure 6.10

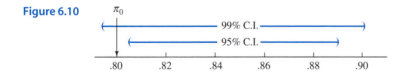

The test of a population proportion is summarized in the box:

Large-Sample Test of a Population Proportion

Application: Bernoulli Population
Assumption: $np > 5$ and $n(1-p) > 5$

Left-Tailed Test	Right-Tailed Test	Two-Tailed Test
$H_0: \pi \geq \pi_0$	$H_0: \pi \leq \pi_0$	$H_0: \pi = \pi_0$
$H_a: \pi < \pi_0$	$H_a: \pi > \pi_0$	$H_a: \pi \neq \pi_0$

continued

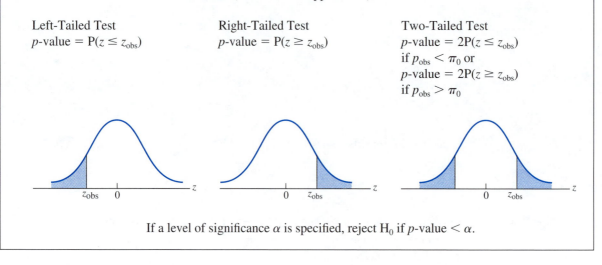

Standardized Test Statistic

$$z = \frac{p - \pi_0}{\sqrt{\pi_0(1 - \pi_0)/n}}$$

From the z-table (Table B.2 in Appendix B), find

Left-Tailed Test
p-value $= P(z \leq z_{obs})$

Right-Tailed Test
p-value $= P(z \geq z_{obs})$

Two-Tailed Test
p-value $= 2P(z \leq z_{obs})$
if $p_{obs} < \pi_0$ or
p-value $= 2P(z \geq z_{obs})$
if $p_{obs} > \pi_0$

If a level of significance α is specified, reject H_0 if p-value $< \alpha$.

EXERCISES 6.2 The Basics

6.16 A school psychologist suspects that more than 10% of all students experience some type of emotional tragedy during their first two years of college. She states the hypotheses as

$$H_0: \pi = .10 \quad \text{versus} \quad H_a: \pi > .10$$

a. Is this a left-tailed, right-tailed, or two-tailed test?
b. What is the correct test statistic to use to test H_0?
c. If the test statistic falls on the left tail of the sampling distribution, is it possible to reject H_0? Explain your answer.

6.17 Following up on Exercise 6.16, the psychologist selected a random sample of 1600 students that had completed two years of college. She determined that 180 of the 1600 had experienced some type of emotional tragedy in their first two years of college. Is there sufficient evidence to verify her claim?

6.18 In Exercise 6.3 in Section 6.1 a claim was made that 60% of the adult population thinks that there is too much violence on television. You thought that figure was too high and were asked to set up null and alternative hypotheses to evaluate the claim. The hypotheses should be stated as

$$H_0: \pi = .60 \quad \text{versus} \quad H_a: \pi < .60$$

a. Is this a left-tail, right-tail, or two-tail test?
b. What is the correct test statistic to use to test H_0?
c. If the test statistic falls on the right tail of the sampling distribution, is it possible to reject H_0? Explain your answer.

6.19 As a follow-up to Exercise 6.18, a random sample of 200 adults found that 110 thought that there is too much violence on television. Is this enough evidence to reject the claim?

6.20   The government believes that no more than 25% of all college students would favor reducing penalties for the use of marijuana. A sample of 2400 college students revealed that 750 favor reducing the penalties.

    a. Set up null and alternative hypotheses to evaluate the government's claim.

    b. Give the form of the standardized test statistic and calculate its observed value.

    c. Compute the p-value and determine whether there is sufficient statistical evidence to reject the government's claim.

    d. State your conclusion so that a nonstatistician could understand it.

6.21   In Exercise 6.20, if you use the government's estimate of 25%, how large a sample is needed to estimate the percent to within 1%, using a 95% confidence interval?

6.22   We wish to determine whether a greater proportion of students than local citizens favor the sale of beer in the county. We know that half of the local citizens oppose the sale of beer. In a random sample of 400 students, we find that 230 favor the sale of beer. Is there sufficient evidence to suggest that a greater proportion of students than local citizens favor the sale of beer in the county?

    a. Set up null and alternative hypotheses to determine whether a greater proportion of students than local citizens favor the sale of beer in the county.

    b. Give the form of the standardized test statistic and calculate its observed value.

    c. Compute the p-value and determine whether there is sufficient statistical evidence to support the alternative hypothesis.

    d. State your conclusion so that a nonstatistician could understand it.

6.23   Juries in the 75 largest U.S. counties disposed of 12,026 tort, contract, and real property cases (Bureau of Justice Statistics, *Civil Jury Cases and Verdicts in Large Counties*, July 1995). Of all those cases, 33% were classified as automobile accident suits. In Cook County, Illinois, there were 600 jury cases. With a test of significance, determine whether the percent of automobile accident suits in Cook County exceeds the national average if there were 210 automobile accident suits out of the 600 cases. State the null and alternative hypotheses and give the z statistic and the p-value. State your conclusion so that a person with no statistical training could understand it.

6.24   In Exercise 6.23, the Bureau of Justice Statistics reported that 1362 medical malpractice suits were decided by a jury. In 413 of those cases, the plaintiff won. In this sample, the plaintiff won only 30.3% of the cases. Is the evidence from this sample strong enough to conclude that less than one-third of all medical malpractice suits are won by the plaintiff? State the null and alternative hypotheses and give the z statistic and the p-value. State your conclusion so that a person with no statistical training could understand it.

6.25   In a *USA Today*/CNN/Gallup nationwide telephone poll of 1022 randomly selected adults, 470 said that they were pleased with the way the president was dealing with the economy. A presidential consultant stated that at least half of all American adults support the president.

    a. State the null and alternative hypotheses to test the consultant's claim.

    b. Calculate the standardized test statistic, z.

    c. A 10% level of significance is generally accepted as the dividing line between statistical significance and insignificance. What must the standard z test statistic be in order for there to be significant evidence to reject the consultant's claim?

    d. Does the z statistic found in part **b** satisfy the condition in part **c**?

    e. Calculate the p-value. Is it less than a 10% level of significance?

    f. Should the null hypothesis be rejected?

    g. State your conclusion.

6.26   A manufacturer of automobiles purchases machine bolts from a supplier who claims that no more than 5% of the bolts are defective. From a random sample of 400 bolts, it is found that 28 are defective. Is there sufficient evidence to reject the supplier's claim?

6.27   A psychologist has developed a new aptitude test and believes that 80% of the public should score above 50 on the test. From a sample of 200 people, 164 scored above 50. Is there statistical evidence that the claim made by the psychologist is not valid? For the results to be significant at the 5% level of significance, how many out of 200 will have to score above 50 on the aptitude test?

6.28   A presidential aide said that at least half of the nation agreed with the U.S. invasion of Grenada in 1984. However, a Roper poll of 2000 Americans found that only 34% supported the invasion. Did the 34% happen simply by chance, or is there statistical evidence that the aide was in error?

6.29   One-third of all ex-convicts return to jail within three years. One state is trying a new rehabilitation system on its prisoners. After the release of 200 prisoners who had participated in the new system, only 48 returned within three years. Is there statistical evidence that the new system is reducing the proportion of repeat offenders? What is the smallest level of significance at which these results will still be significant?

6.30   A doctor claims that at least 80% of his patients never have significant arthritic pain after a six-month stay in his Caribbean treatment center. Out of a random sample of 75 of his former patients, 45 said that his treatment was ineffective and that they now have arthritic pain. Is there statistical evidence to refute the doctor's claim?

6.3   Testing a Population Mean

Learning Objectives for this section:

☐ Be able to test a population mean with a t-test, based on $\bar{x}$.

☐ Know the assumptions required for the t-test.

☐ Know the applications for the t-test.

☐ Know the difference between the z-test and the t-test.

In Section 6.1 we introduced hypothesis-testing problems that involve the mean of a measurement population. The z-test was presented to test H_0: $\mu = \mu_0$. Recall, however, that the z-score

$$z = \frac{\bar{x} - \mu_0}{\sigma/\sqrt{n}}$$

depends on the population standard deviation σ. Therefore, to find the p-value, we need to know the numerical value of σ. In most applications, it is unreasonable to assume that we would know the population standard deviation when we are conducting inferences about the population mean.

Testing μ When σ Is Unknown: The t-Test

To develop the procedure for testing H_0: $\mu = \mu_0$ when the population standard deviation σ is unknown, we will, as in the development of the confidence interval for μ, modify the z-score by estimating σ with the sample standard deviation s.

As in the z-test, the evidence against H_0 is measured by the distance the test statistic $\bar{x}$ is from μ_0, the value of μ being tested. This distance is evaluated by calculating the observed value of the standardized test statistic

$$t = \frac{\bar{x} - \mu_0}{s/\sqrt{n}}$$

Notice that σ in the z-score has been replaced by s.

When the parent population is *normally distributed*, recall that t is distributed as Student's t with $n - 1$ degrees of freedom (see Section 5.4). Therefore, the p-value associated with the observed value of t is found in the t-table (Table B.3 in Appendix B) with $n - 1$ degrees of freedom. Computer programs print the p-value; however, if you must look it up in a table, you should be cautioned that the exact *p*-value cannot be found as with a *z*-table. With the *t*-table you will only be able to find bounds for the *p*-value as described later in the summary and illustrated in Examples 6.8 and 6.9.

Like the *t*-interval, the *t*-test is robust against violations of the normality assumption; that is, modest departures from normality are permissible. We must, however, guard against severely skewed distributions or symmetric distributions with long tails. Again, boxplots and normal probability plots should be used to identify outliers and departures from normality. The same rules of thumb for using the *t*-interval procedure apply to using the *t*-test:

Rules for Applying the *t*-Test

- If the sample size is less than 15, the population distribution should be close to normal.
- If the sample size is between 15 and 30, the *t*-test can be used when the population distribution is reasonably symmetric and has no outliers.
- If the sample size is greater than 30, the *t*-test procedure is permissible except when there are extreme outliers or strong skewness.

If these criteria are not met, refer to Section 6.4 for an alternative to the *t*-test.

The *t*-test is summarized in the box:

Test of μ Based on $\bar{x}$ When σ Is Unknown

Application: Symmetric population distributions with tails that are not unusually long. If the sample size is small, the parent distribution should not deviate substantially from normality.

Assumption: σ is unknown

Left-Tailed Test	Right-Tailed Test	Two-Tailed Test
$H_0: \mu \geq \mu_0$	$H_0: \mu \leq \mu_0$	$H_0: \mu = \mu_0$
$H_a: \mu < \mu_0$	$H_a: \mu > \mu_0$	$H_a: \mu \neq \mu_0$

Standardized test statistic:

$$t = \frac{\bar{x} - \mu_0}{s/\sqrt{n}}$$

p-value $= P(t \leq t_{obs})$ p-value $= P(t \geq t_{obs})$ p-value is same as one tailed and doubled to account for both tails.

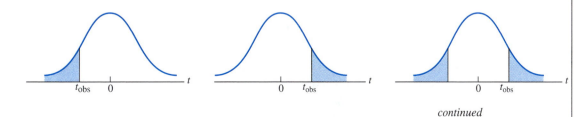

continued

From the t-table, with $n - 1$ degrees of freedom, find the p-value most closely associated with t_{obs}. If t_{obs} falls between two table values, give the two associated probabilities as bounds for the p-value. The p-value for the two-tailed test is calculated as in a one-tailed test and then doubled to account for both tails.

If a level of significance α is specified, reject H_0 if p-value $< \alpha$.

Example 6.8   To justify raising its rates, an insurance company claims that the mean medical expense for all middle-class families is at least $700 per year. A survey of 100 randomly selected middle-class families found that their mean medical expense for the year was $670 and the standard deviation was $140. Assuming that the distribution of medical expenses is symmetric without unusually long tails, is there evidence that the insurance company is misinformed?

Solution   The insurance company claims that the mean medical expense is at least $700. We doubt that, however, and believe that it is less than $700. This becomes the research hypothesis. Denoting the mean medical expense for all middle-class families as μ, we can state the null and alternate hypotheses as

$$H_0: \mu \geq 700 \quad \text{versus} \quad H_a: \mu < 700$$

Because the underlying distribution of medical expenses is symmetric without excessively long tails, the choice for a test statistic is $\bar{x}$. The standardized test statistic is

$$t = \frac{\bar{x} - 700}{s/\sqrt{n}}$$

and has 99 degrees of freedom ($n = 100$).

The observed value of t becomes

$$t_{obs} = \frac{670 - 700}{140/\sqrt{100}} = -2.14$$

The degrees of freedom in the t-table jump from 80 to 120 because there is very little difference between the corresponding critical values. Because the degrees of freedom here are 99, we will use the table values associated with 80 degrees of freedom.

The absolute value of t_{obs}, 2.14, falls between table values 1.99 (upper tail probability .025) and 2.374 (upper tail probability .01). Therefore, we have

$$.01 < p\text{-value} < .025$$

which suggests that the null hypothesis should be rejected. There is significant evidence to indicate that the insurance company's claim is inaccurate. We might suggest that the insurance company check its sources.   ■

Comparing the z-Test and the t-Test

Because of the large number of degrees of freedom (df $= 99$) in Example 6.8, there is very little difference between the z distribution and the t distribution. For example, if we had used the z distribution to find the p-value in the example, we would have found

$$p\text{-value} = P(z \leq -2.14) = 1.0 - .9838 = .0162$$

This is in complete agreement with the *t*-test, which found

$$.01 < p\text{-value} < .025$$

(With a computer we can find *p*-value $= P(t_{99} \le -2.14) = .0174$.)

The difference between a *z*-test and a *t*-test begins to diminish when the sample size is relatively large; say, $n > 30$. When the sample size is small, however, care must be exercised in applying the *t*-test.

Example 6.9   A chemist measures the haptoglobin concentration (in grams per liter) in the blood serum taken from a random sample of eight healthy adults. These are the values:

WORKSHEET: Haptoglo.mtw

1.82 3.32 1.07 1.27 0.49 3.79 0.15 1.98

Source: J. C. Miller and J. N. Miller, (1988), *Statistics for Analytical Chemistry*, 2nd Ed. (New York: Halsted Press).

Is there statistical evidence that the mean haptoglobin concentration in adults is less than 2 grams per liter?

Solution   Let μ denote the population mean haptoglobin concentration in adults. The research hypothesis is that the concentration is less than 2; therefore, we have

$$H_0: \mu \ge 2 \quad \text{versus} \quad H_a: \mu < 2$$

The normal probability plot of the data in Figure 6.11 does not rule out normality; therefore, we will assume that the parent distribution is normally distributed and use $\bar{x}$ as the test statistic.

Figure 6.11

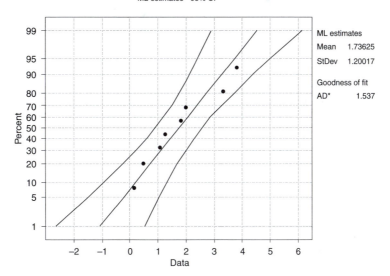

Normal Probability Plot for Concent
ML estimates - 95% CI

Because we are sampling from a normally distributed population, the standardized form of the test statistic is

$$t = \frac{\bar{x} - 2}{s/\sqrt{n}}$$

and has a t distribution with $n - 1 = 7$ degrees of freedom. The observed value is

$$t_{obs} = \frac{1.736 - 2}{1.283/\sqrt{8}} = -0.58$$

Thus, we have

$$p\text{-value} = P(t \leq -0.58)$$

Going to the t-table with 7 degrees of freedom, we see that a tail probability of .20 (20%) corresponds to a t-score of -0.896 and the value of t_{obs} is only -0.58. Consequently, we can say that

$$p\text{-value} > .20$$

which means that we do not have evidence to reject the null hypothesis. In conclusion, we can say that there is insufficient evidence that the mean haptoglobin concentration in adults is less than 2 grams per liter. ■

COMPUTER TIP

Testing a Population Mean

Just as in the one-sample confidence interval procedures, there are two alternatives for the one-sample test of hypothesis about a population mean. They are the one-sample z-test and the one-sample t-test. To execute either one, from the **Stat** menu select **Basic Statistics** and click on **1-Sample z** or **1-Sample t**. When the next window appears, **Select** the **Variable** containing the data. In the **Test Mean** box, type in the value of the mean being tested. (For the one-sample z-test, you must give a value for the population standard deviation, σ, in the **Sigma** box.) Next choose **Options** and select the **Alternative** depending on whether it is a one-tailed or two-tailed test. Finally, click **OK**.

Example 6.10 Psychologists wish to investigate the learning ability of schizophrenic people after they have taken a specified dose of a tranquilizer. Thirteen patients were given the drug, and one hour later they were given a standardized exam. Their scores are listed here:

WORKSHEET: Schizoph.mtw

15, 20, 30, 27, 24, 22, 22, 17, 21, 25, 23, 27, 25

Generally, patients score around 20 on the exam. Is there statistical evidence that taking the tranquilizer has affected their scores?

Solution Let μ denote the mean score on the standardized exam by all schizophrenics who could be administered the tranquilizer. The null and alternate hypotheses are:

$$H_0: \mu = 20 \quad \text{versus} \quad H_a: \mu \neq 20$$

The dotplot in Figure 6.12 shows that only three patients scored 20 or below on the exam. Is it conceivable that we would observe this pattern of variability if the population mean were indeed 20?

Figure 6.12

Dotplot for Score

Descriptive Statistics: score

Variable	N	Mean	Median	TrMean	StDev	SE Mean
score	13	22.92	23.00	23.00	4.13	1.15

Variable	Minimum	Maximum	Q1	Q3
score	15.00	30.00	20.50	26.00

From the descriptive statistics, we see that the sample mean is 22.92. Is the difference between $\bar{x} = 22.92$ and $\mu_0 = 20$ statistically significant? To answer this, we must first verify that conditions for the t-test are met. The boxplot and normal probability plot of the data in Figure 6.13 reveal no outliers or significant departures from normality. There appears to be no unusual behavior in the data that would prohibit using a t-test.

Figure 6.13

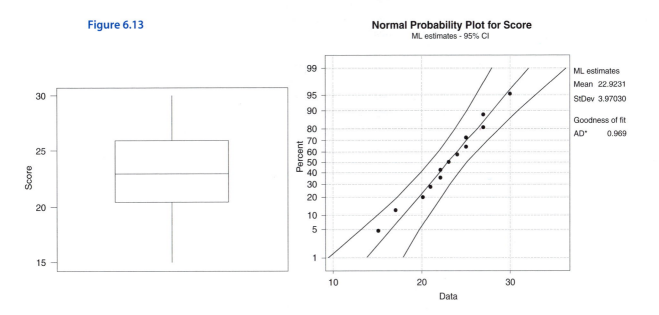

Figure 6.14 **One-Sample T: score**

Test of mu = 20 vs mu not = 20

Variable	N	Mean	StDev	SE Mean
score	13	22.92	4.13	1.15

Variable	95.0% CI		T	P
score	(20.43,	25.42)	2.55	0.025

Figure 6.14 gives the results of the t-test applied to the data in worksheet: **Schizoph.mtw**. The p-value $= 0.025$ is computed exactly by the computer instead of finding bounds as we did using the t-table in the previous example. In conclusion, there is significant evidence to indicate that the mean score is not 20 after the tranquilizer has been taken. ■

EXERCISES 6.3 The Basics

6.31 The chamber of commerce of a particular city claims that the mean carbon dioxide level of air pollution is no greater than 4.9 parts per million (ppm). A random sample of 16 readings resulted in $\bar{x} = 5.6$ ppm and $s = 2.1$ ppm. If we assume that the carbon dioxide level is normally distributed, is there sufficient evidence against the chamber of commerce's claim? State null and alternative hypotheses to evaluate the conjecture. Calculate the t statistic and its p-value and state your conclusion to the test. Why is it important to assume that the parent distribution is normal?

6.32 A manufacturer of a device to decrease gas consumption claims that when installed on a particular make of car, autos will consume, on average, no more than 5.5 gallons of gas for each 100 miles traveled. A consumer group tests 40 of the cars and finds an average consumption of 5.65 gallons per 100 miles and a standard deviation of 1.52 gallons. Do these results cast doubt on the claim made by the manufacturer? Is it necessary that the parent distribution be approximately normally distributed? Explain.

6.33 The average power consumption of 25 randomly selected families in a community for a given period is 125.6 kilowatt-hours and the standard deviation is 20.3 kilowatt-hours. If we assume that kilowatt usage is normally distributed, is there evidence that the mean usage for the entire community exceeds 120 kilowatt hours? State null and alternative hypotheses to evaluate the conjecture. Calculate the t statistic and its p-value and state your conclusion to the test.

6.34 The average length of time required to complete a certain aptitude test is claimed to be 80 minutes. A sample of 25 students yielded an average of 86.5 minutes and a standard deviation of 15.4 minutes. If we assume normality of the parent distribution, is there evidence to support the claim? State null and alternative hypotheses to evaluate the conjecture. Calculate the t statistic and its p-value and state your conclusion to the test. Why is it important to assume that the parent distribution is normal?

6.35 Past experience indicates that the scores on the first exam in a beginning history class are normally distributed with a mean of 72. This semester's class of 16 students had an average of 76 and a standard deviation of 14.4. Is this class "better than usual"? Evaluate this claim with a test of significance. Why is it important to assume that the scores in the past have been normally distributed?

6.36 To test the claim that the average home in a certain town is within 5.5 miles of the nearest fire department, an insurance company measured the distance from 25 randomly selected homes to the nearest fire department and found $\bar{x} = 5.8$ miles and $s = 2.4$ miles. Determine what the insurance company found out with a test of significance. What assumption did you make about the underlying parent distribution?

6.37 An instructor gave his class an examination that he knows from years of experience has a mean of 78. His present class of 20 students obtained an average score of 82 and a standard deviation of 7. Is he correct in assuming that this is a superior class? What assumption did you make about the underlying parent distribution?

6.38 A class of 50 eighth-graders took a standardized reading test. Their scores had a mean of 107.5 and a standard deviation of 10.5. The national mean score on the test is 100. Do we have sufficient statistical evidence to proclaim that this class is superior in reading ability?

Interpreting Computer Output

6.39 Here are the daily price returns (in pence) of Abbey National shares between 7/31/91 and 10/8/91:

WORKSHEET: Abbey.mtw

296 296 300 302 300 304 303 299 293 294 294 293 295 287 288 297 305 307 307 304
303 304 304 309 309 309 307 306 304 300 296 301 298 295 295 293 292 297 294 293
306 303 301 303 308 305 302 301 297 299

Source: Buckle, D. (1995), Bayesian Inference for Stable Distributions, *Journal of the American Statistical Association, 90,* 605–613.

a. From the histogram and boxplot, do you notice any unusual behavior in the data that would prohibit using a *t*-test to test the population mean?

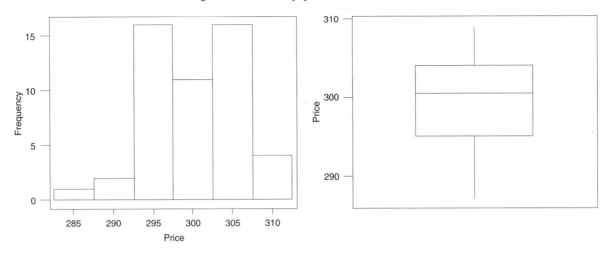

b. Using the descriptive statistics given here, determine with a test of significance whether the mean price differs from 300 pence.

Descriptive Statistics

Variable	N	Mean	Median	TrMean	StDev	SEMean
price	50	299.96	300.50	300.09	5.61	0.79

Variable	Min	Max	Q1	Q3
price	287.00	309.00	295.00	304.00

6.40 Water pH levels for 75 water samples in the Great Smoky Mountains were collected by Kaufman and colleagues (1988) as part of an Environmental Protection Agency (EPA) study. The data listed in worksheet: **Smokyph.mtw** are summarized here.

Character Stem-and-Leaf Display

```
Stem-and-leaf of waterphN = 75
Leaf Unit = 0.10

    1      6 3
    4      6 555
   11      6 6677777
   32      6 88888888899999999999
  (19)     7 0000000000000001111
   24      7 2222222233333
   13      7 444
   10      7 6
    9      7 899
    6      8 01
    4      8 223
    1      8 4
```

Descriptive Statistics

Variable	N	Mean	Median	TrMean	StDev	SEMean
waterph	75	7.1396	7.0300	7.1052	0.4394	0.0507

Variable	Min	Max	Q1	Q3
waterph	6.3800	8.4300	6.8700	7.2400

Source: Schmoyer, R. L. (1994), Permutation Tests for Correlation in Regression Errors, *Journal of the American Statistical Association, 89,* 1507–1516.

a. From the stem-and-leaf plot of the water pH levels, do you detect any unusual behavior in the data?
b. Would the *t*-test be appropriate for testing a hypothesis about the population mean? Explain.
c. State the null and alternative hypotheses necessary for determining whether the mean pH level is 7.
d. Test the hypothesis stated in part **c**. Be sure to calculate the *t* statistic and the *p*-value. State your conclusion so that a nonstatistican could understand it.

6.41 In Chesapeake Bay, complex changes in salinity are caused by the mixture of fresh water and sea-water during the diurnal tidal cycle. The fresh water from the Chesapeake River floats across the denser brine in the bay and during low tide it travels farther down the estuary. There is a counter-flow, however, along the bottom that carries the dense marine water up the bay during the waning tide. Here are the surface salinity measurements (in parts per thousand) taken at station 11, offshore from Annapolis, Maryland, on July 3–4, 1927.

WORKSHEET: Chesapea.mtw
salinity

6.97 6.20 5.93 6.32 6.36 6.72 6.80 6.90 7.14 6.91 6.76 6.74 7.20 7.45 7.47 7.47

Source: J. Davis, Statistics and Data Analysis in Geology, 2nd Ed., John Wiley and Sons, New York.

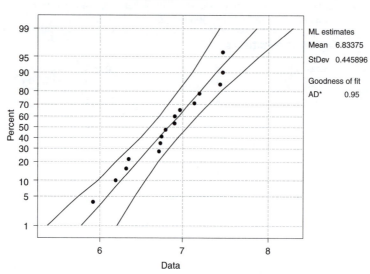

Normal Probability Plot for Salinity
ML estimates - 95% CI

From the normal probability plot, is there evidence that the normality assumption has been violated by the data? Is it appropriate to test the hypothesis using the *t*-test?

6.42 The computer printout below is for a test of hypothesis about the mean surface salinity levels at station 11 offshore from Annapolis, Maryland (see Exercise 6.41).

t-Test of the Mean

```
Test of mu = 7.000 vs mu not = 7.000

Variable    N     Mean    StDev    SE Mean        T    P-Value
salinity    16    6.834   0.461      0.115    −1.44       0.17
```

a. State the null and alternative hypotheses.
b. Is it a right-tailed, left-tailed, or two-tailed test?
c. Is there statistical evidence to reject the null hypothesis?
d. Give a conclusion for the test.

6.43 The charges for a coronary bypass at 17 hospitals in North Carolina are given in worksheet: **Bypass.mtw**. The normal probability plot in Exercise 5.105 in Section 5.6 strongly supports the normality of the data. Use these descriptive statistics to test the hypothesis that the mean charge for a coronary bypass in North Carolina is $35,000.

Descriptive Statistics

Variable	N	Mean	Median	TrMean	StDev	SEMean
charge	17	32406	32428	32501	3591	871

Variable	Min	Max	Q1	Q3
charge	24810	38578	29359	34919

Computer Exercises

6.44 These data are the survival times in weeks for 20 male rats that were exposed to a high level of radiation (see Exercise 1.52 in Section 1.3). It was determined in Exercise 5.62 that the parent distribution is close to normally distributed.

WORKSHEET: Rat.mtw

152 152 115 109 137 88 94 77 160 165 125 40 128 123 136 101 62 153 83 69

Source: Lawless, J. (1982), *Statistical Models and Methods for Lifetime Data*, John Wiley, New York.

a. State the null and alternative hypotheses to determine whether there is statistical evidence that the mean survival time is greater than 100 weeks.
b. What is the value of the t statistic? What is its p-value?
c. Is there evidence to reject the null hypothesis?
d. State your conclusion to this test of significance.

6.45 The Eocene lake deposits of the Rocky Mountains consist of thinly laminated dolomitic oil shales hundreds of feet thick. It is generally accepted that the laminations are varves, or layered deposits caused by seasonal climatic changes in the lake basins. By measuring the thickness of these laminations, scientists record annual changes in the rate of deposition throughout the lake's history. These data are the thicknesses (in millimeters) of a varved section of the Green River oil shale deposit near the western shore of one of the major lakes:

WORKSHEET: Grnriv2.mtw

6.0 7.2 7.1 7.1 7.2 7.4 8.0 8.6 10.0 11.4 12.0 11.0 9.6 8.7 7.6 7.2 7.2 7.8 8.1 7.8 7.1
7.2 7.1 7.0 7.0 7.7 8.6 9.0 12.0 13.7 14.0 13.6 12.1 12.9 12.8 11.1 9.0 7.5 7.5 8.4 8.4
7.9 7.0 6.7 6.8 7.3 7.3 7.2 8.1 9.8 11.0 10.8 9.5 8.1 7.2 7.1 6.8 7.0 7.1 5.6 3.8 3.4 4.2
4.8 4.5 3.6 3.0 2.8 4.1 6.8 8.1 7.8 6.4 4.6 3.7 4.0 4.2 4.5 5.9 7.3 7.3 6.7 6.0 5.8 5.7
6.5 8.2 10.2 12.3 13.2 13.2 12.4 9.7 9.2 9.3 8.3 6.0 5.7 6.1 6.3 6.3

Source: J. Davis, *Statistics and Data Analysis in Geology*, 2nd Ed., John Wiley and Sons, New York.

a. Does a stem-and-leaf plot or boxplot show any unusual characteristics in the data?
b. Check the shape of the distribution with a normal probablility plot.
c. Does a test of the population mean using $\bar{x}$ as the test statistic seem appropriate?
d. Perform a test of significance to determine whether the mean thickness of the varves is less than 8 millimeters. Be sure to state the null and alternative hypotheses and calculate the t statistic and the p-value. State your conclusion so that a geologist without much statistical training could understand it.

6.46 According to a recent report, after five years on the job, American workers get an average of 24 days of paid holidays and vacation leave each year. These are the numbers of days of paid holidays and vacation leave taken by a random sample of 35 workers in the American textile industry.

WORKSHEET: Vacation.mtw

23 12 10 34 25 16 27 18 28 13 14 20 8 21 23 33 30 13 16 14 38 19 6 11 15 21
10 39 42 25 12 17 19 26 20

a. Construct a histogram and boxplot of the data. Are there any unusual observations in the data set?
b. Would the t-test be appropriate for testing a hypothesis about the population mean? Explain.

c. Test the null hypothesis that the mean vacation leave for textile workers is no different from the mean for all American workers.

6.47 An instructor believes that students who score below 80 on their first test most likely will fail developmental mathematics. In order to determine whether this is a valid predictor, she records the following scores on the first test for all students who failed developmental mathematics in the fall semester of 1995:

WORKSHEET: Devmath.mtw

84, 88, 96, 87, 65, 98, 41, 92, 78, 70, 93, 77, 39, 73, 62, 74, 51, 88, 100, 89, 100, 79, 69, 74, 69, 84, 49, 65, 77, 48, 61, 86, 68, 90, 68, 76, 67, 40, 84, 77

Source: Data provided by Dr. Anita Kitchens.

Is there statistical evidence that the mean score for the students who failed developmental mathematics is less than 80?

6.48 To determine the flow characteristics of oil through a valve, the inlet oil temperature is measured in degrees Fahrenheit. Here is a sample of 12 readings:

WORKSHEET: Inletoil.mtw

93, 99, 97, 99, 94, 91, 93, 90, 89, 92, 90, 93

We want to know whether there is evidence that the mean inlet oil temperature is significantly less than 98 degrees.
a. Based on a normal probability plot, can we assume that the parent distribution of oil temperatures is normally distributed?
b. What are the null and alternative hypotheses to evaluate the claim that the mean inlet oil temperature is significantly less than 98 degrees?
c. Perform the test of significance suggested in part **b**. Be sure to give the standardized test statistic, its *p*-value, and a conclusion.

6.49 In the manufacture of integrated circuits (chips) used in computers, approximately 200 chips are typically processed as part of a wafer, which is a thin disk about 20 centimeters in diameter. The production of wafers (which can take months) is handled in lots of approximately 20 wafers. In one part of the process, a thin layer of silicon oxide is placed on the surface of a wafer. The thickness of the oxide layer is critical to the performance of the resulting chips. Two wafers are randomly selected from each of 30 lots. Four measurements of the thickness of the oxide layer on each wafer are then taken. The data set gives the averages of the eight measurements obtained from each set of two wafers:

WORKSHEET: Chipavg.mtw

962.50	1047.50	966.25	998.75	951.25	927.50	1096.25	1076.25	1052.50	992.50
1035.00	1058.75	1033.75	1050.00	986.25	1001.25	1101.25	966.25	1001.25	1003.75
1086.25	1067.50	1026.25	1058.75	1030.00	1035.00	1038.75	967.50	865.00	1006.25

Source: E. Yashchin, Likelihood Ratio Methods for Monitoring Parameters of a Nested Random Effect Model, *Journal of the American Statistical Association*, 90 (1995), 729–738.

a. From a normal probability plot, is there evidence that the normality assumption has been violated by the data? Is it appropriate to test the population mean using the *t*-test?
b. Give the computer printout necessary to test the claim that the mean thickness is 1000. State the hypotheses and give your conclusion.

6.4 Testing a Population Median

Learning Objectives for this section:

❏ Be able to test a population median with a large or small sample.

❏ Know the applications for a test of the population median.

❏ Know the assumptions required for the test of a population median.

When the parent population distribution is highly skewed, we learned in confidence interval estimation that the population median is often a better indicator of the center of the distribution than is the population mean. Also, in Section 5.2 it was pointed out that we should identify the population median as the measure of center when the parent distribution is symmetric with long tails. Thus, there are important situations where the population median is the parameter to estimate and/or test. In this section, we describe a procedure commonly known as the *sign test* for testing the population median, θ.

Small-Sample Test of the Population Median

Suppose we are interested in investigating the study habits of students in a beginning college math class. We suspect that the number of hours of study outside class each week is skewed right. Therefore, the parameter of concern is the median number of hours of study outside class each week. We believe that the median number of hours is more than 5, and we decide to test

$$H_0: \theta \leq 5 \quad \text{versus} \quad H_a: \theta > 5$$

If the median is 5 hours, one would expect approximately half of the sample observations to be below 5 and half to be above 5. On the other hand, if the population median is more than 5 (the research hypothesis claim), one would expect an abundance of observations greater than 5. Thus, it seems reasonable to reject the null hypothesis if a "large" number of observations are more than 5.

If, on the other hand, the alternative had been

$$H_a: \theta < 5$$

that is, a left-tailed test, we would reject the null hypothesis if a very "small" number of observations were more than 5.

To test the hypothesis, we choose the test statistic

$$G = \text{number of observations} > 5$$

Clearly, the range of values for G is

$$\{0, 1, 2, \ldots, n\}$$

where n is the sample size. Moreover, if the null hypothesis is true, the proportion of successes (observations greater than 5) should be 50%. The sampling distribution of G is a binomial distribution with parameters n and $\pi = .5$.

Suppose that from a sample of 15 students, we find that 9 study more than 5 hours per week; that is, the observed value of G is $G_{\text{obs}} = 9$. Is this evidence sufficient to reject H_0 and conclude that the median exceeds 5 hours per week? The *p*-value is the probability of observing a value of the test statistic at least as large as G_{obs} when the null hypothesis is true—namely,

$$p\text{-value} = P(G \geq G_{\text{obs}}) = P(G \geq 9)$$

where G is binomial with $n = 15$ and $\pi = .5$.

Going to the binomial table (Table B.1 in Appendix B), we find

$$
\begin{aligned}
p\text{-value} &= P(G \geq 9) \\
&= 1 - P(G \leq 8) \\
&= 1 - .696 \\
&= .304
\end{aligned}
$$

Clearly, this *p*-value is insignificant, and therefore there is insufficient evidence to indicate that the true median is more than 5 hours per week. We see from the binomial table that it would have taken 12 or more observations (out of 15) greater than 5 to reject H_0 with a *p*-value of .014 + .003 = .017.

One problem that may occur when this test is conducted is that some observations may equal the value of the median being tested. Those observations that are "tied" with the median value are discarded, and the sample size *n* is reduced by that amount. For example, in the preceding example, if 2 of the 15 observations were equal to 5, the sample size would be reduced to 13.

The complete small-sample ($n \leq 20$) test of the population median is summarized in the box.

Small-Sample Test of the Population Median

Application: Skewed or symmetric distributions with long tails.

Assumption: Parent distribution is continuous and $n \leq 20$.

Left-Tailed Test	Right-Tailed Test	Two-Tailed Test
$H_0: \theta \geq \theta_0$	$H_0: \theta \leq \theta_0$	$H_0: \theta = \theta_0$
$H_a: \theta < \theta_0$	$H_a: \theta > \theta_0$	$H_a: \theta \neq \theta_0$

Test Statistic

$$G = \text{no. of observations} > \theta_0$$

From the binomial probability table with n and $\pi = .5$ find

p-value = $P(G \leq G_{obs})$	*p*-value = $P(G \geq G_{obs})$	*p*-value = $2P(G \leq G_{obs})$ if $G_{obs} < n/2$ or $= 2P(G \geq G_{obs})$ if $G_{obs} > n/2$

If a level of significance α is specified, reject H_0 if *p*-value $< \alpha$.

Example 6.11   A psychologist measures the threshold reaction time (in seconds) for persons subjected to emotional stress and obtains these results:

WORKSHEET: Reaction.mtw

14.3, 13.7, 15.4, 14.7, 12.4, 13.1, 9.2, 14.2, 14.4, 15.8, 11.3, 15.0

Is there evidence that the median threshold reaction time is less than 15 seconds?

Solution   The stem-and-leaf plot of the data in Figure 6.15 indicates that the distribution is skewed left, and therefore the median, θ, is the appropriate measure to investigate.

Figure 6.15

```
Stem-and-leaf of time    N = 12
Leaf Unit = 0.10
       1        9 2
       1       10
       2       11 3
       3       12 4
       5       13 17
     (4)       14 2347
       3       15 048
```

We wish to obtain evidence that θ is less than 15 (the research hypothesis); thus, the null and the alternative hypotheses are

$$H_0\!: \theta \geq 15 \quad \text{versus} \quad H_a\!: \theta < 15$$

From the ordered stem-and-leaf plot, we find that the test statistic, G = no. of observations > 15, has an observed value of

$$G_{obs} = 2$$

One observation is 15, the value being tested, so we must reduce the sample size by one. Going to the binomial table with $n = 11$ and $\pi = .5$ we find (see Figure 6.16)

$$p\text{-value} = P(G \leq 2) = .032^+$$

Figure 6.16

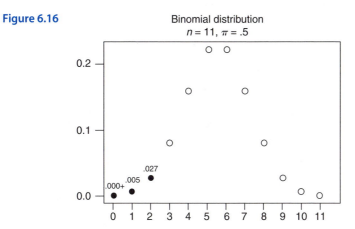

Binomial distribution
$n = 11$, $\pi = .5$

We have sufficient evidence to reject the null hypothesis and conclude that the median threshold reaction time is less than 15 seconds. ■

Large-Sample Test of the Population Median

For large samples, we can no longer use the binomial table, so we use a normal approximation of the binomial distribution. Recall from Chapter 3 that the mean of the binomial distribution is $n\pi$ and the standard deviation is $\sqrt{n\pi(1 - \pi)}$. Therefore, the standardized form of the test statistic G becomes

$$z = \frac{G - n\pi}{\sqrt{n\pi(1 - \pi)}}$$

It can be shown that the sampling distribution of z is approximately a standard normal distribution. Therefore, the p-value can be found in the normal table (Table B.2 in Appendix B). For the median test, π is always equal to .5. Thus, the test statistic, z, simplifies to

$$z = \frac{G - n/2}{\sqrt{n/4}} = \frac{2G - n}{\sqrt{n}}$$

A summary of the test is given in the box.

Large-Sample Test of the Population Median

Application: Skewed or symmetric distributions with long tails.
Assumption: Parent distribution is continuous and $n > 20$.

Left-Tailed Test
$H_0: \theta \geq \theta_0$
$H_a: \theta < \theta_0$

Right-Tailed Test
$H_0: \theta \leq \theta_0$
$H_a: \theta > \theta_0$

Two-Tailed Test
$H_0: \theta = \theta_0$
$H_a: \theta \neq \theta_0$

Standardized test statistic:

$$z = \frac{2G - n}{\sqrt{n}}$$

where G = no. of observations $> \theta_0$. From the z-table find the

p-value = $P(z \leq z_{obs})$

p-value = $P(z \geq z_{obs})$

p-value = $2P(z \leq z_{obs})$ if $z_{obs} < n/2$
or = $2P(z \geq z_{obs})$ if $z_{obs} > n/2$

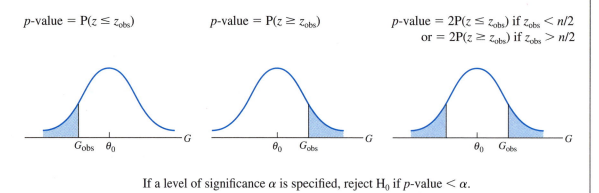

If a level of significance α is specified, reject H_0 if p-value $< \alpha$.

Example 6.12 An experimental lethal drug is injected into mice, and the survival times (in seconds) are recorded. The researcher suspects that the median survival time is less than 45 seconds. From these data, see whether there is evidence to support her theory.

WORKSHEET: Lethal.mtw

32, 37, 44, 41, 65, 27, 35, 29, 54, 42, 38, 57, 40, 24, 39,
43, 82, 52, 74, 25, 34, 25, 47, 34, 24, 38, 91, 30, 35, 31

Solution The boxplot of the data in Figure 6.17 point out that the data are skewed right.

Figure 6.17

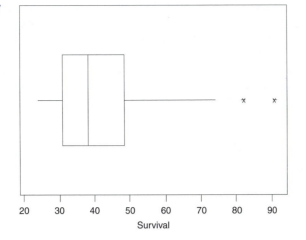

Thus, the investigator will test the population median. The null and alternative hypotheses are

$$H_0: \theta \geq 45 \quad \text{versus} \quad H_a: \theta < 45$$

The sample size is large, and consequently the test statistic is

$$z = \frac{2G - n}{\sqrt{n}}$$

where G is the number of observations greater than 45.

The number of observations greater than 45 is 8, so the observed value of z is

$$z_{\text{obs}} = \frac{(2)(8) - 30}{\sqrt{30}} = -2.56$$

From the normal table (Table B.2) we have

$$p\text{-value} = P(z \leq -2.56) = .0052$$

which is highly significant. The researcher has established statistical evidence that the median survival time is less than 45 seconds. ■

The test procedures for the population median described in this section are intended for skewed distributions or symmetric distributions with long tails. They are appropriate, however, for testing the median of any shape of continuous distribution. If the median best describes the population characteristic of concern, these procedures (large or small) should be used. In fact, if the sample size is small and the normality requirement for the t-test described in Section 6.3 is not met, this test is recommended.

C O M P U T E R   T I P

Testing a Population Median

To perform the Minitab test for a population median, from the **Stat** menu choose **Nonparametrics** and click on **1-Sample Sign** test. When the Sign test window appears, **Select** the variable to be tested, click on **Test median**, and type in the value of the median being tested in the null hypothesis. Choose an **Alternative** depending on whether it is a left-, right-, or two-tailed test. Then click on **OK**.

Figure 6.18 shows the results obtained by applying the sign test to the data in Example 6.11. Notice that the *p*-value is slightly different from the value we obtained in Example 6.11. This is because the binomial probabilities in Table B.1 have been rounded to three decimal places. If more significant decimal places are taken, we would obtain the same *p*-value as the computer. Also, notice that the computer gives the number of observations that are below the value being tested and the number of observations equal to the value of the median being tested.

Figure 6.18   **Sign Test for Median: time**

```
Sign test of median = 15.00  versus  < 15.00

          N     Below    Equal    Above         P    Median
time     12         9        1        2    0.0327     14.25
```

EXERCISES 6.4   **The Basics**

6.50   Test the hypotheses using the sample information. Be sure to find the *p*-value and interpret the results.

a. $H_0: \theta \geq 50$ versus $H_a: \theta < 50$
    $n = 14$ G = no. of observations $> 50 = 5$.

b. $H_0: \theta \leq 3.5$ versus $H_a: \theta > 3.5$
    $n = 8$ G = no. of observations $> 3.5 = 7$.

c. $H_0: \theta = 100$ versus $H_a: \theta \neq 100$
    $n = 18$ G = no. of observations $> 100 = 4$.

6.51   Test the hypotheses using the sample information. Be sure to find the *p*-value and interpret the results.

a. $H_0: \theta \geq 400$ versus $H_a: \theta < 400$
    $n = 50$ G = no. of observations $> 400 = 14$.

b. $H_0: \theta \leq 12.5$ versus $H_a: \theta > 12.5$
    $n = 100$ G = no. of observations $> 12.5 = 56$.

c. $H_0: \theta = 75$ versus $H_a: \theta \neq 75$
    $n = 200$ G = no. of observations $> 75 = 78$.

6.52   According to test theory, the median mental age of the 16 girls with scores listed below should be 100. Does the median mental age differ significantly from 100? Set up null and alternative hypotheses and perform a test of significance. Be sure to calculate the test statistic and its *p*-value and state a conclusion.

WORKSHEET: Mental.mtw

Mental	Ages	87	89	93	93	93	95	95	99	99
		102	108	108	113	113	114	114		

6.53   A retail merchant is considering opening a new department store that caters to college students in the community. One concern is the age of students in the freshmen class. She obtained the following ages from a random sample of 30 college freshmen:

WORKSHEET: Freshman.mtw

19, 18, 19, 22, 18, 21, 20, 19, 19, 28, 19, 20, 19, 18, 20, 15, 33, 19, 19, 20, 20, 19, 18, 20, 19, 20, 19, 26, 21, 19

Test the hypothesis that the median age for all college freshmen at the university is 19.

6.54 A supplier of storm windows claims that the median leakage from a 50-mph wind is not more than 12.5% (.125). A sample of nine windows yields these results:

WORKSHEET: Window.mtw

Leakage   .13   .17   .13   .18   .14   .12   .11   .14   .20

Is there statistical evidence that the supplier's claim is untrue? Conduct a test of significance and state your conclusion so that the supplier will understand it.

Interpreting Computer Output

6.55 Double Nobel prize–winner Linus Pauling strongly advocated the use of vitamin C as a treatment for cancer. The survival times (in days) of terminal stomach cancer patients who were treated with vitamin C are given here:

WORKSHEET: Cancer.mtw

column 1
Survival time in days
124 42 25 45 412 51 1112 46 103 876 146 340 396

Source: Cameron, E. and Pauling, L. (1978), Supplemental Ascorbate in the Supportive Treatment of Cancer, *Proceedings of the National Academy of Science,* USA, *75,* 4538–4542.

a. What do the boxplot and normal probability plot of the data tell about the underlying parent distribution?

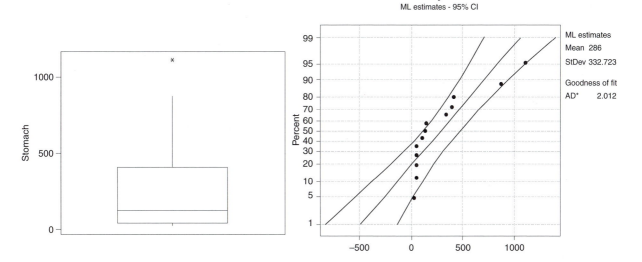

b. Does the computer output indicate that the median survival time exceeds 100 days? Explain by setting up the null and alternative hypotheses and performing a test of significance. Be sure to state a conclusion.

Sign Test for Median

Sign test of median = 100.0 versus > 100.0

	N	BELOW	EQUAL	ABOVE	P-value	MEDIAN
stomach	13	5	0	8	0.2905	124.0

6.56 A researcher claims that the average self-criticism score on the Tennessee Self-Concept Scale is at least 30 for all gifted high school students at one school. A sample of 20 students had these scores:

WORKSHEET: Tenness.mtw

29.8, 30.4, 27.5, 29.8, 31.0, 30.2, 29.5, 29.0, 27.0, 35.8, 25.4, 34.2, 30.7, 31.5, 28.2, 28.9, 24.2, 32.4, 29.2, 28.5

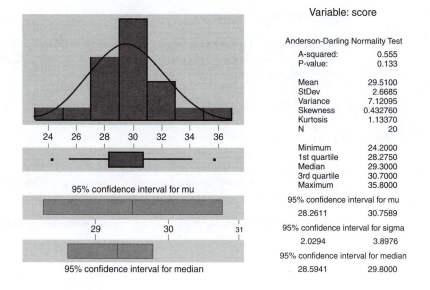

Descriptive Statistics

Variable: score

Anderson-Darling Normality Test

A-squared:	0.555
P-value:	0.133
Mean	29.5100
StDev	2.6685
Variance	7.12095
Skewness	0.432760
Kurtosis	1.13370
N	20
Minimum	24.2000
1st quartile	28.2750
Median	29.3000
3rd quartile	30.7000
Maximum	35.8000

95% confidence interval for mu

28.2611   30.7589

95% confidence interval for sigma

2.0294   3.8976

95% confidence interval for median

28.5941   29.8000

95% confidence interval for mu

95% confidence interval for median

a. The histogram and boxplot indicate that the parent distribution may be a long-tailed distribution. Does this create a problem for this *t*-test? Explain.

One-Sample T: score

```
Test of mu = 30 vs mu < 30

Variable          N         Mean       StDev     SE Mean
score            20       29.510       2.669       0.597

Variable      95.0% Upper Bound          T           P
score                     30.542      -0.82       0.211
```

Based on the computer printout, is there evidence to refute the researcher's claim? Be sure to state the null and alternative hypotheses and state a conclusion.

b. Because of the sample size and the shape of the distribution, do you think that a test of the population median might be better? Remember, a test of the median does not require that the parent distribution be normally distributed. State the hypotheses and give a conclusion based on the following printout.

Sign Test for Median: score

```
Sign test of median = 30.00  versus  < 30.00

              N     Below    Equal    Above          P     Median
score        20        15        0        5     0.0207      29.30
```

Do these results differ from the results in part **a**? Which is the better test? Explain why.

6.57   The stem-and-leaf plot shows the varve thickness from a sequence through an Eocene lake deposit in the Rocky Mountains. The deposit, part of the Green River Formation, consists of thinly laminated dolomitic oil shales hundreds of feet thick. In Exercise 6.45 in Section 6.3, you analyzed a similar oil shale deposit that is located only 10 miles from this deposit. In that exercise, you learned that the median varve thickness was 7.3 millimeters. Is there statistical evidence that the median varve thickness for this deposit exceeds 7.3 millimeters? Set up hypotheses, perform the test of significance, calculate the test statistic and *p*-value, and state your conclusion.

WORKSHEET Greenriv.mtw
Stem-and-Leaf Display: thick

```
Stem-and-leaf of thick    N = 37
Leaf Unit = 0.10

    1        8 9
    8        9 0012244
   11        9 899
   17       10 001233
  (5)       10 67888
   15       11 0013
   11       11 7
   10       12 03
    8       12 6
    7       13 04
    5       13 5
    4       14 2
    3       14 6
    2       15 0
    1       15 6
```

Is the apparent skewness in the data also demonstrated in the normal probability plot? Is the population median the correct parameter to investigate in this problem, or should we conduct a test of the population mean?

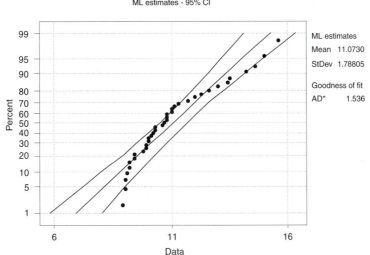

Normal Probability Plot for Thick
ML estimates - 95% CI

ML estimates
Mean 11.0730
StDev 1.78805

Goodness of fit
AD*   1.536

6.58 Exercise 6.40 in Section 6.3 looked at the pH levels of water samples collected in the Great Smoky Mountains. You performed a test of hypothesis on the population mean using $\bar{x}$ as the test statistic. Do the boxplot and normal probability plot of the pH levels support the decision to test the population mean, or would a test of the population median be more appropriate? Explain why.

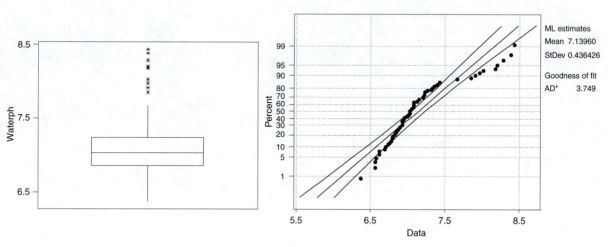

Normal Probability Plot for Waterph
ML estimates - 95% CI

6.59 These are the results of a test of the median pH levels of the water samples collected in the Great Smoky Mountains (worksheet: **Smokyph.mtw**).

Sign Test for Median: waterph

```
Sign test of median = 7.000  versus  not = 7.000
```

	N	Below	Equal	Above	P	Median
waterph	75	32	1	42	0.2955	7.030

a. State the null and alternative hypotheses.
b. Is the test a one-tailed or two-tailed test?
c. Compute the value of the standardized test statistic.
d. What is the *p*-value? Should the null hypothesis be rejected?
e. State your conclusion.

6.60 Of critical concern in AIDS research is the "incubation" time of the HIV virus—that is, the time from HIV infection to the clinical manifestation of full-blown AIDS. In column 1 of worksheet **Aids.mtw**, the incubation times (duration in months) for 295 patients who were thought to be infected with HIV by a blood transfusion are presented.

Source: Kalbsleisch, J. and Lawless, J., (1989), An analysis of the data on transfusion-related AIDS, *Journal of the American Statistical Association, 84,* 360–372.

a. Examine the boxplot and normal probability plot and comment on the general shape of the distribution of the incubation time.
b. Should the investigator draw inferences about the mean or the median incubation time?

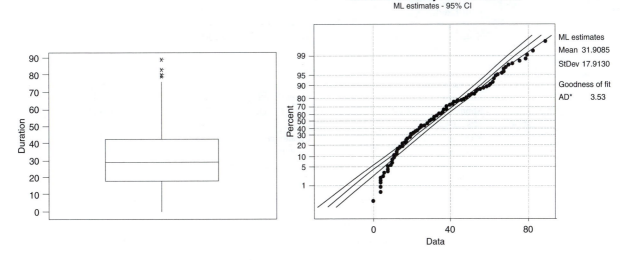

c. Using $\bar{x}$ as your test statistic, test the hypothesis that the mean incubation period exceeds 30 months. Be sure to state the hypotheses, give the p-value, and provide a conclusion.

d. Perform a test of significance to determine whether the median incubation time exceeds 2 years.

e. Which test of significance, **c** or **d**, do you recommend as the appropriate analysis of the data?

Computer Exercises

6.61 To assess the predictive validity of Klopfer's Prognostic Rating Scale (PRS) for people who have received behavior modification therapy, these data were obtained following psychotherapy:

WORKSHEET: Prognost.mtw

11.9 8.2 6.9 11.7 7.4 6.5 9.5 7.4 6.3 9.4 7.3 6.8 8.7 7.1 4.9

Source: Newmark, C., et al. (1973), Predictive Validity of the Rorschach Prognostic Rating Scale with Behavior Modification Techniques, *Journal of Clinical Psychology*, 29, 246–248.

We want to know whether there is evidence that a typical rating score differs significantly from a standard score of 9.

a. Construct a normal probability plot of the data. Does it appear that the assumptions for the t-test are satisfied, or should we test the population median?

b. State the hypotheses for the test.

c. Perform the test and state your conclusion.

6.62 What does it cost to operate a state law enforcement agency? Listed are the operating expenditures (dollars) per resident per year for each of the states in 1993.

Source: Bureau of Justice Statistics, *Law Enforcement Management and Administrative Statistics, 1993*, NCJ-148825, September 1995, page 84.

WORKSHEET: Statelaw.mtw

State	cost	State	cost	State	cost
Alabama	5	Connecticut	9	Illinois	17
Alaska	4	Delaware	3	Indiana	7
Arizona	8	Florida	11	Iowa	5
Arkansas	3	Georgia	8	Kansas	3
California	59	Hawaii	*	Kentucky	8
Colorado	4	Idaho	1	Louisiana	4

continued

WORKSHEET: Statelaw.mtw
continued

State	cost	State	cost	State	cost
Maine	2	New Jersey	11	South Dakota	1
Maryland	16	New Mexico	1	Tennessee	8
Massachusetts	11	New York	23	Texas	33
Michigan	24	North Carolina	9	Utah	*
Minnesota	4	North Dakota	1	Vermont	2
Mississippi	*	Ohio	13	Virginia	11
Missouri	8	Oklahoma	6	Washington	11
Montana	1	Oregon	7	West Virginia	3
Nebraska	3	Pennsylvania	32	Wisconsin	3
Nevada	3	Rhode Island	2	Wyoming	1
New Hampshire	*	South Carolina	3		

a. Notice in the descriptive statistics that the mean cost is $8.96 per resident and the median cost is $5.50 per resident. Why is there such a difference between the two measures of the center of the distribution?

Descriptive Statistics

Variable	N	N*	Mean	Median	TrMean	StDev	SEMean
cost	46	4	8.96	5.50	7.57	10.70	1.58

Variable	Min	Max	Q1	Q3
cost	1.00	59.00	3.00	11.00

b. Construct a stem-and-leaf plot of the data; does it explain the difference between the mean and median cost of operating the agencies?
c. To conduct a test of hypothesis, should we test the population mean or the population median? Explain your answer.
d. From the computer printout, state the null and alternative hypotheses being tested.

Sign Test for Median

Sign test of median = 8.000 versus < 8.000

	N	N*	BELOW	EQUAL	ABOVE	P-VALUE	MEDIAN
cost	46	4	26	5	15	0.0586	5.500

e. Is there statistical evidence to reject the null hypothesis? Provide a conclusion to this test of significance.

6.63 It is reported that the median annual income for social workers with less than five years' experience is $27,500. A random sample of 25 social workers from North Carolina (with less than five years' experience) had these incomes:

WORKSHEET: Social.mtw

25200, 26500, 26700, 27900, 28000, 25500, 22200, 24700, 27000, 26500, 27700, 21500, 26000, 23500, 28500, 27000, 24500, 27600, 28100, 26400, 27900, 25900, 27600, 26100, 27400

Perform a test of significance to determine whether there is evidence that the North Carolina social workers are underpaid.

6.64 Column 2 of worksheet: **Cancer.mtw** contains survival times of terminal bronchus cancer patients. Perform an analysis on these data similar to that given to the data in Exercise 6.55. With a test of significance, determine whether the median survival time for the bronchus patients exceeds 100 days.

6.65 The parallax of the sun is the angle subtended by the earth's radius, as if viewed from the surface of the sun. Knowing this angle and the radius of the earth, it is possible to determine the distance between the earth and the sun. Astronomers in the 18th century were concerned with the absolute dimensions of the solar system and were eager to determine the parallax of the sun. Astronomer Edmund Halley (1656–1742) suggested that the parallax of the sun could be determined by observing a "transit of Venus," which is the time it takes the planet Venus to travel across the face of the sun, as viewed from the earth. Transits of Venus are quite rare. James Short used the one in 1761 to calculate several determinations of the parallax of the sun. His measurements are stored in worksheet: **Short.mtw**. Columns C1 through C8 in the worksheet represent Short's calculations from different pairs of observations of the transit. Column C9 gives all 158 measurements made by Short and C10 indicates the sample number.

Source: Stigler, S., (1977), Do Robust Estimators Work with Real Data?, *The Annals of Statistics, 5*, 1055–1078; and Short, J. (1763), Second Paper Concerning the Parallax of the Sun Determined from the Observations of the Late Transit of Venus, *Philosophical Trans. Royal Society*, London, *53*, 300–345.

  a. Construct a histogram of Short's 158 measurements. Do you observe any unusual observations? Would you classify the distribution as being "normally" distributed or symmetrical with long tails?
  b. Construct a boxplot and a normal probability of Short's 158 measurements of the parallax of the sun. Based on the boxplot and the normal probability plot, would you classify the distribution as being close to normal or symmetrical with long tails? Does this agree with your analysis in part **a**?
  c. What parameter would you suggest as a measure of the "true" parallax of the sun?
  d. What estimator of the "true" parallax of the sun do you recommend, $\bar{x}$ or M?
  e. The "true" parallax of the sun has been determined to be 8.798 seconds of a degree. Short's 158 measurements are stored in column 9 of worksheet: **Short.mtw**. Based on your recommendations in parts **c** and **d**, test the hypothesis that Short's measurements were consistent with the "true" parallax of the sun.

6.5 Summary and Review Exercises

Key Concepts

- The *null hypothesis* is the hypothesis being tested in a *test of significance*. The *alternative hypothesis*, what is believed by the researcher to be true, is also called the *research hypothesis*.

- A *Type I error* is rejecting a true null hypothesis. A *Type II error* is failing to reject a false null hypothesis. The probability of a Type I error is denoted by α and is called the *level of significance*. The probability of a Type II error is denoted by β.

- A *test statistic* is the statistic that is used to test the null hypothesis. The null hypothesis will be rejected if the test statistic differs significantly from the hypothesized value of the parameter.

- The *p-value* is the probability of observing a value of the test statistic at least as extreme as that given by the observed data under the assumption that the null hypothesis is true. The null hypothesis is rejected if the *p*-value is smaller than the level of significance.

- A *right-tailed test* is a test of hypothesis where the *p*-value is calculated on the right tail of the sampling distribution of the test statistic. A *left-tailed test* is a test where the *p*-value is calculated on the left tail of the sampling distribution of the test statistic. A *two-tailed test* is one where one-half the *p*-value is on either tail of the sampling distribution of the test statistic.

- These are the steps to follow in testing hypotheses:

  - Formulate the null and alternative hypotheses.
  - Decide on an appropriate test statistic.
  - Determine the sampling distribution of the test statistic.
  - Collect the data and calculate the *p*-value.
  - Interpret the results.

- The test of a population mean is based on $\bar{x}$ and is either a *z*-test or a *t*-test, depending on whether we know the population standard deviation.

■ If the parent distribution is skewed or symmetric with long tails, the inference should be concerned with the population median and the test based on the sample median.

■ The test of a population proportion is a z-test and is similar to the test of the population mean based on $\bar{x}$.

**Statistical
Insight
Revisited**

The Statistical Insight given at the beginning of the chapter was concerned with the EPA gas mileage estimates for a Honda Civic. The EPA says that one should average at least 40 miles per gallon in highway driving. Based on the sample of 35 different occasions given in worksheet: **Honda.mtw**, is there statistical evidence to deny the EPA claim?

a. This histogram of the data shows no unusual behavior and does not deviate substantially from normality.

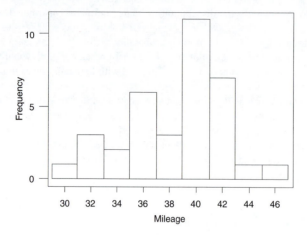

b. The boxplot shows a somewhat symmetric distribution without any outliers. The sample size is 35, so the conditions for a t-test are satisfied.

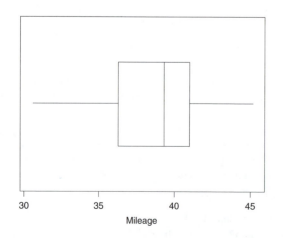

c. The formulation of the null and alternative hypotheses necessary for evaluating the EPA prediction is

$$H_o: \mu \geq 40 \quad \text{versus} \quad H_a: \mu < 40$$

d. Based on the results found in part **b**, the test statistic is $\bar{x}$ and we should perform a t-test.

e. Here are the results of a t-test for the hypotheses stated in part **c**:

One-Sample T: mileage

```
Test of mu = 40 vs mu < 40

Variable            N         Mean       StDev      SE Mean
mileage            35        38.480      3.561       0.602

Variable        95.0% Upper Bound          T          P
mileage                   39.498        -2.53      0.008
```

f. With a test statistic $T = -2.53$ and a p-value $= .008$ there is highly significant evidence to reject the null hypothesis. Based on these data, we have reason to doubt the EPA ratings for this vehicle.

Questions for Review

To test your skills, answer the following questions.

The Basics

Multiple Choice (Exercises 6.66–6.72)

6.66 In a test of significance
a. the null hypothesis is believed to be true and we are trying to find evidence to support it.
b. the alternative hypothesis is believed to be true and we are trying to find evidence to support it.
c. neither of the above.

6.67 If the null hypothesis is rejected,
a. only a Type I error is possible.
b. only a Type II error is possible.
c. both Type I and Type II errors are possible.
d. neither a Type I nor Type II error is possible.

6.68 If the probability of a Type I error is α, the probability of a Type II error is
a. also α b. $1 - \alpha$ c. 0   d. unknown

6.69 The null hypothesis should be rejected if
a. the p-value is smaller than the level of significance.
b. the p-value is greater than the level of significance.
c. the p-value is greater than 90%.
d. the p-value is greater than 10%.

6.70 In testing H_0: $\mu = 50$ versus H_a: $\mu > 50$ with $n = 100$, $\bar{x} = 52.4$, which of the following values of s leads to the rejection of H_0?
a. 200   b. 20   c. 2   d. none of the above

6.71 If the parent distribution is normal, the choice of estimator for the population mean is
a. $\bar{x}$ b. p c. M   d. s

6.72 In a test of significance
a. it is possible to prove the alternative hypothesis.
b. it is possible to find evidence to support the alternative.
c. it is not possible to find evidence to support the alternative.
d. none of the above.

6.73 Why is the alternate hypothesis also called the research hypothesis?

6.74 A study showed that more than 70% of the women who undergo breast biopsies to remove benign lumps face no unusual risk of later developing breast cancer. Formulate null and alternative hypotheses to test the validity of this claim.

6.75   The pH factor measures a substance's acidity or alkalinity. A soil pH of 5.5 is recommended for raising Christmas trees. Formulate the null and alternative hypotheses to test that the soil pH is at the desired level.

6.76   The county agent believes that the farmland in his county needs more than 2000 pounds of lime per acre. Out of 60 random soil samples, it was found that the average required lime per acre was 2080 pounds and the standard deviation was 240 pounds. Set up the null and alternative hypotheses to determine whether the data support the agent's claim. Test the hypothesis by determining the value of the test statistic and its p-value; then state your conclusion.

6.77   The mean self-concept score on a standardized test is 15. A group of 24 politicians was given the exam and scored an average of 17.8 with a standard deviation of 6.2. Generally, scores on standardized tests are close to being normally distributed. With a test of significance, determine whether the politicians scored significantly above the norm.

6.78   It is believed that the percent of convicted felons with a history of juvenile delinquency is 70%. Is there evidence to contradict this claim if out of 200 convicted felons, we find that 154 have a history of juvenile delinquency?

6.79   Do the data given below on STEP science test scores for a class of ability-grouped students provide sufficient evidence to conclude that they fall significantly below the national median of 80? First, verify that the data are close to normal with a normal probability plot.

WORKSHEET: Step.mtw

58, 60, 82, 80, 67, 70, 65, 73, 75, 77, 82, 68

6.80   Of all Ph.D.s awarded in English departments in 1997, 58% went to women. Of the 32 Ph.D.s awarded by the university system of a particular state, 17 were awarded to women. Was this an unusually low percentage to be awarded to women in this state? Evaluate with a test of significance.

6.81   A manufacturer claims that the average life span of its washing machines is at least 4 years. It is assumed that the distribution of the life spans is approximately normal. A random sample of 32 of these machines had an average life span of 3.6 years and a standard deviation of 1.5 years. Is there statistical evidence to dispute the manufacturer's claim?

6.82   A psychologist wishes to evaluate a new technique that he developed to improve rote memorization. Subjects are asked to memorize 50 word phrases using his technique. He thinks his technique is successful if the subjects average more than 40 correct phrases. How should he set up his null and alternative hypotheses to evaluate his theory?

6.83   Interpret the Type I and II errors for Exercise 6.82.

6.84   A city government claims that the average residence tax is no higher than $800 per year. We wish to test this claim; therefore, we randomly select 45 property owners and determine the average residence tax to be $848. Can we reject the claim made by the city government if it is known that the distribution of residence taxes is near normal with a standard deviation of $200?

6.85   According to the norms published for a certain intelligence test, the average for college freshmen is expected to be 65 points and the standard deviation is 10 points. A sample of 81 freshmen at State U. averaged 68.2 points on the test. Is there sufficient evidence to say that State U. freshmen are more intelligent, as measured by this test, than the average college freshman?

6.86   To study the effect of birth control pills on exercise capacity, a physiologist measures the maximum oxygen uptake during a treadmill session. It is known that the average maximal oxygen uptake for females not on the Pill is 36 milliliters per kilogram of body weight. A random sample of 36 females on the Pill had an average maximal oxygen uptake of 33.6 ml/kg of body weight and a standard deviation of 4.8 ml/kg. Is there statistical evidence that women on the Pill have a significantly smaller maximum oxygen uptake than women not on the Pill?

6.87 A group of nine children in a slow ability group was given a standardized exam that has a national mean score of 50. Their mean was 42 and the standard deviation was 15. Are these children really slow ability as measured by this test? What assumption must be made about the distribution of the scores on the standardized exam?

6.88 The height of adults in a certain town has a mean of 65.42 inches. A sample of 144 adults living in a depressed section of town is found to have a mean height of 64.82 inches and a standard deviation of 2.32 inches. Does this indicate that the adult height of residents in the depressed area is significantly less than that of all residents of the town?

6.89 A utility company claims that its customers' heating bills average no more than $250 during the winter period of December, January, and February. A consumer group sampled 100 accounts for the three-month period and found an average cost of $263 and a standard deviation of $72. Is there statistical evidence to cast doubt on the utility company's claim?

6.90 A manufacturer of string has established from several years of experience that the breaking strength of her string has a mean of 15.9 pounds and a standard deviation of 2.4 pounds. A change is made in the manufacturing process, after which a sample of 64 pieces is taken; the mean breaking strength is found to be 16.2 pounds. Assuming breaking strength is normally distributed and the new process has the same standard deviation as the old, can we say that the average breaking strength of the string from the new process is greater than the average from the old process? Is the normality assumption really required in this exercise? Explain.

6.91 A spokesperson for a statewide organization for raising the legal drinking age claims that at least 70% of the state's population thinks that the legal drinking age should be raised from 18 to at least 20. A random sample of 200 people was asked whether they approve of raising the drinking age to at least 20. Of the 200, 132 favored raising the limit. Is there statistical evidence to doubt the spokesperson's claim?

6.92 A factory makes a certain computer part that, according to specifications, must have a mean length of 1.5 centimeters. In a random sample of 16 parts from a shipment, the average length was found to be 1.56 centimeters and the standard deviation was .09 centimeters. Should this shipment be rejected?

6.93 Suppose you own a large racing stable and are contemplating the purchase of a two-year-old stallion. Based on past experience, you think a horse of this age should be able to run the mile in less than 98.5 seconds. You race the horse and obtain these times (in seconds):

WORKSHEET: Stable.mtw

104.6, 98.8, 101.4, 98.2, 99.7, 102.5, 103.6, 98.7, 101.5

You decide that you will reject the horse if you feel that his lifetime median time on the mile will exceed 98.5 seconds. Based on the data, should you reject the horse? Three years later, it is determined that his lifetime median time is 100 seconds. Did you make an error? If so, what type?

6.94 Suppose a preadmission algebra exam has a standardized mean of 200 and a standard deviation of 50. A sample of 100 students from a large high school takes the college admission exam. The average score turns out to be 215. Is this unusual? Explain with a test of significance.

Interpreting Computer Output

6.95   A study of the length of long-distance phone calls for a small business firm found this random sample of times (in minutes):

WORKSHEET: Phone.mtw

12.8   3.5   2.9   9.4   8.7   3.5   4.8   7.7   5.9   6.2   2.8   4.7   1.7   2.6   1.9   6.5   2.8   3.1   7.2   15.8

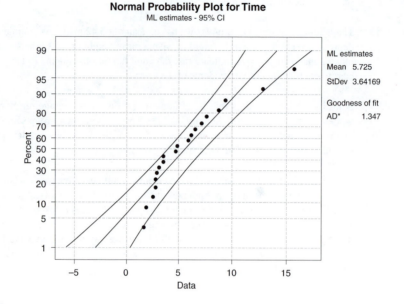

Normal Probability Plot for Time
ML estimates - 95% CI

ML estimates
Mean 5.725
StDev 3.64169

Goodness of fit
AD*   1.347

The normal probability plot indicates a degree of skewness in the data; therefore, we test the population median.

Sign Test for Median

```
Sign test of median = 5.000  versus  > 5.000

              N     BELOW     EQUAL     ABOVE     P-VALUE     MEDIAN
time         20        11         0         9      0.7483      4.750
```

Based on the computer output:
a. State the null and alternative hypotheses.
b. Is it a one-tailed or two-tailed test?
c. What is the value of the test statistic, G?
d. Is this a large-sample or small-sample test?
e. Why is the p-value so large?
f. Is there evidence to reject the null hypothesis?
g. State your conclusion to the test.

6.96   A standardized psychology exam has a mean score of 70. A research psychologist wished to see whether a counseling process has an effect on performance on the exam. She administered the exam to 18 volunteers who had participated in the counseling process, and obtained these scores:

WORKSHEET: Counsel.mtw

68,   71,   75,   65,   61,   70,   70,   64,   71,   73,   62,   78,   70,   69,   76,   67,   69,   72

From the boxplot and normal probability plot of the data, assess the assumptions required for the *t*-test.

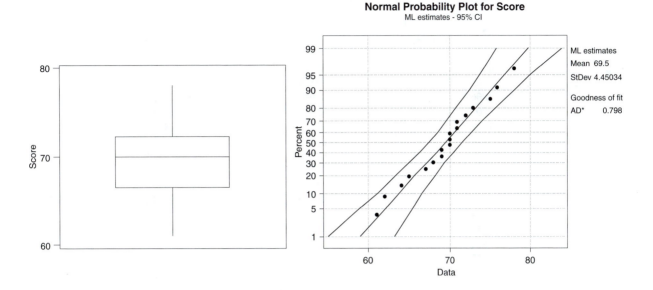

Descriptive Statistics

Variable	N	Mean	Median	TrMean	StDev	SEMean
score	18	69.50	70.00	69.50	4.58	1.08

Variable	Min	Max	Q1	Q3
score	61.00	78.00	66.50	72.25

Use the summary statistics to determine whether there is evidence that the counseling process reduces one's score on the exam.

Computer Exercises   **6.97**   The table gives a 100-year history (1770–1869) of the worldwide seismic activity based on the annual incidence of severe earthquakes. Examine these data to determine whether the mean severity exceeds 100.

WORKSHEET: Earthqk.mtw

Earthquake severity from 1770 to 1869

year	severity	year	severity	year	severity
1770	66	1781	50	1792	70
1771	62	1782	122	1793	46
1772	66	1783	127	1794	96
1773	197	1784	152	1795	78
1774	63	1785	216	1796	110
1775	0	1786	171	1797	79
1776	121	1787	70	1798	85
1777	0	1788	141	1799	113
1778	113	1789	69	1800	59
1779	27	1790	160	1801	86
1780	107	1791	92	1802	199

continued

WORKSHEET: Earthqk.mtw
continued

Earthquake severity from 1770 to 1869

year	severity	year	severity	year	severity
1803	53	1826	60	1848	76
1804	81	1827	118	1849	64
1805	81	1828	206	1850	31
1806	156	1829	122	1851	138
1807	27	1830	134	1852	163
1808	81	1831	131	1853	98
1809	107	1832	84	1854	70
1810	152	1833	100	1855	155
1811	99	1834	99	1856	97
1812	177	1835	99	1857	82
1813	48	1836	69	1858	90
1814	70	1837	67	1859	122
1815	158	1838	26	1860	70
1816	22	1839	106	1861	96
1817	43	1840	108	1862	111
1818	102	1841	155	1863	42
1819	111	1842	40	1864	97
1820	90	1843	75	1865	91
1821	86	1844	99	1866	64
1822	119	1845	86	1867	81
1823	82	1846	127	1868	162
1824	79	1847	201	1869	137
1825	111				

Source: Quenouille, M. H., (1952), *Associated Measurements*, Butterworth, London, p 279.

a. Construct a histogram of the severity measurements. What is the general shape of the histogram?

b. Construct a boxplot and a normal probability plot of the data. Are they consistent with the shape given in part **a**?

c. Do you think it is more important to study the mean severity or the median severity?

d. Using $\bar{x}$ as your test statistic, test the hypothesis that the mean severity over this period of time is no more than 100. Be sure to state the hypotheses, give the *p*-value, and form a conclusion.

6.98 In Exercise 6.45 we investigated varve thicknesses of oil shale deposits in Eocene Lake in the Rocky Mountains. There we learned that the distribution (Worksheet: **Grnriv2.mtw**) was skewed right. Because the sample size was large ($n = 101$), however, we chose to use a *t*-test to determine whether the mean varve thickness is less than eight millimeters.

a. Verify that the distribution is skewed with a histogram and boxplot.

b. From the *t*-test in Exercise 6.45 we found insufficient evidence (*p*-value = 0.26) that the mean varve thickness is less than eight millimeters. Because of the skewness, do you think that a test of the median thickness is more appropriate?

c. Test the hypothesis that the median varve thickness is less than eight millimeters. Be sure to state the hypotheses, give your *p*-value, and form a conclusion. Compare these results with those found in Exercise 6.45.

 6.99 As a follow-up to Example 6.10 in Section 6.3, a psychologist examines the amount of learning exhibited by patients with schizophrenia after they take a specified dose of a tranquilizer. One hour after taking the drug, 17 patients were given a second exam on which schizophrenics normally score an average of 22. Their scores are listed here:

WORKSHEET: Schizop2.mtw

36 29 30 32 37 15 34 23 32 5 24 10 25 34 13 30 33

a. Construct a histogram of the data. What is the general shape of the histogram?
b. Construct a boxplot and a normal probability plot of the data. Are they consistent with the shape given in part **a**?
c. For a typical or representative measure, should the psychologist investigate the mean exam score or the median exam score?
d. Based on your answer to part **c**, formulate the null and alternative hypotheses to determine whether the tranquilizer significantly improved the amount of learning exhibited by all schizophrenics.
e. Test the hypothesis stated in part **d**. Give the test statistic and its *p*-value and state your conclusion.

6.100   In most states, police departments have a certain percentage of unmarked police cars. These are the percentages of marked cars found in 65 police departments in Florida:

WORKSHEET: Marked.mtw

61 74 49 65 62 62 73 71 37 54 79 64 54 75 92 63 74 63 55 58 56 78 63
74 65 52 53 55 57 79 56 69 51 54 56 74 51 68 61 61 60 64 58 56 58 51
72 51 63 62 62 50 58 60 51 50 68 73 56 60 52 41 51 75 52

Source: Law Enforcement Management and Administrative Statistics, 1993, Bureau of Justice Statistics, NCJ-148825, September 1995, p 147–148.

a. Construct a histogram of the data. What is the general shape of the histogram?
b. Construct a boxplot and a normal probability plot of the data. Are they consistent with the shape given in part **a**? Are there any outliers in the data?
c. Based on the results found in part **b**, should the parameter of interest be μ or θ?
d. A criminal justice expert is interested in whether the average percent of unmarked police cars in Florida exceeds 60%. Formulate the null and alternative hypotheses for evaluating the question.
e. Test the hypothesis stated in part **d**. Give the test statistic and its *p*-value, and state your conclusion.

7

Comparing Distributions

In the preceding two chapters, we studied inference procedures for a single population. Now we investigate problems that involve two populations. Samples from each are used to make inferences about two population parameters. The inferences from the two samples may take the form of an interval estimate or a test of hypothesis.

Tests that compare two population means are among the most widely used statistical procedures. In addition to comparing two population means, we study the problem of comparing two population proportions. You will see that a simple extension of the concepts developed for the single population can enormously expand the realm of statistical analysis.

Contents

■ **STATISTICAL INSIGHT**
7.1 Introduction
7.2 Comparing Two Population Proportions
7.3 Comparing Two Population Means
7.4 Comparing Two Population Centers: Variances Equal
7.5 Comparing Two Population Centers: Matched Samples
7.6 Choosing the Right Procedure
7.7 Summary and Review Exercises

Are Vitamins Helpful in Warding Off Cancer, Heart Disease, and the Ravages of Aging?

In the early 1970s, Nobel Prize–winner Linus Pauling was ridiculed for his claim that large doses of vitamin C could cure everything from the common cold to cancer. A Canadian study, however, suggests that high doses of vitamins C and E appear to reduce the risk of cataracts by at least 50%. Other studies have linked increased doses of vitamin C with reduced risk of cancer and heart disease.

All the evidence is not positive, however. A Mayo Clinic study found that 51 terminally ill cancer patients given massive doses of vitamin C for up to 1 year fared no better than 49 similar patients who were given useless placebo pills. About equal percentages of both groups died within 1 year—49% of the vitamin C group and 47% of the placebo group. And 96 of the total 100 patients, all of whom had advanced colorectal cancer, experienced a worsening of the disease during the study.

In other studies, beta-carotene, which is turned into vitamin A by the body, has appeared to reduce the risk of heart disease, stroke, and certain cancers. Doctors at Harvard Medical School studied 22,000 male physi-

cians over a 10-year period and discovered that men with a history of cardiac disease who were given beta-carotene supplements of 50 milligrams every other day suffered half as many heart attacks, strokes, and deaths as those who took placebo pills. A University of Arizona Cancer Center study found that daily beta-carotene pills taken for 3 to 6 months dramatically reduced precancerous mouth lesions in 70% of patients.

The common theme in each of the studies is the attempt to compare an experimental group with a control group. In some studies we compare mean measurements, and in others we compare percentages. In the Mayo Clinic study, for example, 49% of the terminally ill cancer patients in the vitamin C group, as compared with 47% in the placebo group, died within 1 year. This is a comparison of Bernoulli proportions. Here the data speak for themselves; the vitamin C group did not fare any better than the placebo group, and thus there is no need for a formal analysis of the data. Other studies, however, require a formal test of significance.

Source: The New Scoop on Vitamins, *Time* Magazine, April 6, 1992.

7.1   Introduction

Learning Objectives for this section:

- ❏ Understand the basic properties of a comparative experiment.
- ❏ Know about extraneous variables and how their effects may be confounded with a treatment.
- ❏ Understand the importance of randomization and how it helps reduce bias in an experiment.
- ❏ Understand the basic properties of the matched-pairs design.
- ❏ Be able to distinguish between an experiment and an observational study.
- ❏ Know that an observational study cannot establish causation.

A widely used method of statistical analysis is the comparison of two populations. We often wish to compare a new procedure with an old established procedure, or to compare one product with another, or one treatment with another treatment. The key word is *compare*; we wish to compare the characteristics of one population with those of another population. To compare the two populations, random samples from each must be chosen and then used to make inferences about the parameters of the two populations. The observations in the two samples often come from an *experimental design*.

Comparative Experiment

In the *comparative experiment*, subjects are randomly divided into two equivalent groups, with one group receiving a *treatment* and the other serving as a *control group*. Then measurements of a *response variable* are taken on the subjects in the two groups. A configuration of the design is

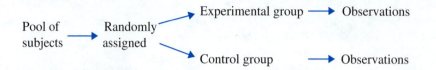

Confounding Variables

The comparative experiment is used to avoid the *confounding* of an *extraneous* variable with the treatment.

> A **confounding variable** is one whose effect on the response variable is mixed up with the effect of the treatment.

Example 7.1   To study the effect that a particular drug might have on blood pressure, a physician selects a random sample of her patients, administers the drug, and then measures their blood pressures. Any number of extraneous variables such as age, gender, and weight could be confounded with the treatment on the blood pressures of the patients. She cannot determine to what extent the experimental drug (treatment) affects a given patient's blood pressure, and to what extent their age, gender, and weight affect it.   ■

In the comparative experiment, however, the control group is treated the same as the experimental group except that it does not receive the treatment. Extraneous factors should affect both groups in the same way, and thus their effects are nullified. Any observed difference between the two groups should be due to the treatment.

It is possible to have a comparative experiment without a control group. Suppose we were comparing two drugs. We could assign drug A to one group and drug B to the other group. The two experimental groups serve as controls for each other. Any observed difference in the two groups should be due to the effects of the different drugs.

Randomization

To reduce possible bias in the comparative experiment, subjects are randomly assigned to the groups. It is hoped that *randomization* will create equivalent groups prior to the experiment. For example, to study the effect of alcohol on reaction time, we certainly would not want our subjects to choose which group (control or experimental) they would enter. Even if the subjects are homogeneous (alike in all characteristics relevant to the experiment), they do have some differences. Random allocation tends to distribute these differences evenly between the two groups.

Matched-Pairs Design

Randomization tends to average out the differences and produce nearly equivalent groups for comparison; however, there is no guarantee that it will. For example, suppose 20 subjects are to be randomly assigned to two experimental groups of ten subjects each. It is possible that a randomized procedure would put ten subjects of high intelligence in one group and ten subjects of low intelligence in the other group. If it is important that the two groups be equivalent as far as intelligence is concerned, the randomization procedure has failed. The problem can be avoided by assigning subjects according to the *matched-pairs design*.

In the matched-pairs design, the 20 subjects are divided into ten intelligence groups (pairs) of two subjects each. The two subjects in each pair should be as much alike as possible prior to the experiment and then randomly assigned to the two experimental groups. The end result is that each of the groups has subjects from each of the ten intelligence groups, and thus the groups are equivalent as far as intelligence of the subjects is concerned. A configuration of the design is shown here:

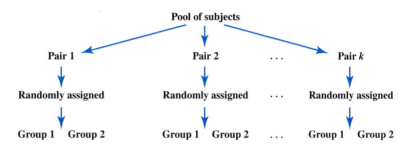

Note that each pair consists of two matched subjects who are randomly assigned to the two treatment groups. Because each pair is similar prior to the experiment, any observed difference is then due to the treatment.

Example 7.2 A clothing manufacturer would like to compare the durability of a newly designed line of children's clothing with that of its existing line of clothing. Ten sets of identical twin children are chosen for the experiment. One of each set of twins is randomly selected to wear the new clothing; the other will wear the old type. After a period of time, the durability of the two types of clothing will be evaluated. In this example, the sets of twins make up the matched pairs and the new and old lines of clothing are the two groups. A configuration of the design follows:

Twin set	1	2	3	4	5	6	7	8	9	10
Clothing type A	x	x	x	x	x	x	x	x	x	x
Clothing type B	y	y	y	y	y	y	y	y	y	y

A value of x denotes a measure of durability for type A, and a value of y denotes the measure of durability for type B. Identical twins are used as matched pairs because it is assumed that they will give approximately the same treatment to the two sets of clothing. ∎

In a modification of this design, the *before–after design*, a group of subjects serve as their own match and observations are taken both before and after the treatment:

Pool of subjects ⇒ Observation ⇒ Treatment ⇒ Observation

The *randomized block design* (covered in more advanced textbooks) is a generalization of the matched-pairs design where there are b blocks and g groups and the blocks can represent any variable that has a confounding effect on the response variable.

Observational Studies

In many situations it is not possible to randomly allocate subjects to groups to study the effects of a treatment. For example, if we wish to study the effects of income on one's spending habits, it is not possible to assign subjects randomly to different income groups. However, we can observe the spending habits of people who already fall into different income groups and then make statistical generalizations about how income affects one's spending habits.

An **observational study** (also called a quasi experiment) is an experiment in which one observes how a treatment has already affected the subjects.

A major distinction between an observational study and a designed experiment is that an observational study can only reveal an *association* between the treatment and response variable, whereas designed experiments can establish *causation* (cause and effect). The association that we see in an observational study may be due to confounding variables whose effects cannot be reduced by randomization as they can in the designed experiment.

Example 7.3   A university would like to compare the salaries of male and female faculty members. It is not possible to assign the subjects randomly to a gender. Therefore, the two groups (male and female) must be compared, as they are, in an observational study.

Suppose that a difference is observed between the salaries of men and women. Does this mean that the university should be accused of gender discrimination? Possibly not. It is possible, and very likely, that the years of experience differ for the two groups, which could account for the discrepancy in the salaries. We would say that the variable, years of experience, is confounded with gender on the professors' salaries. It is possible, however, to control for the confounding variable by comparing the salaries of males and females with similar years of experience. ∎

We close this section with three examples that illustrate the type of problems we will see in the remainder of this chapter.

Example 7.4   A study published in the *Journal of the American Medical Association* by Swedish doctors at Sahlgren's Hospital in Goteborg suggests that nicotine-laced gum helps smokers to stop smoking. The study showed that 29 of 106 smokers who chewed nicotine gum while trying to quit remained smoke free for 1 year. In a similar group, 16 of 100 smokers who chewed regular gum remained smoke free for 1 year. Both groups participated in counseling sessions to help them quit.

In this example, we assume that the two groups of smokers are independent of each other, and we are interested in comparing two Bernoulli proportions. For instance, we can let π_1 denote the proportion of smokers who use nicotine-laced gum and remain smoke free for 1 year, and let π_2 denote the proportion of smokers who chew regular gum and remain smoke free for 1 year. The problem, then, is to compare π_1 and π_2. The quantity $\pi_1 - \pi_2$ represents the difference between the proportions of the two groups. An inference problem can involve finding a confidence interval estimate of $\pi_1 - \pi_2$ or a test of hypothesis about the difference between the two proportions. If the research hypothesis is that the nicotine-laced gum is beneficial in helping smokers to remain smoke free, the null and alternative hypotheses are

$$H_0: \pi_1 = \pi_2 \quad \text{versus} \quad H_a: \pi_1 > \pi_2$$

The analysis for this type of problem is explained in Section 7.2. ∎

Example 7.5   A team of physicians would like to compare the recovery times of two different techniques after a certain operation. Subjects are to be randomly and independently assigned to the two techniques, and after the operation their recovery times are to be recorded. If μ_1 denotes the mean recovery time for one technique and μ_2 denotes the mean recovery time for the other technique, then $\mu_1 - \mu_2$ represents the difference in mean recovery times for the two techniques. An inference can be one of finding a confidence interval estimate of $\mu_1 - \mu_2$ or a test of hypothesis about the difference between the mean recovery times for the two techniques. If the null hypothesis says that there is no difference between the two techniques, the hypotheses are

$$H_0: \mu_1 = \mu_2 \quad \text{versus} \quad H_a: \mu_1 \neq \mu_2$$

The computations for the confidence interval and the hypothesis-testing procedures are presented in Section 7.3 when information about the population variances is unknown, and in Section 7.4 if we can assume that the population variances are equal.   ∎

Example 7.6   To examine the effects of jogging on resting pulse rate, a researcher divided a sample of 100 men and women into an experimental and a control group, with each member of one group paired with a member of the other group according to gender, age, weight, and height. All subjects were asked to continue their everyday activities except that the members of the experimental group were asked to jog 4 days a week. They began jogging a quarter-mile each day and built up, over a 6-month period, to 2 miles a day. After 6 months, the resting pulse rate was measured for both members of each pair of subjects.

The pulse rate of a member of the experimental group is related to the pulse rate of the corresponding member of the control group because the subjects were paired according to gender, age, weight, and height. If we let μ_d denote the mean of the population of differences between the experimental and control groups, the test of significance that there is no difference between the groups can be stated as

$$H_0: \mu_d = 0 \quad \text{versus} \quad H_a: \mu_d \neq 0$$

The analysis of the matched-pairs experiment is described in Section 7.5.   ∎

In each example, the aim of the study is to compare the two groups by evaluating the difference between specific parameters in the respective populations. In the following sections, we develop confidence interval estimates and test of hypothesis procedures for comparing the difference between these parameters.

EXERCISES 7.1   The Basics

7.1   In each situation, identify the parameters under study and state the null and alternative hypotheses.

   a. An economist for the Census Bureau wishes to compare the costs associated with renting a home in the West versus the East. Realizing that the distribution of rents is typically skewed, she focuses on the median gross rents for the two regions. She has no information to suggest that one region is generally more expensive than the other.

   b. The British bioscience company Zeneca Group, PLC, has harnessed microorganisms that ingest corn sugar to make plastic (*Wall Street Journal*, March 23, 1994). The goal is to make a plastic product that is biodegradable. Mean biodegradation rates for plastic and ordinary paper grocery bags in a compost pile were compared. The experimenter wishes to show that the plastic biodegrades faster than the paper.

   c. Has the percent of persons not covered by health insurance changed from last year? One researcher believes that it has increased.

7.2 Does Example 7.5 describe a controlled experiment or an observational study?

7.3 Does Example 7.6 describe a controlled experiment or an observational study?

7.4 In Example 7.5, suppose that the results showed a statistically significant difference (p-value $< .05$) between the mean recovery times for the two techniques. Explain what this means using terminology that a nonstatistician would understand. Do not use statistical wording, such as "statistically significant."

7.5 In Example 7.6, suppose that the statistical analysis of data produced a p-value $= .014$. What does this mean in terms of a test of significance? Should H_0 be rejected? If so, what does this mean? Explain your conclusion so that someone who is not knowledgeable about statistics could understand it.

7.6 Example 7.4 does not specifically state whether the study is a controlled experiment or an observational study. It appears on the surface that there is a significant difference between the two proportions of smokers who remained smoke free for 1 year. If this is an observational study and the difference is statistically significant, can we say that the nicotine-laced gum caused the smokers to remain smoke free for 1 year? Explain your answer.

7.7 Is in-line skating more dangerous than skateboarding? According to the Consumer Product Safety Commission, in 1993 in-line skating caused 37,000 injuries and skateboarding caused 27,700 injuries. What additional information is needed to compare the relative danger of the two sports? Define the parameters of interest and state the null and alternative hypotheses to evaluate whether in-line skating is more dangerous than skateboarding.

7.8 An 18-month Duke University study of 413 patients in a rural North Carolina clinic suggests that family stress is linked to increased health problems (AP wire release, March 15, 1995). The report states that those patients with high family stress had more follow-up visits to the clinic, more referrals to specialists, more hospitalizations, and more severe illnesses. Can we conclude from this study that family stress causes increased health problems? Is this an observational study or a designed experiment?

7.9 A 1994 study in the *Journal of the National Cancer Institute* showed a possible link between exposure to electricity and breast cancer. The researchers compared the death records of 267 female electrical workers and 138,564 other working women. They found that the electrical workers were 38% more likely to die of breast cancer. The report also stated that the link was strongest for telephone installers and line workers but that there was no link among telephone operators, computer operators, and others in traditionally female jobs with potential electrical exposure. Does this mean that unusually high doses of electricity cause breast cancer? Is this an observational study or a designed experiment?

Computer Exercises

7.10 Water pH levels and elevations (in miles) for 75 water samples in the Great Smoky Mountains were collected by Kaufman et al. (1988) as part of an Environmental Protection Agency (EPA) study. The data listed in worksheet: **Smokyph.mtw** also appear in Schmoyer (1994). The elevations are coded as 0 (low) for elevations below .60 miles and 1 (high) for elevations above or equal to .60 miles. State the null and alternative hypotheses necessary for determining whether the pH level is greater at the lower elevations. From the data, construct side-by-side dotplots of the water pH levels for low and high elevations. Comment on the shapes of the two distributions. Are they similar? Are they symmetric? Does either distribution appear to be long tailed? Construct normal probability plots of the two samples. Do they indicate anything about the lengths of the tails of the distributions?

7.11 Worksheet: **Indy500.mtw** gives the qualifying speeds and the number of previous starts for drivers in the 79th Indianapolis 500 car race held May 28, 1995. In an effort to determine whether the qualifying speed is related to the number of previous starts, the drivers were separated into two groups, those with fewer than four previous Indy 500 starts and those with five or more previous Indy 500 starts. State the null and alternative hypotheses for determining whether there is a difference between the qualifying speeds for experienced and inexperienced drivers. From the data, construct side-by-side dotplots and boxplots of the two samples. Comment on the shapes of the two distributions. Are the two samples equally variable? Does normality seem reasonable in either distribution? Does it appear that the more experienced drivers had higher qualifying speeds?

7.2 Comparing Two Population Proportions

Learning Objectives for this section:

❑ Be able to find a confidence interval for the difference between two proportions.

❑ Be able to test hypotheses about the difference between two proportions.

❑ Know the applications and assumptions for the inference procedures concerning the difference between two proportions.

In Chapters 5 and 6 we learned how to statistically evaluate a single Bernoulli proportion with a confidence interval estimate (Section 5.3) and with a test of hypothesis (Section 6.2). For example, when a CNN/Gallup nationwide telephone poll of 1022 adults reports a 54% approval rate for the president's health care policy, we know how to interpret this information.

Now suppose the results of the poll indicate that 56% of Democrats and 47% of Republicans approve of the president's health care policy. How do we deal with two proportions? How do we compare the results? Are the percentages significantly different? In this section, you will learn how to statistically compare two Bernoulli proportions. Recall that a Bernoulli proportion is the probability of a success in a situation where the outcome is classified as either a success or a failure. For example, if the respondent favored the president's health care policy, we classify the outcome as a success; otherwise, it is classified as a failure. In the example, let π_1 denote the Bernoulli proportion of Democrats who approve of the president's health care policy and let π_2 denote the Bernoulli proportion of Republicans who approve of the president's health care policy. The statistics problem, then, is to compare π_1 and π_2. The procedures developed previously to estimate and test hypotheses about a single Bernoulli proportion are modified to make inferences about the *difference* between two Bernoulli proportions.

As suggested, let π_1 and π_2 denote the proportions of successes in the two Bernoulli populations. From random samples of size n_1 and n_2 that are drawn *independently* from the two populations, we calculate the two sample proportions, p_1 and p_2. From the two sample proportions, we make inferences about the difference between the population proportions, $\pi_1 - \pi_2$.

As in the one-sample inference procedures for π, we need an estimator of the parameter $\pi_1 - \pi_2$ and we need to know its sampling distribution. Extending the Central Limit Theorem from one sample to two samples, we have

---

Sampling Distribution of $p_1 - p_2$

When n_1 and n_2 are sufficiently large, the sampling distribution of $p_1 - p_2$ is

1. approximately normally distributed
2. centered at $\pi_1 - \pi_2$
3. and has a standard deviation of $\sqrt{\dfrac{\pi_1(1 - \pi_1)}{n_1} + \dfrac{\pi_2(1 - \pi_2)}{n_2}}$

Sufficiently large means that if the sample proportions p_1 and p_2 are somewhere between .2 and .8, the sample sizes need to be at least 25. If, on the other hand, either sample proportion is smaller than .2 or larger than .8, the sample sizes should perhaps be as large as 50. Technically speaking, for the normal approximation the accepted criteria is $n_i p_i > 5$ and $n_i(1 - p_i) > 5$ for $i = 1, 2$.

Thus, the statistic $p_1 - p_2$ is an unbiased estimator of $\pi_1 - \pi_2$ and is approximately normally distributed (when n_1 and n_2 are sufficiently large). From its standard deviation we estimate π_1 and π_2 to obtain the standard error

$$SE(p_1 - p_2) = \sqrt{\frac{p_1(1 - p_1)}{n_1} + \frac{p_2(1 - p_2)}{n_2}}$$

Confidence Interval for $\pi_1 - \pi_2$

The confidence interval for $\pi_1 - \pi_2$ is constructed, as before, using the form

$$\text{Estimator} \pm \text{Margin of Error}$$

Because of the normality of the sampling distribution, the margin of error is the critical value z^*, found in the standard normal probability table, times the standard error.

The final form of the confidence interval for $\pi_1 - \pi_2$ becomes:

$$(p_1 - p_2) \pm z^* \sqrt{\frac{p_1(1 - p_1)}{n_1} + \frac{p_2(1 - p_2)}{n_2}}$$

Example 7.7   Example 7.4 reported that 29 out of 106 smokers who chewed nicotine-laced gum remained smoke free for 1 year. Of the 100 smokers who chewed regular gum, only 16 remained smoke free for 1 year. Use this information to find a 98% confidence interval for the difference in the proportions of smokers who successfully use nicotine-laced gum and those who successfully use regular gum.

Solution   Following the suggestion in Example 7.4, we let π_1 denote the proportion of smokers who use nicotine-laced gum and remain smoke free for 1 year, and let π_2 denote the proportion of smokers who chew regular gum and remain smoke free for 1 year. The quantity $\pi_1 - \pi_2$ then represents the difference between the proportions from the two groups. We wish to find a 98% confidence interval for $\pi_1 - \pi_2$.

From the sample data, we have

$$p_1 = \frac{29}{106} = .274 \quad \text{and} \quad p_2 = \frac{16}{100} = .16$$

and

$$SE(p_1 - p_2) = \sqrt{\frac{(.274)(.726)}{106} + \frac{(.16)(.84)}{100}} = .0568$$

For a 98% confidence interval, we have $z^* = 2.33$. The confidence interval for $\pi_1 - \pi_2$ becomes

$$(.274 - .16) \pm 2.33(.0568) \quad \text{or} \quad .114 \pm .132$$

so the interval is $(-.018, .246)$.

Note that the confidence interval for the difference between the proportions contains both positive and negative values. This indicates that a test of hypothesis would not find a significant difference between the two proportions when the level of significance is 2%. Thus, although the percents (27.4% and 16%) appear significantly different, it is likely that the perceived difference between these two sample proportions happened purely by chance. Repeating the experiment may yield percents in the opposite direction. ∎

Large-Sample Hypothesis Test of $\pi_1 - \pi_2$

When we compare two proportions, as in Example 7.7, it may be that the sample proportions are considerably different and yet there is no difference between the population proportions. With a test of hypothesis, we can determine how likely it is that we will see a difference in the sample proportions when there is no difference in the population proportions.

Following the procedures outlined in the one-sample theory, the null and alternative hypotheses to evaluate claims involving π_1 and π_2 can take one of three forms:

$$\text{Left tailed---H}_0: \pi_1 - \pi_2 \geq 0 \quad \text{versus} \quad \text{H}_a: \pi_1 - \pi_2 < 0$$

$$\text{Right tailed---H}_0: \pi_1 - \pi_2 \leq 0 \quad \text{versus} \quad \text{H}_a: \pi_1 - \pi_2 > 0$$

$$\text{Two tailed---H}_0: \pi_1 - \pi_2 = 0 \quad \text{versus} \quad \text{H}_a: \pi_1 - \pi_2 \neq 0$$

Recall that in the development of the hypothesis test, we assume that the null hypothesis is true and then attempt to find statistical evidence to the contrary. More specifically, when calculating the p-value, we assume the equality part of the null hypothesis so as to maximize the p-value; that is, we assume that $\pi_1 - \pi_2 = 0$. Under this assumption, the standard deviation of $p_1 - p_2$ is

$$\sqrt{\pi(1-\pi)}\sqrt{\frac{1}{n_1} + \frac{1}{n_2}}$$

where π is the common value of π_1 and π_2.

Pooling together the information in the two samples, we obtain p, a pooled estimate of π:

$$p = \frac{x_1 + x_2}{n_1 + n_2}$$

where x_1 is the number of successes in sample 1 and x_2 is the number of successes in sample 2. Using this estimate of π, we have the standard error

$$\text{SE}(p_1 - p_2) = \sqrt{p(1-p)}\sqrt{\frac{1}{n_1} + \frac{1}{n_2}}$$

Because H_0 is assumed to be true (until evidence to the contrary is found), the standardized test statistic is

$$z = \frac{p_1 - p_2}{\sqrt{p(1-p)}\sqrt{1/n_1 + 1/n_2}}$$

which has an approximate *standard* normal sampling distribution when n_1 and n_2 are sufficiently large. At this point, it is easy to see that the remainder of the test is just like the z-test for a single proportion presented in Section 6.2.

Example 7.8 The campaign manager for a presidential candidate wishes to test the claim that the proportion of Ohio voters who favor the candidate is at least as large as the proportion of California voters who favor her. Given these data, test the manager's claim at a 5% level of significance.

	n	Number in favor
Ohio	100	36
California	200	84

Solution   Let π_1 denote the proportion of Ohio voters who favor the candidate, and let π_2 denote the proportion of California voters who favor her. The claim of the campaign manager is that $\pi_1 \geq \pi_2$. To test this claim, the null and alternative hypotheses are

$$H_0: \pi_1 \geq \pi_2 \quad \text{versus} \quad H_a: \pi_1 < \pi_2$$

Pooling the sample information, we find

$$p = \frac{36 + 84}{100 + 200} = \frac{120}{300} = .4$$

Thus, we have

$$SE(p_1 - p_2) = \sqrt{(.4)(.6)} \sqrt{\frac{1}{100} + \frac{1}{200}} = (.4899)(.12247) = .06$$

The observed value of the test statistic is

$$z_{obs} = \frac{.36 - .42}{.06} = -1.0$$

From the z-table we find

$$p\text{-value} = P(z < -1.0) = .1587 > \alpha \, (= .05)$$

Consequently, there is insufficient evidence to reject the null hypothesis. There is no statistical information to refute the campaign manager's claim. ∎

Occasionally the evidence in a statistical study is such that a formal test of significance is not required. A case in point is the Mayo Clinic study described in the Statistical Insight at the beginning of this chapter. The objective of the study was to determine whether large doses of vitamin C prolonged the lives of patients who had cancer. The hypotheses of interest are

$$H_0: \pi_1 \geq \pi_2 \quad \text{versus} \quad H_a: \pi_1 < \pi_2$$

where π_1 is the proportion of patients in the experimental group who die within 1 year and π_2 is the proportion of patients in the control group who die within 1 year. The sample results of the study showed that a greater percentage of the experimental group died within 1 year. Without completing a test, we realize that the null hypothesis is supported by the sample data; there is no statistical evidence to support the research hypothesis. In fact, the p-value of the test would exceed .50 because the test is left tailed and the test statistic falls on the right tail of the sampling distribution. Only in those cases where the evidence is in question is a formal test of significance required.

The inference procedures for $\pi_1 - \pi_2$ are summarized in the box.

Inference Procedures for $\pi_1 - \pi_2$

Application: Bernoulli populations.

Assumptions: Independent samples and sufficiently large for a normal approximation.

Confidence Interval

A $(1 - \alpha)100\%$ confidence interval for $\pi_1 - \pi_2$ is given by the limits

$$(p_1 - p_2) \pm z^* \sqrt{\frac{p_1(1 - p_1)}{n_1} + \frac{p_2(1 - p_2)}{n_2}}$$

where z^* is the upper $\alpha/2$ critical value found in the standard normal probability table.

Large-Sample Hypothesis Test of $\pi_1 - \pi_2$

Left-Tailed Test	Right-Tailed Test	Two-Tailed Test
$H_0: \pi_1 - \pi_2 \geq 0$	$H_0: \pi_1 - \pi_2 \leq 0$	$H_0: \pi_1 - \pi_2 = 0$
$H_a: \pi_1 - \pi_2 < 0$	$H_a: \pi_1 - \pi_2 > 0$	$H_a: \pi_1 - \pi_2 \neq 0$

Standardized test statistic

$$z = \frac{p_1 - p_2}{\sqrt{p(1 - p)}\sqrt{1/n_1 + 1/n_2}}$$

We are given that

$$p_1 = \frac{x_1}{n_1}, \quad p_2 = \frac{x_2}{n_2}, \quad \text{and} \quad p = \frac{x_1 + x_2}{n_1 + n_2}$$

where x_1 is the number of successes in sample 1 and x_2 is the number of successes in sample 2. From the z-table we find

Left-Tailed Test	Right-Tailed Test	Two-Tailed Test
p-value $= P(z < z_{obs})$	p-value $= P(z > z_{obs})$	Find the p-value as in a one-tailed test and then double it.

If a level of significance α is specified, reject H_0 if p-value $< \alpha$.

EXERCISES 7.2 The Basics

7.12 Independent random samples of size 200 each were selected from two Bernoulli populations. The number of successes in sample 1 and sample 2 were 65 and 74, respectively.

a. Calculate individual estimates of the two Bernoulli proportions, π_1 and π_2. If $\pi_1 = \pi_2$, what is the pooled estimate of π, the common value of π_1 and π_2?

b. Calculate the standard error of $p_1 - p_2$ if we cannot assume that $\pi_1 = \pi_2$.

c. Calculate the standard error of $p_1 - p_2$ if we can assume that $\pi_1 = \pi_2$.

d. To test the null hypothesis that $\pi_1 = \pi_2$ versus the alternative that $\pi_1 \neq \pi_2$, should the standard error of $p_1 - p_2$ be calculated as in part **b** or part **c**?

e. Test the null hypothesis stated in part **d**. Be sure to perform all calculations, including the z-statistic and the p-value.

f. Suppose the two Bernoulli proportions represent the percentages of female and male students who have cars registered on campus. How does this information affect your analysis of the data? How does it affect your conclusion regarding the test of significance?

7.13   In a survey taken to help understand emotions, 22 of 70 persons under 18 years of age expressed a fear of meeting people and 23 of 90 persons 18 years old and over expressed the same fear.

a. Formulate the null and alternative hypotheses necessary to determine whether there is a statistical difference between the proportions of those who fear meeting people in the two age groups.

b. What test statistic is used to test the hypotheses in part **a**?

c. To test the hypotheses, should the standard error of the test statistic be computed with a pooled estimate of π?

d. Complete the test. Be sure to compute the test statistic and its p-value and state your conclusion.

7.14   Suppose 250 castings produced in mold A contained 19 defectives and 300 castings produced in mold B contained 27 defectives.

a. Calculate individual estimates of the proportions of defectives produced in mold A and mold B.

b. To find a 99% confidence interval for the difference between the proportions of defectives produced by the two molds, should the two sample proportions be pooled together to calculate the standard error of $p_1 - p_2$?

c. Calculate the confidence interval from part **b**.

d. Does the interval contain both positive and negative values? What does this signify about the two proportions of defectives?

7.15   The FDA requires that the mean antibody strength for a measles vaccine exceed 1.6 before it can be placed on the market. In a sample of size 55, firm A's vaccine exceeded the requirements 37 times. In a sample of size 46, firm B's vaccine exceeded the requirements 33 times. Is there a significant difference between the success rates for the two firms?

a. Formulate null and alternative hypotheses to evaluate the difference.

b. What test statistic is used to test the hypotheses?

c. To test the hypotheses, should the standard error of the test statistic be computed with a pooled estimate of π?

d. Complete the test. Be sure to compute the test statistic and its p-value and state your conclusion.

7.16   In hypothesis testing, why do we calculate a pooled estimate of π when we calculate the standard error of $p_1 - p_2$? Why do we not pool in confidence interval estimation?

7.17   In a particular occupation, a random sample of 210 men found that 65 smoked. In a random sample of 240 women, 87 smoked. Are the percents of men and women in this occupation who smoke significantly different? Formulate null and alternative hypotheses to evaluate the difference. Complete the test and be sure to give the p-value of the test statistic as well as a summary.

7.18   To test the effectiveness of a vaccine against a certain disease, 120 experimental animals were given the vaccine and 180 were not. All 300 animals were then infected with the disease. Among those vaccinated, l5 died as a result of the disease. In the control group, 36 died. Can we conclude that the vaccine was effective in reducing the mortality rate? Formulate the null and alternative hypotheses if we believe that the vaccine was effective in reducing the mortality rate. Test the hypotheses and be sure to give the p-value of the test statistic as well as a summary. Is this a controlled experiment or an observational study?

7.19   A sample of 500 adults ranging in age from 24 through 54 was divided randomly into two groups. Those in group 1 consumed 1000 milligrams of vitamin C each day for 6 months. Those in group 2 consumed a placebo each day for the same 6 months. Of the 238 (12 dropped out of the experiment for various reasons) in group 1, 42 had at least one cold during the period. Of the 241 in group 2, 61 had at least one cold during the period. Is there a significant difference between the percents that had at least one cold in each group? Formulate the hypotheses and conduct the test. Is this a controlled experiment or an observational study?

7.20   To be judged proficient in reading at a certain university, a student must score 85 or above on the Nelson-Denny reading test. Of the 1250 entering freshmen, 190 were randomly selected and given a special 2-week course in reading. After the course, 146 of the 190 were judged proficient in reading.

Of the remaining 1060 entering freshmen, 762 were judged proficient in reading. Did the reading course increase reading scores? To evaluate with a test of significance, should the test be a one-tailed or a two-tailed test? Formulate the hypotheses and conduct the test. Be sure to give the *p*-value and state your conclusion.

7.21 To study the effectiveness of the drug AZT for preventing AIDS among their offspring, 164 pregnant HIV-positive women were randomly assigned to receive AZT and a like group of 160 were randomly assigned to receive a placebo (*Newsweek*, March 7, 1994). Among the 164 births in the AZT group, 13 children were diagnosed as HIV-positive, and among the 160 births in the placebo group, 40 children were diagnosed as HIV-positive. Is there statistical evidence that AZT reduces the chances of children of HIV-positive mothers developing AIDS? Evaluate with a test of significance. Should the test be a one-tailed or a two-tailed test? Formulate the hypotheses and conduct the test. Be sure to give the *p*-value and state your conclusion.

7.22 Do you worry about being a victim of crime? Do people who live in cities feel more vulnerable than those who live in the suburbs? A telephone poll of 500 adult Americans conducted for *TIME*/CNN on August 12, 1993, found that 59% of those living in cities versus 57% of those living in the suburbs were worried about being a victim of crime. Are these figures statistically significantly different? Assume there were 250 adults in each sample; that is, 250 lived in cities and 250 lived in the suburbs. Test the hypothesis that there is no difference in the population proportions.

7.23 A survey by the American Association of University Women reported that 64% of third-grade girls and 66% of third-grade boys said they were good in math. Assume that the survey involved 150 girls and 200 boys. Are the percentages statistically different? Formulate null and alternative hypotheses to test the equality of the two proportions. Conduct a test of significance to evaluate the findings. Report the *p*-value and write a summary of the results.

7.24 In the same study as that reported in Exercise 7.23, the American Association of University Women determined that only 48% of 11th-grade girls versus 60% of 11th-grade boys felt they were good in math. Assume again that the sample sizes were 150 and 200, respectively. Conduct the same test of significance for the 11th-graders as you did for the third-graders in Exercise 7.23. Are the results different? What can you conclude about girls' and boys' feelings about math as they get older?

7.3 Comparing Two Population Means

Learning Objectives for this section:

❏ Be able to find a confidence interval for the difference between two population means.

❏ Be able to test hypotheses about the difference between two population means.

❏ Know the applications and assumptions for the inference procedures concerning the difference between two population means.

For a certain medical operation, patients are randomly and independently assigned to two different techniques to determine whether the recovery times differ for the two methods (see Example 7.5). A manufacturer of plastic bags for grocery stores wishes to compare the mean tensile strengths of bags produced by two different manufacturing processes. A commercial fishery compares the haddock catches from a small-mesh net versus a large-mesh net. Is the mean pH level of water in the Great Smoky Mountains the same for high and low elevations? In each of these problems, we wish to compare the means of two independent samples selected from two different populations. Initially, we should examine the two samples with dotplots that have a common scale and/or side-by-side boxplots. If the graphical information indicates that the distributions are somewhat symmetric with tails that are not excessively long, we can proceed to make inferences about the means of the two populations much like the single-population inference procedures given in Section 6.3. As in the inference procedures for proportions given in the preceding section, the procedures used to estimate and test hypotheses about a single population mean are modified to apply to the problem of comparing two population means.

Two-Sample t-Procedures

Let μ_1 and σ_1 denote the mean and standard deviation of population 1 and let μ_2 and σ_2 denote the mean and standard deviation of population 2, respectively. Suppose we have two *independent* samples of sizes n_1 and n_2 from the two populations. From the two samples we can make inferences about the difference, $\mu_1 - \mu_2$, between the two population means.

We first need a good estimator of the parameter $\mu_1 - \mu_2$. As in single-parameter estimation, the choice of an estimator depends on the distribution of the underlying parent populations. Initially, we will assume that both populations are normally distributed. Because of the robustness of the inference procedures, however, we can relax the condition and apply the results to situations where the two parent distributions are symmetric with tails that are not excessively long. In later sections we will consider skewed and long-tailed distributions.

Following the single-parameter estimation problem, a reasonable estimator of $\mu_1 - \mu_2$ is the difference between the two sample means, $\bar{x}_1 - \bar{x}_2$. As before, to form a confidence interval or construct a test of significance, we need the sampling distribution and standard error of the estimator. Extending the results from one sample to two samples we have these results.

Sampling Distribution of $\bar{x}_1 - \bar{x}_2$

Independent random samples of size n_1 and n_2 are selected from two populations that have means μ_1 and μ_2 and standard deviations σ_1 and σ_2, respectively. If the parent distributions are close to normal, or if n_1 and n_2 are each large, then the sampling distribution of the difference between the sample means, $\bar{x}_1 - \bar{x}_2$, is

1. approximately normally distributed
2. centered at $\mu_1 - \mu_2$
3. and has a standard deviation of $\sqrt{\sigma_1^2/n_1 + \sigma_2^2/n_2}$

Note: If the two populations are normally distributed, the sampling distribution of $\bar{x}_1 - \bar{x}_2$ is also normally distributed regardless of the sample sizes. It is only when the population distributions deviate from normality that the sample sizes must be large. Unless there is strong skewness or symmetry with long tails, a sample size of 30 or more is considered large.

Because the sampling distribution of $\bar{x}_1 - \bar{x}_2$ is approximately normal with a mean of $\mu_1 - \mu_2$ and a standard deviation of $\sqrt{\sigma_1^2/n_1 + \sigma_2^2/n_2}$ we have that the standardized form

$$z = \frac{(\bar{x}_1 - \bar{x}_2) - (\mu_1 - \mu_2)}{\sqrt{\dfrac{\sigma_1^2}{n_1} + \dfrac{\sigma_2^2}{n_2}}}$$

has an approximate *standard* normal sampling distribution.

In practice, however, we rarely ever know the numerical values of σ_1^2 and σ_2^2. Consequently, they are replaced with their respective estimators, s_1^2 and s_2^2. Then we have the standard error

$$\mathrm{SE}(\bar{x}_1 - \bar{x}_2) = \sqrt{\dfrac{s_1^2}{n_1} + \dfrac{s_2^2}{n_2}}$$

Because we have estimated population variances with sample variances, the sampling distribution of

$$t = \frac{(\bar{x}_1 - \bar{x}_2) - (\mu_1 - \mu_2)}{\sqrt{\dfrac{s_1^2}{n_1} + \dfrac{s_2^2}{n_2}}}$$

is no longer standard normal. The distribution is, however, approximated with a t-distribution that has degrees of freedom given by this formula:

$$df = \frac{(s_1^2/n_1 + s_2^2/n_2)^2}{\dfrac{(s_1^2/n_1)^2}{n_1 - 1} + \dfrac{(s_2^2/n_2)^2}{n_2 - 1}}$$

If the formula yields a decimal number, round it down to the nearest whole number.

Confidence Interval for $\mu_1 - \mu_2$ Based on $\bar{x}_1 - \bar{x}_2$

For a $(1 - \alpha)100\%$ confidence interval for $\mu_1 - \mu_2$, the estimator is $\bar{x}_1 - \bar{x}_2$, the standard error is $\sqrt{s_1^2/n_1 + s_2^2/n_2}$ and the critical value, t^*, is found in the t-table with degrees of freedom given by the above formula. The final form of the confidence interval for $\mu_1 - \mu_2$ becomes:

$$(\bar{x}_1 - \bar{x}_2) \pm t^* \sqrt{\frac{s_1^2}{n_1} + \frac{s_2^2}{n_2}}$$

Example 7.9   Wind speed data were gathered during January and July at the site proposed for a wind generator to determine whether the production of electricity by the wind generator will be different in the two months. Assume that the distribution of wind speeds (in knots per hour) at this location is symmetric without long tails. From the summary data, construct a 99% confidence interval for the difference between the mean wind speeds in January and July.

	n	$\bar{x}$	s^2
January	32	23.4	26.42
July	35	16.2	23.875

Solution   Let μ_1 denote the mean wind speed in January and μ_2 denote the mean wind speed in July. The problem is to construct a confidence interval for $\mu_1 - \mu_2$. Because we can assume that the distribution of wind speeds at this location is symmetric with tails that are not unusually long, we use $\bar{x}_1 - \bar{x}_2$ as the estimate of $\mu_1 - \mu_2$. From the summary data, we have

$$\bar{x}_1 - \bar{x}_2 = 23.4 - 16.2 = 7.2$$

and

$$SE(\bar{x}_1 - \bar{x}_2) = \sqrt{\frac{26.42}{32} + \frac{23.875}{35}} = 1.228$$

Because a 99% confidence interval is desired, t^* is the upper .005 ($1 - \alpha = .99$; $\alpha = .01$; $\alpha/2 = .005$) critical value found in the t-table with

$$df = \frac{[26.42/32 + 23.875/35]^2}{\dfrac{(26.42/32)^2}{31} + \dfrac{(23.875/35)^2}{34}} \approx 63$$

From the t-table with 63 degrees of freedom we find the upper .005 critical value to be $t^* = 2.66$. (We take the closest value, which corresponds to 60 degrees of freedom.)

Therefore, the 99% confidence interval for $\mu_1 - \mu_2$ is

$$7.2 \pm (2.66)(1.228) \quad \text{or} \quad 7.2 \pm 3.266$$

Thus, we are reasonably sure that the difference between the mean wind speeds in the two months is somewhere in the interval (3.934, 10.466).

The length of this interval is rather large and may not provide much useful information. To decrease the length, we can reduce the confidence level from 99% to, say, 90%, but this may not be desirable. The other choice is to take larger samples. Given that the equipment is already in place to measure the wind speed, collecting more data would be a reasonable next step. ∎

Hypothesis Test for $\mu_1 - \mu_2$ Based on $\bar{x}_1 - \bar{x}_2$

We next develop the hypothesis testing procedures for the difference between two population means. The null and alternate hypotheses can take one of three forms:

$$\text{Left Tailed—H}_0\text{: } \mu_1 - \mu_2 \geq 0 \quad \text{versus} \quad \text{H}_a\text{: } \mu_1 - \mu_2 < 0$$

$$\text{Right Tailed—H}_0\text{: } \mu_1 - \mu_2 \leq 0 \quad \text{versus} \quad \text{H}_a\text{: } \mu_1 - \mu_2 > 0$$

$$\text{Two Tailed—H}_0\text{: } \mu_1 - \mu_2 = 0 \quad \text{versus} \quad \text{H}_a\text{: } \mu_1 - \mu_2 \neq 0$$

In the left-tailed test, for example, testing H_0: $\mu_1 - \mu_2 \geq 0$ is the same as testing H_0: $\mu_1 \geq \mu_2$. Therefore, the test of the difference between two population means is appropriate when we wish to determine whether the mean response under one condition is greater than (or less than in the case of a right-tailed test) the mean response under another condition. Also, it is easy to see that the inference is for the difference between two population means, $\mu_1 - \mu_2$. Given that the population distributions are symmetric and do not have long tails, the choice for the test statistic is the estimator, $\bar{x}_1 - \bar{x}_2$.

In the test of H_0: $\mu_1 - \mu_2 = 0$, the problem is to determine whether $\bar{x}_1 - \bar{x}_2$ is significantly different from zero. As in single-parameter inference, significant departures of a test statistic from a given value are measured by first computing the "standardized" test statistic and then finding the associated p-value. In this case, the standardized test statistic is

$$t = \frac{(\bar{x}_1 - \bar{x}_2) - (\mu_1 - \mu_2)}{\sqrt{\dfrac{s_1^2}{n_1} + \dfrac{s_2^2}{n_2}}}$$

which has an approximate t distribution.

In testing H_0: $\mu_1 - \mu_2 \geq 0$ versus H_a: $\mu_1 - \mu_2 < 0$, for example, we have

$$p\text{-value} = P(t < t_{obs})$$

Recall that testing a null hypothesis, such as H_0: $\mu_1 - \mu_2 \geq 0$, is equivalent to testing H_0: $\mu_1 - \mu_2 = 0$, because the p-value is maximized when the null is stated with an equal sign.

where t_{obs} is found by substituting the observed values of n_1, n_2, $\bar{x}_1$, $\bar{x}_2$, s_1^2, and s_2^2 into the formula for t.

Example 7.10   Plastic grocery bags have almost completely replaced the standard brown paper bags at the supermarket. When plastic bags were introduced, problems with tearing and ripping resulted in some very upset customers. One particular company was trying to increase the tensile strength of the bags and still hold down production costs by adjusting temperature and pressure in the production runs. These summary data are from two independent random samples and give the tensile strengths of plastic bags from two different production runs:

$$\text{Sample 1:} \quad n_1 = 32 \quad \bar{x}_1 = 102.33 \quad s_1 = 14.06$$

$$\text{Sample 2:} \quad n_2 = 40 \quad \bar{x}_2 = 118.19 \quad s_2 = 24.44$$

Assume the two population distributions that produced the data are symmetric with tails that are not excessively long. Determine whether there is a significant difference between the mean tensile strengths from the two production runs.

Solution   Let μ_1 and μ_2 denote the mean tensile strengths from the two different production runs. The null and alternative hypotheses are:

$$H_0: \mu_1 - \mu_2 = 0 \quad \text{versus} \quad H_a: \mu_1 - \mu_2 \neq 0$$

Under the assumption that the null hypothesis is true (that is, $\mu_1 - \mu_2 = 0$), the standardized test statistic becomes

$$t = \frac{(\bar{x}_1 - \bar{x}_2)}{\sqrt{\dfrac{s_1^2}{n_1} + \dfrac{s_2^2}{n_2}}}$$

From the sample data we find

$$t_{obs} = \frac{102.33 - 118.19}{\sqrt{(14.06)^2/32 + (24.44)^2/40}} = -3.45$$

$$df = \frac{[(14.06)^2/32 + (24.44)^2/40]^2}{\dfrac{((14.06)^2/32))^2}{31} + \dfrac{((2.44)^2/40))^2}{39}} \approx 64$$

Because the test is two tailed,

$$p\text{-value} = 2P(t < -3.45)$$

From the t-table with 60 degrees of freedom (the closest value to 64) we see that 3.46 corresponds to a tail probability of .0005. Because this is so close to 3.45, we have

$$p\text{-value} \approx 2(.0005) = .001$$

This small p-value indicates that the difference between the mean tensile strengths of the plastic bags from the two different production runs is highly significant.   ∎

The highly significant difference in mean tensile strength of plastic bags from the two different production runs in the preceding example was not at all apparent from the summary data. A graphical illustration of the data, such as boxplots or dotplots, prior to any formal test may reveal differences that might indicate that a formal test is not required.

Example 7.11 The actual data for the tensile strengths of plastic bags for the two production runs described in Example 7.10 are given in worksheet: **Tensile.mtw**. Complete side-by-side boxplots of the data and comment on any differences.

Solution Figure 7.1 shows side-by-side boxplots of the two production runs.

Figure 7.1

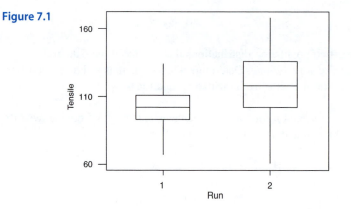

Both distributions appear symmetric without excessively long tails. The boxplots show that the median for run 1 is very near the first quartile of run 2 and the median of run 2 is greater than the third quartile of run 1. Notice also that the data in run 2 are considerably more variable than in run 1. Thus, run 2, on average, produces a higher tensile strength than run 1, but it has considerably more variability. In all, the boxplots clearly show a difference in the two production runs. The test of significance given in Example 7.10 is for means and not medians. Because of the symmetry, however, there is very little difference between the two, and the test of significance simply confirms what we see in the boxplots. ■

The inference procedures for $\mu_1 - \mu_2$ are summarized in the box.

Two-Sample t-Procedures for $\mu_1 - \mu_2$

Application: Symmetric distributions whose tails are not excessively long.

Assumption: The two samples are independent of each other. If the parent distributions deviate substantially from normality, the sample sizes should be large. In most cases, larger than 30 will suffice.

Confidence Interval

A $(1 - \alpha)100\%$ confidence interval for $\mu_1 - \mu_2$ is given by the limits

$$(\bar{x}_1 - \bar{x}_2) \pm t^* \sqrt{\frac{s_1^2}{n_1} + \frac{s_2^2}{n_2}}$$

where t^* is the upper $\alpha/2$ critical value found in the t-table.

continued

Two-Sample t-Test

Left-Tailed Test	Right-Tailed Test	Two-Tailed Test
H_0: $\mu_1 - \mu_2 \geq 0$	H_0: $\mu_1 - \mu_2 \leq 0$	H_0: $\mu_1 - \mu_2 = 0$
H_a: $\mu_1 - \mu_2 < 0$	H_a: $\mu_1 - \mu_2 > 0$	H_a: $\mu_1 - \mu_2 \neq 0$

Standardized test statistic

$$t = \frac{(\bar{x}_1 - \bar{x}_2)}{\sqrt{\dfrac{s_1^2}{n_1} + \dfrac{s_2^2}{n_2}}}$$

The degrees of freedom for the confidence interval and the test of hypothesis are given by

$$df = \frac{(s_1^2/n_1 + s_2^2/n_2)^2}{\dfrac{(s_1^2/n_1)^2}{n_1 - 1} + \dfrac{(s_2^2/n_2)^2}{n_2 - 1}} \quad \text{(round down to the nearest integer)}$$

The p-value is the tail probability most closely associated with the observed value of t, t_{obs}. If t_{obs} falls between two table values, give the two associated probabilities as bounds for the p-value. The p-value for the two-tailed test is calculated as above in the single-tailed test and then doubled to account for both tails.

If a level of significance is specified, reject H_0 if p-value $< \alpha$.

Computer Analysis

As demonstrated in Examples 7.9 and 7.10, the computations for the two-sample t-test are rather complicated, especially for calculating the degrees of freedom. Therefore, it is highly recommended that you use a computer to analyze data involving two or more samples. As you learn more about the analysis of statistical data, you will realize that the computer is almost indispensable. This is especially true in the later chapters of this book on regression analysis and the analysis of variance.

Figure 7.2 is the computer output of the two-sample t-test applied to the data in worksheet: **Tensile.mtw**. Notice that the summary statistics (after rounding) are as they were in Example 7.10. The procedure has computed a 95% confidence interval for the difference between the means and has confirmed the test statistic as $t = -3.45$. To find the p-value, the computer used 64 degrees of freedom where we used 60 (our table does not have 64 degrees of freedom). We got the same p-value because when the degrees of freedom become large, there is very little difference between the t critical values.

Figure 7.2   Two-Sample t-Test and CI: for Tensile, Run

```
Two-sample T for Tensile

Run        N        Mean        StDev        SE Mean
1         32        102.3        14.1            2.5
2         40        118.2        24.4            3.9

Difference = mu (1) - mu (2)
Estimate for difference: -15.85
95% CI for difference: (-25.03, -6.67)
T-test of difference = 0 (vs not =): T-Value = -3.45 P-Value = 0.001 DF = 64
```

Robustness of the Two-Sample *t*-Procedures

The development of the two-sample *t*-procedures relied on the assumption that the two parent populations are normally distributed. As was pointed out in the one-sample case, rarely will we ever see distributions that are exactly normally distributed. As it turns out, this is not a serious problem, however, because of the robustness of the *t*-procedures. In fact, if the sample sizes are reasonably close and the distributions are similar in shape, the two-sample *t*-procedures are more robust than the one-sample *t*-procedures. And as before, when we have large samples, the normality of the parent distributions is not as important because of the Central Limit Theorem. Therefore, the *t*-procedures are valid over a large spectrum of underlying parent distributions. This does not mean all distributions, however. As pointed out previously, the statistical properties of $\bar{x}$ as an estimator of the population mean tend to deteriorate when there are outliers. The result is a loss of precision in estimating μ with the usual confidence interval based on $\bar{x}$. The same is true when estimating or testing the difference in population means. If the parent distributions are symmetric with long tails or heavily skewed, the confidence interval and hypothesis-testing problems should not be based on *t*-procedures. The Wilcoxon rank-sum test, covered in Section 7.4, gives an alternative for the *t*-procedures.

A Conservative Approach for the Two-Sample *t*-Procedures

As previously suggested, a computer is recommended for the analysis of data for the two-sample *t*-procedures. If we are performing calculations without a computer, however, the formula for calculating degrees of freedom is somewhat complicated. In those cases it is recommended that we use the following conservative approach for the two-sample *t*-procedures.

For a conservative approach to the two-sample *t*-procedures, the degrees of freedom are given by

$$\text{df} = \text{smaller of } n_1 - 1 \text{ and } n_2 - 1$$

This conservative approach gives a confidence interval whose confidence level is slightly higher than what is reported, and in hypothesis testing gives a *p*-value larger than its actual value. As the sample sizes increase, the confidence level of the interval becomes closer to its stated level and the *p*-value for a test becomes more accurate.

EXERCISES 7.3   The Basics

7.25   The standardized norm for a science test for tenth-graders has a mean of 75 and a standard deviation of 20. In the Grover County school district, 123 tenth-grade students were randomly divided into two groups. One group of 61 students received instruction through a traditional lecture class, and the other group of 62 students received instruction in an experimental class. The average grade on the science test in the traditional group was 77.2 and the standard deviation was 19.6; the average in the experimental group was 78.6 and the standard deviation was 42.4. Is there statistical evidence of a difference between the two groups? Formulate null and alternative hypotheses to evaluate the difference between the two groups. Test the hypothesis and be sure to give the observed value of the test statistic and its *p*-value as well as a summary report. Is this a controlled experiment or an observational study?

7.26   The FDA tests tobaccos in two different types of cigars for nicotine content and obtains these results (in milligrams):

Brand A	$n_1 = 23$	$\bar{x}_1 = 85.3$	$s_1 = 12.44$
Brand B	$n_2 = 25$	$\bar{x}_2 = 89.8$	$s_2 = 26.67$

Do these results indicate that there is a significant difference between the mean nicotine content of the two types of cigars? Assume that the population distributions satisfy the conditions for the two-sample *t*-test.

7.27   In an experiment to study the effects of a particular drug on the number of errors in maze-learning behavior of rats, these results were obtained:

Drug Group	$n_1 = 12$	$\bar{x}_1 = 18.67$	$s_1 = 11.21$
Placebo Group	$n_2 = 16$	$\bar{x}_2 = 16.3$	$s_2 = 4.13$

a. State the null hypothesis that the drug has no effect on the errors.
b. What assumptions are required to test the hypothesis in part **a** using the two-sample *t*-test?
c. Assume that the assumptions in part **b** are met and proceed with the test.
d. Calculate the *p*-value and state your conclusion from the test.

7.28   A store owner wants to know whether an advertising campaign has increased mean daily income. The daily incomes for the 2 weeks prior to the campaign were recorded as well as the income for a 2-week period after the campaign.

Before campaign	$n_1 = 12$	$\bar{x}_1 = \$2277$	$s_1 = \$375$
After campaign	$n_2 = 12$	$\bar{x}_2 = \$2564$	$s_2 = \$938$

a. State the null and alternative hypotheses to determine whether the campaign increased mean daily income.
b. What assumptions are required to test the hypotheses in part **a** using the two-sample *t*-test?
c. Based on the values of the two standard deviations, do you think that the population variances are homogeneous? Is this a requirement for the two-sample *t*-test?
d. Assume that the assumptions for the two-sample *t*-test in part **b** are met and proceed with the test. Be sure to calculate the *p*-value and state your conclusion from the test.

7.29   Worksheet **Fish.mtw** contains the lengths and numbers of fish caught with a small-mesh (35-mm diamond-mesh) and a large-mesh (87-mm diamond-mesh) codend (see Exercise 5.83 in Section 5.5). The data produced these summary statistics:

Small-mesh	$n_1 = 739$	$\bar{x}_1 = 33.4235$	$s_1 = 3.41653$
Large-mesh	$n_2 = 767$	$\bar{x}_2 = 34.6154$	$s_2 = 3.16754$

Source: R. Millar (1992), Estimating the Size-Selectivity of Fishing Gear by Conditioning on the Total Catch," *Journal of the American Statistical Association,* 87, 962–968.

a. Formulate null and alternative hypotheses to examine the difference between the lengths of fish caught by the two different size codends.
b. Based on the sample sizes, do you think that the Central Limit Theorem applies to these data?
c. From the summary statistics, calculate the standardized test statistic and its *p*-value and make a decision.
d. Based on your analysis, does the codend size make a difference in the size of fish caught?

Interpreting Computer Output

7.30   The back-to-back stem-and-leaf plot gives the grades from two introductory statistics classes that meet at different times:

WORKSHEET: Statclas.mtw

```
               9am                      2pm
                     0 | 6 | 0
             8 7 7 6 6 6 | 6 |
                   4 2 | 7 | 1 4
             8 8 8 7 7 | 7 | 6 6 6 6 6 9 9
         4 4 3 3 3 2 1 | 8 | 0 1 2 2 2 2 2 4
     9 9 8 7 6 6 5 5 5 | 8 | 5 5 5 6 6 7 8 9 9
           4 4 3 2 1 | 9 | 0 1 2 3
                   8 | 9 | 6
```

a. Do the data satisfy the requirements for the two-sample *t*-procedures?
b. State the null and alternative hypotheses necessary for determining whether there is a significant difference between the mean grades for the two classes.
c. Use this computer printout to determine the observed value of the standardized test statistic that is used to test the hypotheses.

Two Sample *t*-Test and Confidence Interval

```
Twosample T for 9am vs 2pm
             N       Mean      StDev      SE Mean
9am         36      81.00       9.56          1.6
2pm         32      82.50       7.29          1.3

95% C.I. for mu 9am — mu 2pm: ( −5.6, 2.6)
T-Test mu 9am = mu 2pm (vs not =): T= −0.73 P=0.47 DF= 64
```

d. Are the sample sizes large enough for the Central Limit Theorem to apply to these data?
e. From the *p*-value, determine whether to reject the null hypothesis. State your conclusion.

7.31 At a large industrial plant, employees were classified according to age and given a leadership exam. The data are the scores on the exam:

WORKSHEET: Leader.mtw

	Age								
	Under 35				Over 35				
	25	13	9	46	24	31	43	23	13
	25	30	17	20	23	21	42	34	14
	17	20	37	25	15	38	30	14	
Leadership	26	23	20	17	45	19	20	27	
Exam	18	26	11	36	26	38	29	9	
	30	12	32	54	50	41	13	15	
	24	20	16	8	16	68	32	7	
	21	37	31	26	9	28	30	51	

a. State the null and alternative hypotheses to evaluate the claim that the mean score for the over-35 age group exceeds the mean score for the under-35 group.

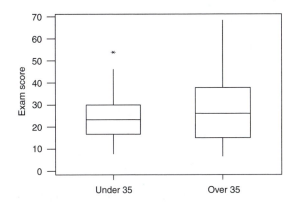

b. Based on the boxplots, do you think the data meet the requirements for the two-sample t-procedures?

c. Use these descriptive statistics to complete the test of hypothesis given in part **a**. Summarize your results.

Descriptive Statistics

Variable	N	Mean	Median	TrMean	StDev	SEMean
under35	32	24.13	23.50	23.39	10.32	1.82
over35	34	27.59	26.50	26.77	14.04	2.41

Variable	Min	Max	Q1	Q3
under35	8.00	54.00	17.00	30.00
over35	7.00	68.00	15.00	38.00

7.32 Do the following boxplots indicate a difference between group A and group B? Give a verbal description of what you see in the boxplots. Is a test of significance required? Why, or why not?

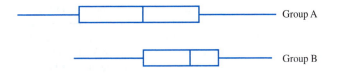

7.33 Do the following boxplots indicate a difference between group A and group B? Give a verbal description of what you see in the boxplots. Is a test of significance required? Why, or why not?

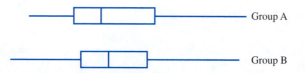

7.34 Patients with small cell lung cancer (SCLC) received one of two types of treatments: arm A or arm B. Their survival times are recorded in worksheet **Censored.mtw**.

Source: Ying, Z., Jung, S., Wei, L. (1995), Survival Analysis with Median Regression Models, *Journal of the American Statistical Association*, 90, 178–184.

a. Based on the side-by-side boxplots, do you think that the data meet the requirements for the two-sample *t*-procedures?

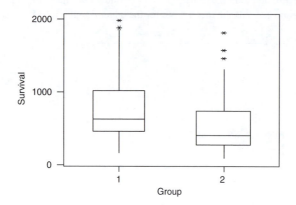

b. Notice from the descriptive statistics that there is a large difference between the mean and median of each group. What would cause such a large difference between a mean and a median?

Descriptive Statistics

Variable	N	Mean	Median	TrMean	StDev	SEMean
survivA	62	766.1	623.0	737.1	456.8	58.0
survivB	59	546.5	395.0	508.9	391.1	50.9

Variable	Min	Max	Q1	Q3
survivA	152.0	1980.0	461.8	1016.7
survivB	83.0	1820.0	265.0	731.0

c. Is it better to compare the population means or the population medians for these data? Explain your answer.

d. Without formally analyzing these data, do you think one treatment is better than the other in increasing survival time?

Computer Exercises

7.35 Exercise 7.10 in Section 7.1 describes worksheet: **Smokyph.mtw**, which gives the water pH values and elevations for 75 water samples taken from the Great Smoky Mountains. Use the Descriptive Statistics command to confirm that the mean pH for the 42 low elevations is 7.2181 and the standard deviation is .5005, and the mean pH for the 33 high elevations is 7.0397 and the standard deviation is .3273. Construct a 90% confidence interval for the difference between the mean pH levels at the low and high elevations. Does the interval contain both positive and negative values?

What does this mean? If the confidence level is increased to 99%, what is the effect on the resulting confidence interval? Is it possible for zero to be inside a 99% confidence interval and not inside a 90% interval?

7.36 Exercise 7.11 in Section 7.1 describes worksheet: **Indy500.mtw**, which gives the qualifying speeds and the numbers of previous starts for drivers in the 1995 Indy 500 race. In an effort to determine whether the qualifying speed is related to the number of previous starts, recall that the drivers were separated into two groups: those with fewer than four previous Indy 500 starts and those with five or more previous Indy 500 starts. In Exercise 7.11 in Section 7.1 you were to evaluate the distributions and state the null and alternative hypotheses to determine whether there is a difference between the qualifying speeds for experienced and inexperienced drivers.

a. Based on your results in Exercise 7.11, do you think that a two-sample t-test is appropriate for these data?

b. Complete the test and determine whether the more experienced drivers have faster qualifying speeds.

7.4 Comparing Two Population Centers: Variances Equal

Learning Objectives for this section:

❏ Be able to find a confidence interval for the difference between two population means using the pooled t-procedures.

❏ Be able to test hypotheses about the difference between two population means using the pooled t-test.

❏ Know the applications and assumptions for the inference procedures concerning the difference between two population means using the pooled t-procedures.

❏ Know how to compare population centers with the two-sample Wilcoxon rank-sum test.

In this section, we present inference procedures for comparing the centers of two population distributions when the population variances are homogeneous; that is, when σ_1^2 and σ_2^2 are equal. For those situations when the parent distributions are symmetric without unusually long tails, we give the *pooled t-procedures* based on $\bar{x}_1 - \bar{x}_2$. When the parent distributions are symmetric with long tails or heavily skewed, we give a nonparametric test called the *Wilcoxon rank-sum test*.

Pooled *t*-Procedures

In certain situations, such as some randomized experiments, it is safe to assume that the population variances are equal. In Example 7.9, we compared wind speed data in the months of January and July at a site proposed for a wind generator. From the summary data, we see that there is very little difference between the two sample variances: $s_{\text{January}}^2 = 26.42$ and $s_{\text{July}}^2 = 23.875$. In this problem, assuming that the population variances are equal is probably justified. In Example 7.10, however, the sample variances of the tensile strengths of plastic bags from two production runs are $s_1^2 = (14.06)^2 = 197.6836$ and $s_2^2 = (24.44)^2 = 597.3136$. The larger sample variance is more than three times greater than the smaller sample variance. There is some doubt whether the two population variances are equal. Some statisticians feel comfortable in assuming homogeneous population variances as long as one sample variance is not more than four times greater than the other, whereas others say not more than three times greater. We will take a closer look at this issue later.

To develop the *pooled t-procedures*, let us assume that the population variances are equal. Then we can write the standard deviation of $\bar{x}_1 - \bar{x}_2$ as

$$\sqrt{\frac{\sigma_1^2}{n_1} + \frac{\sigma_2^2}{n_2}} = \sigma \sqrt{\frac{1}{n_1} + \frac{1}{n_2}}$$

Because $\sigma_1^2 = \sigma_2^2$, the common value can be factored out of the two terms and then taken out from under the square root.

where σ is the common standard deviation. Because $\sigma_1^2 = \sigma_2^2 = \sigma^2$, we have that both s_1^2 and s_2^2 are unbiased estimators of σ^2. We can combine the two estimators using a weighted or pooled average to obtain a single unbiased estimator of σ^2:

$$s_p^2 = \frac{(n_1 - 1)s_1^2 + (n_2 - 1)s_2^2}{n_1 + n_2 - 2}$$

Notice that if $n_1 = n_2$ then $s_p^2 = (s_1^2 + s_2^2)/2$.

Observe that s_1^2 is weighted by its degrees of freedom, $n_1 - 1$, and s_2^2 is weighted by its degrees of freedom, $n_2 - 1$. The *pooled variance*, s_p^2, then has $n_1 - 1 + n_2 - 1 = n_1 + n_2 - 2$ degrees of freedom. Using $s_p = \sqrt{s_p^2}$ as an estimate of σ, we have the standard error:

$$SE(\bar{x}_1 - \bar{x}_2) = s_p \sqrt{\frac{1}{n_1} + \frac{1}{n_2}}$$

If the parent distributions are normal, it follows that

$$t = \frac{(\bar{x}_1 - \bar{x}_2) - (\mu_1 - \mu_2)}{s_p \sqrt{\frac{1}{n_1} + \frac{1}{n_2}}}$$

has a Student's t distribution with $n_1 + n_2 - 2$ degrees of freedom. Consequently, the critical value, t^*, for the confidence interval and the p-value for the test of hypothesis are found in the t-table (Table B.3 in Appendix B). As before, the normality assumption can be relaxed somewhat because of the robustness of the t-procedures. We should, however, find an alternative to the t-procedures when there are outliers or severe skewness.

The pooled t inference procedures for $\mu_1 - \mu_2$ are given in the box.

Pooled t-Procedures for $\mu_1 - \mu_2$

Application: Symmetric distributions whose tails are not excessively long.

Assumptions: The two samples are independent of each other and the population variances are equal. If the parent distributions deviate substantially from normality, the sample sizes should be large. In most cases, larger than 30 will suffice.

Confidence Interval

A $(1 - \alpha)100\%$ confidence interval for $\mu_1 - \mu_2$ is given by the limits

$$(\bar{x}_1 - \bar{x}_2) \pm t^* s_p \sqrt{\frac{1}{n_1} + \frac{1}{n_2}}$$

where

$$s_p = \sqrt{\frac{(n_1 - 1)s_1^2 + (n_2 - 1)s_2^2}{n_1 + n_2 - 2}}$$

and t^* is the upper $\alpha/2$ critical point of the t distribution with $n_1 + n_2 - 2$ degrees of freedom.

continued

Pooled t-Test

Left-Tailed Test	Right-Tailed Test	Two-Tailed Test
$H_0: \mu_1 - \mu_2 \geq 0$	$H_0: \mu_1 - \mu_2 \leq 0$	$H_0: \mu_1 - \mu_2 = 0$
$H_a: \mu_1 - \mu_2 < 0$	$H_a: \mu_1 - \mu_2 > 0$	$H_a: \mu_1 - \mu_2 \neq 0$

Standardized test statistic

$$t = \frac{\bar{x}_1 - \bar{x}_2}{s_p \sqrt{1/n_1 + 1/n_2}}$$

From the t-table with $n_1 + n_2 - 2$ degrees of freedom, the p-value is the tail probability most closely associated with the observed value of t, t_{obs}. If t_{obs} falls between two table values, give the two associated probabilities as bounds for the p-value. The p-value for the two-tailed test is calculated as in the single-tailed test and then doubled to account for both tails.

If a level of significance is specified, then reject H_0 if p-value $< \alpha$.

Example 7.12 **Confidence Interval for $\mu_1 - \mu_2$**

Polychlorinated biphenyls (PCBs) are classified as health hazards. To study the concentration of PCBs in a river, a group of environmentalists measured the PCB levels (in parts per million) in fish at two locations on the river. Their results are given here:

	n	$\bar{x}$	s
Location A:	8	25.2	3.8
Location B:	8	23.1	4.2

Assuming that the variability of the concentration is the same in the two areas and that the distributions are close to normal, construct a 95% confidence interval to estimate the difference between the mean concentration levels at the two sites.

Solution Because the sample sizes $n_1 = n_2 = 8$ are rather small, it is important that the parent populations be reasonably close to being normally distributed. Furthermore, the closeness of the standard deviations suggests homogeneous variances and thus the pooled t-procedures are appropriate. The pooled degrees of freedom are $n_1 + n_2 - 2 = 14$, and for a 95% confidence interval we have that $\alpha = .05$, so that $\alpha/2 = .025$. We then find from the t-table with 14 degrees of freedom that $t^* = 2.145$. The pooled variance is found as follows:

$$s_p^2 = \frac{(n_1 - 1)s_1^2 + (n_2 - 1)s_2^2}{n_1 + n_2 - 2} = \frac{7(3.8)^2 + 7(4.2)^2}{14} = 16.04$$

Therefore, the standard error is

$$SE(\bar{x}_1 - \bar{x}_2) = s_p \sqrt{\frac{1}{n_1} + \frac{1}{n_2}} = \sqrt{16.04} \sqrt{\frac{1}{8} + \frac{1}{8}} = 2.0025$$

The confidence interval becomes

$$(25.2 - 23.1) \pm 2.145 \, (2.0025) \qquad \text{or} \qquad 2.1 \pm 4.295$$

which in turn gives the interval $(-2.195, 6.395)$.

Notice that the confidence interval runs from about -2.2 to $+6.4$, which includes both positive and negative values. This means that a test of hypothesis would not detect a significant difference (with $\alpha = .05$) between the two population means. ∎

The hypothesis test for comparing two population means based on the pooled t-test is developed in a manner similar to the development of the confidence interval.

Example 7.13 **Test for $\mu_1 - \mu_2$**

A random sample of 14 fourth-graders took a standardized achievement test immediately after an hour of recess. A second random sample of 16 fourth-graders took the same test after a 1-hour rest period. From these summary data, test the hypothesis of no difference between the mean scores for the two groups.

	n	$\bar{x}$	s
Recess group	14	56.5	6.2
Rest group	16	62.2	9.8

Solution Because the scores on most standardized tests are normally distributed, the test of hypothesis is based on the difference between the sample means. The null and alternate hypotheses are:

$$H_0: \mu_1 = \mu_2 \quad \text{versus} \quad H_a: \mu_1 \neq \mu_2$$

where μ_1 denotes the mean test score for the recess group and μ_2 denotes the mean test score for the rest group. Normality and the closeness of the standard deviations suggest that the pooled t-test is justified. The test statistic is

$$t = \frac{\bar{x}_1 - \bar{x}_2}{s_p\sqrt{\dfrac{1}{n_1} + \dfrac{1}{n_2}}}$$

The sampling distribution is t with degrees of freedom

$$n_1 + n_2 - 2 = 14 + 16 - 2 = 28$$

From the sample data we have for the observed value of t

$$t_{obs} = \frac{56.5 - 62.2}{\sqrt{[13(6.2)^2 + 15(9.8)^2]/28}\;\sqrt{1/14 + 1/16}}$$

$$= \frac{-5.7}{(8.3245)(.366)}$$

$$= -1.87$$

From the t-table with 28 degrees of freedom, we see that

$$1.701 < 1.87 < 2.048$$

and therefore

The *p*-value is doubled because this is a two-tailed test.

$$2(.025) < p\text{-value} < 2(.05) \quad \text{or} \quad .05 < p\text{-value} < .10$$

Thus, we can say that there is a mildly significant difference between the rest group and the recess group on the achievement test. ∎

Pooled *t*-Procedures

The pooled *t*-procedures are executed in the same manner as the two-sample *t*-test given in the preceding section. The only difference is that for the pooled *t*-procedures, you must click on the **Assume equal variances** box and then click on **OK**.

Example 7.14   A math achievement test is given to a random sample of 25 high school students. The scores and gender (coded as 1 for girls and 2 for boys) are given here:

W O R K S H E E T : Achieve.mtw

Gender	1	2	2	1	2	1	1	1	1	2	2	1	2
Score	87	68	87	91	67	78	81	72	95	74	81	89	93
Gender	2	2	1	2	1	1	2	1	2	2	1	1	
Score	60	78	93	74	83	74	92	75	81	62	85	95	

Is there a significant difference in the scores for boys and girls?

Solution   Let μ_1 denote the mean score for girls and μ_2 denote the mean score for boys. Then the null and alternative hypotheses are

$$H_0: \mu_1 = \mu_2 \quad \text{versus} \quad H_a: \mu_1 \neq \mu_2$$

Rejection of the null hypothesis indicates a difference in the scores for boys and girls.

To conduct a pooled *t*-test, we should first check assumptions. The normal probability plots in Figure 7.3 show no unusual deviations from normality.

Figure 7.3

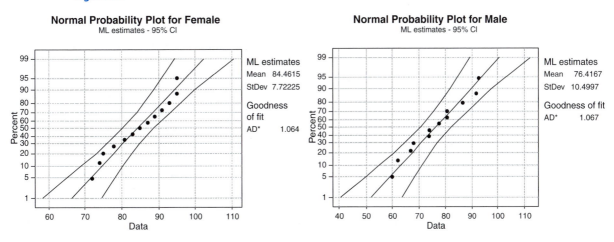

The side-by-side boxplots in Figure 7.4 show that the variances are reasonably homogeneous. The boxplots also show that girls generally scored higher than boys. To determine whether the difference is significant, we will conduct a pooled *t*-test.

Figure 7.4

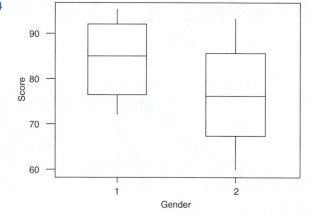

Figure 7.5 is the computer output. Notice that the two standard deviations are close in value. The ratio of the sample variances is $(11.0/8.04)^2 = 1.87$, which suggests that the population variances are homogeneous. Based on a p-value = .047, there is evidence of a significant difference between the scores for boys and girls.

Figure 7.5   Two-Sample t-Test and CI: Female, Male

```
Two-sample T for Female vs Male

                 N        Mean       StDev       SE Mean
Female          13       84.46        8.04           2.2
Male            12       76.4        11.0            3.2

Difference = mu Female − mu Male
Estimate for difference: 8.04
95% CI for difference: (0.14, 15.95)
T-test of difference = 0 (vs not =): T-Value = 2.10 P-Value = 0.047 DF = 23
Both use Pooled StDev = 9.55
```

When to Use the Pooled t-Procedures

The pooled t-procedures assume that the populations are reasonably close to normally distributed and their variances are homogeneous. The normal probability plot is excellent for detecting deviations from normality. Although there is a formal test for the equality of population variances, we have chosen not to present it because it is so sensitive to moderate departures from normality (extremely nonrobust). Moser and Stevens (1992) suggest not to conduct any formal test for homogeneity and to use the pooled t-test only when the ratio of population variances is *known* to be near 1.0. In practice, the ratio of the population variances is rarely known; therefore, their suggestion may be too restrictive. Instead, we suggest that you informally compare the variability of the two samples by viewing side-by-side boxplots of the data and look at the ratio of the sample variances before applying either the pooled t- or the two-sample t-procedures. If the ratio of the sample variances is greater than 4, we should not assume homogeneous variances; if the ratio is less than 3, it is probably safe to assume homogeneous variances. If the ratio is between 3 and 4, as in Example 7.10, we should use the pooled t-procedure if the sample sizes are close; otherwise, use the two-sample t-procedure.

The Wilcoxon Rank-Sum Test

If the assumptions for the pooled t-procedures are not met, there are certain situations when the nonparametric *Wilcoxon rank-sum test* is appropriate. For the two-sample Wilcoxon rank-sum test, no specific distribution assumptions are made except that if a difference exists between the distributions, it is in their locations. This means that the two distributions should be similar in shape and have rather homogeneous variances.

Remember that the boxplot is an excellent tool for detecting a distribution's shape. Side-by-side boxplots are useful for comparing shapes.

For this test, the population medians are the preferred measures of location to study. Inferences, then, are concerned with the difference between the population medians. We give a modified version of the test in which the ordinary pooled t-test is applied to the *rank-transformed* data (Conover and Iman, 1981). The resulting test gives a good approximation to the Wilcoxon test when both sample sizes are at least 10.

To rank-transform a data set, arrange the data in increasing order and assign ranks of 1, 2, 3, and so on to the data values. The pooled t-test is then conducted on the ranks. If two or more observations have the same value (tied), then average ranks are given. For example, this data set

$$8.6, \quad 11.2, \quad 9.3, \quad 7.4, \quad 11.2, \quad 10.4, \quad 9.1, \quad 12.1$$

has two entries of 11.2. They would normally be assigned ranks 6 and 7, but because they are tied, each is assigned a rank of $(6 + 7)/2 = 6.5$. The completed transformed ranking (in the same order as the given data) is

$$2, \quad 6.5, \quad 4, \quad 1, \quad 6.5, \quad 5, \quad 3, \quad 8$$

To conduct the two-sample Wilcoxon rank-sum test, the two samples are combined into one data set and then ranks are assigned. Then the pooled t-test is applied to the two sets of ranks.

The Two-Sample Wilcoxon Rank-Sum Test

Application: Comparing centers of general populations.

Assumptions: Population distributions are similar except for possibly different centers. Samples are independent; $n_1 \geq 10$ and $n_2 \geq 10$.

The hypothesis-testing procedures are the same as the pooled t-test applied to the rank-transformed data.

The next example illustrates the test.

Example 7.15   **Skewed Distributions**

A study was undertaken to determine whether taking a particular drug influences one's acquisition of a response. The dependent measurement is the number of trials required to master a given task. A group of 28 subjects are randomly assigned to the experimental and control groups. Those assigned to the experimental group are given the drug, and those assigned to the control group are given a placebo. After a certain period of time, the numbers of trials to master the task are recorded here:

WORKSHEET: Drug.mtw

Experimental group	17	15	5	14	18	3	16	13	15	16	17	8	19
Control group		14	10	1	12	11	14	8	10	2	12	16	12	15	4	12

The null hypothesis for the Wilcoxon rank-sum test can be stated more generally as H_0: The population distributions are identical. In this example, the distributions differ at most in their medians; hence the null hypothesis is stated as an equality of population medians.

Test the hypothesis to see whether the drug has an effect on the dependent measurement.

Solution Figure 7.6 gives back-to-back stem-and-leaf plots, and Figure 7.7 shows side-by-side boxplots of the data. These graphs indicate that both distributions are similarly skewed. Because the distributions are not symmetric (but have the same shape), we shall use the two-sample Wilcoxon rank-sum test to test the equality of the population medians.

Given that the drug may have a positive or a negative effect on the number of trials to complete the task, we conduct a two-tailed test.

Let θ_1 = the median number of trials to complete the task with the drug and θ_2 = the median number without the drug. The null and alternative hypotheses are:

$$H_0: \theta_1 - \theta_2 = 0 \quad \text{versus} \quad H_a: \theta_1 - \theta_2 \neq 0$$

Figure 7.6

Figure 7.7

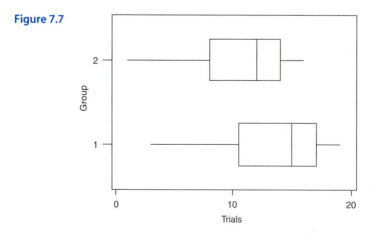

From the back-to-back stem-and-leaf plot, it is easy to rank the combined samples. Figure 7.8 is the same stem-and-leaf plot with the observations replaced by their ranks.

Figure 7.8

```
             Experimental                    Control

                                      0 | 1
                              3       0 | 2
                              5       0 | 4
                                      0 |
                            6.5       0 | 6.5
                                      1 | 8.5,  8.5,  10
                             15       1 | 12.5,  12.5,  12.5,  12.5
                     20, 20, 17       1 | 17,  17,  20
          25.5, 25.5, 23, 23          1 | 23
                     28, 27           1 |
```

The averages and standard deviations for the ranks in the two groups are

$$\text{Experimental:} \quad \bar{r}_1 = 18.346 \qquad s_{r_1} = 8.594$$

$$\text{Control:} \qquad \bar{r}_2 = 11.167 \qquad s_{r_2} = 6.377$$

$$s_p = \sqrt{\frac{12(8.594)^2 + 14(6.377)^2}{26}} = 7.482$$

The test statistic is the usual t-test applied to the rank-transformed data. Thus, we have

$$t_{\text{obs}} = \frac{18.346 - 11.167}{7.482\sqrt{1/13 + 1/15}} = \frac{7.179}{2.835} = 2.53$$

From the t-table with 26 degrees of freedom, we find that $2(.005) < p\text{-value} < 2(.01)$ which gives $.01 < p\text{-value} < .02$. Thus, we can declare a significant difference between the two groups; that is, there is sufficient evidence to say that the drug has an effect on the number of trials needed to master the task. ∎

To investigate the difference between population centers when the parent distributions are symmetric with long tails, it has been suggested that inferences not be based on t-procedures. It is recommended that we use the Wilcoxon rank-sum test when the t-test is in doubt and, in particular, when the parent distributions are symmetric with long tails.

Example 7.16 **Symmetric Distributions with Long Tails**

These are social adjustment scores for a rural group and a city group of children. Test to see whether there is a significant difference in the social adjustment scores for the two groups of children.

WORKSHEET: Rural.mtw

Rural	55	57	62	58	34	52	63	84	50	56	98	58	54	60	55	51
City	61	59	64	42	58	65	81	67	69	23	63	51	65	68	53	61	68

Solution   The null and alternative hypotheses are:

$$H_0: \mu_1 = \mu_2 \quad \text{versus} \quad H_a: \mu_1 \neq \mu_2$$

where μ_1 denotes the mean social adjustment score for the rural group and μ_2 denotes the mean social adjustment score for the city group. Figure 7.9 gives back-to-back stem-and-leaf plots and Figure 7.10 shows side-by-side boxplots for the two groups. Because of the long tails, we should not conduct a *t*-test. Noting that the two distributions are symmetrically distributed and similar in shape, we compare the population means using the Wilcoxon rank-sum test.

Figure 7.9

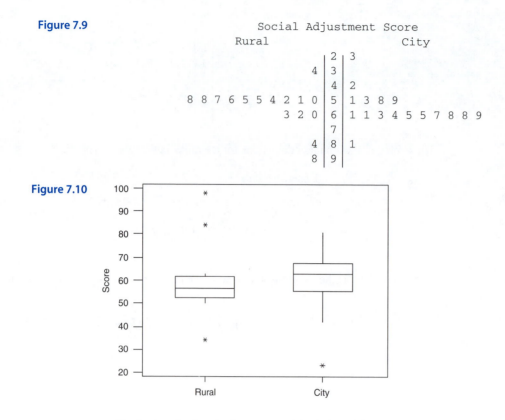

```
                             Social Adjustment Score
                   Rural                           City
                                     |2| 3
                                  4  |3|
                                  4  |4| 2
            8 8 7 6 5 5 4 2 1 0   |5| 1 3 8 9
                        3 2 0     |6| 1 1 3 4 5 5 7 8 8 9
                                  |7|
                                  4  |8| 1
                                  8  |9|
```

Figure 7.10

Replacing the data in the stem-and-leaf plot with ranks, we get Figure 7.11.

Figure 7.11

```
                        Rural                          City
                                          |2| 1
                                    2     |3|
                                    4     |4| 3
  15, 15, 13, 12, 10.5, 10.5, 9, 7, 5.5, 4 |5| 5.5, 8, 15, 17
                       22.5, 21, 18 |6| 19.5, 19.5, 22.5, 24, 25.5, 25.5, 27, 28.5, 28.5, 30
                                          |7|
                                 32  |8| 31
                                 33  |9|
```

From the ranks, we get these averages and standard deviations:

Rural $n_1 = 16$ $\bar{r}_1 = 14.375$ $s_{r_1} = 9.099$ $s_p = 9.458$

City $n_2 = 17$ $\bar{r}_2 = 19.471$ $s_{r_2} = 9.783$

The test statistic is

$$t = \frac{\bar{r}_1 - \bar{r}_2}{s_p\sqrt{1/n_1 + 1/n_2}}$$

$$t_{obs} = \frac{14.375 - 19.471}{9.458\sqrt{1/16 + 1/17}} = -1.55$$

From the t-table and $16 + 17 - 2 = 31$ degrees of freedom, we find that

$$2(.05) < p\text{-value} < 2(.10) \text{ or } .10 < p\text{-value} < .20$$

which indicates no significant difference in the rural and city social adjustment scores. ∎

For the Wilcoxon rank-sum test, the two samples must be combined and ranked before proceeding with a pooled t-test on the ranks. For computer analysis, this is best accomplished if the data are in one column and the groups identified are in another column.

Example 7.17   A new method of making concrete blocks has been proposed. To test whether the new method increases compressive strength, ten sample blocks are made by each method. The compressive strengths in pounds per square inch are listed here:

WORKSHEET: Concrete.mtw

New Method	152	147	134	146	138	156	145	137	157	160
Old Method	132	146	151	127	137	125	138	141	127	137

The data are stored in column C1 with a group identifier (1 for the new method and 2 for the old method) in column C2. Test the hypothesis that the new method is no better than the old using the two-sample Wilcoxon rank-sum test.

Solution   If we let μ_1 denote the mean compressive strength with the new method and μ_2 denote the mean compressive strength with the old method, the null and alternative hypotheses are

$$H_0: \mu_1 \leq \mu_2 \quad \text{versus} \quad H_a: \mu_1 > \mu_2$$

To perform the Wilcoxon rank-sum test we must rank the data in column C1.

COMPUTER TIP

Ranking Data

To rank the data in C1, choose the **Manip** menu and select **Rank**. When the Rank window appears, **Select** C1 for the **Rank data in** box and choose C3 for the **Store ranks in** box and select **OK**.

Now we have the ranks in C3 and the group identifier for the methods in C2. To complete the Wilcoxon rank-sum test, simply apply the two-sample t-test to the ranks in C3 with the subscripts in C2. Because this is a right-tailed test, we must choose the *greater than* alternative in the 2-sample t Options window. Figure 7.12 is the computer output. Based on this p-value, there is highly significant evidence that the new method has increased compressive strength. ∎

Figure 7.12 Two-Sample *t*-Test and CI: Ranks, Meth

```
Two-sample T for Ranks

Method        N          Mean        StDev        SE Mean
1            10         13.60         5.16            1.6
2            10          7.40         5.05            1.6

Difference = mu (1) - mu (2)
Estimate for difference: 6.20
95% CI for difference: (1.40, 11.00)
t-test of difference = 0 (vs not =): T-Value = 2.72 P-Value = 0.014 DF = 18
Both use Pooled StDev = 5.11
```

EXERCISES 7.4 The Basics

7.37 An instructor of an introductory history course is interested in knowing, in general, whether the average grades on the final exam differ from the fall semester to the spring semester. From the statistics, construct a 90% confidence interval based on $\bar{x}_1 - \bar{x}_2$ for the difference between the two population means.

	n	$\bar{x}$	s
Fall	150	82.4	11.56
Spring	150	84.2	11.44

Should we use pooled or unpooled *t*-procedures? Does the interval contain both positive and negative values? What does this signify about the average grades for the fall and spring classes?

7.38 With $\bar{x}_1 - \bar{x}_2$ as the test statistic and the following summary statistics, test the null hypothesis:

$$H_0: \mu_1 - \mu_2 \geq 0 \quad \text{versus} \quad H_a: \mu_1 - \mu_2 < 0$$

$$n_1 = 45 \ \ \bar{x}_1 = 115.3 \ \ s_1 = 15.92$$
$$n_2 = 45 \ \ \bar{x}_2 = 123.6 \ \ s_2 = 17.65$$

a. Check the homogeneous variances assumption.
b. Calculate the standard error of the test statistic.
c. Calculate the observed value of the standardized test statistic.
d. Is this a one-tailed or two-tailed test? Calculate the *p*-value.
e. Based on the *p*-value, should the null hypothesis be rejected?

7.39 It is believed that the mean amount of coffee dispensed by vending machine A in the student lounge is less than that of machine B in the cafeteria. These summary statistics were obtained from samples of each machine:

	n	$\bar{x}$	s
Machine A	10	9.8	1.4
Machine B	12	10.1	.8

a. State the null and alternative hypotheses to determine whether the mean amount of coffee dispensed by vending machine A in the student lounge is less than that of machine B in the cafeteria.
b. To test the claim with the pooled *t*-test, we must make certain assumptions. What are two of those assumptions?
c. Assume that the assumptions in part **b** are met and proceed with the test. Is there statistical evidence to support the conjecture?

7.40   These data are arranged in a back-to-back stem-and-leaf plot:

WORKSHEET: BacktoBack.mtw

```
            0 │ 14 │ 1
              │ 15 │
            5 │ 16 │
              │ 17 │ 3
        8   4 │ 18 │ 2 7
      8 5   3 │ 19 │ 0 8
        2   0 │ 20 │ 1 3 7
        9   7 │ 21 │ 2 8 9
            3 │ 22 │
```

a. If these data are independent random samples from two populations, do you think that inferences should be about $\mu_1 - \mu_2$ or $\theta_1 - \theta_2$? Explain your reasoning.

b. Based on your answer to part **a**, either conduct a pooled t-test or the two-sample Wilcoxon rank-sum test.

7.41   Two independent random samples were selected from two populations with the following values:

WORKSHEET: Independent.mtw

```
            Sample A                    Sample B
      6  │                        4  │
      6  │ 7                      9  │
      7  │ 2                      0  │
      7  │                        5  │
      8  │ 2 3                    0  │ 1 4
      8  │ 6 7                    5  │ 6 7 7 8 9 9
      9  │ 0 1 2 2 3 3 4          0  │ 1 3 4
      9  │ 5 5 6 8 8 9            5  │ 8
     10  │                        2  │
     10  │ 5 7
     11  │                        3  │
     11  │ 8                      8  │
     12  │
     12  │ 5
```

a. Do the samples appear to come from normal populations?

b. Do the variances appear to be homogeneous?

c. Would boxplots for the two samples help in answering the questions in parts **a** and **b**? Explain.

d. It is quite possible that the two samples came from long-tailed distributions. Is it advisable to test the hypothesis that $\mu_1 = \mu_2$ using the pooled t-test?

e. Complete an analysis of the data by completing the two-sample Wilcoxon rank-sum test.

Interpreting Computer Output

7.42   Strength tests on two types of wool fabric produced these results:

WORKSHEET: Wool.mtw

Type 1	138	127	148	134	125	136	152	110	137	160
Type 2	134	137	135	140	130	134	120	157	162	114

a. State the null and alternative hypotheses to see whether there is a significant difference between the mean strengths of the two types of wool.

b. If we want to conduct a pooled *t*-test with small samples, the data should come from normal populations with homogeneous variances. Based on these normal probability plots of the two samples, is the normality assumption met?

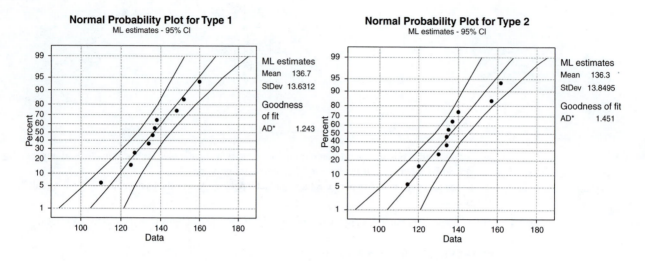

c. Based on the side-by-side boxplots of the two samples, do the variances of the two samples appear to be homogeneous?

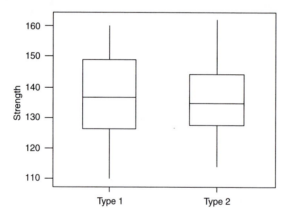

d. Given this computer printout, complete the pooled *t*-test and state your conclusion.

Two-Sample *t*-Test and CI: Type 1, Type 2

```
Two-sample T for Type 1 vs Type 2

              N      Mean     StDev     SE Mean
Type 1       10     136.7      14.4        4.5
Type 2       10     136.3      14.6        4.6

Difference = mu Type 1 - mu Type 2
Estimate for difference: 0.40
95% CI for difference: (-13.21, 14.01)
t-test of difference = 0 (vs not =): T-Value = 0.06 P-Value = 0.951 DF = 18
Both use Pooled StDev = 14.5
```

7.43 Nationally, 25% of all college freshmen enroll in some type of remedial math course. To determine whether men and women differ in their preenrollment ability, a math placement exam (scored from 0 to 40) was given to a sample of 35 incoming freshmen women and a sample of 42 incoming freshmen men. These are their scores:

WORKSHEET: Remedial.mtw

Females: 28, 21, 4, 20, 16, 19, 39, 22, 5, 18, 17, 21, 19, 3, 11, 22, 19, 18, 17, 35, 21, 18, 16, 5, 19, 21, 20, 20, 40, 21, 19, 20, 22, 38, 3

Males: 18, 22, 16, 14, 2, 16, 18, 20, 22, 19, 11, 35, 18, 22, 4, 21, 20, 15, 19, 17, 38, 18, 16, 4, 19, 18, 16, 17, 33, 16, 19, 18, 21, 20, 17, 17, 13, 14, 19, 15, 38, 40

a. Based on side-by-side boxplots of the two samples, does there appear to be a significant difference between the mean abilities of men and women?

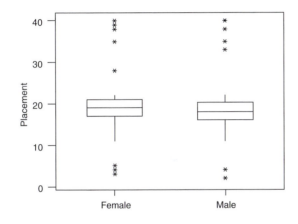

b. Use the descriptive statistics for the two samples to construct a 98% confidence interval based on $\bar{x}_1 - \bar{x}_2$ for the mean difference in ability for men and women.

Descriptive Statistics

Variable	N	Mean	Median	TrMean	StDev	SEMean
female	35	19.34	19.00	19.10	9.09	1.54
male	42	18.93	18.00	18.71	8.03	1.24

Variable	Min	Max	Q1	Q3
female	3.00	40.00	17.00	21.00
male	2.00	40.00	16.00	20.25

c. Does the interval from part **b** contain both positive and negative values? Does this suggest a significant difference between the abilities of men and women?

d. The confidence interval in part **b** is based on $\bar{x}_1 - \bar{x}_2$. From the appearance of the boxplots, can you think of an alternative procedure for finding a confidence interval for $\mu_1 - \mu_2$?

7.44 Measurements of viscosity for a certain substance were taken on two days:

WORKSHEET: Viscosit.mtw

First day	35.4 38.3 34.0 36.2 37.2 32.8 36.0 35.2 36.0 31.3 35.7
Second day	37.0 38.6 35.1 37.1 36.2 36.8 37.6 34.8 35.8 38.2 32.1

a. From the boxplots of the two samples, do you think that the populations are sufficiently skewed that inferences should be made about the population medians instead of the population means?

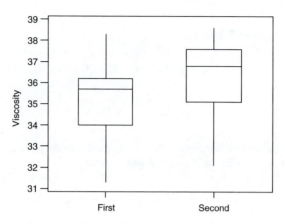

b. Investigate the normal probability plots of the two samples. Does it appear that the normality assumption for the pooled *t*-test has been violated?

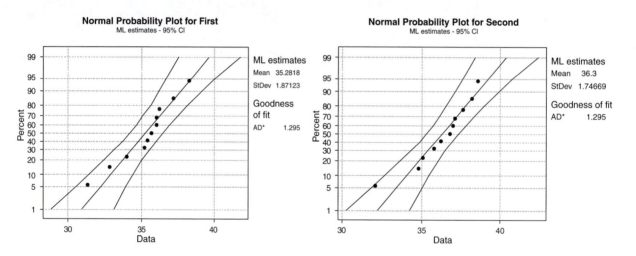

c. In terms of the population means, state the null and alternative hypotheses to determine whether the population has changed from one day to the next.

d. Using the computer printout, complete the pooled *t*-test and state your conclusion.

Two-Sample *t*-Test and CI: first, second

```
Two-sample T for first vs second

             N      Mean     StDev    SE Mean
first        11     35.28    1.96     0.59
second       11     36.30    1.83     0.55

Difference = mu first – mu second
Estimate for difference: –1.018
95% CI for difference: (–2.707, 0.670)
t-test of difference = 0 (vs not =): T-Value = –1.26 P-Value = 0.223 DF = 20
Both use Pooled StDev = 1.90
```

7.45 A manufacturer feels that the amount of carbon monoxide emitted by its smokestacks is less than the smoke from its competitor's stacks. The EPA released these readings:

WORKSHEET: Monoxide.mtw

Manufacturer	2.7 3.1 3.1 2.9 2.5 3.4 3.4 3.4 2.4
Competitor	3.7 3.0 3.5 3.8 2.8 3.5 3.4 3.6 2.7 3.7

a. State the null and alternative hypotheses to test the manufacturer's claim.
b. If we want to conduct a pooled t-test with small sample sizes, the data should come from a population that is approximately normally distributed with homogeneous variances. From the box-plots of the two samples, evaluate the normality assumption.

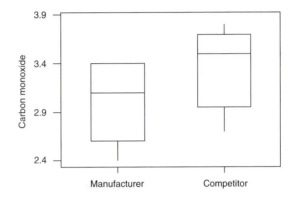

c. Do the variances of the two samples appear to be homogeneous?
d. The pooled t-test is robust except when the parent distributions are severely skewed or have unusually long tails. In light of your answers to parts **b** and **c**, do you think that the pooled t-test can be applied to these data?
e. From the descriptive statistics, complete the pooled t-test and state your conclusion.

Descriptive Statistics

Variable	N	Mean	Median	TrMean	StDev	SEMean
manufac	9	2.989	3.100	2.989	0.389	0.130
compet	10	3.370	3.500	3.400	0.395	0.125

Variable	Min	Max	Q1	Q3
manufac	2.400	3.400	2.600	3.400
compet	2.700	3.800	2.950	3.700

7.46 A distributor of auto gears wishes to determine with which of two manufacturers of gears he should do business. He obtained, from the two manufacturers, these numbers of defective gears in the production of 100 gears per day for 20 consecutive days.

WORKSHEET: Autogear.mtw

Manufacturer A	16 25 15 26 21 22 17 26 23 20 28 32 43 18 16 36 21 16 29 26
Manufacturer B	28 24 28 42 17 31 26 33 26 24 32 36 38 26 25 30 21 36 18 34

a. State the null and alternative hypotheses to see whether there is a significant difference in the mean numbers of defective gears per day from the two manufacturers.

b. If we wish to conduct a pooled *t*-test with small sample sizes, the data should come from approximately normally distributed populations with homogeneous variances. Do the boxplots of the two samples indicate that the normality assumption is in question? Do the variances of the two samples appear to be homogeneous?

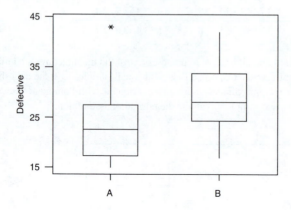

c. From the descriptive statistics, complete the pooled *t*-test and state your conclusion.

Descriptive Statistics

Variable	N	Mean	Median	TrMean	StDev	SEMean
A	20	23.80	22.50	23.22	7.32	1.64
B	20	28.75	28.00	28.67	6.59	1.47

Variable	Min	Max	Q1	Q3
A	15.00	43.00	17.25	27.50
B	17.00	42.00	24.25	33.75

Computer Exercises

7.47 Exercise 6.49 in Section 6.3 describes the process of manufacturing integrated circuits (chips) used in computers. Recall that in one part of the process, a thin layer of silicon oxide is placed on the surface of a wafer. The thickness of the oxide layer is critical to the performance of the resulting chips. Two wafers are randomly selected from each of 30 lots. Four measurements of the thickness of the oxide layer on each wafer are then taken. Columns 1 and 2 of worksheet: **Chipavg.mtw** contain the average of the four measurements for each of the two wafers selected from the 30 lots.

WORKSHEET: Chipavg.mtw

wafer1
940.0 1042.5 942.5 1007.5 985.0 925.0 1072.5 1070.0 1047.5 982.5 1050.0 1057.5 1057.5
1070.0 970.0 997.5 1107.5 932.5 960.0 1000.0 1105.0 1037.5 1037.5 1015.0 1002.5 972.5
1070.0 1005.0 842.5 1057.5

wafer2
985.0 1052.5 990.0 990.0 917.5 930.0 1120.0 1082.5 1057.5 1002.5 1020.0 1060.0 1010.0
1030.0 1002.5 1005.0 1095.0 1000.0 1042.5 1007.5 1067.5 1097.5 1015.0 1102.5 1057.5 1097.5
1007.5 930.0 887.5 955.0

Source: Yashchin, E. (1995), Likelihood Ratio Methods for Monitoring Parameters of a Nested Random Effect Model, *Journal of the American Statistical Association,* 90, 729–738.

a. From the data set, construct side-by-side boxplots of the two samples. Comment on the shapes of the two distributions. Are the two samples equally variable?
b. Construct normal probability plots of the two samples. Does normality seem reasonable in either distribution?

c. Based on your results in parts **a** and **b**, complete the pooled *t*-test, the two-sample *t*-test, or the two-sample Wilcoxon rank-sum test to determine whether the mean thickness of the oxide layer is different for the two sets of wafers.

7.48 Patients with small cell lung cancer (SCLC) received one of two types of treatments: Arm A or Arm B. Their survival times are recorded in worksheet: **Censored.mtw**.

Source: Ying, Z., Jung, S., and Wei, L. (1995), Survival Analysis with Median Regression Models, *Journal of the American Statistical Association*, 90, 178–184.

a. In Exercise 7.34 in Section 7.3, you determined that the two distributions are similarly skewed right. Does this mean that inferences should be about the population means or the population medians?
b. State null and alternative hypotheses for determining whether the median survival time is greater for Arm A.
c. Test the hypotheses from part **b**. Be sure to give the *p*-value and state your conclusion.

7.49 In Exercise 5.14 in Section 5.1, 25 schizophrenic patients were classified as psychotic or nonpsychotic after being treated with an antipsychotic drug. Cerebrospinal fluid was take from each patient and assayed for dopamine b-hydroxylase (DBH) activity. We wish to compare the DBH activity (units are nmol/(ml)(h)/(mg) of protein) for the two groups.

WORKSHEET: Dopamine.mtw

nonpsych (group 1)
.0104, .0105, .0112, .0116, .0130, .0145, .0154, .0156, .0170, .0180, .0200, .0200, .0210, .0230, .0252

psychotic (group 2)
.0150, .0204, .0208, .0222, .0226, .0245, .0270, .0275, .0306, .0320

(For ease of calculations the data values in the worksheet were multiplied by 10,000 so that .0104, for example, becomes 104.)

Source: Sternberg, D. E., Van Kammen, D. P., and Bunney, W. E. (1982), Schizophrenia: Dopamine b-hydroxylase Activity and Treatment Response, *Science,* 216, 1423–1425.

Previously, boxplots and normal probability plots of the two samples concluded that both groups have approximately normally distributed parent populations with homogeneous variances. Does one group tend to have higher DBH activity? Conduct a formal test of significance to evaluate the antipsychotic drug. Be sure to identify the test you are using, give the *p*-value, and state your conclusion.

7.50 Exercise 1.53 in Section 1.3 described a special diet mixed with a drug compound designed to reduce low-density lipoproteins (LDL) cholesterol that was fed to a treatment group of quail. A placebo group of quail was fed the same special diet for the same period of time but without the drug compound.

WORKSHEET: Quail.mtw

placebo
64 49 54 64 97 66 76 44 71 89 70 72 71 55 60 62 46 77 86 71

treatment
40 31 50 48 152 44 74 38 81 64

Source: McKean, J., and Vidmar, T. (1994), A Comparison of Two Rank-Based Methods for the Analysis of Linear Models, *The American Statistician*, 48, 220–229.

In Exercises 5.66 and 5.67 in Section 5.4, you analyzed these data and determined that if we remove the one outlier in the treatment group (152), then both groups are reasonably close to being normally distributed.

a. Remove the outlier and conduct a formal test of significance to evaluate the drug compound. Be sure to identify the test you are using, give the *p*-value, and state your conclusion.
b. Conduct the same test without removing the outlier and compare to the results found in part **a**.

7.5   Comparing Two Population Centers: Matched Samples

Learning Objectives for this section:

❑ Be able to find a confidence interval for the difference between two population means using matched samples.

❑ Be able to test hypotheses about the difference between two population means using matched samples.

❑ Be able to apply the matched-pairs design in the before–after experiment.

❑ Be able to apply the Wilcoxon signed-rank test to the matched-pairs experiment.

Suppose we wish to evaluate a new method of reading instruction for students with low ability. An important part of the evaluation is a comparison of reading achievement scores for students using the experimental method with those using the standard method. For the comparison, one possibility is to independently and randomly assign students to the two methods and compare their reading achievement scores at the end of the project. The difference in the mean achievement test scores, $\mu_1 - \mu_2$, can be investigated using the methods of Sections 7.3 or 7.4.

Although we are dealing with students who have low ability, they will still have varying levels of ability prior to exposure to the method of instruction. Given that they are independently and randomly assigned to the two groups, it is likely that the students assigned to the two groups will be comparable in reading skills; however, there is no guarantee. A difference in reading achievement at the end of the project might be due only to their different levels of skill prior to their instruction. The *matched-pairs experiment* is a method of assigning subjects to groups that prevents this from happening and guarantees equivalent groups prior to the instruction.

Matched-Pairs *t*-Test

For the study, a reading skills test is given to all students *before* they are assigned to the groups. Next, the students are paired according to reading skill. One of each pair is assigned to the experimental group, and the other is assigned to the control group. A comparison of *matched-pairs* of achievement test scores then gives a fair evaluation of the new method of instruction.

The matched-pairs test is also called the paired-difference test.

Analysis of the matched-pairs experiment is straightforward. We reduce the matched-pairs to a single sample by taking the difference between the two observations in each pair. These differences form a single sample that can be analyzed using the one-sample techniques of Chapters 5 and 6.

Suppose that the matched-pairs of observations are

$$(x_1,y_1), (x_2,y_2), \cdots, (x_n,y_n)$$

where x_i is the score for the ith subject in group 1 and y_i is the score for the matching ith subject in group 2. Let $d_i = y_i - x_i$ for $i = 1, 2, \cdots, n$ be the difference in the scores. Then $d_1, d_2, \cdots, d_n$ forms a single sample from which the mean, $\bar{d}$, and the standard deviation, s_d, can be calculated.

If we let $\mu_d = \mu_1 - \mu_2$ be the mean of the population of differences, the null hypothesis can be stated as

$$H_0: \mu_d = 0$$

If the population of differences is normally distributed, the test statistic

$$t = \frac{\bar{d}}{s_d/\sqrt{n}}$$

is distributed as Student's t with $n - 1$ degrees of freedom. The remainder of the test proceeds as in the one-sample t-test found in Chapter 6.

The confidence interval for μ_d and the test of hypothesis are summarized in the box.

Inferences for $\mu_1 - \mu_2$ based on Matched Samples

Application: Matched samples.

Assumptions: If the sample size is small, the distribution of difference scores should be approximately normally distributed.

Confidence Interval

A $(1 - \alpha)100\%$ confidence interval for $\mu_1 - \mu_2$ based on matched samples is

$$\bar{d} \pm t^* \frac{s_d}{\sqrt{n}}$$

Matched-pairs t-test

Left-Tailed Test	Right-Tailed Test	Two-Tailed Test
$H_0: \mu_d \geq 0$	$H_0: \mu_d \leq 0$	$H_0: \mu_d = 0$
$H_a: \mu_d < 0$	$H_a: \mu_d > 0$	$H_a: \mu_d \neq 0$

Standardized Test Statistic

$$t = \frac{\bar{d}}{s_d/\sqrt{n}}$$

From the t-table with $n - 1$ degrees of freedom, find the p-value most closely associated with t_{obs}. If t_{obs} falls between two table values, give the two associated probabilities as bounds for the p-value.

The p-value for the two-tailed test is calculated as in the one-tailed test and then doubled to account for both tails.

If a level of significance α is specified, then reject H_0 if p-value $< \alpha$.

Example 7.18   A group of 24 low-ability students of the same age were tested for reading skills and then paired according to ability. One member of each pair was assigned to an experimental reading group and the other to a control group. The experimental group was taught using a new method of reading instruction, and the control group was taught using the standard method that normally is used with low-ability students. After the period of instruction, all students were given a reading achievement test with these results:

WORKSHEET: Lowabil.mtw

					Reading Achievement Test Scores							
Pair	1	2	3	4	5	6	7	8	9	10	11	12
Experimental	82	65	63	71	48	74	61	65	73	92	57	66
Control	78	60	64	66	51	68	61	59	71	88	50	57

Perform a test of significance to determine whether the experimental group performed significantly better than the control group.

Solution Because the data are matched, we investigate the difference scores:

Pair	1	2	3	4	5	6	7	8	9	10	11	12
Difference	4	5	−1	5	−3	6	0	6	2	4	7	9

The normal probability plot of the difference scores in Figure 7.13 indicates that normality is a reasonable assumption.

Figure 7.13

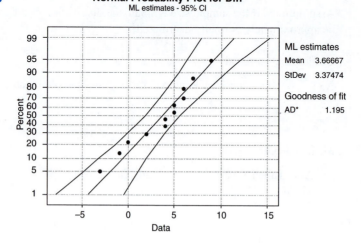

Normal Probability Plot for Diff
ML estimates - 95% CI

ML estimates
Mean   3.66667
StDev   3.37474

Goodness of fit
AD*   1.195

To determine whether the experimental group scored higher than the control group, the matched-pairs t-test is performed with these hypotheses:

$$H_0: \mu_d \leq 0 \quad \text{versus} \quad H_a: \mu_d > 0$$

where $\mu_d = \mu_1 - \mu_2$, μ_1 = mean response for the experimental group, and μ_2 = mean response for the control group.

From the differences, we easily find

When calculating $\bar{d}$, be sure that the negative values are added in as negatives.

$$\bar{d} = 3.667 \quad s_d = 3.525$$

and therefore,

$$t_{\text{obs}} = \frac{3.667}{3.525/\sqrt{12}} = 3.604$$

From the t-table with $n - 1 = 11$ degrees of freedom, we find that the t-score corresponding to .005 is 3.106. Because

$$t_{\text{obs}} = 3.604 > 3.106$$

we have the p-value $< .005$. With such a small p-value, the evidence is highly significant that the new method of reading instruction for low-ability students is better than the old one. ∎

Before–After Experiment

Another application of the matched-pairs design is the *before–after experiment*, in which the same subjects serve in both groups. Each subject is measured both before and after the treatment.

Example 7.19 A group of women in a large city were given instructions on self-defense. Prior to the course, they were tested to determine their self-confidence. After the course they were given the same test. A high score on the test indicates a high degree of self-confidence. Do these self-confidence scores indicate that the course significantly increased the women's self-confidence?

W O R K S H E E T : Selfdefe.mtw

Woman	1	2	3	4	5	6	7	8	9
Before course	6	10	8	6	5	4	3	8	5
After course	7	12	7	5	8	6	5	8	6

Solution The data are matched, so difference scores are found by

$$\text{Difference} = \text{after score} - \text{before score}$$

The results are

Woman	1	2	3	4	5	6	7	8	9
Difference	1	2	-1	-1	3	2	2	0	1

The normal probability plot of the differences in Figure 7.14 shows no serious departures from normality, so a *t*-test on the differences is justified.

Figure 7.14

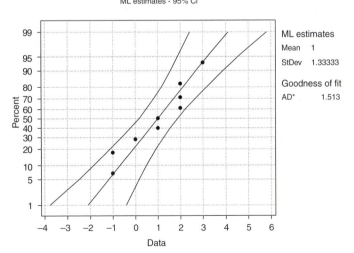

Normal Probability Plot for Difference
ML estimates - 95% CI

ML estimates
Mean   1
StDev   1.33333

Goodness of fit
AD*   1.513

The null and alternate hypotheses are

$$H_0: \mu_d \leq 0 \quad \text{versus} \quad H_a: \mu_d > 0$$

where $\mu_d = \mu_1 - \mu_2$, $\mu_1 =$ mean self-confidence score after the course, and $\mu_2 =$ mean self-confidence score before the course. Rejection of the null hypothesis indicates that $\mu_1 > \mu_2$, which says that the self-defense course increased the women's self-confidence.

From the differences we find

$$\bar{d} = 1.0 \text{ and } s_d = 1.414$$

The test statistic is

$$t_{\text{obs}} = \frac{1.0}{1.414/\sqrt{9}} = 2.122$$

From the t-table with 8 degrees of freedom, we find

$$.025 < p\text{-value} < .05$$

because $1.860 < 2.122 < 2.306$. There is evidence to reject H_0. Therefore, we can say there is significant statistical evidence that the self-defense course was effective in increasing self-confidence.   ∎

Computer Analysis

In Chapters 5 and 6, we introduced the one-sample t procedures in Minitab. These commands also can be used to analyze the difference between two population means when the samples are matched. We first compute the difference between the two observations in each pair and then perform the usual one-sample t-test or t-confidence interval.

Example 7.20   A psychologist who is interested in testing the relationship between stress and short-term memory administered a test to 12 subjects prior to their exposure to a stressful situation, and then retested them after the situation. From the following data can we conclude that the stressful situation decreases one's performance on a test that measures short-term memory?

W O R K S H E E T : Stress.mtw

Prestress	13	15	9	13	15	17	13	16	11	13	9	12
Poststress	10	14	7	15	11	14	13	14	9	14	9	10

Compute the difference in scores and perform a one-tailed t-test.

Solution   Let μ_1 denote the mean prestress test score and μ_2 the mean poststress test score. Also let $\mu_d = \mu_2 - \mu_1$. Then the null and alternative hypotheses are:

$$H_0: \mu_d \geq 0 \quad \text{versus} \quad H_a: \mu_d < 0$$

Rejection of H_0 establishes the alternative, which says that the mean poststress test score is significantly lower than the mean prestress test score.

C O M P U T E R   T I P

Calculating Differences

Having stored the prestress scores in C1 and the poststress scores in C2 of the Minitab worksheet, we can compute the difference scores and store them in a third column in the following manner.

Under the **Calc** menu, select **Calculator**. In the **Store result in variable** box of the Calculator window, identify the column where you wish to store the difference scores; for example, type **C3**. In the **Expression** box, type the mathematical expression **C2 − C1**, and then click on **OK**.

A normal probability plot of the differences will verify that the conditions for the paired t-test are met, so we apply the one-sample t-test as before. Figure 7.15 gives the output from the one-sample t-test with a left-tailed alternative. The p-value $= .012$ is significant, and thus we reject the null hypothesis. There is statistical evidence that the mean post-stress test score is significantly lower than the mean prestress test score.

Figure 7.15   **One-Sample T: Difference**

```
Test of mu = 0 vs mu < 0

Variable                 N       Mean         StDev      SE Mean
Difference              12     -1.333         1.775        0.512

Variable           95.0% Upper Bound              T          P
Difference                      -0.413          -2.60      0.012     ■
```

Wilcoxon Signed-Rank Test for the Matched-Pairs Experiment

A sufficient condition for the difference distribution to be symmetric is for the two distributions to be identical in shape and spread.

It is possible that the distribution of difference scores will deviate substantially from normality. In those cases we have the *Wilcoxon Signed-Rank Test*. The null hypothesis for the Wilcoxon signed-rank test states that the population distributions from which the matched-pairs came are identical. If the two distributions differ only in location, the test is a test of equality of means (as in the matched-pairs t-test) or medians, depending on the measure of location that is used. Although this test does not require normality of the difference scores, it does assume that the distribution is symmetric.

As in the Wilcoxon rank-sum test, we will give the Conover-Iman (1981) version of the Wilcoxon signed-rank test, whereby the usual t-test is applied to the ranks. This rank-transform test provides a good approximation to the Wilcoxon signed-rank test when the sample size is at least 10.

To conduct a test of hypothesis, the absolute values of the d_i's are ranked and if the original d_i was negative, the corresponding rank is given a negative sign. If a d_i is 0 (tied observation), then d_i is discarded from the analysis and the sample size is reduced by 1. The remainder of the test is applying the ordinary t-test to the signed ranks. The test statistic is

$$t = \frac{\bar{r}}{s_r / \sqrt{n}}$$

where $\bar{r}$ and s_r are the mean and standard deviation of the signed ranks, respectively. It is closely approximated by Student's t-distribution with $n - 1$ degrees of freedom when $n \geq 10$. The test is illustrated in the next example.

Example 7.21   To evaluate a new fabric softener, a consumer organization purchased two of each type of ten different garments for children. Ten pieces of clothing were washed 15 times in the experimental softener. The like ten pieces were washed 15 times without a softener. After the washings, the two groups of clothing were measured for softness. (A high number indicates that the garment is judged more soft than one with a lower number.)

WORKSHEET: Fabric.mtw

Type garment	1	2	3	4	5	6	7	8	9	10
With softener	12	3	12	16	4	24	11	17	19	8
Without softener	8	4	15	14	6	21	10	15	22	7

Test to see whether the softener significantly increases the softness.

Solution   Clearly, the data are matched pairs, so we find the differences:

Garment	1	2	3	4	5	6	7	8	9	10
Difference	4	−1	−3	2	−2	3	1	2	−3	1

Figure 7.16

The dotplot of the differences in Figure 7.16 indicates that the normality assumption might not be met, yet symmetry seems plausible. From these observations, it seems reasonable to use the Wilcoxon signed-rank test. The null and alternate hypotheses are

$$H_0\text{: } \mu_d \leq 0 \quad \text{versus} \quad H_a\text{: } \mu_d > 0$$

where $\mu_d = \mu_1 - \mu_2$, μ_1 = mean softness measurement with the softener, and μ_2 = mean softness measurement without the softener. Table 7.1 is useful in calculating the test statistic:

Table 7.1

d_i	$|d_i|$	rank$|d_i|$	signed ranks
4	4	10	10
−1	1	2	−2
−3	3	8	−8
2	2	5	5
−2	2	5	−5
3	3	8	8
1	1	2	2
2	2	5	5
−3	3	8	−8
1	1	2	2

Remember that tied observations are given average ranks. From the signed ranks, we find

$$\bar{r} = .9 \text{ and } s_r = 6.42$$

Thus,

$$t_{\text{obs}} = \frac{.9}{6.42/\sqrt{10}} = .443$$

From the t-table with 9 degrees of freedom, we see that

$$p\text{-value} > .10$$

which indicates that there is insufficient evidence to claim that the softener is effective in making the clothing more soft. ∎

COMPUTER TIP

Calculating Signed Ranks

With the Minitab **Calculator** we can easily calculate the signed ranks from the original data.

As before, calculate the difference and store them in a column such as C3. Then in the **Store result in variable** box of the Calculator window, identify a column to store the signed ranks, such as **C4**. In the **Expression** box, type the following expression and then click on **OK**:

```
signs(C3)*rank(absolute(C3))
```

Having stored the signed ranks in C4, you execute the ordinary one-sample t-test on column C4.

As pointed out in Section 7.1, the matched-pairs experiment is a special case of the randomized block experiment. In the randomized block design, subjects are separated into blocks based on some extraneous variable that has a confounding effect on the dependent variable. After the blocking is complete, the subjects within each block are randomly assigned to the treatments. In the matched-pairs experiment, each pair of subjects is a block, with one of the pair being randomly assigned to group 1 and the other to group 2.

EXERCISES 7.5 The Basics

7.51 Explain why it would not be possible to conduct a before–after experiment in Example 7.18.

7.52 To test the effectiveness of a drug to relieve asthma, a group of subjects was randomly given a drug and a placebo on two different occasions. After 1 hour an asthmatic relief index was obtained for each subject, with these results:

WORKSHEET: Asthmati.mtw

Subject	1	2	3	4	5	6	7	8	9
Drug	28	31	17	22	12	32	24	18	25
Placebo	32	33	19	26	17	30	26	19	25

a. To conduct a matched-pairs t-test or confidence interval, the samples should be matched. In what way are these samples matched?

b. To conduct a matched-pairs *t*-test, what is assumed about the distribution of the difference scores? Does the dotplot of the difference scores indicate that the assumption is met?

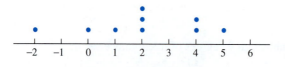

c. Assume that all assumptions are met. Do the data indicate that the drug significantly reduced the asthmatic relief index? Use the matched-pairs *t*-test.

7.53 A study was designed to determine the effect of a certain movie on the moral attitude of young children. The scores are the rating from 0 to 20 on a moral attitude scale recorded before and after the children viewed the film. A high score is associated with a high morality.

WORKSHEET: Movie.mtw

Before	14	16	15	18	15	17	19	17	17	16	19	15
After	14	18	16	17	16	19	20	18	19	15	18	16

a. In what way are these samples matched?
b. To conduct a matched-pairs *t*-test, what is assumed about the distribution of the difference scores?
c. Construct a dotplot of the difference scores. Based on the dotplot, can you tell whether the assumption is met?
d. Assume that all assumptions are met and construct a 99% confidence interval for μ_d, the mean of the population of difference scores.
e. Based on the interval in part **d**, can you conclude that the movie had an effect on the moral attitude of the children?

7.54 In a study using identical twins, one twin was given a drug and then given an intelligence test while under the influence of the drug. The other twin was given the same intelligence test under normal drug-free conditions. Here are their test scores:

WORKSHEET: Twin.mtw

Twin A (no drug)	83	74	67	64	70	67	81	64	72
Twin B (drug)	78	74	63	66	68	63	77	65	70

a. In what way are the two samples matched?
b. For the matched-pairs *t*-test, the difference scores should be approximately normally distributed. Construct a dotplot of the difference scores. Is normality a reasonable assumption in this case?
c. With so few scores it may be difficult to verify the normality assumption. Assume that it is met and complete the matched-pairs *t*-test to determine whether the drug had an influence on the test scores.

7.55 A company wished to study the effectiveness of a coffee break on the productivity of its workers. The productivity of nine randomly selected workers was measured on a day with no coffee break and later on a day when the workers were given a 10-minute coffee break. The scores measuring productivity are listed here:

WORKSHEET: Coffee.mtw

Worker	1	2	3	4	5	6	7	8	9
Without coffee break	23	35	29	33	43	32	41	38	40
With coffee break	28	38	29	37	42	30	43	37	39

a. In what way are the samples matched?
b. Construct a dotplot of the difference scores to see whether the assumptions for the matched-pairs *t*-test are met.
c. Use the Wilcoxon signed-rank test to determine whether a coffee break increases productivity.

Interpreting Computer Output

7.56 As a test of learning ability, nine randomly selected eighth-graders were given a spelling test. After a 2-week course of instruction, they were given a similar test. Here are the test scores:

WORKSHEET: Spelling.mtw

Before	90	72	80	57	64	70	98	76	59
After	95	79	90	60	62	70	99	80	58

a. In what way are the two samples matched?

b. For the matched-pairs *t*-test, the difference scores should be approximately normally distributed. Does this normal probability plot of the difference scores indicate that normality is a reasonable assumption in this case?

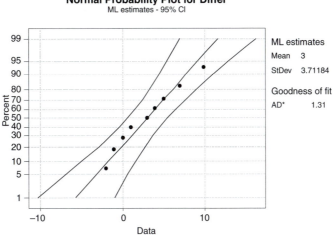

Normal Probability Plot for Differ
ML estimates - 95% CI

ML estimates
Mean 3
StDev 3.71184

Goodness of fit
AD* 1.31

c. With so few scores it may be difficult to verify the normality assumption. Assume that it is met and complete the matched-pairs *t*-test to determine whether there was improvement in the spelling scores.

7.57 The superintendent of school district A believes that her students, on the whole, have better study habits than the students of school district B. Eleven students from each district were paired according to IQ and then an independent party scored their study habits. The results are given here:

WORKSHEET: Habits.mtw

A	105	109	115	112	124	107	121	112	104	101	114
B	115	103	110	125	99	121	119	106	100	97	105

a. To conduct the matched-pairs *t*-test, the samples should be matched. In what way are these samples matched?

b. To conduct a matched-pairs *t*-test, what is assumed about the distribution of the difference scores? Based on the dotplot of the difference scores, do you think that the assumption is met?

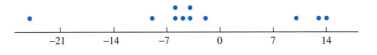

c. Listed here are the difference scores and the signed ranks. If you think that all the assumptions from part **b** are met, complete a matched-pairs *t*-test of the hypothesis H_0: $\mu_d = 0$; otherwise, conduct the Wilcoxon signed-rank test.

```
differ    10    -6    -5    13    -25    14    -2    -6    -4    -4    -9
signrks 8.0  -5.5 -4.0  9.0 -11.0 10.0 -1.0 -5.5 -2.5 -2.5 -7.0
```

d. Based on your analysis, can you conclude that there is a difference in the study habits of students in the two districts?

7.58 To evaluate a speed-reading course, a group of 15 subjects was asked to read two comparable articles before and after the course. These are their scores on a reading comprehension test:

WORKSHEET: Speed.mtw

Before course	57	80	64	72	90	59	76	98	70	57	94	77	46	71	89
After course	60	90	62	79	95	58	80	99	75	64	63	80	79	73	91

a. Outliers may have an adverse effect on the matched-pairs *t*-test. Examine the boxplot and normal probability plot of the difference scores to check the normality assumption. Are there any outliers?

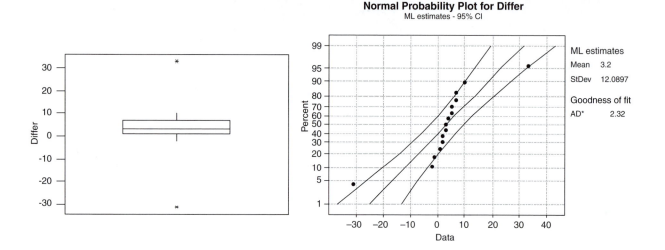

b. Based on your findings in part **a**, should we conduct the matched-pairs *t*-test or the Wilcoxon signed-rank test to determine whether the course was beneficial?

c. Examine the two computer printouts of the data. The first is the matched-pairs *t*-test and the second is the Wilcoxon signed-rank test. What are the null and alternative hypotheses? What are the two *p*-values for the tests? What conclusion can you draw from these two different procedures? Which is the proper analysis of the data?

t-Test of the Mean

```
Test of mu = 0.00 vs mu > 0.00

Variable     N    Mean    StDev    SE Mean      T    P-Value
differ      15    3.20    12.51       3.23    0.99       0.17
```

t-Test of the Mean

Test of mu = 0.00 vs mu > 0.00

Variable	N	Mean	StDev	SE Mean	T	P-Value
signrnks	15	5.40	7.55	1.95	2.77	0.0075

d. Based on your analysis, did the speed-reading course increase comprehension?

Computer Exercises

7.59 Twelve sets of identical twins were taught music recognition by two techniques (each twin was taught with a different method). At the end of the course, their improvement scores were recorded.

WORKSHEET: Music.mtw

Twin set	1	2	3	4	5	6	7	8	9	10	11	12
Method 1	7	4	6	1	5	1	6	3	4	6	5	3
Method 2	2	4	3	2	3	4	2	4	4	3	5	2

a. Construct a dotplot and normal probability plot of the difference scores. Is normality a reasonable assumption in this case?

b. Assume that the normality assumption is met. Complete the matched-pairs *t*-test to determine whether there is a statistical difference between the improvement scores for the two methods.

7.60 A sample of ten students enrolled in a course in German were asked to copy a passage written in German. After experimental instruction in German, the same ten students were asked to copy the same passage. The numbers of errors are listed here:

WORKSHEET: German.mtw

# errors before	10	6	8	7	7	12	4	0	7	10
# errors after	6	4	5	3	6	8	0	1	8	5

a. Construct a dotplot and normal probability plot of the difference scores to see whether the assumptions for the matched-pairs *t*-test are met.

c. Use the Wilcoxon signed-rank test to test the hypothesis that the mean numbers of errors made before and after the course are not significantly different.

7.61 Two psychiatrists were asked to rate each of 20 prison inmates on their rehabilitative potential using a scale from 0 to 15. The larger the rating, the greater the potential for rehabilitation.

WORKSHEET: Rehab.mtw

Inmate	1	2	3	4	5	6	7	8	9	10
Psychiatrist 1	7	12	7	5	8	6	5	8	5	9
Psychiatrist 2	5	10	8	6	5	4	5	6	5	8

Inmate	11	12	13	14	15	16	17	18	19	20
Psychiatrist 1	6	7	3	5	7	9	5	11	4	10
Psychiatrist 2	9	7	5	4	8	4	8	9	3	12

a. Construct a dotplot, boxplot, and normal probability plot of the difference scores. Does it appear that the assumptions for the matched-pairs *t*-test are met?

b. Complete a matched-pairs *t*-test to determine whether there is a significant difference in the mean rating scores.

7.62 A popular data set to be analyzed by statisticians is from Charles Darwin's study of cross-fertilized and self-fertilized plants. In his experiment, Darwin planted a pair of seedlings, one produced by cross-fertilization and the other by self-fertilization, on opposite sides of a pot to determine which grew the fastest. The data are the heights of the plants after a fixed period of time.

WORKSHEET: Darwin.mtw

pot	cross	self
1	23.500	17.375
1	12.000	20.375
1	21.000	20.000
2	22.000	20.000
2	19.125	18.375
2	21.500	18.625
3	22.125	18.625
3	20.375	15.250
3	18.250	16.500
3	21.625	18.000
3	23.250	16.250
4	21.000	18.000
4	22.125	12.750
4	23.000	15.500
4	12.000	18.000

Source: Darwin, C. (1876), *The Effect of Cross- and Self-Fertilization in the Vegetable Kingdom*, 2nd edition, London.

a. Construct a stem-and-leaf plot of each sample. Are the data symmetric or skewed? Are the samples matched?
b. Create a column of difference scores, and construct a stem-and-leaf plot of the resulting data. Do you think the data are more suited for analysis by the paired *t*-test or the Wilcoxon signed-rank test? Explain.
c. Perform the Wilcoxon signed-rank test. Is there evidence of a significant difference in the heights of the plants from the two methods of fertilization?

7.6   Choosing the Right Procedure

Learning Objectives for this section:

☐ Be able to distinguish between independent and matched samples.

☐ Know how to assess the shape of a distribution so that the correct test procedure is chosen.

☐ Understand the characteristics of the three test procedures for independent samples.

☐ Understand the characteristics of the two test procedures for matched samples.

Several procedures for comparing two population parameters are presented in this chapter. You have learned how to compare two Bernoulli proportions, two population means, and two population medians. For proportions, only the large-sample case is presented. The small-sample comparison case is not presented because most applications use large samples. A typical application might involve an analysis of survey results, and rarely is a survey conducted with a small sample. This is not to say that small-sample analysis of proportions is unimportant, however. In some situations, such as clinical trial studies, the sample sizes are extremely small. These are special situations, however, and most likely you will cover the details of the comparison of proportions with small samples in a second course in statistics.

The remainder of this section is devoted to comparing the centers of two population distributions. Choosing the right procedure depends on certain characteristics of the two samples and on the shapes of their parent distributions. Are the samples independent? Samples are independent when two separate random samples are chosen from two separate populations. If the samples are not independent, they may be matched in such a way that each observation in one sample is naturally paired with a corresponding observation in the other sample. For example, if the same subject is measured both before and after a treatment, then obviously the data are matched because the two measurements are taken on the same subject.

Independent Samples

For independent samples, we presented four inference procedures. There is considerable overlap in the applications of these procedures, so deciding which to use in a particular situation can be confusing. Some procedures are very restrictive in their applications (such as the pooled t-test) and others apply to a wide variety of situations. To simplify the selection process, you should understand the basic conditions that apply to each procedure.

POOLED t-TEST   As mentioned, the pooled t-test, presented in Section 7.4, is the most restrictive procedure of all. For small samples, it requires that

- the parent distributions are normally distributed.
- the population variances are equal.

Because of the robustness of the t-procedure, however, these conditions can be relaxed somewhat and the test (or confidence interval) will still be valid. The conditions for which they are not valid are long tails and heavy skewness. In those cases, you should avoid the t-test and use one of the other procedures.

TWO-SAMPLE t-TEST   The two-sample t-test, presented in Section 7.3, is similar to the pooled t-test except that it does not require that the population variances be equal. It should be used to test equality of population means when the parent distributions are reasonably close to normal. It can be used in place of the pooled t-test as well. Recent statistical studies have shown that it is almost as powerful as the pooled t-test even when the population variances are equal. Therefore, it is recommended that we use the pooled t-test only in those cases where we are reasonably sure that the population variances are equal.

If the sample sizes are large, the Central Limit Theorem applies and says that the sampling distribution of $\bar{x}_1 - \bar{x}_2$ is approximately normally distributed. Because the Central Limit Theorem applies regardless of the shape of the underlying parent distribution, there are hardly any restrictions on using the two-sample t-test when the sample sizes are large. On the other hand, recall that the t-test is used for making inferences about $\mu_1 - \mu_2$. If the parent distributions are skewed, it has been suggested that we make inferences about the difference between the population medians, $\theta_1 - \theta_2$, instead of the difference between the population means. If this is the case, you should use the Wilcoxon procedure instead of the t-test. Furthermore, you should use the Wilcoxon procedure when the parent distributions are symmetric with long tails. Thus, the t-test is best suited for situations where the parent distributions are symmetrical without excessively long tails.

WILCOXON TEST   The two-sample Wilcoxon rank-sum test does not require any specific shape for the parent distributions, but it does require that the distributions be similarly shaped. For example, if population 1 is skewed left, then population 2 should be skewed left. In fact, the test requires that if a difference exists between the two populations, it is reflected only in the centers of the two distributions. On the surface, it appears that the applications for the Wilcoxon test are very limited. In reality, however, there are numerous comparison problems where the two population distributions are similar in shape. For example, the distributions of the weights of men and women are similar in shape; it is just that men weigh more than women, which means that the distributions differ in location but not in shape. So the applications for the Wilcoxon test are numerous, and in fact, many statisticians suggest that it should be used in all cases except when the parent distributions are known to be normally distributed. For small samples, this is good advice. For large samples, however, the Wilcoxon test should be used in those situations where the parent distributions are similarly skewed or symmetric with unusually long tails.

Example 7.22 shows that the Wilcoxon rank-sum test is better than the two-sample *t*-test in detecting a difference between population means when the parent distributions are symmetric with long tails.

Example 7.22   Do physically active women have stronger bones than nonactive women as they grow older? Research by doctors at the University of North Carolina at Chapel Hill indicate that they do, which suggests that active women are less likely to have fractures as they grow older. A study of 300 women reported in the February 1985 issue of the *Journal of Orthopedic Research* indicates that older athletic women, aged 55 to 75, have arm and spine bone measurements in the same range as younger athletic women. Measurements of bone density were taken from a random sample of 70 women, 35 of whom are physically active; the rest are considered nonactive:

WORKSHEET: Bones.mtw

Active women:
213, 227, 211, 208, 155, 204, 216, 219, 224, 202, 207, 212, 184,
214, 245, 210, 192, 218, 219, 163, 230, 214, 209, 210, 226, 203, 208, 207, 217,
257, 212, 203, 232, 221, 215
Nonactive women:
201, 205, 187, 208, 203, 265, 201, 210, 219, 205, 173, 202, 199, 216, 192, 207,
243, 194, 201, 217, 270, 209, 185, 176, 202, 211, 213, 208, 236, 214, 209, 162,
204, 206, 213

Is there statistical evidence to support the hypothesis that active women have greater bone density?

Solution   Figure 7.17 is an ordered back-to-back stem-and-leaf plot, and Figure 7.18 shows side-by-side boxplots for the two data sets. Both distributions appear symmetric with long tails, and because of the symmetry we will test the equality of the population means. The research hypothesis is that the mean bone density for active women is greater than the mean bone density for nonactive women. Thus, we have

$$H_0: \mu_1 - \mu_2 \le 0 \quad \text{versus} \quad H_a: \mu_1 - \mu_2 > 0$$

Figure 7.17

```
                 Active                                      Non-active
                          5 | 15 |
                          3 | 16 | 2
                            | 17 | 3 6
                          4 | 18 | 5 7
                          2 | 19 | 2 4 9
          9 8 8 7 7 4 3 3 2 | 20 | 1 1 1 2 2 3 4 5 5 6 7 8 8 9 9
    9 9 8 7 6 5 4 4 3 2 2 1 0 0 | 21 | 0 1 3 3 4 6 7 9
                    7 6 4 1 | 22 |
                        2 0 | 23 | 6
                          5 | 24 | 3
                          7 | 25 |
                            | 26 | 5
                            | 27 | 0
```

Figure 7.18

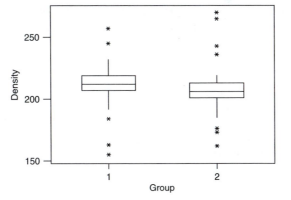

Figure 7.19 show the results of applying the two-sample *t*-test to the data. The *p*-value = .203 indicates that we cannot reject the null hypothesis. Based on these data and the two-sample *t*-test, there is an insignificant difference between the mean bone density of active and nonactive women.

Figure 7.19 Two-Sample *t*-Test and CI: Active, Nonactive

```
Two-sample T for Active vs Nonactive

                N       Mean     StDev    SE Mean
Active          35      211.6    18.7        3.2
Nonactive       35      207.6    21.4        3.6

Difference = mu Active − mu Nonactive
Estimate for difference: 4.03
95% lower bound for difference: −4.00
T-Test of difference = 0 (vs −): T-Value = 0.84 P-Value = 0.203 DF = 66
```

Figure 7.20 shows the results of applying the two-sample *t*-test to the ranks, which is equivalent to the Wilcoxon rank-sum test. Here, the *p*-value = .016 shows that the mean bone density of active women is significantly greater than the mean bone density of nonactive women.

Figure 7.20 Two-Sample *t*-Test and CI: Ranks, group

```
Two-sample T for Ranks

group           N       Mean     StDev    SE Mean
1               35      40.7     19.4        3.3
2               35      30.3     20.2        3.4

Difference = mu (1) − mu (2)
Estimate for difference: 10.34
95% lower bound for difference: 2.45
T-Test of difference = 0 (vs −): T-Value = 2.18 P-Value = 0.016 DF = 67
```

Thus, the Wilcoxon rank-sum test has detected a significant difference between population means that the ordinary two-sample *t*-test overlooked. ∎

Matched Samples

For matched samples, we presented two procedures: the matched-pairs *t*-test and the Wilcoxon signed-rank test.

MATCHED-PAIRS *t*-TEST   As in all previous *t*-tests, the matched-pairs *t*-test is derived assuming that the underlying distribution (in this case the distribution of difference scores) is normally distributed. Again, however, robustness of the *t* distribution allows us to relax this condition, especially when the sample size is increased, and apply it to those situations where the distribution of difference scores does not have excessively long tails. In fact, for large samples, the *t* distribution approaches the standard normal distribution, and thus the test can be thought of as a *z*-test for large samples and applied to a wide variety of cases.

WILCOXON SIGNED-RANK TEST   For small samples, when we have reservations about using the *t*-test, we can apply the Wilcoxon signed-rank test. The only condition is that the difference scores be symmetrically distributed.

Summary

Clearly, all cases that you will confront in the practice of statistics have not been addressed in this chapter. Numerous other testing procedures can be applied in special cases. What you have been exposed to are the standard tests that you will see in routine comparison problems. As implied in the discussion, two or more tests may be applied to a particular problem; your job is to apply the procedure you think best. An exploratory approach to data analysis is the only way you can critically appraise the situation and come up with a reasonable plan of attack.

To help in your decision, consider the following steps when trying to decide which procedure is best for a particular situation.

Choosing the Right Procedure

1. Determine whether you have independent or matched samples.
2. In the case of independent samples, analyze the samples to determine the shapes of the two parent distributions. In the case of matched samples, analyze the difference scores to determine the shape of the difference population.
3. Based on the shape(s) of the parent distribution(s), decide whether the inference should be for means or for medians. In other words, are the parent distributions symmetric or skewed? If they are symmetric, are they normally distributed, symmetric with short tails, or symmetric with long tails?
4. If the inference is for the population medians, the nonparametric Wilcoxon procedures are to be used.
5. If the inference is for the population means, determine whether the standard *t*-procedures apply. That is, if your sample sizes are small, do the normality and homogeneous variances assumptions seem reasonable? If not, apply the Wilcoxon procedure. If the sample sizes are large, examine the length of the tails. If the distributions are symmetric without excessively long tails, use the *t*-procedures; otherwise, use the Wilcoxon procedure.

7.7   Summary and Review Exercises

Key Concepts   ■   Three basic inference problems are addressed in this chapter: the comparison of two population proportions, the comparison of two population centers using independent samples, and the comparison of two population centers using matched samples.

- Only the large-sample inference procedures for the difference between two population proportions are given. The confidence interval is

$$(p_1 - p_2) \pm z^* \sqrt{\frac{p_1(1 - p_1)}{n_1} + \frac{p_2(1 - p_2)}{n_2}}$$

and the test statistic for testing the hypothesis that $\pi_1 = \pi_2$ is

$$z = \frac{p_1 - p_2}{\sqrt{p(1 - p)}\sqrt{1/n_1 + 1/n_2}}$$

where

$$p = \frac{x_1 + x_2}{n_1 + n_2}$$

is the pooled estimate of π.

- The confidence interval for $\mu_1 - \mu_2$ based on $\bar{x}_1 - \bar{x}_2$ is

$$(\bar{x}_1 - \bar{x}_2) \pm t^* \, SE(\bar{x}_1 - \bar{x}_2)$$

- The test statistic for testing $\mu_1 - \mu_2$ is

$$t = \frac{(\bar{x}_1 - \bar{x}_2)}{SE(\bar{x}_1 - \bar{x}_2)}$$

If the population variances are equal, then

$$SE(\bar{x}_1 - \bar{x}_2) = s_p \sqrt{1/n_1 + 1/n_2}$$

and

$$t = \frac{(\bar{x}_1 - \bar{x}_2) - (\mu_1 - \mu_2)}{s_p \sqrt{1/n_1 + 1/n_2}}$$

has a Student's t distribution with $n_1 + n_2 - 2$ degrees of freedom.

- If the population variances are unequal, then

$$SE(\bar{x}_1 - \bar{x}_2) = \sqrt{s_1^2/n_1 + s_2^2/n_2}$$

and

$$t = \frac{(\bar{x}_1 - \bar{x}_2) - (\mu_1 - \mu_2)}{\sqrt{s_1^2/n_1 + s_2^2/n_2}}$$

has an approximate t-distribution with degrees of freedom given by

$$df = \frac{[s_1^2/n_1 + s_2^2/n_2]^2}{\dfrac{(s_1^2/n_1)^2}{n_1 - 1} + \dfrac{(s_2^2/n_2)^2}{n_2 - 1}}$$

- The *Wilcoxon rank-sum test* is an alternative to the two-sample t-test when the assumption of normality is not met. For the Wilcoxon rank-sum test, the usual t-test is applied to the rank-transformed data.

- When the samples are matched, the test of the difference between population means is called the *matched pairs t-test*. The test statistic is

$$t = \frac{\bar{d}}{s_d/\sqrt{n}}$$

which has a Student's t distribution with $n - 1$ degrees of freedom. Again, the assumption of normality must be met for this t-test unless the sample size is large. If the assumption is not met, then the *Wilcoxon signed-rank test* is recommended. Here the usual t-test is applied to the signed ranks.

Use the following problems to test your skills.

The Basics

7.63   The U.S. Department of Agriculture lists the average size of the farms (in acres) for each state. In a comparison of the farms in two regions of Iowa, these data were obtained:

Region 1: $n_1 = 65$ $\bar{x}_1 = 315$ $s_1 = 32.7$
Region 2: $n_2 = 60$ $\bar{x}_2 = 302$ $s_2 = 28.4$

a. Are the standard deviations close enough that we should pool them together to estimate the standard error of $\bar{x}_1 - \bar{x}_2$?
b. Construct a 90% confidence interval for $\mu_1 - \mu_2$ based on $\bar{x}_1 - \bar{x}_2$.
c. Does the interval contain both positive and negative values? What does this tell us about the sizes of the farms in the two regions of Iowa?

7.64   Life expectancy tables indicate that women live longer than men. To examine this theory, a statistics class, over a 1-month period, randomly selected from the local newspaper obituaries of ten men and ten women. An analysis of the data gave these values:

Women: $n_1 = 10$ $\bar{x}_1 = 79.7$ $s_1 = 11.32$
Men: $n_2 = 10$ $\bar{x}_2 = 76.4$ $s_2 = 9.64$

With $\bar{x}_1 - \bar{x}_2$ as the test statistic, test

$H_0: \mu_1 - \mu_2 \leq 0$ versus $H_a: \mu_1 - \mu_2 > 0$

where μ_1 = mean life expectancy of women and μ_2 = mean life expectancy of men
a. In words, what does the null hypothesis say?
b. Should we pool the standard deviations together to estimate the standard error of $\bar{x}_1 - \bar{x}_2$?
c. What is the p-value of the test statistic?
d. State your conclusion.

7.65   Test the hypothesis that there is no significant difference between the population medians using the following data that are arranged in a back-to-back stem-and-leaf plot. Because both distributions appear to be skewed, use the Wilcoxon rank-sum test. Why do we suggest a test of the equality of the population medians instead of the population means?

WORKSHEET: Skewed.mtm

	21	0 4 7
3 6 9	22	5 6 9 9
5 8	23	1 4 8
2 6 9	24	5 8
0 3 5 8	25	0 7
3 7 7	26	3
0 6	27	0 1
8	28	
7	29	2
2	30	7
	31	
4	32	

7.66   In a random sample of 200 youth drivers, 54 were judged to be careless drivers. In a random sample of 200 adult drivers, 38 were judged to be careless drivers. Find a 99% confidence interval for the difference between the percents of youth and adult drivers judged to be careless drivers.

7.67   Do more women choose diet soft drinks than men? In a random sample of 160 women, 94 preferred diet soft drinks to regular soft drinks. In a random sample of 135 men, 71 preferred diet to regular soft drinks. Do the data imply that significantly more women than men prefer diet soft drinks? Perform the test of significance.

7.68 Sixteen children were selected from a first-grade class and paired according to IQ such that one member of each pair had attended kindergarten and the other had not. A reading test was given to all 16 children, resulting in these scores:

WORKSHEET: Kinder.mtw

Pair	1	2	3	4	5	6	7	8
Kindergarten	83	74	67	64	70	67	81	64
No kindergarten	78	74	63	66	68	63	77	65

Is there evidence that kindergarten is beneficial for reading skills?

7.69 A 1984 study by the Centers for Disease Control in Atlanta indicates that giving aspirin to children who have the chicken pox or the flu can cause Reye's syndrome. Dr. Sidney M. Wolfe of the Public Citizen Health Research Group deemed it the most convincing study to date to link aspirin and the life-threatening illness. The study traced 29 cases of Reye's syndrome and 143 control cases of children who did not develop the illness. Twenty-eight of the 29 children who contracted Reye's syndrome took aspirin, whereas 64 of those in the control group took aspirin. Construct a 95% confidence interval for the difference between the proportions of those who took aspirin in the Reye's syndrome group and those who took aspirin in the control group.

7.70 Can computers help high school students prepare for the SAT exam? An educational consultant, in Milton, Florida, George Hopmeier, studied 90 Florida high school students. Half of the students, without computer coaching, averaged 370 on the SAT math portion and had a standard deviation of 125. The other half, who used computer coaching, averaged 407 and had a standard deviation of 80. Is there evidence that computer coaching improved SAT math scores? To answer this question, you need to formulate null and alternative hypotheses and conduct a test of significance. Be sure to give the *p*-value and state your conclusion.

7.71 To test the effect of a physical fitness course on one's physical ability, the number of sit-ups that a person could do in one minute, both before and after the course, were recorded. Nine randomly selected participants scored as follows:

WORKSHEET: Fitness.mtw

Before	28	31	17	22	12	32	24	18	25
After	32	33	19	26	17	30	26	19	25

Test the hypothesis that there was an increase in the number of sit-ups.

7.72 Sixty-four individuals randomly selected from a metropolitan area were asked to indicate their preference for the candidates for a political office. A week later, one of the candidates visited the area. Following the visit, a second sample of 64 individuals was asked to indicate their preference for the same list of candidates. Thirty-two of the 64 respondents favored the candidate before his visit, whereas 40 of the 64 favored him after his visit. Is there evidence to indicate that personal contact has a positive impact?

7.73 A study by the American Association for Counseling and Development found that in 1983, 59% of all 11th-graders were interested in business and tech jobs. Suppose the study was based on a random sample of 400 11th-graders and we wish to determine whether the percent has changed significantly since then. A random sample of 400 11th-graders this year found that 251 were interested in similar jobs. With a test of hypothesis, determine whether significantly more are interested now than in the past.

7.74 According to a survey by the Road Information Program, Ohio is the pothole capital of the United States with an estimated 6.8 million potholes. Potholes are caused by water seeping into cracks, freezing and swelling the pavement. Suppose we wish to compare two types of material for making pavement. A measurement is devised to test their resistance to water. From the following data, determine whether there is a significant difference in the resistance to water for the two types of materials.

	Material A	Material B
n	25	25
$\bar{x}$	63.8	87.2
s	16.55	35.75

7.75 Carolyn S. Hartsough, a University of California, Berkeley, educational psychologist, found that out of 301 hyperactive children, 26% of the mothers had poor health during pregnancy. This compares with 16% of mothers of 191 normal children. Is the percentage for hyperactive children significantly greater than the percentage for normal children? Does this study show that poor health during pregnancy results in hyperactive children?

7.76 High levels of blood fat contribute to atherosclerosis (hardening of the arteries), which is an underlying cause of heart attacks and strokes. Dr. Robert Knopp of the University of Washington School of Medicine in Seattle analyzed 381 diabetics and found that women had higher levels of LDL, or "bad cholesterol," than men, and lower levels of HDL, or "good cholesterol." From the following simulated data, determine whether the level of LDL in women is greater than that in men.

Women	Men
$n_1 = 201$	$n_2 = 180$
$\bar{x}_1 = 52.35$	$\bar{x}_2 = 47.28$
$s_1 = 6.55$	$s_2 = 17.29$

Interpreting Computer Output

7.77 In an experiment to evaluate a new variety of corn, 12 plots of land were divided in half, with the new variety planted on one half and a standard variety planted on the other half. Here are the yields obtained on the 12 plots of land:

WORKSHEET: Corn.mtw

						Yield per acre						
Plot	1	2	3	4	5	6	7	8	9	10	11	12
New variety	110	103	95	94	87	119	102	93	87	98	105	117
Standard	102	86	88	75	89	102	105	88	83	89	100	110

a. The difference between the yields for the new variety and the standard variety are shown in the boxplot. Does it appear that the assumptions for an inference based on the t distribution are violated?

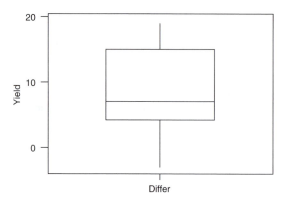

b. To test the hypothesis that the new variety has a significantly higher yield per acre than the standard variety, what testing procedure do you recommend?

c. State the null and alternative hypotheses necessary to conduct the test of significance suggested in part **b**.

d. Using these descriptive statistics, calculate the test statistic and its *p*-value.

Descriptive Statistics: differ

Variable	N	Mean	Median	TrMean	StDev	SE Mean
differ	12	7.75	7.00	7.70	7.01	2.02

Variable	Minimum	Maximum	Q1	Q3
differ	−3.00	19.00	4.25	15.00

e. Write your conclusion.

7.78 Do students who attend private high schools spend more time on homework than students who attend public high schools? Here are the numbers of hours per week spent on homework for a random sample of 15 private high school students and a random sample of 15 public high school students:

WORKSHEET: Homework.mtw

Private	21.3 16.8 8.5 12.6 15.8 19.3 18.5 24.6 18.3 12.9 15.7 18.4 18.7 22.6 20.5
Public	15.3 17.4 12.3 10.7 16.4 11.3 17.6 13.9 20.2 16.8 23.6 14.2 5.7 18.8 9.4

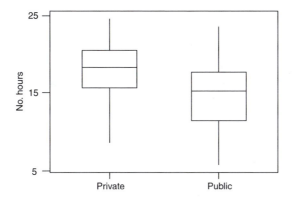

a. Do the boxplots indicate that the assumptions for a *t*-interval or a *t*-test have been violated?

b. Should an inference about the difference between the population means be based on independent or matched samples?

c. Using these descriptive statistics, find a 98% confidence interval for the difference between the mean numbers of hours per week spent on homework for private and public high school students.

Descriptive Statistics: Private, Public

Variable	N	Mean	Median	TrMean	StDev	SE Mean
Private	15	17.63	18.40	17.80	4.14	1.07
Public	15	14.91	15.30	14.95	4.57	1.18

Variable	Minimum	Maximum	Q1	Q3
Private	8.50	24.60	15.70	20.50
Public	5.70	23.60	11.30	17.60

d. Based on the confidence interval from part **c**, what can be said about the number of hours per week spent on homework?

7.79   Use the data in Exercise 7.78 to test the hypothesis that private high school students spend significantly more time on homework per week than public high school students. Be sure to state the hypotheses and give your conclusion. Compare these results to the conclusion given in part **d** of Exercise 7.78.

Computer Exercises

7.80   Fourteen employees who have not completed high school were given a reading test. Afterward, they were given formal vocabulary training and then retested on their reading skills. From the following test scores, determine whether there is any difference between the scores before and after the vocabulary training.

WORKSHEET: Vocab.mtw

First test	84	55	43	64	72	65	72	52	49	80	38	93	77	60
Second test	86	52	50	72	70	67	80	50	62	81	56	90	78	64

Store the data in two columns of your worksheet and have the computer take the differences and conduct a *t*-test. What is your justification for applying a *t*-test to the difference scores?

7.81   The results of quality control tests on two manufacturing processes are as follows:

WORKSHEET: Quality.mtw

Process I	1.5	2.5	3.4	2.3	3.2	2.8	1.9	
Process II	2.5	3.0	2.7	4.0	3.5	2.0	1.8	3.7

Use a computer to determine whether the mean results of the two processes are equivalent.

7.82   A fourth-grade teacher thinks that his students are on the whole better spellers than his colleague's students. Ten students were randomly selected from each teacher's class and given a standardized spelling test. Use the following results and a computer to determine whether the fourth-grade teacher is justified in his claim.

WORKSHEET: Spellers.mtw

Fourth-grade class	105	109	115	112	124	107	121	112	104	119
Colleague's class	115	103	110	125	99	121	119	106	100	123

7.83   A developer of housing projects would like to reduce his costs on kitchen cabinets. He obtained these cost estimates from two suppliers in 20 prospective homes:

WORKSHEET: Cabinets.mtw

Home	1	2	3	4	5	6	7	8	9	10
Supplier A	380	560	425	389	568	651	595	455	540	520
Supplier B	325	470	420	375	574	595	570	475	560	500
Home	11	12	13	14	15	16	17	18	19	20
Supplier A	375	468	492	510	379	750	520	480	394	624
Supplier B	362	465	445	490	350	780	512	465	382	614

Do the data indicate a significant difference between the estimates from the two suppliers?

7.84   Dr. Marvin Moser of Yale University School of Medicine says that automated blood pressure machines in airports and shopping malls are generally unreliable. He suggests having your blood pressure checked regularly by an expert or using an at-home device. Suppose 15 adult men check their blood pressure on an automated machine, and then they have an expert check it.

From the following data, determine whether the mean difference in diastolic blood pressure is significant.

WORKSHEET: Blood.mtw

Machine	68	82	94	106	92	80	76	74	110	93	86	65	74	84	100
Expert	72	84	89	100	97	88	84	70	103	84	86	63	69	87	93

Note that this test may not evaluate accuracy. It will tell you whether the machine tends to give high readings or tends to give low readings. However, the machine could be unreliable, with some readings unusually high and some unusually low. The differences may cancel each other out, in which case the test statistic would be insignificant. Can you think of a better way to check the accuracy of the machines?

7.85 Worksheet: **Commute.mtw** contains commuting times to work in urban areas for 1980 and 1990. The data are based on a 1994 study by the Federal Highway Administration. (See Exercise 1.121 in Section 1.6.) Here are side-by-side boxplots of the two samples.

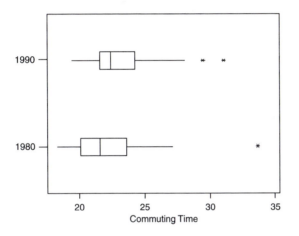

a. Are the boxplots sufficiently skewed that we should compare the median commuting times instead of the mean commuting times?
b. Stack the two samples into one column of the worksheet and store the codes in an adjacent column. Rank the stacked column and then apply the pooled t-test to the ranks.
c. What is the name of the procedure of applying the ordinary pooled t-test to the ranked data?
d. Interpret the results of the analysis. Is there evidence that the median commuting time has changed from 1980 to 1990?

7.86 Here are the mean arterial blood pressures of 11 subjects before and after they received oxytocin. Does oxytocin affect arterial blood pressure?

WORKSHEET: Oxytocin.mtw

Subject	1	2	3	4	5	6	7	8	9	10	11
Before	95	173	94	97	81	100	97	104	72	101	83
After	55	90	36	59	46	46	49	92	23	55	49

7.87 Microparticles, small quantities of micron-sized bits of dust, have been injected into the atmosphere by volcanic eruptions, dust storms, and micrometeorite infall. The particles are extremely small and would remain suspended indefinitely, except that they are scrubbed from the atmosphere by snow because they are nuclei for tiny ice crystals. It has been hypothesized that the concentration of microparticles in snow should be uniform over the earth. Samples of snow were obtained from the permanent snowfields of the Greenland ice cap and from Antarctica. The snow was melted and the volume concentration (in parts per billion) are given here:

WORKSHEET: Snow.mtw

antarc
3.7 2.0 1.3 3.9 0.2 1.4 4.2 4.9 0.6 1.4 4.4 3.2 1.7 2.1 4.2 3.5

greenld
3.7 7.8 1.9 2.0 1.1 1.3 1.9 3.7 3.4 1.6 2.4 1.3 2.6 3.7 2.2 1.8 1.2 0.8

Source: Davis, J., *Statistics and Data Analysis in Geology,* John Wiley, New York.

a. Construct side-by-side boxplots of the two samples. Do you notice any outliers? Do you think the conditions for the *t*-test are satisfied?
b. Remove any outliers by placing an * in the cell that contains the outlier. Repeat part **a**. Do you now think that the conditions for the *t*-test are satisfied?
c. If we conduct a *t*-test, should we pool the standard deviations together or conduct an unpooled *t*-test?
d. Based on your answer to part **c**, complete the test. Be sure to state your hypotheses and give a conclusion.

8

Analysis of Categorical Data

In most problems we have considered up to this point, the data have been the result of some numerical measurement. Statistical inference procedures for numerical variables such as IQ, reaction time, and earnings per share were presented in Chapters 5 through 7. The results of surveys and some experiments, however, are classified into categories. A public opinion poll, for example, records the opinions of respondents on a particular issue; the data are the frequency counts for the categories: for, against, and no opinion.

An experiment that studies childhood leukemia records the cancer history of the parents; the data are frequency counts for categorical variables that indicate the parents' history of cancer. An admission officer at a university records the gender of the applicants and their intended major. Each is a categorical variable, and the data are frequency counts for the categories of the variables. The objective of this chapter is to present methods of analyzing data of this type.

Contents

■ STATISTICAL INSIGHT
8.1 The Chi-Square Goodness-of-Fit Test
8.2 The Chi-Square Test of Independence
8.3 The Chi-Square Test of Homogeneity
8.4 Summary and Review Exercises

How Does Race Relate to Type of Drug Offense for Federal Prisoners?

Federal inmates in prison for a drug offense are usually sentenced for possession or trafficking in heroin, crack, cocaine, or marijuana. A 1994 report, *Comparing Federal and State Prison Inmates*, by the U.S. Department of Justice classified 1991 federal inmates by type of drug offense and race. The results in Table 8.1 suggest that blacks, for example, are more likely to be convicted for dealing in crack than either whites or Hispanics. On the other hand, very few blacks are convicted for dealing in marijuana as compared with whites and Hispanics. Whites are more likely to be convicted for dealing in cocaine and marijuana. There also seems to be a very high conviction rate for Hispanics dealing in cocaine. The data seem to suggest that the variables "type of drug offense" and "race" are dependent. In this chapter, we develop the chi-square distribution, which allows us to statistically evaluate the relationship between categorical variables such as these. (See Section 8.4 for a discussion of this Statistical Insight.)

Table 8.1
Numbers of federal inmates convicted for dealing in illegal drugs in 1991

Race	Type of Drug Offense			
	heroin	crack	cocaine	marijuana
white	407	106	4525	2825
black	1156	2513	4439	442
hispanic	1314	348	7297	2675

Source: C. Wolf Harlow (1994), *Comparing Federal and State Prison Inmates*, NCJ-145864, U.S. Department of Justice, Bureau of Justice Statistics.

8.1 The Chi-Square Goodness-of-Fit Test

Learning Objectives for this section:

❏ Understand the basic properties of the multinomial experiment.

❏ Know how to calculate the expected number of outcomes to fall in categories of a multinomial experiment.

❏ Know the assumptions required for the chi-square goodness-of-fit test.

❏ Know how to conduct a chi-square goodness-of-fit test.

Sampling from a Bernoulli population can be thought of as classifying data from a random sample into one of two categories: success or failure. The resulting categorical data are then used to make inferences about the proportion of successes in the Bernoulli population. Now we generalize to categorical data having more than two categories. For example, a Time Warner survey of 425 randomly selected 8- to 12-year-old schoolchildren found that 130 aspired to be teachers, 105 doctors, 80 lawyers, 60 police officers, and 50 firefighters. These data, shown in the following table, are classified into five different categories corresponding to the different occupations.

Occupation	Teacher	Doctor	Lawyer	Police	Firefighter	Total
Number of children	130	105	80	60	50	425

In another example, the Red Cross classifies a group of potential donors according to blood type. The categories are the different blood types, and the data are the numbers of donors who have the various blood types. Categories can also be defined by ranges of values of a numerical variable, such as income level (low, medium, or high).

Categorical data, as described above, are statistically analyzed with the chi-square (χ^2) *statistic* introduced by Karl Pearson in 1900. Pearson developed the chi-square *goodness-of-fit test* to evaluate probabilities associated with the categories in a *multinomial experiment*. The multinomial experiment, which is an extension of the binomial experiment, satisfies these conditions.

The Multinomial Experiment

1. The experiment consists of n identical, independent trials.
2. The outcome of each trial falls in one of k categories.
3. The probabilities associated with the k outcomes, denoted by $\pi_1, \pi_2, \ldots, \pi_k$, remain the same from trial to trial. Since there are only k possible outcomes, we have

$$\pi_1 + \pi_2 + \cdots + \pi_k = 1$$

4. The experimenter records the values of $o_1, o_2, \cdots, o_k$, where o_j ($j = 1, 2, \ldots, k$) is equal to the observed number of trials in which the outcome is in category j. Note that

$$n = o_1 + o_2 + \cdots + o_k$$

Let us determine whether the Time Warner survey example fits the description of a multinomial experiment.

1. The n identical trials are the 425 randomly chosen children, and because the choice of any one child does not depend on the choice of any other child, the trials are independent.
2. Each child chooses one of the $k = 5$ occupations. (In the actual survey, the children were able to choose any occupation, but for simplicity we are assuming that each child must choose one of the five occupations.)
3. The probabilities with which the children choose the five occupations are denoted as $\pi_1, \pi_2, \pi_3, \pi_4$ and π_5, and they remain the same from one child to the next.
4. The observed counts are

$$o_1 = 130 \quad o_2 = 105 \quad o_3 = 80 \quad o_4 = 60 \quad o_5 = 50$$

and sum to $n = 425$.

The four conditions are satisfied; thus, the example describes a multinomial experiment.

The goodness-of-fit test consists of determining whether the observed counts in the categories of the variable agree with (fit) the probabilities specified in the multinomial experiment. For example, do children choose the occupations with equal probabilities? If they are chosen equally, each occupation has a probability of 1/5 of being chosen and the null hypothesis

$$H_0: \pi_1 = \pi_2 = \pi_3 = \pi_4 = \pi_5 = \frac{1}{5}$$

should be true. The alternative hypothesis is stated as

$$H_a: \text{at least one } \pi_i \neq \frac{1}{5}$$

To develop a test statistic to test the hypothesis, we compare the observed values, o_i, with the values we expect if the hypothesis is true. If H_0 is true, then out of 425 children, we *expect* each occupation to be chosen $(1/5)(425) = 85$ times.

Expected Number of Outcomes in a Multinomial Experiment

Out of n trials of a multinomial experiment, the expected number of outcomes to fall in category j is $e_j = n\pi_j$. Remember that the expected cell count is calculated assuming H_0 is true.

We realize, however, that the choices could be evenly distributed and not all $o_j = 85$ exactly. There could be some discrepancies in a random sample of only 425 children.

The χ^2 Test Statistic

The chi-square (χ^2) statistic, proposed by Pearson, measures the amount of disagreement between the observed data and the expected data.

Pearson χ^2 Test Statistic

$$\chi^2 = \sum \frac{(o_j - e_j)^2}{e_j}$$

where the sum is over all categories, with o_j being the observed frequency count and e_j the expected frequency count in category j.

If there are large differences between o_j and e_j, χ^2 will be large, which in turn suggests that the null hypothesis should be rejected. But how large does it have to get for us to have significant evidence to reject the null hypothesis? This can be answered only by investigating the sampling distribution of the χ^2 statistic.

When n is large and H_0 is true, the sampling distribution of χ^2 is known to be approximately *chi-square with $k - 1$ degrees of freedom*. Figure 8.1 illustrates a typical chi-square density curve. The curve begins at 0 and is skewed right. As the degrees of freedom increase, the distribution stretches out along the horizontal axis.

Remember that k is the number of categories. There are five occupations to choose from; therefore, $k = 5$.

Figure 8.1
Chi-square distribution

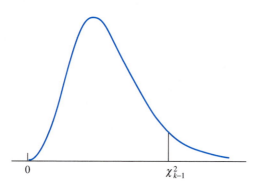

Table B.4 in Appendix B gives the upper-tail probabilities of the χ^2 distribution for various degrees of freedom. For example, when the degrees of freedom are 16, 10% of the area lies above 23.5, 5% lies above 26.3, and 1% lies above 32.0. The table also gives lower-tail probabilities. For 16 degrees of freedom, we see that 95% of the area lies above 7.96; consequently, 5% lies below 7.96.

Goodness-of-Fit Test

For the χ^2 goodness-of-fit test, we are interested only in large values of χ^2 and the upper-tail probabilities. Returning to our example, we have

$$\chi^2_{\text{obs}} = \frac{(130 - 85)^2}{85} + \frac{(105 - 85)^2}{85} + \frac{(80 - 85)^2}{85} + \frac{(60 - 85)^2}{85} + \frac{(50 - 85)^2}{85}$$

$$= \frac{1}{85}[(45)^2 + (20)^2 + (-5)^2 + (-25)^2 + (-35)^2] = 50.59$$

From the χ^2 table with $k - 1 = 5 - 1 = 4$ degrees of freedom, we see that 0.5% of the area lies above 14.9. Thus, less than 0.5% of the area lies above 50.59; that is, p-value $<$ .005, and therefore there is sufficient evidence to reject H_0. Based on these sample results, there is highly significant evidence that the occupations are not chosen equally by schoolchildren between 8 and 12 years old.

In this example, we are investigating statistically how well the observed data fit the conjectured probabilities in the null hypothesis. This is why the test is called the χ^2 goodness-of-fit test. The general form of the test is given in the box.

Chi-Square Goodness-of-Fit Test

Application: Multinomial experiments.

Assumptions:

1. The experiment satisfies the properties of a multinomial experiment.
2. No expected cell count, e_j, is less than 1, and no more than 20% of the e_j's are less than 5.
   (This is so that the χ^2 approximation will be accurate.)

$$H_0: \pi_1 = p_1, \pi_2 = p_2, \ldots \pi_k = p_k$$

where $p_1, p_2, \ldots, p_k$ are the hypothesized values of the multinomial probabilities.

H_a: At least one of the multinomial probabilities does not equal the hypothesized value.

Test statistic:

$$\chi^2 = \sum \frac{(o_j - e_j)^2}{e_j}$$

where $e_j = np_j$

The test is a right-tailed test where the p-value is found in the χ^2 table with $k - 1$ degrees of freedom. Usually the exact value cannot be found, but bounds for it can be found by finding the closest values to the observed value of the χ^2 statistic.

Example 8.1 A local grocery store stocks four brands of cola. Suppose that, nationally, brand A commands 40% of the market, brand B has 35%, brand C has 20%, and brand D has 5%. Of 2000 colas sold during 1 week in the store, 615 were brand A, 804 were brand B, 383 were brand C, and 198 were brand D. Do the data collected at the local grocery store fit the national percentages?

Solution Let

π_1 = proportion of people who buy brand A at the grocery store

π_2 = proportion of people who buy brand B at the grocery store

π_3 = proportion of people who buy brand C at the grocery store

π_4 = proportion of people who buy brand D at the grocery store

The null hypothesis, which says that the local percentages are the same as the national percentages, is

$$H_0: \pi_1 = .40, \pi_2 = .35, \pi_3 = .20, \pi_4 = .05$$

and the alternative is

$$H_a: \text{At least one of the proportions differs from its hypothesized value}$$

The test statistic is

$$\chi^2 = \sum \frac{(o_j - e_j)^2}{e_j}$$

where $e_1 = (2000)(.40) = 800$
$e_2 = (2000)(.35) = 700$
$e_3 = (2000)(.20) = 400$
$e_4 = (2000)(.05) = 100$

Because all of these values are greater than 5 and the conditions of a multinomial experiment are met, we conduct a χ^2 goodness-of-fit test.

The observed value of χ^2 is

$$\chi^2_{obs} = \frac{(615 - 800)^2}{800} + \frac{(804 - 700)^2}{700} + \frac{(383 - 400)^2}{400} + \frac{(198 - 100)^2}{100}$$

$$= 154.995$$

From the χ^2 table with 3 degrees of freedom, we see, as depicted in Figure 8.2, that p-value $< .005$ because any χ^2 value greater than 12.84 has a tail probability less than .005.

Figure 8.2
Chi-square distribution

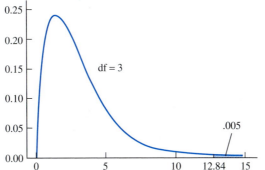

This is highly significant evidence that H_0 should be rejected. Clearly, the shoppers at the local grocery store do not choose the same brands of cola as the rest of the nation. The manager of the grocery store is well advised not to follow the national percentages when stocking cola. ∎

The sampling distribution of χ^2 is only approximately chi-square with $k - 1$ degrees of freedom. The approximation becomes better as the value of n gets larger. The value of n must be large enough to guarantee that no e_j ($e_j = np_j$) is less than 1 and no more than 20% of the e_j's are less than 5 (Cochran, 1954). If these conditions are not met, you should combine cells so that the assumption is satisfied.

EXERCISES 8.1   The Basics

8.1 A multinomial experiment with $k = 5$ and $n = 500$ yielded these results:

Category	1	2	3	4	5
o_j	92	97	106	85	120

Is there evidence to reject the hypothesis that the categories are equally likely? Prior to conducting the test, verify that the assumptions for the goodness-of-fit test are met.

8.2 A multinomial experiment with four possible outcomes and 100 trials produced these data:

Category	1	2	3	4
o_j	44	29	21	6

Is there evidence to reject this hypothesis?

$$H_0: \pi_1 = 40\%, \quad \pi_2 = 30\%, \quad \pi_3 = 20\%, \quad \pi_4 = 10\%$$

Prior to conducting the test, verify that the assumptions for the goodness-of-fit test are met.

8.3 A local official claims that of all voters in the district, 40% are Democrats, 45% are Republican, 7% are conservative, 5% are liberal, and the remaining 3% are classified as other. The following party preferences were found from a random sample of 1200 voters:

Party	Democrat	Republican	Conservative	Liberal	Other
Number of voters	504	523	72	70	31

Are the sample results consistent with the claim made by the official? Prior to conducting the test, verify that the assumptions for the goodness-of-fit test are met.

8.4 In 1882, R. Wolf tossed a die 20,000 times and recorded the number of times each side faced up. Was the die fair?

Die	1	2	3	4	5	6
Frequency	3407	3631	3176	2916	3448	3422

Source: R. Wolf (1882), *Vierteljahresschrift Naturforschl Ges*, Zurich, 207, 242.

8.5 A large jar has red, black, blue, and white marbles in it. One hundred marbles are drawn from the jar with these results:

Color	red	black	blue	white
Frequency	28	19	22	31

Is there statistical evidence that the proportions of marbles of the four colors are the same? Prior to conducting the test, verify that the assumptions for the goodness-of-fit test are met.

8.6 The numbers of books borrowed from a public library for a particular week are given here:

Day	Monday	Tuesday	Wednesday	Thursday	Friday
Books	125	105	120	114	136

Determine whether the number of books borrowed depends on the day of the week.

8.7 Five different strains of flies were tested for their resistance to a particular chemical agent. A large resort area was sprayed with the chemical. Afterward, 1000 dead flies were randomly selected and classified according to their strain, as follows:

Strain	1	2	3	4	5	total
Number killed	265	178	301	115	141	1000

Assuming that each strain is equally prevalent in the resort area, is there evidence that some strains are more resistant to the chemical agent than others?

8.8 According to Information Resources, these are the market shares for different cereal companies

Kellogg	General Mills	Post	Quaker Oats	Other
35%	29%	12%	7%	17%

A survey of 740 shoppers selected randomly from grocery stores in California were asked to name their favorite brand of cereal. From the following data, determine whether the cereal companies have the same market share in California as nationwide.

Kellogg	General Mills	Post	Quaker Oats	Other
220	163	109	95	153

8.9 Should companies be self-insured and provide their own group health plan for their employees? A survey of 461,208 employees of private companies by Medstat Systems at the request of the *Wall Street Journal* (June 17, 1994) found that 31% of the employees had no medical claims, 50% had claims less than $1000, 14% had claims between $1000 and $5000, and 5% had claims that exceeded $5000. A small company obtained the health records of its 250 employees and found that 48 had no claims, 88 had claims less than $1000, 55 had claims between $1000 and $5000, and the rest had claims in excess of $5000. Do these data suggest that the claims from the employees of this company follow the same distribution as that found by the *Wall Street Journal*?

8.10 Are coal-mining disasters more likely to happen in certain months? These data were obtained on 191 coal-mining disasters that occurred from March 15, 1851, to March 22, 1962, inclusive:

Number of disasters

Jan	Feb	Mar	Apr	May	Jun	Jul	Aug	Sep	Oct	Nov	Dec
14	20	20	13	14	10	18	15	11	16	16	24

Source: R. Jarrett (1979), A Note on the Intervals between Coal Mining Disasters, *Biometrika*, 66, 191–193.

Can you conclude from the data that accidents are equally likely in all 12 months of the year? Prior to conducting the test, verify that the assumptions for the goodness-of-fit test are met. State your hypothesis, give the test statistic and its *p*-value, and provide your conclusion.

8.11 The following results were obtained in a 1962 national survey of 452 college students on the issue of student academic integrity:

Cheating behavior
(1962 survey)

Copied from another student during a test without his/her knowledge	20%
Copied from another student during a test with his/her knowledge	12%
Used unpermitted crib notes during a test	16%
Used unfair methods to learn what was on a test before it was given	35%
Helped someone else cheat on a test	22%

The same questions were used in a 1993 survey of 1782 students, with the following results:

Cheating behavior
(1993 survey)

Copied from another student during a test without his/her knowledge	45%
Copied from another student during a test with his/her knowledge	32%
Used unpermitted crib notes during a test	27%
Used unfair methods to learn what was on a test before it was given	29%
Helped someone else cheat on a test	38%

Source: Academic Integrity, Student Life and Learning Report, Vol. 4 No. 1, October 25, 1994, Appalachian State University.

Can the chi-square goodness-of-fit test be used to compare the results of the 1993 survey with the results found in 1962? Why, or why not? If yes, complete the test.

8.12 Here are the percents of total deaths in 1990 from various causes:

Cause	Percent
Heart Disease	33.5
Cancer	23.4
Stroke	6.7
Accidents	4.3
Lung Disease	4.1
Other	28.0

Source: U.S. Department of Health and Human Service (August 28, 1991), *Monthly Vital Statistics Report*, 39 (13).

In 1998, there were 2,338,075 deaths distributed as follows:

Cause	Number
Heart Disease	724,269
Cancer	538,947
Stroke	158,060
Accidents	93,207
Lung Disease	114,381
Other	709,211

Source: World Almanac and Book of Facts, 2000.

With a test of significance, determine whether the 1998 figures are consistent with the 1990 percents.

8.2 The Chi-Square Test of Independence

Learning Objectives for this section:

❏ Understand the basic principles of the chi-square test of independence.

❏ Know how to calculate the expected cell counts in a contingency table.

❏ Know the assumptions required for the chi-square test of independence.

❏ Be able to conduct a chi-square test of independence.

When frequency count data are classified according to two or more variables or populations, they are called *cross-tabulated* data and are displayed in a contingency table (introduced in Chapter 2). For example, subjects might be classified in a contingency table according to their religious preferences and their attitudes about abortion. College students might be classified according to class rank and study habits (good, average, bad).

In this section, we investigate the chi-square test of independence where subjects in a single population are classified according to two different categorical variables. In Section 8.3, we investigate the chi-square test of homogeneity, where subjects from several different populations are classified according to a single categorical variable. In both cases, the chi-square statistic measures the discrepancy between actual counts and expected counts.

Test of Independence

For the test of independence, the aim is to determine whether there is any dependency between the classifying variables. Suppose that 200 randomly selected people are asked about their views on gun control and their preferred political party. Are their views on gun control affected by their political party preference? Table 8.2 is a contingency table of the distribution of the frequency counts from the sample of 200 people.

Table 8.2

		Opinion on gun control			
		Favor	Oppose	No opinion	Total
Political Party	Democrat	44	48	18	110
	Republican	32	48	10	90
	Total	76	96	28	200

Note that we have a single sample that is classified two ways.

The test is constructed in such a way that the classifying variables are assumed to be independent until the data prove otherwise. The null hypothesis is

H$_0$: Political party and opinion on gun control are independent

and the alternative hypothesis is

H$_a$: Political party and opinion on gun control are dependent

A cell is the intersection of a specific row and a specific column.

To utilize the χ^2 test statistic, we must find the expected cell counts and then compare them to the observed cell counts. We will not go into the details, but under the assumption that the classifying variables are independent, we have the following.

The Expected Cell Count

For cell j, the expected cell count is

$$e_j = \frac{RC}{N}$$

where R is the total for the row containing cell j, C is the total for the column containing cell j, and N is the grand total.

Thus, for the first cell

$$\text{Expected[Democrat and favor gun control]} \quad = \frac{(110)(76)}{200} = 41.8$$

In a similar fashion,

$$\text{Expected[Democrat and oppose gun control]} \quad = \frac{(110)(96)}{200} = 52.8$$

$$\text{Expected[Democrat and no opinion]} \quad = \frac{(110)(28)}{200} = 15.4$$

$$\text{Expected[Republican and favor gun control]} \quad = \frac{(90)(76)}{200} = 34.2$$

$$\text{Expected[Republican and oppose gun control]} = \frac{(90)(96)}{200} = 43.2$$

$$\text{Expected[Republican and no opinion]} \quad = \frac{(90)(28)}{200} = 12.6$$

Table 8.3 gives the observed cell counts and the expected cell counts in parentheses.

Table 8.3

		Opinion on gun control			
		Favor	Oppose	No opinion	Total
	Democrat	44 (41.8)	48 (52.8)	18 (15.4)	110
Political Party	Republican	32 (34.2)	48 (43.2)	10 (12.6)	90
	Total	76	96	28	200

Because the expected cell counts were calculated under the assumption that H_0 is true (political party and opinion are independent), a large discrepancy between the observed and expected cell counts should lead to the rejection of H_0 and thus establish that the two classifications are dependent. To compare the differences between observed and expected counts, we calculate the χ^2 statistic as in the previous section:

$$\chi^2_{obs} = \frac{(44 - 41.8)^2}{41.8} + \frac{(48 - 52.8)^2}{52.8} + \frac{(18 - 15.4)^2}{15.4} + \frac{(32 - 34.2)^2}{34.2}$$

$$+ \frac{(48 - 43.2)^2}{43.2} + \frac{(10 - 12.6)^2}{12.6}$$

$$= 2.2025$$

To statistically evaluate the observed value of χ^2, we must determine its p-value from the χ^2 table. But first we must determine the number of degrees of freedom for the test statistic. We ask, of the six expected cell counts, how many are free to vary?

Degrees of Freedom

For data arranged in a contingency table with r rows and c columns, the number of degrees of freedom are

$$df = (r - 1)(c - 1)$$

The example has two rows and three columns; thus, $df = (2 - 1)(3 - 1) = 2$. From the χ^2 table with 2 degrees of freedom, we see that the observed χ^2 statistic of 2.2025 falls below 4.61, which is the value associated with a tail probability of .10. Thus, we have p-value $> .10$, which says that there is insufficient evidence to reject H_0. We have failed to show a dependence between political party and opinion on gun control.

We now summarize the test.

Chi-Square Test of Independence

Application: Test the independence of two classifying variables.

Assumptions:

1. The experiment satisfies the properties of a multinomial experiment.
2. No expected cell count is less than 1, and no more than 20% of the expected cell counts are less than 5.

H_0: The two classifications are independent
H_a: The two classifications are dependent

Test Statistic:
$$\chi^2 = \sum \frac{(o_j - e_j)^2}{e_j}$$

where o_j represents the observed cell frequencies and e_j represents the expected cell frequencies given by

$$e_j = \frac{RC}{N}$$

continued

where R = row total for cell j; C = column total for cell j; and N = grand total or the total number of subjects.

The test is a right-tailed test, where the p-value is found in the χ^2 table with $(r - 1)(c - 1)$ degrees of freedom (r denotes the number of rows and c denotes the number of columns).

Example 8.2   A random sample of 400 undergraduate college students was classified according to class and study habits. From the data in Table 8.4, test to see whether the two classifications are independent.

Table 8.4

	Study Habits			
	Good	Average	Bad	Total
Freshmen	20	42	58	120
Sophomore	25	48	32	105
Junior	31	28	35	94
Senior	24	27	30	81
Total	100	145	155	400

Solution   The null hypothesis is

H_0: class rank and study habits are independent

and the alternative hypothesis is

H_a: class rank and study habits are dependent.

To find the value of the test statistic, we must first find the expected cell counts. For the first cell in the first row and first column, we have

$$e_1 = \frac{(120)(100)}{400} = 30$$

The next cell in the first row has expected count

$$e_2 = \frac{(120)(145)}{400} = 43.5$$

And the expected count in the third cell in the first row is

$$e_3 = \frac{(120)(155)}{400} = 46.5$$

Continuing, we get the expected cell counts shown in Table 8.5 at the top of page 432. Note that all expected cell counts exceed 5, so the conditions for the χ^2 approximation are satisfied. Substituting the observed and expected values into the formula for the χ^2 statistic, we find

$$\chi^2_{obs} = \frac{(20 - 30)^2}{30} + \frac{(42 - 43.5)^2}{43.5} + \cdots + \frac{(30 - 31.39)^2}{31.39}$$

$$= 15.221$$

Table 8.5

	Study Habits			
	Good	Average	Bad	Total
Freshmen	30	43.5	46.5	120
Sophomore	26.26	38.06	40.69	105
Junior	23.5	34.08	36.42	94
Senior	20.25	29.36	31.39	81
Total	100	145	155	400

The degrees of freedom for the χ^2 statistic are

$$df = (r - 1)(c - 1) = (4 - 1)(3 - 1) = 6$$

From the χ^2 tables with 6 degrees of freedom, we see that

$$14.45 < 15.221 < 16.81$$

and thus $.01 < p\text{-value} < .025$. There is statistical evidence to reject H_0. The data suggest that study habits of students are related to their class rank. ∎

COMPUTER TIP

Chi-square Test

To conduct a chi-square analysis of data in a contingency table, first key the observed counts of the table into the worksheet. From the **Stat** menu, select **Tables** and then select **Chisquare Test**. In the Chisquare Test window, **Select** the columns and click on **OK**.

Figure 8.3 shows the Minitab output for the above example. Note that there is a slight disagreement between the results found in Example 8.2 and the results found by Minitab. These differences are due to round-off error. Also note that Minitab computes the actual p-value of .019.

Figure 8.3

```
Expected counts are printed below observed counts

                C1        C2        C3      Total
      1         20        42        58        120
             30.00     43.50     46.50

      2         25        48        32        105
             26.25     38.06     40.69

      3         31        28        35         94
             23.50     34.08     36.42

      4         24        27        30         81
             20.25     29.36     31.39

  Total        100       145       155        400

ChiSq =    3.333 + 0.052 + 2.844 +
           0.060 + 2.595 + 1.855 +
           2.394 + 1.083 + 0.056 +
           0.695 + 0.190 + 0.061 = 15.216
  df = 6, p = 0.019
```

In the preceding example, the frequencies for the contingency table were already given. In the next example, we show how Minitab can construct the table of frequencies from the data.

Example 8.3   A survey was conducted in a voting district to record (among several other variables) the political party and gender of the respondents. The political party and gender were coded as follows and stored in the Minitab worksheet: **Politic.mtw:**

Party	Gender
1—democrat	1—female
2—republican	2—male
3—other	

Organize the data in a contingency table according to the variables *political party* and *gender*. Perform a chi-square test of independence on the two variables.

Solution   The null and alternative hypotheses are

H_0: Political party and gender are independent.

H_a: Political party and gender are dependent.

COMPUTER TIP

Cross Tabulation

From the Stat menu select **Tables** followed by **Cross Tabulation**. In the Cross Tabulation window, **Select** the two columns, C1 and C2. (If more than two columns are selected, a separate chi-square test is done for each two-way table.) Check **Chisquare analysis**, click on **Above and expected count**, and click **OK**.

Figure 8.4 gives the Minitab output.

Figure 8.4   **Tabulated Statistics: Party, Gender**

```
Rows:      Party     Columns:      Gender

              1            2          All

 1           49           55          104
           49.50        54.50       104.00

 2           64           73          137
           65.21        71.79       137.00

 3            6            3            9
            4.28         4.72         9.00

All         119          131          250
          119.00       131.00       250.00
```

Chi-square = 1.365, DF = 2, P-Value = 0.505
2 Cells with expected counts less than 5.0

```
     Cell Contents —
                  Count
                  Exp Freq
```

Notice that Minitab has warned that two cells have expected counts less than 5.0. This is not a serious violation of assumptions here because the expected counts are very close to 5.0 and the criteria suggest that no more than 20% of the cell counts be less than 5. Because the p-value $= 0.505 > .10$, there is insufficient evidence to reject H_0. The two classifications are assumed to be independent. ■

EXERCISES 8.2 The Basics

8.13 College students were classified as being either left- or right-handed and as being either mathematically inclined or not. From the data in the table, test the hypothesis that the two classifications are independent.

	Predominant Hand	
	Left	Right
Mathematically inclined	12	93
Not considered mathematically gifted	7	108

8.14 Suppose a number of patients were treated for cancer with these results:

		Tumor Regression?	
		Yes	No
Toxic Reaction	Yes	15	5
	No	4	22

Determine whether there is a relationship between the presence of a toxic reaction and tumor regression. Are the expected cell counts greater than 5? Why is this important?

8.15 In Table 2.3 in Section 2.5 (reproduced here), automobile dealers were classified according to the type of dealership and the service rendered to customers on their 15,000-mile checkups.

	Service Rendered		
Type Dealership	Number that replace parts before they are needed	Number that only perform services recommended by manufacturer	Total
Honda Accord	19 (9.8)	2 (11.2)	21
Saturn	4 (8.9)	15 (10.1)	19
Ford Taurus	8 (9.8)	13 (11.2)	21
Dodge Caravan	11 (9.8)	10 (11.2)	21
Mazda Miata	12 (9.8)	9 (11.2)	21
Toyota Lexus	3 (8.9)	16 (10.1)	19
Total	57	65	122

The numbers in parentheses are the expected counts if the two classifications are independent. Use the chi-square test to determine whether the discrepancies between the actual counts and the expected counts indicate that there is dependency between the two classifications.

8.16 An experimental psychologist wishes to study the effects that three different drugs have on one's ability to learn a list of nonsense syllables. Sixty subjects were categorized as to the type of drug they had been receiving for the past 3 months and their ability to memorize the list of nonsense syllables.

		Drug		
		A	B	C
Ability to memorize the list of syllables	Low	12	6	6
	Medium	8	5	11
	High	2	7	3

Determine whether the type drug is related to the subject's ability to memorize the list of nonsense syllables. Are all of the required assumptions for the χ^2 test met?

8.17 Does the desire to participate in class projects relate to a child's academic achievement? To study this issue, a sample of 80 third-grade students were asked if they wished to participate in the science project program. The students were then classified according to their academic standing. From these data, is there evidence of a relationship between the desire to participate in the science project program and their academic standing? (Expected cell counts are in parentheses.)

		Desire to participate in science project	
		Yes	No
Academic standing	Below Average	17 (17.23)	9 (8.77)
	Average	14 (17.89)	13 (9.11)
	Above Average	22 (17.89)	5 (9.11)
Total		53	27

8.18 *USA Today*/CNN/Gallup national telephone polls found the following opinions on an embargo against Cuba:

	Embargo Against Cuba	
	Dec., 1993	June, 1994
Favor	660	700
Oppose	260	240

The headline stated "Embargo against Cuba gains favor."
a. Conduct a chi-square test of independence on the data.
b. Is the change of opinion from December 1993 to June 1994 large enough to justify the claim made in the headline?

Interpreting Computer Output

8.19   Based on the computer output, test the independence of the two variables displayed in this contingency table:

	Variable A		
	Level 1	Level 2	Level 3
Variable B Level 1	65	39	16
Level 2	133	156	61

Chi-Square Test

Expected counts are printed below observed counts

	level1	level2	level3	Total
1	65	39	16	120
	50.55	49.79	19.66	
2	133	156	61	350
	147.45	145.21	57.34	
Total	198	195	77	470

ChiSq =   4.129 +   2.337 +   0.681 +
           1.415 +   0.801 +   0.234 =   9.597

df = 2, p = 0.008

8.20   A political pollster would like to determine whether the voters' feelings on a local referendum are related to their views on freedom of the press. A sample of 237 voters were asked how they plan to vote on the referendum, and they were asked to choose one of the following that most closely represents their view on freedom of the press.

A. The press is at liberty to report anything it sees fit.

B. The press should not report anything that would jeopardize anyone's life.

C. The press should not report anything that would jeopardize anyone's life or reputation.

The contingency table lists the number of voters in each category:

WORKSHEET: Referend.mtw

Response to question	Opinion on referendum			
	For	Against	Undecided	Total
A	24	29	7	60
B	68	39	12	119
C	47	8	3	58

Using the accompanying computer printout, determine whether there is a relationship between opinion on the referendum and view on freedom of the press. Are any assumptions for the chi-square test in question?

Chi-Square Test: For, Against, Undecided

```
Expected counts are printed below observed counts

              For    Against  Undecided    Total
     1         24        29          7         60
            35.19      19.24       5.57

     2         68        39         12        119
            69.79      38.16      11.05

     3         47         8          3         58
            34.02      18.60       5.38

Total        139        76         22        237

Chi-Sq =    3.558 +   4.950 +    0.367 +
            0.046 +   0.018 +    0.082 +
            4.955 +   6.040 +    1.056 = 21.074
DF = 4, P-Value = 0.000
```

8.21 Are snoring and heart disease related? The accompanying data are the results of a study reported in the *British Medical Journal*. Subjects were classified according to the amount they snored (as reported by their spouses) and whether they had a history of heart disease.

WORKSHEET: Snore.mtw

		Nonsnorer	Occasional snorer	Nearly every night	Snores every night	Total
Heart	Yes	24	35	21	30	110
Disease	No	1355	603	192	224	2374

Source: Norton, P. and Dunn, E. (1985), Snoring as a Risk Factor for Disease, *British Medical Journal*, *291*, 630–632.

Using the accompanying computer printout, determine whether snoring and heart disease are statistically related. Are any assumptions for the chi-square test in question?

Chi-Square Test: Non, occasion, nearly, every

```
Expected counts are printed below observed counts

            Non    occasion    nearly    every    Total
    1        24         35        21        30      110
           61.07      28.25      9.43     11.25

    2      1355        603       192       224     2374
         1317.93     609.75    203.57    242.75

Total    1379        638       213       254     2484

Chi-Sq =  22.499 +   1.611 +  14.186 +  31.262 +
           1.043 +   0.075 +   0.657 +   1.449 = 72.782
DF = 3, P-Value = 0.000
```

Computer Exercises

8.22 Numerous seasonal allergy medicines are advertised to relieve sneezing, runny nose, watery eyes, and congestion. In most cases, however, there are associated side effects. In double-blind, controlled clinical trials, Marion Merrell Dow, Inc., the maker of Seldane-D, reported the following results in June 1991:

WORKSHEET: Allergy.mtw

		Medication		
		Seldane-D	Pseudoephedrine	Placebo
Adverse Event	Insomnia	97	77	12
	Headache	65	49	43
	Drowsiness	27	14	22

Source: Marion Merrell Dow, Inc. Kansas City, Mo 64114.

a. Is there statistical dependence between the medication and the adverse events?
b. Discard the insomnia data and determine, as in part **a**, whether there is statistical dependence between the medication and the adverse events headache and drowsiness.
c. In light of the results found in part **b**, what is causing the dependence found in part **a**?
d. Marion Merrell Dow reported that there was no significant difference in drowsiness between those who took Seldane-D and those who took a placebo. Do the data support this claim?

8.23 In 1984, history was made when Geraldine Ferraro was nominated as the first woman vice presidential candidate. For the first time ever, the gender of the candidate was an issue. On August 7 and 9, 1984, *Time* magazine conducted a telephone survey of 1000 voters. In each contingency table, determine whether the gender of the respondent is independent of his or her choice.
a. If the election were tomorrow, for whom would you vote?

WORKSHEET: Ferraro1.mtw

	Reagan/Bush	Mondale/Ferraro	Undecided
Men	245	140	115
Women	205	160	135

b. Who would be a better vice president?

WORKSHEET: Ferraro2.mtw

	Bush	Ferraro	Undecided
Men	245	155	100
Women	185	235	80

8.24 Prior to the enactment of seat belt legislation in the province of Alberta, Canada, data were collected on 86,769 automobile accident reports to determine the effectiveness of seat belts in preventing injury. The table gives the injury level of the driver and whether he or she was wearing a seatbelt:

WORKSHEET: Seatbelt.mtw

	Injury level			
Seatbelt	None	Minimal	Minor	Major/Fatal
Yes	12,813	647	359	42
No	65,963	4000	2642	303

Source: Jobson, J. (1992), *Applied Multivariate Data Analysis*, Springer-Verlag, New York, p. 18.

Is there statistical evidence that seat belts help prevent injury? Be sure to formulate null and alternative hypotheses, give the *p*-value, and state your conclusion.

8.25   A study of the absenteeism of bus drivers in the transit system for the city of Edmonton, Alberta, produced these data:

WORKSHEET: Bus.mtw

	Type of Shift				
Attendance	AM	Noon	PM	Swing	Split
Absent	454	208	491	160	1599
Present	5806	2112	3989	3790	10754

Source: Jobson, J. (1992), *Applied Multivariate Data Analysis*, Springer-Verlag, New York, p. 67.

Based on the data, is the attendance of bus drivers dependent on the shift? Be sure to state the null and alternative hypotheses. Give the *p*-value of the test statistic and state your conclusion.

8.3   The Chi-Square Test of Homogeneity

Learning Objectives for this section:

❏ Understand the basic principles of the chi-square test of homogeneity.

❏ Know the difference between the chi-square test of homogeneity and the chi-square test of independence.

❏ Know the assumptions required for the chi-square test of homogeneity.

❏ Be able to conduct a chi-square test of homogeneity.

In the test of independence, we attempt to determine whether two characteristics (variables) associated with the subjects in a *single* population are independent. For example, subjects randomly selected from a population may be classified according to their views on gun control and their preferred political party. We then determine whether their views on gun control are independent of their preferred political party. The chi-square test of homogeneity, presented in this section, attempts to determine whether *several* populations are similar or *homogeneous* with respect to some variable. With the variable as one classification and the populations as the other, the data form a two-way contingency table. The assumptions and statements of the null and alternative hypotheses are different from those for the test of independence, but the details of the analysis are the same. We illustrate with an example.

Example 8.4   Suppose that over a 2-year period, 120 patients with heart disease were treated with one of two drugs (A or B). After a period of time each patient's condition was rated as no change, improved, or greatly improved. Table 8.6, a contingency table, gives the distribution of frequency counts. Determine whether the patients' conditions are similar with respect to the two drugs.

Table 8.6

		Patient's Condition			
		No change	Improved	Greatly Improved	Total
Drug	A	15	22	33	70
	B	20	18	12	50
Total		35	40	45	120

Solution   The null and alternative hypotheses are:

H_0: The proportions of patients falling in the three categories are the same for drug A and drug B.

H_a: The proportions of patients falling in the three categories are not the same for drug A and drug B.

If we denote the probabilities of falling in the three categories (cells) for drug A as π_{A1}, π_{A2}, and π_{A3} and for drug B as π_{B1}, π_{B2}, and π_{B3}, the null hypothesis can be stated as:

$$H_0: \pi_{A1} = \pi_{B1}, \pi_{A2} = \pi_{B2}, \pi_{A3} = \pi_{B3}$$

From the two samples, one of size 70 from the population of patients who received drug A and the other of size 50 from the population of patients who received drug B, we wish to determine whether the two populations are homogeneous with respect to the cell probabilities.

Because the sample sizes are determined before the data are collected, the row (or column) marginal totals in the test of homogeneity are fixed quantities whose values are the sizes of the samples taken from the populations. In the test of independence, however, the grand total is the only fixed quantity with a value that is the size of the sample taken from the single population. Thus, a feature that distinguishes the test of homogeneity from the test of independence is whether the marginal totals are fixed or random quantities.

Under the assumptions that the marginal totals are fixed and the null hypothesis is true, the expected cell counts for the test of homogeneity are calculated just as they were in the test for independence; namely, the expected cell counts are

$$e_j = \frac{RC}{N}$$

where R is the row total for cell j, C is the column total for cell j, and N is the grand total. Thus, in the example, we have

$$E[A \text{ and no change}] \quad = \frac{(70)(35)}{120} = 20.42$$

$$E[A \text{ and improved}] \quad = \frac{(70)(40)}{120} = 23.33$$

$$E[A \text{ and greatly improved}] = \frac{(70)(45)}{120} = 26.25$$

$$E[B \text{ and no change}] \quad = \frac{(50)(35)}{120} = 14.58$$

$$E[B \text{ and improved}] \quad = \frac{(50)(40)}{120} = 16.67$$

$$E[B \text{ and greatly improved}] = \frac{(50)(45)}{120} = 18.75$$

Table 8.7 gives the observed cell counts and the expected cell counts in parentheses.

Table 8.7

		Patient's condition			
		No Change	Improved	Greatly Improved	Total
Drug	A	15 (20.42)	22 (23.33)	33 (26.25)	70
	B	20 (14.58)	18 (16.67)	12 (18.75)	50
Total		35	40	45	120

The test statistic is computed exactly as in the preceding section; that is,

$$\chi^2 = \sum \frac{(o_j - e_j)^2}{e_j}$$

From the data we have

$$\chi^2_{obs} = \frac{(15 - 20.42)^2}{20.42} + \frac{(22 - 23.33)^2}{23.33} + \frac{(33 - 26.25)^2}{26.25}$$

$$+ \frac{(20 - 14.58)^2}{14.58} + \frac{(18 - 16.67)^2}{16.67} + \frac{(12 - 18.75)^2}{18.75}$$

$$= 7.798$$

Because there are two rows and three columns, the degrees of freedom are df = $(2 - 1)(3 - 1) = 2$. From the χ^2 table (Table B.4 in Appendix B), we see that the observed χ^2 statistic of 7.801 falls between 7.38 and 9.21, the values associated with tail probabilities of .025 and .01, respectively. Thus, we have

$$.01 < p\text{-value} < .025$$

which indicates that there is evidence to reject H_0. We conclude that the patients' conditions depend on the drug received. ∎

Here is a summary of the test procedure.

Chi-Square Test of Homogeneity

Application: Contingency table with fixed marginal totals.

Assumptions:

1. A random sample is selected from each of the row category populations. The sample sizes (row marginal totals) are fixed prior to sampling.
2. No expected cell count is less than 1, and no more than 20% of the cell counts are less than 5.

H_0: The populations are homogeneous with respect to the variable of classification
H_a: The populations are not homogeneous

Test statistic:
$$\chi^2 = \sum \frac{(o_j - e_j)^2}{e_j}$$

where o_j represents the observed cell frequencies and e_j represents the expected cell frequencies given by

$$e_j = \frac{RC}{N}$$

where R = row total for cell j, C = column total for cell j, N = grand total or the total number of subjects.

   The test is a right-tailed test, where the p-value is found in the χ^2 table with $(r - 1)(c - 1)$ degrees of freedom (r denotes the number of rows and c denotes the number of columns).

EXERCISES 8.3 The Basics

8.26 A pollster sampled 200 voters, 100 from District 1 and 100 from District 2, to determine their opinions on an upcoming referendum. The results of the survey are given in the contingency table, with expected counts in parentheses.

	Opinion on referendum			
	Favor	Against	Undecided	Total
District 1	72 (66)	21 (27.5)	7 (6.5)	100
District 2	60 (66)	34 (27.5)	6 (6.5)	100

Is there evidence that the two districts will vote differently on the referendum?
a. Identify the variable of classification.
b. Identify the populations.
c. Formulate your hypotheses.
d. Calculate the test statistic and its *p*-value and state your conclusion.
e. Are any assumptions for the chi-square test in question?

8.27 A study was conducted to compare two treatments for smokers who wish to stop smoking. Three hundred smokers were divided between the two methods with the following results:

	Stopped Smoking	Smoke less	Smoke the same
Treatment A	44	38	68
Treatment B	33	42	75

Do the data suggest that the two treatments are equally effective in helping smokers stop smoking?
a. Identify the variable of classification.
b. Identify the populations.
c. Formulate your hypotheses.
d. Calculate the test statistic and its *p*-value and state your conclusion.
e. Are any assumptions for the chi-square test in question?

8.28 The collegiate record for the single-season average rushing yards per game is held by Marcus Allen, who in 1981 with USC, rushed for an average of 212.9 yards per game. Generally when a record is broken, the back who carries the ball receives the credit in the press release; many believe, however, that the linemen should receive the credit. Samples of 50 football players and 70 members of the press were asked who should receive the most credit: the back or the linemen. From these data, determine whether their views are similar.

	Who should receive the most credit?			
	Linemen	Back	Both	Total
Players	22	14	14	50
Press	6	48	16	70

a. Identify the variable of classification.
b. Identify the populations.
c. Conduct the chi-square test. Be sure to state the hypotheses, give the *p*-value, and present your conclusion.

8.29 It is reported that a Vietnam veteran is much more likely to commit suicide than a nonveteran. From the list of those eligible for the draft in 1970, a sample of 100 Vietnam veterans and a sample of 100 nonveterans were selected. The 200 individuals were asked whether they had ever contemplated suicide. The results are recorded in the contingency table.

		Vietnam Veteran	
		Yes	No
Contemplated suicide	Yes	32	11
	No	68	89

Determine with a test of significance whether there is statistical evidence to suggest that the proportion of veterans who have contemplated suicide is different from the proportion of nonveterans. What is the variable of classification, and what are the populations?

8.30 An experiment is designed to study the side effects of two drugs used as treatments for a certain ailment. A group of 90 subjects are assigned to two drug groups. After the specified drug is given, the side effects are classified as follows:

	Side Effects			
	Major	Minor	None	Total
Drug A	13	15	17	45
Drug B	8	21	16	45

Are the side effects distributed the same for the two drugs? State your hypotheses accordingly. Are the marginal totals fixed? Complete the test, give the p-value, and state your conclusion.

Interpreting Computer Output

8.31 From the contingency table data and the computer analysis, test the hypothesis that the proportions that fall into the three categories are the same for the three populations:

	Category			
	1	2	3	Total
Population 1	35	16	29	80
Population 2	29	21	30	80
Population 3	25	27	28	80

Chi-Square Test

```
Expected counts are printed below observed counts

          Cat1       Cat2       Cat3      Total
    1       35         16         29         80
          29.67      21.33      29.00

    2       29         21         30         80
          29.67      21.33      29.00

    3       25         27         28         80
          29.67      21.33      29.00

Total       89         64         87        240

ChiSq =   0.959 +    1.333 +    0.000 +
          0.015 +    0.005 +    0.034 +
          0.734 +    1.505 +    0.034 =    4.621
df = 4,  p = 0.328
```

Be sure to state the hypotheses and give the *p*-value. Does it appear that the row marginal totals are fixed? Do the expected cell frequencies exceed 5? Is there evidence to reject the null hypothesis?

8.32 Do male and female college students differ in their favorite sport? Random samples of 100 college women and 100 college men were asked to name their favorite sport. The results are recorded in the contingency table:

WORKSHEET: Sports.mtw

	Football	Basketball	Baseball	Tennis
		Favorite Sport		
Male	33	38	24	5
Female	38	21	15	26

Are the favorite sports distributed the same for men and women?
a. Identify the variable of classification.
b. Identify the populations.
c. Formulate your hypotheses.
d. From the provided computer printout, calculate the test statistic and its *p*-value and state your conclusion.
e. Are any assumptions for the chi-square test in question?

Chi-Square Test: football, basketbl, baseball, tennis

```
Expected counts are printed below observed counts

          football    basketbl    baseball    tennis     Total
   1            33          38          24         5       100
            35.50       29.50       19.50     15.50

   2            38          21          15        26       100
            35.50       29.50       19.50     15.50

Total           71          59          39        31       200

Chi-Sq =    0.176 +    2.449 +    1.038 +   7.113 +
            0.176 +    2.449 +    1.038 +   7.113 = 21.553
DF = 3, P-Value = 0.000
```

Computer Exercises

8.33 The justice system in the 75 largest counties in the United States disposed of 540 spouse-murder cases in 1988. Can we conclude from the following data that the outcomes of murder cases are different for husband and wife defendants?

WORKSHEET: Spouse.mtw

Result	Husband	Wife
	Defendant	
Not prosecuted	35	35
Pleaded guilty	146	87
Convicted at trial	130	69
Acquitted at trial	7	31

Source: Bureau of Justice Statistics (September 1995), *Spouse Murder Defendants in Large Urban Counties*, Executive Summary, NCJ-156831.

a. Identify the variable of classification.
b. Identify the populations.
c. Conduct the chi-square test. Be sure to state the hypotheses, give the *p*-value, and present your conclusion.

8.34   Are foreign cars safer than domestic cars? The Insurance Institute for Highway Safety (IIHS) and the Highway Loss Data Institute (HLDI) collect vehicle loss data from major insurers and the federal government to produce fatality ratings for most makes of automobiles. The fatality ratings are based on actual occupant deaths per registered vehicle. These data are based on foreign and domestic vehicles (pickup trucks included) for the 1988–92 model years.

WORKSHEET: Vehicle.mtw

	Fatality rating				
	Much better than average	Above average	Average	Below average	Much worse than average
Foreign	11	0	10	4	12
Domestic	30	9	38	7	30

Source: Insurance Institute for Highway Safety and the Highway Loss Data Institute, 1995.

Based on these data, are the fatality ratings similar for foreign and domestic vehicles?

a. Identify the variable of classification.
b. Identify the populations.
c. Formulate your hypotheses.
d. Calculate the test statistic and its *p*-value and state your conclusion.
e. Are any assumptions for the chi-square test in question?

8.35   Determine whether the insurance injury ratings for Chevrolet vehicles improved from 1990 to 1993 by analyzing these data.

WORKSHEET: Chevy.mtw

	Chevrolet Injury Frequency				
	Much better than average	Above average	Average	Below average	Much worse than average
1988–90	16	5	5	3	4
1991–93	12	2	12	2	6

Source: Insurance Institute for Highway Safety and the Highway Loss Data Institute, 1995.

a. Identify the variable of classification.
b. Identify the populations. Are the row marginal totals fixed? Explain.
c. Are any assumptions for the chi-square test in question?
d. Adjust the data so that the assumptions from part **c** are met.
e. Formulate your hypotheses.
f. Calculate the test statistic and its *p*-value and state your conclusion.

8.4   Summary and Review Exercises

Key Concepts   ■   When data consist of frequency counts and satisfy the properties of a multinomial experiment the statistic

$$\chi^2 = \sum \frac{(o_j - e_j)^2}{e_j}$$

is used to test hypotheses about the data. We discussed three applications of this statistic: the χ^2 goodness-of-fit test, the χ^2 test of independence, and the χ^2 test of homogeneity. To use the χ^2 statistic in any of the three cases, the expected cell counts must be found. For the goodness-of-fit test, the expected counts are

$$e_j = n \pi_j$$

where n is the number of subjects and π_j is the hypothesized probability of an observation falling into category j.

- For the test of independence and test of homogeneity, the expected cell counts are found by

$$e_j = \frac{RC}{N}$$

where

R = row total for cell j, C = column total for cell j, and N = total number of subjects

- The test statistic is approximately distributed as χ^2 with the degrees of freedom being $k - 1$ in the goodness-of-fit and $(r - 1)(c - 1)$ in the test of independence and the test of homogeneity. In all cases, the approximation is valid if no expected cell count is less than 1 and no more than 20% are less than 5.

Statistical Insight Revisited

The Statistical Insight in the beginning of the chapter was concerned with whether the type of drug offense for federal prisoners is related to their race. The data, obtained from the 1994 report by the Department of Justice, are in worksheet: **Inmate.mtw**.

a. The null and alternative hypotheses to evaluate the relationship between race and type of drug offense are

H_0: The type of drug offense is independent of race.
H_a: The type of drug offense is dependent on race.

b. Conducting the chi-square test of independence we have

Chi-Square Test: heroin, crack, cocaine, marijuana

Expected counts are printed below observed counts

	heroin	crack	cocaine	marijuan	Total
white	407	106	4525	2825	7863
	806.57	831.80	4558.79	1665.84	
black	1156	2513	4439	442	8550
	877.04	904.48	4957.09	1811.39	
hispanic	1314	348	7297	2675	11634
	1193.39	1230.72	6745.12	2464.76	
Total	2877	2967	16261	5942	28047

```
Chi-Sq = 197.944 +  633.309 +   0.250 +  806.582 +
          88.729 + 2.9E+03 +  54.148 +  1.0E+03 +
          12.189 +  633.123 +  45.154 +   17.932 = 6385.210
DF =  6,  P-Value = 0.000
```

The chi-square statistic is extremely large (p-value = 0.000) so the null hypothesis should be rejected. Based on these data there is a strong relationship between race and the type of drug offense.

c. The largest differences between actual and expected counts occur in cells with rows intersecting with crack and marijuana. Many more blacks are convicted for a crack drug offense than either whites or Hispanics. Very few blacks are convicted for a marijuana drug offense. A small number of whites and Hispanics are convicted on crack offenses, but a very larger number of whites are convicted for marijuana offenses.

Questions for Review

Use the following problems to test your skills.

The Basics

8.36 A multinomial experiment with three categories resulted in this table:

Category		
1	2	3
48	102	150

Test the hypothesis that $\pi_1 = 1/6$, $\pi_2 = 2/6$, $\pi_3 = 3/6$.

8.37 From these data, test the hypothesis that the proportions that fall into the three categories are the same for the two populations. (Expected cell counts are in parentheses.) Is this a test of independence or a test of homogeneity?

	Category			
	1	2	3	Total
Population 1	22 (25.5)	33 (30.5)	45 (44)	100
Population 2	29 (25.5)	28 (30.5)	43 (44)	100

8.38 Teenage suicide is a serious problem in the United States. The family economic status of a sample of 38 teenagers who committed suicide was distributed as follows:

Family Status			
Upper class	Upper middle class	Lower middle class	Lower class
14	10	8	6

Is there statistical evidence that the proportions in the four classes differ?

8.39 People who are said to have a Type A personality are aggressive and always on the go. A Type B personality is the opposite. Suppose 50 Type A and 50 Type B persons were classified according to their risk of a heart attack with these results:

	Risk of a heart attack			
	Low	Average	High	Total
Type A	9	27	14	50
Type B	12	31	7	50

Determine whether the risk of a heart attack is distributed the same for Type A and Type B personalities. Is this a test of independence or a test of homogeneity?

8.40 As of December 31, 1998, there were 3320 prisoners on death row in state and federal prisons. Here is the distribution of the prisoners according to regions of the country.

Region	Number on death row	Percent of U.S. resident population age ≥ 18
Northeast	232	20%
Midwest	481	24%
South	1838	35%
West	769	21%

Are the prisoners on death row distributed similar to the general population as far as region of the country is concerned? State your null and alternative hypotheses to examine this conjecture. Give the value of the test statistic and its p-value and state your conclusion.

8.41 A multinomial experiment with four categories resulted in this table:

	Category			
1	2	3	4	Total
29	74	32	65	200

Test the hypothesis that $\pi_1 = \pi_3 = .20$, $\pi_2 = \pi_4 = .30$.

8.42 Mutual funds continually revise their holdings, hoping to improve their portfolios. At one point, a certain mutual fund classified its holdings as follows:

Income Stocks	Growth Stocks	High-Risk Stocks
48	39	43

Is there statistical evidence that its holdings are equally divided among the three types of stocks?

8.43 A sample of 120 diamonds from a diamond mine were classified as low grade, medium grade, and high grade:

	Low grade	Medium grade	High grade	Total
Number of diamonds	52	46	22	120

It is suspected that the percents of low- and medium-grade diamonds are the same and each is twice that of high grade. Is there evidence to support this conjecture?

8.44 A company sales manager wishes to know whether all of his salespeople are contributing the same effort to the company. He records the number of sales made by each of his six salespeople over a given time. Are the sales equally distributed among the salespeople?

Salesperson	A	B	C	D	E	F	Total
Number of sales	34	21	44	52	37	42	230

8.45 The contingency table shows the results of an experiment designed to study the effects of vaccinating laboratory animals against a particular disease.

	Got the disease	Did not get the disease
Vaccinated	12	38
Not vaccinated	21	29

Test the independence of disease and vaccination.

8.46 Executives of small, medium, and large corporations were polled on the issue of the economy. From the data, determine whether the size of the corporation is related to outlook on the economy.

	Economic Outlook		
	Optimistic	Cautious	Pessimistic
Small corporation	38	22	14
Medium corporation	25	20	11
Large corporation	20	8	12

8.47 In a 14-year study, 200 female college graduates were divided into two groups according to their careers. Those in male-dominated fields such as law, government, business, science, and medicine were classified in the professional group. Those who had never been employed or who had careers in teaching, nursing, secretarial, and social work were classified in the traditional group. From the data, determine whether marital status is independent of career.

	Married	
	Yes	No
Professional	65	35
Traditional	85	15

8.48 It is believed that in many species of mammals and birds, a greater proportion of young males than females die when resources are scarce. For example, due to its greater size, the domestic ram requires 15% more food than the ewe. To study this further, a group of scientists observed three populations of red deer in three areas where the food supply was considered low, medium, and high, to record the number of deer that die before their first birthday. From the data, which gives the number of red deer that died before their first birthday, determine whether there is any association between the gender of the dead deer and the food supply.

		Number of deaths of red deer before their first birthday	
		Males	Females
Food supply	Low	23	17
	Medium	12	10
	High	7	8

Interpreting
Computer
Output

8.49 From the computer printout, determine whether the row and column classifications in the contingency table are independent. Is this a chi-square test of independence or a test of homogeneity?

```
                 Column
Row      1       2       3       4
   1     15      26      18      31
   2     19      21      25      32
```

Chi-Square Test

```
Expected counts are printed below observed counts

           C1       C2       C3       C4     Total
   1       15       26       18       31        90
        16.36    22.62    20.70    30.32

   2       19       21       25       32        97
        17.64    24.38    22.30    32.68

Total      34       47       43       63       187

ChiSq =  0.114 +  0.505 +  0.351 +  0.015 +
         0.105 +  0.469 +  0.326 +  0.014 =   1.899
df = 3, p = 0.594
```

8.50 Do the demographic characteristics of developmental students differ at 2-year colleges and 4-year colleges? The results given here are from a study commissioned by the Exxon Education Foundation and conducted by the National Center for Developmental Education in 1992. Based on the computer printout, is there statistical evidence that the distribution of race is the same for the two types of colleges? Be sure to state the hypotheses, give the p-value, and write a conclusion.

WORKSHEET: Develop.mtw

Race	Two-year	Four-year
African Am	545	986
Am Indian	24	66
Asian	71	66
Latino	142	230
White	1587	1939

Source: Research in Developmental Education (1994), V. 11, 2.

Chi-Square Test

```
Expected counts are printed below observed counts

         two-year    four-yr     Total
   1          545        986      1531
          641.26     889.74

   2           24         66        90
           37.70      52.30

   3           71         66       137
           57.38      79.62

   4          142        230       372
          155.81     216.19

   5         1587       1939      3526
         1476.86    2049.14

Total       2369       3287      5656
```

```
ChiSq =    14.448 + 10.413 +
            4.976 +  3.586 +
            3.232 +  2.329 +
            1.224 +  0.882 +
            8.215 +  5.920 = 55.227
df = 4, p = 0.000
```

8.51 Is the abortion rate higher in the Northeast and West than it is in the Midwest and South? Worksheet: **Abortion.mtw** contains abortion rates for each state and the District of Columbia. In the table, the 1996 abortion rate is coded as low if it is less than or equal to 20 abortions per 1000 women and high if it is greater than 20 per 1000 women. From the data and the chi-square test, is there evidence that the abortion rate in 1996 is dependent on the region of the country?

Tabulated Statistics: region, lowhigh

```
                    Rate

Region      Low         High        All

midwest      10            2          12
            7.29         4.71       12.00

northeast     4            7          11
            6.69         4.31       11.00

south        10            5          15
            9.12         5.88       15.00

west          7            6          13
            7.90         5.10       13.00

All          31           20          51
            31.00        20.00       51.00

Chi-Square = 5.792, DF = 3, P-Value = 0.122
2 cells with expected counts less than 5.0
```

8.52 In a study related to the one in Exercise 8.51, the number of facilities performing abortions in 1988 was compared to the number in 1992. From the data and the accompanying computer printout, is there evidence that the facilities performing abortions have changed from 1988 to 1992? State the null and alternative hypotheses, give the *p*-value, and write your conclusion.

	1988	1992
Public hospital	230	180
Private hospital	810	675
Clinic	885	889
Physician's office	657	636

Source: *USA Today*, June 16, 1994.

Chi-Square Test

Expected counts are printed below observed counts

	C1	C2	Total
1	230	180	410
	213.35	196.65	
2	810	675	1485
	772.73	712.27	
3	885	889	1774
	923.11	850.89	
4	657	636	1293
	672.82	620.18	
Total	2582	2380	4962

ChiSq = 1.300 + 1.410 +
       1.798 + 1.951 +
       1.573 + 1.707 +
       0.372 + 0.403 = 10.515
df = 3, p = 0.015

Computer Exercises

8.53 In a local referendum to legalize the sale of beer, voters were surveyed about their opinions on the referendum and their views on the moral issue of selling alcoholic beverages. From the contingency table, determine whether their moral values are related to their opinions on the referendum:

WORKSHEET: Drink.mtw

	Opinion on referendum		
Drinking is	For	Against	Undecided
A. ok	95	83	21
B. tolerated	73	71	18
C. immoral	12	46	8

8.54 The average cost of a funeral ranges from around $2000 in the West to as much as $3000 to $4000 in the central states. One hundred people in each region who had buried a loved one were surveyed and asked to complete a questionnaire regarding the cost of the funeral. From the contingency table, determine whether the perceived cost of a funeral is the same across the regions of the country.

WORKSHEET: Funeral.mtw

Region	Less than expected	Cost about what expected	Cost more than expected
West	15	60	25
Central	20	38	42
South	34	44	22
East	12	40	48

8.55 An article in the *Wall Street Journal* (March 8, 1994) reported that black women are outperforming black men in corporate America. For example, in 1982, there were 1.2 black female professionals for every black male professional, compared to 1.8 in 1994. Is the difference because black women are better educated than black men? Examine these data (random samples of 2140 black women and 1660 black men) to compare the educational levels of black women to those of black men.

WORKSHEET: Blackedu.mtw

	High school dropout	High school graduate	Some college	Bachelor's degree	Graduate degree
Female	486	659	691	208	96
Male	496	530	435	134	65

Counts based on Bureau of Census data.

a. State the null and alternative hypotheses to determine whether there is a difference in the education levels of black women and men.
b. Is this a test of independence or a test of homogeneity?
c. What is the test statistic? What is its observed value?
d. What is the p-value? Should the null hypothesis be rejected?
e. Write your conclusion.

8.56 A summary of the birthday selections for the 1970 draft lottery is given here. The data are the numbers of selections in each month that were for the first half of potential draftees—that is, those with draft numbers less than or equal to 183. If the draft were truly random, each month should have been selected with a probability proportional to the number of days in the month. Test the hypothesis that the months were selected with probabilities proportional to the number of days in the months.

Month	Number of selections in the first 183
Jan	13
Feb	12
Mar	9
Apr	11
May	14
Jun	14
Jly	14
Aug	18
Sep	19
Oct	13
Nov	21
Dec	25

8.57 Compared with boys, girls begin to decline in academic achievement at about age 13 and are below boys in math ability by age 17 (based on interviews of 285 educators and students in ten cities by the National Coalition of Advocates for Students). Many believe that the decline in math ability of girls is due to their own lack of confidence. Samples of 200 13-year-old girls and 200 13-year-old boys were asked how they perceived their math ability, with these results:

WORKSHEET: Ability.mtw

	How do you perceive your math ability?				
	Hopeless	Below average	Average	Above average	Superior
Girls	56	61	54	21	8
Boys	35	43	61	42	19

Is there evidence that 13-year-old girls have less self-confidence than boys as far as math ability is concerned?

8.58 The number of hazardous waste sites and the number located in communities with above-average minority populations are given in worksheet: **Toxic.mtw**. From the following summarized data, determine whether the number of hazardous waste sites located in communities with a large number of minorities is dependent on the region of the country.

	Minority population	
Region	above-average number of minorities	below-average number of minorities
Northeast	29	50
Midwest	133	43
South	63	99
West	85	28

8.59 When you take your car for its annual inspection, what type of inspection station should you choose? In Mecklenburg County (Charlotte), North Carolina, 183 inspection stations inspected at least 500 vehicles in 1992. The table indicates the type of inspection station and the percent of vehicles that passed inspection with perfect scores. Is there statistical evidence that the percent of vehicles that pass with perfect scores depends on the type of inspection station?

WORKSHEET: Inspect.mtw

	Percent that passed		
Type of station	< 70% passed	between 70% and 84%	≥ 85%
Auto inspection station	1	8	4
Auto repair dealer	16	13	7
Tire dealer	16	11	6
Gas station	19	21	6
Car care center	4	4	4
New car dealer	1	5	28

Source: The Charlotte Observer, December 13, 1992.

a. State your hypotheses.
b. Give the value of the test statistic and its *p*-value and state your conclusion.
c. Are any of the assumptions for the chi-square test violated?
d. Why might you remove the new-car dealers from the analysis?
e. Remove the new-car dealers and recalculate the chi-square statistic. What happened?

9

Regression Analysis

In Chapter 2, we introduced bivariate data and the study of association. We now revisit the association problem with the idea of using a statistical model to relate two or more numerical variables in a single population. For example, the grade point averages of students in general are most likely related to their intelligence as measured by their IQ scores, study habits, and motivation. The statistical model is the formula that shows how the variables are related. A wide variety of statistical models relate a response variable to a set of predictor variables; we restrict our attention, however, to models that are linear and involve only one predictor variable.

In some situations it is not clear which variable is the response and which is the predictor. For example, a criminologist studies the relationship between poverty level and crime rate. Can we say crime rate depends on poverty level or vice versa? With *correlation analysis* it is possible to study relationships between two variables without declaring one as the response and the other as the predictor. In this chapter, however, we study simple linear regression models that relate a response variable, *y*, to a predictor variable, *x*. We view the model as the mechanism that produces data that relate the variables. Using inference tools, we statistically evaluate the data with the goal of finding the linear model that best fits the data.

Contents

- **STATISTICAL INSIGHT**
9.1 The Linear Regression Model
9.2 Inference about the Regression Model

9.3 Checking Model Adequacy
9.4 Summary and Review Exercises

Scholastic Assessment Test Scores

Scholastic Assessment Test (SAT) scores are often thought of as a barometer that measures how well high school seniors will do in college. To study trends and further analyze the data, SAT scores were recentered* in 1995. From a high of 509 in 1972 on the recentered scale, the national mean SAT math score declined through the 1970s to 492 in 1980. Since 1981, however, it has steadily increased to 514 in 2000. The mean SAT verbal score, on the other hand, has not come close to its 530 average in 1972. It declined to 502 in 1980, rose to 509 in 1985, and has fluctuated between 499 and 509 ever since. The 2000 SAT verbal mean was 505.

Boys generally perform better than girls on the SAT, and some states have significantly higher averages than others. In 2000, the gap between boys and girls narrowed to 3 points on the verbal section but remained some 35 points on the math section. Some argue that boys outperform girls in math because they generally take more math and science courses in high school. The differences among the states are difficult to explain. In fact, it is suggested that state averages not be compared because the data are not based on random samples of test takers from the states. The students are self-selected, and the participation rate varies from 4% in some states to 80% in others. Some also argue that the differences are linked directly to the states' per-student funding. They claim that in states where funding is high, the scores are high, and where funding is low, the scores are low.

Worksheet: **Sat.mtw** contains verbal, math and total SAT averages by state for years 1994 (not recentered), 1995, and 1999. Also included in the worksheet are the percents of students taking the exam and the state expenditures per student. The table here lists the information for 1999. (To investigate some of the above issues, see Section 5.6 for an analysis of these data.)

*For 1972–1986, a formula was applied to the original mean and standard deviation to convert the mean to the recentered scale. For 1987–1995, individual student scores were converted to the recentered scale and then the mean was recomputed. For 1996–2000 students received scores on the recentered scale.

WORKSHEET: Sat.mtw 1999 SAT Scores

State	Verbal	Math	Total	Percent	Expend
Alabama	561	555	1116	9	4903
Alaska	516	514	1030	50	9097
Arizona	524	525	1049	34	4940
Arkansas	563	556	1119	6	4840
California	497	514	1011	49	5414

WORKSHEET: Sat.mtw 1999 SAT Scores

State	Verbal	Math	Total	Percent	Expend
Colorado	536	540	1076	32	5728
Connecticut	510	509	1019	80	8901
Delaware	503	497	1000	67	7804
Dist of Columbia	494	478	972	77	9019
Florida	499	498	997	53	5986
Georgia	487	482	969	63	5708
Hawaii	482	513	995	52	6144
Idaho	542	540	1082	16	4732
Illinois	569	585	1154	12	6557
Indiana	496	498	994	60	6605
Iowa	594	598	1192	5	6047
Kansas	578	576	1154	9	6158
Kentucky	547	547	1094	12	5929
Louisiana	561	558	1119	8	5201
Maine	507	503	1010	68	6774
Maryland	507	507	1014	65	7543
Massachusetts	511	511	1022	78	7818
Michigan	557	565	1122	11	7568
Minnesota	586	598	1184	9	6371
Mississippi	563	548	1111	4	4312
Missouri	572	572	1144	8	5823
Montana	545	546	1091	21	6112
Nebraska	568	571	1139	8	6472
Nevada	512	517	1029	34	5541
New Hampshire	520	518	1038	72	6236
New Jersey	498	510	1008	80	10211
New Mexico	549	542	1091	12	4674
New York	495	502	997	76	9658
North Carolina	493	493	986	61	5315
North Dakota	594	605	1199	5	5198
Ohio	534	568	1102	25	6517
Oklahoma	567	560	1127	8	5150
Oregon	525	525	1050	53	6792
Pennsylvania	498	495	993	70	7686
Rhode Island	504	499	1003	70	8307
South Carolina	479	475	954	61	5371
South Dakota	585	588	1173	4	4924
Tennessee	559	553	1112	13	5011
Texas	494	499	993	50	5736
Utah	570	568	1138	5	4045
Vermont	514	506	1020	70	7171
Virginia	508	499	1007	65	5677
Washington	525	526	1051	52	6182
West Virginia	527	512	1039	18	6519
Wisconsin	584	595	1179	7	7398
Wyoming	546	551	1097	10	6448

Source: The 2000 World Almanac and Book of Facts, Funk and Wagnalls Corporation, New Jersey.

9.1 The Linear Regression Model

Learning Objectives for this section:

- ❏ Be able to describe the linear regression model.
- ❏ Know the assumptions for the linear regression model.
- ❏ Know how to find the least squares regression equation.
- ❏ Know how to find residuals and obtain the standard error about the regression line.
- ❏ Understand the main tasks in a regression analysis.

In Chapter 2, the scatterplot helped us visually examine the extent to which two numerical variables are related. You learned that although the points of the scatterplot may not lie in a perfect straight line, often a straight line will adequately describe the relationship between the variables. The method of least squares was used to find the equation of the *regression* line that best fits the data.

The Population Regression Line

We now extend the study of regression analysis by examining *statistical models*. In our study of univariate statistics, we viewed the parent population as the population that produces the univariate data in the sample. We now ask the question, Is there a *population regression line* that produces the sample data in a scatterplot? If so, we will refer to the model that relates the response variable y to the predictor variable x as the *linear regression model*.

Linear Regression Model

Given the response variable y and the predictor variable x, the linear regression model that relates the two is

$$y = \beta_0 + \beta_1 x + \epsilon$$

where β_0 and β_1 are the regression coefficients and ϵ is the error term.

The Error Term

Error is used here in the sense that ϵ encompasses all of the variation in the response variable y that is *not* modeled by the linear expression $\beta_0 + \beta_1 x$. Without ϵ, the model would be

$$y = \beta_0 + \beta_1 x$$

and the data, (x_i, y_i), exhibited in a scatterplot would fall in a perfect straight line. The error is then the quantity that causes the points to fall away from the line. The smaller the error, the closer the data are modeled with the straight line. To make inferences about the regression model, we must make these assumptions.

> ## Assumptions for the Linear Regression Model $y = \beta_0 + \beta_1 x + \epsilon$
>
> 1. The mean value of y is
>
> $$\mu_y = \beta_0 + \beta_1 x \qquad \text{(the population regression line)}$$
>
> 2. The error ϵ is normally distributed with a mean 0 and a standard deviation σ.
> 3. Realizations of ϵ (and hence y) are independent.

Implications of the Assumptions

The assumptions about the linear regression model have certain implications that are illustrated in Figure 9.1.

- The first assumption states that the means of the distribution curves drawn in Figure 9.1 are connected with the population regression line.
- Because of the second assumption, all the distribution curves in Figure 9.1 are drawn as normal curves and have the same standard deviation σ for each value of x.

Figure 9.1

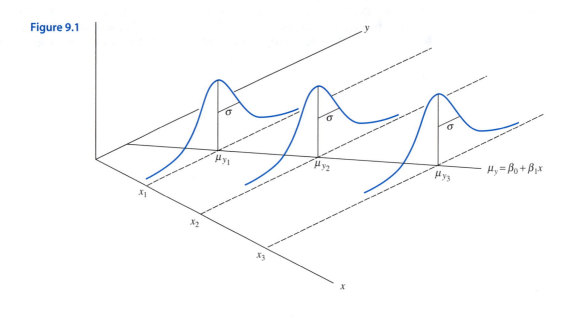

Example 9.1   Selected states were given the chance to have their eighth-grade math proficiency scores compared on a state-by-state basis. Here are the proficiency scores for a group of states that chose to participate.

WORKSHEET: Mathpro.mtw

Conn	23.1	Del	16.1	D.C.	2.8	Ga	14.7	Hawaii	12.5
Ind	17.5	Md	17.1	N.H.	22.5	N.J.	22.8	N.Y.	16.2
N.C.	9.2	Ore	22.6	Pa	19.1	R.I.	15.3	Va	18.9

Source: National Assessment of Educational Progress.

(Thirty-five states actually participated in the study. These are the ones that participated and also where a majority of the students took the SAT exam.) Four years later, when the eighth-graders were seniors in high school, these SAT math scores resulted for the same states:

WORKSHEET: Mathpro.mtw

Conn	472	Del	464	D.C.	443	Ga	446	Hawaii	480
Ind	466	Md	479	N.H.	486	N.J.	475	N.Y.	472
N.C.	455	Ore	491	Pa	462	R.I.	462	Va	469

Source: The College Board.

Suppose that the linear regression model

$$y = \beta_0 + \beta_1 x + \epsilon$$

with $\beta_0 = 440$, $\beta_1 = 1.7$, and $\sigma = 10$ describes the relationship between the SAT math scores and the proficiency scores. Explain how the eight-grade proficiency scores can be used to predict SAT math scores 4 years later by giving the equation for the population regression line and graphing it. For a given value of x, describe the possibilities for y.

Solution   The population regression line is

$$\mu_y = 440 + 1.7x$$

where x represents the proficiency score and μ_y is the mean SAT math score. Figure 9.2 is a graph of the equation.

Figure 9.2

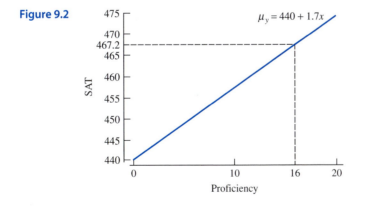

For any x, y is a normally distributed random variable with mean

$$\mu_y = \beta_0 + \beta_1 x = 440 + 1.7x$$

and standard deviation $\sigma_y = 10$. We see that the mean depends on the value of x; in fact, the means are connected by the population regression line.

When a particular state has a proficiency score of, say 16, the mean SAT math score is

$$\mu_y = 440 + 1.7(16) = 467.2$$

(See Figure 9.2.) Furthermore, using the empirical rule when $x = 16$, we are reasonably sure (95% confident) that y will be between 447.2 and 487.2, which is found from

467.2 ± 2(10). Notice that Delaware and New York had proficiency scores close to 16 (16.1 and 16.2, respectively). Their SAT math scores were 464 and 472, both of which are inside the given interval. ■

If the linear model is the correct model in the example, and if indeed $\beta_0 = 440$, $\beta_1 = 1.7$, and $\sigma = 10$, then we know what values to expect for y for a given value of x. In practice, however, we do not know for sure that the linear model is the correct model relating x and y, and certainly, we do not know the numerical values of β_0, β_1, and σ. We must use sample data to verify that the linear model is correct and to estimate β_0, β_1, and σ.

Estimating the Values in the Population Regression Line

Having collected sample data, (x_1,y_1), (x_2,y_2), $\cdots (x_n,y_n)$, we graph them in a scatterplot to see whether the linear regression model seems plausible. If a straight line appears to serve our purposes, we use the sample data to estimate the regression parameters.

Estimating β_0 and β_1

The least squares method, presented in Chapter 2, gives estimates of β_0 and β_1.

Example 9.2   A study at the Stanford University School of Medicine suggests that coffee drinking may be linked to heart disease. In particular, the report says that coffee drinking is strongly related to elevated levels of apolipoprotein B, a cholesterol-associated protein linked to heart disease. From the following data on 15 adult men over the age of 35 who drink from 1 to 5 cups of coffee per day, create a scatterplot and calculate the least squares estimates of β_0 and β_1 for the model

$$y = \beta_0 + \beta_1 x + \epsilon$$

WORKSHEET: Apolipop.mtw

Subject	# of cups of coffee per day	level of apolipoprotein B
1	1	23
2	1	19
3	1	13
4	2	21
5	2	18
6	2	25
7	3	26
8	3	32
9	3	28
10	4	35
11	4	27
12	4	33
13	5	33
14	5	37
15	5	38

Solution   The scatterplot in Figure 9.3 indicates that the linear model is a reasonable model to relate the level of apolipoprotein B to coffee drinking.

Figure 9.3

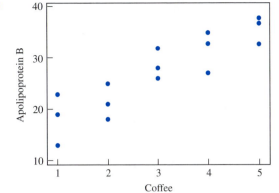

Figure 9.4 is a partial regression printout associated with the data in worksheet: **Apolipop.mtw**. The least squares regression coefficients appear in two places. They are given in the regression equation, and then they are summarized in the second column of the table. The first column of the table gives the predictors, which are the constant associated with b_0, and the dependent variable coffee associated with b_1. In the column Coef are the numerical values of b_0 and b_1.

Figure 9.4   Regression Analysis

```
The regression equation is
apolipB = 13.5 + 4.57 coffee

Predictor          Coef        SE Coef            T              p
Constant         13.500         2.089           6.46         0.000
coffee           4.5667         0.6300          7.25         0.000

s = 3.450         R-sq = 80.2%          R-sq(adj) = 78.6%
```

The slope of the regression line, 4.5667, gives the average increase in apolipoprotein B for each additional cup of coffee per day. This does not mean that if you normally drink 2 cups of coffee per day and then one day you drink an additional cup, your apolipoprotein B level will increase 4.5667 units. For those who normally drink 2 cups per day, the average apolipoprotein B level is estimated to be

$$\hat{y} = 13.5 + 4.5667(2) = 22.6334$$

and for those who normally drink 3 cups per day, the average is estimated to be

$$\hat{y} = 13.5 + 4.5667(3) = 27.2001$$

The difference between those who normally drink 3 cups and those who drink 2 cups per day is the slope = 4.5667. If the linear trend continues, then for each additional cup of coffee, we expect that the average apolipoprotein B level will increase by 4.5667 units. ■

Estimating σ

A residual is the difference between the actual y_i and the predicted y_i found by the equation

$$\hat{y}_i = b_0 + b_1 x_i$$

That is, the residual associated with the data point (x_i, y_i) is

$$e_i = y_i - \hat{y}_i$$

The magnitude of a residual is directly related to the magnitude of σ. In fact, the residuals are realizations of the random variable ϵ, which is normally distributed with mean 0 and standard deviation σ. A large value of σ translates into large residuals. Thus, the standard deviation of the residuals will provide an estimate of σ.

Recall that the sum of squares due to error (also called the residual sum of squares) is given by

$$\text{SSE} = \Sigma e_i^2 = \Sigma(y_i - \hat{y}_i)^2$$

When divided by its degrees of freedom, $n - 2$, it gives an unbiased estimate (does not tend to underestimate or overestimate) of σ^2. Taking a square root gives an estimate of σ.

The **standard error about the regression line,** given by

$$s = \sqrt{\frac{\text{SSE}}{n - 2}}$$

is an estimate of σ.

WHY $n - 2$ DEGREES OF FREEDOM?   Sums of squares such as SSE have degrees of freedom associated with them. For a random sample $x_1, x_2, \ldots x_n$, the ordinary sum of squares, $\Sigma(x_i - \bar{x})^2$, has $n - 1$ degrees of freedom because one of the deviations from the mean is determined by the other $n - 1$ deviations from the mean. We say that one degree of freedom is lost when the sample mean, $\bar{x}$, is used to estimate the population mean μ. Substituting for $\hat{y}_i$ in the formula for SSE we have

$$\text{SSE} = \Sigma(y_i - b_0 - b_1 x_i)^2$$

We see that the two quantities, β_0 and β_1, have been estimated by b_0 and b_1, resulting in a loss of two degrees of freedom. Thus, the degrees of freedom for SSE are $n - 2$. Dividing SSE by $n - 2$ gives an unbiased estimate of σ^2. Its square root, s, then provides an estimate of σ.

The formula for SSE involves calculating all of the predicted y's and then subtracting them from the actual y's to obtain the residuals, which are in turn squared and summed. If you are using a calculator this is very tedious and time consuming. There is an equivalent formula for SSE that is easier to use; but, we will rely on computer output for all numerical calculations. We see in Figure 9.4 that $s = 3.450$.

C O M P U T E R   T I P

Regression Analysis

To perform a regression analysis using Minitab, first enter the bivariate data, (x_i, y_i), in the Minitab worksheet. For example, store the x values in C1 and the y values in C2. From the **Stat** menu, click on **Regression** followed by **Regression**. From the Regression window, **Select** the **Response** variable (y) and the **Predictor** variable (x). Also select **Storage** if you wish to store any quantities such as Residuals, Fits, Coefficients, and so on, that you need. (Later we will discuss these options; at this point, you need not select any of them.) Finally, click on **OK**.

From our discussion of the linear regression model, we extract six main tasks associated with regression analysis.

Main Tasks in a Regression Analysis

1. Construct a scatterplot of the bivariate data.
2. Propose a statistical model that relates the response and predictor variables.
3. Estimate the parameters in the model.
4. Make inferences about the parameters in the model.
5. Check the assumptions of the model.
6. Use the model for making predictions and for generally describing the relationship between the two variables.

In the next section, we will focus on task 4, making inferences about the parameters in the linear regression model.

EXERCISES 9.1 The Basics

9.1 Given $y = 17 - 4x + \epsilon$, find the residual for each data point:
a. (4,3) b. (−1,20) c. (3,5)

9.2 Given $y = 92.88 + 12.93x + \epsilon$, find the residual for each data point:
a. (2,100) b. (−1,72) c. (10,195)

9.3 For a given geographic region, a proposed model relating the price of a new home, y, to the square feet of heated floor space, x, is

$$y = \beta_0 + \beta_1 x + \epsilon$$

with $\beta_0 = 65{,}280$, $\beta_1 = 24.3$, and $\sigma = 2300$.
a. What is the estimated mean price for homes that are 1500 square feet?
b. Using the empirical rule, when $x = 1500$, give an interval that contains 95 percent of the prices. Out of 100 randomly selected homes in this region, how many do you think will have prices outside this interval?
c. How much do you expect the price to increase if the area is increased from 1500 to 1600 square feet?
d. A new home with 1500 square feet was purchased for $97,500. Is this price consistent with the interval found in part **b**?

9.4 The linear model relating state SAT scores, y, to eighth-grade proficiency scores, x, proposed in Example 9.1 was for states that use the SAT as their primary college admissions test. For those states where the SAT is not primary, consider the model $y = \beta_0 + \beta_1 x + \epsilon$, with $\beta_0 = 482$, $\beta_1 = 2.48$, and $\sigma = 25$.
a. What is the estimated mean SAT math score when the proficiency score is 15?
b. Using the empirical rule, when $x = 15$, give an interval that contains 95 percent of all SAT math scores.
c. How much do you expect the SAT score to increase if the proficiency score increases from 15 to 16 points?
d. Michigan and Ohio had proficiency scores very close to 15. Their SAT math scores were 537 and 510, respectively. Are these scores consistent with the interval found in part **b**?

9.5 As soon as a new personal computer hits the market, its price begins to drop. For example, when computers with Intel's Pentium processor were first introduced, the average price exceeded $5000.

Six months later, the average price had dropped to around $3000. For a new model to be released, a proposed model relating its cost (y) to the time on the market (x, in months) for the first 6 months is $y = \beta_0 + \beta_1 x + \epsilon$, with $\beta_0 = 3500$, $\beta_1 = -275$, and $\sigma = 45$.

a. What is the estimated mean cost 3 months after the computer is introduced? What is the initial cost?

b. After a computer is on the market 3 months, use the empirical rule to give an interval that contains the cost 95% of the time. Out of 100 randomly selected computer stores that sell this particular computer, how many do you think have prices outside this interval?

c. How much do you expect the cost to decrease each month?

d. An individual purchased a computer after it had been on the market 3 months for $2750. Is this cost consistent with the interval found in part **b**?

e. Is it reasonable to use this model to predict the cost after the computer has been on the market for 1 year? Explain your answer.

9.6 A Swedish research study claims to have shown a relationship between the risk of heart disease and the waist-to-hip ratio in men and women. The study suggests that as the waist-to-hip ratio increases, so does the risk of heart disease. The 13-year study of about 1500 Swedish men with a high waist-to-hip ratio showed their risk of coronary heart disease, stroke, and death was four times that for men of more normal proportions. A similar study of 1500 women showed that those with a high waist-to-hip ratio had a risk of heart attack eight times greater than slimmer women. From the provided information, do you think that this describes an association problem or a comparison problem? In other words, are we trying to discover a relationship between two variables, or are we comparing the risks of heart disease for two groups of subjects? If it is an association problem, describe the variables that are of interest in the study. If it is a comparison problem, describe the two groups of subjects.

Interpreting Computer Output

9.7 To study how bone density changes as children grow, researchers at Citrus Hill Plus Calcium compiled data for five prominent research journals. They found these percents of peak bone density for different-aged children:

WORKSHEET: Citrus.mtw

Age	2	4	6	8	10	12	14	16	18
Percent of peak bone density	43	49	51	56	63	71	82	91	95

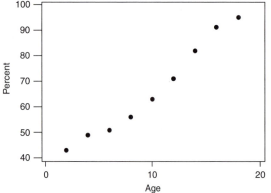

Regression Analysis: percent versus age

```
The regression equation is
percent = 32.5 + 3.42 age

Predictor          Coef       SE Coef           T            P
Constant         32.528         2.428       13.40        0.000
age              3.4250        0.2158       15.87        0.000

S = 3.342          R-Sq = 97.3%          R-Sq(adj) = 96.9%
```

a. What are the least squares estimates of β_0 and β_1 for the model $y = \beta_0 + \beta_1 x + \epsilon$?
b. Graph the regression equation on the scatterplot. Does it fit the data?
c. What is the standard error about the regression line?
d. Is it reasonable to extrapolate the percent of peak bone density to age 20?

9.8 The ability to predict rainfall in an agricultural state such as Kansas would be invaluable for farmers. In an effort to predict the average monthly (May through September) rainfall (in inches) for an area in west central Kansas, researchers measured the rainfall in four surrounding counties. The data are for 35 consecutive years ending in 1970.

WORKSHEET: Rainks.mtw

rain	x1	x2	x3	x4
2.97	1.85	2.36	2.17	2.77
1.88	1.72	1.93	1.25	1.50
2.19	1.94	3.35	2.30	2.64
1.68	0.78	1.23	0.97	1.66
2.40	2.51	2.04	3.19	2.17
3.40	3.55	4.46	3.82	4.06
2.05	2.08	1.67	1.94	2.22
2.11	1.40	1.83	1.94	2.55
2.66	2.36	1.88	2.78	3.18
1.49	2.25	2.34	1.82	2.12
2.45	2.77	3.21	2.57	3.18
2.45	1.70	2.33	2.18	1.95
2.41	2.69	3.22	2.62	3.28
4.10	3.71	3.81	3.30	4.13
3.82	2.98	3.50	3.40	4.28
4.78	4.66	4.45	4.46	4.79
1.25	0.95	1.16	0.90	1.25
1.88	1.45	1.76	1.33	1.98
1.49	1.36	1.95	2.50	1.50
2.02	2.27	2.28	3.21	2.24
1.36	1.05	1.35	0.86	1.39
3.38	2.73	2.92	2.40	4.30
3.16	3.94	2.70	4.30	2.99
2.23	1.94	2.14	2.27	3.52
1.76	1.88	1.71	2.25	2.65
3.33	2.84	2.84	2.62	5.44
2.30	2.16	3.07	2.91	3.46
2.59	3.96	2.46	3.02	4.14
2.28	1.90	2.11	1.84	2.75
3.41	3.94	3.89	3.70	3.86
2.56	1.49	2.71	1.44	2.49
2.60	2.92	2.63	3.32	2.78
2.19	2.21	2.40	2.27	3.76
2.48	3.57	2.62	3.18	2.44
2.23	2.63	2.13	2.12	2.55

Source: R. Picard, K. Berk (1990), Data Splitting, *The American Statistician, 44,* (2), 140–147.

a. Here is a correlation matrix of all pairwise correlations among the five variables. The intersection of a row and a column gives the correlation between the two identified variables. For example, the correlation between rain and $x3$ is 0.748, and the correlation between $x1$ and $x3$ is 0.883.

Correlations (Pearson)

	rain	x1	x2	x3
x1	0.793			
x2	0.818	0.769		
x3	0.748	0.883	0.746	
x4	0.818	0.720	0.734	0.638

Examine the correlations and determine which variable is the best predictor variable for rainfall.

b. Examine the regression computer printout. What is the response variable? What is the predictor variable? Is this the predictor variable that you chose in part **a**? What is the regression equation? What is the standard error about the regression line?

Regression Analysis: rain versus x2

```
The regression equation is
rain = 0.532 + 0.777 x2

Predictor          Coef      SE Coef          T          P
Constant         0.5322       0.2522       2.11      0.043
x2              0.77696      0.09494       8.18      0.000

S = 0.4608         R-Sq = 67.0%       R-Sq(adj) = 66.0%
```

c. This is a scatterplot that accompanies the regression output in part **b**. Graph the regression line on the scatterplot. Does it appear that the line fits the data?

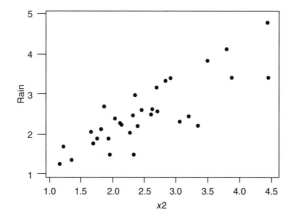

9.9 The data given here are from analyses of the magnesium concentration in stream samples that were collected along a river. Sampling locations were identified on an aerial photograph, and later the distances between samples were measured.

WORKSHEET: Magnesiu.mtw

distance	0	1820	2542	2889	3460	4586	6020	6841	7232	10903	
magnesium	6.44	8.61	5.24	5.73	3.81	4.05	2.95	2.57	3.37	3.84	

distance	11098	11922	12530	14065	14937	16244	17632	19002	20860	22471
magnesium	2.86	1.22	1.09	2.36	2.24	2.05	2.23	0.42	0.87	1.26

Source: Davis, J. (1986), *Statistics and Data Analysis in Geology*, 2d. Ed., John Wiley and Sons, New York, p. 146.

a. From the fitted line plot does it appear that the straight line fits the data? Are there any unusual observations in the scatterplot?

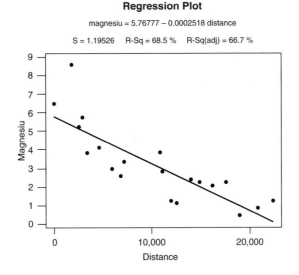

Regression Plot

magnesiu = 5.76777 − 0.0002518 distance

S = 1.19526   R-Sq = 68.5 %   R-Sq(adj) = 66.7 %

b. A regression computer printout is given here. What is the regression equation? Does it fit the data reasonably well? What is the standard error about the regression line?

Regression Analysis: magnesiu versus distance

```
The regression equation is
magnesiu = 5.77 − 0.000252 distance

Predictor            Coef        SE Coef          T          P
Constant           5.7678         0.4952      11.65      0.000
distance       −0.00025184     0.00004027      −6.25      0.000

S = 1.195            R-Sq = 68.5%            R-Sq(adj) = 66.7%
```

Computer Exercises

9.10   In the National Football League draft held each spring, professional teams pick new players from the college ranks. A great deal of effort goes into rating the players to determine who the best prospects are at each position. Probably the most widely used rating tool is the time in which the player can run 40 yards. Here are ratings (y) provided by National Scouting and *USA Today* research (*USA Today*, April 20, 1994) and the times (x, in seconds) for the 40-yard dash for the top offensive linemen in the 1994 draft.

WORKSHEET: NFLdraft.mtw

Rating	6.5	6.1	6.0	5.7	5.5	5.2	5.0	7.6	7.2	7.0	6.5
40-yd.	4.94	5.27	5.27	5.14	5.09	5.23	5.30	5.15	5.20	5.20	5.18

Rating	6.2	6.1	6.0	5.9	5.5	5.3	7.1	7.0	6.4	6.4	6.3
40-yd.	5.23	5.35	5.29	5.20	5.26	5.18	5.06	5.36	5.06	5.26	5.36

Rating	6.2	6.0	5.9	5.7	5.5	5.4	5.2
40-yd.	5.29	5.03	5.25	5.29	5.56	5.29	5.26

Do you think that the time for the 40-yard dash is important in determining the player's rating? To answer this question, graph the data in a scatterplot and calculate the correlation coefficient. Obtain the least squares estimates of β_0 and β_1 for the model $y = \beta_0 + \beta_1 x + \epsilon$ and calculate SSE and the standard error about the regression line. (In the next section, we will statistically evaluate the model.)

9.11 The Children's Defense Fund, a private advocacy group, claims that more than 25% of all children who live in cities with populations over 100,000 were impoverished in 1989 (*USA Today*, August 12, 1992). Listed here are the 20 cities with populations over 100,000 that were identified by the Children's Defense Fund as having the highest rates of poverty:

WORKSHEET: Poverty.mtw

City	Poverty percent	Crime rate
Detroit	46.6	13
Laredo	46.4	4
New Orleans	46.3	9
Flint, MI	44.6	12
Miami	44.1	17
Hartford, CT	43.8	12
Gary, IN	43.0	8
Cleveland	43.0	7
Atlanta	42.9	18
Dayton, OH	40.9	6
St. Louis	39.7	18
Buffalo	38.8	9
Rochester, NY	38.4	5
Milwaukee	37.8	5
Newark, NJ	37.6	19
Cincinnati	37.4	7
Fresno, CA	36.9	8
Shreveport	36.7	6
Waco, TX	36.6	8
Macon, GA	36.1	4

Source: Children's Defense Fund and the Bureau of Justice Statistics.

In addition to the percent of children who live in poverty, the crime rate (per 1000) is given for each city. Does it appear that there is a linear relationship between the poverty rate and the crime rate? What tools do you have to examine the relationship between these two variables? Perform some preliminary analysis of the data to determine whether it is worthwhile to pursue the linear relationship.

9.2 Inference about the Regression Model

Learning Objectives for this section:

❏ Know the assumptions needed to make inferences about the linear regression model.

❏ Be able to test hypotheses about the regression coefficient β_1.

❏ Be able to find a confidence interval for β_1.

❏ Be able to find a confidence interval estimate of μ_y for any given value of x.

The parameter β_0 in the linear regression model

$$y = \beta_0 + \beta_1 x + \epsilon$$

is the *intercept* or *constant term* that gives the expected value of y when x is zero. In many applications, however, β_0 is not of interest because zero is out of the range of reasonable values for x. In those cases, we simply view β_0, along with its estimate b_0, as an initializing quantity that gives the mean value of y when x is not present.

On the other hand, β_1 is extremely important. It is the slope of the regression line (often referred to as the *regression coefficient*) and gives the increase in y when x is increased by one unit. This was revealed in Example 9.2, where it was pointed out that men who drink 3 cups of coffee a day have an estimated apolipoprotein B level that is 4.5667 units higher than those who normally drink 2 cups a day. Remember, however, that 4.5667 is the value of b_1, an estimate of β_1. We have yet to show that it is a good estimator; in fact, we have not shown, in this example, that the linear model is our final choice that relates y to x.

Hypothesis Test for β_1

To statistically evaluate β_1, and the linear model, we begin by testing the null hypothesis

$$H_0: \beta_1 = 0$$

If indeed, $\beta_1 = 0$, then we can say that y is *not* linearly related to x. On the other hand, rejection of H_0 indicates a linear relationship between x and y.

To proceed with the inference about β_1, we use b_1 as the test statistic. As in all previous tests of hypotheses, we need to know the sampling distribution of the test statistic. Using the linear regression assumptions given in Section 9.1, it can be shown that the standardized test statistic

$$t = \frac{b_1 - \beta_1}{\text{SE}(b_1)}$$

has a t-distribution with $n - 2$ degrees of freedom.

The test statistic, as presented, allows us to test any reasonable value for β_1. But because we are testing $\beta_1 = 0$, the test statistic simplifies to

$$t = \frac{b_1}{\text{SE}(b_1)}$$

It can be shown that the standard deviation of b_1 is

$$\frac{\sigma}{\sqrt{\Sigma(x_i - \bar{x})^2}}$$

Using s to estimate σ, we have

$$\text{SE}(b_1) = \frac{s}{\sqrt{\Sigma(x_i - \bar{x})^2}}$$

Because s has $n - 2$ degrees of freedom, the t-distribution has $n - 2$ degrees of freedom. The test of hypothesis about β_1 is summarized here:

Test of Hypothesis about β_1

Assumptions: The assumptions are listed in Section 9.1.

Left-Tailed Test	Right-Tailed Test	Two-Tailed Test
$H_0: \beta_1 \geq 0$	$H_0: \beta_1 \leq 0$	$H_0: \beta_1 = 0$
$H_a: \beta_1 < 0$	$H_a: \beta_1 > 0$	$H_a: \beta_1 \neq 0$

Standardized test statistic

$$t = \frac{b_1}{SE(b_1)}$$

where $SE(b_1) = \dfrac{s}{\sqrt{(x_i - \bar{x})^2}}$.

From the t-table with $n - 2$ degrees of freedom, find the p-value most closely associated with t_{obs}. If t_{obs} falls between two table values, give the two associated probabilities as bounds for the p-value.

The p-value for the two-tailed test is calculated as above and then doubled to account for both tails.

If a level of significance α is specified, reject H_0 if the p-value $< \alpha$.

We illustrate the test in the next example.

Example 9.3   A study to relate the number of years of experience of New York City street patrolpersons with the average number of tickets they write per week found these data:

WORKSHEET: Patrol.mtw

Patrolperson	1	2	3	4	5	6	7	8	9	10
years experience	3	8	2	15	5	20	1	10	7	12
tickets per week	42	30	54	12	32	8	75	28	20	15

Find the least squares estimates of the regression coefficients in the model $y = \beta_0 + \beta_1 x + \epsilon$, and test the hypothesis that $\beta_1 = 0$.

Solution   Figure 9.5 is a partial computer printout from a regression routine.

Figure 9.5
```
The regression equation is
tickets = 55.9 - 2.93 years

Predictor          Coef      SE Coef              T            p
Constant         55.938        6.215           9.00        0.000
years           -2.9322        0.6150          -4.77        0.001

s = 11.21          R-sq = 74.0%         R-sq(adj) = 70.7%
```

We see the regression equation followed by a detailed analysis for the two regression co-efficients. We find

$$b_0 = 55.938 \quad \text{and} \quad b_1 = -2.9322$$

Under the column Stdev we find the standard errors of b_0 and b_1. In particular,

$$SE(b_1) = 0.6150$$

To test the hypothesis $\quad H_0: \beta_1 = 0 \quad \text{versus} \quad H_a: \beta_1 \neq 0$

we need the observed value of the test statistic

$$t = \frac{b_1}{SE(b_1)}$$

We can calculate it as follows:

$$t_{obs} = \frac{-2.9322}{0.6150} = -4.77$$

However, we see it computed in the T column of the computer printout.

The t-ratio has 8 degrees of freedom, and the computer calculates the p-value as .001. Using Table B.3 in Appendix B, we find p-value $< 2(.001) = .002$, which is consistent with the computer printout. We reject the null hypothesis and conclude that there is a significant linear relationship between the number of years of experience and the number of tickets written per week. The regression equation $\hat{y} = 55.94 - 2.93x$ can now be used to predict the number of tickets based on the number of years of experience of a patrolper-son. For example, someone with 5 years of experience is expected to give

$$\hat{y} = 55.94 - 2.93(5) = 41.29 \text{ tickets}$$

We could leave this problem at this point and conclude that we have found the model that best describes the relationship between the two variables under study. However, a *residual analysis* of the data, covered in the next section, reveals that a curvilinear model better fits the data. We will use residual analysis to check model assumptions; however, fitting data to more advanced models is beyond the scope of this text. ∎

Confidence Interval for β_1

In addition to the test of hypothesis, a confidence interval for the regression coefficient can be found. Following the construction of the t-interval for the population mean, μ, we have that the form of the confidence interval for β_1 is

$$b_1 \pm (t^* \text{ critical value})SE(b_1)$$

As in the test of hypothesis, the t^* critical value is found in the t-table with $n - 2$ degrees of freedom.

A $(1 - \alpha)$ 100% **confidence interval for β_1** is given by the limits

$$b_1 \pm t^*SE(b_1)$$

where t^* is the upper critical value associated with an upper $\alpha/2$ probability found in the t-table with $n - 2$ degrees of freedom.

Example 9.4   Returning to Example 9.3, find a 98% confidence interval for β_1.

Solution   From the printout in Figure 9.5 we find that $b_1 = -2.9322$, and the standard error of b_1 is found under the Stdev column to be 0.6150. For a 98% confidence interval and 8 ($n = 10$) degrees of freedom, we find from the t-table that $t^* = 2.896$. Therefore, the confidence interval for β_1 is

$$-2.9322 \pm (2.896)(0.615) \quad \text{or} \quad -2.9322 \pm 1.781$$

The interval is $(-4.713, -1.151)$, and we are 98% confident that β_1 is somewhere inside this interval. Notice that the interval contains only negative values and does not include zero; this indicates that a test of H_0: $\beta_1 = 0$ would be rejected at a level of significance of 2% (the complement of the 98% confidence). This is consistent with the test results obtained in Example 9.3.   ∎

Confidence Interval Estimate of μ_y

The previous confidence interval was for the regression coefficient β_1. A more interesting problem, however, is finding a confidence interval for the mean value of y, $\mu_y = \beta_0 + \beta_1 x$, for any given value of x.

For the linear regression model,

$$y = \beta_0 + \beta_1 x + \epsilon$$

The expected y for a given x—say x'—is

$$\mu_y = \beta_0 + \beta_1 x'$$

This expected y is estimated by

$$\hat{y} = b_0 + b_1 x'$$

With the basic assumptions of independent, normally distributed errors with common variance, the confidence interval for μ_y becomes

$$\hat{y} \pm t^* \text{SE}(\hat{y})$$

The formula for SE $(\hat{y})$ is rather complicated and is given in the box.

For a given x', a $(1 - \alpha)$ 100% **confidence interval for the expected response μ_y** is given by the limits

$$b_0 + b_1 x' \pm t^* \, \text{SE} \, (\hat{y})$$

where

$$\text{SE} \, (\hat{y}) = s \sqrt{\frac{1}{n} + \frac{(x' - \bar{x})^2}{\text{SS}_x}}$$

is the standard error of the estimate of μ_y and t^* is the upper $\alpha/2$ critical value found in the t-table with $n - 2$ degrees of freedom.

Example 9.5   A manufacturer of thermal pane windows wished to analyze the heat loss (in BTUs) through the windows at different outside temperatures. In a controlled environment, the heat loss through three windows was measured at four different outside temperatures (in degrees centigrade). From these data, obtain the estimated regression line.

WORKSHEET: Thermal.mtw

outside temp (C)	−10	−10	−10	0	0	0	10	10	10	20	20	20
heat loss (BTU)	58	62	54	39	41	36	22	26	20	10	13	15

Graph the data in a scatterplot and find a 95% confidence interval for the expected heat loss when the outside temperature is 10°C.

Solution   Figure 9.6 is a scatterplot of the data with the prediction line graphed.

Figure 9.6

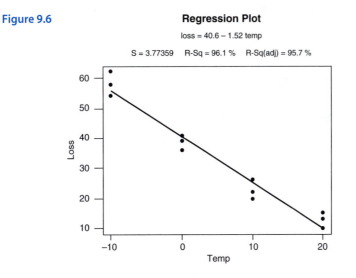

Regression Plot

loss = 40.6 − 1.52 temp

S = 3.77359   R-Sq = 96.1 %   R-Sq(adj) = 95.7 %

The scatterplot (and the fact that R–Sq = 96.1%) shows that a straight line accurately describes the relationship between the two variables. We see that the regression equation is

$$\text{loss} = 40.6 - 1.52 \text{ temp}$$

When the outside temperature is 10°C, we would expect the heat loss to be

$$\hat{y} = 40.6 - 1.52(10) = 25.4 \text{ BTUs}$$

For a 95% confidence interval estimate of the mean heat loss, we add and subtract the margin of error given by

$$t^* \text{SE}(\hat{y}) = t^* \, s \sqrt{\frac{1}{n} + \frac{(x' - \bar{x})^2}{\text{SS}_x}}$$

From the t-table with $n - 2 = 10$ degrees of freedom, we have $t^* = 2.228$. Also, $\bar{x} = 5$, $\text{SS}_x = 1500$, $s = 3.774$.

Substituting in all known values, we have that the 95% confidence interval for the expected heat loss when the outside temperature is 10°C is

$$25.4 \pm 2.228(3.774)\sqrt{\frac{1}{12} + \frac{(10-5)^2}{1500}} = 25.4 \pm 2.659$$

which gives the interval (22.74, 28.06). When the outside temperature is 10°C, we are confident that the mean heat loss is somewhere between 22.74 and 28.06 BTUs. ∎

We complete this section with a detailed analysis of a regression computer printout.

Example 9.6   In Example 9.1 the problem was to relate state SAT math scores to eighth-grade proficiency scores on tests taken 4 years earlier. The data have been stored in the Minitab worksheet: **Mathpro.mtw**. Give an interpretation of the regression output shown in Figure 9.7.

Figure 9.7
Regression Analysis:
Sat-M1 versus
Profic1

```
The regression equation is
Sat-M1 = 439 + 1.73 Profic1

15 cases used 4 cases contain missing values

Predictor          Coef      SE Coef          T          P
Constant        439.270        8.371      52.47      0.000
Profic1          1.7291        0.4774       3.62      0.003

S = 9.919           R-Sq = 50.2%        R-Sq(adj) = 46.4%

Analysis of Variance

Source             DF           SS          MS          F          P
Regression          1       1290.6      1290.6      13.12      0.003
Residual Error     13       1279.1        98.4
Total              14       2569.7

Unusual Observations
Obs    Profic1    Sat-M1        Fit    SE Fit    Residual    St Resid
  3        2.8    443.00     444.11      7.11       -1.11       -0.16 X
  5       12.5    480.00     460.88      3.25       19.12        2.04R

R denotes an observation with a large standardized residual
X denotes an observation whose X value gives it large influence.
```

Solution   From the output in Figure 9.7 we first see the least squares regression line

```
Sat-M1 = 439 + 1.73 Profic1
```

The intercept is $b_0 = 439$, and the slope is $b_1 = 1.73$. In the data set there were 19 states, but only 15 contained usable information. Next, a table is printed that gives more accurate values for b_0 and b_1, with their standard deviations (standard errors), t statistics, and the p-value. The t-ratio is obtained by dividing the coefficient by its standard error. As explained previously, this is the t-ratio for testing the significance of the regression coefficient. For example, to test

$$\text{H}_0: \beta_1 = 0 \quad \text{versus} \quad \text{H}_a: \beta_1 \neq 0$$

the test statistic is $t_{\text{obs}} = 3.62$, which is significant with a p-value $= .003$. This means that there is a linear relationship between the two variables.

If we need to test

$$\text{H}_0: \beta_0 = 0 \quad \text{versus} \quad \text{H}_a: \beta_0 \neq 0$$

the computer output has the necessary information. The test statistic $b_0 = 439.270$, and t-ratio $= 52.47$, is highly significant with a p-value $< .001$. This means that the regression line does not go through the origin.

Next, we see $s = 9.919$. This is the standard error of the regression line that is used as an estimate of σ in the previous t-tests for β_0 and β_1. The coefficient of determination, R–Sq $= 50.2\%$, was introduced in Chapter 2. It measures the percent of variability in the dependent variable that is explained by the regression equation. In this case, we can say that 50.2% of the variability in SAT scores is explained by proficiency scores. The *adjusted* R–Sq $= 46.4\%$ is an adjustment to the value of R^2 based upon the number of independent variables in the equation. Here there is only one independent variable in the equation; hence, there is very little difference between R–Sq and R–Sq(adj). In multiple regression studies (more than one independent variable) the value of R–Sq(adj) becomes relevant.

Analysis of Variance Table

Next, in Figure 9.7 an analysis of variance (ANOVA) table is displayed. An ANOVA table is a summary table of the variability in the data. It is very useful in tabulating the results for hypothesis testing in more complicated statistical analysis problems. We will investigate its properties in the next chapter.

Unusual Observations

To complete the analysis of the computer output in Figure 9.7, we describe the section on Unusual Observations. Two unusual observations are identified in the printout. One is identified as having a large standardized residual, and the other is identified because its x value gives it large influence. An observation with a large standardized residual naturally lies far from the fitted regression line and thus is classified as a regression outlier. As always, you should investigate outliers to determine their cause. If it can be determined that the outlier does not belong to the data set, it should be removed and the regression equation recomputed. Notice in the regression plot in Figure 9.8 that the identified outlier lies well above the regression line with a standardized residual of 2.04. There appears to be another outlier that lies well below the regression line with a standardized residual of -1.96. It is not identified by the computer as being an unusual observation because its standardized residual is less than 2.0 in absolute value. It probably should also be classified as an unusual observation because it cancels the effect of the other outlier. If either observation is removed from the data set, the regression equation will be altered substantially. If both are removed, the regression equation will not be affected very much. An important lesson is that you should visually examine the scatterplot for unusual behavior in addition to examining the computer printout.

Figure 9.8

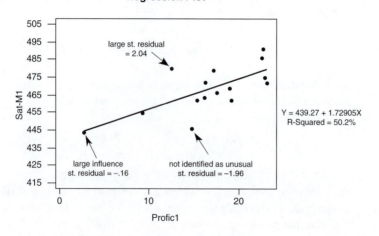

Regression Plot

Influential Observations

The other unusual observation identified by the computer is categorized as an *influential observation*. In regression analysis, an influential observation is an observation whose removal might have a substantial effect on the regression equation. Notice in this case that the observation is not a regression outlier in the sense that it has a large standardized residual ($= -0.16$) but rather is an outlier because its x value is far removed from the remaining x values. Least squares regression is not resistant to outliers and thus may be influenced by observations with large x values. Because of the influence of a large x value, it will pull the regression line toward it and thus will have a small residual. Having a small residual, it will not be detected with the usual residual analysis. It is for this reason that we use a computer with a regression diagnostic procedure in conjunction with a residual analysis. ■

You will learn how to conduct a residual analysis in Section 9.3 when we check the model specifications.

EXERCISES 9.2 ## The Basics/Interpreting Computer Output

9.12 Examine the table to determine whether the model $y = \beta_0 + \beta_1 x + \epsilon$ fits the data reasonably well.

x	6	9	11	15	11
y	5	10	8	14	12

a. Recall that $\Sigma(x_i - \bar{x})^2 = (n - 1)s_x^2$. From these descriptive statistics for x and y, find $\Sigma(x_i - \bar{x})^2$.

Descriptive Statistics

Variable	N	Mean	Median	TrMean	StDev	SEMean
x	5	10.40	11.00	10.40	3.29	1.47
y	5	9.80	10.00	9.80	3.49	1.56

Variable	Min	Max	Q1	Q3
x	6.00	15.00	7.50	13.00
y	5.00	14.00	6.50	13.00

b. From the regression printout, find b_1, $SE(b_1)$, and the standard error about the regression line.

Regression Analysis

```
The regression equation is
y = 0.07 + 0.935 x
```

Predictor	Coef	SE Coef	T	p
Constant	0.074	3.151	0.02	0.983
x	0.9352	0.2916	3.21	0.049

```
s = 1.916      R-sq = 77.4%      R-sq(adj) = 69.9%
```

c. From your answer to part **a** and the standard error about the regression line found in part **b**, verify that $SE(b_1) = s/\sqrt{\Sigma(x_i - \bar{x})^2}$.
d. Find a 95% confidence interval for β_1. Is 0 inside the interval?
e. Examine the computer printout and conduct a t-test to test the hypothesis that $\beta_1 = 0$. What is the p-value? State your conclusion.

9.13 Use the following data and the computer printout to test the significance of the slope of the regression line: $\mu_y = \beta_0 + \beta_1 x$.

x	11	6	11	15	14	15	8	16
y	8	5	10	12	12	9	5	11

Regression Analysis

```
The regression equation is
y = 0.78 + 0.685 x
```

Predictor	Coef	SE Coef	T	p
Constant	0.783	1.903	0.41	0.695
x	0.6848	0.1526	4.49	0.004

```
s = 1.464      R-sq = 77.0%      R-sq(adj) = 73.2%
```

a. What is the standard error about the regression line?
b. What is the regression equation?
c. Verify that the t-ratio to test the hypothesis that $\beta_1 = 0$ is $b_1/SE(b_1)$.
d. Complete the t-test to test the hypothesis that $\beta_1 = 0$. Be sure to give the p-value and state your conclusion.
e. Based on the computer printout, is there evidence to reject the hypothesis H_0: $\beta_0 = 0$? Does this mean that the regression line should go through the origin (0,0)?

9.14 These data match a child's age with the number of gymnastic activities he or she was able to successfully complete:

WORKSHEET: Gym.mtw

Age	2	3	4	4	5	6	7	7
Activities	5	5	6	3	10	9	11	13

Regression Analysis

```
The regression equation is
number = - 0.03 + 1.64 age
```

Predictor	Coef	SE Coef	T	p
Constant	−0.032	2.014	−0.02	0.988
age	1.6383	0.3988	4.11	0.006

```
s = 1.933      R-sq = 73.8%      R-sq(adj) = 69.4%
```

a. Graph the data in a scatterplot. Does it appear that a straight line fits the data?
b. Based on the regression printout, does it appear that the straight line fits the data?

c. Should the regression line go through the origin? Explain. Is there evidence in the computer printout to support this conjecture?

d. To find 95% confidence intervals for the expected y, these values of x' have been stored in column 3 of the worksheet: 2, 4, 6, 8, and 10. These confidence intervals were obtained using an Options routine. What is given in the Fit column? What value of x' produces the narrowest interval?

```
    Fit    Stdev.Fit              95.0% C.I.
  3.245       1.292      (    0.082,    6.408)
  6.521       0.746      (    4.695,    8.347)
  9.798       0.846      (    7.727,   11.869)
 13.074       1.465      (    9.488,   16.661)
 16.351       2.202      (   10.960,   21.742)
```

9.15 A retailer of satellite dishes would like to know the impact of advertising on her sales. For 6 months she records the number of ads run in the newspaper and the number of sales:

WORKSHEET: Adsales.mtw

	Mar	Apr	May	Jun	Jly	Aug
Number of ads	0	3	5	8	10	9
Number of sales	5	7	12	15	17	15

Regression Analysis

```
The regression equation is
sales = 4.67 + 1.23 ads

Predictor        Coef      SE Coef            T            p
Constant       4.6748       0.8018         5.83        0.004
ads            1.2272       0.1176        10.44        0.000

s = 1.017        R-sq = 96.5%        R-sq(adj) = 95.6%
```

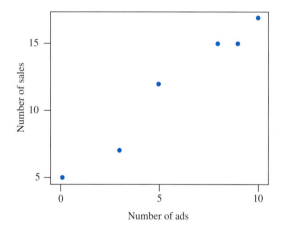

a. Examine the scatterplot. Does there appear to be a linear trend in the data?

b. Obtain the least square estimates of β_0 and β_1 for the linear model $y = \beta_0 + \beta_1 x + \epsilon$.

c. Test the hypothesis that $\beta_1 = 0$. Do the results of this test indicate that the linear model is a reasonable model? What is the value of the coefficient of determination?

d. Predict the number of sales if the retailer runs six ads in June.

e. With the additional information, $SS_x = 74.83$ and $\bar{x} = 5.83$, construct a 98% confidence interval for the expected sales when six ads are placed.

f. The collected data are for March through August. What impact would this have on using this model to predict the number of sales in January?

9.16   In Exercise 9.10 in Section 9.1, a linear model was proposed to relate the ratings of top offensive linemen for the NFL draft to their times in the 40-yard dash. Here is the Minitab output from the Regression routine.

WORKSHEET: NFLdraft.mtw

Regression Analysis

```
The regression equation is
Rating = 13.7 - 1.47 forty
```

Predictor	Coef	SE Coef	T	p
Constant	13.744	5.291	2.60	0.015
forty	-1.466	1.012	-1.45	0.159

```
s = 0.6449      R-sq = 7.2%        R-sq(adj) = 3.8%
```

Analysis of Variance

SOURCE	DF	SS	MS	F	p
Regression	1	0.8724	0.8724	2.10	0.159
Error	27	11.2290	0.4159		
Total	28	12.1014			

Unusual Observations

Obs.	forty	Rating	Fit	Stdev.Fit	Residual	St.Resid
1	4.94	6.500	6.502	0.313	-0.002	-0.00 X
8	5.15	7.600	6.194	0.142	1.406	2.24R
27	5.56	5.500	5.593	0.359	-0.093	-0.17 X

```
R denotes an obs. with a large st. resid.
X denotes an obs. whose X value gives it large influence.
```

a. Find the regression equation. Does it agree with the answer you found in Exercise 9.10?
b. Locate the t-statistic for testing H_0: $\beta_1 = 0$. What is its p-value? Should you reject or accept the null hypothesis? What does this signify about the proposed linear model?
c. If the rating score is a valid measure of the potential of a future offensive lineman in the NFL, should scouts be concerned with his time in the 40-yard dash?
d. Identify any unusual observations and explain why they are classified as unusual.

9.17   The Minitab printout is for the Regression routine for the data in Exercise 9.11 in Section 9.1. The problem was to determine whether the poverty level of children in large cities can be predicted with a linear model based on the city's crime rate.

WORKSHEET: Poverty.mtw

Regression Analysis

```
The regression equation is
Poverty = 39.1 + 0.181 Crime
```

Predictor	Coef	SE Coef	T	p
Constant	39.116	1.858	21.05	0.000
Crime	0.1810	0.1710	1.06	0.304

```
s = 3.667      R-sq = 5.9%        R-sq(adj) = 0.6%
```

Analysis of Variance

Source	DF	SS	MS	F	p
Regression	1	15.06	15.06	1.12	0.304
Error	18	242.02	13.45		
Total	19	257.07			

a. What is the numerical value of R^2? What percent of the variability in poverty rate is explained by the linear model? Is it significant?

b. Locate the *t*-statistic for testing H_0: $\beta_1 = 0$. What is its *p*-value? Should we reject or accept the null hypothesis? What does this signify about the proposed linear model?

c. Should we use this regression equation to predict children's poverty levels from the crime rates? Explain.

9.18 In the early 1960s, tremendous amounts of wastewater from the Rocky Mountain Arsenal (a manufacturing plant for producing various military compounds) were forced, under high pressure, into a deep injection well near Denver, Colorado. Unfortunately, the well, drilled into basement rocks, penetrated the shear zone of a major fault along the Rocky Mountains. There is evidence that the high-pressure injection of waste fluids lubricated and mobilized the fault and triggered earthquakes. The data are the month-by-month volumes of injected water (in million gallons) and the number of earthquakes detected in Denver each month between March 1962 and October 1965.

WORKSHEET: Wastewat.mtw

gallons	4.2	7.2	8.4	8.0	5.2	6.0	5.0	5.6	4.0	3.6	6.0	7.6
number		*	2	12	35	23	29	24	8	6	20	25	22

7.8	6.4	3.6	4.0	3.4	2.4	3.9	0.0	0.0	0.0	0.0	0.0	0.0	0.0
21	42	21	8	6	10	11	12	4	2	5	2	9	9

0.0	0.0	0.0	0.0	0.6	1.8	2.4	2.0	2.0	1.7	1.6	3.6	4.0	6.4
2	4	4	5	2	14	2	7	1	30	9	19	11	38

8.9	5.4	6.4	3.8
62	48	87	5

Source: Davis, J. C. (1986), *Statistics and Data Analysis in Geology*, 2 ed., John Wiley and Sons, New York, p. 228, and Bardwell, G. E. (1970), Some Statistical Features of the Relationship between Rocky Mountain Arsenal Waste Disposal and Frequency of Earthquakes, *Geological Society of America, Engineering Geology Case Histories*, *8*, 33–37.

a. Bardwell presented the two accompanying times series graphs as evidence of a relationship between the volume of injected water (first graph) and the number of earthquakes (second graph). Examine the peaks and valleys of the two graphs. How many prominent peaks do you see? Do they appear in approximately the same place on the two graphs? Near the middle of the data, no wastewater was injected into the well. Do you see this in the first graph? What do you see in the second graph during this same period? Do you think that Bardwell demonstrated a valid relationship between the two variables?

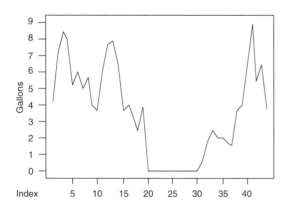

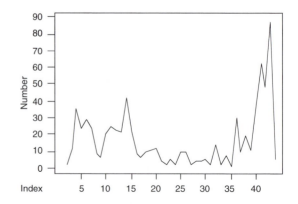

b. Does the scatterplot demonstrate a relationship between the two variables?

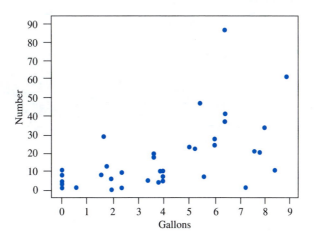

c. Examine the computer printout. What is the correlation between the two variables? What is the regression equation? Is the straight line a good fit for the data?
d. The computer has identified two unusual observations. Explain why they are called unusual. Identify them on this scatterplot.

Correlations (Pearson)

Correlation of gallons and number = 0.602

Regression Analysis

The regression equation is
number = 3.59 + 3.79 gallons

43 cases used 1 cases contain missing values

Predictor	Coef	Stdev	t-ratio	p
Constant	3.590	3.483	1.03	0.309
gallons	3.7905	0.7843	4.83	0.000

s = 14.33   R-sq = 36.3%   R-sq(adj) = 34.7%

Analysis of Variance

SOURCE	DF	SS	MS	F	p
Regression	1	4800.2	4800.2	23.36	0.000
Error	41	8424.9	205.5		
Total	42	13225.1			

Unusual Observations

Obs.	gallons	number	Fit	Stdev.Fit	Residual	St.Resid
2	7.20	2.00	30.88	3.66	−28.88	−2.08R
43	6.40	87.00	27.85	3.18	59.15	4.23R

R denotes an obs. with a large st. resid.

9.19 This regression printout is related to the problem, presented in Examples 2.8 and 2.16 in Chapter 2, of associating the value of a company's brand name with its revenue.

WORKSHEET: Name.mtw

Regression Analysis

```
The regression equation is
value = - 0.889 + 2.02 revenue
```

Predictor	Coef	Stdev	t-ratio	p
Constant	−0.8889	0.4174	−2.13	0.039
revenue	2.0244	0.1158	17.49	0.000

```
s = 2.096        R-sq = 88.4%        R-sq(adj) = 88.1%
```

Analysis of Variance

SOURCE	DF	SS	MS	F	p
Regression	1	1344.1	1344.1	305.82	0.000
Error	40	175.8	4.4		
Total	41	1519.9			

Unusual Observations

Obs.	revenue	value	Fit	Stdev.Fit	Residual	St.Resid
1	15.4	31.200	30.287	1.553	0.913	0.65 X
2	8.4	24.400	16.116	0.779	8.284	4.26R
11	6.0	3.700	11.257	0.539	−7.557	−3.73R

```
R denotes an obs. with a large st. resid.
X denotes an obs. whose X value gives it large influence.
```

a. Is there sufficient evidence to reject the hypothesis H_0: $\beta_1 = 0$? What is the p-value?
b. What evidence in the printout suggests that the straight line provides a strong relationship between revenue and value?
c. The computer has identified three unusual observations. Explain why they are called unusual. Identify them on this scatterplot.

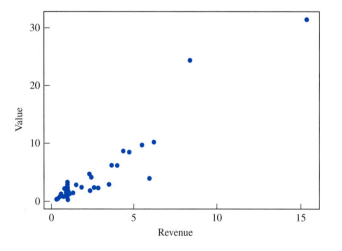

d. The following three confidence intervals are for the expected value of brand names when the revenues are $5, $10, and $20 billion.

Fit	Stdev.Fit	95.0% C.I.
9.233	0.452	(8.320, 10.146)
19.355	0.951	(17.434, 21.276)
39.599	2.077	(35.402, 43.796)

What is the predicted value of a company's brand name when its revenue is $5 billion? What is a 95% confidence interval estimate of the mean brand name value? What do you observe about the interval when the revenue is $20 billion?

Computer Exercises

9.20 In an attempt to cut down on traffic through the city, a free shuttle service from the suburban areas to the downtown area was provided. At the beginning very few people took advantage of the shuttle service, but by the end of 1 year, a decision had to be made whether to add more shuttles. Here are the numbers of people who rode on the shuttle and the numbers of automobiles in the downtown area on 15 different occasions throughout the year:

WORKSHEET: Shuttle.mtw

No. using shuttle	160	180	240	280	440	370	490	620
No. of autos downtown	2460	2730	2560	2600	2290	2370	2410	2040

No. using shuttle		850	840	970	1230	1140	1290	1350
No. of autos downtown		1820	1950	1870	1460	1330	1390	1250

a. Which variable is the dependent variable: the number of people who rode the shuttle or the number of automobiles?

b. Construct a scatterplot of the data. Does there appear to be a linear trend?

c. Obtain the least square estimates of β_0 and β_1 for the model $y = \beta_0 + \beta_1 x + \epsilon$.

d. Test the hypothesis that $\beta_1 = 0$. Do the results of this test indicate that a linear trend is significant?

e. For each additional person who rides the shuttle, how many automobiles are removed from the downtown area?

f. Predict the number of automobiles downtown if 1000 people ride the shuttle.

g. Construct a 99% confidence interval for the expected number of autos when 1000 people ride the shuttle.

h. Would this model be useful in predicting the number of automobiles downtown if 3000 people ride the shuttle? Explain.

9.21 A cigarette manufacturer recorded the nicotine contents (in milligrams) and sales figures (in $100,000) for eight major brands of cigarettes.

WORKSHEET: Nicotine.mtw

Nicotine	.86	1.38	1.67	.25	.59	1.17	1.00	.66
Sales	24	65	83	34	59	62	85	38

a. Which variable is the dependent variable: nicotine content or sales?

b. Construct a scatterplot of the data. Does there appear to be a linear relationship between the two variables?

c. Obtain the least square estimates of β_0 and β_1 for the model $y = \beta_0 + \beta_1 x + \epsilon$.

d. Test the hypothesis that $\beta_1 = 0$. Do the results of this test indicate that sales increase when the nicotine content is increased?

e. How much do sales increase for each additional unit of nicotine?

f. Predict the sales when the nicotine content is 1.0.

g. Construct a 99% confidence interval for the expected sales when the nicotine content is 1.0.

9.22 An owner of a chicken farm would like to test a new feed supplement that is supposed to increase egg production. He randomly selects 12 groups of 100 chickens each and feeds them various levels of the feed supplement. Listed are the amounts of feed supplement and the number of eggs per day.

Group	1	2	3	4	5	6	7	8	9	10	11	12
Feed supplement	10	10	10	15	15	15	20	20	20	25	25	25
Number of eggs	78	84	81	85	79	95	98	96	89	84	93	87

a. Construct a scatterplot of the data. Does there appear to be an upward trend that would predict the number of eggs?

b. From the scatterplot in part **a**, does it appear that a linear trend would continue upward if the feed supplement is increased to 30 units?

c. Obtain the least square estimates of β_0 and β_1 for the model $y = \beta_0 + \beta_1 x + \epsilon$.

d. Test the hypothesis that $\beta_1 = 0$. Do the results of this test indicate that a linear trend is significant?

e. How many additional eggs are expected to be produced for each additional unit of feed supplement?

f. Predict the number of eggs if the feed supplement is set at 20 units.

g. Would this model be useful in predicting the number of eggs if the feed supplement is increased to 40 units? Explain.

9.23 How significant is the earned run average (ERA) for pitchers in major league baseball in determining the number of wins for the team? To investigate, consider the numbers of wins (out of 162 games) and the earned run averages for the National League teams at the end of the 1990 season.

WORKSHEET: Wins.mtw

Team	Wins	ERA
Atlanta	65	4.58
Chicago	77	4.34
Cincinnati	91	3.39
Houston	75	3.61
Los Angeles	86	3.72
Montreal	85	3.37
New York	91	3.43
Philadelphia	77	4.07
Pittsburgh	95	3.40
St. Louis	70	3.87
San Diego	75	3.68
San Francisco	85	4.08

a. Construct a scatterplot of the data. Does it appear that a linear model would describe the relationship between the number of wins and the earned run average?

b. Obtain a regression computer printout. Find the regression equation. What is the change in the number of wins for a unit change in the earned run average?

c. Locate the t-statistic for testing H_0: $\beta_1 = 0$. What is its p-value? Should we reject or accept the null hypothesis? What does this signify about the proposed linear model?

d. Should we use this regression equation to predict the number of wins based on the pitchers' earned run average?

e. Did the computer identify any unusual observations?

f. Construct a 99% confidence interval for the expected number of wins when the earned run average is 3.8.

9.24   Example 9.6 gave the computer output relating state SAT math scores and eighth-grade proficiency scores. The analysis was based on data from states where a majority of students take the SAT exam. In the Minitab worksheet: **Mathpro.mtw**, the same data are given for states where a majority do not take the SAT exam. The data are listed under the variables Sat-M2 and Profic2. Conduct an analysis of these data similar to the analysis given in Example 9.6.

a. Compare your output with that given in Example 9.6. What is R^2 for your data? Is the t-statistic significant?

b. Of the two data sets, which has a linear model that better fits the data?

c. Create side-by-side boxplots of the SAT math scores for the two groups. Is one more variable than the other? Are the medians similar?

9.3   Checking Model Adequacy

Learning Objectives for this section:

☐ Be able to describe the linear regression model.

☐ Know the assumptions for the linear regression model.

☐ Know that the assumptions can be checked by a careful examination of the residuals.

☐ Know how to construct and use residual plots.

☐ Know how to check the assumptions of linearity, constant standard deviation, and normality.

☐ Be able to check for outliers and know how to deal with them.

In the preceding sections, you learned how to find estimates and make inferences about the linear regression model

$$y = \beta_0 + \beta_1 x + \epsilon$$

You should be cautioned, however, that the conclusions drawn from the inference procedures can be seriously misleading if the assumptions made about the model (see Section 9.1) are invalid. A regression study is not complete without an investigation of these assumptions.

Evaluating the Residuals

Recall that the residual associated with data point (x_i, y_i) is $e_i = y_i - \hat{y}_i$ where $\hat{y}_i$ is the predicted y_i. Because it is the difference between the actual observed y_i and the predicted y_i obtained from fitting a line to the data, the residual e_i contains information that is *not* described by the linear relationship. If the model is correct, e_i can be viewed as an observed value of the random error ϵ. As such, the residuals, $e_1, e_2, \ldots, e_n$, that are found when we collect data, should distribute themselves randomly about 0 and appear to have come from a normal population with standard deviation σ. When investigating residuals, however, it may be more convenient to investigate *standardized residuals*, which are found by e_i/s where s is the standard error of the regression line that estimates σ. Under normality, the standardized residuals should appear to have come from a *standard* normal distribution. Hence, most of the standardized residuals should be between -2 and $+2$. Because most computer programs for linear regression print standardized residuals, it is easy to see whether they lie between -2 and $+2$.

Computer Caution

When you request standardized residuals with the Minitab regression routine, you actually get *Studentized residuals*. Studentized residuals are very similar to standardized residuals except that the divisor is a modification of s that is adjusted for each individual residual. Residuals associated with observations that have unusually large x values are adjusted the most. For our purposes, however, we do not distinguish between standardized and Studentized residuals.

Residual Plots

A plot of residuals against the predicted values, $\hat{y}_i$, amounts to an examination of y after the linear dependence on the predictor variable is removed. Thus, if the model is correct, the scatter of the residuals should form a horizontal band about 0, as in Figure 9.9.

Figure 9.9
Residual Plot that shows no abnormalities

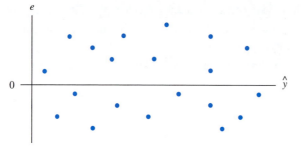

CHECKING LINEARITY A scatterplot is used to visually check the linearity assumption by observing whether the data exhibit straight-line behavior. Pearson's correlation gives a measure of the degree of linearity and its square, the coefficient of determination, R^2, measures the amount of variability in y explained by the predictor variable. The remaining variability is measured by the residuals. Thus, any systematic pattern revealed in a plot of the residuals indicates a departure from linearity. The residual plot in Figure 9.10 reveals a systematic pattern that has not been explained by the linear equation. Although beyond the scope of this book, this residual plot indicates that a nonlinear term should be considered in the model.

Figure 9.10
Residual plot that shows nonlinearity

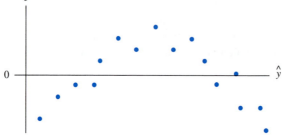

Example 9.7   In an effort to determine the effective duration of a tranquilizer for animals, the concentration of the substance in blood samples taken at various times after the injection are measured. From the data, determine the least squares equation that relates the concentration to the elapsed time after the injection, and conduct a residual analysis.

Elapsed time (hours)	1	2	3	6	12	18
Concentration (mg/ml)	1.8	1.4	1.2	0.9	0.5	0.1

Solution   From the sample data, we find the least squares equation:

$$\text{concentration} = 1.60521 - 0.0888393 \text{ time}$$

The coefficient of determination is R–Sq = 92.6%, which suggests a strong linear relationship. Figure 9.11 is a scatterplot with the equation graphed.

Figure 9.11

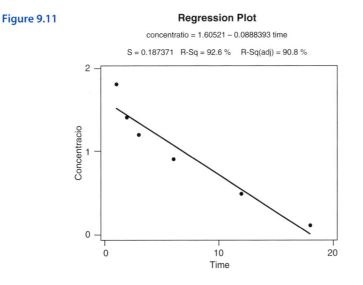

Regression Plot

concentratio = 1.60521 − 0.0888393 time

S = 0.187371   R-Sq = 92.6 %   R-Sq(adj) = 90.8 %

Using the equation, we find the predicted values of y and the residuals listed in Table 9.1. A plot of the residuals against the predicted values in Figure 9.12 shows a definite curvilinear pattern. Although a straight line fits the data adequately, this suggests that a higher-ordered x term in the model will improve the model.

Table 9.1

x_i	y_i	$\hat{y}_1$	e_i
1	1.8	1.5164	.2836
2	1.4	1.4275	−.0275
3	1.2	1.3387	−.1387
6	.9	1.0722	−.1722
12	.5	.5391	−.0391
18	.1	.0061	.0939

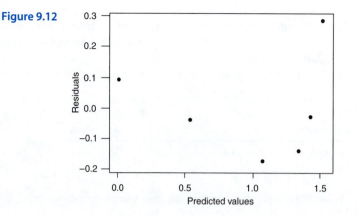

Figure 9.12

CHECKING CONSTANT STANDARD DEVIATION In Figure 9.13 the width of the band increases as the predicted y increases. This indicates that the standard deviation is not constant. Again, it is beyond the scope of this book, but often a transformation of data will correct this disorder.

Figure 9.13
Residual plot that shows nonconstant standard deviation

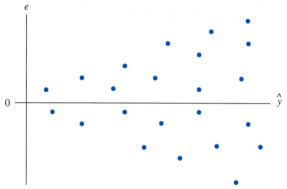

CHECKING NORMALITY The shape of the distribution of the residuals can be checked in the same way that distributions were checked in previous chapters. By using the histogram, stem-and-leaf plot, boxplot, or a simple dot diagram, we can observe whether the data have a normal curve appearance. Also, the normal probability plot is excellent for checking for departures from normality such as skewness and long tails.

CHECKING FOR OUTLIERS As might be expected, outliers tend to have an adverse effect on the least squares regression. Not only do they affect the precision of the inference procedure, but they also affect the actual equation of the straight line. Because the least squares procedure minimizes the sum of squares of the distances of all points to the line, an outlier tends to pull the line toward it in an unusual fashion. A large residual indicates an outlier. In fact, observations with standardized residuals greater than 2 in absolute value are classified as outliers. Also, a boxplot of the residuals is helpful in identifying outliers.

Example 9.8 The data are the research and development (R & D) expenditures and the corresponding sales for a large company. (All numbers are in million dollars.)

WORKSHEET: RandD.mtw

R & D	Sales
1.2	55
2.4	48
3.1	32
4.0	21
4.9	10
6.3	41
1.8	43
2.7	44
3.3	21
3.7	32
4.2	12
5.4	8

Determine whether a linear relationship between the two variables seems plausible by evaluating the regression assumptions.

Solution The least squares regression equation that relates R & D expenditures to sales is given in the Minitab output in Figure 9.14.

Figure 9.14
Regression output
from worksheet:
RandD.mtw

```
The regression equation is
sales = 55.0 - 6.82 rd

Predictor        Coef      SE Coef         T          P
Constant        55.01        10.13      5.43      0.000
rd              -6.816        2.626     -2.60      0.027

S = 12.95          R-Sq = 40.3%       R-Sq(adj) = 34.3%
```

The coefficient of determination R–Sq = 40.3% suggests that approximately 40% of the variability in sales is explained by the R & D expenditures.

Using the regression equation, we find the predicted y_i's and then the residuals in Table 9.2. Dividing the residuals by $s = 12.95$, we get the standardized residuals in the last column.

Table 9.2

x_i	y_i	$\hat{y}_i$	e_i	e_i/s
1.2	55	46.829	8.171	0.631
2.4	48	38.649	9.351	0.722
3.1	32	33.878	-1.878	-0.145
4.0	21	27.743	-6.743	-0.521
4.9	10	21.609	-11.609	-0.896
6.3	41	12.066	28.934	2.234
1.8	43	42.739	0.261	0.020
2.7	44	36.604	7.396	0.571
3.3	21	32.515	-11.515	-0.889
3.7	32	29.788	2.212	0.171
4.2	12	26.380	-14.380	-1.110
5.4	8	18.201	-10.201	-0.788

A normal probability plot of the residuals in Figure 9.15 indicates no serious departures from normality.

Figure 9.15

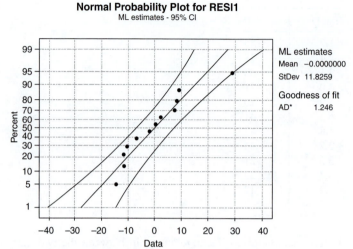

On closer examination of the residuals, however, we notice one large standardized residual (2.234) corresponding to (6.3,41). From the scatterplot with the equation graphed in Figure 9.16, we see that (6.3,41) is an outlier, which explains the large residual. Notice how the regression line is tilted towards the outlier.

Figure 9.16

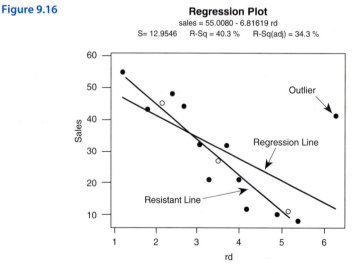

Because of the sensitivity of the least squares method to outliers, robust methods, such as the *resistant line*, have been developed to fit data to a line. In Figure 9.16, we see that the resistant line, which is fit to the medians (identified by o) of the three groups of four observations each, is a more reasonable fit to the data than the least squares line. (The equation for the resistant line is $y = 68.952 - 11.475x$.)   ∎

Often an outlier is the result of an erroneous measurement, in which case it should be corrected or eliminated from the study. If, on the other hand, it is a valid measurement, we should fit a resistant line or conduct a least squares analysis both with and without the outlier. This will reveal the true effect of the outlier.

Figure 9.17 presents the standardized residuals plotted against the predicted y_i's, for the regression equation found in Example 9.8. Notice that the residual corresponding to (6.3,41) shows up clearly as an outlier. Furthermore, the remaining residuals have a systematic pattern that needs to be rectified.

Figure 9.17

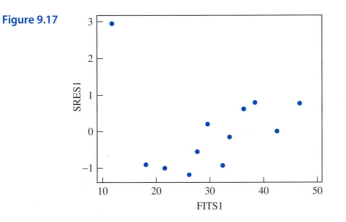

Removal of the outlier, (6.3,41), results in a significant increase in R^2 (from 40.3% to 88.4%) and the much more reasonable residual plot, shown in Figure 9.18. Notice that the residuals appear to be randomly distributed about 0, as they should be. Whether the outlier should be removed from the data set is a decision that the experimenter will have to make. Without the outlier we have a reasonable model relating sales to research and development expenditures. If we choose to retain the outlier, the model needs improvement.

Figure 9.18

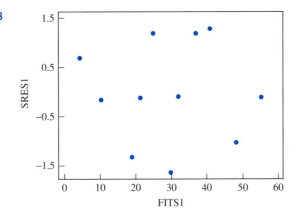

CHECKING INDEPENDENCE   Often, data for a regression study are collected sequentially over a period of time. When the observations are correlated with time, the residuals are dependent. A plot with time on the horizontal axis and the residuals on the vertical axis will often detect violations of the assumption of independence of the residuals. Figure 9.19

shows a definite relationship between the residuals and time. It appears that time should be incorporated in the model.

Figure 9.19
Residual plot against time

A well-known statistical test called the Durbin-Watson test is used to check for correlated data. A discussion of this test and the general topic of time series models can be found in *Time Series Analysis, Forecasting and Control* (San Francisco: Holden-Day, 1970) by G. E. P. Box and G. M. Jenkins.

C O M P U T E R   T I P

Residual Analysis

After entering the data in the worksheet, perform a regression analysis as before. Make sure that in the **Regression-Storage** window you have checked **Standardized residuals** and **Fits**. By checking these two boxes, you store the standardized residuals and predicted values (fits) in the next two columns of the worksheet. Click on **Regression** under the Stat menu again, but this time instead of clicking on Regression a second time, click on **Residual Plots**. In the Residual Plots window, from the list of variables, **Select** the **Fits** and **Residuals** variables, and then click **OK**. Prior to clicking OK, you may want to give a title to the residual plots in the **Title** box.

Example 9.9   The effects of coffee drinking on a cholesterol-associated protein called apolipoprotein B was examined in Example 9.2 in Section 9.1. Perform a residual analysis of these data located in worksheet: **Apolipop.mtw**.

Solution   In worksheet: Apolipop.mtw the standardized residuals and fits have been stored in columns SRES1 and FITS1, respectively. These are the columns that should be selected to produce the residual plot analysis. The residual model diagnostics appear in Figure 9.20. The normal probability plot and the histogram show no serious problems with the normality assumption. The I Chart of the residuals shows random fluctuation about 0, indicating no problems with the independence assumption. Finally, the plot of the residuals versus fits is typical when there are no abnormalities in the model. In summary, the regression equation

$$\text{apolipB} = 13.5 + 4.57 \text{ coffee}$$

which explains approximately 80.2% of the variability in apolipoprotein B level is a very reasonable model. To explain the remaining 20% of variability will require additional predictor variables in a multiple regression setting.   ■

Figure 9.20   **Residual Model Diagnostics**

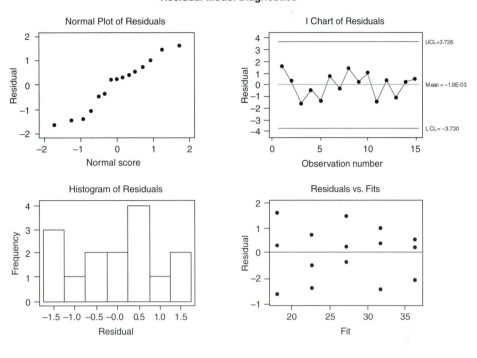

EXERCISES 9.3   The Basics

9.25   The stem-and-leaf plot shows the residuals associated with a simple linear regression equation. Do the residuals appear to violate the assumptions stated for the linear regression model?

```
−3  | 22
−2  | 05,  18
−1  | 22,  35,  51,  74,  81
−0  | 04,  19,  31,  56,  64,  71,  82,  94
 0  | 02,  15,  24,  48,  61,  75,  93
 1  | 12,  34,  47,  63,  74,  92
 2  | 34
 3  | 21,  47
```

9.26   Shown here is a residual plot against the predictor variable in a simple linear regression problem. What pattern do you see in the plot? Does it appear that the assumptions for the linear regression model have been violated?

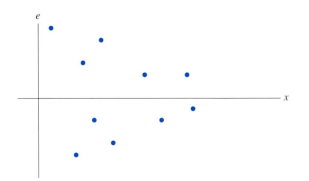

9.27   From the residual plot against time, what observations can you make about the model?

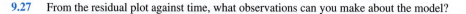

9.28   Here is a residual plot for these data:

x	1	1	1	2	2	2	3	3	3
y	15	5	3	10	8	3	6	5	4

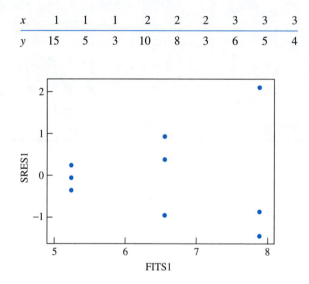

Does it appear that any assumptions for the linear regression model have been violated?

Interpreting Computer Output

9.29   Examine the regression printout for these data:

x	1.2	2.3	3.2	3.9	4.6	6.5	9.8	15.5
y	1.3	2.6	4.8	5.9	7.1	12.1	16.3	24.5

Regression Analysis

```
The regression equation is
y = - 0.423 + 1.66 x

Predictor         Coef        Stdev      t-ratio          p
Constant       -0.4230       0.5160       -0.82      0.444
x              1.65924       0.07023       23.63      0.000

s = 0.8765      R-sq = 98.9%        R-sq(adj)  = 98.8%
```

a. Based on R^2 and the *t*-ratio, does it appear that the straight-line regression equation fits these data?

b. Do you anticipate any significant violation of the assumptions for the linear regression model?

c. Examine the residual plot. Do you detect any violation of assumptions?

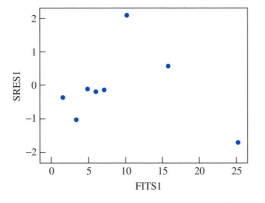

9.30 A school psychologist believes that there is a linear relationship between the verbal test scores for eighth-graders and the number of library books they check out.

WORKSHEET: Verbal.mtw

books	12	15	3	7	10	5	22	9	13	7	25	17	14	19	20
verbal scores	77	85	48	59	75	41	94	72	80	70	98	85	83	96	89

a. Based on the scatterplot, do you think the school psychologist is correct?

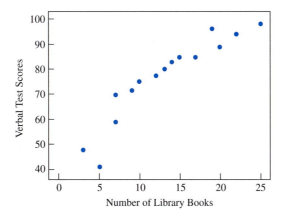

b. Examine the regression printout on the next page. Does it appear that a straight line fits the data? What is the value of R^2? Is the regression coefficient significantly different from 0? What is the *p*-value?

Regression Analysis

```
The regression equation is
verbal = 45.3 + 2.38 number

Predictor        Coef       Stdev      t-ratio          p
Constant       45.346       4.022        11.28      0.000
number         2.3828      0.2751         8.66      0.000

s = 6.695        R-sq = 85.2%        R-sq(adj) = 84.1%

Analysis of Variance

SOURCE            DF         SS         MS          F          p
Regression         1     3363.6     3363.6      75.03      0.000
Error             13      582.8       44.8
Total             14     3946.4

Unusual Observations
Obs.    number    verbal      Fit   Stdev.Fit   Residual    St.Resid
   6       5.0     41.00    57.26        2.84     -16.26      -2.68R
R denotes an obs. with a large st. resid.
```

c. Did the computer identify any unusual observation? If so, what observation is it and why is it classified unusual?

d. Examine the residual plot. Do any unusual observations show up in the residual plot? Is there a random pattern to the residual or should we consider revising the model? What suggestions can you give?

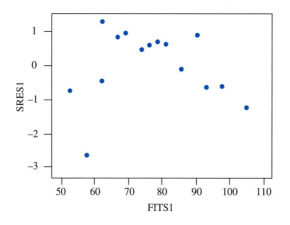

9.31 A residual analysis of Exercise 9.9 in Section 9.1 is shown here. The problem relates the magnesium concentration in stream samples that were collected along a river to the distances between samples.

a. Does it appear that the normality assumption has been violated?

b. Is there a random pattern to the Residuals vs. Fits plot?

c. What adjustments to the model should we consider?

Residual Model Diagnostics

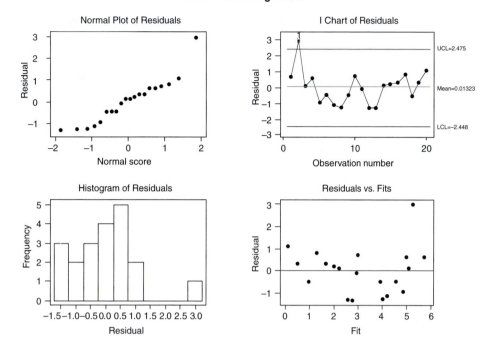

9.32 Of interest to geologists in eastern Louisiana is the moisture content of marine muds that accumulate in small inlets on the Gulf Coast. The following measurements of moisture contents of core samples were obtained by comparing the weight of the sample immediately after its removal from the core barrel with its weight after forced drying. The moisture content is expressed as grams of water per 100 grams of dried sediment. We wish to relate the moisture content to the depth of the core sample.

WORKSHEET: Moisture.mtw

depth	moisture
0	124
5	78
10	54
15	35
20	30
25	21
30	22
35	18
0	137
5	84
10	50
15	32
20	28
25	24
30	23
35	20

Source: Davis, J. C. (1986),
*Statistics and Data Analysis in
Geology,* 2d ed., John Wiley
and Sons, New York, pp. 177, 185.

a. Here is the correlation between depth and moisture content.

$$\text{Correlation of depth and moisture} = -0.870$$

Does this correlation indicate that a straight line will fit the data?

b. Examine this scatterplot. Does it appear that a straight line fits the data?

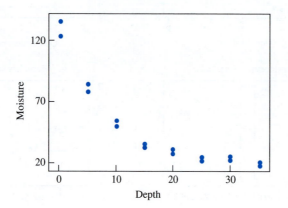

c. Examine the regression printout. What is the regression equation? Do the values of R^2 and the t- statistic indicate that the equation fits the data?

Regression Analysis

```
The regression equation is
moisture = 97.3 - 2.78 depth
```

Predictor	Coef	Stdev	t-ratio	p
Constant	97.333	8.778	11.09	0.000
depth	-2.7762	0.4197	-6.62	0.000

```
s = 19.23      R-sq = 75.8%      R-sq(adj) = 74.0%
```

Analysis of Variance

SOURCE	DF	SS	MS	F	p
Regression	1	16185	16185	43.76	0.000
Error	14	5178	370		
Total	15	21363			

Unusual Observations

Obs.	depth	moisture	Fit	Stdev.Fit	Residual	St.Resid
9	0.0	137.00	97.33	8.78	39.67	2.32R

```
R denotes an obs. with a large st. resid.
```

d. Examine this residual plot associated with the regression equation in part **c**. What does the residual plot tell about the simple linear regression equation? Should the equation be revised? If so, in what way?

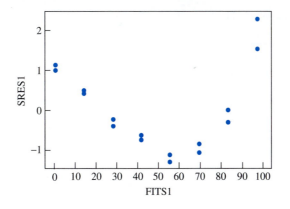

Computer Exercises

9.33   Conduct a residual analysis for the least squares solution given in Exercise 9.14 in Section 9.2. The data, which are repeated here, match a child's age with the number of gymnastic activities he or she is able to complete successfully.

WORKSHEET: Gym.mtw

Age	2	3	4	4	5	6	7	7
Activities	5	5	6	3	10	9	11	13

9.34   The tensile strength of kraft paper (in pounds per square inch) was measured for different percentages of hardwood in the batch of pulp that was used to produce the paper.

WORKSHEET: Hardwood.mtw

tensile	6.3	11.1	20.0	24.0	26.1	30.0	33.8	34.0	38.1	39.9
hardwood	1.0	1.5	2.0	3.0	4.0	4.5	5.0	5.5	6.0	6.5

tensile	42.0	46.1	53.1	52.0	52.5	48.0	42.8	27.8	21.9
hardwood	7.0	8.0	9.0	10.0	11.0	12.0	13.0	14.0	15.0

Source: Joglekar, et al. (1989), Lack-of-Fit Testing when Replicates Are Not Available, *The American Statistician, 43,* (3) p. 135–143.

a. Construct a scatterplot of tensile strength versus hardwood content.
b. Does it appear that a straight line will fit the data?
c. Fit a simple linear regression model to the data. What is the value of R^2? Is the *t*-statistic significant?
d. Complete a residual analysis of the model. Is it clear that the simple linear regression model is inappropriate? What reasons can you give for its inadequacy?

9.35   Joglekar and colleagues (1989) considered the following windmill data that record the direct current (in volts) produced by given wind velocities (in miles per hour).

WORKSHEET: Windmill.mtw

velocity(mph)	output(dc volts)
2.45	0.123
2.70	0.500
2.90	0.653

continued

WORKSHEET: Windmill.mtw
continued

velocity(mph)	output(dc volts)
3.05	0.558
3.40	1.057
3.60	1.137
3.95	1.144
4.10	1.194
4.60	1.562
5.00	1.582
5.45	1.501
5.80	1.737
6.00	1.822
6.20	1.866
6.35	1.930
7.00	1.800
7.40	2.088
7.85	2.179
8.15	2.166
8.80	2.112
9.10	2.303
9.55	2.294
9.70	2.386
10.00	2.236
10.20	2.310

Source: Joglekar, et al. (1989), Lack of Fit Testing when Replicates Are Not Available, *The American Statistician, 43,* (3), 135–143.

Regression Analysis

The regression equation is
output = 0.131 + 0.241 velocity

Predictor	Coef	Stdev	t-ratio	p
Constant	0.1309	0.1260	1.04	0.310
velocity	0.24115	0.01905	12.66	0.000

s = 0.2361   R-sq = 87.4%   R-sq(adj) = 86.9%

Analysis of Variance

SOURCE	DF	SS	MS	F	p
Regression	1	8.9296	8.9296	160.26	0.000
Error	23	1.2816	0.0557		
Total	24	10.2112			

Unusual Observations

Obs.	velocity	output	Fit	Stdev.Fit	Residual	St.Resid
1	2.5	0.1230	0.7217	0.0845	-0.5987	-2.72R

R denotes an obs. with a large st. resid.

The simple linear model seems adequate, with R^2 = 87.4% and t = 12.66, which is highly significant. Examine a residual plot, however, and determine whether there is still a distinct pattern in the data.

9.36   *Consumer Reports* (October 1994) rated these toaster ovens.

WORKSHEET: Toaster.mtw

toaster	score	cost
Black&D T660G	85	85
Toastmaster 336V	77	50
DeLonghi XU20L	75	130
Proctor-Silex 03030	75	48
Black&D SO2500G	75	92
Toastmaster 342	72	60
Munsey M88	70	56
Sears Kenmore 48216	70	70
Proctor-Silex 03010	70	41
Panasonic NT855U	68	70
DeLonghi XU14	65	69
Black&D TRO510	63	55
Black&D TRO400	61	50
Hamilton Beach 336	59	40
Toastmaster 319V	77	39
Black&D TRO200	60	40
Proctor-Silex 03008	59	35

a. Is there a significant linear relationship between the rating given by *Consumer Reports* and the cost of the toaster?

b. What is the value of R^2? Is its value acceptable or should the model be revised in an effort to increase R^2?

c. Conduct a residual analysis of the model. Does the pattern of residuals seem random? What recommendations can you make about the model?

9.37   In Example 9.8 we observed a single outlier that produced an unusually large residual. One way to deal with the outlier is to evaluate the regression line both with and without the outlier. Retrieve the data (worksheet: **RandD.mtw**) into the computer.

a. Perform a regression analysis with all the data. What is the regression equation?

b. What is the value of R^2? What percent of the variability in sales is explained by R & D expenditures?

c. What is the *p*-value associated with the *t*-statistic used to evaluate the regression coefficient? Is it significantly different from zero?

d. Are there any unusual observations? What are they, and why are they identified as unusual?

e. Remove the one outlier and repeat parts **a–d.**

f. With the outlier removed, is the *p*-value for the *t*-statistic more significant? Did R^2 improve? Are there any unusual observations?

9.4   Summary and Review Exercises

Key Concepts

- Regression is the study of the relationship between a *response variable y* and a *predictor variable x*. The regression equation is of the form

$$y = \beta_0 + \beta_1 x + \epsilon$$

- The *error term* ϵ is assumed to be normally distributed with a mean of 0 and a standard deviation of σ.

- From the data $(x_1,y_1), (x_2,y_2), \ldots, (x_n,y_n)$, the *least squares estimates* of β_0 and β_1 are found.

- From the least squares equation, the *predicted value of y* is found by the equation

$$\hat{y}_i = b_0 + b_1 x_i$$

where b_0 and b_1 are the least squares estimates.

■ The *residual* associated with the data point (x_i, y_i) is $e_i = y_i - \hat{y}_i$. All the residuals are squared and summed to get SSE, the *sum of squares due to error*. SSE is then used to find s, the *standard error about the regression line*, which estimates σ.

■ The *coefficient of determination* is the square of the correlation coefficient and gives the percent of the variability in the response variable that is explained by the predictor variable.

■ A test of hypothesis about the regression coefficient, β_1, can be performed. The test statistic for testing $H_0: \beta_1 = 0$ is given by

$$t = \frac{b_1}{\text{SE}(b_1)}$$

where $\text{SE}(b_1) = s/\sqrt{\Sigma(x_i - \bar{x})^2}$. It is distributed as a t-distribution with $n - 2$ degrees of freedom when the basic assumptions are satisfied.

■ Confidence intervals for β_1 can be constructed with the formula

$$b_1 \pm t^* \, \text{SE}(b_1)$$

■ A confidence interval for the expected response μ_y for a given x' is found with the formula

$$b_0 + b_1 x' + t^* \, s \sqrt{\frac{1}{n} + \frac{(x' - \bar{x})^2}{\text{SS}_x}}$$

■ An *analysis of the residuals* makes it possible to check out the assumptions that are made in a regression analysis. It also can point out possible deficiencies in the regression model and suggest alternatives.

Statistical Insight Revisited

The SAT data described in the Statistical Insight at the beginning of the chapter is stored in worksheet **Sat.mtw**. To better understand the data, as usual, we first construct a stem-and-leaf plot (or histogram). Here is a stem-and-leaf plot of the SAT total scores for 1999:

Stem-and-Leaf Display: total99

```
Stem-and-leaf of total99   N = 51
Leaf Unit = 10
    1        9 5
    3        9 67
   10        9 8999999
   18       10 00001111
   24       10 222333
  (3)       10 455
   24       10 7
   23       10 89999
   18       11 011111
   12       11 2233
    8       11 455
    5       11 77
    3       11 899
```

Notice that the distribution shape is bimodal. In Section 1.5, it was pointed out that a bimodal shape indicates that some extraneous variable is acting on the data. If possible, we should identify that variable. Returning to the data set, is it possible to associate the SAT scores with the percent of the student population in each state taking the test? Below is a scatterplot of the total SAT versus the percent taking the test. There is a definite downward trend with a noticeable break around 40 percent. The correlation between the two variables is a highly significant −0.886. This clearly demonstrates that SAT scores become lower as the percent of students taking the test increases.

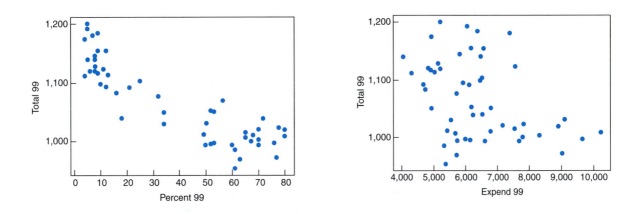

Correlations: total99, percent99

```
Pearson correlation of total99 and percent99 = -0.886
P-Value = 0.000
```

Correlations: total99, expend99

```
Pearson correlation of total99 and expend99 = -0.389
P-Value = 0.005
```

To study the relationship between SAT scores and the state per-student funding of education, we show its scatterplot on the right and the correlation coefficient. The correlation is significant (−0.389) and the scatterplot shows a downward trend. This is the opposite of what we expect to see. As funding increases we expect the SAT scores to also increase, but they decrease! If we look closer, however, we see when the funding is between \$4000 and \$7500, there is no trend; the SAT scores are randomly distributed between 950 and 1200. When funding is between \$7500 and \$10,000, the SAT scores fluctuate randomly about 1000. The scatterplot and correlation coefficient are misleading. In the following scatterplot of expenditures versus percent, we see that funding was less than \$7700 in those states where fewer than 40% took the test.

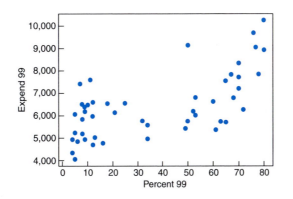

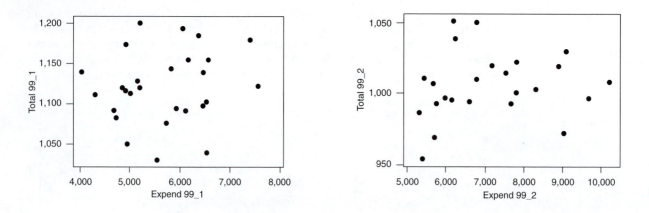

Continuing, we separate the SAT scores into two groups—those where fewer than 40% took the test and those where more than 40% took the test. This is accomplished by creating a code variable that is 1 if the percent is less than .4 and 2 if the percent is greater than .4. From these two scatterplots of total SAT scores versus expenditures for groups 1 and 2, we see no pattern in the data. That is, in those states where fewer than 40% took the test there is no relationship between SAT scores and expenditures per student. The same is true in states where more than 40% took the test. In summary, state funding does not explain the differences among state SAT scores, but clearly the scores are related to percent taking the test.

Questions for Review

Use the following questions to test your skills.

The Basics

9.38 Identify the predictor and response variables in each case:
a. A study of the relationship between robbery rates and population density.
b. A study of the relationship between attitude scores and academic achievement scores.
c. A study of the relationship between the growth rate of rainbow trout and the number of fish per cubic yard of water.
d. A study of the relationship between expenditures per student and teacher's salary.
e. A study that matches the time necessary for a subject to react with the number of alternatives he or she is given.

9.39 Graph these linear equations:
a. $y = 3.1 + 4.7x$ b. $y = -7.3 + 5.5x$ c. $y = -8.3 - 2.1x$

9.40 Identify the slope and y-intercept and graph each equation:
a. $2x + 3y = 6$ b. $-3.1y + 7.4x = 12$ c. $-5.6x - 4.1y = 10$

9.41 Construct a scatterplot for each data set and comment on the general relationship between x and y.

a.
x	100	110	120	130	140	150
y	7.8	6.1	5.4	5.2	4.7	3.5

b.
x	2.5	3.4	4.7	5.2	6.8	7.6
y	10.3	14.2	17.5	22.6	24.8	29.0

9.42 Using the summary information, find the least squares estimates of β_0 and β_1 in the linear model $y = \beta_0 + \beta_1 x + \epsilon$ for the data in Exercise 9.41.
a.

Predictor	Coef	SE Coef	T	P
Constant	14.700	1.342	10.95	0.000
x	-0.07400	0.01064	-6.96	0.002

S = 0.4450   R-Sq = 92.4%   R-Sq(adj) = 90.5%

b.

Predictor	Coef	SE Coef	T	P
Constant	1.950	1.736	1.12	0.324
x	3.5331	0.3253	10.86	0.000

S = 1.415   R-Sq = 96.7%   R-Sq(adj) = 95.9%

9.43   For the data in Exercise 9.41 and the solutions found in Exercise 9.42, find the predicted y for these values of x.
a. $x = 100$ and $x = 160$ b. $x = 1.5$ and $x = 5.0$

9.44   Using the solution to Exercise 9.42, find the standardized residuals associated with these data points:
a. (120,5.4) and (150,3.5). Is either identified as an outlier?
b. (2.5,10.3) and (5.2,22.6). Is either identified as an outlier?

9.45   Graph the equation $y = -7.8 + 6.3x$. Find the residuals for (5,18) and (5,20).

9.46   Given the least squares equation $\hat{y} = 3.5 - 6.8x$, find the predicted y when $x = 3.5$. What is the residual for $(3.5, -20)$?

9.47   A proposed model that relates the failure time and the operating temperature of a piece of electronic equipment is

$$y = \beta_0 + \beta_1 x + \epsilon$$

with $\beta_0 = 200$, $\beta_1 = -.35$, and $\sigma = 1.5$.
a. What is the expected mean failure time when the operating temperature is 45?
b. When the operating temperature is 45, use the empirical rule to give an interval that would contain 95% of all failure times.
c. If operating temperature changes by one unit, how much do you expect the failure time to change?

Interpreting Computer Output

9.48   From the scatterplot and regression output for these data, comment on the general relationship between x and y. Find the predicted y when $x = 3$ and when $x = 6$.

x	2.3	3.4	1.6	6.4	4.2	3.1	5.6	4.9
y	.65	.82	.47	1.23	.92	.74	1.08	1.01

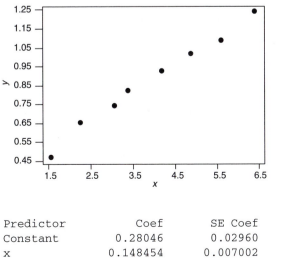

Predictor	Coef	SE Coef	T	P
Constant	0.28046	0.02960	9.47	0.000
x	0.148454	0.007002	21.20	0.000

S = 0.03049   R-Sq = 98.7%   R-Sq(adj) = 98.5%

9.49   From the scatterplot and regression output for these data, comment on the general relationship between x and y. Find the predicted y when $x = 30$ and when $x = 60$.

x	41	52	37	26	45	32	49	55	22	30
y	1.2	2.8	0.6	-1.7	2.9	-1.1	3.1	2.7	-2.4	-1.3

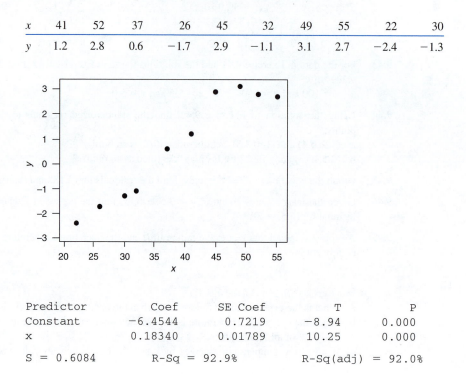

Predictor	Coef	SE Coef	T	P
Constant	-6.4544	0.7219	-8.94	0.000
x	0.18340	0.01789	10.25	0.000

S = 0.6084   R-Sq = 92.9%   R-Sq(adj) = 92.0%

9.50   A manufacturer of a new insulation medium recorded the heat loss (in BTUs) through the insulation as the outside temperature (in degrees centigrade) dropped:

WORKSHEET Insulate.mtw

Outside temp (C°)	-10	-10	0	0	10	10	20	20	30	30
Heat loss	96	91	84	82	68	75	49	51	28	24

Predictor	Coef	SE Coef	T	P
Constant	81.600	2.177	37.48	0.000
temp	-1.6800	0.1257	-13.37	0.000

S = 5.621   R-Sq = 95.7%   R-Sq(adj) = 95.2%

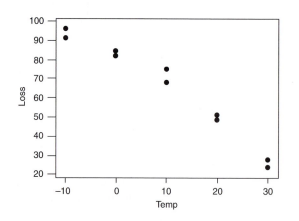

a. Examine the scatterplot for any unusual behavior.
b. Find the least squares estimates of β_0 and β_1 in the model $y = \beta_0 + \beta_1 x + \epsilon$.
c. For each degree increase of outside temperature, how much does the heat loss change?
d. Find the standard error about the regression line.
e. Find the predicted heat loss when the outside temperature is $15°$ C. Should we predict when the outside temperature is $45°$ C?
f. Find a 95% confidence interval for β_1. Is 0 inside the interval? If it is, what does that signify about the relationship between the two variables?

9.51   These data are IQ and GPA scores for a random sample of 12 students:

WORKSHEET: IqGpa.mtw

IQ(x)	115	132	125	120	119	132	105	114	106	139	127	118
GPA(y)	2.2	3.3	3.0	2.6	2.9	3.5	2.2	2.7	3.7	1.8	3.7	2.4

Predictor	Coef	SE Coef	T	P
Constant	2.530	2.305	1.10	0.298
IQ	0.00250	0.01899	0.13	0.898

S = 0.6572   R-Sq = 0.2%   R-Sq(adj) = 0.0%

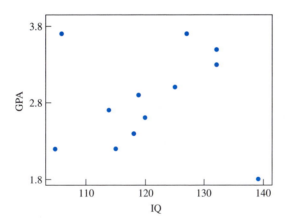

a. Examine the scatterplot for any unusual behavior. Are there any bivariate outliers? Will they have an adverse effect on the least squares regression line?
b. Find the least squares estimates of β_0 and β_1 in the model $y = \beta_0 + \beta_1 x + \epsilon$.
c. What is the value of R^2? What does it tell you about the proposed model?
d. What is the p-value associated with the t-ratio? What does this signify?
e. Should we use this model to predict GPA from one's IQ?

Computer Exercises

9.52   Remove the two bivariate outliers from the worksheet: **IqGpa.mtw** given in Exercise 9.51.
a. Reexamine the scatterplot for any unusual behavior. Does it appear that a straight line will fit the data?
b. Find the least squares estimates of β_0 and β_1 in the model $y = \beta_0 + \beta_1 x + \epsilon$.
c. What is the value of R^2? Has the removal of outliers improved the value of R^2?
d. What is the p-value associated with the t-ratio? What does this signify?
e. Should we use this model to predict GPA from one's IQ?

9.53 To investigate the relationship between the numbers of books read during the term by third-graders and their final spelling scores, an educator collected the following data on 17 randomly selected students:

WORKSHEET: Books.mtw

Number of books read	final spelling score
27	85
11	81
32	98
5	61
17	92
0	36
8	59
15	84
24	90
6	70
4	72
23	95
41	99
7	78
2	58
19	87
13	80

Conduct a simple linear regression analysis of the data. Examine a scatterplot. Construct a residual plot analysis based on the simple linear regression equation.

a. Does the scatterplot show a straight-line pattern?
b. Is the regression coefficient significantly different from 0? What is the p-value associated with the t-test? What is your conclusion?
c. Calculate R^2. How much of the variability in the response variable is explained by the predictor variable?
d. Does the residual plot show a random pattern or do you think that the model should be revised? If so, what revision do you recommend?

9.54 Use the solution to Exercise 9.53 to find the predicted spelling scores for students who have read 10, 15, 20, 25, and 30 books. For these numbers, obtain 95% confidence intervals for the expected responses and graph them on the scatterplot.

9.55 How much confidence do we have in the press to report the facts accurately? One study suggests that one's confidence level is related to his or her years of education. Listed are the number of years of education and the degree of confidence in the press (the higher the score, the more confidence) for 20 randomly selected persons:

WORKSHEET: Press.mtw

Education(yrs)x	12	12	14	8	10	12	11	12	16	
Confidence y	28	36	22	58	41	32	30	62	14	

Education(yrs)x	14	8	12	15	12	9	12	10	14	12	16
Confidence y	21	42	48	25	31	40	42	57	28	16	18

a. Construct a scatterplot and find the least squares estimates of β_0 and β_1 in the model $y = \beta_0 + \beta_1 x + \epsilon$.
b. Calculate the standard error about the regression line.
c. Test the hypothesis that $\beta_1 = 0$.
d. Calculate R^2. How much of the variability in confidence is explained by years of education?
e. Based on your responses, do you think the model in part **a** adequately describes the relationship between the two variables?

9.56   In Exercise 9.55, find the predicted degree of confidence for someone who has had 12 years of education. Repeat for someone who has had 16 years of education.

9.57   Find the standardized residuals associated with the least squares equation found in Exercise 9.55. Construct a normal probability plot of the residuals. Construct a residual plot and comment on the results. Do you think that any assumptions have been violated?

9.58   The owner of a department store decided to investigate the relationship between the amount lost due to shoplifting and the number of customers in the store. Unable to get a count on the number of shoppers, she examined the sales receipts:

WORKSHEET: Shoplift.mtw

Sales receipts ($1000)	8.4	7.1	9.3	12.3	10.8	8.1	6.5	7.8
Shoplifting loss ($100)	16.2	12.3	19.8	18.4	14.6	15.8	11.4	13.1

a. Construct a scatterplot and find the least squares estimates of β_0 and β_1 in the model $y = \beta_0 + \beta_1 x + \epsilon$. Evaluate the model.
b. Test the hypothesis that $\beta_1 = 0$. Is there statistical evidence to reject the hypothesis? What is the p-value?
c. Find the residuals associated with the least squares equation found in part **a**. Plot the residuals and comment on the results.
d. Find the predicted loss due to shoplifting if the sales receipts are $10,000.

9.59   An owner of a large retail store thinks that the monthly gross sales figures of her employees are related to their length of employment. Listed here are July gross sales figures and the number of months of employment for ten employees:

WORKSHEET: Retail.mtw

Months	July gross sales
10	3860
22	4230
8	2650
16	5170
31	4970
2	4780
13	3120
36	4690
18	4920
6	2150

a. What are the least square estimates of β_0, β_1 in the linear regression model?
b. Interpret the coefficient of determination.
c. Test the hypothesis that $\beta_1 = 0$. Is there statistical evidence to reject the hypothesis? What is the p-value?
d. Perform a residual analysis. Does the residual plot indicate any violations of assumptions?

9.60   In an article in *Education*, G. W. Sutton et al. report a negative correlation between stress level and job satisfaction, suggesting a tendency for teachers to report a higher level of job satisfaction when stress levels are low. (WSPT is the Wilson Stress Profile for Teachers.)

WORKSHEET: Jobsat.mtw

WSPT	90	78	85	65	94	82	96	79	80
Job satisfaction	3.6	5.3	4.7	8.9	3.2	4.0	3.8	6.2	6.5

Source: Sutton, G. W. et al., An Evaluation of Teacher Stress and Satisfaction, *Education*, *105* (2).

a. Construct a scatterplot and find the least squares estimates of β_0 and β_1 in the model $y = \beta_0 + \beta_1 x + \epsilon$.

b. What is the value of R^2? How much of the variability in stress level is explained by job satisfaction? Based on R^2, do you think the proposed model adequately describes the relationship between the two variables?

c. Test the hypothesis that $\beta_1 = 0$. Is there statistical evidence to reject the hypothesis? What is the *p*-value?

9.61   A report recently stated that the prices of oranges were expected to rise sharply due to low projected harvests. Here are the average prices California growers charged for a 75-pound box of navel oranges and the size of the harvest (in millions of boxes) for six consecutive years:

WORKSHEET: Orange.mtw

Harvest	72	69	58	70	65	54
Price per box	$5.40	6.10	9.30	6.50	7.20	13.40

a. Construct a scatterplot and find the least squares estimates of β_0 and β_1 in the model $y = \beta_0 + \beta_1 x + \epsilon$.

b. Test the hypothesis that $\beta_1 = 0$. Is there statistical evidence to reject the hypothesis? What is the *p*-value?

c. Calculate R^2. How much of the variability in price is explained by the size of the harvest?

d. Find the residuals associated with the least squares equation. Plot the residuals and comment on the results. Do you recommend revising the model?

e. Using the model found in part **a**, find the predicted price for a box of oranges if the projected harvest is 60 million boxes; then find it for 30 million boxes. Which prediction do you think is more reliable? Why?

9.62   The *IAAF/ATFS Track and Field Statistics Handbook* for the 1984 Los Angeles Olympics gives the national records for women in several races (worksheet: **track.mtw**).

Source: Dawkins, B. (1989), Multivariate Analysis of National Track Records, *The American Statistician, 43,* (2), 110–115.

a. From the data set, create a correlation matrix of all races. Are there any strong correlations?

b. Using the 100-meter, 200-meter, and 400-meter races, produce all possible scatterplots of pairs of columns. Does it appear that a straight line would adequately fit any two variables?

c. Would you normally expect a straight-line relationship between the times for the 100-meter and 200-meter races? Explain.

9.63   In Exercises 9.10 and 9.16, a linear model was proposed to relate the ratings of top offensive linemen for the NFL draft to their times to run the 40-yard dash. We learned that the rating was not linearly related to the race time ($R^2 = 7.2\%$, $t = -1.45$ with *p*-value $= .159$). Worksheet: **NFLdraf2.mtw** contains similar data on the top defensive linemen in the 1994 draft. Analyze these data in the same way; that is, determine whether a simple linear regression equation adequately relates the ratings to the 40-yard dash times.

a. What is R^2? What percent of rating is explained by race time?

b. Is the regression coefficient significantly different from 0? What is the *p*-value?

c. If the model needs improvement, try using the weight of the player. What percent of the rating is explained by weight?

d. Is the regression coefficient significantly different from 0? What about the *t*-test? What is the *p*-value?

Analysis of Variance

This chapter extends the methods of Chapters 6 and 7 to the comparison of more than two population distributions. This is one of the most widely used statistical procedures available.

The approach is somewhat different from the one-sample and two-sample problems, however, because we introduce the idea of comparing between-sample and within-sample variability. For example, subjects treated for some ailment with different types of drugs have different recovery rates, but is the *variability* in their recovery rates due to the different drugs or is it simply because they are different people? Analysis of variance is a procedure that attempts to determine how much of the variability is due to the treatment and how much is due to all other factors.

This chapter on analysis of variance is an elementary introduction to the vast field of experimental design. A second course in statistical methods would begin where this chapter ends.

Contents

■ STATISTICAL INSIGHT
10.1 Introduction
10.2 Comparing Several Means: The One-Way Analysis of Variance
10.3 The Kruskal–Wallis Test
10.4 Summary and Review Exercises

The Supercar Olympics

Which country makes the fastest car? *Car and Driver* (July 1995) conducted tests of five supercars from five different countries: the United States (Dodge Viper RT/10), Germany (Porsche 911 Turbo), Great Britain (Lotus Esprit S4S), Italy (Ferrari F355), and Japan (Acura NSXT).

Among the tests conducted by *Car and Driver* is a comparison of the top speeds achieved by the cars using as much distance as necessary and without exceeding the engine's redline. From six runs, three in each direction to cancel any grade or wind factors at the test facility, can we determine whether there is a difference in the top speeds attained by these five supercars? The table gives the speeds in miles per hour.

Our natural reaction is to look at the average of the six runs for each of the five cars. This is a good approach, but remember that these are sample averages and they only estimate their respective population

WORKSHEET: Supercar.mtw

Acura (1)	Ferrari (2)	Lotus (3)	Porsche (4)	Viper (5)
159.7	179.6	167.4	173.5	172.3
161.5	173.9	163.0	182.4	168.9
163.7	180.2	160.3	171.3	169.5
166.0	183.9	164.9	175.7	174.6
157.7	176.7	160.5	179.1	161.1
161.7	178.4	158.3	175.0	164.2

means. The problem is to determine whether the observed differences between the sample means are *statistically significant*, and thus leads to the conclusion that the corresponding population means are different. In this chapter, we investigate the equality of several population means with a procedure called *analysis of variance*. (See Section 10.4 for a further discussion of the Supercar example.)

10.1 Introduction

In Chapter 7, we presented the methods for comparing two population means. There are situations, however, as in the Statistical Insight example, where we wish to compare more than two means. A college administrator wishes to compare the mean grade point averages of freshmen, sophomores, juniors, and seniors. A building contractor wants to compare the effectiveness of five different types of insulation. A physician wishes to compare the effectiveness of three different drugs that she is using to treat patients. An agricultural experiment station would like to compare the yields of crops treated with four different types of fertilizer. There is an unending list of experiments in which we wish to compare more than two means.

Having learned the two-sample *t*-test in Chapter 7, it is quite natural to suggest that we simply compare all possible pairs of means when we wish to evaluate the equality of several population means. Certainly with a computer and the different samples stored in the columns of a worksheet, it is very easy to execute a *t*-test on any two columns of data. There is a basic flaw in this procedure, however. When conducting many *t*-tests, it is difficult to determine the overall risk of a Type I error. Moreover, the more tests that you conduct, the greater is the risk of a Type I error on at least one of the tests. When comparing five population means with a 5% level of significance, for example, the overall risk of at least one Type I error can exceed 40%. *Analysis of variance* provides us with the tools to compare the means of several populations with a single test where the overall risk of a Type I error is controlled.

As always, any numerical analysis of data should be preceded by exploratory graphs. Graphs such as histograms and boxplots help us formulate good questions and interpret the information contained in the samples. For example, consider Figure 10.1, side-by-side boxplots of the data comparing the top speeds of the supercars. It appears that the Ferrari and Porsche (cars 2 and 4) are superior to the other three cars. But is the Ferrari faster than the Porsche? Is there a significant difference between the Acura and Lotus (cars 1 and 3)? And where does the Dodge Viper (car 5) fit in the comparison?

Figure 10.1

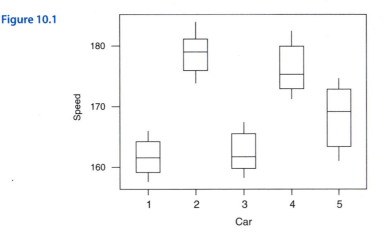

The side-by-side boxplots in Figure 10.1 show differences, but are the observed differences statistically significant? The role of analysis of variance is to perform a numerical test of significance that will test the equality of all the means. That is, the null hypothesis states that there is no difference in the top speeds of all five cars. If this null hypothesis is rejected, we will attempt to determine where the differences actually exist. For example, if there is a difference in the means, is there a significant difference between the Ferrari and the Porsche or between the Acura and the Lotus? (See the Statistical Insight Revisited in Section 10.4 for a complete discussion of these data.)

Homogeneous standard deviations means that the standard deviations of the different populations are close in value.

As in previous inference procedures, the numerical analysis performed depends on the characteristics of the underlying parent distributions. When it can be assumed that the parent distributions are normally distributed with homogeneous standard deviations, an F procedure (presented in Section 10.2) is performed. If the assumptions for the F procedure are in doubt, we have the nonparametric Kruskal–Wallis test in Section 10.3. This procedure is recommended when the distributions are similar in shape but possibly have different centers.

10.2 Comparing Several Means: The One-Way Analysis of Variance

Learning Objectives for this section:

❑ Know the basic outline of the one-way analysis of variance.

❑ Be able to state the null and alternative hypotheses for the one-way analysis of variance.

❑ Be able to describe within- and between-sample variability.

❑ Know how to formulate the F statistic.

❑ Know how to complete an ANOVA table.

❑ Know the assumptions that are necessary for the F-test.

❑ Be able to check the assumptions for an F-test.

❑ Know how to construct individual confidence intervals for the population means in a one-way design.

The statistical design of comparing several population means using independent random samples is called the *one-way analysis of variance* (ANOVA). The design, which describes the procedure for selecting sample data, is illustrated in Table 10.1.

Table 10.1

Population	1	2	· · ·	k	
Population Mean	μ_1	μ_2	· · ·	μ_k	
Population StDev	σ	σ	· · ·	σ	
Sample Size	n_1	n_2	· · ·	n_k	Grand Sample Size $= N$
Sample Mean	$\bar{x}_1$	$\bar{x}_2$	· · ·	$\bar{x}_k$	Grand Sample Mean $= \bar{x}_G$
Sample StDev	s_1	s_2	· · ·	s_k	Pooled StDev. $= s_p$

In the array, we describe k distinct populations with means μ_1, μ_2, through μ_k. Moreover, all k populations have the same standard deviation, σ. By taking samples of sizes $n_1, n_2, \cdots, n_k$ from the respective populations, we use the sample means and standard deviations to test

$$H_0: \mu_1 = \mu_2 = \cdots = \mu_k$$

versus the alternative hypothesis

$$H_a: \text{at least 2 } \mu_i\text{'s are different}$$

The grand mean is the average of the individual averages *only* when all sample sizes are equal.

Note that the sample sizes need not be the same. The grand sample size is $N = n_1 + n_2 + n_3 + \cdots + n_k$ and the grand mean is found by dividing the grand total by the grand sample size; that is, $\bar{x}_G = T/N$, where T is the total of all observed data. As in the two-sample *t*-test in Chapter 7, we use the sample means and the pooled standard deviation to form a test statistic to test the null hypothesis. We illustrate with a simple example.

Suppose we wish to compare the mean gas mileages of standard four-wheel drive pickup trucks manufactured by Chevrolet, Dodge, and Ford. An experiment is designed in which five vehicles of each type are randomly and independently selected from the population of four-wheel drive trucks. (Choosing the samples randomly and independently satisfies the conditions of a one-way ANOVA design.) Each vehicle is driven in a stationary position for the equivalent of 500 miles. The gasoline consumed is measured and the miles per gallon computed. The results appear in Table 10.2.

Table 10.2   WORKSHEET: Trucks.mtw

	Chevy	Dodge	Ford	
	15.2	14.8	15.1	
	15.4	14.4	14.3	
	14.8	14.3	14.6	
	14.4	14.1	13.9	
	14.7	14.4	14.6	
Sample Mean	14.9	14.4	14.5	Grand Mean $= 219/15 = 14.6$
Sample StDev	.4	.255	.442	

Because the sample sizes are equal, the grand mean is the average of the three averages. This is not the case if the sample sizes are unequal.

Do the data indicate differences between the mean gas mileages of the three types of trucks? The sample means are certainly different (14.9, 14.4, 14.5). We must remember, however, that they are sample means, which could be different estimates of a common population mean. It may be that the population mean miles per gallon for each of the three types of vehicles is 14.6, and we have three different sample means that estimate it. We would say that $\mu_1 = \mu_2 = \mu_3 = 14.6$ where μ_1, μ_2, and μ_3 represent the mean miles per gallon for the different trucks. On the other hand, there may indeed be a difference in the population means, and the sample means are reflecting those differences.

Figure 10.2

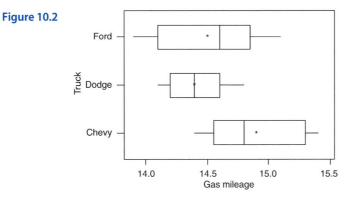

Figure 10.2 shows side-by-side boxplots of the data. Because we are comparing the means of the distributions, we have modified the boxplots by labeling the individual means with asterisks. Notice that the entire boxplot for the Dodge trucks is contained within the middle 50% of the data (box portion of the boxplot) for the Ford trucks. This suggests that there is no difference between the miles per gallon for Dodge and Ford trucks. But what about the Chevrolet trucks? Their boxplot is centered at or above all the data for Dodge trucks and approximately 75% of the data for Ford trucks. Does this mean that the mean miles per gallon for Chevrolet trucks is significantly greater than the other trucks?

Basically, the question is this: "Are the *observed* differences in the *sample* means different enough to conclude that the population means are different?" The question can be answered by testing

$$H_0: \mu_1 = \mu_2 = \mu_3 \quad \text{versus} \quad H_a: \text{at least 2 } \mu_i\text{'s are different}$$

As in all tests of hypotheses, an appropriate test statistic is needed to give a unique numerical measure of how much the sample means differ. To better understand the formulation of the test statistic, consider two hypothetical data sets given in Table 10.3.

Table 10.3
Hypothetical data

	Data Set A			Data Set B		
	Chevy	Dodge	Ford	Chevy	Dodge	Ford
	14.8	14.3	14.4	16.6	13.2	13.5
	14.9	14.4	14.5	16.8	14.7	16.9
	14.9	14.4	14.5	13.2	15.7	15.4
	14.9	14.4	14.5	15.3	12.3	12.8
	15.0	14.5	14.6	12.6	16.1	13.9
Mean	14.9	14.4	14.5	14.9	14.4	14.5
StDev	0.071	0.071	0.071	1.926	1.622	1.645

Notice that the means for the three samples are the same in the two data sets, but the variability of the measurements (standard deviations) are considerably different. This is more clearly demonstrated in the side-by-side boxplots of the data in Figure 10.3.

Figure 10.3

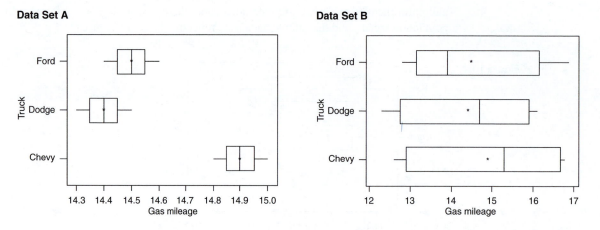

In data set A, there is little variation within each sample as compared to the variation across the samples. The boxplots show a clear distinction between the samples and therefore a clear distinction between the population means. The data suggest that Chevrolets, in general, get 14.9 miles per gallon, Dodges get 14.4 miles per gallon, and Fords get 14.5 miles per gallon. These data lead to the rejection of H_0.

On the other hand, the variability within each sample in data set B overshadows the differences between the means. It is impossible to make a clear distinction between the groups. It is quite possible that all three samples came from a single population with a lot of variability and a mean of 14.6, rather than three samples from three distinct populations. These data certainly would not reject H_0.

The data sets in Tables 10.3 (and Figure 10.3) illustrate *variability between the samples* and *variability within the samples*. In data set A, the variation between the samples is large in comparison to the variation within the samples, and we are able to conclude that there is a difference between the population means. Yet in data set B, the variation between the samples is slight in comparison to the variation within the samples; and consequently, we cannot distinguish between the population means. In other words, by comparing the two measures of variability, we can get an idea of whether or not to reject H_0.

Basic Idea of Analysis of Variance

Analysis of variance is a procedure that compares the variability between the samples to the variability within the samples by computing the ratio

$$F = \frac{\text{variability between the samples}}{\text{variability within the samples}}$$

The F-statistic is a numerical measure of how much the sample means differ. If it becomes unusually large, we should reject H_0.

Within-Sample Variability

The measures of variability in the numerator and denominator of the F-statistic are called *mean squares*. They are extensions of the sample variance, s^2, which measures the variability within a single sample. In the case of multiple samples in the one-way analysis of variance, a reasonable combined measure of variation within the samples is a pooling of all the individual sample variances. Indeed, if $s_1^2, s_2^2, \cdots s_k^2$ represent the sample variances from k samples and $n_1, n_2, \cdots, n_k$ are the associated sample sizes, then

$$\text{MSE} = s_p^2 = \frac{(n_1 - 1)s_1^2 + (n_2 - 1)s_2^2 + \cdots + (n_k - 1)s_k^2}{n_1 + n_2 + \cdots + n_k - k}$$

called the *mean square for error*, is a pooled measure of the variability within the k samples. Notice that it is an extension of s_p^2, the pooled variance used in the pooled t-test in Chapter 7. The numerator of MSE is called the *sum of squares for error* (SSE) because it is a combined measure of errors within each sample. The denominator is the degrees of freedom associated with SSE.

Between-Sample Variability

To measure the variability between the samples, we simply need to calculate the variation across the sample means. If $\bar{x}_1, \bar{x}_2, \cdots, \bar{x}_k$ are the sample means of the k samples and $\bar{x}_G$ is the overall sample mean, then

$$(\bar{x}_1 - \bar{x}_G), (\bar{x}_2 - \bar{x}_G), \cdots, (\bar{x}_k - \bar{x}_G)$$

are the k deviations of the sample means from their grand mean. Summing the squared deviations and dividing by their degrees of freedom, $k - 1$, we obtain the between-sample variation, which is called the mean square for treatments

$$\text{MST} = \frac{\Sigma n_j(\bar{x}_j - \bar{x}_G)^2}{k - 1}$$

The numerator of MST is called the *sum of squares for treatments* (SST) because the various samples arise from the different treatments in an experiment. The denominator is the degrees of freedom associated with SST.

F-Distribution

If the k populations under study in the one-way analysis of variance have the same standard deviation, σ, then MSE, being a pooled estimate of all the individual sample variances, is an unbiased estimator of the common variance σ^2. Additionally, if the null hypothesis is true, MST is also an unbiased estimator of the common variance. Consequently, when H_0 is true, their ratio

$$F = \frac{\text{MST}}{\text{MSE}}$$

should be very close to 1. When H_0 is not true, however, MST tends to overestimate the common variance because it includes between-sample variability. The result is a large F value. Thus, a *significantly* large F value indicates that the variability between the samples

(MST) is significantly larger than the variability within the samples (MSE), which in turn indicates that the null hypothesis, H_0, should be rejected. The values of F that lead to the rejection of H_0 are determined by its sampling distribution.

The sampling distribution of F is not easily obtainable when we are sampling from arbitrary populations. If the null hypothesis is true, however, and if the samples are *independent* and from *normal populations* with *equal standard deviations,* the sampling distribution is the *F-distribution* named after the renowned English statistician, Sir Ronald Fisher (1890–1962).

The F-distribution is similar to the chi-square distribution in that it assumes only non-negative values and is skewed right. Because the F statistic is composed of the ratio of two mean squares, which have their own associated degrees of freedom, we say that the F-distribution has degrees of freedom associated with its numerator and its denominator. Like other sampling distributions, the shape depends on the degrees of freedom. In particular, the amount of skewness depends on the number of degrees of freedom in the numerator and the denominator.

To interpret large F values and assess the statistical evidence against H_0, we need to determine the tail probability (*p*-value) of the F-distribution. Knowing the degrees of freedom of the numerator and denominator and an upper-tail probability of α, the critical value, F* (Figure 10.4) is found in Table B.5 of Appendix B.

Figure 10.4
Density curve for an
F-distribution with
critical value F*

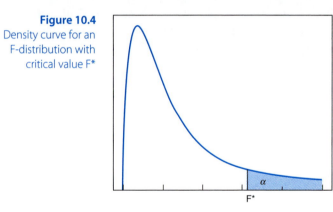

Example 10.1 Suppose degrees of freedom associated with the numerator are 2 and the degrees of freedom associated with the denominator are 12.

a. Find the F critical values associated with upper-tailed probabilities of .10, .05, .025, and .01.
b. Find the approximate *p*-value associated with an F of 2.5, an F of 5.6, and an F of 13.12.

Solution a. From Table B.5 in Appendix B with degrees of freedom (2,12) we find

$$F^*_{.10} = 2.81 \quad F^*_{.05} = 3.89 \quad F^*_{.025} = 5.10 \quad F^*_{.01} = 6.93$$

b. Because 2.5 is less than 2.81, we have that its *p*-value is greater than .10. Because 5.6 is between 5.10 and 6.93, we have that its *p*-value is between .01 and .025. Because 13.12 is greater than 6.93, we have that its *p*-value is less than .01. ■

Here is a summary of the one-way analysis-of-variance F-test.

The One-Way Analysis-of-Variance F-Test

Assumptions:

1. The samples are randomly and independently selected from their respective populations.
2. All of the sampled populations are normally distributed.
3. All of the populations have the same standard deviation. (This is often called the homogeneous standard deviations assumption.)

$$H_0: \mu_1 = \mu_2 = \cdots = \mu_k \quad \text{versus} \quad H_a: \text{at least two } \mu\text{'s differ}$$

$$\text{Test Statistic: } F = \frac{\text{MST}}{\text{MSE}}$$

The test is a right-tailed test where the p-value is found in the F-table with $k - 1$ and $n_1 + n_2 + \cdots + n_k - k$ degrees of freedom in the numerator and denominator, respectively. Unless you are using a computer, the exact p-value cannot be found with the tables, but bounds for it can be found by using the closest value to the observed value of the F-statistic.

If a level of significance α is specified, reject H_0 if p-value $< \alpha$.

Equivalence of F and t^2

If $k = 2$, notice that the assumptions for the F-distribution are the same as the assumptions given for the pooled t-test in Section 7.4. In this case, it can be shown algebraically that the t statistic and the F statistic are related by the equation $t^2 = F$. In other words, the two-tailed pooled t-test is a special case of the F-test when $k = 2$.

Checking Assumptions

Checking the basic assumptions for the F-test and the t-test involves checking the normality of each of the k populations and checking that the standard deviations of the k populations are equal. The normality assumption is best checked with normal probability plots of each of the k samples. A simple way to check the homogeneous standard deviations assumption is to compare the largest sample standard deviation to the smallest sample standard deviation. As long as the largest s is not more than twice as large as the smallest s, we can assume that the population standard deviations are equal.

Checking the Assumptions for the F-Test

1. Examine histograms and boxplots for unusual behavior.
2. Check the normality assumption by constructing normal probability plots for each of the k samples.
3. Check the homogeneous standard deviations assumption by comparing the largest sample standard deviation to the smallest sample standard deviation. The largest standard deviation, s, should be no more than twice as large as the smallest standard deviation.

Example 10.2   Returning to the earlier discussion of the gas mileage of pickup trucks, determine whether there is a significant difference in the mean miles per gallon for the three major brands of trucks. Use the data in Table 10.2 to complete the test.

Solution   Here, μ_1, μ_2, and μ_3 denote the mean miles per gallon for Chevrolet, Dodge, and Ford trucks, respectively. The null hypothesis says that they are the same; that is,

$$H_0: \mu_1 = \mu_2 = \mu_3 \quad \text{versus} \quad H_a: \text{at least two } \mu\text{'s differ}$$

Before proceeding with the F-test, we check assumptions. The boxplots in Figure 10.2 (page 515) show no unusual behavior. The normal probability plots of the three samples in Figure 10.5 support the normality assumption. The largest standard deviation, $s = .442$, is not more than twice as large as the smallest standard deviation, $s = .255$ (see Table 10.4); therefore, it is reasonable to assume that the population standard deviations are equal.

Figure 10.5

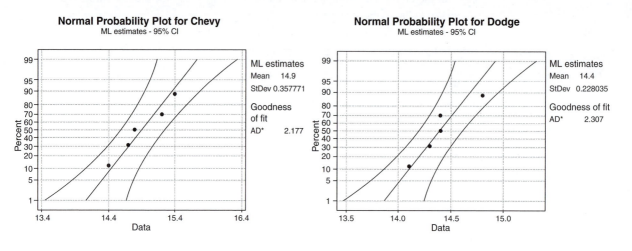

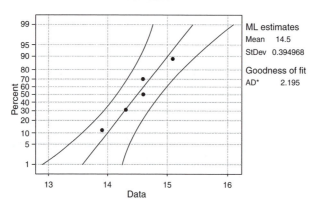

The F-test is summarized in Table 10.4.

Table 10.4
ANOVA table

One-way ANOVA: gas mileage versus truck

```
Analysis of Variance for gas mile
Source    DF       SS      MS      F      P
truck      2    0.700   0.350   2.50   .124
Error     12    1.680   0.140
Total     14    2.380
                                  Individual 95% CIs For Mean
                                  Based on Pooled StDev
Level      N     Mean   StDev   ---------+---------+---------+-------
chevy      5   14.900   0.400                   (----------*---------)
dodge      5   14.400   0.255   (---------*----------)
ford       5   14.500   0.442      (---------*----------)
                                  ---------+---------+---------+-------
Pooled StDev = 0.374                  14.35     14.70     15.05
```

The results of the analysis of variance, shown in Table 10.4, are summarized in an ANOVA table. The first column of the table lists the possible *sources of variation* (Source) in the data. The second column gives the *degrees of freedom* (DF) associated with the sum of squares (SS) given in the third column. The fourth column, which is obtained by dividing the sum of squares by the degrees of freedom, gives the *mean squares* (MS) for treatment and error. The fifth and sixth columns give the *F-ratio* and its *p-value*, respectively. With a *p*-value = 0.124, we fail to reject H_0. Based on the sample information, there is insufficient evidence to say that the trucks differ in gas mileage. ∎

Confidence Intervals

Notice in Table 10.4 that the analysis-of-variance procedure gives confidence intervals for each of the treatment means. The procedure used to construct the intervals is the same as for a one-sample confidence interval based on $\bar{x}$ (given in Section 5.4), except that the estimate of σ is obtained from the pooled standard deviation s_p = Pooled StDev given in the ANOVA table.

A **(1 − α) 100% confidence interval for the mean of treatment *j*** is given by the limits

$$\bar{x}_j \pm t^* \frac{s_p}{\sqrt{n_j}}$$

where t^* is the upper $\alpha/2$ critical value found in the *t*-table with $N - k$ degrees of freedom.

Example 10.3   An experiment was conducted to study the reaction effects of four drugs on a nervous disorder. Twenty-eight subjects with the nervous disorder were independently and randomly assigned to the four drug groups, seven to each group. Unfortunately, two subjects in group 1 and one in group 4 were unable to complete the experiment. The reaction times to an experimental task were recorded for the remaining 25 subjects after they were administered the drug.

WORKSHEET: Nervous.mtw

	Drug Type		
1	2	3	4
3	5	6	2
5	7	5	4
4	3	7	3
6	4	9	4
4	5	6	2
	3	7	5
	6	8	

From the side-by-side boxplot diagram in Figure 10.6, it appears that the mean reaction time for subjects given drug 3 is greater than the means for the other drugs. To investigate further, complete an ANOVA table and test the hypothesis of equality of mean reaction times for the four drugs. Then find individual 95% confidence intervals for the mean reaction time for each of the drugs.

Figure 10.6

Side-by-side boxplots of reaction times to four drugs

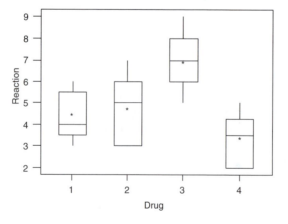

Solution   If we let μ_j denote the mean reaction time to drug j for $j = 1, 2, 3,$ and 4, then the null hypothesis is

$$H_0: \mu_1 = \mu_2 = \mu_3 = \mu_4$$

and the alternative hypothesis is

$$H_a: \text{at least two } \mu\text{'s differ}$$

From the data, we get Table 10.5.

Table 10.5   One-Way ANOVA: reac versus drug

```
Analysis of Variance for reac
Source    DF      SS      MS      F      P
drug       3   43.02   14.34   8.18   0.001
Error     21   36.82    1.75
Total     24   79.84
                               Individual 95% CIs For Mean
                               Based on Pooled StDev
Level   N    Mean   StDev   -------+---------+---------+---------
1       5   4.400   1.140           (-------*------)
2       7   4.714   1.496            (------*-------)
3       7   6.857   1.345                            (-------*------)
4       6   3.333   1.211   (------*------)
                               -------+---------+---------+---------
Pooled StDev = 1.324              3.2       4.8       6.4
```

A p-value $= .001$ means that we should reject H_0 in favor of the alternative. Thus, there is highly significant evidence that the mean reaction effects of the four drugs are different. From the appearance of the boxplots, we speculate that the reaction effect of drug 3 is greater than those of the other three drugs. Can we also conclude that the reaction effect of drug 4 is significantly less than those of the other drugs? Individual confidence intervals may help answer this question. From the ANOVA table, we find

$$s_p = \text{Pooled StDev} = 1.324 \quad \text{and} \quad df = N - k = 21$$

For a 95% confidence interval, we find from the t-table with 21 degrees of freedom, $t^* = 2.08$. The confidence intervals for the mean reaction time for the four drugs are

$$4.400 \pm 2.08*1.324/\sqrt{5} \qquad (3.168, 5.632)$$
$$4.714 \pm 2.08*1.324/\sqrt{7} \qquad (3.673, 5.755)$$
$$6.857 \pm 2.08*1.324/\sqrt{7} \qquad (5.816, 7.898)$$
$$3.333 \pm 2.08*1.324/\sqrt{6} \qquad (2.209, 4.457)$$

Table 10.5 shows the graphs of the four intervals on a common number line. Clearly, the mean reaction time for drug 3 is greater than that of any of the other drugs. The graph also suggests that the mean reaction time for drug 4 is less than that of drugs 1 and 2. To make sure, however, we should conduct a *multiple comparison test*. We will conduct Tukey's Multiple Comparison procedure in Example 10.5. ∎

COMPUTER TIP

One-Way ANOVA

To perform a one-way analysis of variance with Minitab, we have two different forms for the input data. Either all the response data are in one column of the worksheet with a group code value in a separate column, or the response data for the different groups are in separate columns. To execute the ANOVA, from the **Stat** menu select **ANOVA**. If the response data are all in one column, select **One-way**;

continued

Example 10.4   In almost any magazine one can find advertisements for several different brands of cigarettes, each claiming to have the lowest tar and nicotine. To evaluate cigarettes, the Federal Trade Commission (FTC) uses "smoking machines" that measure the tar, nicotine, and carbon monoxide in each cigarette. Carbon monoxide has been linked to heart disease, tar has been linked to cancer, and nicotine is addictive.

Suppose that the amount of tar (measured in milligrams) is recorded for 25 cigarettes randomly selected from each of four brands:

WORKSHEET: Cigar.mtw

brandA	brandB	brandC	brandD
0.41	0.43	0.52	0.43
0.48	0.49	0.48	0.55
0.44	0.52	0.67	0.71
0.37	0.65	0.49	0.65
0.31	0.63	0.38	0.47
0.40	0.55	0.57	0.63
0.53	0.38	0.70	0.40
0.32	0.41	0.45	0.55
0.35	0.58	0.47	0.56
0.52	0.63	0.55	0.32
0.40	0.53	0.49	0.54
0.51	0.57	0.56	0.58
0.53	0.68	0.51	0.42
0.29	0.44	0.60	0.39
0.39	0.35	0.41	0.58
0.48	0.53	0.65	0.46
0.58	0.52	0.57	0.48
0.46	0.43	0.39	0.52
0.59	0.57	0.54	0.38
0.50	0.48	0.61	0.47
0.47	0.55	0.53	0.53
0.51	0.41	0.68	0.61
0.33	0.44	0.50	0.59
0.56	0.53	0.58	0.63
0.61	0.44	0.49	0.44

Perform an analysis of variance on the cigarette data to determine whether one brand is "lowest" in tar.

Solution First, we look at boxplots of the data. The side-by-side boxplots in Figure 10.7 show that all samples are reasonably symmetric and have no outliers. The standard deviations appear to be equal. Normality does not seem out of the question.

Figure 10.7

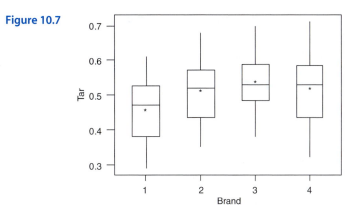

To further assess normality, we construct normal probability plots of each sample in Figure 10.8. In no case should we reject normality.

Figure 10.8

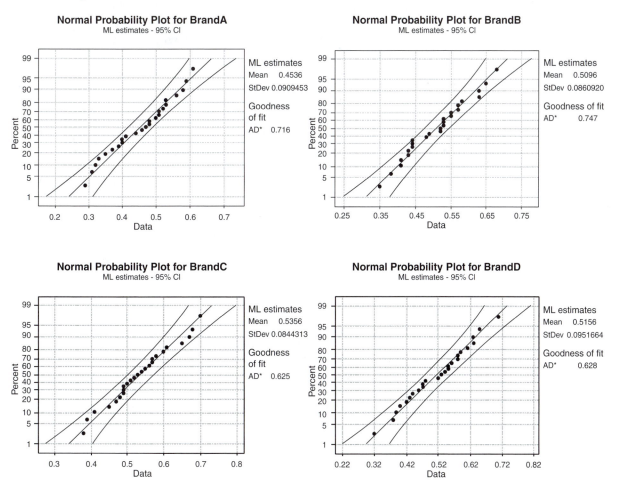

Finally, we complete the one-way analysis of variance in Table 10.6.

Table 10.6

```
One-Way ANOVA: brandA, brandB, brandC, brandD

Analysis of Variance
Source   DF       SS        MS      F       P
Factor    3   0.09260   0.03087   3.72   0.014
Error    96   0.79670   0.00830
Total    99   0.88930
                                Individual 95% CIs For Mean
                                Based on Pooled StDev
Level    N     Mean     StDev   -------+---------+---------+---------
brandA   25   0.45360   0.09282 (-------*------)
brandB   25   0.50960   0.08787             (------*------)
brandC   25   0.53560   0.08617                  (------*------)
brandD   25   0.51560   0.09713              (------*------)
                                -------+---------+---------+---------
Pooled StDev = 0.09110              0.450     0.500     0.550
```

Notice that the standard deviations support the homogeneous standard deviations assumption. The F statistic is 3.72 with p-value $= .014$. We should reject the hypothesis of equal mean tar contents for the four brands of cigarettes. The individual confidence intervals suggest that brand 1 has a mean tar content lower than the other brands. What can be said about the comparison of brands 2, 3, and 4? We look closer at this question with a multiple comparison test in the next example. ∎

The execution of a multiple comparison test is based on the premise that if a confidence interval for the difference between two means includes 0, there is no significant difference between the means. Thus, if an F-test declares a difference between a group of means, we construct confidence intervals for all possible pairs of means and determine which ones include 0. In this manner we are able to determine exactly where differences lie between the means. The next example illustrates Tukey's multiple comparison procedures. A detailed discussion of the procedure can be found in Kitchens: *Exploring Statistics, 2nd Edition,* Duxbury.

Example 10.5   In Example 10.4, the tar contents of four major brands of cigarettes were compared with an analysis-of-variance procedure. The results indicated that a significant difference exists between the different brands. Analyze with Tukey's multiple comparison procedure to determine where differences between means are statistically significant.

Solution   Following the Computer Tip, perform the usual one-way analysis of variance and click on Comparisons. Select Tukey's procedure and then click on OK. Immediately after the ANOVA table and the individual confidence intervals, we get the array shown in Figure 10.9. First, notice that the Family error rate is 0.05 and the Individual error rate is 0.0104. The Individual error rate is the probability that a given confidence interval will not contain the true difference between two group means. The family error rate is the probability that among all the confidence intervals, at least one will not contain the true difference between group means.

Figure 10.9 Tukey's pairwise comparisons

```
     Family error rate = 0.0500
Individual error rate = 0.0103

Critical value = 3.70

Intervals for (column level mean)  −  (row level mean)

                        1              2              3

         2      −0.12341
                 0.01141

         3      −0.14941       −0.09341
                −0.01459        0.04141

         4      −0.12941       −0.07341       −0.04741
                 0.00541        0.06141        0.08741
```

The six groups of numbers in the array are confidence intervals for all possible pairwise comparisons of the means. For example, the group

−0.12341

0.01141

at the intersection of column 1 and row 2 is the confidence interval $(−0.12341, 0.01141)$ for $\mu_1 − \mu_2$. The fact that 0 is inside the interval means that there is no significant difference between μ_1 and μ_2. In fact, the only interval that does not include 0 is the interval $(−0.14941, −0.01459)$ for $\mu_1 − \mu_3$. Consequently, there is a significant difference between μ_1 and μ_3. We conclude that the mean tar contents of Brands 1 and 3 are different.

If we arrange the means in rank order based on the magnitude of the sample means, and then draw a line under those means that are not significantly different, we get the following summary bars:

$\mu_3 \quad \mu_4 \quad \mu_2 \quad \mu_1$

The line from μ_2 to μ_1 is a subset of the line from μ_4 to μ_1 and thus is not needed. Also the lines from μ_4 to μ_2 and from μ_3 to μ_4 are subsets of the line from μ_3 to μ_2 and thus are not needed. Deleting unwanted lines, we get this summary:

$\mu_3 \quad \mu_4 \quad \mu_2 \quad \mu_1$

In other words, the mean tar content of cigarette brand 1 is significantly less than that of brand 3. ■

EXERCISES 10.2 The Basics/Interpreting Computer Output

10.1 Examine the following stem-and-leaf plots and descriptive statistics:

WORKSHEET: ABC.mtw

Group A		Group B		Group C	
0		0	8	0	
1	0 2 4	1	2	1	
2	3 5 6 7 7	2	2 4 4 6	2	2 4 6
3	1 3 6 7 8	3	0 3 5 5 8 9	3	5 8 9 9
4	0 2 3	4	1 4 5	4	0 3 5 6
5		5	0 2	5	3 5 7
6	4 5			6	0 2

Descriptive Statistics

Variable	N	Mean	Median	TrMean	StDev	SEMean
GroupA	18	32.94	32.00	32.38	15.12	3.56
GroupB	17	32.82	35.00	33.20	12.39	3.00
GroupC	16	42.75	41.50	42.86	12.44	3.11

Variable	Min	Max	Q1	Q3
GroupA	10.00	65.00	24.50	40.50
GroupB	8.00	52.00	24.00	42.50
GroupC	22.00	62.00	35.75	54.50

a. Judging from the stem-and-leaf plots, do you think it is reasonable to assume that the samples are from normally distributed populations?

b. Judging from the standard deviations of the three groups, do you think the population standard deviations are homogeneous?

c. Examine the interquartile ranges of the three groups. Do their values support the homogeneous standard deviations assumption?

d. From the standard deviations, calculate the mean square for error.

e. Examine the means of the three groups. Do you think that there is evidence to reject the hypothesis that the population means are equal?

f. Can the grand mean be calculated by averaging the means of the three groups given in the descriptive statistics? Explain.

10.2 Here are side-by-side boxplots of the data given in Exercise 10.1.

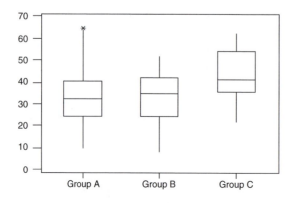

a. Does it appear that the population standard deviations are homogeneous?
b. Do you see evidence that the populations are not normally distributed?
c. Based on the boxplots, do you think there is statistical evidence to reject the hypothesis that the population means are equal?

10.3 This analysis-of-variance table is for the data in Exercise 10.1.

One-Way Analysis of Variance

```
Analysis of Variance
Source     DF        SS        MS          F         p
Factor      2      1069       534       2.96     0.061
Error      48      8662       180
Total      50      9731
                                   Individual 95% CIs For Mean
                                   Based on Pooled StDev
Level      N      Mean     StDev    ---+---------+---------+---------+---
GroupA     18     32.94    15.12    (--------*--------)
GroupB     17     32.82    12.39    (--------*--------)
GroupC     16     42.75    12.44                (---------*---------)
                                    ---+---------+---------+---------+---
Pooled StDev =    13.43            28.0      35.0      42.0      49.0
```

a. What is the sum of squares for error?
b. What is the mean square for error?
c. What is the sum of squares for treatments?
d. What is the mean square for treatments?
e. The pooled standard deviation is 13.43. What is another expression for the square of this figure?
f. State the null and alternative hypotheses for which this analysis is intended.
g. What is the F ratio? What is its p-value? Should the null hypothesis be rejected?
h. Examine the individual confidence intervals. Can you tell which groups are different?

10.4 Summary statistics associated with data calculated from three groups in a one-way ANOVA are given here:

WORKSHEET: Groups.mtw

Descriptive Statistics

Variable	N	Mean	Median	TrMean	StDev	SEMean
GroupA	23	66.35	67.00	66.29	7.67	1.60
GroupB	26	75.69	74.50	75.58	10.08	1.98
GroupC	25	71.64	73.00	71.52	10.06	2.01

Variable	Min	Max	Q1	Q3
GroupA	53.00	81.00	62.00	71.00
GroupB	54.00	100.00	67.75	83.25
GroupC	56.00	90.00	62.00	78.00

a. Can we conclude that the homogeneous standard deviations assumption is met?
b. Give the degrees of freedom associated with the pooled variance, s_p^2.
c. Compute the mean square for error.
d. Examine the means of the three groups. Do you think that there is evidence to reject the hypothesis that the population means are equal?

10.5 This analysis-of-variance table is for the data in Exercise 10.4.

One-Way Analysis of Variance

```
Analysis of Variance
Source     DF        SS        MS          F         p
Factor      2    1067.5     533.7       6.05     0.004
Error      71    6260.5      88.2
Total      73    7328.0
                                   Individual 95% CIs For Mean
                                   Based on Pooled StDev
Level      N      Mean     StDev   ------+---------+---------+---------+
GroupA     23    66.348    7.667   (-------*-------)
GroupB     26    75.692   10.079                        (------*-------)
GroupC     25    71.640   10.058                 (------*-------)
                                   ------+---------+---------+---------+
Pooled StDev =    9.390            65.0      70.0      75.0      80.0
```

a. State the null hypothesis for which this analysis is intended.
b. What is the F ratio? What is its p-value? Should the null hypothesis be rejected?
c. Does this analysis agree with your answer to part **d** of Exercise 10.4?
d. Examine the individual confidence intervals. Can you tell which groups are different?

10.6 The side-by-side boxplots are for the groups in Exercises 10.4 and 10.5. In Exercise 10.5, the null hypothesis was rejected. Do these boxplots illustrate what means are different? Explain.

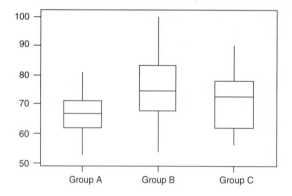

10.7 In Example 7.14 in Section 7.4, a math achievement test was given to a random sample of male and female high school students. A two-tailed, pooled t-test was applied to the data (worksheet: **Achieve.mtw**) with these results:

Two-Sample t-Test and CI: female, male

```
Two-sample T for Female vs Male

           N      Mean     StDev    SE Mean
Female    13     84.46      8.04        2.2
Male      12     76.4       11.0        3.2

Difference = mu Female − mu Male
Estimate for difference: 8.04
95% CI for difference: (0.14, 15.95)
T-test of difference = 0 (vs not =): T-Value = 2.10   P-Value = 0.047   DF = 23
Both use Pooled StDev = 9.55
```

Here is a one-way analysis of variance applied to the same data:

One-Way ANOVA: score versus gender

```
Analysis of Variance for Score
Source          DF          SS          MS          F          P
Gender           1       403.9       403.9       4.43      0.047
Error           23      2098.1        91.2
Total           24      2502.0
```

```
                                    Individual 95% CIs For Mean
                                    Based on Pooled StDev
Level           N        Mean       StDev    ---+---------+---------+---------+---
1              13      84.462       8.038                   (--------*--------)
2              12      76.417      10.967     (---------*---------)
                                             ---+---------+---------+---------+---
Pooled StDev =          9.551              72.0        78.0        84.0        90.0
```

a. Compare the T statistic and its p-value with the F-statistic and its p-value. Is there a relationship between T and F? What about the p-values?

b. Compare the pooled standard deviations for the two test procedures. Are they the same? Find the mean square for error (MSE) and compare it with the pooled standard deviation. What, if any, is the relationship between the two?

c. What general observations can you make about the pooled t-test and the F-test when there are only two groups to compare?

10.8 Can music steady the scalpel? Psychologist Karen Allen, at the State University of New York— Buffalo studied 50 male surgeons under simulated stress. She randomly divided them into three groups: listening to music they chose, listening to her choice of Pachelbel's Canon in D, and no music. She reported in the *Journal of the American Medical Association* that the surgeons stayed relaxed with their own music, their blood pressure rose with her music, and the surgeons jumped to near hypertension with no music.

a. Is this a one-way analysis-of-variance design?

b. What is the response variable?

c. Verbally describe the null hypothesis.

d. What assumptions are necessary to test the hypothesis?

e. What is the test statistic?

f. What is the distribution of the test statistic under the assumption that the null hypothesis is true?

g. Is there a control group?

10.9 Exercise 6.49 in Section 6.3 describes the process of manufacturing integrated circuits (chips) used in computers. In one part of the process, a thin layer of silicon oxide is placed on the surface of a wafer. The thickness of the oxide layer is critical to the performance of the resulting chips. Two wafers are randomly selected from each of 30 lots. Four measurements of the thickness of the oxide layer on each wafer are then taken. In worksheet: **Chips.mtw,** columns C1–C4 contain the four measurements on the first wafer, and columns C5–C8 contain the four measurements on the second of the two wafers selected from the 30 lots.

Source: Yashchin, E. (1995), "Likelihood Ratio Methods for Monitoring Parameters of a Nested Random Effect Model," *Journal of the American Statistical Association, 90,* 729–738.

a. From the side-by-side boxplots, is there evidence of a difference in the measurements of the thicknesses of the two wafers?

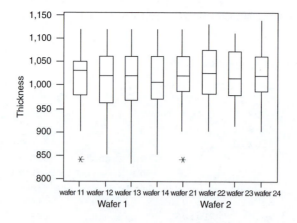

b. Review the assumptions for the F-distribution. Is it appropriate to compare the means of the observations obtained from the eight measurements using a one-way analysis of variance?

10.10 In Example 10.3, an experiment was conducted to study the reaction effects of four drugs on a nervous disorder. The analysis of variance indicates that a significance difference exists between the mean reaction effects. Use the provided Tukey's multiple comparison procedure to determine exactly what means differ.

Tukey's pairwise comparisons

```
     Family error rate = 0.0500
Individual error rate = 0.0111

Critical value = 3.94

Intervals for (column level mean) − (row level mean)

                    1              2              3

        2       −2.474
                 1.846

        3       −4.617         −4.115
                −0.297         −0.171

        4       −1.167         −0.671          1.471
                 3.300          3.433          5.576
```

10.11 A psychologist is studying the effect of drug and electroshock therapy on a subject's ability to solve simple tasks. The number of tasks completed in a 10-minute period is recorded for the subjects, who have been randomly assigned to four treatment groups: drug with electroshock, drug without electroshock, no drug with electroshock, and no drug and no electroshock.

WORKSHEET: Shkdrug.mtw

Drug w/shock	Drug wo/shock	No Drug w/shock	No Drug/no shock
3	2	1	5
1	3	2	3
2	4	2	2
1	0	0	3
4	2	1	4
3	4	2	6
2	2	3	5

continued

WORKSHEET: Shkdrug.mtw
continued

Drug w/shock	Drug wo/shock	No Drug w/shock	No Drug/no shock
4	4	1	4
2	5	0	6
1	6	3	3
3	2	1	4
4	3	3	5
5	5	2	6
3	6	0	4
4	2	1	2
3	3	2	4

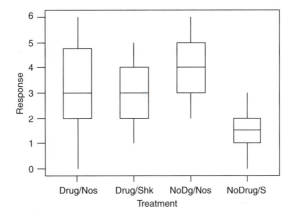

a. State the null and alternative hypotheses to determine whether the drug and electroshock thera-pies have an effect on ability to solve simple tasks.
b. From the appearance of the boxplots, do you think that the null hypothesis should be rejected?
c. Judging from the boxplots, do you think the population standard deviations are homogeneous?
d. Is there evidence in the boxplots that would suggest that the populations are not normally distrib-uted?
e. Do you think that the overall conditions for the ordinary F-test are met, or should we look for an alternative test?

10.12 The analysis of variance is for the data in Exercise 10.11.

One-Way ANOVA: response versus treatment

```
Analysis of Variance for Response
Source      DF       SS       MS        F        P
Treatmen     3    58.13    19.38    11.01    0.000
Error       60   105.63     1.76
Total       63   163.75

                             Individual 95% CIs For Mean
                             Based on Pooled StDev
Level        N     Mean    StDev   ----+---------+---------+---------+--
Drug/NoS    16    3.313    1.662                 (-----*----)
Drug/Shk    16    2.813    1.223            (----*-----)
NoDg/NoS    16    4.125    1.310                      (----*-----)
NoDrug/S    16    1.500    1.033   (-----*----)
                                   ----+---------+---------+---------+--
Pooled StDev =     1.327           1.2       2.4       3.6       4.8
```

```
Tukey's pairwise comparisons

     Family error rate = 0.0500
Individual error rate = 0.0104

Critical value = 3.74

Intervals for (column level mean) - (row level mean)

                Drug/NoS        Drug/Shk        NoDg/NoS

Drug/Shk         -0.741
                  1.741

NoDg/NoS         -2.053          -2.553
                  0.428          -0.072

NoDrug/S          0.572           0.072           1.384
                  3.053           2.553           3.866
```

a. Judging from the standard deviations, do you think the population standard deviations are homogeneous?
b. Should the null hypothesis be rejected? Interpret the decision.
c. Examine the individual confidence intervals. Can you tell which treatments are statistically different?
d. Using the provided Tukey's multiple comparison procedure pinpoint exactly what groups differ.

10.13   The following error scores were obtained for four groups of experimental animals running a maze under different experimental conditions.

WORKSHEET: Maze.mtw

Condition A	Condition B	Condition C	Condition D
16	20	9	15
12	18	11	14
15	22	14	18
13	17	15	20
15	21	8	16
14	19	10	17
15	18	11	17
14	18	10	16

a. State the null and alternative hypotheses for an ANOVA.

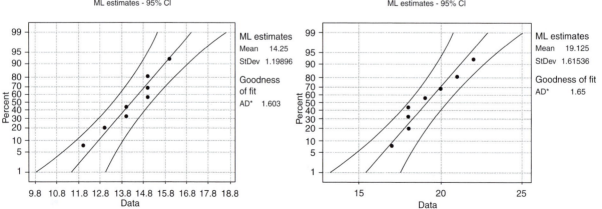

Normal Probability Plot for CondA
ML estimates - 95% CI

ML estimates
Mean   14.25
StDev 1.19896

Goodness of fit
AD*   1.603

Normal Probability Plot for CondB
ML estimates - 95% CI

ML estimates
Mean   19.125
StDev 1.61536

Goodness of fit
AD*   1.65

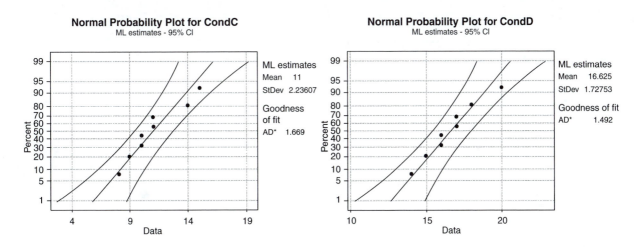

b. Based on the normal probability plots, do you think the normality assumption for an analysis of variance is met?

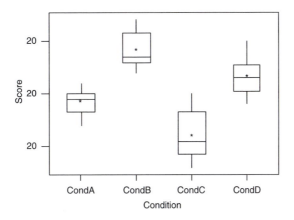

c. Based on the boxplots, does it appear that the population standard deviations are somewhat homogeneous?
d. Do the boxplots indicate that the null hypothesis should be rejected? Explain your answer.

One-Way ANOVA: score versus condition

```
Analysis of Variance for score
Source     DF        SS       MS        F       P
conditio    3    287.75    95.92    27.90    0.000
Error      28     96.25     3.44
Total      31    384.00
                                   Individual 95% CIs For Mean
                                   Based on Pooled StDev
Level       N      Mean    StDev    --------+---------+---------+--------
CondA       8    14.250    1.282                    (----*---)
CondB       8    19.125    1.727                                 (----*---)
CondC       8    11.000    2.390    (----*---)
CondD       8    16.625    1.847                         (---*----)
                                    --------+---------+---------+--------
Pooled StDev =     1.854              12.0      15.0      18.0
```

```
Tukey's pairwise comparisons

    Family error rate = 0.0500
Individual error rate = 0.0108

Critical value = 3.86

Intervals for (column level mean) - (row level mean)

            CondA          CondB          CondC

CondB       -7.405
            -2.345

CondC        0.720          5.595
             5.780         10.655

CondD       -4.905         -0.030         -8.155
             0.155          5.030         -3.095
```

e. From the ANOVA table, test the hypothesis that there is no difference in the mean error scores for the four experimental conditions. Be sure to give the *p*-value and state your conclusion.

f. Pinpoint exactly what experimental conditions differ with Tukey's multiple comparison procedure.

10.14 According to IRS statements, the 1999 salaries (in $1000) of the members of the boards of directors of three different universities are listed here:

WORKSHEET: Board.mtw

University A	University B	University C
70	30	100
120	90	900
85	80	300
200	250	90
60	70	1200
310	55	260
90	180	60

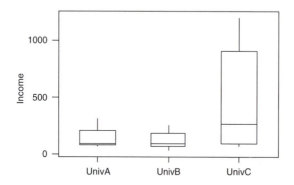

Descriptive Statistics

Variable	N	Mean	Median	TrMean	StDev	SEMean
UnivA	7	133.6	90.0	133.6	90.8	34.3
UnivB	7	107.9	80.0	107.9	78.3	29.6
UnivC	7	416	260	416	451	170

Variable	Min	Max	Q1	Q3
UnivA	60.0	310.0	70.0	200.0
UnivB	30.0	250.0	55.0	180.0
UnivC	60	1200	90	900

a. Based on the side-by-side boxplots, do you think the assumptions for the F-test are satisfied?

b. Based on the standard deviations given in the descriptive statistics, do you think the population standard deviations are homogeneous?

c. Do you recommend an F-test in this situation?

d. Look at the means for the three universities. Is it necessary to conduct a formal test of the hypothesis that the mean salaries for the three groups are the same?

10.15 Three randomly selected groups of chickens are fed three different rations. These are their weight gains during a specified period of time:

WORKSHEET: Chicken.mtw

Ration 1	Ration 2	Ration 3
4	3	6
4	4	7
7	5	7
3	4	7
2	6	6
5	4	8
4	5	5
5	6	6
2	7	7
3	6	6
6	5	7
4	5	5
5	5	6

a. State the null and alternative hypotheses for a one-way analysis-of-variance test.

b. Examine the dotplots. Is there evidence that the normality assumption is violated? Are the population standard deviations homogeneous? In all, does it appear that the assumptions for the F-test are satisfied?

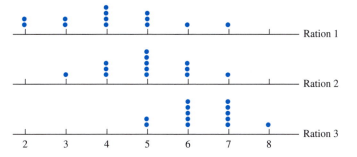

c. From the analysis-of-variance printout, examine the standard deviations. Are the conditions for the F-test satisfied?

d. Test the hypothesis (stated in part **a**) that there is no difference in the average weight gains for the three rations. Be sure to state your conclusion.

One-Way Analysis of Variance

```
Analysis of Variance
Source    DF      SS       MS        F        p
Factor     2    32.97    16.49     12.17    0.000
Error     36    48.77     1.35
Total     38    81.74

                                 Individual 95% CIs For Mean
                                 Based on Pooled StDev
Level      N     Mean    StDev   ---+---------+---------+---------+
Ration1   13    4.154    1.463   (------*-----)
Ration2   13    5.000    1.080             (------*------)
Ration3   13    6.385    0.870                             (------*-----)
                                 ---+---------+---------+---------+
Pooled StDev = 1.164                4.0       5.0       6.0       7.0
```

e. Give an interpretation of the individual confidence intervals.

10.16 An experiment was designed to compare three different cleansing agents. Forty-five subjects with similar skin conditions were randomly assigned to three groups of 15 subjects each. A patch of skin on each individual was exposed to a contaminant and then cleansed with one of the cleansing agents. After 8 hours, the residual contaminants were measured.

WORKSHEET: Clean.mtw

Cleansing Agent

A	B	C
2	6	5
4	7	6
3	9	5
3	8	4
2	6	7
4	6	5
5	8	6
3	6	5
2	7	4
4	8	6
3	5	7
2	6	6
5	7	7
4	8	6
3	8	7

Here are the results of an analysis-of-variance test and Tukey's procedure to compare the mean residual contaminants left by the three cleansing agents:

One-Way Analysis of Variance

```
Analysis of Variance on clean
Source      DF        SS       MS       F         p
agent        2    108.13    54.07    47.44    0.000
Error       42     47.87     1.14
Total       44    156.00
```

```
                                       Individual 95% CIs For Mean
                                       Based on Pooled StDev
Level       N      Mean    StDev    --+---------+---------+---------+----
  1        15     3.267    1.033    (---*--)
  2        15     7.000    1.134                               (---*--)
  3        15     5.733    1.033                     (--*---)
                                    --+---------+---------+---------+---
Pooled StDev =    1.068             3.0       4.5       6.0       7.5
```

Tukey's pairwise comparisons

    Family error rate = 0.0500
Individual error rate = 0.0193

Critical value = 3.44

Intervals for (column level mean) — (row level mean)

```
                     1              2

    2         -4.682
              -2.785

    3         -3.415          0.318
              -1.518          2.215
```

Is there a significant difference in the mean residual contaminants left by the three cleansing agents? Summarize the results of Tukey's multiple comparison procedure.

Computer Exercises

10.17 Suppose that four groups of 11 students each used a different method of programmed learning to study statistics. A standard test was administered to the four groups and graded on a 15-point scale. Given these results, determine whether there was a significant difference in the results of the four methods.

WORKSHEET: Program.mtw

Method I	Method II	Method III	Method IV
3	5	7	4
5	7	5	6
6	7	6	6
8	7	8	7
4	6	7	6
3	6	6	5
5	8	9	5
6	4	8	5
4	6	7	6
6	7	7	5
3	5	8	4

a. State the null and alternative hypotheses for a one-way analysis-of-variance test.
b. Construct side-by-side boxplots. Does it appear that there is a difference among the four methods?
c. Construct normal probability plots for each method. Does the normality assumption seem to be satisfied?
d. Complete an analysis-of-variance printout. Based on the standard deviations, do you think the population standard deviations are homogeneous?
e. In all, do you think that the assumptions for the F-test are satisfied?
f. Test the hypothesis that there is no difference among the four methods. Be sure to state your conclusions.
g. Determine whether a multiple comparison test is necessary. If it is, complete the test and indicate what differences exist among the four methods.

10.18 A laminectomy is the surgical removal of a ruptured disk or all or part of the bony arch of a segment of the spine. Here are the median costs of laminectomies at hospitals across the state of North Carolina in 1992. Each hospital has been classified as a rural, regional, or metropolitan hospital.

WORKSHEET: Laminect.mtw

Rural(1)	Regional(2)	Metropol(3)
5223	7455	5807
5782	4142	6943
8892	6278	4925
7405	9158	4556
8460	7733	2696
4353	4288	6177
9296	7054	5477
5227	8308	5656
6535	11391	5621
10238	3339	4324
4346	6801	5493
4973	3835	6306
11191	7224	5073
5225	5445	8435
4027	5841	5450
6538		

Source: *Consumer's Guide to Hospitalization Charges in North Carolina Hospitals* (August 1994), North Carolina Medical Database Commission, Department of Insurance.

a. Construct side-by-side boxplots. Does it appear that the assumptions for an F-test are satisfied?

b. Construct normal probability plots for each of the three samples. Is normality rejected in any case?

c. Calculate descriptive statistics for each sample. Is the largest standard deviation more than twice as large as the smallest standard deviation? Is the homogeneous standard deviations assumption satisfied?

d. State the null hypothesis that there is no difference in the costs at the three types of hospitals.

e. Test the hypothesis stated in part **d**. Be sure to state your conclusion.

10.19   *El Niño* (Spanish for "the Christ Child") refers to unusually warm ocean currents in the Pacific that appear around Christmastime and lasts for several months. Climatic effects around the globe have been associated with El Niño. Such effects as monsoon rains in the central Pacific and severe droughts and disastrous forest fires in Indonesia and Australia have been linked to El Niño. One hypothesis is that the warm phase of the El Niño tends to suppress hurricanes, whereas a cold phase encourages hurricanes. The following information is the numbers of storms and resulting hurricanes from 1950 through 1995. Each year is classified as a warm, cold, or neutral El Niño year.

WORKSHEET: Hurrican.mtw

year	storms	hurricane	ElNiño	year	storms	hurricane	ElNiño
1950	13	11	cold	1973	7	4	cold
1951	10	8	warm	1974	7	4	cold
1952	7	6	neutral	1975	8	6	cold
1953	14	6	warm	1976	8	6	warm
1954	11	8	cold	1977	6	5	warm
1955	12	9	cold	1978	11	5	neutral
1956	8	4	neutral	1979	8	5	neutral
1957	8	3	warm	1980	11	9	neutral
1958	10	7	neutral	1981	11	7	neutral
1959	11	7	neutral	1982	5	2	warm
1960	7	4	neutral	1983	4	3	warm
1961	11	8	cold	1984	12	5	neutral
1962	5	3	neutral	1985	11	7	cold
1963	9	7	neutral	1986	6	4	warm
1964	12	6	cold	1987	7	3	warm
1965	6	4	warm	1988	12	5	cold
1966	11	7	neutral	1989	11	7	cold
1967	8	6	cold	1990	14	8	neutral
1968	7	4	neutral	1991	8	4	warm
1969	17	12	neutral	1992	6	3	warm
1970	10	5	cold	1993	8	4	warm
1971	13	6	cold	1994	7	3	warm
1972	4	3	warm	1995	17	10	cold

Source: National Hurricane Center.

a. Construct side-by-side boxplots of the numbers of storms associated with cold, warm, and neutral El Niño years. Does it appear that the cold phase encourages storms or that the warm phase suppresses hurricanes?

b. Do the conditions for a one-way analysis of variance seem to be met? That is, is there any unusual behavior in any of the three samples to suggest that the assumptions for the F-test are not satisified?

c. Perform an ANOVA on the number of storms. What is your conclusion?

d. What is the correlation between the number of storms and the number of hurricanes?

e. Perform an ANOVA on the number of hurricanes. How do these results differ from the analysis of the number of storms in part **c**? Based on the correlation found in part **d**, are these results consistent with the results found in part **c**?

10.3   The Kruskal–Wallis Test

Learning Objectives for this section:

- ❏ Know when the Kruskal–Wallis test should be used in place of the ordinary F-test.
- ❏ Know the assumptions for the Kruskal–Wallis test.
- ❏ Be able to conduct a Kruskal–Wallis test.

The one-way analysis of variance presented in Section 10.2 is appropriate to use when the parent distributions are normally distributed and have equal variances. When those assumptions are not met, an alternative to the ordinary F-test is the nonparametric Kruskal–Wallis test. The only assumption made about the parent distributions is that they are similar in shape; then, the Kruskal–Wallis test is used to test equality of location. For example, if the parent distributions are similarly skewed, the Kruskal–Wallis test can be used to test equality of their population medians.

The procedure for the Kruskal–Wallis test that is presented here is the ordinary F-test applied to the rank summary data, n_j, $\bar{x}_{Rj}$, and s_{Rj}. This is an approximation of the original Kruskal–Wallis test, and it is suggested that there be *5 or more observations in each sample*. For samples less than 5, exact tables for the Kruskal–Wallis test can be found in Conover (1980). It is also required that the samples be random and independent and can be ranked. As in the two-sample Wilcoxon test described in Chapter 7, ranks are assigned to the combined samples.

Example 10.6   A study was conducted to compare the hostility levels of high school students in rural, suburban, and urban areas. A psychological test, the Hostility Level Test (HLT), was used to measure the degree of hostility. Fifteen students were randomly selected from each type of school and given the HLT. The data are summarized in the side-by-side ordered stem-and-leaf plots:

W O R K S H E E T : Hostile.mtw

	Rural	Suburban	Urban
1	6		
2	1	2	
3	3	7	3
4			
5	1 3	2	3 4
6	3 4 4 6 7 8 8	3 5 7	
7	2 5 7	0 2 3 4 6 8 9	2 4 6
8		2 3	0 2 3 3 4 6 7 8
9			2

Is there evidence to indicate that the hostility levels of students differ in the three school environments?

Solution   The shapes of the boxplots in Figure 10.10 indicate that all distributions are skewed left. Thus, to compare the hostility levels of students from the different school environments, we suggest that a test of equality of the population *medians* be conducted using the Kruskal–Wallis test.

Figure 10.10

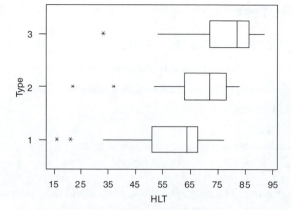

The normality assumption has been violated; hence, the ordinary F-test should not be used. The distributions are similarly skewed, so the Kruskal–Wallis test is suggested.

Let θ_1, θ_2, and θ_3 denote the median hostility levels for students of the three school environments. The null hypothesis is

$$H_0: \theta_1 = \theta_2 = \theta_3 \text{ and the alternative hypothesis is}$$

$$H_a: \text{at least two } \theta_i\text{'s differ}$$

As stated earlier, the form of the Kruskal–Wallis test presented in this text is the ordinary F-test applied to the ranks. From the side-by-side ordered stem-and-leaf plots, the ranks can easily be assigned. Remember that the ranks are assigned to the combined samples, and tied values are given the average of the ranks that normally would have been assigned. In Table 10.7, we denote the ranks for each data value in parentheses.

Table 10.7

	Rural	Suburban	Urban
1	6(1)		
2	1(2)	2(3)	
3	3(4.5)	7(6)	3(4.5)
4			
5	1(7)3(9.5)	2(8)	3(9.5)4(11)
6	3(12.5)4(14.5)4(14.5)6(17)7(18.5)8(20.5)8(20.5)	3(12.5)5(16)7(18.5)	
7	2(24)5(29)7(32)	0(22)2(24)3(26)4(27.5)6(30.5)8(33)9(34)	2(24)4(27.5)6(30.5)
8		2(36.5)3(39)	0(35)2(36.5)3(39)3(39)4(41)6(42)7(43)8(44)
9			2(45)

Figure 10.11 is the complete analysis of variance on the ranks. The p-value is less than 1%, and therefore the null hypothesis is rejected. There is a significant difference in the median hostility scores for the three school environments. ∎

Figure 10.11 One-Way ANOVA: ranks versus type

```
Analysis of Variance for Ranks
Source    DF        SS      MS        F        P
Type       2      2000    1000     7.52    0.002
Error     42      5582     133
Total     44      7582
                                 Individual 95% CIs For Mean
                                 Based on Pooled StDev
Level      N      Mean StDev  ---------+---------+---------+-------
1         15     15.13  9.33  (-------*-------)
2         15     22.43 11.45            (------*-------)
3         15     31.43 13.44                      (------*-------)
                              ---------+---------+---------+-------
Pooled StDev = 11.53                  16.0      24.0      32.0
```

Unlike the ordinary F-test, the Kruskal–Wallis test can be applied to nonnormal distributions. Although no specific distributional shape is required, it is assumed that the distributions are similar in shape. As illustrated in Example 10.6, all three distributions were skewed left. Violations of the assumptions for the ordinary F-test result in a loss of power (the ability to detect significant differences) of the test. In the case of long-tailed distributions (numerous outliers), the loss may be to the point that the test fails to detect differences in populations. The Kruskal–Wallis test, which is based on ranks, is resistant to outliers and thus may detect differences that might otherwise go undetected.

Example 10.7   Three engineering universities wished to compare the salaries of their graduates 10 years after graduation. Seventeen graduates were randomly and independently selected from each of the three universities. Their salaries (in $1000) are recorded in the side-by-side ordered stem-and-leaf plots:

WORKSHEET: Engineer.mtw

	University A	University B	University C
3	0	3	5
4	2 6	9	0
5	0 1 1 4 4 5 6 8 8	3 5 7 7	6
6	0 7 9	0 2 2 4 6 8	0 2 4 4 7 7 9
7		0 1 3 5	0 2 2 8
8	9		0
9		3	5
10	4		1

Is there statistical evidence to indicate that the salaries of the graduates of the three engineering schools differ?

Solution   Because of the apparent symmetry of the distributions, we conduct an ordinary F-test of the equality of the population means. Let μ_1, μ_2, and μ_3 denote the mean salaries 10 years after graduation at universities A, B, and C, respectively. The null hypothesis is

$$H_0: \mu_1 = \mu_2 = \mu_3$$

Having retrieved the data from worksheet: **Engineer.mtw**, we get the complete analysis of variance printout in Figure 10.12. A p-value greater than 10% (0.228) indicates that we should not reject the null hypothesis. Therefore, the ordinary F-test shows no significant difference in mean salaries of the graduates of the three universities.

Figure 10.12   One-Way ANOVA: salary versus university

```
Analysis of Variance for Salary
Source    DF       SS     MS        F         P
universi   2      735    368     1.53     0.228
Error     48    11554    241
Total     50    12289
                              Individual 95% CIs For Mean
                              Based on Pooled StDev
Level      N     Mean   StDev  --------+---------+---------+---------+--------
1         17    58.47   17.08   (----------*---------)
2         17    62.82   12.83            (---------*----------)
3         17    67.76   16.30                    (----------*----------)
                              --------+---------+---------+---------+--------
Pooled StDev = 15.51                  56.0      63.0      70.0
```

The side-by-side boxplots of the data in Figure 10.13, however, show numerous outliers, which indicates that the assumptions for the ordinary F-test have been violated. The distributions appear to be long tailed and similar in shape. We should therefore conduct the Kruskal–Wallis test.

Figure 10.13

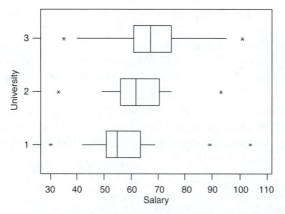

COMPUTER TIP

Ranking Data

There is a Minitab command for the Kruskal–Wallis test; however, the test statistic, though equivalent, is not the same as the one used in this text. If you recall, we simply apply the ordinary F-test to the rank-transformed data. Thus, to perform a Kruskal–Wallis test, we store the data in one column of the worksheet with the levels in a second column, rank the data, and then apply the one-way ANOVA. The data column is ranked by the following procedure: From the **Manip** menu select **Rank**. In the Rank window, select the column to be ranked in the **Rank data in** box, select the column to **Store ranks in**, and click **OK**.

The completed analysis-of-variance table for the Kruskal–Wallis test in Figure 10.14 is obtained by applying the one-way analysis-of-variance procedure to the ranks of the data. The *p*-value is .037, which indicates that H_0 should be rejected. There is statistical evidence that the mean salaries of the three groups of university graduates differ. Because of the long-tailed distributions, the ordinary F-test failed to detect a difference in the means of the three distributions (*p*-value = .228), whereas the Kruskal–Wallis test, the more powerful test in this example, was able to detect the differences. ∎

Figure 10.14 One-Way ANOVA: rank versus university

```
Analysis of Variance for ranks
Source    DF      SS      MS        F         P
universi   2    1413     706      3.52     0.037
Error     48    9625     201
Total     50   11038

                              Individual 95% CIs For Mean
                              Based on Pooled StDev
Level      N    Mean   StDev   -----+---------+---------+---------+--
1         17   19.21   14.64   (--------*--------)
2         17   26.76   13.62           (-------*--------)
3         17   32.03   14.19                   (--------*--------)
                              -----+---------+---------+---------+--
Pooled StDev =        14.16    16.0      24.0      32.0      40.0
```

As in any analysis-of-variance test, if the null hypothesis is rejected, a multiple comparison test should be conducted to further analyze the differences. Tukey's multiple comparison test proceeds just as before, except that it is applied to the ranks.

COMPUTER TIP

Kruskal–Wallis Test

Minitab can perform the standard Kruskal–Wallis test that is found in nonparametric textbooks. Under the **Stat** menu, select **Nonparametrics** and then choose **Kruskal–Wallis**. In the Kruskal–Wallis window, select the **Response** and **Factor** columns and click **OK.**

Figure 10.15 is the output of the Kruskal–Wallis test for Example 10.7. Notice that the p-value $= 0.041$ is very close to the p-value $= 0.037$ that we obtained using the rank-transformed Kruskal–Wallis test.

Figure 10.15   Kruskal–Wallis Test: salary versus university

```
Kruskal-Wallis Test on salary

universi      N      Median     Ave Rank          Z
1            17       55.00        19.2        -2.31
2            17       62.00        26.8         0.26
3            17       67.00        32.0         2.05
Overall      51                    26.0

H = 6.39 DF = 2 P = 0.041
H = 6.40 DF = 2 P = 0.041 (adjusted for ties)
```

It may appear from our discussion that the Kruskal–Wallis test is superior to the ordinary F-test and should be used in most applications. This is far from the truth. The ordinary F-test is a very robust test that can be used in a wide variety of situations. Certainly, if the specified assumptions for the F-test are met, it should be used because, in that case, it is the most powerful test. It is only when the data grossly violate assumptions—specifically, highly skewed or with unusually long tails—that you should resort to the Kruskal–Wallis test.

EXERCISES 10.3   The Basics/Interpreting Computer Output

10.20   What is the Kruskal–Wallis test? Describe how the rank-transformation form of the test is conducted.

10.21   Under what conditions do you choose the Kruskal–Wallis test over the ordinary F-test?

10.22   These summary data were obtained by collectively assigning ranks to three independent samples. Complete an ANOVA table on the ranks for the Kruskal–Wallis test. Include the p-value and state your conclusion.

n	10	10	10	
$\bar{x}_R$	12.45	14.70	19.35	$\bar{x}_G = 15.5$
s_R	8.67	8.43	8.72	

10.23   The following ranks were collectively assigned to three independent samples. Using the descriptive statistics, complete an ANOVA table on the ranks for the Kruskal–Wallis test. Include the p-value.

Sample 1	Sample 2	Sample 3
1	2	5
4	3	9
6	7	12
10	8	18
11	13	20
14	15	22
16	17	23
19	21	24

Descriptive Statistics of Ranks

Variable	N	Mean	Median	TrMean	StDev	SEMean
Sample1	8	10.12	10.50	10.12	6.17	2.18
Sample2	8	10.75	10.50	10.75	6.82	2.41
Sample3	8	16.63	19.00	16.63	7.09	2.51

Variable	Min	Max	Q1	Q3
Sample1	1.00	19.00	4.50	15.50
Sample2	2.00	21.00	4.00	16.50
Sample3	5.00	24.00	9.75	22.75

10.24 Independent random samples of the signal loss per 1000 feet were measured on three types of coaxial cable:

WORKSHEET: Coaxial.mtw

Type A		Type B		Type C	
3	5	3	5 6 9	3	5 6 7 9
4	2 3 4	4	0 1 3 4	4	0 1 2 2 4
4	5 6 6 7 8	4	6 7 8 9	4	5 6 9
5	0 1 4	5	1 3	5	2
5	6	5		5	7
6	4	6	3	6	4
6	8	6	5	6	

a. From the stem-and-leaf plots, comment on the shapes of the distributions. Are they similar?
b. To test the equality of the centers of the distributions, should an ordinary F-test or the Kruskal–Wallis test be conducted?

10.25 Using this partially completed ANOVA table from the data in Exercise 10.24, test the hypothesis that the population medians are equal with the test suggested in part **b** of Exercise 10.24.

One-Way ANOVA: ranks versus cable

```
Analysis of Variance for Ranks
Source      DF       SS       MS        F        P
Cable        2      540
Error       42
Total       44     7574
```

```
                                  Individual 95% CIs For Mean
                                  Based on Pooled StDev
Level      N    Mean   StDev    ----------+---------+---------+------
1         15   27.40   11.76                    (-----------*----------)
2         15   22.67   13.69           (----------*----------)
3         15   18.93   13.29    (-----------*----------)
                                  ----------+---------+---------+------
Pooled StDev =  12.94                18.0       24.0      30.0
```

10.26   Independent random samples of a pollution index were taken from three regions of a county:

WORKSHEET: Region.mtw

West Region		Central Region		East Region	
15	3	15	7	15	3
16	4	16	5 7	16	5 7
17	1 3 6 8	17	0 1 2 5 7 9	17	1 5 6 9
18	0 1 1 2 3 5 9	18	0 2 7 9	18	2 3 4 5 7 9
19	2 4 6	19	4 6	19	2 6
20	0	20		20	0

a. Examine the side-by-side boxplots of the data. Does any boxplot show severe skewness? Does any boxplot show extreme outliers? Do these boxplots suggest that the population standard deviations are homogeneous? Are they similar in shape but possibly centered in different places? What about the normality assumption?

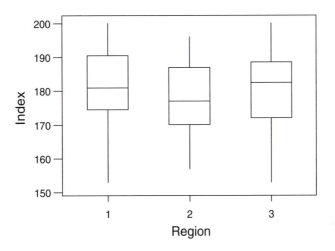

b. Based on your responses to part **a**, are the assumptions for the ordinary F-test satisfied? Are the assumptions for the Kruskal–Wallis test satisfied? Which one is the recommended procedure to test the equality of centers of the three distributions?

10.27   Here are two analysis-of-variance procedures applied to the data in Exercise 10.26. Which one is the ordinary F-test and which one is the Kruskal–Wallis test? Based on your answer to Exercise 10.26, which one is the proper way to analyze these data? Does either test reject the hypothesis that the populations have the same center?

One-Way ANOVA: index versus region

```
Analysis of Variance for Index
Source    DF        SS       MS        F        P
Region     2       115       58     0.42    0.658
Error     45      6150      137
Total     47      6265

                              Individual 95% CIs For Mean
                              Based on Pooled StDev
Level      N      Mean     StDev   --------+---------+---------+--------
1         17    181.06     11.81               (----------*-----------)
2         15    177.40     10.98   (-----------*-----------)
3         16    180.25     12.19         (----------*----------)
                                   --------+---------+---------+--------
Pooled StDev =  11.69                 175.0     180.0     185.0
```

One-Way ANOVA: ranks versus region

```
Analysis of Variance for Ranks
Source    DF      SS      MS        F       P
Region     2     246     123     0.62   0.543
Error     45    8951     199
Total     47    9197

                            Individual 95% CIs For Mean
                            Based on Pooled StDev
Level     N     Mean    StDev   --------+---------+---------+---------
1        17    26.35    13.91                 (----------*-----------)
2        15    21.17    13.84      (-----------*-----------)
3        16    25.66    14.54           (-----------*----------)
                                 --------+---------+---------+---------
Pooled StDev =  14.10              18.0      24.0      30.0
```

10.28 The carbon monoxide level was measured (in parts per million) at three industrial sites at randomly selected times. Is there a significant difference in the carbon monoxide levels at the three sites?

WORKSHEET: Carbon.mtw

Site A	Site B	Site C
.106	.122	.119
.127	.119	.110
.132	.115	.106
.105	.120	.108
.117	.117	.105
.109	.136	.121
.107	.118	.109
.109	.142	.134

a. Here are side-by-side boxplots and normal probability plots of the three samples. Is normality a reasonable assumption in each case? How do you classify the shapes of the distributions? Are the three distributions similar in shape?

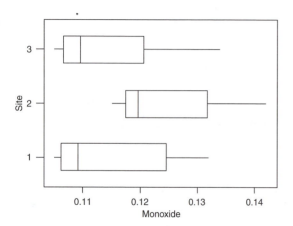

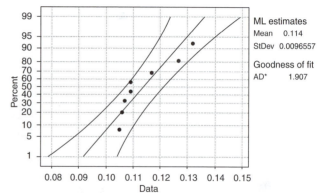

Normal Probability Plot for SiteA
ML estimates - 95% CI

ML estimates
Mean   0.114
StDev 0.0096557

Goodness of fit
AD*   1.907

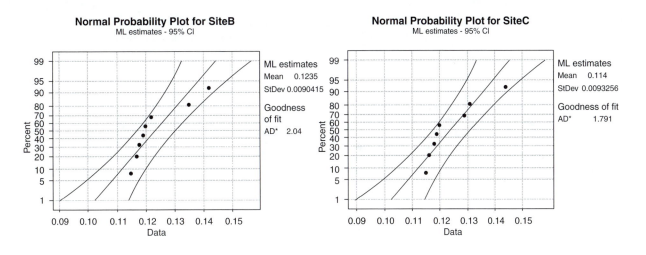

Normal Probability Plot for SiteB
ML estimates - 95% CI

ML estimates
Mean   0.1235
StDev 0.0090415

Goodness
of fit
AD*   2.04

Normal Probability Plot for SiteC
ML estimates - 95% CI

ML estimates
Mean   0.114
StDev 0.0093256

Goodness of fit
AD*   1.791

b. Should the ordinary F-test or Kruskal–Wallis test be used to compare the centers of the three distributions? Explain your reasoning.

10.29 Here are two analysis-of-variance procedures applied to the data in Exercise 10.28. Which one is the ordinary F-test and which one is the Kruskal–Wallis test? Based on your answer to Exercise 10.28, which one is the proper way to analyze these data? Does either test reject the hypothesis that the populations have the same center? For the Kruskal–Wallis test, does the null hypothesis state that the population means are equal or that the population medians are equal? State your conclusion for this problem.

One-Way Analysis of Variance

```
Analysis of Variance on monoxide
Source    DF         SS           MS         F         p
Site       2    0.0004813    0.0002407     2.41     0.114
Error     21    0.0020960    0.0000998
Total     23    0.0025773
                                      Individual 95% CIs For Mean
                                      Based on Pooled StDev
Level      N        Mean        StDev   --------+---------+---------+--------
    1      8     0.11400      0.01032   (---------*---------)
    2      8     0.12350      0.00967                (---------*----------)
    3      8     0.11400      0.00997   (----------*---------)
                                        --------+---------+---------+--------
Pooled StDev =   0.00999                     0.1120    0.1190    0.1260
```

One-Way Analysis of Variance

```
Analysis of Variance on Ranks
Source    DF         SS           MS         F         p
Site       2       244.6        122.3     2.85     0.080
Error     21       901.4         42.9
Total     23      1146.0

                                      Individual 95% CIs For Mean
                                      Based on Pooled StDev
Level      N        Mean        StDev   ---------+---------+---------+------
    1      8       9.938        7.317   (---------*---------)
    2      8      17.000        4.728                (---------*---------)
    3      8      10.563        7.272   (---------*---------)
                                        ---------+---------+---------+------
Pooled StDev =      6.552                    10.0      15.0      20.0
```

10.30 To get extra spending money, students are often eager to sell their used textbooks back to the bookstore at the end of the semester. In one university community, three independent bookstores will buy used books. In an effort to compare the amounts that the bookstores will pay for used books, several students were randomly selected as they exited a bookstore and asked how much they were paid for their used books. The recorded amounts are given in side-by-side stem-and-leaf plots:

WORKSHEET: Bookstor.mtw

Bookstore A		Bookstore B	Bookstore C
1			0
1		6	7
2	3	4	
2	7		6 8
3	1	4	0 0 1 2 3 3 4
3	5 6 6 8 9	5 6 9	5 6 7 7 9
4	0 1 1 2 3 4 4	0 1 1 3 4	0 1 4
4	5 6 6 7 8	5 5 6 7 8 9 9	7
5	0 1 4	1 3	2
5	6		9
6	4	3	4
6	8	5	6

The stem-and-leaf plots suggest that the distributions are long tailed. To compare the centers of the distributions, is the ordinary F-test or the Kruskal–Wallis test recommended?

10.31 Compare the two analysis-of-variance tests that have been applied to the data in Exercise 10.30. Which one is the ordinary F-test and which one is the Kruskal–Wallis test? Using the test you recommended in Exercise 10.30, can we conclude that the mean amount paid for used books is different for the three bookstores? Explain why one test shows a significant difference and the other one does not.

One-Way Analysis of Variance

```
Analysis of Variance on Dollars
Source    DF      SS      MS       F       p
Store      2     573     287     2.19   0.120
Error     69    9023     131
Total     71    9596
```

```
                              Individual 95% CIs For Mean
                              Based on Pooled StDev
Level      N     Mean    StDev   -----+---------+---------+---------+-
    1     26    43.65    10.09                 (--------*--------)
    2     22    43.36    10.90                 (--------*--------)
    3     24    37.54    13.16      (--------*--------)
                                  -----+---------+---------+---------+-
Pooled StDev =  11.44             35.0      40.0      45.0      50.0
```

One-Way Analysis of Variance

```
Analysis of Variance on Ranks
Source    DF      SS      MS       F       p
Store      2    2908    1454     3.56   0.034
Error     69   28153     408
Total     71   31061
```

```
                              Individual 95% CIs For Mean
                              Based on Pooled StDev
Level      N     Mean    StDev   -+---------+---------+---------+-----
    1     26    40.67    19.11                (-------*-------)
    2     22    41.36    19.46                (-------*--------)
    3      4    27.52    21.94     (--------*-------)
                                  -+---------+---------+---------+-----
Pooled StDev =  20.20             20        30        40        50
```

Computer Exercises

10.32 A home heating contractor sells three types of oil heaters. To compare the heating units, these efficiency ratings were obtained on samples of each type of heater:

WORKSHEET: Heating.mtw

Type A	Type B	Type C
75	73	60
71	83	63
74	70	74
86	66	56
77	54	61
84	71	73
76	74	71
75	76	62
57	92	91
96	75	64

a. Construct side-by-side boxplots and normal probability plots of the three samples. Is normality a reasonable assumption in each case? Are there many outliers?

b. Should the ordinary F-test or Kruskal–Wallis test be used to compare the three distributions?

c. Complete the test recommended in part **b** to determine whether there is statistical evidence of a difference in the mean efficiency ratings of the three types of heating units. Be sure to state a conclusion and justify your choice of statistical test.

10.33 The prospective owners of a soon-to-be-built motor lodge wished to evaluate three prospective locations for the business. On randomly selected occasions, they recorded the numbers of vehicles that passed prospective sites in 1-hour periods of time.

WORKSHEET: Lodge.mtw

Site A	Site B	Site C
162	165	165
154	193	160
174	178	155
148	184	168
150	157	140
148	165	151
185	204	163
157	195	175
164	183	182
172	189	150
159	179	139
193	160	164
160	198	181
159	185	176
173	215	177

a. Construct normal probability plots of the three samples. Is normality a reasonable assumption in each case?

b. Construct boxplots of the three samples. Does the variability in the three distributions appear the same?

c. Should the ordinary F-test or Kruskal–Wallis test be used to compare the three distributions?

d. Complete the test recommended in part **c** to determine whether there is a significant difference in the amount of traffic that passes the three points.

10.34 Due to bad debts and risky investments, the U.S. savings and loan industry lost nearly $48 billion from 1987 to 1990. Many of the nation's largest savings and loans (S&Ls) went broke, thus creating a financial crisis for the U.S. government. The Congressional Budget Office estimated it would cost

$215 billion to clean up the mess. Recovery was slow and fragile. Some banks were able to return to profitability on their own; others were taken over by federal regulators. In 1991, S&Ls operating under government control cut losses to $2 billion and those in private hands earned a profit of $1.8 billion. By June 1992, 326 of the nations 2188 S&Ls were still listed as being in financial difficulty. A key measure of the health of a savings and loan company is its problem–asset ratio. Problem assets include bad loans and repossessed property. The ratio measures those assets against tangible capital plus reserves set aside for loan losses. S&Ls with a problem–asset ratio greater than 100% are listed as troubled. A lower number is better, and the national average is 89.91%. Following is the problem-asset ratio for those S&Ls in California, New York, and Texas that were listed as being financially troubled in 1992.

WORKSHEET: Saving.mtw

Calif	NewYork	Texas
103.45	124.23	34033.33
188.29	176.08	417.50
107.55	168.98	416.08
139.72	110.83	564.74
101.15	271.56	127.70
104.98	107.16	162.94
118.80	118.19	534.48
168.02	232.74	335.10
1246.44	163.15	381.59
140.11	120.65	261.91
547.77	187.36	145.67
264.63	207.53	132.29
533.59	278.45	938.13
653.93	474.43	139.62
122.11	123.60	234.63
109.20		367.93
764.61		219.20
112.57		115.70
277.93		141.00
572.66		293.78
378.88		105.41
		100.26
		428.08
		171.92
		812.77
		127.74
		150.91
		131.59
		113.73

a. Construct side-by-side boxplots of the three columns. Describe the boxplots. Can you classify the shapes of the three distributions?

b. Do you recommend the ordinary F-test to compare the means of the three distributions? Why, or why not?

c. Notice that the first value in the Texas column is an extreme outlier. Should the analysis be conducted with or without that observation?

d. Remove the outlier described in part **c** and then construct side-by-side boxplots of the three columns. Can you now classify the shapes of the three distributions?

e. Do you recommend the ordinary F-test to compare the means of the three distributions, or should we compare the medians with the Kruskal–Wallis test?

f. Proceed with the test recommended in part **e**. First analyze with the extreme Texas outlier, and then without it. Does it matter whether the outlier is included?

g. From the analysis in part **f**, determine whether there is a statistically significant difference in the problem-asset ratio for the three states. Be sure to state your conclusion in nonstatistical terms.

10.35 Worksheet: **Toxic.mtw** lists the numbers of hazardous waste sites located in each state. The region of the country for each state is also given.

a. Construct side-by-side boxplots for the number of hazardous waste sites by the region of the country. Compare the variability in the four boxplots. Which region has the least variability in the number of waste sites? Which region has the most variability?

b. Are there any outliers in any region? If so, identify them.

c. Obtain descriptive statistics of the number of sites for each region. Is any standard deviation more than twice as large as any other standard deviation? What does this say about the homogeneous standard deviations assumption?

d. Construct a normal probability plot of the number of sites for each region. Is normality a reasonable assumption in each case?

e. Should the ordinary F-test or the Kruskal–Wallis test be used to compare the four distributions?

f. Complete the ordinary F-test to determine whether there is a significant difference in the mean number of sites in the four regions. Is there a significant difference? State your conclusion.

g. Notice one extreme outlier in the West region. Remove it and conduct the ordinary F-test as in part **f**. What effect did this have on the analysis? Summarize these results.

10.4 Summary and Review Exercises

Key Concepts

- *Analysis of variance* allows one to statistically analyze several population means at one time. *The one-way ANOVA design* is an experiment in which independent random samples are obtained from the several populations. The *total variability* in the data for a one-way design is partitioned into two parts: *within-sample* variability and *between-sample* variability. Within-sample variability is measured by *the sum of squares for error* and has $n_1 + n_2 + \cdots + n_k - k$ degrees of freedom where n_j is the sample size for the sample from the jth population. Between-sample variability is measured by the *sum of squares for treatment* and has $k - 1$ degrees of freedom. Dividing the degrees of freedom into the sum of squares yields a *mean square*. Dividing the mean square for treatments by the mean square for error produces the *F-statistic*, which is used to test the hypothesis of equality of the population means.

- To conduct the F-test, certain assumptions about the populations should be satisfied: the populations should be normally distributed and have homogeneous standard deviations. If the assumptions are not met, the *Kruskal–Wallis* test should be used.

- If an analysis of variance indicates a significant difference in the population means, *Tukey's multiple comparison procedure* is used to further analyze the means.

- If you are using a calculator, these computational formulas for the sum of squares are useful:

$$\text{Total sum of squares} = \text{TSS} = \Sigma\Sigma x_{i,j}^{2} - T^{2}/N$$

$$\text{Sum of squares for treatments} = \text{SST} = \Sigma(T_j^{2}/n_j) - T^{2}/N$$

$$\text{Sum of squares for error} = \text{SSE} = \text{TSS} - \text{SST}$$

where T_j is the total for group j and T is the grand total of all observations. The value $C = T^{2}/N$ appears frequently and is called the *correction term*.

Statistical Insight Revisited

Who is the winner of the Supercar Olympics? The top speeds of five supercars were compared by *Car and Driver* magazine. The data on six runs each are given in worksheet: **Supercar.mtw**.

The side-by-side boxplots shown in Figure 10.1 look reasonably symmetric and have no outliers.

Descriptive Statistics: Acura, Ferrari, Lotus, Porsche, Viper

Variable	N	Mean	Median	TrMean	StDev	SEMean
Acura	6	161.72	161.60	161.72	2.92	1.19
Ferrari	6	178.78	179.00	178.78	3.38	1.38
Lotus	6	162.40	161.75	162.40	3.36	1.37
Porsche	6	176.17	175.35	176.17	4.00	1.63
Viper	6	168.43	169.20	168.43	5.02	2.05

```
Variable     Minimum     Maximum          Q1          Q3
Acura         157.70      166.00      159.20      164.27
Ferrari       173.90      183.90      176.00      181.13
Lotus         158.30      167.40      159.80      165.53
Porsche       171.30      182.40      172.95      179.93
Viper         161.10      174.60      163.42      172.88
```

From these descriptive statistics, we see that no standard deviation is more than twice as large as any other standard deviation, so it is reasonable to accept the homogeneous standard deviation assumption. Furthermore, the normality assumption does not seem out of the question. The ordinary F-test is appropriate for comparing the population means.

The null and alternative hypotheses are

$$H_0: \mu_1 = \mu_2 = \mu_3 = \mu_4 = \mu_5 \text{ versus } H_a: \text{ at least 2 } \mu_i\text{'s are different.}$$

Here is a complete one-way analysis of variance of the data with Tukey's multiple comparison procedure.

One-Way ANOVA: speed versus car

```
Analysis of Variance for speed
Source    DF        SS        MS         F         P
car        4    1456.5     364.1     25.15     0.000
Error     25     362.0      14.5
Total     29    1818.5

                                  Individual 95% CIs For Mean
                                  Based on Pooled StDev
Level     N      Mean     StDev    ----+---------+---------+---------+--
1         6    161.72      2.92    (----*----)
2         6    178.78      3.38                               (---*----)
3         6    162.40      3.36    (----*----)
4         6    176.17      4.00                        (----*---)
5         6    168.43      5.02              (----*---)
                                   ----+---------+---------+---------+--
Pooled StDev =    3.81            161.0     168.0     175.0     182.0

Tukey's pairwise comparisons

     Family error rate =.0500
Individual error rate =.00706

Critical value = 4.15

Intervals for (column level mean) - (row level mean)

                    1          2          3          4
    2         -23.514
              -10.619

    3          -7.131      9.936
               5.764     22.831

    4         -20.897     -3.831    -20.214
              -8.003      9.064     -7.319

    5         -13.164      3.903    -12.481      1.286
              -0.269     16.797      0.414     14.181
```

Based on the F-statistic (25.15) and the p-value $= .000$, there is highly statistical evidence to reject the null hypothesis. The mean speeds of the cars are not the same.

A summary of Tukey's comparison is

$$\mu_1 \qquad \mu_3 \qquad \mu_5 \qquad \mu_4 \qquad \mu_2$$

We see that means 1, 3, and 5 are not significantly different and means 4 and 2 are not significantly different. However, means 2 and 4 are significantly greater than the other three means. In summary, the Acura, Lotus, and Viper are comparable in speed but the Ferrari and Porsche are significantly faster.

Questions for Review

Use the following problems to test your skills.

The Basics

10.36   Describe the basic differences between the pooled t-test and the ordinary F-test.

10.37   Describe the conditions under which you would use each test:
a. The ordinary F-test
b. The Kruskal–Wallis test
c. Tukey's multiple comparison

10.38   Suppose an exploratory analysis of data from four populations suggested that the populations were symmetric with tails that are not excessively long. What mode of analysis would you use: the ordinary F-test or the Kruskal–Wallis test?

10.39   Suppose an exploratory analysis of data from three populations suggested that the populations were similarly skewed. What mode of analysis would you use: the ordinary F-test or the Kruskal–Wallis test?

10.40   From these summary data, construct an analysis of variance table. Compute the p-value associated with the F value and state your conclusion.

n	13	13	13	13
$\bar{x}$	16.5	18.3	14.7	19.5
s	6.72	4.94	4.31	5.82

10.41   This partially completed ANOVA table is for a one-way completely randomized design:

Source	SS	df	MS	F	p-value
Treatment	428	4			
Total	1460	32			

a. How many treatment levels are there?
b. Complete the ANOVA table.
c. Is there a significant difference in the treatment levels?

10.42 From the descriptive statistics for the following data set, construct an analysis-of-variance table. Compute the *p*-value associated with the F value and state your conclusion.

WORKSHEET: GroupABC.mtw

Group A	Group B	Group C
10	10 5	10 6
11 0	11 4	11
12 147	12 2347	12 34
13 03558	13 123588	13 3578
14 136	14 15	14 158
15 02	15 0	15 035
16 4	16	16 0

Descriptive Statistics

Variable	N	Mean	Median	TrMean	StDev	SEMean
GroupA	15	136.60	135.00	136.54	13.58	3.51
GroupB	15	130.53	132.00	131.00	11.80	3.05
GroupC	14	139.14	139.50	140.17	14.58	3.90

Variable	Min	Max	Q1	Q3
GroupA	110.00	164.00	127.00	146.00
GroupB	105.00	150.00	123.00	138.00
GroupC	106.00	160.00	130.75	150.75

10.43 These summary data resulted from an experiment involving a treatment with four levels:

Treatment				
	A1	A2	A3	A4
n	12	12	12	12
Σx_i	135	152	147	184

ANOVA table		
Source	SS	df
Treatment	756.5	3
Error	261.3	44

Complete the ANOVA table and analyze the means using Tukey's multiple comparison procedure with 99% confidence intervals.

10.44 Examine the following data obtained from an experiment involving three treatments.

WORKSHEET: Treatments.mtw

Treatment 1	Treatment 2	Treatment 3
21	41	35
24	44	37
31	38	33
42	37	46
38	42	42
31	48	38
36	39	37
34	32	30

Descriptive Statistics

Variable	N	Mean	Median	TrMean	StDev	SEMean
treat1	8	32.13	32.50	32.13	7.00	2.47
treat2	8	40.13	40.00	40.13	4.82	1.71
treat3	8	37.25	37.00	37.25	5.01	1.77

Variable	Min	Max	Q1	Q3
treat1	21.00	42.00	25.75	37.50
treat2	32.00	48.00	37.25	43.50
treat3	30.00	46.00	33.50	41.00

a. Based on the boxplots, would you say that the distributions are symmetric? Would you rule out normality for any one of the distributions?

b. For all three samples, use the descriptive statistics to calculate the midrange and the midQ and compare them with the mean, median, and trimmed mean. Are they about the same? If so, would you say that the distributions are symmetric?

c. Based on the boxplots, would you say that the standard deviations of the three distributions are homogeneous? Examine the standard deviations given in the descriptive statistics. Do they support your conclusion?

d. Do you think that the assumptions for the ordinary F-test are satisfied? What about the assumptions for the Kruskal–Wallis test?

e. Based on the boxplots, do you think there is significance evidence to reject the hypothesis that the means of the three treatments are the same?

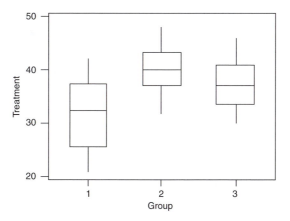

10.45 Independent random samples were selected from three populations:

WORKSHEET: Longtail.mtw

```
 5 |                5 |8           5 |7
 6 |2              6 |3           6 |0
 6 |8              6 |7           6 |
 7 |0              7 |            7 |0 3 4
 7 |6 8            7 |6 8         7 |5 5 6 7 8 8 8 9
 8 |0 1 2 2 3 3 4 4  8 |0 1 3 3 4  8 |0 1 3 4
 8 |5 6 7 9         8 |5 6 7 8 9 9  8 |
 9 |               9 |1 4         9 |3
 9 |8              9 |            9 |8
10 |2             10 |2          10 |0
10 |7             10 |7          10 |
```

a. Use the boxplots and normal probability plots to comment on the shapes of the distributions.
b. To test the equality of the population means, should an ordinary F-test or the Kruskal–Wallis test be conducted? Explain your reasoning.

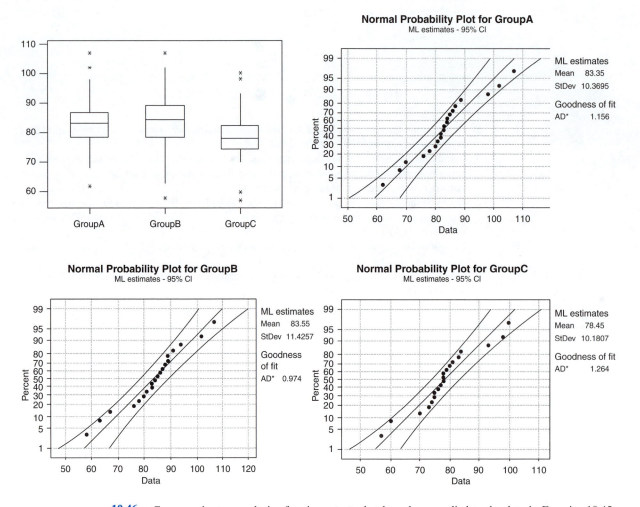

Normal Probability Plot for GroupA
ML estimates - 95% CI

ML estimates
Mean   83.35
StDev 10.3695

Goodness of fit
AD*   1.156

Normal Probability Plot for GroupB
ML estimates - 95% CI

ML estimates
Mean   83.55
StDev 11.4257

Goodness of fit
AD*   0.974

Normal Probability Plot for GroupC
ML estimates - 95% CI

ML estimates
Mean   78.45
StDev 10.1807

Goodness of fit
AD*   1.264

10.46   Compare the two analysis-of-variance tests that have been applied to the data in Exercise 10.45. Which one is the ordinary F-test and which one is the Kruskal–Wallis test? Using the test you recommended in Exercise 10.45, can we conclude that the means of the three populations are different? Explain why one test shows a significant difference and the other one does not.

One-Way ANOVA: score versus group

```
Analysis of Variance for score
Source    DF       SS       MS       F        P
Group      2      334      167    1.39    0.257
Error     57     6834      120
Total     59     7168

                           Individual 95% CIs For Mean
                           Based on Pooled StDev
Level     N     Mean     StDev   ---+---------+---------+---------+---
1        20    83.35     10.64              (---------*---------)
2        20    83.55     11.72              (---------*---------)
3        20    78.45     10.45   (---------*---------)
                                 ---+---------+---------+---------+---
Pooled StDev = 10.95             75.0      80.0      85.0      90.0
```

One-Way ANOVA: ranks versus group

```
Analysis of Variance for Ranks
Source   DF      SS      MS       F       P
Group     2    1693     846    2.97   0.059
Error    57   16265     285
Total    59   17958
```

```
                                Individual 95% CIs For Mean
                                Based on Pooled StDev
Level    N    Mean   StDev   -+---------+---------+---------+-----
1       20   33.23   16.44                     (---------*--------)
2       20   35.20   17.87                       (--------*--------)
3       20   23.08   16.33    (---------*--------)
                             -+---------+---------+---------+-----
Pooled StDev = 16.89         16.0      24.0      32.0      40.0
```

10.47 Radiocarbon dating is a method of determining the age of archaeological sites. The data are the "ages" recorded as years before 1983 (the data were obtained in 1983)—that is, B.C. + 1983—for samples taken from one site at the archaeological excavation of the Danebury Iron Age hill fort. Each of the 60 observations is associated with a pottery shard or fragment that has been classified into one of four phases, referred to as Ceramic Phases 1–4. The phases are thought to be abutting, nonoverlapping periods of stylistically consistent production. Determine whether the mean "ages" for the four phases differ.

WORKSHEET: Archaeo.mtw

phase1	phase2	phase3	phase4
2530	2290	2230	2140
2420	2330	2060	2030
2160	2340	2210	2100
2770	2270	2120	2110
2370	2140	2380	2060
2440	2300	2220	1990
2330	2120	2210	2170
2300	2580	2090	2040
2460	2180	2210	2160
2210		2470	2200
2450		2520	2300
		2330	2060
		2090	2170
		2110	2370
		2250	2120
		2280	2150
		2300	1980
			2260
			2090
			2120
			2000
			2130
			1900

Source: Cunliffe, B. (1984), "Danebury: An Iron-Age Hill Fort in Hampshire," Research Report 2, Council for British Archaeology, London; and Naylor and Smith (1988), "An Archaeological Inference Problem," *Journal of the American Statistical Association, 83,* 588–595.

a. Examine the boxplots. Does it appear that the median ages for the four phases differ? Does anything in the boxplots suggest that the assumptions for the ordinary F-test are not satisfied?

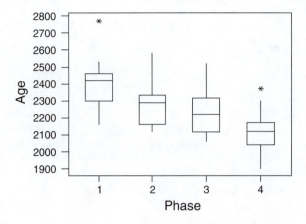

b. Examine the analysis-of-variance table. Is there statistical evidence to reject the null hypothesis that the mean ages are the same? What is the p-value? Based on the confidence intervals, do you think that the periods associated with the four phases are nonoverlapping?

One-Way Analysis of Variance

```
Analysis of Variance for Age
Source    DF       SS        MS        F        p
Phase      3    661865    220622    13.00    0.000
Error     56    950028     16965
Total     59   1611893
                                  Individual 95% CIs For Mean
                                  Based on Pooled StDev
Level     N      Mean     StDev   ---------+---------+---------+-------
phase1   11    2403.6     164.7                         (-----*------)
phase2    9    2283.3     137.7                 (------*-------)
phase3   17    2240.0     131.4              (-----*----)
phase4   23    2115.2     106.8   (---*----)
                                  ---------+---------+---------+-------
Pooled StDev =   130.2                   2160      2280      2400
```

10.48   Examine Tukey's multiple comparison procedure obtained from the analysis of variance in Exercise 10.47. Which pairs of means are not statistically different? What does this say about the periods being nonoverlapping?

Tukey's pairwise comparisons

```
       Family error rate = 0.0500
   Individual error rate = 0.0106

   Critical value = 3.74

   Intervals for (column level mean) - (row level mean)

                      1            2            3

          2        -35
                   275

          3         30          -99
                   297          185

          4        162           33           15
                   415          304          235
```

10.49   Exercise 10.34 in Section 10.3 examined the problem-asset ratio for savings and loan companies in California, New York, and Texas that were listed as being financially troubled in 1992 (worksheet: **Saving.mtw**). An ordinary F-test is applied to the data.

One-Way Analysis of Variance

```
Analysis of Variance for PAR
Source      DF          SS          MS          F        p
State        2     22690724    11345362       0.64    0.532
Error       62    1.103E+09    17784535
Total       64    1.125E+09
                                        Individual 95% CIs For Mean
                                          Based on Pooled StDev
Level       N        Mean       StDev    ----+---------+---------+---------+-
calif       21         322         300         (-----------*-----------)
newyork     15         191          97    (-------------*------------)
texas       29        1452        6270                    (----------*---------)
                                        ----+---------+---------+---------+-
Pooled StDev =        4217             -1500         0      1500      3000
```

Notice that the sample means for the three samples are 322, 191, and 1452, and yet the F-test is insignificant (F = 0.64, p-value = 0.532). With such a large difference between the sample means (the mean for Texas is more than four times larger than the other two), why is the F-test insignificant? Have we violated any assumptions? Explain.

10.50   A psychologist designs an experiment to study the effect of shock treatment on the amount of time one takes to complete a difficult task. Subjects were randomly assigned to three groups. Group 1 received no shock, group 2 received a medium shock, and group 3 received a severe shock. The dependent measure is the number of attempts to complete the task.

WORKSHEET: Shock.mtw

Group 1	Group 2	Group 3
6	11	14
3	9	15
4	7	12
8	14	16
6	10	15
3	9	18
7	13	16
9	11	13
7	10	15

a. State the null and alternate hypotheses to evaluate the effect of shock.
b. What assumptions are necessary for the analysis to be valid?
c. Based on the boxplots, does it appear that the assumptions are satisfied? Do you think shock had a significant effect?

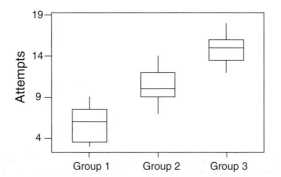

d. Using the accompanying information, complete the ANOVA table and test the hypothesis stated in part **a**.

One-Way Analysis of Variance

```
Analysis of Variance
Source     DF        SS        MS          F        p
Factor      2     364.52
Error      24      98.00
Total      26     462.52
                                   Individual 95% CIs For Mean
                                   Based on Pooled StDev
Level      N      Mean     StDev    --------+---------+---------+--------
Group1     9     5.889     2.147    (---*---)
Group2     9    10.444     2.128                   (---*---)
Group3     9    14.889     1.764                             (---*---)
                                    --------+---------+---------+--------
Pooled StDev =  2.021                   7.0      10.5      14.0
```

10.51   To assess the impact of the level of impurities in a particular ingredient on the solubility of a type of aspirin tablet, a scientist wishes to test the null hypothesis that the mean dissolving time is the same regardless of the impurity level. In test batches, these dissolving times (in seconds) were obtained:

WORKSHEET: Asprin.mtw

Level of impurity		
1%	5%	10%
2.0	1.9	2.3
1.8	2.3	2.3
1.7	2.2	2.2
1.9	1.9	2.1
2.1	2.2	2.6

Descriptive Statistics

Variable	N	Mean	Median	TrMean	StDev	SEMean
1%	5	1.9000	1.9000	1.9000	0.1581	0.0707
5%	5	2.1000	2.2000	2.1000	0.1871	0.0837
10%	5	2.3000	2.3000	2.3000	0.1871	0.0837

Variable	Min	Max	Q1	Q3
	1.7000	2.1000	1.7500	2.0500
5%	1.9000	2.3000	1.9000	2.2500
10%	2.1000	2.6000	2.1500	2.4500

a. State the null and alternative hypotheses for an ANOVA test.
b. Does it appear that the assumptions for the F-test are satisfied?
c. Complete an ANOVA table.
d. Test the hypothesis that there is no difference in the dissolving time for the different impurity levels.

Computer Exercises

10.52   Independent random samples were selected from three populations:

WORKSHEET: Median.mtw

0		0		0	0 1
0		0	5 6 7	0	5 6 8
1	1 2 4 4	1	0 2 4 4	1	0 1 2 3
1	5 6 7 8 9	1	5 6 6 9	1	5 6 8
2	0 2 3	2	0 3	2	4
2	5	2	7	2	5
3	0	3	2	3	1
3	5	3		3	

a. Comment on the shapes of the distributions.
b. To test the equality of the medians of the distributions, should an ordinary F-test or the Kruskal–Wallis test be conducted?
c. State the null and alternative hypotheses and complete the test recommended in part **b**.

10.53   To compare the efficiency of the pit crews of major NASCAR teams, the durations of pit stops were measured for three top crews. These are the times (in seconds) for 12 randomly selected pit stops:

WORKSHEET: NASCAR.mtw

Team A	Team B	Team C
25	25	30
22	30	35
18	24	32
30	26	26
24	22	37
15	15	43
40	32	36
23	46	40
10	20	35
20	28	25
45	35	55
25	25	33

a. Complete side-by-side boxplots and comment on the shapes of the distributions. Are the tails of the distributions unusually long? What about symmetry?
b. Construct normal probability plots of the three samples. Do you detect any significant departures from normality?
c. To test the equality of the centers of the distributions, should an ordinary F-test or the Kruskal–Wallis test be conducted?

10.54   Using the test that you recommended in part **c** of Exercise 10.53, perform an analysis to determine whether a significant difference exists between the times it takes for a pit stop for the three crews.

10.55   The delay times (in minutes) were recorded on 20 flights selected randomly from each of four major air carriers:

WORKSHEET: Delay.mtw

Carrier A	Carrier B	Carrier C	Carrier D
20	15	20	25
14	17	27	17
12	10	22	10

continued

WORKSHEET: Delay.mtw
continued

Carrier A	Carrier B	Carrier C	Carrier D
20	36	35	5
17	18	26	22
30	20	24	35
19	5	15	19
7	16	17	24
22	20	10	3
18	13	25	20
10	42	45	15
15	15	20	40
13	8	16	16
5	17	12	10
19	10	5	9
25	4	21	19
45	19	32	45
14	25	23	15
40	12	10	5
10	5	15	10

a. Construct normal probability plots and boxplots of the four samples.
b. Based on the results of part **a**, should the analysis of the data be an ordinary F-test or the Kruskal–Wallis test?

10.56   Using the test you recommended in part **b** of Exercise 10.55, perform an analysis to determine whether a significant difference exists between the delay times for the four air carriers.

10.57   Perform Tukey's multiple comparison procedure on the means you found in Exercise 10.56, unless you think one is not really needed. If not, why not?

10.58   A study was conducted to evaluate three treatments for arthritis. Arthritis sufferers were randomly assigned to three groups. After their treatment, the times until they experienced relief were measured:

WORKSHEET: Arthriti.mtw

Treatment A	Treatment B	Treatment C
40	73	50
35	32	75
47	47	34
52	52	47
31	34	87
61	60	45
92	77	38
46	42	25
50	20	86
49	81	39
93	75	42
84	35	30
72	25	75
43	40	36
80	90	90
85	33	32
30	30	89

a. Construct normal probability plots and boxplots of the three samples. Comment on the shapes of the distributions.

b. Based on the results of part **a**, should the analysis of the data be an ordinary F-test or the Kruskal–Wallis test?

10.59 Using the test that you recommended in part **b** of Exercise 10.58, perform an analysis to determine whether a significant difference exists among the relief times for the three treatments for arthritis.

10.60 These are the median costs of appendectomies at hospitals across the state of North Carolina in 1992. Each hospital has been classified as a rural, regional, or metropolitan hospital.

WORKSHEET: Appendec.mtw

Rural (1)

| 3821 | 3981 | 3931 | 5582 | 4591 | 3840 | 4053 | 5104 | 4673 | 3935 | 5442 | 5159 | 2861 | 5012 | 4891 |
| 4887 | 3597 | 4210 | 4175 | 3628 | 4519 | 5693 | 8000 | 6439 | 3987 | 3997 | 2671 | 4683 | 4896 |

Regional (2)

| 5498 | 6046 | 4775 | 4844 | 4026 | 4347 | 6389 | 2659 | 4072 | 3441 | 3677 | 5299 | 3822 | 5119 | 4071 |
| 4336 |

Metropol (3)

| 4257 | 6163 | 6266 | 2478 | 2251 | 5143 | 4532 | 4212 | 3556 | 3362 | 4508 | 3366 | 5506 | 4950 |

Source: Consumer's Guide to Hospitalization Charges in North Carolina Hospitals (August 1994), North Carolina Medical Database Commission, Department of Insurance.

a. From the side-by-side boxplots, does it appear that the assumptions for an F-test are satisfied?

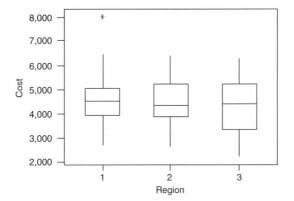

b. Construct normal probability plots for each of the three samples. Is normality rejected in any case?

c. Obtain descriptive statistics for each sample. Is any standard deviation more than twice as large or larger than any other standard deviation? Is the homogeneous standard deviations assumption satisfied?

d. State the null hypothesis that there is no difference in the costs at the three types of hospitals.

e. Test the hypothesis stated in part **d**. Be sure to state your conclusion.

References

Andrews, D. F., and A. M. Herzberg. *Data—A Collection of Problems from Many Fields for the Student and Research Worker.* New York: Springer-Verlag, 1985.

Bellout, G. "Is Vitamin C Really Good for Colds?" *Consumer Reports*, February 1976, pp. 66–70.

Box, G. E. P. "Some Theorems on Quadratic Forms Applied in the Study of Analysis of Variance Problems, I. Effect of Inequality of Variance in the One-Way Classification." *Annals of Mathematical Statistics*, 25 (1954), pp. 290–302.

Box, G. E. P., G. Hunter, and J. S. Hunter. *Statistics for Experimentation.* New York: Wiley, 1978.

Box, G. E. P., and G. M. Jenkins. *Time Series Analysis, Forecasting and Control.* San Francisco: Holden-Day, 1970.

Chambers, J. M., W. S. Cleveland, B. Kleiner, and P. A. Tukey. *Graphical Methods for Data Analysis.* Boston: Duxbury Press and Belmont, Calif.: Wadsworth, 1983.

Cochran, W. G. "Some Methods for Strengthening the Common χ^2 Tests," *Biometrics*, 10 (1954), pp. 417–451.

Conover, W. J. *Practical Nonparametric Statistics*, 2d ed. New York: Wiley, 1980.

Conover, W. J. and R. L. Iman. "Rank Transformation as a Bridge Between Parametric and Nonparametric Statistics," *The American Statistician*, 35, no. 3 (1981).

Cunliffe, B. *Danebury: An Iron-Age Hill Fort in Hampshire.* Research Report 2. London: Council for British Archaeology, 1984.

Daniel, C., and F. Wood. *Fitting Equations to Data.*, 2d ed. New York: Wiley, 1980.

Davis, J. *Statistics and Data Analysis in Geology*, 2d ed. New York: Wiley, 1986.

Dawkins, B. "Multivariate Analysis of National Track Records," *The American Statistician*, 43, no. 2 (1989), pp. 110–115.

Famighetti, R., ed. *The World Almanac & Book of Facts—2000.* New Jersey: Primedia Reference Inc., 1999.

Fienberg, S. E. "Randomization and Social Affairs: The 1970 Draft Lottery," *Science*, 171 (January 22, 1971), pp. 255–261.

Freedman, D., R. Pisani, R. Purves, and A. Adhikari. *Statistics*, 2d ed. New York: W. W. Norton, 1991.

Freeh, L. J., director. *Crime in the United States—Uniform Crime Reports for the United States.* Washington, D.C.: Federal Bureau of Investigation, U.S. Department of Justice, 2000.

Gnanadesikan, R., ed. *Statistical Data Analysis—Proceedings of Symposia in Applied Mathematics—Volume 28.* Providence, R.I.: American Mathematical Society, 1983.

Greenhouse, J. B., and S. M. Greenhouse. "An Aspirin a Day . . . ?" *Chance*, 1, no. 4 (Fall 1988), pp. 24–31.

Gross, A. M. "Confidence Interval Robustness with Long-Tailed Symmetric Distributions," *Journal of the American Statistical Association*, 71, no. 354 (1976).

Hamilton, T. H., I. Rubinoff, R. H. Barth, Jr., and G. L. Bush. "Species Abundance: Natural Regulation of Insular Variation," *Science*, 142 (1963), pp. 1575–1577.

Hand, D. J., et al., eds. *A Handbook of Small Data Sets.* London: Chapman and Hall, 1994.

Hartwig, F., and B. E. Dearing. *Exploratory Data Analysis.* Beverly Hills, Calif.: Sage Publications, 1979.

Hoaglin, D. C., F. Mosteller, and J. W. Tukey. *Understanding Robust and Exploratory Data Analysis.* New York: Wiley, 1983.

Hooke, R. "Basketball, Baseball, and the Null Hypothesis," *Chance*, 2, no. 4 (1989), pp. 35–37.

Hora, S. C., and W. J. Conover. "The *F* Statistic in the Two-Way Layout with Rank-Score Transformed Data," *Journal of the American Statistical Association*, 79, no. 387 (1984).

Huber, P. J. "Robust Statistics: A Review," *The Annals of Mathematical Statistics*, 43, no. 4 (1972).

Iman, R. L., S. C. Hora, and W. J. Conover, "Comparison of Asymptotically Distribution-Free Procedures for the Analysis of Complete Blocks," *Journal of the American Statistical Association*, 79, no. 387 (1984).

Ito, P. K. "Robustness of ANOVA and MANOVA Test Procedures," *Handbook of Statistics*, vol. 1, P. R. Krishnaiah, ed. Amsterdam: North-Holland, 1980, pp. 199–236.

Joglekar, G., J. H. Schuenemeyer, and V. LaRiccia. "Lack-of-Fit Testing When Replicates Are Not Available," *The American Statistician*, 43, no. 3 (1989), pp. 135–143.

Kalbfleisch, J., and J. Lawless. "An Analysis of the Data on Transfusion-Related AIDS," *Journal of the American Statistical Association*, 84 (1989), pp. 360–372.

Kaufman, P., A. Herlihy, J. Elwood, J. Mitch, W. Overton, M. Sale, J. Messer, K. Cougan, D. Pech, K. Reckhow, A. Kinney, S. Christie, D. Brown, C. Hagley, and H. Jager. *Chemical Characteristics of Streams in the Mid-Atlantic and Southeastern U.S., Vol. I: Population Descriptions and Physico-Chemical Relationships*, EPA/600/3088/021a. Washington, D.C.: U.S. Environmental Protection Agency, 1988.

Kempthorne, O. "Teaching of Statistics: Content Versus Form," *The American Statistician*, 31, no. 1 (1980).

Koopmans, L. H. *An Introduction to Contemporary Statistics.* Boston: Duxbury Press, 1981.

McEntire, A., and A. N. Kitchens, "A New Focus for Educational Improvement Through Cognitive and Other Structuring of Subconscious Personal Axioms," *Education*, 105, no. 2 (1984).

Miller, J. C., and J. N. Miller. *Statistics for Analytical Chemistry*, 2d ed. New York: Halsted Press, 1988.

Moser, B. K., and G. R. Stevens. "Homogeneity of Variances in the Two-Sample Means Test," *The American Statistician*, 46 (1992), pp. 19–21.

Mosteller, F. "The Teaching of Statistics: Classroom and Platform Performance," *The American Statistician*, 34, no. 1 (1980).

Mosteller, F., and J. W. Tukey. *Data Analysis and Regression, A Second Course in Statistics.* Reading, Mass.: Addison-Wesley, 1977.

Naylor, J., and A. Smith. "An Archaeological Inference Problem," *Journal of the American Statistical Association*, 83 (1988), pp. 588–595.

Neter, J., M. H. Kutner, C. J. Nachtsheim, and W. Wasserman. *Applied Linear Statistical Models*, 4th ed. Chicago: Irwin, 1996.

Noether, G. E. "The Role of Nonparametrics in Introductory Statistics Courses," *The American Statistician*, 34, no. 1 (1980).

Rousseeuw, P., and A. Leroy. *Robust Regression and Outlier Detection*, New York: Wiley, 1987.

Sacks, J., and D. Ylvisaker, "A Note on Huber's Robust Estimation of a Location Parameter," *The Annals of Mathematical Statistics*, 43, no. 4 (1972).

Schmoyer, R. "Permutation Tests for Correlation in Regression Errors," *Journal of the American Statistical Association*, 89 (1994), pp. 1507–1516.

Soofi, E., N. Ebrahimi, and M. Habibullah. "Information Distinguishability with Application to Analysis of Failure Data," *Journal of the American Statistical Association*, 90 (1995), pp. 657–668.

Staudte, R., and S. Sheather. *Robust Estimation and Testing.* New York: Wiley, 1990.

Stigler, S. M. "Do Robust Estimators Work with 'Real' Data?" *The Annals of Statistics*, 5, no. 6 (1977).

Sutton, G. W., T. J. Huberty, and R. Price. "An Evaluation of Teacher Stress and Job Satisfaction," *Education*, 105, no. 2 (1984).

Tanur, J., et al., eds. *Statistics: A Guide to the Unknown*, 3d ed. San Francisco: Holden-Day, 1989.

Tong, H. *Threshold Models in Nonlinear Time Series Analysis.* New York: Springer-Verlag, 1983.

Tukey, J. W. *Exploratory Data Analysis.* Reading, Mass.: Addison-Wesley, 1977.

Tukey, J. W., and D. H. McLaughlin. "Less Vulnerable Confidence and Significance Procedures for Location Based on a Single Sample: Trimming/Winsorization 3," *Sankhyā Series A*, 25 (1963), pp. 331–352.

Velleman, P. F., and D. C. Hoaglin. *Applications, Basics, and Computing of Exploratory Data Analysis.* North Scituate, Mass.: Duxbury Press, 1981.

vos Savant, M. "Ask Marilyn," *Parade*, September 9, 1990, p. 15.

vos Savant, M. "Ask Marilyn," *Parade*, December 2, 1990, p. 25.

vos Savant, M. "Ask Marilyn," *Parade*, February 17, 1991, p. 12.

vos Savant, M. "Ask Marilyn," *Parade*, July 7, 1991, p. 28.

Wade, N. "IQ and Heredity: Suspicion of Fraud Beclouds Classic Experiment," *Science*, 194 (1976), pp. 916–919.

Wheeler, M. *Lies, Damn Lies, and Statistics.* New York: Liveright, 1976.

Wilner, D. M., et al. *The Housing Environment and Family Life.* Baltimore: Johns Hopkins Press, 1962.

Yashchin, E. "Likelihood Ratio Methods for Monitoring Parameters of a Nested Random Effect Model," *Journal of the American Statistical Association*, 90 (1995), pp. 729–738.

Ying, Z., S. Jung, and L. Wei. "Survival Analysis with Median Regression Models," *Journal of the American Statistical Association*, 90 (1995), pp. 178–184.

Yuen, K., and W. J. Dixon. "The Approximate Behavior and Performance of the Two-Sample Trimmed *t.*" *Biometrika*, 60, no. 2 (1973), p. 369.

Data Sets

ABBEY.MTW	Exercise 6.39	Daily price returns of Abbey National shares
ABC.MTW	Exercise 10.1	Three samples to illustrate analysis of variance
ABILENE.MTW	Exercise 1.23	Crimes reported in Abilene, Texas
	Exercise 2.79	
ABILITY.MTW	Exercise 8.57	Perceived math ability for 13-year-olds by gender
ABORTION.MTW	Exercise 8.51	Abortion rate by region of country
ABSENT.MTW	Exercise 1.28	Number of absent days for 20 employees
ACHIEVE.MTW	Example 7.14	Math achievement test scores by gender for 25 high school students
	Exercise 10.7	
ADSALES.MTW	Exercise 9.15	Number of ads versus number of sales for a retailer of satellite dishes
AGGRESS.MTW	Exercise 1.66	Aggressive tendency scores for a group of teenage members of a street gang
	Exercise 1.81	
AID.MTW	Exercise 1.91	Monthly payments per person for families in the AFDC federal program
	Exercise 3.68	
AIDS.MTW	Exercise 6.60	Incubation times for 295 patients thought to be infected with HIV by a blood transfusion
AIRDISASTERS.MTW	Exercise 1.12	Aircraft disasters in five different decades
AIRLINE.MTW	Example 2.9	Percentage of on-time arrivals and number of complaints for 11 airlines
ALCOHOL.MTW	Exercise 5.79	Ages at which 14 female alcoholics began drinking
ALLERGY.MTW	Exercise 8.22	Allergy medicines by adverse events
ANESTHET.MTW	Exercise 5.58	Recovery times for anesthetized patients
ANXIETY.MTW	Exercise 2.96	Math test scores versus anxiety scores before the test
APOLIPOP.MTW	Example 9.2	Level of apolipoprotein B and number of cups of coffee consumed per day for 15 adult males
	Example 9.9	
APPEND.MTW	Exercise 1.119	Median costs of an appendectomy at 20 hospitals in North Carolina
APPENDEC.MTW	Exercise 10.60	Median costs of appendectomies at three different types of North Carolina hospitals
APTITUDE.MTW	Exercise 2.1	Aptitude test scores versus productivity in a factory
	Exercise 2.26	
	Exercise 2.35	
	Exercise 2.51	
ARCHAEO.MTW	Example 1.16	Radiocarbon ages of observations taken from an archaeological site
	Exercise 5.120	
	Exercise 10.47	
ARTHRITI.MTW	Exercise 10.58	Time of relief for three treatments of arthritis
ARTIFICI.MTW	Exercise 1.107	Duration of operation for 15 artificial heart transplants
ASPRIN.MTW	Exercise 10.51	Dissolving time versus level of impurities in aspirin tables
ASTHMATI.MTW	Exercise 7.52	Asthmatic relief index on nine subjects given a drug and a placebo
ATTORNEY.MTW	Example 2.2	Number of convictions reported by U.S. attorney's offices
	Exercise 2.43	
	Exercise 2.57	
AUTOGEAR.MTW	Exercise 7.46	Number of defective auto gears produced by two manufacturers
BACKTOBACK.MTW	Exercise 7.40	Illustrates inferences based on pooled *t*-test versus Wilcoxon rank sum test
BBSALARIES.MTW	Exercise 1.11	Baseball salaries for members of five major league teams

BIGTEN.MTW	Exercise 1.124 Exercise 2.94	Graduation rates for student athletes and nonathletes in the Big Ten Conference
BIOLOGY.MTW	Exercise 1.49	Test scores on first exam in biology class
BIRTH.MTW	Example 1.10	Live birth rates in 1990 and 1998 for all states
BLACKEDU.MTW	Exercise 8.55	Education level of blacks by gender
BLOOD.MTW	Exercise 7.84	Blood pressure of 15 adult males taken by machine and by an expert
BOARD.MTW	Exercise 10.14	Incomes of board members from three different universities
BONES.MTW	Example 7.22	Bone density measurements of 35 physically active and 35 nonactive women
BOOKS.MTW	Exercise 9.53	Number of books read and final spelling scores for 17 third-graders
BOOKSTOR.MTW	Exercise 10.30–31	Prices paid for used books at three different bookstores
BRAIN.MTW	Example 2.3 Example 2.20 Exercise 2.15 Exercise 2.44 Exercise 2.58	Brain weight versus body weight of 28 animals
BUMPERS.MTW	Exercise 1.73	Repair costs of vehicles crashed into a barrier at five miles per hour
BUS.MTW	Exercise 8.25	Attendance of bus drivers versus type of shift
BYPASS.MTW	Exercise 5.104 Exercise 6.43	Median charges for coronary bypass at 17 hospitals in North Carolina
CABINETS.MTW	Exercise 7.83	Estimates of costs of kitchen cabinets by two suppliers on 20 prospective homes
CANCER.MTW	Exercise 6.55 Exercise 6.64	Survival times of terminal cancer patients treated with vitamin C
CARBON.MTW	Exercise 10.28–29	Carbon monoxide level measured at three industrial sites
CAT.MTW	Exercise 1.116	Reading scores on the California achievement test for a group of third-graders
CENSORED.MTW	Exercise 7.34 Exercise 7.48	Entry age and survival time of patients with small cell lung cancer under two different treatments
CHALLENG.MTW	Example 1.11–13 Example 2.11 Example 5.1	Temperatures and O-ring failures for the launches of the space shuttle Challenger
CHEMIST.MTW	Example 5.3	Starting salaries of 50 chemistry majors
CHESAPEA.MTW	Exercise 6.46	Surface salinity measurements taken offshore from Annapolis, Maryland in 1927
CHEVY.MTW	Exercise 8.35	Insurance injury ratings of Chevrolet vehicles for 1990 and 1993 models
CHICKEN.MTW	Exercise 10.15	Weight gain of chickens fed three different rations
CHIPAVG.MTW	Exercise 6.49 Exercise 7.47	Measurements of the thickness of the oxide layer of manufactured integrated circuits
CHIPS.MTW	Exercise 10.9	Four measurements on a first wafer and four measurements on a second wafer selected from 30 lots
CIGAR.MTW	Example 10.4	Milligrams of tar in 25 cigarettes selected randomly from four different brands
CIGARETT.MTW	Exercise 2.27	Effect of mother's smoking on birth weight of newborn
CITRUS.MTW	Exercise 9.7	Percent of peak bone density of different aged children
CLEAN.MTW	Exercise 10.16	Residual contaminant following the use of three different cleansing agents
COAXIAL.MTW	Exercise 10.24–25	Signal loss from three types of coaxial cable
COFFEE.MTW	Exercise 7.55	Productivity of workers with and without a coffee break
COINS.MTW	Exercise 5.68	Yearly returns on 12 investments
COMMUTE.MTW	Exercise 1.13 Exercise 1.121 Exercise 7.85	Commuting times for selected cities in 1980 and 1990
CONCEPT.MTW	Exercise 1.68 Exercise 1.82	Tennessee self-concept scale scores for a group of teenage boys
CONCRETE.MTW	Example 7.17	Compressive strength of concrete blocks made by two different methods
CORN.MTW	Exercise 7.77	Comparison of the yields of a new variety and a standard variety of corn planted on 12 plots of land

CORRELAT.MTW	Exercise 2.23	Exercise to illustrate correlation
COUNSEL.MTW	Exercise 6.96	Scores of 18 volunteers who participated in a counseling process
CPI.MTW	Exercise 1.34	Consumer price index from 1970 to 1998
CRIME.MTW	Exercise 1.90	Violent crime rates for the states in 1983 and 1993
	Exercise 2.30	
	Exercise 2.32	
	Exercise 3.64	
	Exercise 5.113	
DARWIN.MTW	Exercise 7.62	Charles Darwin's study of cross-fertilized and self-fertilized plants
DEALERS.MTW	Example 2.22	Automobile dealers classified according to type dealership and service rendered to customers
DEALERS.XLS	Example 2.22	Excel version of worksheet dealers
DEFECTIV.MTW	Exercise 1.27	Number of defective items produced by 20 employees
DEGREE.MTW	Exercise 2.75	Percent of bachelor's degrees awarded women in 1970 versus 1990
DELAY.MTW	Exercise 10.55	Delay times on 20 flights from four major air carriers
DEPEND.MTW	Exercise 1.26	Number of dependent children for 50 families
DETROIT.MTW	Exercise 5.21	Educational levels of a sample of 40 auto workers in Detroit
DEVELOP.MTW	Exercise 8.5	Demographic characteristics of developmental students.
DEVMATH.MTW	Exercise 6.47	Test scores for students who failed developmental mathematics in the fall semester 1995
DICE.MTW	Exercise 3.109	Outcomes and probabilities of the roll of a pair of fair dice
DIESEL.MTW	Exercise 2.8	Diesel fuel prices in 1999–2000 in nine regions of the country
DIPLOMAT.MTW	Exercise 1.14	Parking tickets issued to diplomats
	Exercise 1.37	
DISPOSAL.MTW	Exercise 1.127	Toxic intensity for plants producing herbicidal preparations
DOGS.MTW	Exercise 2.88	Rankings of the favorite breeds of dogs
DOMESTIC.MTW	Exercise 1.20	Rates of domestic violence per 1000 women by age groups
DOPAMINE.MTW	Exercise 5.14	Dopamine b-hydroxylase activity of schizophrenic patients treated with an antipsychotic drug
	Exercise 7.49	
DOWJONES.MTW	Exercise 1.35	Closing year-end Dow Jones Industrial averages from 1896 through 2000
DRINK.MTW	Exercise 8.53	Opinion on referendum by view on moral issue of selling alcoholic beverages
DRUG.MTW	Example 7.15	Number of trials to master a task for a group of 28 subjects assigned to a control and an experimental group
DYSLEXIA.MTW	Exercise 2.90	Data on a group of college students diagnosed with dyslexia
EARTHQK.MTW	Exercise 6.97	One hundred year record of worldwide seismic activity (1770–1869)
EDUCAT.MTW	Exercise 2.41	Crime rates versus the percent of the population without a high school degree
EGGS.MTW	Exercise 9.22	Number of eggs versus amounts of feed supplement
ELDERLY.MTW	Exercise 1.92	Percent of the population over the age of 65
	Exercise 2.61	
ENERGY.MTW	Exercise 2.5	Amount of energy consumed by homes versus their sizes
	Exercise 2.24	
	Exercise 2.55	
ENGINEER.MTW	Example 10.7	Salaries after ten years for graduates of three different universities
ENTRANCE.MTW	Example 1.8	College entrance exam scores for 24 high school seniors
EPAMINICOMPAC.MTW	Exercise 1.65	Fuel efficiency ratings for compact vehicles in 2001
EPATWOSEATER.MTW	Exercise 5.8	Fuel efficiency ratings for two-seater vehicles in 2001
EXECUTIV.MTW	Exercise 1.104	Ages of 25 executives
EXERCISE.MTW	Exercise 1.44	Weight loss for 30 members of an exercise program
FABRIC.MTW	Example 7.21	Measures of softness of ten different clothing garments washed with and without a softener
FAITHFUL.MTW	Exercise 5.12	Waiting times between successive eruptions of the Old Faithful geyser
	Exercise 5.111	
FAMILY.MTW	Exercise 2.89	Size of family versus cost per person per week for groceries
FERRARO1.MTW	Exercise 8.23	Choice of presidential ticket in 1984 by gender
FERRARO2.MTW	Exercise 8.23	Choice of vice presidential candidate in 1984 by gender

FERTILITY.MTW	Exercise 1.125	Fertility rates of all 50 states and DC
FIRSTCHI.MTW	Exercise 5.11	Ages of women at the birth of their first child
FISH.MTW	Exercise 5.83	Length and number of fish caught with small and large mesh codend
	Exercise 5.119	
	Exercise 7.29	
FITNESS.MTW	Exercise 7.71	Number of sit-ups before and after a physical fitness course
FLORIDA2000.MTW	Stat Insight Ch 2	Florida voter results in the 2000 presidential election
FLUID.MTW	Exercise 5.76	Breakdown times of an insulating fluid under various levels of voltage stress
FOOD.MTW	Exercise 5.106	Annual food expenditures for 40 single households in Ohio
FRAMINGH.MTW	Exercise 1.55	Cholesterol values of 62 subjects in the Framingham Heart Study
	Exercise 1.75	
	Exercise 3.69	
	Exercise 5.60	
FRESHMAN.MTW	Exercise 6.53	Ages of a random sample of 30 college freshmen
FUNERAL.MTW	Exercise 8.54	Cost of funeral by region of country
GALAXIE.MTW	Example 5.2	Velocities of 82 galaxies in the Corona Borealis region
GALLUP.MTW	Exercise 2.76	Results of a Gallup poll on possession of marijuana as a criminal offense conducted in 1980
GASOLINE.MTW	Exercise 1.45	Price of regular unleaded gasoline obtained from 25 service stations
GERMAN.MTW	Exercise 7.60	Number of errors in copying a german passage before and after an experimental course in German
GOLF.MTW	Exercise 5.24	Distances a golf ball can be driven by 20 professional golfers
GOVERNOR.MTW	Exercise 5.112	Annual salaries for state governors in 1994
GPA.MTW	Example 2.13	High school GPA versus college GPA
GRADES.MTW	Exercise 1.120	Test grades in a beginning statistics class
GRADUATE.MTW	Exercise 1.118	Graduation rates for student athletes in the Southeastern Conference
GREENRIV.MTW	Exercise 6.57	Varve thickness from a sequence through an Eocene lake deposit in the Rocky Mountains
GRNRIV2.MTW	Exercise 6.45	Thickness of a varved section of the Green River oil shale deposit near a major lake in the Rocky Mountains
GROUPABC.MTW	Exercise10.42	Group data to illustrate analysis of variance
GROUPS.MTW	Exercise 10.4	An illustration of analysis of variance
GYM.MTW	Exercise 2.21	Children's age versus number of completed gymnastic activities
	Exercise 9.14	
	Exercise 9.32	
HABITS.MTW	Exercise 7.57	Study habits of students in two matched school districts
HAPTOGLO.MTW	Example 6.9	Haptoglobin concentration in blood serum of eight healthy adults
HARDWARE.MTW	Example 2.18	Daily receipts for a small hardware store for 31 working days
HARDWOOD.MTW	Exercise 9.33	Tensile strength of Kraft paper for different percentages of hardwood in the batches of pulp
HEAT.MTW	Exercise 1.29	Primary heating sources of homes on indian reservations versus all households
HEATING.MTW	Exercise 10.32	Fuel efficiency ratings for three types of oil heaters
HODGKIN.MTW	Exercise 2.77	Results of treatments for Hodgkin's disease
HOMES.MTW	Stat Insight Ch 5	Median prices of single-family homes in 65 metropolitan statistical areas
HOMEWORK.MTW	Exercise 7.78	Number of hours per week spent on homework for private and public high school students
HONDA.MTW	Stat Insight Ch 6	Miles per gallon for a Honda Civic on 35 different occasions
HOSTILE.MTW	Example 10.6	Hostility levels of high school students from rural, suburban, and urban areas
HOUSING.MTW	Exercise 5.82	Median home prices for 1984 and 1993 in 37 markets across the United States
HURRICAN.MTW	Example 1.6	Number of storms, hurricanes, and El Niño effects from 1950 through 1995
	Exercise 1.38	
	Exercise 10.19	

ICEBERG.MTW	Exercise 2.46	Number of icebergs sighted each month south of Newfoundland and south
	Exercise 2.60	of the Grand Banks in 1920
INCOME.MTW	Exercise 1.33	Percent change in personal income from 1st to 2nd quarter in 2000
INDEPENDENT.MTW	Exercise 7.41	Illustrates a comparison problem for long-tailed distributions
INDIAN.MTW	Exercise 2.95	Educational attainment versus per captia income and poverty rate for
		American Indians living on reservations
INDIAPOL.MTW	Exercise 1.128	Average miles per hour for the winners of the Indianapolis 500 race
INDY500.MTW	Exercise 7.11	Qualifying miles per hour and number of previous starts for drivers in 79th
	Exercise 7.36	Indianapolis 500 race
INFLATIO.MTW	Exercise 2.12	Private pay increase of salaried employees versus inflation rate
	Exercise 2.29	
INLETOIL.MTW	Exercise 5.91	Inlet oil temperature through a valve
	Exercise 6.48	
INMATE.MTW	Stat Insight Ch 8	Type of drug offense by race
INSPECT.MTW	Exercise 8.59	Percent of vehicles passing inspection by type inspection station
INSULATE.MTW	Exercise 9.50	Heat loss through a new insulating medium
IQGPA	Exercise 9.51–52	GPA versus IQ for 12 individuals
IRISES.MTW	Example 1.15	R. A. Fisher's famous data on sepal length of a species of Iris Setosa
	Example 5.19	
JDPOWER.MTW	Exercise 2.14	Number of problems reported per 100 cars in 1994 versus 1995
	Exercise 2.17	
	Exercise 2.31	
	Exercise 2.33	
	Exercise 2.40	
JOBSAT.MTW	Exercise 9.60	Job satisfaction and stress level for nine school teachers
KIDSMOKE.MTW	Exercise 4.85	Smoking habits of boys and girls ages 12 to 18
KILOWATT.MTW	Example 5.9	Rates per kilowatt-hour for each of the 50 states and DC
KINDER.MTW	Exercise 7.68	Reading scores for first grade children who attended kindergarten versus
		those who did not
LAMINECT.MTW	Exercise 10.18	Median costs of laminectomies at hospitals across North Carolina in
		1992
LEAD.MTW	Example 1.17	Lead levels in children's blood whose parents worked in a battery factory
LEADER.MTW	Exercise 7.31	Leadership exam scores by age for employees in an industrial plant
LETHAL.MTW	Example 6.12	Survival time of mice injected with an experimental lethal drug
LIFE.MTW	Exercise 1.31	Life expectancy of men and women in U.S.
LIFESPAN.MTW	Exercise 2.4	Life span of electronic components used in a spacecraft versus heat
	Exercise 2.37	
	Exercise 2.49	
LIGNTMONTH.MTW	Exercise 2.6	Relationship between damage reports and deaths caused by lightning
LODGE.MTW	Exercise 10.33	Measured traffic at three prospective locations for a motor lodge
LONGTAIL.MTW	Exercise 10.45	Long-tailed distributions to illustrate Kruskal–Wallis test
LOWABIL.MTW	Example 7.18	Reading skills of 24 matched low ability students
MAGNESIU.MTW	Exercise 9.9	Magnesium concentration and distances between samples
MALPRACT.MTW	Exercise 5.73	Amounts awarded in 17 malpractice cases
MANAGER.MTW	Exercise 5.81	Advertised salaries offered general managers of major corporations in
		1995
MARKED.MTW	Exercise 6.100	Percent of marked cars in 65 police departments in Florida
MATH.MTW	Exercise 1.69	Standardized math test scores for 30 students
MATHCOMP.MTW	Exercise 5.26	Standardized math competency for a group of entering freshmen at a small
		community college
MATHPRO.MTW	Example 9.1	Math proficiency and SAT scores by states
	Example 9.6	
	Exercise 9.24	
MAZE.MTW	Exercise 10.13	Error scores for four groups of experimental animals running a maze
MEDIAN.MTW	Exercise 10.52	Illustrates test of equality of medians with the Kruskal–Wallis test
MENTAL.MTW	Exercise 6.52	Median mental ages of 16 girls

MERCURY.MTW	Example 1.9	Concentration of mercury in 25 lake trout
METRENT.MTW	Exercise 5.117	Monthly rental costs in metro areas with 1 million or more persons
MILLER.MTW	Example 5.7	Miller personality test scores for a group of college students applying for graduate school
MILLER1.MTW	Exercise 1.41	Twenty scores on the Miller personality test
MOISTURE.MTW	Exercise 9.37	Moisture content and depth of core sample for marine muds in eastern Louisiana
MONOXIDE.MTW	Exercise 7.45	Carbon monoxide emitted by smoke stacks of a manufacturer and a competitor
MOVIE.MTW	Exercise 7.53	Moral attitude scale on 15 subjects before and after viewing a movie
MUSIC.MTW	Exercise 7.59	Improvement scores for identical twins taught music recognition by two techniques
NAME.MTW	Example 2.8 Exercise 2.28 Exercise 9.19	Estimated value of a brand name product and the company's revenue
NASCAR.MTW	Exercise 10.53	Efficiency of pit crews for three major NASCAR teams
NERVOUS.MTW	Example 10.3	Reaction effects of four drugs on 25 subjects with a nervous disorder
NEWSTAND.MTW	Exercise 1.43	Daily profits for 20 newsstands
NFLDRAF2.MTW	Exercise 9.63	Rating, time in 40-yard dash, and weight of top defensive linemen in the 1994 NFL draft
NFLDRAFT.MTW	Exercise 9.10 Exercise 9.16	Rating, time in 40-yard dash, and weight of top offensive linemen in the 1994 NFL draft
NICOTINE.MTW	Exercise 9.21	Nicotine content versus sales for eight major brands of cigarettes
ORANGE.MTW	Exercise 9.61	Price of oranges versus size of the harvest
ORIOLES.MTW	Example 1.3	Salaries of members of the Baltimore Orioles baseball team
OXYTOCIN.MTW	Exercise 7.86	Arterial blood pressure of 11 subjects before and after receiving oxytocin
PARENTED.XLS	Exercise 1.32	Excel worksheet of education backgrounds of parents of entering freshmen at a state university
PATROL.MTW	Example 9.3	Years of experience and number of tickets given by patrolpersons in New York City
PEARSON.MTW	Exercise 2.20	Karl Pearson's data on heights of brothers and sisters
PHONE.MTW	Exercise 6.95	Length of long-distance phone calls for a small business firm
POISON.MTW	Exercise 1.113	Number of poisonings reported to 16 poison control centers
POLITIC.MTW	Example 8.3	Political party and gender in a voting district
POLLUTIO.MTW	Exercise 5.59	Air pollution index for 15 randomly selected days for a major western city
POROSITY.MTW	Exercise 5.86	Porosity measurements on 20 samples of Tensleep Sandstone, Pennsylvanian from Bighorn Basin in Wyoming
POVERTY.MTW	Exercise 9.11 Exercise 9.17	Percent poverty and crime rate for selected cities
PRECINCT.MTW	Exercise 2.2 Exercise 2.38	Robbery rates versus percent low income in eight precincts
PREJUDIC.MTW	Example 5.10 Exercise 5.22	Racial prejudice measured on a sample of 25 high school students
PRESIDEN.MTW	Exercise 1.126	Ages at inauguration and death of U.S. presidents
PRESS.MTW	Exercise 9.55	Degree of confidence in the press versus education level for 20 randomly selected persons
PROGNOST.MTW	Exercise 6.61	Klopfer's prognostic rating scale for subjects receiving behavior modification therapy
PROGRAM.MTW	Exercise 10.17	Effects of four different methods of programmed learning for statistics students
PSAT.MTW	Exercise 2.50	PSAT scores versus SAT scores
PSYCH.MTW	Exercise 1.42	Correct responses for 24 students in a psychology experiment
PUERTO.MTW	Exercise 5.22 Exercise 5.65	Weekly incomes of a random sample of 50 Puerto Rican families in Miami

QUAIL.MTW	Exercise 1.53	Plasma LDL levels in two groups of quail
	Exercise 1.77	
	Exercise 1.88	
	Exercise 5.66	
	Exercise 7.50	
QUALITY.MTW	Exercise 7.81	Quality control test scores on two manufacturing processes
RAINKS.MTW	Exercise 9.8	Rainfall in an area of west central Kansas and four surrounding counties
RANDD.MTW	Example 9.8	Research and development expenditures and sales of a large company
	Exercise 9.36	
RAT.MTW	Exercise 1.52	Survival times of 20 rats exposed to high levels of radiation
	Exercise 1.76	
	Exercise 5.62	
	Exercise 6.44	
RATINGS.MTW	Example 2.6	Grade point averages versus teacher's ratings
REACTION.MTW	Exercise 6.11	Threshold reaction time for persons subjected to emotional stress
READING.MTW	Exercise 1.72	Standardized reading scores for 30 fifth-graders
	Exercise 2.10	
READIQ.MTW	Exercise 2.53	Reading scores versus IQ scores
REFEREND.MTW	Exercise 8.20	Opinion on referendum by view on freedom of the press
REGION.MTW	Exercise 10.26	Pollution index taken in three regions of the country
REGISTER.MTW	Exercise 2.3	Maintenance cost versus age of cash registers in a department store
	Exercise 2.39	
	Exercise 2.54	
REHAB.MTW	Exercise 7.61	Rehabilitative potential of 20 prison inmates as judged by two psychiatrists
REMEDIAL.MTW	Exercise 7.43	Math placement test score for 35 freshmen females and 42 freshmen males
RENTALS.MTW	Exercise 1.122	Weekly rentals for 45 apartments
REPAIR.MTW	Exercise 5.77	Recorded times for repairing 22 automobiles involved in wrecks
RETAIL.MTW	Exercise 9.59	Length of employment versus gross sales for ten employees of a large retail store
RONBROWN1.MTW	Exercise 2.9	Oceanography data obtained at site 1 by scientist aboard the ship Ron Brown
RONBROWN2.MTW	Example 2.4	Oceanography data obtained at site 2 by scientist aboard the ship Ron Brown
	Exercise 2.56	
RURAL.MTW	Example 7.16	Social adjustment scores for a rural group and a city group of children
SALARY.MTW	Exercise 3.66	Starting salaries for 25 new PhD psychologists
SALINITY.MTW	Exercise 5.27	Surface-water salinity measurements from Whitewater Bay, Florida
	Exercise 5.64	
SAT.MTW	Stat Insight Ch 9	SAT scores, percent taking exam and state funding per student by state for 1994, 1995, and 1999
SAVING.MTW	Exercise 10.34	Problem asset ratio for savings and loan companies in California, New York, and Texas
	Exercise 10.49	
SCALES.MTW	Exercise 1.89	Readings obtained from a 100-pound weight placed on four brands of bathroom scales
SCHIZOP2.MTW	Exercise 6.99	Exam scores for 17 patients to assess the learning ability of schizophrenics after taking a specified dose of a tranquilizer
SCHIZOPH.MTW	Example 6.10	Standardized exam scores for 13 patients to investigate the learning ability of schizophrenics after a specified dose of a tranquilizer
SEATBELT.MTW	Exercise 8.24	Injury level versus seatbelt usage
SELFDEFE.MTW	Example 7.19	Self-confidence scores for nine women before and after instructions on self-defense
SENIOR.MTW	Exercise 1.83	Reaction times of 30 senior citizens applying for drivers license renewals
	Exercise 3.67	
SENTENCE.MTW	Exercise 1.123	Sentences of 41 prisoners convicted of a homicide offense
SHKDRUG.MTW	Exercise 10.11–12	Effects of a drug and electroshock therapy on the ability to solve simple tasks

SHOCK.MTW	Exercise 10.50	Effect of experimental shock on time to complete difficult task
SHOPLIFT.MTW	Exercise 9.58	Sales receipts versus shoplifting losses for a department store
SHORT.MTW	Exercise 6.65	James Short's measurements of the parallax of the sun
SHUTTLE.MTW	Exercise 9.20	Number of people riding shuttle versus number of automobiles in the downtown area
SIMPSON.MTW	Example 1.18	Grade point averages of men and women participating in various sports—an illustration of Simpson's paradox
SITUP.MTW	Exercise 1.47	Maximum number of situps by participants in an exercise class
SKEWED.MTW	Exercise 7.65	Illustrates the Wilcoxon Rank Sum test
SKIN.MTW	Exercise 5.20	Survival times of closely and poorly matched skin grafts on burn patients
SLC.MTW	Exercise 5.116	Sodium-lithium countertransport activity on 190 individuals from six large English kindred
SMOKYPH.MTW	Exercise 6.40	Water pH levels of 75 water samples taken in the Great Smoky Mountains
	Exercise 6.59	
	Exercise 7.10	
	Exercise 7.35	
SNORE.MTW	Exercise 8.21	Snoring versus heart disease
SNOW.MTW	Exercise 7.87	Concentration of microparticles in snowfields of Greenland and Antarctica
SOCCER.MTW	Exercise 1.46	Weights of 25 soccer players
SOCIAL.MTW	Exercise 6.63	Median income level for 25 social workers from North Carolina
SOPHMOR.MTW	Exercise 2.42	Grade point averages, SAT scores, and final grade in college algebra for 20 sophomores
SOUTH.MTW	Exercise 1.84	Murder rates for 30 cities in the South
SPEED.MTW	Exercise 7.58	Speed reading scores before and after a course on speed reading
SPELLERS.MTW	Exercise 7.82	Standardized spelling test scores for two fourth-grade classes
SPELLING.MTW	Exercise 7.56	Spelling scores for nine eighth-graders before and after a two-week course of instruction
SPORTS.MTW	Exercise 8.32	Favorite sport by gender
SPOUSE.MTW	Exercise 8.33	Convictions in spouse murder cases by gender
STABLE.MTW	Exercise 6.93	Times of a two-year old stallion on a one mile run
STAMP.MTW	Stat Insight Ch 1	Thicknesses of 1872 Hidalgo stamps issued in Mexico
	Exercise 5.110	
STATCLAS.MTW	Exercise 7.30	Grades for two introductory statistics classes
STATELAW.MTW	Exercise 6.62	Operating expenditures per resident for each of the state law enforcement agencies
STATISTI.MTW	Exercise 1.70	Test scores for two beginning statistics classes
	Exercise 1.87	
STEP.MTW	Exercise 6.79	STEP science test scores for a class of ability-grouped students
STRESS.MTW	Example 7.20	Short-term memory test scores on 12 subjects before and after a stressful situation
STUDY.MTW	Exercise 5.25	Number of hours studied per week by a sample of 50 freshmen
SUBMARIN.MTW	Exercise 2.16	Number of German submarines sunk by U.S. Navy in World War II
	Exercise 2.45	
	Exercise 2.59	
SUBWAY.MTW	Exercise 5.19	Time it takes a subway to travel from the airport to downtown
SUNSPOT.MTW	Example 1.7	Wolfer sunspot numbers from 1700 through 2000
SUPERBOWL.MTW	Exercise 1.54	Margin of victory in Superbowls I to XXXV
SUPERCAR.MTW	Stat Insight Ch10	Top speeds attained by five makes of supercars
TABLROCK.MTW	Exercise 5.63	Ozone concentrations at Mt. Mitchell, North Carolina
TEACHER.MTW	Exercise 5.114	Average teacher's salaries across the states in the 70s, 80s, and 90s
TENNESS.MTW	Exercise 6.56	Tennessee self-concept scores for 20 gifted high school students
TENSILE.MTW	Example 7.11	Tensile strength of plastic bags from two production runs
TEST1.MTW	Exercise 5.80	Grades on the first test in a statistics class
THERMAL.MTW	Example 9.5	Heat loss of thermal pane windows versus outside temperature
TICKET.MTW	Exercise 5.18	Time to complete an airline ticket reservation
TOASTER.MTW	Exercise 9.35	Consumer Reports (Oct 94) rating of toaster ovens versus the cost

TONSILS.MTW	Exercise 2.78	Size of tonsils collected from 1398 children
TORT.MTW	Exercise 5.13	The number of torts, average number of months to process a tort, and county population from the court files of the nation's largest counties
TOXIC.MTW	Exercise 1.55	Hazardous waste sites near minority communities
	Exercise 5.108	
	Exercise 5.109	
	Exercise 8.58	
	Exercise 10.35	
TRACK.MTW	Exercise 2.97	National Olympic records for women in several races
	Exercise 5.115	
	Exercise 9.62	
TRACK15.MTW	Exercise 1.36	Olympic winning times for the men's 1500-meter run
TREES.MTW	Exercise 1.50	Number of trees in 20 grids
TREATMENTS.MTW	Exercise 10.44	Illustrates analysis of variance for three treatment groups
TRUCKS.MTW	Table 10.2	Miles per gallon for standard four-wheel drive trucks manufactured by Chevrolet, Dodge, and Ford
	Example 10.2	
TV.MTW	Example 2.1	Percent of students that watch more than six hours of TV per day versus national math test scores
TWIN.MTW	Exercise 7.54	Intelligence test scores for identical twins in which one twin is given a drug
UNDERGRAD.MTW	Table 1.1	Data set describing a sample of undergraduate students
	Exercise 1.15	
VACATION.MTW	Exercise 6.46	Number of days of paid holidays and vacation leave for sample of 35 textile workers
	Exercise 6.98	
VACCINE.MTW	Exercise 1.111	Reported serious reactions due to vaccines in 11 southern states
VEHICLE.MTW	Exercise 8.34	Fatality ratings for foreign and domestic vehicles
VERBAL.MTW	Exercise 9.30	Verbal test scores and number of library books checked out for 15 eighth-graders
VICTORIA.MTW	Exercise 2.98	Number of sunspots versus mean annual level of Lake Victoria Nyanza from 1902 to 1921
VISCOSIT.MTW	Exercise 7.44	Viscosity measurements of a substance on two different days
VISUAL.MTW	Exercise 5.6	Visual acuity of a group of subjects tested under a specified dose of a drug
VOCAB.MTW	Exercise 7.80	Reading scores before and after vocabulary training for 14 employees who did not complete high school
WASTEWAT.MTW	Exercise 9.18	Volume of injected waste water from Rocky Mountain Arsenal and number of earthquakes near Denver
WEATHER94.XLS	Exercise 1.30	Excel worksheet of weather casualties in 1994
WHEAT.MTW	Exercise 2.11	Price of a bushel of wheat versus the national weekly earnings of production workers
WINDMILL.MTW	Exercise 9.34	Direct current produced by different wind velocities
WINDOW.MTW	Exercise 6.54	Wind leakage for storm windows exposed to a 50 mph wind
WINS.MTW	Exercise 9.23	Baseball team wins versus seven independent variables for National League teams in 1990
WOOL.MTW	Exercise 7.42	Strength tests of two types of wool fabric
YEARSUNSPOT.MTW	Exercise 2.7	Monthly sunspot activity from 1974 to 2000

Tables

B.1   Binomial Probabilities

B.2   Probabilities for the Standard Normal Distribution

B.3   Critical Values of Students' *t* Distribution

B.4   Critical Values of the Chi-Square Distribution

B.5   Critical Values of the F-Distribution

TABLE B.1   Binomial Probabilities

n	k	.01	.05	.10	.20	.30	.40	.50	.60	.70	.80	.90	.95	.99
								π						
2	0	.980	.902	.810	.640	.490	.360	.250	.160	.090	.040	.010	.002	.0+
	1	.020	.095	.180	.320	.420	.480	.500	.480	.420	.320	.180	.095	.020
	2	.0+	.002	.010	.040	.090	.160	.250	.360	.490	.640	.810	.902	.980
3	0	.970	.857	.729	.512	.343	.216	.125	.064	.027	.008	.001	.0+	.0+
	1	.029	.135	.243	.384	.441	.432	.375	.288	.189	.096	.027	.007	.0+
	2	.0+	.007	.027	.096	.189	.288	.375	.432	.441	.384	.243	.135	.029
	3	.0+	.0+	.001	.008	.027	.064	.125	.216	.343	.512	.729	.857	.970
4	0	.961	.815	.656	.410	.240	.130	.062	.026	.008	.002	.0+	.0+	.0+
	1	.039	.171	.292	.410	.412	.346	.250	.154	.076	.026	.004	.0+	.0+
	2	.001	.014	.049	.154	.265	.346	.375	.346	.265	.154	.049	.014	.001
	3	.0+	.0+	.004	.026	.076	.154	.250	.346	.412	.410	.292	.171	.039
	4	.0+	.0+	.0+	.002	.008	.026	.062	.130	.240	.410	.656	.815	.961
5	0	.951	.774	.590	.328	.168	.078	.031	.010	.002	.0+	.0+	.0+	.0+
	1	.048	.204	.328	.410	.360	.259	.156	.077	.028	.006	.0+	.0+	.0+
	2	.001	.021	.073	.205	.309	.346	.312	.230	.132	.051	.008	.001	.0+
	3	.0+	.001	.008	.051	.132	.230	.312	.346	.309	.205	.073	.021	.001
	4	.0+	.0+	.0+	.006	.028	.077	.156	.259	.360	.410	.328	.204	.048
	5	.0+	.0+	.0+	.0+	.002	.010	.031	.078	.168	.328	.590	.774	.951
6	0	.941	.735	.531	.262	.118	.047	.016	.004	.001	.0+	.0+	.0+	.0+
	1	.057	.232	.354	.393	.303	.187	.094	.037	.010	.002	.0+	.0+	.0+
	2	.001	.031	.098	.246	.324	.311	.234	.138	.060	.015	.001	.0+	.0+
	3	.0+	.002	.015	.082	.185	.276	.312	.276	.185	.082	.015	.002	.0+
	4	.0+	.0+	.001	.015	.060	.138	.234	.311	.324	.246	.098	.031	.001
	5	.0+	.0+	.0+	.002	.010	.037	.094	.187	.303	.393	.354	.232	.057
	6	.0+	.0+	.0+	.0+	.001	.004	.016	.047	.118	.262	.531	.735	.941
7	0	.932	.698	.478	.210	.082	.028	.008	.002	.0+	.0+	.0+	.0+	.0+
	1	.066	.257	.372	.367	.247	.131	.055	.017	.004	.0+	.0+	.0+	.0+
	2	.002	.041	.124	.275	.318	.261	.164	.077	.025	.004	.0+	.0+	.0+
	3	.0+	.004	.023	.115	.227	.290	.273	.194	.097	.029	.003	.0+	.0+
	4	.0+	.0+	.003	.029	.097	.194	.273	.290	.227	.115	.023	.004	.0+
	5	.0+	.0+	.0+	.004	.025	.077	.164	.261	.318	.275	.124	.041	.002
	6	.0+	.0+	.0+	.0+	.004	.017	.055	.131	.247	.367	.372	.257	.066
	7	.0+	.0+	.0+	.0+	.0+	.002	.008	.028	.082	.210	.478	.698	.932
8	0	.923	.663	.430	.168	.058	.017	.004	.001	.0+	.0+	.0+	.0+	.0+
	1	.075	.279	.383	.336	.198	.090	.031	.008	.001	.0+	.0+	.0+	.0+
	2	.003	.051	.149	.294	.296	.209	.109	.041	.010	.001	.0+	.0+	.0+
	3	.0+	.005	.033	.147	.254	.279	.219	.124	.047	.009	.0+	.0+	.0+
	4	.0+	.0+	.005	.046	.136	.232	.273	.232	.136	.046	.005	.0+	.0+
	5	.0+	.0+	.0+	.009	.047	.124	.219	.279	.254	.147	.033	.005	.0+
	6	.0+	.0+	.0+	.001	.010	.041	.109	.209	.296	.294	.149	.051	.003
	7	.0+	.0+	.0+	.0+	.001	.008	.031	.090	.198	.336	.383	.279	.075
	8	.0+	.0+	.0+	.0+	.0+	.001	.004	.017	.058	.168	.430	.663	.923
9	0	.914	.630	.387	.134	.040	.010	.002	.0+	.0+	.0+	.0+	.0+	.0+
	1	.083	.299	.387	.302	.156	.060	.018	.004	.0+	.0+	.0+	.0+	.0+
	2	.003	.063	.172	.302	.267	.161	.070	.021	.004	.0+	.0+	.0+	.0+
	3	.0+	.008	.045	.176	.267	.251	.164	.074	.021	.003	.0+	.0+	.0+
	4	.0+	.001	.007	.066	.172	.251	.246	.167	.074	.017	.001	.0+	.0+

TABLE B.1 Binomial Probabilities (continued)

n	k	.01	.05	.10	.20	.30	.40	.50	.60	.70	.80	.90	.95	.99
	5	.0+	.0+	.001	.017	.074	.167	.246	.251	.172	.066	.007	.001	.0+
	6	.0+	.0+	.0+	.003	.021	.074	.164	.251	.267	.176	.045	.008	.0+
	7	.0+	.0+	.0+	.0+	.004	.021	.070	.161	.267	.302	.172	.063	.003
	8	.0+	.0+	.0+	.0+	.0+	.004	.018	.060	.156	.302	.387	.299	.083
	9	.0+	.0+	.0+	.0+	.0+	.0+	.002	.010	.040	.134	.387	.630	.914
10	0	.904	.599	.349	.107	.028	.006	.001	.0+	.0+	.0+	.0+	.0+	.0+
	1	.091	.315	.387	.268	.121	.040	.010	.002	.0+	.0+	.0+	.0+	.0+
	2	.004	.075	.194	.302	.233	.121	.044	.011	.001	.0+	.0+	.0+	.0+
	3	.0+	.010	.057	.201	.267	.215	.117	.042	.009	.001	.0+	.0+	.0+
	4	.0+	.001	.011	.088	.200	.251	.205	.111	.037	.006	.0+	.0+	.0+
	5	.0+	.0+	.001	.026	.103	.201	.246	.201	.103	.026	.001	.0+	.0+
	6	.0+	.0+	.0+	.006	.037	.111	.205	.251	.200	.088	.011	.001	.0+
	7	.0+	.0+	.0÷	.001	.009	.042	.117	.215	.267	.201	.057	.010	.0+
	8	.0+	.0+	.0+	.0+	.001	.011	.044	.121	.233	.302	.194	.075	.004
	9	.0+	.0+	.0+	.0+	.0+	.002	.010	.040	.121	.268	.387	.315	.091
	10	.0+	.0+	.0+	.0+	.0+	.0+	.001	.006	.028	.107	.349	.599	.904
11	0	.895	.569	.314	.086	.020	.004	.0+	.0+	.0+	.0+	.0+	.0+	.0+
	1	.099	.329	.384	.236	.093	.027	.005	.001	.0+	.0+	.0+	.0+	.0+
	2	.005	.087	.213	.295	.200	.089	.027	.005	.001	.0+	.0+	.0+	.0+
	3	.0+	.014	.071	.221	.257	.177	.081	.023	.004	.0+	.0+	.0+	.0+
	4	.0+	.001	.016	.111	.220	.236	.161	.070	.017	.002	.0+	.0+	.0+
	5	.0+	.0+	.002	.039	.132	.221	.226	.147	.057	.010	.0+	.0+	.0+
	6	.0+	.0+	.0+	.010	.057	.147	.226	.221	.132	.039	.002	.0+	.0+
	7	.0+	.0+	.0+	.002	.017	.070	.161	.236	.220	.111	.016	.001	.0+
	8	.0+	.0+	.0+	.0+	.004	.023	.081	.177	.257	.221	.071	.014	.0+
	9	.0+	.0+	.0+	.0+	.001	.005	.027	.089	.200	.295	.213	.087	.005
	10	.0+	.0+	.0+	.0+	.0+	.001	.005	.027	.093	.236	.384	.329	.099
	11	.0+	.0+	.0+	.0+	.0+	.0+	.0+	.004	.020	.086	.314	.569	.895
12	0	.886	.540	.282	.069	.014	.002	.0+	.0+	.0+	.0+	.0+	.0+	.0+
	1	.107	.341	.377	.206	.071	.017	.003	.0+	.0+	.0+	.0+	.0+	.0+
	2	.006	.099	.230	.283	.168	.064	.016	.002	.0+	.0+	.0+	.0+	.0+
	3	.0+	.017	.085	.236	.240	.142	.054	.012	.001	.0+	.0+	.0+	.0+
	4	.0+	.002	.021	.133	.231	.213	.121	.042	.008	.001	.0+	.0+	.0+
	5	.0+	.0+	.004	.053	.158	.227	.193	.101	.029	.003	.0+	.0+	.0+
	6	.0+	.0+	.0+	.016	.079	.177	.226	.177	.079	.016	.0+	.0+	.0+
	7	.0+	.0+	.0+	.003	.029	.101	.193	.227	.158	.053	.004	.0+	.0+
	8	.0+	.0+	.0+	.001	.008	.042	.121	.213	.231	.133	.021	.002	.0+
	9	.0+	.0+	.0+	.0+	.001	.012	.054	.142	.240	.236	.085	.017	.0+
	10	.0+	.0+	.0+	.0+	.0+	.002	.016	.064	.168	.283	.230	.099	.006
	11	.0+	.0+	.0+	.0+	.0+	.0+	.003	.017	.071	.206	.377	.341	.107
	12	.0+	.0+	.0+	.0+	.0+	.0+	.0+	.002	.014	.069	.282	.540	.886
13	0	.878	.513	.254	.055	.010	.001	.0+	.0+	.0+	.0+	.0+	.0+	.0+
	1	.115	.351	.367	.179	.054	.011	.002	.0+	.0+	.0+	.0+	.0+	.0+
	2	.007	.111	.245	.268	.139	.045	.010	.001	.0+	.0+	.0+	.0+	.0+
	3	.0+	.021	.100	.246	.218	.111	.035	.006	.001	.0+	.0+	.0+	.0+
	4	.0+	.003	.028	.154	.234	.184	.087	.024	.003	.0+	.0+	.0+	.0+

TABLE B.1   Binomial Probabilities (continued)

n	k	.01	.05	.10	.20	.30	.40	.50	.60	.70	.80	.90	.95	.99
	5	.0+	.0+	.006	.069	.180	.221	.157	.066	.014	.001	.0+	.0+	.0+
	6	.0+	.0+	.001	.023	.103	.197	.209	.131	.044	.006	.0+	.0+	.0+
	7	.0+	.0+	.0+	.006	.044	.131	.209	.197	.103	.023	.001	.0+	.0+
	8	.0+	.0+	.0+	.001	.014	.066	.157	.221	.180	.069	.006	.0+	.0+
	9	.0+	.0+	.0+	.0+	.003	.024	.087	.184	.234	.154	.028	.003	.0+
	10	.0+	.0+	.0+	.0+	.001	.006	.035	.111	.218	.246	.100	.021	.0+
	11	.0+	.0+	.0+	.0+	.0+	.001	.010	.045	.139	.268	.245	.111	.007
	12	.0+	.0+	.0+	.0+	.0+	.0+	.002	.011	.054	.179	.367	.351	.115
	13	.0+	.0+	.0+	.0+	.0+	.0+	.0+	.001	.010	.055	.254	.513	.878
14	0	.869	.488	.229	.044	.007	.001	.0+	.0+	.0+	.0+	.0+	.0+	.0+
	1	.123	.359	.356	.154	.041	.007	.001	.0+	.0+	.0+	.0+	.0+	.0+
	2	.008	.123	.257	.250	.113	.032	.006	.001	.0+	.0+	.0+	.0+	.0+
	3	.0+	.026	.114	.250	.194	.085	.022	.003	.0+	.0+	.0+	.0+	.0+
	4	.0+	.004	.035	.172	.229	.155	.061	.014	.001	.0+	.0+	.0+	.0+
	5	.0+	.0+	.008	.086	.196	.207	.122	.041	.007	.0+	.0+	.0+	.0+
	6	.0+	.0+	.001	.032	.126	.207	.183	.092	.023	.002	.0+	.0+	.0+
	7	.0+	.0+	.0+	.009	.062	.157	.209	.157	.062	.009	.0+	.0+	.0+
	8	.0+	.0+	.0+	.002	.023	.092	.183	.207	.126	.032	.001	.0+	.0+
	9	.0+	.0+	.0+	.0+	.007	.041	.122	.207	.196	.086	.008	.0+	.0+
	10	.0+	.0+	.0+	.0+	.001	.014	.061	.155	.229	.172	.035	.004	.0+
	11	.0+	.0+	.0+	.0+	.0+	.003	.022	.085	.194	.250	.114	.026	.0+
	12	.0+	.0+	.0+	.0+	.0+	.001	.006	.032	.113	.250	.257	.123	.008
	13	.0+	.0+	.0+	.0+	.0+	.0+	.001	.007	.041	.154	.356	.359	.123
	14	.0+	.0+	.0+	.0+	.0+	.0+	.0+	.001	.007	.044	.229	.488	.869
15	0	.860	.463	.206	.035	.005	.0+	.0+	.0+	.0+	.0+	.0+	.0+	.0+
	1	.130	.366	.343	.132	.031	.005	.0+	.0+	.0+	.0+	.0+	.0+	.0+
	2	.009	.135	.267	.231	.092	.022	.003	.0+	.0+	.0+	.0+	.0+	.0+
	3	.0+	.031	.129	.250	.170	.063	.014	.002	.0+	.0+	.0+	.0+	.0+
	4	.0+	.005	.043	.188	.219	.127	.042	.007	.001	.0+	.0+	.0+	.0+
	5	.0+	.001	.010	.103	.206	.186	.092	.024	.003	.0+	.0+	.0+	.0+
	6	.0+	.0+	.002	.043	.147	.207	.153	.061	.012	.001	.0+	.0+	.0+
	7	.0+	.0+	.0+	.014	.081	.177	.196	.118	.035	.003	.0+	.0+	.0+
	8	.0+	.0+	.0+	.003	.035	.118	.196	.177	.081	.014	.0+	.0+	.0+
	9	.0+	.0+	.0+	.001	.012	.061	.153	.207	.147	.043	.002	.0+	.0+
	10	.0+	.0+	.0+	.0+	.003	.024	.092	.186	.206	.103	.010	.001	.0+
	11	.0+	.0+	.0+	.0+	.001	.007	.042	.127	.219	.188	.043	.005	.0+
	12	.0+	.0+	.0+	.0+	.0+	.002	.014	.063	.170	.250	.129	.031	.0+
	13	.0+	.0+	.0+	.0+	.0+	.0+	.003	.022	.092	.231	.267	.135	.009
	14	.0+	.0+	.0+	.0+	.0+	.0+	.0+	.005	.031	.132	.343	.366	.130
	15	.0+	.0+	.0+	.0+	.0+	.0+	.0+	.0+	.005	.035	.206	.463	.860
16	0	.851	.440	.185	.028	.003	.0+	.0+	.0+	.0+	.0+	.0+	.0+	.0+
	1	.138	.371	.329	.113	.023	.003	.0+	.0+	.0+	.0+	.0+	.0+	.0+
	2	.010	.146	.274	.211	.073	.015	.002	.0+	.0+	.0+	.0+	.0+	.0+
	3	.0+	.036	.142	.246	.146	.047	.008	.001	.0+	.0+	.0+	.0+	.0+
	4	.0+	.006	.051	.200	.204	.101	.028	.004	.0+	.0+	.0+	.0+	.0+
	5	.0+	.001	.014	.120	.210	.162	.067	.014	.001	.0+	.0+	.0+	.0+
	6	.0+	.0+	.003	.055	.165	.198	.122	.039	.006	.0+	.0+	.0+	.0+
	7	.0+	.0+	.0+	.020	.101	.189	.175	.084	.018	.001	.0+	.0+	.0+
	8	.0+	.0+	.0+	.006	.049	.142	.196	.142	.049	.006	.0+	.0+	.0+

TABLE B.1 Binomial Probabilities (continued)

								π							
n	k	.01	.05	.10	.20	.30	.40	.50	.60	.70	.80	.90	.95	.99
	9	.0+	.0+	.0+	.001	.018	.084	.175	.189	.101	.020	.0+	.0+	.0+
	10	.0+	.0+	.0+	.0+	.006	.039	.122	.198	.165	.055	.003	.0+	.0+
	11	.0+	.0+	.0+	.0+	.001	.014	.067	.162	.210	.120	.014	.001	.0+
	12	.0+	.0+	.0+	.0+	.0+	.004	.028	.101	.204	.200	.051	.006	.0+
	13	.0+	.0+	.0+	.0+	.0+	.001	.008	.047	.146	.246	.142	.036	.0+
	14	.0+	.0+	.0+	.0+	.0+	.0+	.002	.015	.073	.211	.274	.146	.010
	15	.0+	.0+	.0+	.0+	.0+	.0+	.0+	.003	.023	.113	.329	.371	.138
	16	.0+	.0+	.0+	.0+	.0+	.0+	.0+	.0+	.003	.028	.185	.440	.851
17	0	.843	.418	.167	.022	.002	.0+	.0+	.0+	.0+	.0+	.0+	.0+	.0+
	1	.145	.374	.315	.096	.017	.002	.0+	.0+	.0+	.0+	.0+	.0+	.0+
	2	.012	.158	.280	.191	.058	.010	.001	.0+	.0+	.0+	.0+	.0+	.0+
	3	.001	.042	.156	.239	.124	.034	.005	.0+	.0+	.0+	.0+	.0+	.0+
	4	.0+	.008	.060	.209	.187	.080	.018	.002	.0+	.0+	.0+	.0+	.0+
	5	.0+	.001	.018	.136	.208	.138	.047	.008	.001	.0+	.0+	.0+	.0+
	6	.0+	.0+	.004	.068	.178	.184	.094	.024	.003	.0+	.0+	.0+	.0+
	7	.0+	.0+	.001	.027	.120	.193	.148	.057	.010	.0+	.0+	.0+	.0+
	8	.0+	.0+	.0+	.008	.064	.161	.186	.107	.028	.002	.0+	.0+	.0+
	9	.0+	.0+	.0+	.002	.028	.107	.186	.161	.064	.008	.0+	.0+	.0+
	10	.0+	.0+	.0+	.0+	.010	.057	.148	.193	.120	.027	.001	.0+	.0+
	11	.0+	.0+	.0+	.0+	.003	.024	.094	.184	.178	.068	.004	.0+	.0+
	12	.0+	.0+	.0+	.0+	.001	.008	.047	.138	.208	.136	.018	.001	.0+
	13	.0+	.0+	.0+	.0+	.0+	.002	.018	.080	.187	.209	.060	.008	.0+
	14	.0+	.0+	.0+	.0+	.0+	.0+	.005	.034	.124	.239	.156	.042	.001
	15	.0+	.0+	.0+	.0+	.0+	.0+	.001	.010	.058	.191	.280	.158	.012
	16	.0+	.0+	.0+	.0+	.0+	.0+	.0+	.002	.017	.096	.315	.374	.145
	17	.0+	.0+	.0+	.0+	.0+	.0+	.0+	.0+	.002	.022	.167	.418	.843
18	0	.835	.397	.150	.018	.002	.0+	.0+	.0+	.0+	.0+	.0+	.0+	.0+
	1	.152	.376	.300	.081	.013	.001	.0+	.0+	.0+	.0+	.0+	.0+	.0+
	2	.013	.168	.284	.172	.046	.007	.001	.0+	.0+	.0+	.0+	.0+	.0+
	3	.001	.047	.168	.230	.105	.025	.003	.0+	.0+	.0+	.0+	.0+	.0+
	4	.0+	.009	.070	.215	.168	.061	.012	.001	.0+	.0+	.0+	.0+	.0+
	5	.0+	.001	.002	.151	.202	.115	.033	.004	.0+	.0+	.0+	.0+	.0+
	6	.0+	.0+	.005	.082	.187	.166	.071	.014	.001	.0+	.0+	.0+	.0+
	7	.0+	.0+	.001	.035	.138	.189	.121	.037	.005	.0+	.0+	.0+	.0+
	8	.0+	.0+	.0+	.012	.081	.173	.167	.077	.015	.001	.0+	.0+	.0+
	9	.0+	.0+	.0+	.003	.038	.128	.186	.128	.038	.003	.0+	.0+	.0+
	10	.0+	.0+	.0+	.001	.015	.077	.167	.173	.081	.012	.0+	.0+	.0+
	11	.0+	.0+	.0+	.0+	.005	.037	.121	.189	.138	.035	.001	.0+	.0+
	12	.0+	.0+	.0+	.0+	.001	.014	.071	.166	.187	.082	.005	.0+	.0+
	13	.0+	.0+	.0+	.0+	.0+	.004	.033	.115	.202	.151	.022	.001	.0+
	14	.0+	.0+	.0+	.0+	.0+	.001	.012	.061	.168	.215	.070	.009	.0+
	15	.0+	.0+	.0+	.0+	.0+	.0+	.003	.025	.105	.230	.168	.097	.001
	16	.0+	.0+	.0+	.0+	.0+	.0+	.001	.007	.046	.172	.284	.168	.013
	17	.0+	.0+	.0+	.0+	.0+	.0+	.0+	.001	.013	.081	.300	.376	.152
	18	.0+	.0+	.0+	.0+	.0+	.0+	.0+	.0+	.002	.018	.150	.397	.835

TABLE B.1   Binomial Probabilities (continued)

n	k	.01	.05	.10	.20	.30	.40	.50	.60	.70	.80	.90	.95	.99
19	0	.826	.377	.135	.014	.001	.0+	.0+	.0+	.0+	.0+	.0+	.0+	.0+
	1	.159	.377	.285	.069	.009	.001	.0+	.0+	.0+	.0+	.0+	.0+	.0+
	2	.014	.179	.285	.154	.036	.005	.0+	.0+	.0+	.0+	.0+	.0+	.0+
	3	.001	.053	.180	.218	.087	.018	.002	.0+	.0+	.0+	.0+	.0+	.0+
	4	.0+	.011	.080	.218	.149	.047	.007	.0+	.0+	.0+	.0+	.0+	.0+
	5	.0+	.002	.027	.164	.192	.093	.022	.002	.0+	.0+	.0+	.0+	.0+
	6	.0+	.0+	.007	.096	.192	.145	.052	.008	.0+	.0+	.0+	.0+	.0+
	7	.0+	.0+	.001	.044	.152	.180	.096	.024	.002	.0+	.0+	.0+	.0+
	8	.0+	.0+	.0+	.017	.098	.180	.144	.053	.008	.0+	.0+	.0+	.0+
	9	.0+	.0+	.0+	.005	.051	.146	.176	.098	.022	.001	.0+	.0+	.0+
	10	.0+	.0+	.0+	.001	.022	.098	.176	.146	.051	.005	.0+	.0+	.0+
	11	.0+	.0+	.0+	.0+	.008	.053	.144	.180	.098	.017	.0+	.0+	.0+
	12	.0+	.0+	.0+	.0+	.002	.024	.096	.180	.152	.044	.001	.0+	.0+
	13	.0+	.0+	.0+	.0+	.0+	.008	.052	.145	.192	.096	.007	.0+	.0+
	14	.0+	.0+	.0+	.0+	.0+	.002	.022	.093	.192	.164	.027	.002	.0+
	15	.0+	.0+	.0+	.0+	.0+	.0+	.007	.047	.149	.218	.080	.011	.0+
	16	.0+	.0+	.0+	.0+	.0+	.0+	.002	.018	.087	.218	.180	.053	.001
	17	.0+	.0+	.0+	.0+	.0+	.0+	.0+	.005	.036	.154	.285	.179	.014
	18	.0+	.0+	.0+	.0+	.0+	.0+	.0+	.001	.009	.069	.285	.377	.159
	19	.0+	.0+	.0+	.0+	.0+	.0+	.0+	.0+	.001	.014	.135	.377	.826
20	0	.818	.358	.122	.012	.001	.0+	.0+	.0+	.0+	.0+	.0+	.0+	.0+
	1	.165	.377	.270	.058	.007	.0+	.0+	.0+	.0+	.0+	.0+	.0+	.0+
	2	.016	.189	.285	.137	.028	.003	.0+	.0+	.0+	.0+	.0+	.0+	.0+
	3	.001	.060	.190	.205	.072	.012	.001	.0+	.0+	.0+	.0+	.0+	.0+
	4	.0+	.013	.090	.218	.130	.035	.005	.0+	.0+	.0+	.0+	.0+	.0+
	5	.0+	.002	.032	.175	.179	.075	.015	.001	.0+	.0+	.0+	.0+	.0+
	6	.0+	.0+	.009	.109	.192	.124	.037	.005	.0+	.0+	.0+	.0+	.0+
	7	.0+	.0+	.002	.054	.164	.166	.074	.015	.001	.0+	.0+	.0+	.0+
	8	.0+	.0+	.0+	.022	.114	.180	.120	.036	.004	.0+	.0+	.0+	.0+
	9	.0+	.0+	.0+	.007	.065	.160	.160	.071	.012	.0+	.0+	.0+	.0+
	10	.0+	.0+	.0+	.002	.031	.117	.176	.117	.031	.002	.0+	.0+	.0+
	11	.0+	.0+	.0+	.0+	.012	.071	.160	.160	.065	.007	.0+	.0+	.0+
	12	.0+	.0+	.0+	.0+	.004	.036	.120	.180	.114	.022	.0+	.0+	.0+
	13	.0+	.0+	.0+	.0+	.001	.015	.074	.166	.164	.054	.002	.0+	.0+
	14	.0+	.0+	.0+	.0+	.0+	.005	.037	.124	.192	.109	.009	.0+	.0+
	15	.0+	.0+	.0+	.0+	.0+	.001	.015	.075	.179	.175	.032	.002	.0+
	16	.0+	.0+	.0+	.0+	.0+	.0+	.005	.035	.130	.218	.090	.013	.0+
	17	.0+	.0+	.0+	.0+	.0+	.0+	.001	.012	.072	.205	.190	.060	.001
	18	.0+	.0+	.0+	.0+	.0+	.0+	.0+	.003	.028	.137	.285	.189	.016
	19	.0+	.0+	.0+	.0+	.0+	.0+	.0+	.0+	.007	.058	.270	.377	.165
	20	.0+	.0+	.0+	.0+	.0+	.0+	.0+	.0+	.001	.012	.122	.358	.818

TABLE B.2   Probabilities for the Standard Normal Distribution

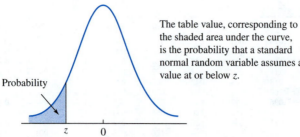

The table value, corresponding to the shaded area under the curve, is the probability that a standard normal random variable assumes a value at or below z.

Second decimal place in z

z	.00	.01	.02	.03	.04	.05	.06	.07	.08	.09
−5.0	.0000003									
−4.5	.000003									
−4.0	.00003									
−3.5	.0002									
−3.4	.0003	.0003	.0003	.0003	.0003	.0003	.0003	.0003	.0003	.0002
−3.3	.0005	.0005	.0005	.0004	.0004	.0004	.0004	.0004	.0004	.0003
−3.2	.0007	.0007	.0006	.0006	.0006	.0006	.0006	.0005	.0005	.0005
−3.1	.0010	.0009	.0009	.0009	.0008	.0008	.0008	.0008	.0007	.0007
−3.0	.0013	.0013	.0013	.0012	.0012	.0011	.0011	.0011	.0010	.0010
−2.9	.0019	.0018	.0018	.0017	.0016	.0016	.0015	.0015	.0014	.0014
−2.8	.0026	.0025	.0024	.0023	.0023	.0022	.0021	.0021	.0020	.0019
−2.7	.0035	.0034	.0033	.0032	.0031	.0030	.0029	.0028	.0027	.0026
−2.6	.0047	.0045	.0044	.0043	.0041	.0040	.0039	.0038	.0037	.0036
−2.5	.0062	.0060	.0059	.0057	.0055	.0054	.0052	.0051	.0049	.0048
−2.4	.0082	.0080	.0078	.0075	.0073	.0071	.0069	.0068	.0066	.0064
−2.3	.0107	.0104	.0102	.0099	.0096	.0094	.0091	.0089	.0087	.0084
−2.2	.0139	.0136	.0132	.0129	.0125	.0122	.0119	.0116	.0113	.0110
−2.1	.0179	.0174	.0170	.0166	.0162	.0158	.0154	.0150	.0146	.0143
−2.0	.0228	.0222	.0217	.0212	.0207	.0202	.0197	.0192	.0188	.0183
−1.9	.0287	.0281	.0274	.0268	.0262	.0256	.0250	.0244	.0239	.0233
−1.8	.0359	.0351	.0344	.0336	.0329	.0322	.0314	.0307	.0301	.0294
−1.7	.0446	.0436	.0427	.0418	.0409	.0401	.0392	.0384	.0375	.0367
−1.6	.0548	.0537	.0526	.0516	.0505	.0495	.0485	.0475	.0465	.0455
−1.5	.0668	.0655	.0643	.0630	.0618	.0606	.0594	.0582	.0571	.0559
−1.4	.0808	.0793	.0778	.0764	.0749	.0735	.0721	.0708	.0694	.0681
−1.3	.0968	.0951	.0934	.0918	.0901	.0885	.0869	.0853	.0838	.0823
−1.2	.1151	.1131	.1112	.1093	.1075	.1056	.1038	.1020	.1003	.0985
−1.1	.1357	.1335	.1314	.1292	.1271	.1251	.1230	.1210	.1190	.1170
−1.0	.1587	.1562	.1539	.1515	.1492	.1469	.1446	.1423	.1401	.1379
−0.9	.1841	.1814	.1788	.1762	.1736	.1711	.1685	.1660	.1635	.1611
−0.8	.2119	.2090	.2061	.2033	.2005	.1977	.1949	.1922	.1894	.1867
−0.7	.2420	.2389	.2358	.2327	.2296	.2266	.2236	.2206	.2177	.2148
−0.6	.2743	.2709	.2676	.2643	.2611	.2578	.2546	.2514	.2483	.2451
−0.5	.3085	.3050	.3015	.2981	.2946	.2912	.2877	.2843	.2810	.2776
−0.4	.3446	.3409	.3372	.3336	.3300	.3264	.3228	.3192	.3156	.3121
−0.3	.3821	.3783	.3745	.3707	.3669	.3632	.3594	.3557	.3520	.3483
−0.2	.4207	.4168	.4129	.4090	.4052	.4013	.3974	.3936	.3897	.3859
−0.1	.4602	.4562	.4522	.4483	.4443	.4404	.4364	.4325	.4286	.4247
−0.0	.5000	.4960	.4920	.4880	.4840	.4801	.4761	.4721	.4681	.4641

TABLE B.2 Probabilities for the Standard Normal Distribution (continued)

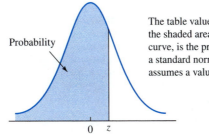

Probability

The table value, corresponding to the shaded area under the curve, is the probability that a standard normal random variable assumes a value at or below z.

Second decimal place in z

z	.00	.01	.02	.03	.04	.05	.06	.07	.08	.09
0.0	.5000	.5040	.5080	.5120	.5160	.5199	.5239	.5279	.5319	.5359
0.1	.5398	.5438	.5478	.5517	.5557	.5596	.5636	.5675	.5714	.5753
0.2	.5793	.5832	.5871	.5910	.5948	.5987	.6026	.6064	.6103	.6141
0.3	.6179	.6217	.6255	.6293	.6331	.6368	.6406	.6443	.6480	.6517
0.4	.6554	.6591	.6628	.6664	.6700	.6736	.6772	.6808	.6844	.6879
0.5	.6915	.6950	.6985	.7019	.7054	.7088	.7123	.7157	.7190	.7224
0.6	.7257	.7291	.7324	.7357	.7389	.7422	.7454	.7486	.7517	.7549
0.7	.7580	.7611	.7642	.7673	.7704	.7734	.7764	.7794	.7823	.7852
0.8	.7881	.7910	.7939	.7967	.7995	.8023	.8051	.8078	.8106	.8133
0.9	.8159	.8186	.8212	.8238	.8264	.8289	.8315	.8340	.8365	.8389
1.0	.8413	.8438	.8461	.8485	.8508	.8531	.8554	.8577	.8599	.8621
1.1	.8643	.8665	.8686	.8708	.8729	.8749	.8770	.8790	.8810	.8830
1.2	.8849	.8869	.8888	.8907	.8925	.8944	.8962	.8980	.8997	.9015
1.3	.9032	.9049	.9066	.9082	.9099	.9115	.9131	.9147	.9162	.9177
1.4	.9192	.9207	.9222	.9236	.9251	.9265	.9279	.9292	.9306	.9319
1.5	.9332	.9345	.9357	.9370	.9382	.9394	.9406	.9418	.9429	.9441
1.6	.9452	.9463	.9474	.9484	.9495	.9505	.9515	.9525	.9535	.9545
1.7	.9554	.9564	.9573	.9582	.9591	.9599	.9608	.9616	.9625	.9633
1.8	.9641	.9649	.9656	.9664	.9671	.9678	.9686	.9693	.9699	.9706
1.9	.9713	.9719	.9726	.9732	.9738	.9744	.9750	.9756	.9761	.9767
2.0	.9772	.9778	.9783	.9788	.9793	.9798	.9803	.9808	.9812	.9817
2.1	.9821	.9826	.9830	.9834	.9838	.9842	.9846	.9850	.9854	.9857
2.2	.9861	.9864	.9868	.9871	.9875	.9878	.9881	.9884	.9887	.9890
2.3	.9893	.9896	.9898	.9901	.9904	.9906	.9909	.9911	.9913	.9916
2.4	.9918	.9920	.9922	.9925	.9927	.9929	.9931	.9932	.9934	.9936
2.5	.9938	.9940	.9941	.9943	.9945	.9946	.9948	.9949	.9951	.9952
2.6	.9953	.9955	.9956	.9957	.9959	.9960	.9961	.9962	.9963	.9964
2.7	.9965	.9966	.9967	.9968	.9969	.9970	.9971	.9972	.9973	.9974
2.8	.9974	.9975	.9976	.9977	.9977	.9978	.9979	.9979	.9980	.9981
2.9	.9981	.9982	.9982	.9983	.9984	.9984	.9985	.9985	.9986	.9986
3.0	.9987	.9987	.9987	.9988	.9988	.9989	.9989	.9989	.9990	.9990
3.1	.9990	.9991	.9991	.9991	.9992	.9992	.9992	.9992	.9993	.9993
3.2	.9993	.9993	.9994	.9994	.9994	.9994	.9994	.9995	.9995	.9995
3.3	.9995	.9995	.9995	.9996	.9996	.9996	.9996	.9996	.9996	.9997
3.4	.9997	.9997	.9997	.9997	.9997	.9997	.9997	.9997	.9997	.9998
3.5	.9998									
4.0	.99997									
4.5	.999997									
5.0	.9999997									

TABLE B.3   Critical Values of Students' t Distribution

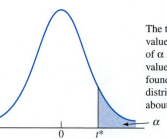

The table values are the critical values for students' t for an area of α in the right-hand tail. Critical values for the left-hand tail are found by knowing that the distribution is symmetric about zero.

Amount of α in one-tail

Degrees of freedom	.2	.1	.05	.025	.01	.005	.0025	.001
1	1.376	3.078	6.314	12.706	31.821	63.657	127.3	318.3
2	1.061	1.886	2.920	4.303	6.965	9.925	14.09	22.33
3	.978	1.638	2.353	3.182	4.541	5.841	7.453	10.21
4	.941	1.533	2.132	2.776	3.747	4.604	5.598	7.173
5	.920	1.476	2.015	2.571	3.365	4.032	4.773	5.893
6	.906	1.440	1.943	2.447	3.143	3.707	4.317	5.208
7	.896	1.415	1.895	2.365	2.998	3.499	4.029	4.785
8	.889	1.397	1.860	2.306	2.896	3.355	3.833	4.501
9	.883	1.383	1.833	2.262	2.821	3.250	3.690	4.297
10	.879	1.372	1.812	2.228	2.764	3.169	3.581	4.144
11	.876	1.363	1.796	2.201	2.718	3.106	3.497	4.025
12	.873	1.356	1.782	2.179	2.681	3.055	3.428	3.930
13	.870	1.350	1.771	2.160	2.650	3.012	3.372	3.852
14	.868	1.345	1.761	2.145	2.624	2.977	3.326	3.787
15	.866	1.341	1.753	2.131	2.602	2.947	3.286	3.733
16	.865	1.337	1.746	2.120	2.583	2.921	3.252	3.686
17	.863	1.333	1.740	2.110	2.567	2.898	3.222	3.646
18	.862	1.330	1.734	2.101	2.552	2.878	3.197	3.611
19	.861	1.328	1.729	2.093	2.539	2.861	3.174	3.579
20	.860	1.325	1.725	2.086	2.528	2.845	3.153	3.552
21	.859	1.323	1.721	2.080	2.518	2.831	3.135	3.527
22	.858	1.321	1.717	2.074	2.508	2.819	3.119	3.505
23	.858	1.319	1.714	2.069	2.500	2.807	3.104	3.485
24	.857	1.318	1.711	2.064	2.492	2.797	3.091	3.467
25	.856	1.316	1.708	2.060	2.485	2.787	3.078	3.450
26	.856	1.315	1.706	2.056	2.479	2.779	3.067	3.435
27	.855	1.314	1.703	2.052	2.473	2.771	3.057	3.421
28	.855	1.313	1.701	2.048	2.467	2.763	3.047	3.408
29	.854	1.311	1.699	2.045	2.462	2.756	3.038	3.396
30	.854	1.310	1.697	2.042	2.457	2.750	3.030	3.385
40	.851	1.303	1.684	2.021	2.423	2.704	2.971	3.307
50	.849	1.299	1.676	2.009	2.403	2.678	2.937	3.261
60	.848	1.296	1.671	2.000	2.390	2.660	2.915	3.232
80	.846	1.292	1.664	1.990	2.374	2.639	2.887	3.195
120	.845	1.289	1.658	1.980	2.358	2.617	2.860	3.160
∞	.842	1.282	1.645	1.960	2.326	2.576	2.807	3.090

TABLE B.4   **Critical Values of the Chi-Square Distribution**

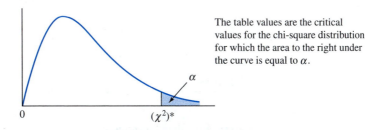

The table values are the critical values for the chi-square distribution for which the area to the right under the curve is equal to α.

Amount of α in right-hand tail

Degrees of freedom	.995	.990	.975	.950	.900	.200	.100	.050	.025	.010	.005	.0025	.001
1	0.0000393	0.000157	0.000982	0.00393	0.0158	1.64	2.71	3.84	5.02	6.63	7.88	9.14	10.8
2	0.0100	0.0201	0.0506	0.103	0.211	3.22	4.61	5.99	7.38	9.21	10.6	12.0	13.8
3	0.0717	0.115	0.216	0.352	0.584	4.64	6.25	7.81	9.35	11.3	12.8	14.3	16.3
4	0.207	0.297	0.484	0.711	1.0636	5.99	7.78	9.49	11.1	13.3	14.9	16.4	18.5
5	0.412	0.554	0.831	1.15	1.61	7.29	9.24	11.1	12.8	15.1	16.7	18.4	20.5
6	0.676	0.872	1.24	1.64	2.20	8.56	10.6	12.6	14.5	16.8	18.5	20.3	22.5
7	0.989	1.24	1.69	2.17	2.83	9.80	12.0	14.1	16.0	18.5	20.3	22.0	24.3
8	1.34	1.65	2.18	2.73	3.49	11.0	13.4	15.5	17.5	20.1	22.0	23.8	26.1
9	1.73	2.09	2.70	3.33	4.17	12.2	14.7	16.9	19.0	21.7	23.6	25.5	27.9
10	2.16	2.56	3.25	3.94	4.87	13.4	16.0	18.3	20.5	23.2	25.2	27.1	29.6
11	2.60	3.05	3.82	4.58	5.58	14.6	17.3	19.7	21.9	24.7	26.8	28.7	31.3
12	3.07	3.57	4.40	5.23	6.30	15.8	18.5	21.0	23.3	26.2	28.3	30.3	32.9
13	3.57	4.11	5.01	5.90	7.04	17.0	19.8	22.4	24.7	27.7	29.8	31.9	34.5
14	4.07	4.66	5.63	6.57	7.79	18.2	21.1	23.7	26.1	29.1	31.3	33.4	36.1
15	4.60	5.23	6.26	7.26	8.55	19.3	22.3	25.0	27.5	30.6	32.8	35.0	37.7
16	5.14	5.81	6.91	7.96	9.31	20.5	23.5	26.3	28.8	32.0	34.3	36.5	39.3
17	5.70	6.41	7.56	8.67	10.1	21.6	24.8	27.6	30.2	33.4	35.7	38.0	40.8
18	6.26	7.01	8.23	9.39	10.9	22.8	26.0	28.9	31.5	34.8	37.2	39.4	42.3
19	6.84	7.63	8.91	10.1	11.7	23.9	27.2	30.1	32.9	36.2	38.6	40.9	43.8
20	7.43	8.26	9.59	10.9	12.4	25.0	28.4	31.4	34.2	37.6	40.0	42.3	45.3
21	8.03	8.90	10.3	11.6	13.2	26.2	29.6	32.7	35.5	38.9	41.4	43.8	46.8
22	8.64	9.54	11.0	12.3	14.0	27.3	30.8	33.9	36.8	40.3	42.8	45.2	48.3
23	9.26	10.2	11.7	13.1	14.8	28.4	32.0	35.2	38.1	41.6	44.2	46.6	49.7
24	9.89	10.9	12.4	13.8	15.7	29.6	33.2	36.4	39.4	43.0	45.6	48.0	51.2
25	10.5	11.5	13.1	14.6	16.5	30.7	34.4	37.7	40.6	44.3	46.9	49.4	52.6
26	11.2	12.2	13.8	15.4	17.3	31.8	35.6	38.9	41.9	45.6	48.3	50.8	54.1
27	11.8	12.9	14.6	16.2	18.1	32.9	36.7	40.1	43.2	47.0	49.6	52.2	55.5
28	12.5	13.6	15.3	16.9	18.9	34.0	37.9	41.3	44.5	48.3	51.0	53.6	56.9
29	13.1	14.3	16.0	17.7	19.8	35.1	39.1	42.6	45.7	49.6	52.3	55.0	58.3
30	13.8	15.0	16.8	18.5	20.6	36.3	40.3	43.8	47.0	50.9	53.7	56.3	59.7
40	20.7	22.2	24.4	26.5	29.1	47.3	51.8	55.8	59.3	63.7	66.8	69.7	73.4
50	28.0	29.7	32.4	34.8	37.7	58.2	63.2	67.5	71.4	76.2	79.5	82.7	86.7
60	35.5	37.5	40.5	43.2	46.5	69.0	74.4	79.1	83.3	88.4	92.0	95.3	99.6
70	43.3	45.4	48.8	51.7	55.3	79.7	85.5	90.5	95.0	100.0	104.2	107.8	112.3
80	51.2	53.5	57.2	60.4	64.3	90.4	96.6	101.9	106.6	112.3	116.3	120.1	124.8
90	59.2	61.8	65.6	69.1	73.3	101.1	107.6	113.1	118.1	124.1	128.3	132.2	137.2
100	67.3	70.1	74.2	77.9	82.4	111.7	118.5	124.3	129.6	135.8	140.2	144.3	149.4

TABLE B.5   Critical Values of the F-Distribution ($\alpha = .10$)

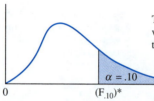

The table values are the critical values of F for which the area under the curve to the right is equal to .10.

$\alpha = .10$

$(F_{.10})^*$

Degrees of freedom for numerator

	1	2	3	4	5	6	7	8	9
1	39.86	49.50	53.59	55.83	57.24	58.20	58.91	59.44	59.86
2	8.53	9.00	9.16	9.24	9.29	9.33	9.35	9.37	9.38
3	5.54	5.46	5.39	5.34	5.31	5.28	5.27	5.25	5.24
4	4.54	4.32	4.19	4.11	4.05	4.01	3.98	3.95	3.94
5	4.06	3.78	3.62	3.52	3.45	3.40	3.37	3.34	3.32
6	3.78	3.46	3.29	3.18	3.11	3.05	3.01	2.98	2.96
7	3.59	3.26	3.07	2.96	2.88	2.83	2.78	2.75	2.72
8	3.46	3.11	2.92	2.81	2.73	2.67	2.62	2.59	2.56
9	3.36	3.01	2.81	2.69	2.61	2.55	2.51	2.47	2.44
10	3.29	2.92	2.73	2.61	2.52	2.46	2.41	2.38	2.35
11	3.23	2.86	2.66	2.54	2.45	2.39	2.34	2.30	2.27
12	3.18	2.81	2.61	2.48	2.39	2.33	2.28	2.24	2.21
13	3.14	2.76	2.56	2.43	2.35	2.28	2.23	2.20	2.16
14	3.10	2.73	2.52	2.39	2.31	2.24	2.19	2.15	2.12
15	3.07	2.70	2.49	2.36	2.27	2.21	2.16	2.12	2.09
16	3.05	2.67	2.46	2.33	2.24	2.18	2.13	2.09	2.06
17	3.03	2.64	2.44	2.31	2.22	2.15	2.10	2.06	2.03
18	3.01	2.62	2.42	2.29	2.20	2.13	2.08	2.04	2.00
19	2.99	2.61	2.40	2.27	2.18	2.11	2.06	2.02	1.98
20	2.97	2.59	2.38	2.25	2.16	2.09	2.04	2.00	1.96
21	2.96	2.57	2.36	2.23	2.14	2.08	2.02	1.98	1.95
22	2.95	2.56	2.35	2.22	2.13	2.06	2.01	1.97	1.93
23	2.94	2.55	2.34	2.21	2.11	2.05	1.99	1.95	1.92
24	2.93	2.54	2.33	2.19	2.10	2.04	1.98	1.94	1.91
25	2.92	2.53	2.32	2.18	2.09	2.02	1.97	1.93	1.89
26	2.91	2.52	2.31	2.17	2.08	2.01	1.96	1.92	1.88
27	2.90	2.51	2.30	2.17	2.07	2.00	1.95	1.91	1.87
28	2.89	2.50	2.29	2.16	2.06	2.00	1.94	1.90	1.87
29	2.89	2.50	2.28	2.15	2.06	1.99	1.93	1.89	1.86
30	2.88	2.49	2.28	2.14	2.05	1.98	1.93	1.88	1.85
40	2.84	2.44	2.23	2.09	2.00	1.93	1.87	1.83	1.79
60	2.79	2.39	2.18	2.04	1.95	1.87	1.82	1.77	1.74
120	2.75	2.35	2.13	1.99	1.90	1.82	1.77	1.72	1.68
∞	2.71	2.30	2.08	1.94	1.85	1.77	1.72	1.67	1.63

Degrees of freedom for denominator

TABLE B.5   Critical Values of the F-Distribution ($\alpha = .10$) (continued)

		Degrees of freedom for numerator								
	10	**12**	**15**	**20**	**24**	**30**	**40**	**60**	**120**	**∞**
1	60.19	60.71	61.22	61.74	62.00	62.26	62.53	62.79	63.06	63.33
2	9.39	9.41	9.42	9.44	9.45	9.46	9.47	9.47	9.48	9.49
3	5.23	5.22	5.20	5.18	5.18	5.17	5.16	5.15	5.14	5.13
4	3.92	3.90	3.87	3.84	3.83	3.82	3.80	3.79	3.78	3.76
5	3.30	3.27	3.24	3.21	3.19	3.17	3.16	3.14	3.12	3.10
6	2.94	2.90	2.87	2.84	2.82	2.80	2.78	2.76	2.74	2.72
7	2.70	2.67	2.63	2.59	2.58	2.56	2.54	2.51	2.49	2.47
8	2.54	2.50	2.46	2.42	2.40	2.38	2.36	2.34	2.32	2.29
9	2.42	2.38	2.34	2.30	2.28	2.25	2.23	2.21	2.18	2.16
10	2.32	2.28	2.24	2.20	2.18	2.16	2.13	2.11	2.08	2.06
11	2.25	2.21	2.17	2.12	2.10	2.08	2.05	2.03	2.00	1.97
12	2.19	2.15	2.10	2.06	2.04	2.01	1.99	1.96	1.93	1.90
13	2.14	2.10	2.05	2.01	1.98	1.96	1.93	1.90	1.88	1.85
14	2.10	2.05	2.01	1.96	1.94	1.91	1.89	1.86	1.83	1.80
15	2.06	2.02	1.97	1.92	1.90	1.87	1.85	1.82	1.79	1.76
16	2.03	1.99	1.94	1.89	1.87	1.84	1.81	1.78	1.75	1.72
17	2.00	1.96	1.91	1.86	1.84	1.81	1.78	1.75	1.72	1.69
18	1.98	1.93	1.89	1.84	1.81	1.78	1.75	1.72	1.69	1.66
19	1.96	1.91	1.86	1.81	1.79	1.76	1.73	1.70	1.67	1.63
20	1.94	1.89	1.84	1.79	1.77	1.74	1.71	1.68	1.64	1.61
21	1.92	1.87	1.83	1.78	1.75	1.72	1.69	1.66	1.62	1.59
22	1.90	1.86	1.81	1.76	1.73	1.70	1.67	1.64	1.60	1.57
23	1.89	1.84	1.80	1.74	1.72	1.69	1.66	1.62	1.59	1.55
24	1.88	1.83	1.78	1.73	1.70	1.67	1.64	1.61	1.57	1.53
25	1.87	1.82	1.77	1.72	1.69	1.66	1.63	1.59	1.56	1.52
26	1.86	1.81	1.76	1.71	1.68	1.65	1.61	1.58	1.54	1.50
27	1.85	1.80	1.75	1.70	1.67	1.64	1.60	1.57	1.53	1.49
28	1.84	1.79	1.74	1.69	1.66	1.63	1.59	1.56	1.52	1.48
29	1.83	1.78	1.73	1.68	1.65	1.62	1.58	1.55	1.51	1.47
30	1.82	1.77	1.72	1.67	1.64	1.61	1.57	1.54	1.50	1.46
40	1.76	1.71	1.66	1.61	1.57	1.54	1.51	1.47	1.42	1.38
60	1.71	1.66	1.60	1.54	1.51	1.48	1.44	1.40	1.35	1.29
120	1.65	1.60	1.55	1.48	1.45	1.41	1.37	1.32	1.26	1.19
∞	1.60	1.55	1.49	1.42	1.38	1.34	1.30	1.24	1.17	1.00

Degrees of freedom for denominator

TABLE B.5   Critical Values of the F-Distribution ($\alpha = .05$)

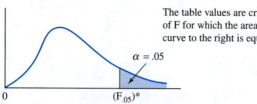

The table values are critical values of F for which the area under the curve to the right is equal to .05.

$\alpha = .05$

0 $(F_{.05})^*$

	Degrees of freedom for numerator								
	1	**2**	**3**	**4**	**5**	**6**	**7**	**8**	**9**
1	161.4	199.5	215.7	224.6	230.2	234.0	236.8	238.9	240.5
2	18.51	19.00	19.16	19.25	19.30	19.33	19.35	19.37	19.38
3	10.13	9.55	9.28	9.12	9.01	8.94	8.89	8.85	8.81
4	7.71	6.94	6.59	6.39	6.26	6.16	6.09	6.04	6.00
5	6.61	5.79	5.41	5.19	5.05	4.95	4.88	4.82	4.77
6	5.99	5.14	4.76	4.53	4.39	4.28	4.21	4.15	4.10
7	5.59	4.74	4.35	4.12	3.97	3.87	3.79	3.73	3.68
8	5.32	4.46	4.07	3.84	3.69	3.58	3.50	3.44	3.39
9	5.12	4.26	3.86	3.63	3.48	3.37	3.29	3.23	3.18
10	4.96	4.10	3.71	3.48	3.33	3.22	3.14	3.07	3.02
11	4.84	3.98	3.59	3.36	3.20	3.09	3.01	2.95	2.90
12	4.75	3.89	3.49	3.26	3.11	3.00	2.91	2.85	2.80
13	4.67	3.81	3.41	3.18	3.03	2.92	2.83	2.77	2.71
14	4.60	3.74	3.34	3.11	2.96	2.85	2.76	2.70	2.65
15	4.54	3.68	3.29	3.06	2.90	2.79	2.71	2.64	2.59
16	4.49	3.63	3.24	3.01	2.85	2.74	2.66	2.59	2.54
17	4.45	3.59	3.20	2.96	2.81	2.70	2.61	2.55	2.49
18	4.41	3.55	3.16	2.93	2.77	2.66	2.58	2.51	2.46
19	4.38	3.52	3.13	2.90	2.74	2.63	2.54	2.48	2.42
20	4.35	3.49	3.10	2.87	2.71	2.60	2.51	2.45	2.39
21	4.32	3.47	3.07	2.84	2.68	2.57	2.49	2.42	2.37
22	4.30	3.44	3.05	2.82	2.66	2.55	2.46	2.40	2.34
23	4.28	3.42	3.03	2.80	2.64	2.53	2.44	2.37	2.32
24	4.26	3.40	3.01	2.78	2.62	2.51	2.42	2.36	2.30
25	4.24	3.39	2.99	2.76	2.60	2.49	2.40	2.34	2.28
30	4.17	3.32	2.92	2.69	2.53	2.42	2.33	2.27	2.21
40	4.08	3.23	2.84	2.61	2.45	2.34	2.25	2.18	2.12
60	4.00	3.15	2.76	2.53	2.37	2.25	2.17	2.10	2.04
120	3.92	3.07	2.68	2.45	2.29	2.17	2.09	2.02	1.96
∞	3.84	3.00	2.60	2.37	2.21	2.10	2.01	1.94	1.88

Degrees of freedom for denominator

TABLE B.5   Critical Values of the F-Distribution ($\alpha = .05$) (continued)

		10	12	15	20	24	30	40	60	120	∞
						Degrees of freedom for numerator					
	1	241.9	243.9	245.9	248.0	249.1	250.1	251.1	252.2	253.3	254.3
	2	19.40	19.41	19.43	19.45	19.45	19.46	19.47	19.48	19.49	19.50
	3	8.79	8.74	8.70	8.66	8.64	8.62	8.59	8.57	8.55	8.53
	4	5.96	5.91	5.86	5.80	5.77	5.75	5.72	5.69	5.66	5.63
	5	4.74	4.68	4.62	4.56	4.53	4.50	4.46	4.43	4.40	4.36
	6	4.06	4.00	3.94	3.87	3.84	3.81	3.77	3.74	3.70	3.67
	7	3.64	3.57	3.51	3.44	3.41	3.38	3.34	3.30	3.27	3.23
	8	3.35	3.28	3.22	3.15	3.12	3.08	3.04	3.01	2.97	2.93
	9	3.14	3.07	3.01	2.94	2.90	2.86	2.83	2.79	2.75	2.71
	10	2.98	2.91	2.85	2.77	2.74	2.70	2.66	2.62	2.58	2.54
Degrees of freedom for denominator	11	2.85	2.79	2.72	2.65	2.61	2.57	2.53	2.49	2.45	2.40
	12	2.75	2.69	2.62	2.54	2.51	2.47	2.43	2.38	2.34	2.30
	13	2.67	2.60	2.53	2.46	2.42	2.38	2.34	2.30	2.25	2.21
	14	2.60	2.53	2.46	2.39	2.35	2.31	2.27	2.22	2.18	2.13
	15	2.54	2.48	2.40	2.33	2.29	2.25	2.20	2.16	2.11	2.07
	16	2.49	2.42	2.35	2.28	2.24	2.19	2.15	2.11	2.06	2.01
	17	2.45	2.38	2.31	2.23	2.19	2.15	2.10	2.06	2.01	1.96
	18	2.41	2.34	2.27	2.19	2.15	2.11	2.06	2.02	1.97	1.92
	19	2.38	2.31	2.23	2.16	2.11	2.07	2.03	1.98	1.93	1.88
	20	2.35	2.28	2.20	2.12	2.08	2.04	1.99	1.95	1.90	1.84
	21	2.32	2.25	2.18	2.10	2.05	2.01	1.96	1.92	1.87	1.81
	22	2.30	2.23	2.15	2.07	2.03	1.98	1.94	1.89	1.84	1.78
	23	2.27	2.20	2.13	2.05	2.01	1.96	1.91	1.86	1.81	1.76
	24	2.25	2.18	2.11	2.03	1.98	1.94	1.89	1.84	1.79	1.73
	25	2.24	2.16	2.09	2.01	1.96	1.92	1.87	1.82	1.77	1.71
	30	2.16	2.09	2.01	1.93	1.89	1.84	1.79	1.74	1.68	1.62
	40	2.08	2.00	1.92	1.84	1.79	1.74	1.69	1.64	1.58	1.51
	60	1.99	1.92	1.84	1.75	1.70	1.65	1.59	1.53	1.47	1.39
	120	1.91	1.83	1.75	1.66	1.61	1.55	1.50	1.43	1.35	1.25
	∞	1.83	1.75	1.67	1.57	1.52	1.46	1.39	1.32	1.22	1.00

TABLE B.5 Critical Values of the F-Distribution ($\alpha = .025$)

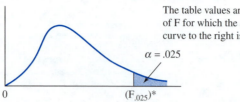

The table values are critical values of F for which the area under the curve to the right is equal to .025.

$\alpha = .025$

$(F_{.025})^*$

Degrees of freedom for numerator

		1	2	3	4	5	6	7	8	9
	1	648	800	864	900	922	937	948	957	963
	2	38.51	39.00	39.17	39.25	39.30	39.33	39.36	39.37	39.39
	3	17.44	16.04	15.44	15.10	14.88	14.73	14.62	14.54	14.47
	4	12.22	10.65	9.98	9.60	9.36	9.20	9.07	8.98	8.90
	5	10.01	8.43	7.76	7.39	7.15	6.98	6.85	6.76	6.68
	6	8.81	7.26	6.60	6.23	5.99	5.82	5.70	5.60	5.52
	7	8.07	6.54	5.89	5.52	5.29	5.12	4.99	4.90	4.82
	8	7.57	6.06	5.42	5.05	4.82	4.65	4.53	4.43	4.36
	9	7.21	5.71	5.08	4.72	4.48	4.32	4.20	4.10	4.03
Degrees of freedom for denominator	10	6.94	5.46	4.83	4.47	4.24	4.07	3.95	3.85	3.78
	11	6.72	5.26	4.63	4.28	4.04	3.88	3.76	3.66	3.59
	12	6.55	5.10	4.47	4.12	3.89	3.73	3.61	3.51	3.44
	13	6.41	4.97	4.35	4.00	3.77	3.60	3.48	3.39	3.31
	14	6.30	4.86	4.24	3.89	3.66	3.50	3.38	3.29	3.21
	15	6.20	4.77	4.15	3.80	3.58	3.41	3.29	3.20	3.12
	16	6.12	4.69	4.08	3.73	3.50	3.34	3.22	3.12	3.05
	17	6.04	4.62	4.01	3.66	3.44	3.28	3.16	3.06	2.98
	18	5.98	4.56	3.95	3.61	3.38	3.22	3.10	3.01	2.93
	19	5.92	4.51	3.90	3.56	3.33	3.17	3.05	2.96	2.88
	20	5.87	4.46	3.86	3.51	3.29	3.13	3.01	2.91	2.84
	21	5.83	4.42	3.82	3.48	3.25	3.09	2.97	2.87	2.80
	22	5.79	4.38	3.78	3.44	3.22	3.05	2.93	2.84	2.76
	23	5.75	4.35	3.75	3.41	3.18	3.02	2.90	2.81	2.73
	24	5.72	4.32	3.72	3.38	3.15	2.99	2.87	2.78	2.70
	25	5.69	4.29	3.69	3.35	3.13	2.97	2.85	2.75	2.68
	30	5.57	4.18	3.59	3.25	3.03	2.87	2.75	2.65	2.57
	40	5.42	4.05	3.46	3.13	2.90	2.74	2.62	2.53	2.45
	60	5.29	3.93	3.34	3.01	2.79	2.63	2.51	2.41	2.33
	120	5.15	3.80	3.23	2.89	2.67	2.52	2.39	2.30	2.22
	∞	5.02	3.69	3.12	2.79	2.57	2.41	2.29	2.19	2.11

TABLE B.5   Critical Values of the F-Distribution (α = .025) (continued)

	Degrees of freedom for numerator									
	10	**12**	**15**	**20**	**24**	**30**	**40**	**60**	**120**	**∞**
1	969	977	985	993	997	1,001	1,006	1,010	1,014	1,018
2	39.40	39.41	39.43	39.45	39.46	39.46	39.47	39.48	39.49	39.50
3	14.42	14.34	14.25	14.17	14.12	14.08	14.04	13.99	13.95	13.90
4	8.84	8.75	8.66	8.56	8.51	8.46	8.41	8.36	8.31	8.26
5	6.62	6.52	6.43	6.33	6.28	6.23	6.18	6.12	6.07	6.02
6	5.46	5.37	5.27	5.17	5.12	5.07	5.01	4.96	4.90	4.85
7	4.76	4.67	4.57	4.47	4.42	4.36	4.31	4.25	4.20	4.14
8	4.30	4.20	4.10	4.00	3.95	3.89	3.84	3.78	3.73	3.67
9	3.96	3.87	3.77	3.67	3.61	3.56	3.51	3.45	3.39	3.33
10	3.72	3.62	3.52	3.42	3.37	3.31	3.26	3.20	3.14	3.08
11	3.53	3.43	3.33	3.23	3.17	3.12	3.06	3.00	2.94	2.88
12	3.37	3.28	3.18	3.07	3.02	2.96	2.91	2.85	2.79	2.72
13	3.25	3.15	3.05	2.95	2.89	2.84	2.78	2.72	2.66	2.60
14	3.15	3.05	2.95	2.84	2.79	2.73	2.67	2.61	2.55	2.49
15	3.06	2.96	2.86	2.76	2.70	2.64	2.59	2.52	2.46	2.40
16	2.99	2.89	2.79	2.68	2.63	2.57	2.51	2.45	2.38	2.32
17	2.92	2.82	2.72	2.62	2.56	2.50	2.44	2.38	2.32	2.25
18	2.87	2.77	2.67	2.56	2.50	2.44	2.38	2.32	2.26	2.19
19	2.82	2.72	2.62	2.51	2.45	2.39	2.33	2.27	2.20	2.13
20	2.77	2.68	2.57	2.46	2.41	2.35	2.29	2.22	2.16	2.09
21	2.73	2.64	2.53	2.42	2.37	2.31	2.25	2.18	2.11	2.04
22	2.70	2.60	2.50	2.39	2.33	2.27	2.21	2.14	2.08	2.00
23	2.67	2.57	2.47	2.36	2.30	2.24	2.18	2.11	2.04	1.97
24	2.64	2.54	2.44	2.33	2.27	2.21	2.15	2.08	2.01	1.94
25	2.61	2.51	2.41	2.30	2.24	2.18	2.12	2.05	1.98	1.91
30	2.51	2.41	2.31	2.20	2.14	2.07	2.01	1.94	1.87	1.79
40	2.39	2.29	2.18	2.07	2.01	1.94	1.88	1.80	1.72	1.64
60	2.27	2.17	2.06	1.94	1.88	1.82	1.74	1.67	1.58	1.48
120	2.16	2.05	1.94	1.82	1.76	1.69	1.61	1.53	1.43	1.31
∞	2.05	1.94	1.83	1.71	1.64	1.57	1.48	1.39	1.27	1.00

Degrees of freedom for denominator

TABLE B.5   Critical Values of the F-Distribution ($\alpha = .01$)

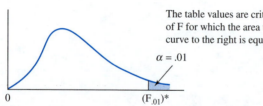

The table values are critical values of F for which the area under the curve to the right is equal to .01.

$\alpha = .01$

0

$(F_{.01})^*$

Degrees of freedom for numerator

	1	2	3	4	5	6	7	8	9
1	4,052	5,000	5,403	5,625	5,764	5,859	5,928	5,981	6,022
2	98.50	99.00	99.17	99.25	99.30	99.33	99.36	99.37	99.39
3	34.12	30.82	29.46	28.71	28.24	27.91	27.67	27.49	27.35
4	21.20	18.00	16.69	15.98	15.52	15.21	14.98	14.80	14.66
5	16.26	13.27	12.06	11.39	10.97	10.67	10.46	10.29	10.16
6	13.75	10.92	9.78	9.15	8.75	8.47	8.26	8.10	7.98
7	12.25	9.55	8.45	7.85	7.46	7.19	6.99	6.84	6.72
8	11.26	8.65	7.59	7.01	6.63	6.37	6.18	6.03	5.91
9	10.56	8.02	6.99	6.42	6.06	5.80	5.61	5.47	5.35
10	10.04	7.56	6.55	5.99	5.64	5.39	5.20	5.06	4.94
11	9.65	7.21	6.22	5.67	5.32	5.07	4.89	4.74	4.63
12	9.33	6.93	5.95	5.41	5.06	4.82	4.64	4.50	4.39
13	9.07	6.70	5.74	5.21	4.86	4.62	4.44	4.30	4.19
14	8.86	6.51	5.56	5.04	4.69	4.46	4.28	4.14	4.03
15	8.68	6.36	5.42	4.89	4.56	4.32	4.14	4.00	3.89
16	8.53	6.23	5.29	4.77	4.44	4.20	4.03	3.89	3.78
17	8.40	6.11	5.18	4.67	4.34	4.10	3.93	3.79	3.68
18	8.29	6.01	5.09	4.58	4.25	4.01	3.84	3.71	3.60
19	8.18	5.93	5.01	4.50	4.17	3.94	3.77	3.63	3.52
20	8.10	5.85	4.94	4.43	4.10	3.87	3.70	3.56	3.46
21	8.02	5.78	4.87	4.37	4.04	3.81	3.64	3.51	3.40
22	7.95	5.72	4.82	4.31	3.99	3.76	3.59	3.45	3.35
23	7.88	5.66	4.76	4.26	3.94	3.71	3.54	3.41	3.30
24	7.82	5.61	4.72	4.22	3.90	3.67	3.50	3.36	3.26
25	7.77	5.57	4.68	4.18	3.85	3.63	3.46	3.32	3.22
30	7.56	5.39	4.51	4.02	3.70	3.47	3.30	3.17	3.07
40	7.31	5.18	4.31	3.83	3.51	3.29	3.12	2.99	2.89
60	7.08	4.98	4.13	3.65	3.34	3.12	2.95	2.82	2.72
120	6.85	4.79	3.95	3.48	3.17	2.96	2.79	2.66	2.56
∞	6.63	4.61	3.78	3.32	3.02	2.80	2.64	2.51	2.41

Degrees of freedom for denominator

TABLE B.5   Critical Values of the F-Distribution ($\alpha = .01$) (continued)

		\multicolumn{10}{c}{Degrees of freedom for numerator}
		10	**12**	**15**	**20**	**24**	**30**	**40**	**60**	**120**	**∞**
	1	6,056	6,106	6,157	6,209	6,235	6,261	6,287	6,313	6,339	6,366
	2	99.40	99.42	99.43	99.45	99.46	99.47	99.47	99.48	99.49	99.50
	3	27.23	27.05	26.87	26.69	26.60	26.50	26.41	26.32	26.22	26.13
	4	14.55	14.37	14.20	14.02	13.93	13.84	13.75	13.65	13.56	13.46
	5	10.05	9.89	9.72	9.55	9.47	9.38	9.29	9.20	9.11	9.02
	6	7.87	7.72	7.56	7.40	7.31	7.23	7.14	7.06	6.97	6.88
	7	6.62	6.47	6.31	6.16	6.07	5.99	5.91	5.82	5.74	5.65
	8	5.81	5.67	5.52	5.36	5.28	5.20	5.12	5.03	4.95	4.86
	9	5.26	5.11	4.96	4.81	4.73	4.65	4.57	4.48	4.40	4.31
	10	4.85	4.71	4.56	4.41	4.33	4.25	4.17	4.08	4.00	3.91
	11	4.54	4.40	4.25	4.10	4.02	3.94	3.86	3.78	3.69	3.60
	12	4.30	4.16	4.01	3.86	3.78	3.70	3.62	3.54	3.45	3.36
	13	4.10	3.96	3.82	3.66	3.59	3.51	3.43	3.34	3.25	3.17
	14	3.94	3.80	3.66	3.51	3.43	3.35	3.27	3.18	3.09	3.00
	15	3.80	3.67	3.52	3.37	3.29	3.21	3.13	3.05	2.96	2.87
	16	3.69	3.55	3.41	3.26	3.18	3.10	3.02	2.93	2.84	2.75
	17	3.59	3.46	3.31	3.16	3.08	3.00	2.92	2.83	2.75	2.65
	18	3.51	3.37	3.23	3.08	3.00	2.92	2.84	2.75	2.66	2.57
	19	3.43	3.30	3.15	3.00	2.92	2.84	2.76	2.67	2.58	2.49
	20	3.37	3.23	3.09	2.94	2.86	2.78	2.69	2.61	2.52	2.42
	21	3.31	3.17	3.03	2.88	2.80	2.72	2.64	2.55	2.46	2.36
	22	3.26	3.12	2.98	2.83	2.75	2.67	2.58	2.50	2.40	2.31
	23	3.21	3.07	2.93	2.78	2.70	2.62	2.54	2.45	2.35	2.26
	24	3.17	3.03	2.89	2.74	2.66	2.58	2.49	2.40	2.31	2.21
	25	3.13	2.99	2.85	2.70	2.62	2.54	2.45	2.36	2.27	2.17
	30	2.98	2.84	2.70	2.55	2.47	2.39	2.30	2.21	2.11	2.01
	40	2.80	2.66	2.52	2.37	2.29	2.20	2.11	2.02	1.92	1.80
	60	2.63	2.50	2.35	2.20	2.12	2.03	1.94	1.84	1.73	1.60
	120	2.47	2.34	2.19	2.03	1.95	1.86	1.76	1.66	1.53	1.38
	∞	2.32	2.18	2.04	1.88	1.79	1.70	1.59	1.47	1.32	1.00

Degrees of freedom for denominator

Answers to Selected Odd-Numbered Exercises

Your answers may not agree completely because of round-off error. Most solutions were found using the Minitab computer package.

Chapter 1

1.1 The population of units is all students at the university, but the data in the sample would not generalize because it is taken from a specific psychology class.

1.3 Individual unit: An accident
The population of units is all accidents.
Categorical variables: Alcohol involved (Yes/No), Make and model of vehicles, Gender of driver
Numerical variables: How many vehicles involved, How many occupants, Age of driver
The sample size is 1000.

1.5 **a.** It would be impossible to test the drug on **all** potential users of the product.
b. If we distributed free samples to everyone then no one would be required to purchase the product.
c. It would be impossible to count every tree in the forest.
d. If we determined the life of every battery we would deplete our inventory.

1.7 Gender—categorical, Major—categorical, Grade point average—numerical, Number of times using placement service—numerical, Type of employment—categorical. The data set is multivariate because several variables are measured.

1.9 **a.** Threshold reaction time is numerical and continuous. The population of interest is persons subject to emotional stress.
b. Occupation is categorical and Salary is numerical. The population of interest is Vietnam veterans discharged prior to 1974.

1.11 All distributions have salaries that extend out on the right tail, some more than others. The salaries for the White Sox are less spread out than the other distributions. The majority of their players make near the minimum salary.

1.13 **a.** A city is an individual unit.
b. Two variables are measured: 1980 commuting time and 1990 commuting time. The values of the two variables measured on Charlotte, NC are 19.9 and 21.6.
c. The numerical values are estimates of commuting times obtained from 1990 census data.

d. Houston had the greatest change in commute time from 1980 to 1990 of 3.6 minutes.
g. Generally commuting times have gone up from 1980 to 1990.

1.15 **a.** The variability of the grade point averages for the Junior class is slightly less than the other three classes. The Junior class distribution is centered lower than the other three classes. The Sophomore class distribution is centered slightly higher than the Freshman class.
b. There does not appear to be a difference between the distributions for men and women.
c. Those who drop 0 courses generally have higher SAT scores. Those who drop 4 courses generally have lower SAT scores.

1.19

Eye color	Frequency	Percent
blue	124	31.6%
brown	150	38.3
green	15	3.8
hazel	103	26.3
Total	392	100.0

1.23 **a.** Larceny theft had the highest percent of 64.1785%.
b. Larceny theft and Vehicle theft amounts to 69.28% of all crimes.

1.27

number	Count	Percent
6	3	15.00
7	4	20.00
8	6	30.00
9	4	20.00
10	3	15.00 The data are
N=	20	perfectly symmetric.

1.29 **a.** The first column sums to 101%. It does not sum to 100% because of roundoff error.
b. The heating sources for American Indians not on reservations are very similar to the heating sources for all U.S. households. The heating sources for American Indians living on reservations, however, differ substantially from those for all U.S. households. Most American Indians living on reservations still heat with wood. The differences are much easier to see in the comparison bar graph.
c. American Indians on reservations heat mostly with wood.
d. All U.S. households heat mostly with utility gas.

1.31 The life expectancies of men and women were about the same in 1920. Each decade after that, however, women tend to live longer than men. The graphs grew farther apart each decade until 1970. Since 1970 the difference has remained about the same.

1.33 Percent change in personal income from first quarter 2000 to second quarter 2000.

		Freq	Percent
Class 1	% change ≤ 0.5	2	3.92
Class 2	0.5 <% change ≤ 1.0	5	9.80
Class 3	1.0 <% change ≤ 1.5	13	25.49
Class 4	1.5 <% change ≤ 2.0	22	43.14
Class 5	% change < 2.0	9	17.65
Total		51	100.00

1.35 Until about 1980 the Dow-Jones Industrial Average stayed rather constant. In the 1990s there was a sharp increase in the average price.

1.37 The diplomats from the Ukraine are the worst offenders and the diplomats from S. Korea are the least offenders.

1.41 Stem-and-leaf of miller N = 20
Leaf Unit = 1.0

```
 2    1 67
 3    1 8
 5    2 01
10    2 22233
10    2 5555
 6    2 67
 4    2 9
 3    3 01
 1    3 3
```

This is a five-stem stem-and-leaf plot. The data are symmetric.

1.43 Stem-and-leaf of profit N = 20
Leaf Unit = 1.0

```
 2    5 89
 3    6 1
 3    6
 7    6 4455
 7    6
 8    6 8
 9    7 0
10    7 3
10    7 4455
 6    7 66667
 1    7
 1    8 1
```

These data appear to have two modes.

1.45 Stem-and-leaf of price N = 25
Leaf Unit = 0.010

```
 1     14 6
 9     14 88899999
(9)    15 000000011
 7     15 222
 4     15
 4     15
 4     15 9
 3     16 001
```

There is a mode at $1.50 and another mode at $1.60.

1.47 Stem-and-leaf of number N = 20
Leaf Unit = 1.0

```
 1     1 8
 2     2 4
 5     2 789
 7     3 12
(5)    3 56779
 8     4 12
 6     4
 6     5 4
 5     5 5568
 1     6 2
```

These data have a gap between 42 and 54. We should classify the distributed as bimodal.

1.51 The data are slightly skewed right. By recording the day of the week the hits were taken, the designer may see that some days are more likely to be hit than other days.

1.53 The observation 152 is clearly an outlier. It will increase the mean of the treatment group substantially. If we choose to discard it, we see that remaining treatment data are centered below the placebo group. It appears that the treatment drug has effectively reduced the LDL cholesterol in quail.

1.55 Both distributions are severely skewed right.

1.61 Standard deviation measures variability. Adding ten to each observation only shifts the data to the right ten points. It does not affect the variability so the standard deviation is not changed. The IQR also measures variability and thus is not affected by adding 10 to each observation.

1.63 Multiplying each observation by a constant results in the standard deviation and IQR being multiplied by that constant.

1.65 There is very little difference between the mean (= 18.273) and median (= 18.5). Because the data are reasonably symmetric a trimmed mean would be similar in value to the mean and median.

1.67

	Low	Q_1	M	Q_3	High
Original data	11	28.25	34.5	39.75	46
After subtracting ten	1	18.25	24.5	29.75	36

Each value in the five-number summary diagram is reduced by ten.

1.69 **a.** The data are close to being bell shaped so we apply the empirical rule. Approximately 68% of the data should be within one standard deviation of the mean, that is, between the limits 45.15 to 64.65. The actual percent is 18/30 = 60%. Approximately 95% of the data should be within two standard deviations of the mean, that is, between the limits 35.4 and 74.4. The actual percent is 29/30 = 96.67%. All the data are within three standard deviations of the mean.

   b. The z-score for 62 is $z = (62 - 54.9)/9.75 = .728$, so 62 is less than one standard deviation above the mean.

1.71 $\bar{x}_1 = 78.58$, $s_1 = 6.73$ $\bar{x}_2 = 80.81$, $s_2 = 9.06$ A score of 82 in Class 1 is $z = (82 - 78.58)/6.73 = .51$ standard deviations above the mean. A score of 82 in Class 2 is $z = (82 - 80.81)/9.06 = .13$ standard deviations above the mean. The score of 82 is higher in Class 1.

1.73 $\bar{x} = 2122$, $s = 798$ According to the empirical rule approximately 68% of the data should be between 1324 and 2920, that is within one standard deviation of the mean. For these data we find $13/23 = 56.5\%$ of the observations. This does not agree with the empirical rule because the distribution is not bell shaped. From the stem-and-leaf plot we see that the distribution of the data has possibly four different peaks and has observations heavily concentrated on the ends.

1.75 **a.** Stem-and-leaf of cholest N = 62
        Leaf Unit = 10

```
    1      1 6
    4      1 899
   13      2 001111111
   30      2 22223333333333333
  (11)     2 44444555555
   21      2 66666677777
   10      2 88
    8      3 000
    5      3 233
    2      3 5
    1      3
    1      3 9
```

  b. Variable   N   Mean Median StDev
        cholest   62 250.03 241.50 41.44

        Minimum   Maximum   Q1   Q3
         167.00   393.00 225.00 268.50

  c. One standard deviation above and below the mean gives the range 208.59 to 291.47. There are 48 out of 62 or 77.42% of the observations between these two values. Two standard deviations above and below the mean gives the range 167.15 to 332.91. There are 57 out of 62 or 91.94% of the observations between these two values. The empirical rule says that these percents should be 68% and 95% respectively. This is inconsistent with the empirical rule because the data are not bell shaped. It is skewed right.

  d. The largest cholesterol level is $z = (393 - 250.03)/41.44 = 3.45$ standard deviations above the mean. It most definitely is an outlier.

1.77 Variable   N   Mean Median   StDev
         placebo   20   67.20   68.00   13.97
         treatmen   10   62.2   49.0   35.4

         Minimum   Maximum   Q1   Q3
          44.00   97.00   56.25   75.00
          31.0   152.0   39.5   75.8

  a. The treatment group has a mean plasma LDL level 5 units lower than the mean for the placebo group.

  b. The treatment group has a median plasma LDL level 19 units lower than the median for the placebo group.

  c. The difference between the medians is large because the median of the treatment group is not affected by the extreme outlier at 152. The outlier, on the other hand, increases the mean of the treatment group so that it is comparable to the mean of the placebo group.

  d. The standard deviation of the treatment group is much greater than the standard deviation of the placebo group because of the extreme outlier at 152.

  e. The observation 152 is $z = (152 - 62.2)/35.4 = 2.54$ standard deviations above the mean. It is an outlier.

  f. Because the observation is so unusually large in comparison to the other treatment group observations, we are probably justified in removing it for further investigation of the difference between the two groups.

  g. Without the outlier the difference between means is 14.98 which is more in line with the observed difference between the medians.

  h. Without the outlier we see a clear difference between the plasma LDL levels of the two groups. It appears that the drug compound significantly reduced the LDL levels.

1.81 **a.** Low Q_1 M Q_3 High
         11   28.25   34.5   39.75   46

  b. $Q_1 - 1.5(IQR) = 28.25 - 1.5(11.5) = 11$
        $Q_3 + 1.5(IQR) = 39.75 + 1.5(11.5) = 57$
        There are no observations smaller than 11 nor larger than 57 so there are no outliers.

  c. The distribution is slightly skewed left.

  d. The median and interquartile range are measures of the center and spread of the distribution.

1.83 **a.** Stem-and-leaf of reaction N = 31
        Leaf Unit = 1.0

```
    1       3 1
    2       4 2
    3       5 5
    7       6 0446
   10       7 179
   14       8 0456
  (11)      9 00344667889
    6      10 1579
    2      11 0
    1      12 8
```

  b. Low Q_1 M Q_3 High
        31   71   90   98   128

  c. $Q_1 - 1.5(IQR) = 71 - 1.5(27) = 30.5$
        $Q_3 + 1.5(IQR) = 98 + 1.5(27) = 138.5$
        There are no observations smaller than 30.5 and no observation larger than 138.5 so there are no outliers.

  d. There is slight skewness to the left.

1.85 a. $Q_1 - 1.5(IQR) = 1456 - 1.5(1546) = -863$
$Q_3 + 1.5(IQR) = 3002 + 1.5(1546) = 5321$
There are no observations smaller than -863 and no observations larger than 5321 so there are no outliers.

b. Based on the boxplot, the distribution appears symmetric.

1.87 The distribution for Class 2 has greater variability. Students in Class 2 tend to score higher than students in Class 1, however, the median is greater in Class 1.

1.89 Brands A and C have greater variability than brands B and D. Brand B has little variability but underestimates the weight. Brands C and D give accurate estimates but C is too variable. Brand D gives a very precise accurate estimate of the 100 pounds.

1.91 a. The histogram shows a skewed right distribution.

b.
```
Variable   N    Mean   Median  StDev
payment   51  123.59  114.23  45.95

   Minimum  Maximum     Q1      Q3
    42.29   253.54    91.20  144.74
```
The percent of data that fall within one standard deviation of the mean is $35/51 = 68.6\%$ and the percent within two standard deviations of the mean is $50/51 = 98\%$. These percents are in close agreement with the empirical rule.

c. There is one outlier on the right tail.

d. Without the outlier the data are close to normally distributed. The mean and standard deviation are measures of the center and spread.

1.97 a. Population of interest: All oil stocks
b. Sample: Ten selected oil stocks
c. Variables: Price/earning ratio, Current price

1.99 a. Numerical— **b.** Numerical—
Continuous   Continous
c. Categorical **d.** Numerical—Discrete
e. Categorical **f.** Numerical—Discrete

1.103 $\bar{x} = 10$ Deviations from the mean are $-2, 2, 8, 6,$ $-1, 0, -8, -2, -3$, and they sum to 0.
$SS = (-2)^2 + (2)^2 + (8)^2 + (6)^2 + (-1)^2 + (0)^2 + (-8)^2 + (-2)^2 + (-3)^2 = 186$
$s = \sqrt{186/8} = 4.822$

1.105
Legal Drinking Age	frequency
21	720
20	50
19	60
18	110
undecided	60
Total	1000

1.107 a.
```
Stem-and-leaf of duration  N = 15
Leaf Unit = 0.10

  1      1 8
  4      2 134
 (4)     2 5568
  7      3 011
  4      3 58
  2      4
  2      4
  2      5
  2      5
  2      6
  2      6 5
  1      7 0
```
Aside from the two lengthy operations, the data appears bell shaped.

b. $\bar{x} = 3.267$, $M = 2.8$ After removing the two extreme observations: $\bar{x} = 2.731$, $M = 2.6$ By removing the two extreme observations, the mean is decreased by a large amount and the median is barely affected.

c. A new patient should be told that the surgery should take about 2.7 hours. That is, the two extreme observations should not be considered in the calculations.

1.109 Center—Mean, median, trimmed mean
Variability—Range, standard deviation, IQR

1.111 b.
Low	Q_1	M	Q_3	High
3.9	9.2	24.8	54.9	138.8

$IQR = 45.7$
$Q_1 - 1.5(IQR) = 9.2 - 1.5(45.7) = -59.35$
$Q_3 + 1.5(IQR) = 54.9 + 1.5(45.7) = 123.45$
There are no observations smaller than -59.35 and one observation larger than 123.45 so 138.8 is an outlier.

1.115 a. Categorical **b.** Numerical **c.** Categorical
d. Categorical **e.** Numerical

1.117
Quality	number
excellent	17
good	22
fair	13
needs improv	5
redo	8

1.119 a. There is only 101 dollars difference between the mean and median. The estimated cost is 4733 dollars.

b. The limits that are within two standard deviations of the mean are 2701 to 6765. There is only one observation, 2478, outside these limits.

1.121 The outlier in the 1980 data set is New York. The outliers in the 1990 data set are New York, Washington, and Houston. The median commuting time in 1990 was greater than in 1980. The variability of commuting times as measured with the range, interquartile range, or standard deviation is greater in 1980. Commuting times increased from 1980 to 1990 and became less variable.

1.123 a. The stem-and-leaf plot shows a skewed right distribution.

b. There is a difference between the mean and median but it is not large, only 5.07.

c. From the empirical rule, 95% of the prison sentences should be between $155.07 - 2(25.86) = 103.35$ months and $155.07 + 2(25.86) = 206.79$ months.

1.125 a. Stem-and-leaf of rate   N = 51
Leaf Unit = 0.10

```
    4      1 5555
   15      1 66666677777
  (18)     1 888888888899999999
   18      2 0000011111
    8      2 2223
    4      2 445
    1      2
    1      2
    1      3
    1      3 2
```

There is one unusual observation, Utah's birth rate of 3.2.

b. The mean is slightly larger than the median. The outlier, 3.2, will increase the mean.

Variable	N	Mean	Median	StDev
rate	51	1.9059	1.9000	0.3069

Minimum	Maximum	Q1	Q3
1.5000	3.2000	1.7000	2.1000

c.

Low	Q_1	M	Q_3	High
1.5	1.7	1.9	2.1	3.2

$IQR = 0.4$
$Q_1 - 1.5(IQR) = 1.7 - 1.5(0.4) = 1.1$
$Q_3 + 1.5(IQR) = 2.1 + 1.5(0.4) = 2.7$
There is one observation greater than 2.7 and no observation smaller than 1.1 so there is one outlier.

e. The distribution is skewed right but without the outlier it is close to symmetric.

f. The best estimate of the average fertility is the median 1.9. That rate will not maintain constant population growth.

1.127 a. Stem-and-leaf of pounds   N = 29
Leaf Unit = 0.10

```
    4      0 2348
   11      1 0233467
   (7)     2 0224469
   11      3 34459
    6      4 333
    3      5 46
    1      6
    1      7
    1      8
    1      9 1
```

The distribution has one unusually large observation (9.12).

b. There is one outlier. Without the outlier the distribution is close to symmetric.

c.

Variable	N	Mean	Median	StDev
pounds	29	2.757	2.430	1.913

Minimum	Maximum	Q1	Q3
0.280	9.120	1.385	3.765

Because of the outlier, the mean is substantially larger than the median. The median is the better of the two.

d. The measure of center is the median and the measure of spread is the interquartile range.

Chapter 2

2.1 There appears to be a strong linear trend between the two variables.

2.3 The relationship between the age of the cash register and the maintenance cost for the first five years is almost a perfect straight line. The observation at year seven may suggest a sharp upturn in maintenance cost for older machines; however, we cannot say for sure because there is only one observation. If that trend continues then the relationship is nonlinear.

2.5 Up to a size of 2000 sq ft, the energy consumption follows a very definite straight line positive trend. After 2000 sq ft, however, the energy consumption tends to level off, again in a straight line. Because the amount of energy consumed depends on the size of the home, we should graph kilowatt-hours/month on the vertical axis and the size of the home on the horizontal (x) axis. A straight line will fit the data but perhaps a better model would be to use one straight line for homes between 1000 and 2000 sq ft and use a second straight line (with different slope) for homes with more than 2000 sq ft.

2.7 There is no positive or negative trend in these data. The monthly sunspot activity does, however, have a very predictable cyclical behavior.

2.9 The temperature is somewhat constant for shallow depths but begins a very steep decline for depths between about 50 meters and 200 meters where it reaches the temperature of 15 degrees centigrade. After about 250 meters the temperature begins to level off but still declining to about 5 degrees centigrade at 1000 meters.

2.11 a. Weekly earnings have gone up every year since 1980. The positive trend is very predictable and rather dramatic.

b. The trend in wheat prices prior to 1987 was a definite negative linear trend.

c. In 1988 the price of wheat jumped significantly from $2.57/bushel to $3.72/bushel and remained at $3.72/bushel in 1989. In 1990 the price plummeted to $2.61/bushel. There may have been government supports in 1988 and 1989 to explain the high prices.

d. Between 1990 and 1995 the price of wheat increased with a substantial positive linear trend.

e. Since 1995 the price has plummeted to the level it was in the late 1980s.

f. Two variables that affect wheat price are supply/demand and weather.

2.13 b. When the inflation rate is low and medium there is a lot of variability in the pay increase. There is very little variability in pay increase when the inflation rate is high.

2.15 a. The outlier in body weight causes the horizontal scale of the scatterplot to be out of proportion with respect to the rest of the data. We cannot see a relationship that might exist between the variables.

b. The scatterplot of log (brain wt) to log (body wt) shows a definite positive linear trend. We see three bivariate outliers corresponding to the dinosaurs.

2.21 b. $r = 0.859$ This is consistent with the scatterplot.

c. The relationship is positive. Normally we would expect that as a child's age increases so do the number of gymnastic activities.

d. Strength, coordination, training

2.23 a. There is a negative linear trend.

b. There is more variability in y when x is large. This would cause predicted y values to be less reliable when x is large.

c. There are no outliers.

d. Because the slope is negative, the correlation will be negative.

e. Except for a shift along the x and y axes, the scatter will not change.

f. The correlation ($r = -0.814$) is not affected by shifts in the x or y axes.

2.25 The scatterplot indicates that a straight line will not fit the data. The correlation is low, $r = -0.35$. The relationship exhibited by the scatterplot is curvilinear if we ignore the data for 1988 and 1989. We should not be looking for a linear trend unless we divide the data in two parts, previous to 1988 and after 1990.

2.27 b. There is a negative linear trend.

c. $r = -0.884$ The correlation is negative because of the negative trend. As the average number of cigarettes smoked per day goes up the birth weight of the child goes down.

2.29 The correlation of increase and inflation $= 0.815$ suggest that a straight line should fit the data reasonably well; the scatterplot, however, suggest otherwise. As indicated in the solution of Exercise 2.12, there is a sharp increase in the pay increase when the inflation rate jumps from 5% to 7% but after 7% the pay increase is very steady even for inflation rates up to 14%. We should not use a straight line to describe the relationship over the entire period.

2.31 Correlation of 1994 and 1995 $= 0.822$. Based on the correlation and the scatterplot, a straight line will fit the data reasonably well.

2.35 a. The regression equation is product $= 12.44 + 1.207$ aptitude.

b. The magnitude of the correlation suggests that a straight line will fit the data.

c. The residual for (9,23) is -0.303 and the residual for (20,40) is 3.42.

d. The predicted productivity for an aptitude of 15 is $12.44 + 1.207(15) = 30.545$.

e. Predicting outside the range of x is extrapolation and 40 is outside the range of x. The prediction would be questionable.

2.37 a. The regression equation is life $= 1136 - 3.22$ heat.

b. $-100.143, 69.714, 108.571, -68.571, 33.286, -42.857$

c. SSE $= 34323$

d. The predicted life span when the expected heat is 180 is $1136 - 3.22(180) = 556.4$.

2.39 a. The regression equation is cost $= -16.125 + 23.0034$ age

b. $r^2 = 85.9\%$. So 86% of the variability in cost is explained by age. The line provides a good fit to the data.

c. We can use the equation to predict cost when age is five years but not ten years. Ten years is extrapolation beyond available data.

d. The predicted maintenance cost when the age is five years is $-16.125 + 23.0034(5) = 98.892$.

e. The residual for (1,23) is 16.1216 and the residual for (5,86) is -12.892.

2.41 a. There is no trend in the data. It appears that the crime rate is not related to the percent without high school degrees.

b. The outlier is an outlier with respect to the crime rate and is a bivariate outlier. It is not an outlier with respect to nodegree.

c. Because it is such a high crime rate it is probably for the District of Columbia.

d. We should not fit a straight line to these data. The crime rate varies from 0 to 1000 regardless of what the value is of nodegree.

2.43 a. The correlation with outlier is 0.703 and without outlier is 0.617. The outlier increases the correlation and should be classified as an influential observation.

b. Because the attorneys' office in Washington, DC is so different from the other offices, we should remove it from the data set before describing the relationship between convictions and staff.

c. Without the outlier the regression equation is Convict $= 45.9 + 9.18$ Staff.

2.45 The regression equation is actual $= 0.762768 + 1.03899$ reported. The scatterplot shows a linear pattern. The value of $r^2 = 80.7\%$ indicates a good linear fit.

2.49 The residual plot shows a random pattern. A straight line adequately describes the relationship between the variables.

2.51
  a. FITS 23.3033, 32.9567, 28.1300, 35.3700, 36.5767, 40.1967, 26.9233, 30.5433
     RESI -0.30333, 2.04333, 0.87000, -2.37000, 3.42333, -2.19667, -1.92333, 0.45667
  b. The residual plot shows a random pattern.
  c. The coefficient of determination is $r^2 = (.936)^2 = 87.6\%$.
  d. Indications are that **a** straight line will fit the data.

2.53
  a. The regression line is Reading $= -45.1166 + 0.817459$ IQ.
  b. The line seems to fit the data.
  c. The coefficient of determination is $r^2 = 76.2\%$. 76.2% of the variability in reading scores is explained by variability in IQ.
  d. The equation in part **a** can be used to predict reading scores.
  e. The predicted reading score of an IQ of 100 is 36.63 and for an IQ of 120 is 52.98. The residual for (100, 46) is 9.37 and the residual for (120, 59) is 6.02.

2.55 The residual plot does not show a random pattern. The model can be improved by further analysis.

2.57
  a. We cannot interpret the pattern in the residual plot because of the one outlier.
  b. Without the one outlier, the residual plot shows a random pattern.

2.59
  a. RESI1 -0.87974, -0.84075, 1.08127, 0.15925, -1.95772, -2.95772, 0.88633, -4.23064, 0.92532, 1.73037, -2.30862, 1.12026, 1.08127, 4.73037, 3.84734, -2.38660
     The residual plot shows a random pattern.
  b. The coefficient of determination is $r^2 = 80.7\%$. This indicates that a straight line will explain almost 81% of the variability in the actual count.
  c. A straight line describes the relationship between the actual count and the reported count.

2.61
  a. The pattern shown in the scatterplot is a positive linear trend.
  b. There are two bivariate outliers. Because their x-values are extreme they are influential observations.
  c. The coefficient of determination is $r^2 = 79.6\%$. Almost 80% of the variability in the 1998 data is explained by the variability in the 1985 data. A straight line adequately describes the relationship between the variables.
  d. The removal of the outliers will decrease the coefficient of determination.
  e. The regression equation is 98 percent $= 2.85 + 0.619\ 85$ percent.
     Because the slope of the regression equation is less than one the percent over the age of 65 has decreased in the 13-year period.

2.71
  a. $50/228 = 21.9\%$
  b. $2/64 = 3.125\%$
  c. A convicted black defendant, when the victim is white, is seven times more likely to be given the death penalty than a convicted white defendant, when the victim is black.

2.73
  a. $425/630 = 67.5\%$
  b. $360/500 = 72\%$
  c. $60/630 = 9.5\%$
  d. The figures in the graph reflect frequencies and not percents.
  e. The graph should give percents.

2.75
  a. The areas of business and engineering showed the most dramatic change from 1970 to 1990.
  b. Those areas that remained about the same from 1970 to 1990 include education and foreign languages.
  c. 15.7%
  d. Because there are so many categories and we are making a comparison, a comparison bar graph would illustrate the information better than a comparison pie chart.
  e. The table illustrates the information better than a graph because there are so many categories.

2.77
  a. $74/104 = 71.15\%$
  b. $74/314 = 23.57\%$
  c. The percentages in parts **a** and **b** are different because they are computed relative to different marginal totals. In part **a** the percent is relative to all histological type LP. In part **b** the percent is relative to all positive responses.
  d. $126/538 = 23.42\%$

2.79
  a. Crimes that increased from 1992 to 1999 include Murder, Larceny theft, and Vehicle theft.
  b. Crimes that decreased from 1992 to 1999 include Forcible rape, Aggravated assault, Burglary, and Arson.
  c. Robbery stayed about the same.
  d. Larceny theft is the most prevalent crime. Murder is the least prevalent crime.

2.83
  a. Response variable—Yield, Predictor—Rainfall
  b. Response—Health cost, Predictor—Number of new AIDS cases
  c. Response—Crime rate, Predictor—Poverty rate

2.89
  b. Correlation of Number and Cost $= -0.868$
  c. There is a negative linear trend.
  d. The regression equation is Cost $= 88.6 - 4.09$ Number
  e. Cost per person for a family of 4 $= 88.6 - 4.09$ (4) $= 72.24$

2.91
  b. $210/500 = 42\%$
  c. $142/238 = 59.7\%$
  d. $96/290 = 33.1\%$
  e. $142/500 = 28.4\%$
  f. Of accidents with at least one fatality almost 60% involved alcohol and of the no alcohol accidents only 33% had at least one fatality. We would have

to say that there is an association between alcohol and fatalities.

2.93 **a.** 46/152

b. 46/125

c. 46/400

d. 46/88 = .523, 46/152 = .303, 28/132 = .212, 5/28 = .179 The Single group is most happy with their jobs.

2.95 **a.** There is a moderate linear trend between per capita income and percent with a high school diploma.

b. There is no relationship between poverty and percent with a high school diploma.

c. Only in part **a** is there a linear relationship, but it is not strong.

d. The correlation between percent with high school diploma and per capita income is modest. Poverty rate and per capita income are highly correlated with a negative trend.

2.97 **c** The scatterplot of the 100m versus the 400m has the greater variablity. There is less variability in the 100m versus 200m because in many cases the same runners run both races. Rarely however would a runner run both the 100m and the 400m race.

d. The correlation between 100m and 200m will be higher because the scatter is more tightly clustered about a straight line.

e. Correlation of 100m and 200m = 0.953
Correlation of 100m and 400m = 0.835

f. In both cases a straight line will adequately describe the relationship.

Chapter 3

3.1 **a.** 2/6 = 1/3   **b.** 1/200   **c.** 3/10   **d.** 26/52 = 1/2

3.3 **a.** 1/3

b. No, there are many more residents of California than either Washington or Oregon. The chances are great that the selected resident would be from California.

3.5 **a.** 18/38

b. 2/38

c. $(18/38)^3$

3.7 S = {(H_1,M), (H_1,F), (H_2,M), (H_2,F)}

3.9 **a.** {3,4,5,6,7}   **b.** {5}   **c.** A = {3,5,7}   **d.** {2,8}

e. {4,6}   **f.** {3,4,5,6,7}   **g.** {5}   **h.**{5}

i. No, see b.

3.11 **a.** If A and B are mutually exclusive, then P(A and B) = 0. So P(A or B) = .4 + .3 = .7.

b. If A and B are independent, then P(A and B) = (.4)(.3) = .12. So P(A or B) = .4 + .3 − .12 = .58.

3.13 S = {(MMM),(MMF),(MFM),(FMM),(MFF), (FMF),(FFM),(FFF)}

If 60% of the student body is female then the outcomes are not equally likely. If 50% of the student body is female then the outcomes are equally likely, and P({(MMM)})= . . . = P({(FFF)}) = 1/8.

3.15 S = {(AAA),(AAD),(ADA),(DAA),(ADD),(DAD), (DDA),(DDD)}

If the probability one application is approved is .6, then the outcomes are not equal. If the probability one application is approved is .5, then the outcomes are equal to 1/8. The probability that at least two loans are approved is P({(AAA)}) + P({(AAD)}) + P({(ADA)}) + P({(DAA)}) = 4/8.

3.17 If Team A finding the hiker is independent of Team B finding the hiker then P(A and B) = (.3)(.4) = .12, and P(A or B) = .3 + .4 − .12 = .58.

3.19 S = {WW,WL,LW,LL}, P({WW}) = $(.8)^2$ = .64, P({LL}) = $(.2)^2$ = .04, P({WL,LW}) = 2(.8)(.2) = .32

3.21 S ={BB, BN, NB, NN} Because P(B) = .3 the outcomes are not equally likely. From the simulation P(BB) = P(2) = .077 and P(BN, NB) = P(1) = .401 Assuming independent selections for her suits, P({NN}) = (.7)(.7) = .49, P({NB}) = (.7)(.3) = .21, P({BN}) = (.3)(.7) = .21, and P({BB}) = (.3)(.3) = .09. P(two days in a row) = P(BB) = .09 (compared to .077) and P(BN, NB) = .21 + .21 = .42 (compared to .401). The simulated probabilities are very close to the theoretical probabilities.

3.23 **a.** 17%

b. 10%

c. No, conditional probabilities are not additive.

d. 1 − .10 − .18 = .72 = 72%

3.25 **a.** Yes

b. 31%

c. 42%, 64%

d. Yes, less crime is committed in February than October.

e. Overall feeling about crime was better in 1998 because those indicating "more crime" dropped from 54% to 31% and those indicating "less crime" increased from 19% to 48%.

3.29 Yes, **b** gives the number of successes out of 15 independent Bernoulli trials.

3.31 **a.** 9   **b.** 0   **c.** 1/15 + 2/15 + 3/15 = 6/15   **d.** No, a binomial random variable assumes values {0, 1, 2, 3, ... , n}.

3.33 **a.** S = {WWW,WWL,WLW,LWW,WLL, LWL,LLW,LLL}

b. $(.4)^3$, $(.4)^2(.6)$, $(.4)^2(.6)$, $(.4)^2(.6)$, $(.4)(.6)^2$, $(.4)(.6)^2$, $(.4)(.6)^2$, $(.6)^3$

c. S_x = {0, 1, 2, 3}

d.

x	P(x)
0	$(.6)^3$
1	$3(.4)(.6)^2$
2	$3(.4)^2(.6)$
3	$(.4)^3$

e. P(at least 2 wins) = $3(.4)^2(.6)$ + $(.4)^3$ = .352

3.35 Using the Binomial probability tables

a. .028 + .121 + .233 + .267 + .200 = .849

b. .246 + .393 + .262 =.901

c. .343 + .206 = .549   **d.** .328 + .410 + .205 = .943

3.37 Let Y be the number of successful shows out of eight new shows. Then Y is a binomial random variable with $n = 8$ and $\pi = .2$
   a. Using Table B2, $.168 + .336 + .294 = .798$
   b. $\mu = (8)(.2) = 1.6$

3.39 **a.** Having an accident during the week could be identified as a "success"; however, the number of trials is not specified. So the number of accidents per week is not a binomial random variable.
   b. A violent crime being committed during a month could be identified as a success; however, the number of trials is not specified. So the number of crimes committed per month is not a binomial random variable.
   c. If we assume that the probability of a successful heart transplant is the same for all five patients, then the number of successful heart transplant is a binomial random variable with $n = 5$.
   d. The length of a prison term is continuous and cannot be a binomial random variable.
   e. The number approved for food stamps is binomial with $n = 50$.

3.41 $\mu \pm 2\sigma = -0.5$ to 6.3 which according to the empirical rule should contain approximately 95% of the distribution. In this case it contains 100%.

3.43
y	$P(y)$
0	$(.75)^3 = .42187$
1	$3(.25)(.75)^2 = .421875$
2	$3(.25)^2(.75) = .140625$
3	$(.25)^3 = .015625$

$$\mu = (3)(.25) = .75,$$
$$\sigma = \sqrt{(3)(.25)(.75)} = .75$$

$\mu \pm 2\sigma = -.75$ to 2.25 which contains 98.44% of the distribution. According to the empirical rule should contain approximately 95% of the distribution.

3.45 Let Y be the number who experience relief out of 16. Y is binomial with $n = 16$ and $\pi = .9$. Using the binomial probability tables,
   a. $P(Y = 16) = .185$
   b. $P(Y \geq 14) = .274 + .329 + .185 = .788$
   c. $\mu = 16(.9) = 14.4$

3.47 $P(x \geq 35) = 1 - P(x \leq 34) = 1 - .9696 = .0304$
From the simulation we find $P(x \geq 35) = 14/200 = .07$
$P(x < 20) = P(x \leq 19) = .0079$
From the simulation we find $P(x < 20) = 2/200 = .01$
If 56% are white then out of a sample of 50 we expect $n * \pi = (50)(.56) = 28$ to be white. The number that came up most frequently in the simulation is 27.

3.53 **a.** .9773 or 97.73% **b.** .9953 or 99.53%
   c. .9131 or 91.31% **d.** .0228 or 2.28%
   e. .8166 or 81.66% **f.** .2786 or 27.86%

3.55 $Z = (T - 30)/10\ P(T > 60) = P(Z > 3) =$
$1 - .9987 = .0013$

3.57 By finding the z-score $= (\text{score} - \mu)/\sigma$

3.59 Let Y be the score on the test. Y is normal with $\mu = 450$ and $\sigma = 100$, $Z = (Y - 450)/100$, and $Y = 450 + Z(100)$.
   a. $P(350 < Y < 550) = P(-1 < Z < 1) = .8413 - .1587 = .6826$
   b. $P(Y > 400) = P(Z > -.5) = 1 - .3085 = .6915$
   c. The upper 5% would be the 95th percentile. So $z = 1.64$ and $y = 450 + 1.64(100) = 614$.

3.61 **a.** $P(Y < 9000) = P(Z < 0.5) = .6915$
   b. $P(Y < 5000) = P(Z < -1.5) = .0668$
   c. $P(Y > 13,000) = P(Z > 2.5) = 1 - .9938 = .0062$
   d. $P(3000 < Y < 6000) = P(-2.5 < Z < -1.0) = .1587 - .0062 = .1525$

3.63 Let Y be the score on the test. Y is normal with $\mu = 100$ and $\sigma = 5$ and $Y = 100 + Z(5)$.
The value of Z with 5% of the total area above it is $z = 1.64$, so $Y = 100 + 1.64(5) = 108.2$.

3.65 The normal probability plot supports the conjecture that the data are coming from a normally distributed population.

3.67 The boxplot in Exercise 1.83 gives an indication that the data are coming from a population that is slightly skewed left. This normal probability plot, however, indicates that normality is a reasonable conjecture.

3.69 The pattern of the normal probability plot is that of a distribution that is skewed right which agrees with the previous assessment given in Exercise 1.56.

3.71 d **3.73** d **3.75** a

3.77 **a.** $S = \{rrr, rrn, rnr, nrr, rnn, nrn, nnr, nnn\}$
   b. No, only 25% return for an advanced degree. If 50% return then the outcomes are equally likely.
   c. $(.75)^3 = .422$ **d.** $3(.25)^2(.75) + (.25)^3 = .156$

3.79 $1/3, 0$

3.81 **a.** $3/5$ **b.** $3/7$

3.83 **a.** BC, BM, BP, BS, CM, CP, CS, MP, MS, PS
   b. $1/10$ **c.** $4/10$ **d.** $3/10$

3.85 $.4452, .2119$

3.87 **a.** T **b.** F **c.** F **d.** T **e.** F **f.** F

3.89 **a.** $.0359$ **b.** 3.14

3.91 **a.** $z = 3.33, .0004$ **b** $z = .83, .7967$
   c. $z = 1.67$ and $z = 5$, $1 - .9525 = .0475$

3.93 **a.** $z = -1.33$ and $z = .6$, $.7257 - .0918 = .6339$
   b. $z = 1.6$, $1 - .9452 = .0548$
   c. $50 + 1.28(15) = 69.2$

3.95 **a.** $L = \{(1,1),(6,6)\}; P(L) = 2/36$ **b.** $1 - 2/36 - 8/36 = 26/36$

3.97 $S = \{RR, RG, RB, GR, GG, GB, BR, BG, BB\}$
$A = \{RR\}$ $B = \{GG, GB, BG, BB\}$
$C = \{RG, RB, GR, BR\}$

3.99 **a.** $S = \{(1,H),(1,T),(2,1),(2,2),(2,3),(2,4),(2,5),(2,6)$
   $(3,H),(3,T),(4,1),(4,2),(4,3),(4,4),(5,5),(4,6)$
   $(5,H),(5,T),(6,1),(6,2),(6,3),(6,4),(6,5),(6,6)\}$
Let O be the event an odd number occurs, E be the event an even number occurs, H be the event a head occurs, and T be the event a tail occurs.
$P\{(O,H)\} = P(O)P(H) = (1/6)(1/2) = 1/12$
for $O = 1,3,5$

$P\{(O,T)\} = P(O)P(T) = (1/6)(1/2) = 1/12$
for $O = 1,3,5$
$P\{(E,x))\} = P(E)P(\{x\}) = (1/6)(1/6) = 1/36$,
for $E = 2,4\ 6$ and $x = 1,2,3,4,5,6$
  b. P(head on coin) = $3(1/12) = 1/4$

3.101 Let Pi = Pair i, for $i = 1,2,3,4$.
$S = \{\{P1,P2\},\{P1,P3\},\{P1,P4\},\{P2,P3\},$
$\{P2,P4\},\{P3,P4\}\}$
Because she randomly chooses two pairs, the outcomes will be equally likely and each will have probability 1/6.

3.103 Let Y be the number of completed passes out of the next ten. Y is binomial with $n = 10$ and $\pi = .6$
  a. $P(Y = 7,8,$ or $9) = .215 + .121 + .040 = .376$
  b. $\mu = 10(.6) = 6$
  c. $\sigma = \sqrt{10(.6)(.4)} = 1.55$

3.105 Let Y be the number of trees that survive out of 14. Y is binomial with $n = 14$ and $\pi = .9$.
$P(Y \geq 10) = .035 + .114 + .257 + .356 + .229 = .991$

3.107 a. $S = \{SSS, SSF, SFS, FSS, SFF, FSF, FFS, FFF\}$
  b. $(1/3)^3, (1/3)^2(2/3), (1/3)^2(2/3), (1/3)^2(2/3),$
  $(1/3)(2/3)^2, (1/3)(2/3)^2, (1/3)(2/3)^2, (2/3)^3$
  c. $3(1/3)^2(2/3) + (1/3)^3$
  d. $(2/3)^3$
  e. $3(1/3) = 1$

Chapter 4

4.1 The study would probably only generalize to readers of Ann Landers who would volunteer to return the questionnaire. Results from volunteer surveys such as this very rarely generalize to any reasonable population.

4.3 The population consists of potential customers for the new shopping center. Taking every tenth name is okay as long as the list is random with respect to the questions addressed in the questionnaire.

4.5 a. The population of interest is the student body at this university.
  b. A random sample could be obtained by using a list of random numbers to select students from a frame of the entire student body.
  c. Convenience sampling
  d. Cluster sampling
  e. Systematic sampling

4.7 This is convenience sampling because no scientific method is used to select the respondents.

4.9 Systematic sampling would be the easiest and because there is no order other than being listed alphabetically, the sample would be representative of all charge accounts.

4.11 The *Digest* prediction (the worst ever made in a national poll) was inaccurate for several reasons. First, it was a voluntary survey; only 23% of those receiving cards returned them. Generally voluntary surveys are biased because only those who feel strongly about the issue will bother to respond. Second, the sample of names was taken from subscription lists of magazines,

telephone directories, and automobile registrations. This was clearly biased toward those with higher incomes and better education. This is especially important because this poll was made during the depression era. Simply stated, the sample was not representative of the population. It had an overrepresentation of the upper middle class and the more highly educated. There were many laborers and farm workers at that time, and it is likely that they were not properly represented in the sample. (George Gallup observed the bias in the *Digest*'s poll and predicted that Roosevelt would win the election with 56% of the popular vote. This was the beginning of the nationally recognized Gallup Poll.)

4.13 Enumerate the subjects in the sampling frame from 1 to 10,000. Select those subjects whose numbers correspond to the 500 random numbers.

4.15 a. $p_1 - p_2$, the difference between two sample proportions is a statistic.
  b. The figure $1.22 is a statistic because it was found from a sample of 6000 service stations.
  c. The figure 50% is a parameter because it pertains to the population of all Americans.

4.17 $72 \pm 3\sigma/\sqrt{n} = 66$ to 78

4.19 $Z = (1900 - 1800)/(400/\sqrt{90}) = 2.37$
$P(\bar{x} > 1900) = P(Z > 2.37) = 1 - .9911 = .0089$

4.21 a. $Z = (34 - 31)/2 = 1.5, Z = (28 - 31)/2 = -1.5$
  probability $= .9332 - .0668 = .8664$
  b. $Z = (34 - 31)/(2/\sqrt{25}) = 7.5, Z = (28 - 31)/$
  $(2/\sqrt{25}) = -7.5$ probability $= 1^- - 0 = 1^-$

4.23 $12.63 \pm 2(5/\sqrt{236}) = 11.98$ to 13.28

4.25 The sampling distribution appears to be normally distributed. The value of μ is 160. The standard deviation appears to be two units. Approximately 95% of the distribution lies between 156 and 164.

4.27 The histogram in Exercise 4.26 is close to being normally distributed, centered at 30, and ranges from about 25 to 35. The histogram in this exercise is also close to normal, centered at 30, but ranges from about 27.5 to 32.5. The normal probability plot here also supports the conjecture that the sampling distribution of $\bar{x}$ is normally distributed when the population is also normally distributed. The only difference between the two exercises is that this distribution is less variable. This is a result of the fact that the standard deviation of the sampling distribution of $\bar{x}$ is $\sigma/\sqrt{n} = 8/\sqrt{64} = 1$, whereas for the previous exercise, when $n = 16$, the standard deviation of the sampling distribution of $\bar{x}$ was $8/\sqrt{16} = 2$.

Variable	N	Mean	Median	StDev
C65	2000	30.016	30.021	0.982

Minimum	Maximum	Q1	Q3
26.898	33.059	29.326	30.679

Notice that the mean of the distribution of sample means is 30.016, very close to 30, and the standard deviation is 0.982, very close to 1.

4.29 The histogram appears much closer to a normal distribution than in Exercise 4.28. The normal probability plot does not support the conjecture that the sampling distribution of $\bar{x}$ is closely approximated by a normal distribution. It does, however, appear more normally distributed than in Exercise 4.28. It seems, in the case of the exponential distribution, that the sample size should be larger than 64 in order for the Central Limit Theorem to guarantee that the sampling distribution of $\bar{x}$ is normally distributed.

Variable	N	Mean	Median	StDev
C65	2000	5.0104	4.9917	0.6408

Minimum	Maximum	Q1	Q3
3.0610	7.2078	4.5731	5.4168

The mean of the distribution of sample means is 5.0104 which is very close, as it should be, to the population mean of 5. The standard deviation of the distribution of sample means, 0.6408, is smaller than it was in Exercise 5.34, 1.2396, when the sample size was only 16.

In comparing the results of this exercise and the results found in Exercise 4.28 we see that when the sample size is only 16, the sampling distribution of $\bar{x}$ is not approximately normally distributed. When the sample size is 64, however, the sampling distribution of $\bar{x}$ is closer to being normally distributed. Recall that the Central Limit Theorem says that it will be approximately normally distributed when the sample size is relatively large. For the heavily skewed exponential distribution the sample size should be larger than 64.

4.31 **a.** mean $= .7$, standard deviation $= \sqrt{(.7)(.3)/100} = .0458$
   b. mean $= .7$, standard deviation $= \sqrt{(.7)(.3)/400} = .0229$
   c. mean $= .7$, standard deviation $= \sqrt{(.7)(.3)/1000} = .0145$
   d. mean $= .7$, standard deviation $= \sqrt{(.7)(.3)/1600} = .0115$

4.33 The maximum standard deviation of p occurs when $\sigma = .5$.
   a. $\sqrt{(.5).5)/25} = .1$ **b.** $\sqrt{(.5).5)/100} = .05$
   c. $\sqrt{(.5).5)/200} = .0354$ **d.** $\sqrt{(.5).5)/500} = .0224$
   e. $\sqrt{(.5).5)/1000} = .0158$ **f.** $\sqrt{(.5).5)/2000} = .0112$

4.35 In a random sample of 1000, by the Central Limit Theorem p is approximately normal with mean π and standard deviation $\sqrt{\pi(1-\pi)/1000}$. Using $p = .23$ as an estimate of π, we would estimate the standard deviation of p to be $\sqrt{(.23)(.77)/1000} = .0133$ and we would expect p to be no more than two standard deviations from the true value of π, i.e., we would expect p to be within $\pm 2(.0133) = \pm .0266$ of the true value of π.

4.37 $\pi = .7$ and $n = 200$. By the Central Limit Theorem p is approximately normal with mean $= .7$ and standard deviation $= \sqrt{(.7)(.3)/200} = .0324$ and $Z = (p - .7)/.0324$.
P($p > .65$) = P($Z > -1.54$) = $1 - .0618 = .9382$

4.39 $\pi = .59$ and $n = 400$. By the Central Limit Theorem p is approximately normal with mean $= .59$ and standard deviation $= \sqrt{(.59)(.41)/400} = .0246$ and $Z = (p - .59)/.0246$.
   a. P($p > .6$) = P($Z > .41$) = $1 - .6591 = .3409$
   b. $\pi a = .43$, $Z = (p - .43)/.\sqrt{(.43)(.57)/400}$, P($p > .6$) = P($Z > 6.87$) = 0^+
   c. The answer to part **b** is so small because with only 43% voting for Clinton, it is very unlikely that more than 60% of a random sample of 400 voters would vote for him. The difference between 60% and 43% is much greater than the difference between 60% and 59%.

4.41 Assuming $\pi = .5$, with $n = 1008$, by the Central Limit Theorem p is approximately normal with mean $= .5$ and standard deviation $= \sqrt{(.5)(.5)/1008} = .016$ and $Z = (p - .5)/.016$.
P($p \leq .44$) = P($Z \leq -3.75$) = 0^+ Because the probability that it happened by chance is so small, there is strong evidence that the aide is in error.

4.43 $\pi = .25$ and $n = 36$, $\sqrt{(.25)(.75)/36} = .0722$ and $Z = (p - .25)/.0722$
P($p > .3056$) = P($Z > .77$) = $1 - .7794 = .2206$
These results are not unusual and therefore are not substantially better than what would normally be expected.

4.45 The sampling distribution appears close to normal with a mean of .2. About 95% of the distribution falls between .16 and .24 so the standard deviation is close to .02.

4.51 If we are only interested in automobile accidents in 1993 then it is a parameter. If the population is all automobile accidents then these data are assumed to be a sample and then 15.9 is a statistic.

4.53 $p = 40/200 = .2$, $Z = (.2 - .176)/\sqrt{(.176)(.824)/200} = .89$ P($p > .2$) = P($Z > .89$) = $1 - .8133 = .1867$

4.55 **a.** 10, .2   **b.** 10, .4   **c.** 50, 1.2   **d.** 50, 2.4

4.57 The sampling distribution of $\bar{x}$ should be approximately normal, so the standard deviation of $\bar{x}$ can be approximated with Range/4. i.e., standard deviation is approximately $800/4 = 200$.

4.59 $Z = (60 - 56.7)/(9.3/\sqrt{36}) = 2.13$ P($\bar{x} > 60$) = P($Z > 2.13$) = $1 - .9834 = .0166$

4.61 For sufficiently large n, the sampling distribution of p is approximately normally distributed with a mean of π and a standard deviation of $\sqrt{\pi(1-\pi)/n}$.

4.63 $\sqrt{(.5)(.5)/200} = .0354$

4.65 **a.** $Z = (145 - 149)/(40/\sqrt{100}) = -1$
      P($\bar{x} > 145$) = P($Z > -1$) = $1 - .1587 = .8413$
   b. $Z = (155 - 149)/(40/\sqrt{100}) = 1.5$
      P($\bar{x} < 155$) = P($Z < 1.5$) = $.9332$
   c. $Z = (159.6 - 149)/(40/\sqrt{100}) = 2.65$
      $Z = (151.2 - 149)/(40/\sqrt{100}) = .55$
      P($151.2 < \bar{x} < 159.6$) = P($.55 < Z < 2.65$) = $9960 - .7088 = .2872$

4.67 **a.** $Z = (80 - 75)/15 = .33$ $P(y > 80) =$
$P(Z > .33) = 1 - .6293 = .3707$
b. $Z = (95 - 75)/15 = 1.33$ $Z = (84 - 75)/15 =$
$.6$ $P(84 < y < 95) = P(.6 < Z < 1.33) =$
$.9082 - .7257 = .1825$
c. $75 + (.84)(15) = 87.6$
d. $Z = (80 - 75)/(15/\sqrt{25}) = 1.67$ $P(\bar{x} > 80) =$
$P(Z > 1.67) = 1 - .9525 = .0475$

4.69 standard deviation $= 2.5/\sqrt{n} = .4$, so $n = (2.5/.4)^2 =$
$39.0625 \approx 40$

4.71 $\bar{x}$ is approximately normal and $Z = (\bar{x} - 45)/1.35$
$P(\bar{x} < 43) = P(Z < -1.48) = .0694$

4.73 $\pi = .12$, p is approximately normal and $Z =$
$(p - .12)/.0257$ $P(p < .094) = P(Z < -1.01) =$
$.1562$

4.75 $Z = (p - .29)/\sqrt{(.29(.71)/200}$ If $p = 65/200 = .325$
then $Z = (.325 - .29)/\sqrt{(.29(.71)/200} = 1.09$ which
is not unusual at all. It is very conceivable that more
than 65 out of a random sample of 200 criminal
victimizations would involve firearms.

4.77 standard deviation $= 15/\sqrt{n} = 2$ thus $n = (15/2)^2 =$
$56.25 \approx 57$

4.79 $Z = (9.2 - 8)/(3/\sqrt{100}) = 4$. The amount these
students study is a full four standard deviations above
what typical college students study. This almost
certainly did not happen by chance.

4.81 **b.** Both sampling distributions are centered at 2.0.
c. Both distributions are symmetrical and appear to
be normally distributed.
d. The distribution of the median is more variable.
e. The values for the sample mean are more tightly
clustered about the population mean than are the
values of the sample median when the population
is normally distributed.

4.83 **b.**

Variable	N	Mean	Median	StDev
mean	2000	4.9921	4.7123	2.2211
median	2000	3.9173	3.4456	2.3330

Minimum	Maximum	Q1	Q3
0.6723	14.9218	3.3403	6.2418
0.1589	16.6560	2.1848	5.1848

Using the describe command we see that the
sampling distribution of the mean is centered
around 5 (the mean of the population) and the
sampling distribution of the median is centered
around 3.9.
c. Both distributions are skewed right.
d. From the descriptive statistics we see that the
distribution of the median is a little more variable
than is the distribution of the mean.
e. The sampling distributions are skewed right
because the original population (exponential) is
skewed right. These sampling distributions are based
on averages of five observations from the
population, consequently they are not as skewed as
the population.

4.85 **a.** Using the Cross Tabulation command we find that
$207/1000 = 20.7\%$ of the kids smoke.
b. From the tabulated statistics we find that
$105/480 = 21.875\%$ of the 480 females smoke.
c. We find that 102 of the 520 males smoke. The
percent is $102/480 = 21.25\%$.
d. There are 520 males and 480 females in the study.
e. Slightly more females (21.875%) smoke than
males (21.25%).
f. These percents are considerably higher than those
found in the NCHS survey.

Chapter 5

5.1 **a.** μ **b.** σ **c.** θ **d.** π

5.3 Because some animals are very slow and may never
respond, the distribution is probably skewed right in
which case the median is the preferred parameter.

5.5 No, 6.2 is only the average time. This sample is
randomly selected from a particular large community
and the results would not necessarily generalize to all
large communities. Nothing can be said about small or
rural communities from a sample taken from a large
community.

5.7 **a.** Stem-and-leaf of miller   N = 25
Leaf Unit = 1.0

1	1 4
2	1 6
5	1 889
7	2 01
2	2 22233
(5)	2 45555
8	2 67
6	2 9
5	3 01
3	3 3
2	3 45

b.

Low	Q$_1$	Median	Q$_3$	High
14	21	24	27	35

d. The boxplot is symmetric. We should classify the
parent distribution as symmetric.

5.9 **a.** The dotplot reveals a multimodal distribution that
was not apparent from the boxplot.
b. There is a group of two-seaters, probably sports
cars, that do not get very good gas mileage and
then there is a large group that get average gas
mileage and finally there is one car that gets an
extremely high gas mileage.
c. Because the distribution is symmetric with long
tails a trimmed mean would give a representative
measure of the fuel efficiency of the vehicles.

5.11 The boxplot and normal probability plot give strong
evidence that the distribution is heavily skewed right.
Because of the appearance of the normal probability
plot it is not reasonable to say that the distribution is
close to normal.

5.13 **a.** Based on a histogram and boxplot we should classify the shape of the distribution as skewed right.

b.
```
Variable   N    Mean    Median   StDev
months     45  18.933   19.000   6.489

    Min         Max        Q1       Q3
 10.000      42.000    13.500   22.000
```

For skewed right distributions the mean is greater than the median. In these data, however, the median is greater than the mean. Further analysis is necessary.

c. There are three modes if we count the extreme outlier as a mode. This is a multimodal distribution, which explains why the median is greater than the mean. The size of the county may account for the different modes in the months to process a tort.

d. There is a large concentration of data around the first mode. This will cause the mean to be smaller than it would have been had the distribution been only skewed right.

5.15 p is an unbiased estimator of π because its sampling distribution is centered at π.

5.19 Parameter of interest is the mean or median time of the subway trip. Possible estimators: $\bar{x}$ or M. $\bar{x}$ = 39.743 M = 39.950 A stem-and-leaf plot shows the distribution to be skewed left hence, the median M = 39.95 is the preferred statistic.

5.21 The boxplot and normal probability plot strongly suggest that the parent population of educational levels is a symmetric long-tailed distribution. The 5% trimmed mean $\bar{x}_{T.05}$ = 11.306, given in the descriptive statistics, is an estimate of the mean of the distribution.

5.23 An estimate of the percent of registered voters in Miami who are Puerto Rican is $p = 45/225$ = .20.

5.25 **a.**
```
Stem-and-leaf of hours   N = 50
Leaf Unit = 1.0

   1      0 5
   2      1 0
   3      1 5
   6      2 000
  13      2 5558889
  23      3 0000022344
 (13)     3 5555556778899
  14      4 000011234
   5      4 55
   3      5 0
   2      5
   2      6 0
   1      6 5
```
The distribution appears to be symmetrical with long tails.

b. The normal probability plot gives the appearance of a long-tailed distribution. This agrees with the assessment given in part **a**.

c. The mean and trimmed mean are equal because of the almost perfect symmetry in the data. For symmetric distribution all standard measures of the center agree.

d. For long-tailed distributions the upper and lower outliers may not balance each other out exactly so we generally prefer a trimmed mean as an estimate of the center. Thus our estimate is $\bar{x}_{T.05}$ = 34.05.

e. Because of the long tailed nature of the distribution, the standard deviation is not a good measure of variability. Perhaps a better measure of variability is the IQR.

5.27 **a.**
```
Stem-and-leaf of salinity  N = 48
Leaf Unit = 1.0

    1      3 4
    6      3 56679
   14      4 00002234
  (13)     4 6666777888999
   21      5 00112333
   13      5 68899
    8      6 000123
    2      6 7
    1      7
    1      7 8
```
The stem-and-leaf plot appears bell shaped with one large value at 78.

b. The normal probability plot does not rule out normality.

c. Aside from the one observation at 78 the data is very close to being normally distributed. The recommended estimator of the mean salinity level is the sample mean. The estimate is $\bar{x}$ = 49.54.

5.29 The margin of error will decrease by a factor of 2.

5.31 Margin of Error = .0366

$.03 = 2.58\sqrt{(.61)(.39)}/\sqrt{n}$, so $n = 1760$

5.33 $p = 260/1000 = .26$ $.26 \pm 2.58\sqrt{(.26)(.74)/1000}$
The confidence interval is .224 to .296

$.03 = 2.58\sqrt{(.26)(.74)}/\sqrt{n}$, so $n = 1423$

5.35
```
State            confidence interval
Arizona          22.436, 25.964
Colorado         13.532, 16.668
Florida          16.520, 18.480
Montana          17.836, 21.364
New Jersey       15.224, 17.576
New Mexico       19.336, 22.864
```

5.37
```
White, non-Hispanic     50.292, 63.708
Black, non-Hispanic     47.228, 64.772
Hispanic                45.066, 56.934
Other races             35.616, 60.384
```
The white and black groups had approximately the same percent witness a crime. There is a lot of overlap in the intervals; it is difficult to detect a difference between the groups.

5.39 $.02 = 1.96\sqrt{(.3)(.7)}/\sqrt{n}$, so $n = 2017$

5.41 $.02 = 1.96\sqrt{(.5)(.5)}/\sqrt{n}$, so $n = 2401$

5.43 The following intervals do not contain .3: (.212, .288), (.316, .400), (.306, .390), (.314, .398), (.208, .284), and (.216, .292). Ninety-four of the 100 intervals contain .3. This is a reasonable number given that they are 95% confidence intervals. Out of 100 different intervals we expect approximately 95 to contain .3.

5.47 **a.** $465.8 \pm 1.96*50/\sqrt{60}$ The confidence interval is 453.15 to 478.45.
  b. $465.8 \pm 2.58*50/\sqrt{60}$ The confidence interval is 449.15 to 482.45.
  c. The 99% confidence interval has greater validity.
  d. The 95% confidence interval has greater precision.

5.49 Interval A is the 95% interval because a 95% confidence interval is generally shorter than a 99% confidence interval.

5.51 $1325 \pm 1.96*42.4/\sqrt{100}$ The confidence interval is 1316.69 to 1333.31. We cannot say that there is a 95% chance that the mean living area is inside the interval. Either the mean is inside the interval or it is not. Probability (95% chance) is only relevant prior to sampling.

5.53 $560 \pm 1.645*35/\sqrt{100}$ The confidence interval is 554.24 to 565.76. A 95% confidence interval would be wider because the margin of error would be multiplied by 1.96 instead of 1.645.

5.55 $2418 \pm 2.58*325/\sqrt{100}$ The confidence interval is 2334.15 to 2501.85. The interval does contain 2500. The advertiser's claim is a valid claim.

5.57 $69.2 \pm 2.776*3.27/\sqrt{5}$ The confidence interval is 65.1 to 73.3.

5.59 The stem-and-leaf display suggests that the distribution is symmetric. With a sample size of 15 it is a good idea to check for normality. This is accomplished with a normal probability plot. Assuming the conditions for a t-interval we have $57.17 \pm 2.624*8.21/\sqrt{15}$. The confidence interval is 51.6 to 62.7

5.61 $250.03 \pm 2.000*41.44/\sqrt{62}$ The confidence interval is 239.50 to 260.56. Because the sample size is 62 the normality condition is not a serious requirement for maintaining the validity of the t-interval.

5.63 **a.** The boxplot and normal probability plot indicate strong right skewness.
  b. The sample size is 629 with 90 missing observations. With such a large sample size it is permissible to find a confidence interval for the mean ozone using a t-interval.
  c. $23.693 \pm 2.58*19.276/\sqrt{629}$ The confidence interval is 21.71 to 25.68.
  d. With such a large sample size the validity is not affected by the shape of the population distribution.

5.65 The t-interval is robust except in the cases of long tails and severe skewness. With a sample size of 50, a t-interval is certainly appropriate for a short-tailed distribution. In fact, for a short-tailed distribution a t-interval is appropriate for a sample size as small as 20.

```
Confidence Intervals

Variable      N        Mean    StDev
   income     49      271.94    65.61

SE Mean      90.0 % C.I.
   9.37    (256.21,  287.66)
```

5.67 In the first set of intervals the placebo interval is completely contained within the treatment interval. In this case it is impossible to detect a difference between the two groups. In the second set of intervals the placebo interval lies sufficiently above the treatment interval to detect a difference between the two groups. If the outliers are typical and included in the data, we cannot detect a difference between the mean responses with the treatment and with the placebo. For these data a comparison of population medians may be more appropriate.

5.69 **a.** 2 **b.** 3 **c.** 4 **d.** 6

5.71 **a.** $n = 16$, Position of $C = 5$; $C_L = 52$, $C_H = 62$
  b. $n = 18$, Position of $C = 5$; $C_L = 59$, $C_H = 79$
  c. $n = 13$, Position of $C = 3$; $C_L = 59$, $C_H = 92$

5.73 $n = 17$, Position of $C = 6$; $C_L = 275$, $C_H = 850$

```
Confidence Interval based on x̄

Variable      N        Mean    StDev
   award      17         915      941

SE Mean      90.0 % C.I.
   228      (517,       1314)
```

5.75 **a.** Position of $C = (25 - 1.96*\sqrt{25})/2 \approx 8$
    Confidence Interval (27,45)
  b. Position of $C = (25 - 1.96*\sqrt{25}/2 \approx 8$
    Confidence Interval (15,36)
  c. Position of $C = (24 - 1.96*\sqrt{24}/2 = 7.2 \approx 8$
    Confidence Interval (12,36)

5.77 Position of $C = (22 - 2.33*\sqrt{22}/2 = 6$
  Confidence Interval (4.1, 9.6)

5.79 The normal probability plot indicates that the data are skewed.
  $n = 14$, Position of $C = 3$; $C_L = 16$, $C_H = 24$

5.81 Position of $C = (26 - 1.96*\sqrt{26}/2 = 9$
  Confidence Interval (78,000, 99,900)

5.83 **a.** Small mesh: Position of median $(739 + 1)/2 = 370$, M = 33
    Large mesh: Position of median $(767 + 1)/2 = 384$, M = 34
  b. Small mesh: Position of quartile $(370 + 1)/2 = 185.5$, $Q_1 = 31$ $Q_3 = 35$, IQR = 4.
    Large mesh: Position of quartile $(384 + 1)/2 = 192.5$, $Q_1 = 33$ $Q_3 = 36$, IQR = 3.
  c. It seems that the length of fish caught with the large mesh codend is slightly larger than with the small mesh codend.
  d. Small mesh: Position of $C = (739 - 2.58*\sqrt{739})/2 = 335$, (33, 34)
    Large mesh: Position of $C = (767 - 2.58*\sqrt{767})/2 = 348$, (34, 35)

There is no overlap in the two intervals. The length of fish caught with a small mesh codend is somewhere between 33 and 34 cm and with the large mesh codend is somewhere between 34 and 35 cm.

5.85 **a.** π **b.** μ **c.** θ **d.** π

5.87 $.03 = 1.96*\sqrt{(.5)(.5)}/\sqrt{n}$, so $n = 1068$

5.89 $160 \pm 1.96*32/\sqrt{64}$ The confidence interval is $(152.16, 167.84)$.

5.91 A 95% confidence interval is $(91.177, 95.490)$ For the t-interval to be valid with this small of a sample it is important that the parent distribution be close to normal.

5.93 $30 = 1.645*120/\sqrt{n}$, so $n = 44$

5.95 **a.** $.04 = 2.33\sqrt{(.4)(.6)}/\sqrt{n}$, so $n = 815$
  b. $.32 \pm 2.33\sqrt{(.32)(.68)}/815$ The confidence interval is .282 to .358.

5.97 $.8 \pm 2.33\sqrt{(.8)(.2)}/400$ The confidence interval is .75 to .85.

5.99 **a.** $.03 = 1.96\sqrt{(.5)(.5)}/\sqrt{n}$, so $n = 1068$
  b. $.75 \pm 1.96\sqrt{(.75)(.25)}/400$ The confidence interval is .71 to .79.

5.101 $p = 220/549 = .4$
  $.4 \pm 1.645\sqrt{(.4)(.6)}/549$ The confidence interval is .366 to .434.

5.103 The normal probability plot indicates that the data are from a normally distributed population. The conditions for a t-interval are satisfied. The 90% confidence interval is $(18.09, 22.11)$.

5.105 The normal probability plot shows no departures from normality. The conditions for the t-interval are satisfied. A 90% confidence interval is $(30886, 33927)$.

5.107 The normal probability plot shows a considerable departure from normality. The appearance is that of a skewed right distribution; this agrees with the assessment given in Exercise 5.106.

5.109 Position of $C = (51 - 2.33*\sqrt{51})/2 \approx 18$. The confidence interval is $(3, 12)$

5.111 Waitime 1 $57.109 \pm 1.9822*8.148/\sqrt{110}$ The confidence interval is $(55.569, 58.649)$
  Waitime 2 $81.164 \pm 1.9727*7.303/\sqrt{189}$ The confidence interval is $(80.116, 82.212)$

```
Wait 1 x----x
Wait 2                        x--x
-----|-----|-----|-----|-----|-----|------|--
    55    60    65    70    75    80    90
```

The precision of the waitime 2 interval is greater (narrower interval) because it is based on a larger sample size. Also the standard deviation of the waittime 2 data is slightly smaller. This causes the interval to be shorter but the real difference in precision of the two intervals is because of the sample sizes. There is a clear distinction between the wait times for the two types of eruptions.

5.113 With outlier the 95% confidence interval is $(485.9, 727.7)$.

Without outlier the 95% confidence interval is $(479.3, 645.3)$.
The mean crime rate is 44.5 units smaller without the outlier. The standard deviation is 137.7 units smaller without the outlier. The one outlier has inflated both the mean and the standard deviation. Because the crime rate in Washington, DC is so much out of line with the crime rates in the states it is a reasonable approach to construct the confidence after deleting the outlier. The confidence interval $(479.3, 645.3)$ is more representative of the data.

5.115 The 200m data shows some slight skewness but otherwise the distributions are close to being normally distributed. The median times are obviously different because the races are longer, but there is more variability in the times for the longer races. This is probably understandable to those who are familiar with track events.

Variable	N	Mean	Median	StDev
100m	55	11.619	11.600	0.452
200m	55	23.642	23.540	1.111
400m	55	53.406	53.300	2.678

Min	Max	Q1	Q3
10.790	12.900	11.250	11.950
21.710	27.100	22.820	24.440
47.990	60.400	51.500	55.080

Notice how the standard deviations increase for the longer races. Also notice that the mean (or median) time for the 400m race is considerably more than just twice the time for a 200m race. Because of the symmetry, the mean, median and trimmed mean are very close in value for all three races. Any one of the three would suffice as a measure of the center of the distribution. We have learned, however, that the sample mean is the best estimate of the center of normally distributed populations.

5.117 **a.** The data are moderately skewed right.
  b. With a sample size of 46 and the fact that the skewness is not severe, the normality assumption is not crucial.
  c. A 99% confidence interval is $(469.3, 560.5)$.

5.119 The two distributions do not deviate significantly from normality and the sample sizes are large enough that normality is really not an issue. Confidence intervals for the means are certainly appropriate.
  $\bar{x}_{small} = 24700/739 = 33.4235$
  $s_{small} = \sqrt{8614.4/738} = 3.4165$
  $\bar{x}_{large} = 26550/767 = 34.6154$
  $s_{large} = \sqrt{7685.5/766} = 3.1675$
  Small mesh confidence interval: $33.4235 \pm 2.58*3.4165/\sqrt{739}$ $(33.099, 33.748)$
  Large mesh confidence interval: $34.6154 \pm 2.58*3.1675/\sqrt{767}$ $(34.320, 34.910)$
  In comparison the small mesh confidence interval for the median is $(33, 34)$ and the large mesh confidence interval for the median is $(34, 35)$. Both intervals for the mean are centered at approximately the same place

as the intervals for the medians. The mean intervals are more precise than the median intervals. In general, it appears that the large mesh codend yields catches 1 cm larger than small mesh codend.

Chapter 6

6.1   **a.** $H_0: \mu = 50$ versus $H_a: \mu > 50$
     b. $H_0: \mu = 50$ versus $H_a: \mu < 50$
     c. $H_0: \mu = 50$ versus $H_a: \mu > 50$
     d. $H_0: \mu = 50$ versus $H_a: \mu < 50$
     e. $H_0: \mu = 50$ versus $H_a: \mu \neq 50$

6.3 $H_0: \pi = .60$ versus $H_a: \pi < .60$

6.5   **a.** No. The evidence suggests that the alternative hypothesis is true; however, we may have committed a Type I error.
     b. No, there is the possibility of a Type II error.
     c. Yes, because the rejection region is larger for a 5% test.

6.7   No, Type I and II errors are not complementary events.

6.9   **a.** Type I : saying that the patient will not recover when in fact he will.
     Type II : saying that the patient will recover when in fact he will not.
     Committing a Type I error is more serious.
     b. Type I : concluding that the fire is not out and it actually is.
     Type II : concluding that the fire is out and it actually is not out.
     Committing a Type II error is more serious.

6.11 $H_0: \mu = .18$ versus $H_a: \mu > .18$
     A Type II error is committed if the geologist concludes that the mean porosity measurement does not exceed 18% when in fact it does.

6.13   If we set up hypotheses as follows: $H_0: \mu = 78$ versus $H_a: \mu > 78$ we find $z = (81.4 - 78)/(7/\sqrt{35}) = 2.87$ which gives p-value $= .0021$. Based on these results there is highly significant evidence that her class is superior. Because of the sample size of 35 the normality assumption is not that crucial. We should, however, look for severe skewness or extremely long tails.

6.15   **a.** $H_0: \mu = 550$ versus $H_a: \mu > 550$
     b. The population standard deviation was given, $\sigma = 120$.
     c. The difference between $\bar{x} = 582.6$ and the hypothesized value of $\mu = 550$ is 32.6 points which is 2.57 standard errors. This is a significant difference.
     d. p-value $= .0051$ which rejects the null hypothesis.
     f. Based on this random sample of size 90, there is highly significant evidence that this year's freshman class scored unusually high.

6.17 $z_{obs} = (.1125 - .1)/(\sqrt{(.1)(.9)/1600}) = 1.67$
     p-value $= .0475$
     Based on a random sample of 1600 adults there is significant evidence that more than 10% of the adult population is illiterate.

6.19 $z_{obs} = (.55 - .6)/(\sqrt{(.6)(.4)/200}) = -1.44$
     p-value $= .0749$
     Based on a random sample of 200 adults there is moderately significant evidence that less than 60% of the adult population believes that there is too much violence on television.

6.21 $.01 = 1.96*(\sqrt{(.25)(.75)/n})$, so $n = 7203$.

6.23 $H_0: \pi = .33$ versus $H_a: \pi > .33$
     $z_{obs} = (.35 - .33)/(\sqrt{(.33)(.67)/600}) = 1.04$
     p-value $= .1492$
     Based on these 600 cases there is insignificant evidence that Cook County exceeds the national average in the number of automobile accident suits.

6.25   **a.** $H_0: \pi \geq .50$ versus $H_a: \pi < .50$
     b. $z_{obs} = (.46 - .50)/(\sqrt{(.5)(.5)/1022}) = -2.56$
     c. If the z statistic is less than 1.28 in absolute value then the results are insignificant.
     d. The value of z_{obs} is clearly in the rejection region.
     e. The p-value $= .0052$ is less than the 10% level of significance.
     f. The null hypothesis should be rejected.
     g. Based on this random sample of 1022 adults there is highly significant evidence that the president does not have the support of the majority on this issue.

6.27 $H_0: \pi = .8$ versus $H_a: \pi \neq .8$
     $z_{obs} = (.82 - .8)/(\sqrt{(.8)(.2)/200}) = .71$
     p-value $= 2(.2389) = .4778$
     Based on a random sample of 200 people there is insufficient evidence to refute the psychologist's claim. For a two-tailed test the value of z_{obs} must be greater than 1.96 in absolute value for the results to be significant at the 5% level of significance. This means that $1.96 = (p - .8)/(\sqrt{(.8)(.2)/200})$. Solving we find $p = .855$, which is 171 out of 200.

6.29 $H_0: \pi = 1/3$ versus $H_a: \pi < 1/3$
     $z_{obs} = (.24 - .33)/(\sqrt{(.33)(.67)/200}) = -2.71$
     p-value $= .0034$
     Based on this sample of 200 prisoners there is highly significant evidence that less than 1/3 are returning to jail within three years.

6.31 $H_0: \mu \leq 4.9$ versus $H_a: \mu > 4.9$
     $t_{obs} = (5.6 - 4.9)/(2.1/\sqrt{16}) = 1.333$ p-value $> .10$
     Based on a random sample of 16 readings there is insufficient evidence that the mean carbon dioxide level is more than 4.9 ppm. For the t-test to be valid when the sample size is small it is important that the population distribution be close to normal. A sample size of 16 is moderately small and therefore the normality assumption will assure the accuracy of the t-test.

6.33 $H_0: \mu \leq 120$ versus $H_a: \mu > 120$
     $t_{obs} = (125.6 - 120)/(20.3/\sqrt{25}) = 1.379$
     p-value $= .0903$
     Based on 25 randomly selected families there is moderately significant evidence that mean kilowatt usage exceeds 120 kilowatt hours.

6.35 $H_0: \mu \leq 72$ versus $H_a: \mu > 72$
$t_{obs} = (76 - 72)/(14.4/\sqrt{16}) = 1.111$
p-value $= .1420$
Based on the information provided there is insufficient evidence that this is a superior class. Having a normally distributed population insures the accuracy of the t-test. On the other hand, for a sample size of 16 we should insist that the population distribution be at least symmetric without outliers.

6.37 $H_0: \mu \leq 78$ versus $H_a: \mu > 78$
$t_{obs} = (82 - 78)/(7/\sqrt{20}) = 2.556$ p-value $= .0097$
Based on the provided data there is highly significant evidence that this is a superior class. Because we are using a t-test and the sample size is 20 we assume that the distribution of scores on this exam is at least symmetric without outliers.

6.39 **a.** There is no unusual behavior in the histogram or boxplot to suggest that we should not conduct a t-test. With 50 observations there is no reason to base the test on any statistic other than $\bar{x}$.
  b. $t_{obs} = (299.96 - 300)/(5.61/\sqrt{50}) = -0.05$
    p-value $= .96$ There is no evidence whatsoever to conclude that the mean price differs from 300 pence.

6.41 The normal probability plot supports the normality assumption. It is very much appropriate to test the population mean with a t-test.

6.43 $H_0: \mu = 35,000$ versus $H_a: \mu \neq 35,000$
$t_{obs} = (32,406 - 35,000)/(3591/\sqrt{17}) = -2.98$
p-value $= .0089$
Based on the information provided there is highly significant evidence that the mean charge for a coronary bypass in North Carolina is different from $35,000.

6.45 **a.** The stem-and-leaf plot indicates that the data are skewed right and may be bimodal.
  b. A normal probability plot indicates that the data are not from a normally distributed population.
  c. Because the sample size is 101 it is reasonable to test the population mean and use $\bar{x}$ as the test statistic. A better alternative for skewed distributions, however, will be given in the next section.
  d. Test of mu = 8.000 vs mu < 8.000

Variable	N	Mean	StDev
thick	101	7.831	2.587

SE Mean	T	P-value
0.257	−0.66	0.26

There is insufficient evidence that the mean thickness of the varves is less than eight millimeters.

6.47 Test of mu = 80.00 vs mu < 80.00

Variable	N	Mean	StDev
score	40	74.40	16.50

SE Mean	T	P-value
2.61	−2.15	0.019

These data suggest that the mean score on the first test for those who fail developmental math is significantly below 80.

6.49 **a.** The normal probability plot supports the normality assumption. It is very much appropriate to test the population mean with a t-test.
  b. Test of mu = 1000 vs
    mu not = 1000

Variable	N	Mean	StDev
thickness	30	1016.33	52.95

SE Mean	95.0% CI	T	P
9.67	(996.56, 1036.11)	1.69	0.102

There is insufficient evidence to reject the claim that the mean thickness is 1000.

6.51 **a.** $z_{obs} = (2*14 - 50)/\sqrt{50} = -3.11$;
    p-value $= .0009$
    There is highly significant evidence that the median is less than 400.
  b. $z_{obs} = (2*56 - 100)/\sqrt{100} = 1.2$;
    p-value $= .1151$
    There is insufficient evidence that the median is more than 12.5.
  c. $z_{obs} = (2*78 - 200)/\sqrt{200} = -3.11$;
    p-value $= 2(.0009) = .0018$
    There is highly significant evidence that the median is not 75.

6.53 Sign test of median = 19.00 versus N.E. 19.00

	N	BELOW	EQUAL	ABOVE
age	30	5	12	13

P-VALUE	MEDIAN
0.0963	19.00

There is mildly significant evidence that the median age is not 19.

6.55 **a.** The distribution is heavily skewed right.
  b. $H_0: \theta \leq 100$ versus $H_a: \theta > 100$
    Because the p-value $= .2905$ there is insufficient evidence to conclude that the median survival time for terminal stomach cancer patients treated with vitamin C exceeds 100 days.

6.57 $H_0: \theta \leq 7.3$ versus $H_a: \theta > 7.3$
Sign test of median = 7.300
versus G.T. 7.300

	N	BELOW	EQUAL	ABOVE
thick	37	0	0	37

P-VALUE	MEDIAN
0.0000	10.70

All 37 observations are greater than 7.3, that is, $T = 37$. There is highly significant evidence that the median varve thickness exceeds 7.3 mm.
Because the stem-and-leaf plot appears to be skewed right, testing the population median is appropriate. The normal probability plot indicates that the skewness is not severe; thus, testing the population mean would not be out of the question.

6.59 **a.** $H_0: \theta = 7$ versus $H_a: \theta \neq 7$
  b. This is a two-tailed test.

c. The standardized test statistic is
$z = (2T - n)/\sqrt{n} = 1.04$.

d. The *p*-value $= .2955$. The null hypothesis should not be rejected.

e. Based on a random sample of 75 water samples there is insignificant evidence that the median pH level is different from 7.

6.61 a. The normal probability plot shows a moderate departure from normality. A boxplot, however, shows no outliers so, with a sample size of 15, the assumptions for the *t*-test are satisfied.

b. $H_0: \mu = 9$ versus $H_a: \mu \neq 9$

c. T-test of the Mean
Test of mu = 9 vs mu not = 9

Variable	N	Mean	StDev
score	15	8.000	1.947

SE Mean	95.0% CI	T	P
0.503	(6.922, 9.078)	−1.99	0.067

Based on the *p*-value there is statistical evidence that the mean rating score differs significantly from 9.

6.63 $H_0: \theta \geq 27{,}500$ versus $H_a: \theta < 27{,}500$

Sign test of median = 27500
versus L.T. 27500

	N	BELOW	EQUAL	ABOVE
income	25	17	0	8

P-VALUE	MEDIAN
0.0539	26500

There is significant evidence that the median salary for North Carolina social workers is less than $27,500.

6.65 a. The histogram shows no unusual observations. Based on the histogram the distribution is symmetric and possibly close to normal.

b. The boxplot shows symmetry with long tails which was not as apparent in the histogram. The normal probability plot supports this conjecture.

c. The population mean is the suggested measure of the "true" parallax of the sun.

d. Because the median is resistant to outliers, the sample median, M, is the best estimator of the "true" parallax of the sun.

e. Sign test of median = 8.798
versus not = 8.798

	N	Below	Equal	Above
Parallax	158	118	0	40

P-VALUE	Median
0.0000	8.550

There is highly significant evidence that Short's measurements were not consistent with the "true" parallax of the sun.

6.67 a **6.69** a **6.71** a

6.73 Because it is the hypothesis that the researcher wishes to support.

6.75 $H_0: \theta = 5.5$ versus $H_a: \theta \neq 5.5$

6.77 $H_0: \mu = 15$ versus $H_a: \mu > 15$
$t_{obs} = (17.8 - 15)/(6.2/\sqrt{24}) = 2.21$ *p*-value $= .0187$
There is significant evidence that the politicians scored above the norm.

6.79 The normal probability plot shows that the data are close to normal and therefore a *t*-test is permissible.
$H_0: \mu = 80$ versus $H_a: \mu < 80$.
Test of mu = 80.00 vs mu < 80.00

Variable	N	Mean	StDev
score	12	71.42	8.14

SE Mean	T	P-value
2.35	−3.65	0.0019

Based on a random sample of 12 ability-grouped students there is highly significant evidence that the group falls below the national mean.

6.81 $H_0: \mu = 4$ versus $H_a: \mu < 4$
$t_{obs} = (3.6 - 4)/(1.5/\sqrt{32}) = -1.51$, *p*-value $= .0706$
Moderately significant evidence exists that the mean life is less than four years.

6.83 A Type I error will occur if the psychologist decides the subjects average memorizing more than 40 correct phrases when in fact they average 40 or less correct phrases.
A Type II error will occur if the psychologist decides the subjects average memorizing 40 or less phrases correctly when in fact they average more than 40.

6.85 $H_0: \mu = 65$ versus $H_a: \mu > 65$
$z_{obs} = (68.2 - 65)/(10/\sqrt{81}) = 2.88$ *p*-value $= .002$
Based on a sample of 81 freshmen there is highly significant evidence that State University freshmen are more intelligent than the average freshman.

6.87 $H_0: \mu = 50$ versus $H_a: \mu < 50$
$t_{obs} = (42 - 50)/(15/\sqrt{9}) = -1.6$ *p*-value $= .0741$
Based on a random sample of nine slow-ability children there is moderately significant evidence that their mean score is less than 50. For the *t*-test to be valid it must be assumed that the distribution of scores for slow learners is approximately normal.

6.89 $H_0: \mu = 250$ versus $H_a: \mu > 250$
$t_{obs} = (263 - 250)/(72/\sqrt{100}) = 1.81$
p-value $= .0367$
Based on a random sample of 100 accounts there is significant evidence to cast doubt on the utility company's claim.

6.91 $H_0: \pi \geq .7$ versus $H_a: \pi < .7$
$z_{obs} = (.66 - .7)/(\sqrt{(.7)(.3)/200}) = -1.23$,
p-value $= .1093$
Based on a random sample of 200 people there is insufficient evidence to conclude that less than 70% favor raising the drinking age.

6.93 $H_0: \theta \le 98.5$ versus $H_a: \theta > 98.5$
Sign test of median $= 98.50$
versus G.T. 98.50

	N		BELOW	EQUAL	ABOVE
time	9		1	0	8

	P-VALUE	MEDIAN
	0.0195	101.4

$T = 8$. Based on nine trials there is significant evidence that the median time is more than 98.5 seconds, therefore you reject the horse. Because his lifetime median time is 100 seconds you did not make an error.

6.95 **a.** $H_0: \theta \le 5$ versus $H_a: \theta > 5$
b. This is a right-tailed test.
c. The test statistic $T = 9$.
d. This is a small sample test.
e. Because this is a right-tailed test we expect a large number of observations greater than five, the hypothesized value of the median. Less than half, however, are greater than five. In other words the observed value of the test statistic has fallen on the left tail of the sampling distribution instead of the right tail, making the p-value greater than 50%.
f. The null hypothesis should not be rejected.
g. Based on a random sample of 20 times there is insufficient evidence to conclude that the length of long-distance phone calls for the small business firm exceeds five minutes.

6.97 **a.** The histogram looks reasonably symmetric and possibly normally distributed.
b. The boxplot and normal probability plot show that the distribution is slightly skewed right.
c. The deviation from symmetry is not enough to warrant a study of the population median. There should be very little difference between the mean and median.
d. Test of mu $= 100.00$ vs mu > 100.00

Variable	N	Mean	StDev
severity	100	97.97	44.70

SE Mean	T	P-Value
4.47	-0.45	0.67

The sample mean is less than 100 and therefore we certainly would not reject the null hypothesis in favor of the alternative that says that μ is greater than 100. This is also indicated by the p-value $= .67$

6.99 **a.** The histogram is highly skewed left.
b. The boxplot and normal probability plot also show a skewed left distribution.
c. The median exam best represents a typical measurement.
d. $H_0: \theta \le 22$ versus $H_a: \theta > 22$

e. Sign test of median $= 22.00$
versus G.T. 22.00

	N		BELOW	EQUAL	ABOVE
score	17		4	0	13

	P-VALUE	MEDIAN
	0.0245	30.00

Based on the test scores of these 17 patients there is evidence that the tranquilizer significantly improved the amount of learning exhibited by schizophrenics.

Chapter 7

7.1 **a.** Parameters under study are the population medians.
$H_0: \theta_1 = \theta_2$ versus $H_a: \theta_1 \ne \theta_2$
b. Parameters under study are the population means.
$H_0: \mu_1 = \mu_2$ versus $H_a: \mu_1 < \mu_2$
$\mu_1 =$ mean biodegradation rate for plastic and $\mu_2 =$ mean biodegradation rate for paper
c. Parameters under study are the population proportions.
$H_0: \pi_1 = \pi_2$ versus $H_a: \pi_1 < \pi_2$.
$\pi_1 =$ percent covered last year and $\pi_2 =$ percent covered this year

7.3 Controlled experiment

7.5 The null hypothesis should be rejected. Based on 100 men and women assigned to an experimental and control group there is statistical evidence that mean resting pulse of those who jog over a period of six months is different from those who do not jog.

7.7 To compare the relative danger of the two sports we need the number of participants as well as the number of injuries. Parameters under study are the population proportions.
$H_0: \pi_1 = \pi_2$ versus $H_a: \pi_1 > \pi_2$
$\pi_1 =$ proportion of injuries from in-line skating and $\pi_2 =$ proportion of injuries from skateboarding

7.9 Subjects were not randomly assigned to the female electrical workers group; this is an observational study. The study does not prove that high doses of electricity causes breast cancer.

7.11 $H_0: \mu_1 = \mu_2$ versus $H_a: \mu_1 \ne \mu_2$
$\mu_1 =$ the mean qualifying time for inexperienced drivers and $\mu_2 =$ the mean qualifying time for experienced drivers
Group 1 is reasonably symmetric and possibly normally distributed. Group 2 is much more variable and possibly bimodal. A group of experienced drivers did no better than the inexperienced drivers but then a group of experienced drivers did significantly better than the inexperienced drivers.

7.13 **a.** $H_0: \pi_1 = \pi_2$ versus $H_a: \pi_1 \ne \pi_2$

b. $z = \dfrac{p_1 - p_2}{\sqrt{(1-p)} \, \sqrt{1/n_1 + 1/n_2}}$

c. For hypothesis testing the sample proportions should be pooled to estimate the standard error.

d. $z_{obs} = .81$; p-value $= 2(.209) = .418$. There is insufficient evidence to declare a difference in the proportions in the two age groups that fear meeting people.

7.15 a. $H_0: \pi_1 = \pi_2$ versus $H_a: \pi_1 \neq \pi_2$

b. $z = \dfrac{p_1 - p_2}{\sqrt{p(1-p)}\,\sqrt{1/n_1 + 1/n_2}}$

c. For hypothesis testing the sample proportions should be pooled to estimate the standard error.

d. $z_{obs} = -.48$; p-value $= 2(.3156) = .6312$. There is insufficient evidence to conclude that the two firms have different success rates.

7.17 $H_0: \pi_1 = \pi_2$ versus $H_a: \pi_1 \neq \pi_2$
$z_{obs} = -1.12$; p-value $= 2(.1314) = .2628$. There is insufficient evidence that the proportion of men who smoke is different from the proportion of women who smoke.

7.19 $H_0: \pi_1 = \pi_2$ versus $H_a: \pi_1 \neq \pi_2$
$z_{obs} = -2.05$; p-value $= 2(.0202) = .0404$. There is significant evidence that the percent of people who had at least one cold is different for the two groups of adults.

7.21 This should be a one-tailed test.
$H_0: \pi_1 = \pi_2$ versus $H_a: \pi_1 < \pi_2$
$z_{obs} = -4.16$; p-value $< .00003$. Based on these data there is highly significant evidence that AZT reduces the chances of children of HIV-positive mothers developing AIDS.

7.23 $H_0: \pi_1 = \pi_2$ versus $H_a: \pi_1 \neq \pi_2$
$\pi_1 = $ proportion of third-grade girls who say they are good in math and $\pi_2 = $ proportion of third-grade boys who say they are good in math
$z_{obs} = -.39$; p-value $= 2(.3483) = .6966$. Based on random samples of 150 third-grade girls and 200 third-grade boys there is insignificant evidence that the percents of third-grade girls and boys who say they are good in math are different.

7.25 $H_0: \mu_1 = \mu_2$ versus $H_a: \mu_1 \neq \mu_2$
$\mu_1 = $ the mean science test grade for the traditional group and $\mu_2 = $ the mean science test grade for the experimental group
$t_{obs} = -.24$, p-value $> .40$. There is insufficient evidence to conclude that the mean score for the experimental group is different from the mean score of the traditional group.

7.27 a. $H_0: \mu_1 = \mu_2$ versus $H_a: \mu_1 \neq \mu_2$
$\mu_1 = $ the mean number of errors with drug and $\mu_2 = $ the mean number of errors with placebo

b. The parent distributions should not deviate substantially from normality and the samples should be independent.

c. $t_{obs} = .7$

d. df $= 11$, p-value $> .10$ There is insignificant evidence to conclude that the mean number of errors with the drug is different from the mean number of errors with a placebo.

7.29 a. $H_0: \mu_1 = \mu_2$ versus $H_a: \mu_1 \neq \mu_2$
$\mu_1 = $ the mean length of fish caught with a small-mesh codend and $\mu_2 = $ the mean length of fish caught with a large-mesh codend

b. The sample sizes are clearly large enough that the Central Limit Theorem would apply.

c. $z_{obs} = -7.014$ p-value $= 2(0*) = 0^+$ Reject H_0

d. Based on these data the difference between the length of fish caught with small-mesh codend and the length of fish caught with a large-mesh codend is highly significant.

7.31 a. $H_0: \mu_1 = \mu_2$ versus $H_a: \mu_1 > \mu_2$
$\mu_1 = $ the mean score for the over 35 age group and $\mu_2 = $ the mean score for the under 35 age group

b. The boxplots show no significant departures from the basic assumptions for inferences based on the two-sample t-procedures.

c. $t_{obs} = 1.15$. Using the conservative approach we get df $= 31$ and $.10 < p$-value $< .15$. Using the more exact method of calculating degrees of freedom we get df $= 60$ and the same p-value. In either case there is insignificant evidence to conclude that the mean for the over-35 group is greater than the mean for the under-35 group.

7.33 Both distributions are close to symmetric. The variability of the two groups is about the same leading to the conclusion that the two groups are homogeneous. The middle 50% of group B is completely contained within the middle 50% of group A. It is doubtful that there is a significant difference between the two groups.

7.35
```
Two sample T for waterph
code     N     Mean    StDev    SE Mean
0       42    7.218    0.500     0.077
1       33    7.040    0.327     0.057

90% CI for mu (0) - mu (1):
(0.018, 0.338)
T-test mu (0) = mu (1) (vs not =):
T= 1.86 P=0.067 DF= 70
```
The interval does not contain both positive and negative values. The interval goes from .018 to .338 which contains only positive numbers; there are no negative numbers inside the interval. This means that a two-tailed test of $\mu_1 = \mu_2$ will be rejected at the 10% level of significance. In fact, the computer printout shows p-value $= .067$ which will reject at a 10% level of significance but not at a 5% level. If the confidence level is increased to 99% the interval will become wider. Because the 99% interval is wider it may contain negative values.

7.37 Because the sample standard deviations are so close we should use pooled t-procedures. $(82.4 - 84.2) \pm 1.645(11.5)\sqrt{1/150 + 1/150}$ The confidence interval is $(-3.98, .38)$.

The interval contains both positive and negative values. This means that a two-tailed test of the equality of mean grades for fall and spring classes would not be rejected.

7.39 **a.** $H_0: \mu_1 = \mu_2$ versus $H_a: \mu_1 < \mu_2$
μ_1 = the mean amount of coffee dispensed by vending machine A and μ_2 = the mean amount of coffee dispensed by vending machine B
  b. For small samples the parent populations should be close to a normal distribution and the population variances should be homogeneous.
  c. $t_{obs} = -0.63$ p-value $> .10$. Insufficient evidence exists to support the conjecture that machine A dispenses significantly less than machine B.

7.41 **a.** The distributions seem symmetric but normality is difficult to ascertain.
  b. The variances are homogeneous.
  c. Because the tails in the stem-and-leaf plots look long boxplots would help identify possible outliers.
  d. If the distributions are long tailed we should not conduct inferences with the pooled t-test.
  e. The pooled t-test applied to the ranks is equivalent to the Wilcoxon rank sum test. We have

```
Two sample T for ranks
group   N    Mean   StDev   SE Mean
1       23   27.2   12.9    2.7
2       23   19.8   13.2    2.7

95% CI for mu (1) − mu (2):
(−0.4, 15.1)
T-test mu (1) = mu (2) (vs not
=): T= 1.92 P=0.061 DF= 44
Both use Pooled StDev = 13.0
```
Based on a p-value $= .061$ there is mildly significant evidence of a difference between the means of these two distributions.

7.43 **a.** The boxplots show two distributions with symmetric long tails. There does not seem to be a difference between the mean abilities of men and women.
  b. 98% CI for mu female − mu male: $(-4.2, 5.1)$
  c. The interval contains both positive and negative values. This suggests that there is no significant difference between the abilities of men and women.
  d. Because of the long-tailed distributions a confidence interval based on the Wilcoxon rank sum procedure might be more appropriate.

7.45 **a.** $H_0: \mu_1 = \mu_2$ versus $H_a: \mu_1 \neq \mu_2$
μ_1 = the mean carbon monoxide emitted by manufacturer and μ_2 = the mean carbon monoxide emitted by competitor
  b. Both distributions appear to be short tailed.
  c. The variances seem to be homogeneous.

  d. Neither distribution is severely skewed nor has unusually long tails (they are short tailed); the pooled t-test can be applied to these data.
  e.
```
Two sample T for manufac vs compet
           N    Mean    StDev  SE Mean
manufac    9    2.989   0.389   0.13
compet     10   3.370   0.395   0.12

95% CI for mu manufac − mu
compet: (−0.76, −0.00)
T-test mu manufac = mu compet (vs
not =): T= −2.12 P=0.049 DF= 17
Both use Pooled StDev = 0.392
```
Based on these data there is a significant difference between the mean carbon monoxide emitted by the manufacturer and by the competitor.

7.47 **a.** From the boxplots the distributions are symmetric with homogeneous variances.
  b. Normality is a reasonable assumption in both cases.
  c. Pooled t-test
```
Twosample T for wafer1 vs wafer2
          N     Mean    StDev  SE Mean
wafer1    30    1012.1  59.7    11
wafer2    30    1020.6  58.7    11

95% C.I. for mu wafer1 − mu
wafer2: (−39, 22)
T-test mu wafer1 = mu wafer2 (vs
not =): T= −0.56 P=0.58 DF= 58
Both use Pooled StDev = 59.2
```
With a p-value $= .58$ there is an insignificant difference between the mean oxide thicknesses for the two sets of wafers.

7.49 $H_0: \mu_1 = \mu_2$ versus $H_a: \mu_1 \neq \mu_2$
μ_1 = the mean DBH activity for the nonpsychotic group and μ_2 = the mean DBH activity for the psychotic group
Conditions are satisfied for the pooled t-test. We have
```
Twosample T for DBH
group   N     Mean    StDev  SE Mean
1       15    164.3   47.0    12
2       10    242.6   51.4    16

95% C.I. for mu 1 − mu 2: (−120, −37)
T-test mu 1 = mu 2 (vs not =):
T= −3.94 P=0.0007 DF= 23
Both use Pooled StDev = 48.7
```
Based on these results there is highly significant evidence of a difference in the DBH activity for the two groups. The DBH activity for the nonpsychotic group was significantly below that of the psychotic group.

7.51 The experiment is to compare a new method of reading instruction to the old standard method. If a student is first taught the standard method they will gain knowledge that will confound with the knowledge gained if they are then taught the new method. Their reading ability will improve by exposure to either reading method and to then be taught the other method would be meaningless. To compare the methods we need two equivalent homogeneous groups with one being taught the standard method and the other group being taught the new method and then compare their reading abilities.

7.53 **a.** The samples are matched because the same subject is measured before and after viewing the movie.
b. The distribution of difference scores should be close to normal.
c. There may be some departure from normality but the sample size is so small it is difficult to tell. There appears to be no gross departures that would preclude the paired t-test.
d. 99% confidence interval for μ_d is $(-0.369, 1.702)$.
e. The confidence interval contains both positive and negative values; thus, there is no change in moral attitudes after viewing the film.

7.55 **a.** The samples are matched because the same subject is measured both with and without a coffee break.
b. The dotplot indicates that the normality assumption may not be met, but symmetry may be plausible. The Wilcoxon signed rank test would apply.
c. Test of mu = 0.00 vs mu not = 0.00

Variable	N	Mean	StDev
sgnrnks	9	1.67	5.66

SE Mean	T	P-Value
1.89	0.88	0.40

There is insufficient evidence that a coffee break increases productivity.

7.57 **a.** The samples are matched because the subjects are paired according to their IQ.
b. The distribution of difference scores should be close to normal. There may be some departure from normality. The dotplot shows symmetry but the distribution may be long tailed.
c. The Wilcoxon signed rank test may be more appropriate for these data.

Test of mu = 0.00 vs mu not = 0.00

Variable	N	Mean	StDev
signrks	11	−1.09	7.01

SE Mean	T	P-Value
2.11	−0.52	0.62

d. There is insufficient evidence to suggest a difference between the study habits of students of the two school districts.

7.59 **a.** The dotplot and normal probability plot show no significant departures from normality.
b. Test of mu = 0.000 vs mu not = 0.000

Variable	N	Mean	StDev
differ	12	1.083	2.353

SE Mean	T	P-Value
0.679	1.59	0.14

There is insufficient evidence that their improvement scores are different for the two techniques.

7.61 **a.** Based on the dotplot, boxplot, and normal probability plot there appears to be no significant departures from the assumptions for the matched pairs t-test.
b. Test of mu = 0.000 vs mu not = 0.000

Variable	N	Mean	StDe
differ	20	0.400	2.088

SE Mean	T	P-Value
0.467	0.86	0.40

There is insufficient evidence of a difference between the ratings given by the two psychiatrists.

7.63 **a.** The standard deviations are close enough that we can pool them together to estimate the standard error of $\bar{x}_1 - \bar{x}_2$.
b. $13 \pm 1.66*30.71\sqrt{(1/65 + 1/60)}$ The confidence interval is 3.87 to 22.13.
c. The interval contains only positive values. This means that a two-tailed test of the equality of the population means would be rejected at the 10% level of significance. The mean size in Region one is greater than the mean size in Region two.

7.65 $H_0: \theta_1 = \theta_2$ versus $H_a: \theta_1 \neq \theta_2$

Twosample T for ranks

group	N	Mean	StDev	SE Mean
1	21	23.7	10.6	2.3
2	19	17.0	12.1	2.8

95% C.I. for mu 1 − mu 2: (−0.5, 14.0)
T-test mu 1 = mu 2 (vs not =):
T= 1.87 P=0.069 DF= 38
Both use Pooled StDev = 11.3
There is a mildly significant difference between the two population medians.

7.67 $H_0: \pi_1 = \pi_2$ versus $H_a: \pi_1 < \pi_2$
$z_{obs} = 1.06$; p-value $= .1446$. There is insufficient evidence that the females choose diet drinks more frequently than males.

7.69 $(.966 - .448) \pm$
$1.96\sqrt{[(.966)(.034)/29 + (.448)(.552)/143]}$ The confidence interval is .413 to .623.

7.71 $H_0: \mu_d \leq 0$ versus $H_a: \mu_d > 0$ where $\mu_d = \mu_1 - \mu_2$
μ_1 = mean number of sit-ups after the course, μ_2 =

mean number of sit-ups before the course
A matched pairs *t*-test:

```
Test of mu = 0.000 vs mu > 0.000

Variable        N        Mean       StDev
differ          9       2.000       2.179

SE Mean         T       P-value
  0.726       2.75       0.012
```

There is significant evidence that the physical fitness course improved the number of sit-ups by the participants.

7.73 $H_0: \pi_1 = \pi_2$ versus $H_a: \pi_1 > \pi_2$
π_1 = proportion interested in business and tech now and π_2 = proportion interested in business and tech in the past.
$z_{obs} = 1.09$; *p*-value = .1379. There is insignificant evidence that a higher proportion of students are interested in business and tech now than in the past.

7.75 $H_0: \pi_1 = \pi_2$ versus $H_a: \pi_1 > \pi_2$
π_1 = proportion from the hyperactive group and π_2 = proportion from the normal group
$z_{obs} = 2.61$; *p*-value = .0045. There is highly significant evidence that a higher proportion of hyperactive children had mothers with poor health during pregnancy.

7.77 **a.** The assumptions are not violated.
b. Matched pairs *t*-test.
c. $H_0: \mu_d \leq 0$ versus $H_a: \mu_d > 0$
where $\mu_d = \mu_1 - \mu_2$
μ_1 = mean yield for the new variety, μ_2 = mean yield for the standard variety.
d.
```
Test of mu = 0.00 vs mu > 0.00

Variable        N        Mean       StDev
differ         12        7.75        7.01

SE Mean         T       P-value
  2.02        3.83       0.0014
```
e. There is highly significant evidence that the yield per acre for the new variety is higher than the standard variety.

7.79
```
Twosample T for Private vs Public
            N     Mean     StDev    SE Mean
Private    15    17.63      4.14        1.1
Public     15    14.91      4.57        1.2

98% C.I. for mu Private - mu Public:
(-1.2, 6.7)
T-test mu Private = mu Public
(vs >): T= 1.71 P=0.049 DF= 28
Both use Pooled StDev = 4.36
```
The conclusion is that private high school students spend significantly more time on homework per week than do public high school students. These results are different from the results found in part **d** of Exercise 7.87 because this is a one-tailed test whereas the confidence interval is equivalent to a two-tailed test. It generally is easier to reject the null hypothesis in favor of a one-tailed alternative than a two-tailed alternative.

7.81 The following side-by-side boxplots show that the two distributions are symmetric, equally variable, without outliers. A pooled *t*-test can be used to compare the population means.
```
Twosample T for Process1 vs Process2

            N     Mean    StDev   SE Mean
Process1    7    2.514    0.682      0.26
Process2    8    2.900    0.796      0.28

95% C.I. for mu Process1
- mu Process2: (-1.22, 0.45)
T-test mu Process1 = mu Process2
(vs not =): T= -1.00 P=0.34 DF= 13
Both use Pooled StDev = 0.746
```
Based on the *p*-value there is an insignificant difference between the quality scores for the two processes.

7.83 The following boxplot of the difference scores identifies one outlier but otherwise the distribution is symmetric and does not deviate substantially from normality. The matched pairs *t*-test can be used to analyze these data.
```
Test of mu = 0.00 vs mu not = 0.00

Variable        N        Mean       StDev
differ         20       17.30       28.49

SE Mean         T       P-value
  6.37        2.72       0.014
```
There is significant evidence of a difference in the estimates from the two suppliers.

7.85 **a.** The skewness is such that we should compare population medians instead of population means.
b.
```
Twosample T for ranks
year    N    Mean    StDev   SE Mean
1      39    34.7    23.2       3.7
2      39    44.3    21.3       3.4

95% C.I. for mu 1 - mu 2: (-19.7, 0.4)
T-test mu 1 = mu 2 (vs not =):
T= -1.92 P=0.059 DF= 76
Both use Pooled StDev = 22.3
```
c. Applying the pooled *t*-test to the ranks is equivalent to the two-sample Wilcoxon rank sum test.
d. Based on the *p*-value found in part **b** there is mildly significant evidence of a difference between the median commuting times for 1980 and 1990.

7.87 **a.** There is one extreme outlier in the Greenland data. The extreme outlier will distort the value of the mean concentration for the Greenland sample. The resulting *t*-test may be misleading.
b. With the outlier removed both distributions appear to be approximately normal. A *t*-test is appropriate for analysis of the data.
c. The distributions are reasonably homogeneous; a pooled *t*-test seems appropriate.
d. $H_0: \mu_1 = \mu_2$ versus $H_a: \mu_1 \neq \mu_2$
μ_1 = the mean concentration of microparticles from the Antarctica snowfield and μ_2 = the mean

concentration of microparticles from the Greenland ice cap

```
Twosample T for antarc vs greenld
           N    Mean   StDev  SE Mean
antarc    16    2.67    1.49     0.37
greenld   17    2.153  0.964    0.23
```

95% C.I. for mu antarc −
mu greenld: (−0.37, 1.40)
T-test mu antarc = mu greenld (vs
not =): T= 1.19 P=0.244 DF= 31
Both use Pooled StDev = 1.25
Even with the removal of the outlier there is an insignificant difference between the concentration of microparticles from the two snowfields.

Chapter 8

8.1 H_0: $\pi_1 = \pi_2 = \pi_3 = \pi_4 = \pi_5 = 1/5$ versus H_a: at least one is not the same
$\chi^2_{obs} = 7.34$; p-value $> .10$ There is insufficient evidence to conclude that the classes are unequal.

8.3 If H_0: $\pi_1 = .40$, $\pi_2 = .45$, $\pi_3 = .07$, $\pi_4 = .05$, $\pi_5 = .03$ is true, then $e_1 = 480$, $e_2 = 540$, $e_3 = 84$, $e_4 = 60$, $e_5 = 36$ and $\chi^2_{obs} = 5.81$. p-value $> .10$. There is insufficient evidence to conclude that the percentages are different from those specified by the official.

8.5 H_0: $\pi_1 = \pi_2 = \pi_3 = \pi_4 = 1/4$ versus H_a: at least one is not the same
$\chi^2_{obs} = 3.6$; p-value $> .10$. There is insufficient evidence to conclude that the proportions of marbles are different.

8.7 H_0: The strains are equally resistant to the chemical agent.
H_a: The strains are not equally resistant to the chemical agent.
If H_0 is true then $e_1 = e_2 = e_3 = e_4 = e_5 = 200$ and $\chi^2_{obs} = 128.08$; p-value $< .005$. The evidence is highly significant that some strains are more resistant to the chemical than others.

8.9 If H_0: $\pi_1 = .31$, $\pi_2 = .50$, $\pi_3 = .14$, $\pi_4 = .05$ is true, then $e_1 = 77.5$, $e_2 = 125$, $e_3 = 35$, $e_4 = 12.5$ and $\chi^2_{obs} = 206.59$ p-value $< .005$. There is highly significant evidence that the claims from the employees of this company do not follow the same distribution as given in the *Wall Street Journal*.

8.11 The categories on the survey do not constitute a multinomial experiment and therefore the chi-square goodness of fit does not apply. The categories of a multinomial experiment must be independent and the associated probabilities must sum to unity. This is clearly not the case in this study.

8.13 H_0: Being mathematically inclined is independent of being left- or right-handed.
H_a: Being mathematically inclined is dependent on being left- or right-handed.
$\chi^2_{obs} = 0.948 + 0.090 + 0.865 + 0.082 = 1.985$; df $= 1$, $p = 0.159$. There is insufficient evidence to conclude that being mathematically inclined depends on being left- or right-handed.

8.15 H_0: The service rendered is independent of the type dealership.
H_a: The service rendered is dependent on the type dealership.
$\chi^2_{obs} = 30.297$ df $= 5$, $p = 0.000$. There is highly significant evidence that the service rendered is dependent on the type dealership.

8.17 H_0: A student's desire to participate in the science project program is independent of their academic standing.
H_a: A student's desire to participate in the science project program is dependent on their academic standing.
$\chi^2_{obs} = 5.314$; df $= 2$, $p = 0.071$. There is mildly significant evidence that participating in the science project depends on academic standing.

8.19 H_0: Variables A and B are independent.
H_a: Variables A and B are dependent.
$\chi^2_{obs} = 9.597$; df $= 2$, $p = 0.008$. There is highly significant evidence that the two variables are dependent.

8.21 H_0: Snoring and heart disease are independent.
H_a: Snoring and heart disease are dependent.
$\chi^2_{obs} = 72.782$ df $= 3$, $p = 0.000$. There is highly significant evidence that snoring and heart disease are statistically related.

8.23 **a.** H_0: A person's choice for president is independent of their gender.
H_a: A person's choice for president is dependent on their gender.
$\chi^2_{obs} = 6.489$; df $= 2$, $p = 0.039$. There is significant evidence that choice for president and gender of the respondent are dependent.
b. H_0: A person's choice for vice president is independent of their gender.
H_a: A person's choice for vice president is dependent on their gender.
$\chi^2_{obs} = 27.005$; df $= 2$, $p = 0.000$. There is highly significant evidence that choice vice president and gender of the respondent are dependent.

8.25 H_0: Attendance of bus drivers and type of shift are independent.
H_a: Attendance of bus drivers and type of shift are dependent.
$\chi^2_{obs} = 336.432$; df $= 4$, $p = 0.000$. There is highly significant evidence that the attendance of bus drivers and type of shift are statistically dependent.

8.27 **a.** The classification variable is the smoking habit of the respondent.
b. The populations are the two treatments.
c. H_0: The two treatments are equally effective.
H_a: The two treatments are not equally effective.
d. $\chi^2_{obs} = 2.114$; df $= 2$, $p = 0.348$. There is insufficient evidence that the two treatments are not equally effective.
e. No assumptions for the test have been violated.

8.29 The classification variable is whether or not the subject has contemplated suicide. The populations are Vietnam veterans and nonveterans.
H_0: The proportion who have contemplated suicide is the same for both veterans and nonveterans.
H_a: The proportion who have contemplated suicide is the not the same for veterans and nonveterans.
$\chi^2_{obs} = 13.065$; df $= 1$, $p = 0.000$. There is highly significant evidence that the proportion of veterans who have contemplated suicide is different from non-veterans.

8.31 H_0: The proportions falling in these three categories are the same for all three populations.
H_a: The proportions falling in these three categories are different for at least one of these populations.
$\chi^2_{obs} = 4.621$; p-value $= .329$. The row marginal totals are all 80; they are fixed values. All of the expected cell counts exceed 5 (the smallest is 21.33) so the chi-square approximation should be good. Based on the p-value there is insufficient evidence that the proportions falling in the three categories are different.

8.33 **a.** The classification variable is the outcome of the murder case.
   b. The populations are defendant husbands and defendant wives.
   c. H_0: The results of murder cases are the same for husband and wife defendants.
   H_a: The results of murder cases are different for husband and wife defendants.
   $\chi^2_{obs} = 32.765$ df $= 3$, $p = 0.000$. There is highly significant evidence that the outcomes of murder cases are different for husband and wife defendants.

8.35 **a.** The classification variable is the injury frequency.
   b. The populations are 1988–90 and 1991–93 Chevrolet vehicles. It is assumed that samples of vehicles from the two model years are fixed and have been classified into the injury categories.
   c. ChiSq $= 5.326$ df $= 4$, $p = 0.256$
   Five cells with expected counts less than 5.0. From the computer printout we see that five of the ten cells have expected counts less than five. This may cause serious problems with the chi-square approximation.
   d. We adjust the data by combining the first and second columns and combining the fourth and fifth columns. We now have three classifications: Above average, Average, Below average.
   ChiSq $= 4.335$ df $= 2$, $p = 0.115$
   e. H_0: Injury frequency is the same for the two model years.
   H_a: Injury frequency is different for the two model years.
   f. $\chi^2_{obs} = 4.335$; p-value $= .115$ Based on these data the injury frequency distribution is the same for the two model years.

8.37 H_0: The proportions in the categories are the same for both populations.
H_a: The proportions in the categories are different for the populations.
$\chi^2_{obs} = 1.416$; df $= 2$, $p = 0.493$. There is insignificant evidence that the proportions are different for the two populations. This is the chi-square test of homogeneity.

8.39 H_0: The risk of heart attack is distributed the same for both personality types.
H_a: The risk of heart attack is not distributed the same for both personality types.
χ^2_{obs} 3.038; df $= 2$, $p = 0.219$. There is insufficient evidence that risk of heart attack is distributed differently for Type A and Type B personalities. This is the chi-square test of homogeneity.

8.41 $\chi^2_{obs} = 8.31$; $.025 < p$-value $< .05$
There is significant evidence to reject the hypothesis that $\pi_1 = \pi_3 = .20$, and $\pi_2 = \pi_4 = .30$.

8.43 H_0: $\pi_L = \pi_M = .4$, $\pi_H = .2$ H_a: At least one proportion is not as specified in H_0.
$\chi^2_{obs} = .58333$; p-value $> .5$ There is insufficient evidence to conclude the alternative.

8.45 H_0: Disease and vaccination are independent.
H_a: Disease and vaccination are dependent.
$\chi^2_{obs} = 3.664$; df $= 1$, $p = 0.056$. There is significant evidence that disease and vaccination are dependent.

8.47 H_0: Marital status and career are independent.
H_a: Marital status and career are dependent.
$\chi^2_{obs} = 10.667$ df $= 1$, $p = 0.001$. There is highly significant evidence that marital status and career are related.

8.49 H_0: The row and column classifications are independent.
H_a: The row and column classifications are dependent.
$\chi^2_{obs} = 1.899$; p-value $= .594$. There is insufficient evidence to conclude that the row and column classifications are dependent. This is the chi-square test of independence.

8.51 H_0: Abortion rate is independent of the region of the country.
H_a: Abortion rate is dependent on the region of the country.
$\chi^2_{obs} = 5.792$; df $= 3$, p-value $= .122$
There is insignificant evidence that abortion rate is dependent on the region of the country.

8.53 H_0: Moral values and opinion on the referendum are independent.
H_a: Moral values and opinion on the referendum are dependent.
$\chi^2_{obs} = 19.702$ df $= 4$, $p = 0.001$. Their moral values are related to their opinion on the referendum.

8.55 **a.** H_0: The education levels of black women is the same as black men.
   H_a: The education levels of black women is different from black men.
   b. This is a chi-square test of homogeneity.
   c. ChiSq $= 34.195$

d. df = 4, p = 0.000 The null hypothesis should be rejected.

e. There is a highly significant difference between the education levels of black women and men.

8.57 H_0: Perceived math ability is the same for girls and boys. H_a: Perceived math ability is different for girls and boys. ChiSq = 19.869 df = 4, p = 0.001. Based on these data there is highly significant evidence that perceived math ability is different for girls and boys.

8.59 **a.** H_0: Percent passing with perfect scores is independent of the type inspection station. H_a: Percent passing with perfect scores is dependent on the type inspection station.

b. ChiSq = 61.027 df = 10, p = 0.000 Six cells with expected counts less than 5.0 Based on the p-value there is highly significant evidence that the percent of vehicles passing with perfect scores depends on the type inspection station.

c. Six of the 18 cells (33%) have expected cell counts less than five.

d. New car dealers generally inspect new cars. Almost all new cars should pass with a perfect score. The new car dealers should not be considered typical inspection stations.

e. ChiSq = 9.834 df = 8, p = 0.278 Four cells with expected counts less than 5.0 Without the new car dealers the test results are insignificant, i.e., there is no evidence that the percent of vehicles passing with perfect scores depends on the type inspection station.

Chapter 9

9.1 **a.** $e = 3 - 17 + 16 = 2$ **b.** $e = 20 - 17 - 4 = -1$

c. $e = 5 - 17 + 12 = 0$

9.3 **a.** $\mu_y = 65{,}280 + 24.3(1500) = 101{,}730$

b. $101730 \pm 2(2300)$ The interval is (97130, 106330). Out of 100 randomly selected homes we expect five to have prices outside this interval.

c. $24.3(100) = 2430$

d. The cost 97,500 is consistent with the interval found in part **b**.

9.5 **a.** The mean cost three months after being introduced is $\mu_y = 3500 - 275(3) = 2675$. The initial cost is 3500.

b. $2675 \pm 2(45)$ The interval is (2585, 2765). We expect five to have prices outside this interval.

c. 275

d. After three months $2750 is consistent with the interval found in part **b**.

e. The model is proposed for the first six months. There is no reason to believe that the same linear model will apply for a whole year.

9.7 **a.** The regression equation is percent = 32.5 + 3.42 age so the least squares estimates of β_0 and β_1 are 32.5 and 3.42, respectively.

b. A straight line fits the data very well.

c. The standard error about the regression line is s = 3.342.

d. The regression equation is based on ages 2 through 18. Extrapolating to age 20 is acceptable but we would not want to extrapolate much beyond that point.

9.9 **a.** A straight line fits the data reasonably well, but a curvilinear fit seems more appropriate. There are no unusual observations other than the curvilinear pattern.

b. The regression equation is: magnesiu = 5.77 − .000252 distance. The standard error about the regression line is s = 1.195.

9.11 A scatterplot and the correlation coefficient are two tools used to evaluate a linear relationship. The Correlation of Poverty and Crime = 0.242 which is very small. The scatterplot does not show a linear pattern. It is not worthwhile to pursue a linear relationship between the two variables.

9.13 **a.** s = 1.464

b. $\hat{y} = .783 + .6848x$

c. $t_{obs} = b_1/SE(b_1) = .6848/.1526 = 4.49$

d. p-value = .004 There is highly significant evidence that $\beta_1 \neq 0$.

e. p-value = .695 We cannot reject H_0: $\beta_0 = 0$. The population regression line should go through the origin.

9.15 **a.** There is a very definite linear trend.

b. $b_0 = 4.6748$ $b_1 = 1.2272$

c. t_{obs} = 10.44 p-value = .000 The linear relationship is highly significant. The coefficient of determination is R-sq = 96.5%

d. sales = 4.6748 + 1.2272(6) = 12.038 ≈ 12

e. A 98% confidence interval for the expected sales when six ads are placed is (10.480, 13.595).

f. This model should not be used to predict sales for the month of January. Data for the month of January should be used to develop the model.

9.17 **a.** The coefficient of determination is R^2 = .059. The percent of variability in poverty rate explained by the linear model is 5.9%. This is unacceptable.

b. t_{obs} = 1.06, p-value = .304 The null hypothesis should be accepted. This means that the proposed linear model is inadequate for explaining the relationship.

c. This regression equation should not be used to predict children's poverty by the crime rate.

9.19 **a.** t_{obs} = 17.49, p-value = .000 The null hypothesis is rejected.

b. R^2 = 88.4% and t_{obs} = 17.49

c. One observation is unusual because its x value is so large (15.4). The other two are unusual because their standardized residual exceeds 2.0. In the scatterplot they are the two observations in the top right-hand corner and the one observation below the main scatter.

d. The predicted value of a company's brand name is 9.233 when their revenue is five billion dollars. A 95% interval estimate of the mean brand name value is (8.32, 10.146). The interval when the revenue is $20 billion is off the chart. This is extrapolation and should be avoided.

9.21 **a.** The Response variable is sales.
  b. There is a weak upward linear trend.
  c. The regression equation is
    sales $= 24.9 + 33.1$ nicotine
  d. $t_{obs} = 2.24$ p-value $= 0.067$ There is mildly significant evidence of a linear trend. The regression coefficient is positive $(+33.1)$ therefore sales will increase when nicotine content increases
  e. Sales increases 33.1 $(\times\$100{,}000)$ for each additional milligram of nicotine.
  f. Predicted sales when nicotine content is 1.0 is $24.9 + 33.1(1) = 58$ $(\times\$100{,}000)$
  g. For $x^* = 1.0$ a 99% confidence interval for expected sales is (34.34, 81.63).

9.23 **a.** There appears to be a downward linear trend.
  b. The regression equation is wins $= 144 - 16.5$ era. A unit change in era results in a decrease of 16.5 wins.
  c. $t_{obs} = -3.24$, p-value $= .009$ The null hypothesis should be rejected. The proposed linear model is highly significant.
  d. This regression equation can be used to predict the number of wins based on the earned run average of the pitchers.
  e. There are no unusual observations.
  f. For $x^* = 3.8$ a 99% confidence interval is (74.74, 87.10).

9.25 No, the residuals appear normally distributed about 0.

9.27 There appears to be a relationship between the residuals and time. Time should be incorporated in the model.

9.29 **a.** $R^2 = 98.9\%$ and the t-ratio is highly significant. The straight line fits the data very well.
  b. It is doubtful that there would be any significant violations of assumptions.
  c. No distinct pattern is exhibited by the residual plot.

9.31 **a.** One observation causes the normal probability plot to show a nonnormal pattern.
  b. Again, one observation causes a nonrandom pattern in the residual plot.
  c. Adjustments would be to investigate the relationship without the one unusual observations.

9.33 There may be a slight departure from normality but otherwise there is no unusual behavior shown in this residual analysis.

9.35 The residuals show a very distinct curvilinear pattern that indicates that the linear equation is inadequate. The model needs revising to include a quadratic component.

9.37 **a.** The regression equation is sales $= 55.0 - 6.82$ rd
    rd: $t_{obs} = -2.60$, p-value $= 0.027$ The regression coefficient is significantly different from 0.

b. R-sq $= 40.3\%$ which is the percent of variability in sales explained by R & D.
  c. $t = -2.6$, p-value $= 0.027$ The regression coefficient is significantly different from 0.
  d. Unusual Observations

Obs.	rd	sales	Fit	Stdev.Fit
6	6.30	41.00	12.07	8.05

Residual	St.Resid
28.93	2.85R

  R denotes an obs. with a large st. resid. The standardized residual for observation six is unusually large.
  e. The regression equation is sales $= 69.9 - 12.1$ rd
    rd: $t_{obs} = -8.26$, p-value $= 0.000$ The regression coefficient is highly significantly different from 0. R-sq $= 88.4\%$ which is the percent of variability in sales explained by R and D.
  f. R^2 improves from 40.3% to 88.4%. The t-statistic are highly significant. There are no unusual observations.

9.41 **a.** The relationship between x and y appears to be curvilinear.
  b. There appears to be a strong linear relationship between x and y.

9.43 **a.** $x' = 100$, $\hat{y} = 7.3$; $x' = 160 = 2.86$
  b. $x' = 1.5$, $\hat{y} = 7.25$; $x' = 5.0$, $\hat{y} = 19.616$

9.45 $x = 5$, $\hat{y} = 23.7$, $e = -5.7$ $x = 5$, $\hat{y} = 23.7$, $e = -3.7$

9.47 **a.** $\mu_y = 200 - .35(45) = 184.25$
  b. $184.25 \pm 2(1.5)$. The interval is (181.25, 187.25).
  c. $-.35$

9.49 The value of $R^2 = 92.9\%$ is very high, however, the scatterplot shows a linear relationship for x between 20 and 40 and a different linear relationship when x is between 45 and 55. For $x' = 30$, we have $\hat{y} = -.96$ and for $x' = 60$, we have $\hat{y} = 4.53$.

9.51 **a.** There are two bivariate outliers in the upper left-hand and lower right-hand corners. They will have an adverse affect on the regression line that describes the relationship between IQ and GPA.
  b. The regression equation is
    GPA $= 2.53 + 0.0025$ IQ.
  c. The value of $R^2 = 0.2\%$ is extremely small indicating no linear relationship. The model does not describe the relationship between IQ and GPA.
  d. The p-value $= 0.898$ is insignificant. There is no linear relationship.
  e. We should not use this model to predict GPA from IQ.

9.53 The regression equation is
  spelling $= 59.5 + 1.23$ book
  a. A straight line will fit the data, however, there is a slight curvature to the pattern that suggests that a polynomial model might fit the data better.
  b. The t-ratio for testing the regression coefficient is highly significant: $t_{obs} = 5.93$ with a p-value $= .000$.

c. The value of $R^2 = 70.1\%$ is large, indicating a strong linear relationship.

d. The residual analysis shows that normality may be an issue and the Residuals versus Fits plot shows a curvilinear pattern. We should consider a polynomial model.

9.55 a. The regression equation is

confid = 85.5 − 4.23 educat

b. The standard error about the regression line is s = 10.35

c.
```
Predictor  Coef  Stdev  t-ratio      p
educat   −4.231  1.020  −4.15  0.001
```

d. R-sq = 48.9% Thus, 48.9% of the variability in confidence is explained by education.

e. This model adequately describes the relationship between the two variables.

9.57 The normal probability plot does not rule out normality of the residuals. The residual plot appears to have a random pattern. It appears that the assumptions have been met.

9.59 a. The regression equation is

sales = 3246 + 49.9 months

b. R-sq = 25.8% Only 25.8% of the variability in sales is explained by months.

c.
```
Predictor  Coef  Stdev  t-ratio      p
months   49.85  29.89    1.67  0.134
```
There is insignificant evidence to reject that $\beta_1 = 0$. There is no linear relationship between sales and months.

d. A possible violation of assumptions is that the variance does not appear to be constant.

9.61 a. The regression equation is

price = 33.4 − 0.393 harvest

b.
```
Predictor      Coef     Stdev
harvest    −0.39267   0.06303
t-ratio          p
  −6.23      0.003
```

c. R-sq = 90.7% This model explains 90.7% of the variability.

d. The model adequately describes the relationship between the two variables.

e.
```
                Fit    Stdev.Fit
For x* = 60    9.816      0.508
For x* = 30   21.596      2.224
       95.0% C.I.        95.0% P.I.
( 8.405, 11.227)  ( 6.663, 12.968)
(15.419, 27.773)  (14.807, 28.385) XX
```
The prediction for $x^* = 60$ is more reliable because 60 is within the range of values of the collected data. $x^* = 30$ is far removed from the rest of the data. The resulting estimate would be and extrapolation beyond the collected data.

9.63 a. Only R-sq = 4.4% of the variability in rating is explained by the forty time.

b. The regression coefficient is not significant (p-value = .327)

d. The value of R^2 improves to only 12.4% and the regression coefficient is only mildly significant (p-value = .091).

Chapter 10

10.1 a. Based on the stem-and-leaf plots there are no serious departures from normality.

b. The standard deviations of the three samples are very close indicating that the standard deviations are homogeneous.

c. The interquartile ranges of the three samples are comparable indicating that the standard deviations are homogeneous.

d. MSE $= s_p^2 = [17(15.12)^2 + 16(12.39)^2 + 15(12.44)^2]/48 = 180.4988$

e. The mean of group C is considerably larger than the other two means indicating that we should reject the null hypothesis.

f. Averaging the three means will not yield the grand mean because the group sizes are different.

10.3 a. SSE = 8662 **b.** MSE = 180
c. SST = 1069 **d.** MST = 534

e. The square of the pooled standard deviation is the mean square error, MSE.

f. H_0: $\mu_A = \mu_B = \mu_C$ versus H_a: at least 2 μ's differ

g. F = 2.96, p-value = .061 The null hypothesis can be rejected but the results are only mildly significant.

h. Based on individual confidence intervals, group C scored significantly higher than the other two groups.

10.5 a. H_0: $\mu_A = \mu_B = \mu_C$ versus H_a: at least 2 μ's differ

b. F = 6.05, p-value = .004 The null hypothesis should be rejected; the results are highly significant.

c. This agrees with 10.4 part **d**.

d. It is apparent from the individual confidence intervals that groups A and B are significantly different. There may not be a difference between groups B and C.

10.7 a. T = 2.10, p-value = .047 and F = 4.43 , p-value = .047 The relationship between the two is T^2 = F because there are only two groups. Because of the exact relationship between the two their p-values are exactly the same.

b. The pooled standard deviations for the two tests are exactly the same (except for slight round off error) Furthermore, MSE = 91.2 $= s_p^2 = (9.55)^2$.

c. The independent pooled t-test and the F-test are equivalent when we have only two groups to compare.

10.9 a. Based on the boxplots it is doubtful that there is a significant difference between the thickness of the wafers.

b. The measurements are not independent because four measurements of the thickness of the oxide layer are taken on the same wafer.

10.11 a. $H_0: \mu_1 = \mu_2 = \mu_3 = \mu_4$ versus H_a: at least 2 μ_i's differ

b. Based on the boxplots there appears to be a difference between the number of tasks completed by the subjects in the four groups. The null hypothesis should be rejected.

c. The standard deviations may not be homogeneous because the variability in Drug/NoS appears to be greater than NoDrug/S.

d. Based on the boxplots, normality is a reasonable assumption.

e. The homogeneous standard deviations assumption should be investigated further. If it can be verified then the ordinary F-test is appropriate.

10.13 a. $H_0: \mu_A = \mu_B = \mu_C = \mu_D$ versus H_a: at least 2 μ's differ

b. From the normal probability plots all four samples appear to have come from normally distributed populations.

c. The standard deviations seem to be homogeneous.

d. Condition B clearly dominates Conditions A and C; the null hypothesis should be rejected.

e. F = 27.90, p-value = .000. There is a highly significant difference between the error scores for the animals under the different experimental conditions.

f. CondB   CondD   CondA   CondC

Condition C is significantly below the other three conditions.

10.15 a. $H_0: \mu_1 = \mu_2 = \mu_3$ versus H_a: at least 2 μ_i's differ

b. There do not appear to be any violations of the normality assumption. All indications are that the standard deviations are homogeneous. If the samples are random and independent the assumptions for an F-test are satisfied.

c. The sample standard deviations are close enough that the population standard deviations are homogeneous.

d. F = 12.17, p-value = .000 The null hypothesis is rejected. There is a highly significant difference between the average weight gained under the three rations.

e. The weight gain with ration three is significantly greater than the other two rations.

10.17 a. $H_0: \mu_1 = \mu_2 = \mu_3 = \mu_4$ versus H_a: at least 2 μ_i's differ

b. The mean for Method 3 appears significantly greater than the means for Methods 1 and 4.

c. All four samples appear to have come from normally distributed populations.

d. The population standard deviations are homogeneous.

e. The assumptions for the F-test are satisfied.

f. Based on the p-value (=.001) there is a highly significant difference between the test scores for the students given the different methods of programmed learning to study statistics.

g. Because the null hypothesis is rejected we should conduct a multiple comparison test.

Method 3   Method 2   Method 4   Method 1

Method 3 is significantly greater than Methods 4 and 1.

10.19 a. There appears to be a clear distinction between the number of storms for cold and warm El Niño years. The warm phase seems to suppress the number of storms. The number of storms for neutral years seems to be a mixture of the cold and warm years.

b. The only question regarding assumptions for the F-test is normality for the neutral years. It is doubtful, however, that the violation is so severe to cause problems with the F-test.

c. Based on the p-value there is a highly significant difference between the number of storms for the different types of El Niño years. The warm El Niño has suppressed the number of storms.

d. Correlation of storms and hurricanes = 0.817.

e. Based on the p-value there is a highly significant difference between the number of hurricanes for the different types of El Niño years. These results are very similar to the analysis of the number of storms. The warm El Niño has suppressed the number of hurricanes. The correlation found in part **d** is very high indicating that when the number of storms decreases the number of hurricanes decreases. Consequently, if there is a low number of storms in a warm El Niño year then there should be a low number of hurricanes. This is completely consistent with the results of the two F-test.

10.21 The Kruskal–Wallis test would be used over the ordinary F-test when there is evidence that the populations are not normally distributed but similarly shaped.

10.23 SST = $8(10.12 - 12.5)^2 + 8(10.75 - 12.5)^2 + 8(16.63 - 12.5)^2 = 206.27$
SSE = $7(6.17)^2 + 7(6.82)^2 + 7(7.09)^2 = 943.95$

source	df	SS	MS
Treatment	2	206.27	103.135
Error	21	943.95	44.95

F	p-value
2.29	> .10

10.25 The ANOVA applied to the ranks is insignificant. Thus, there is insufficient evidence to declare a difference between the population medians.

10.27 The first ANOVA table is the ordinary F-test, which is the recommended test. Neither test, however, leads to the rejection of the null hypothesis. The difference between the means is insignificant.

10.29 The first ANOVA is the ordinary F-test and the second is the Kruskal–Wallis test (ordinary F applied to ranks). Because of the similar skewness of the distributions the Kruskal–Wallis is the recommended test. The K–W test finds mildly significant evidence to reject the null hypothesis. For the K–W test the null hypothesis is that the population medians are equal. Based on these data there is mildly significant evidence that the median carbon monoxide level is different at the three sites.

10.31 The first ANOVA is the ordinary F-test and the second is the Kruskal–Wallis test (ordinary F applied to ranks). Because the distributions are long-tailed the Kruskal–Wallis is the recommended test. Based on the K–W test there is a significant difference between the mean amounts paid for used books at the three bookstores. The ordinary F-test did not detect the difference because it is sensitive to outliers.

10.33 a. Normality is a reasonable assumption in all three distributions.
  b. The variability in the three distributions appears homogeneous.
  c. It is reasonable to apply the ordinary F-test to these data.
  d. Analysis of Variance on Traffic

```
Source    DF        SS        MS
Site       2      3952      1976
Error     42      9036       215
Total     44     12988
F = 9.18  p-value = 0.000
Level     N      Mean    StDev
  1      15    163.87    13.24
  2      15    183.33    16.62
  3      15    163.07    13.93
Pooled StDev =    14.67
```

```
Individual 95% CIs For Mean
Based on Pooled StDev
-----+---------+---------+---------+-
 (-------*-------)
                   (------*-------)
(-------*-------)
-----+---------+---------+---------+-
   160       170       180       190
```

Based on the p-value the difference in the amount of traffic passing the three points is highly significant.

10.35 a. The midwest is the most variable. Except for the one outlier, the west has the least variability.
  b. The only outlier is in the west region associated with California.

c. The standard deviation for the west region is twice the standard deviation of the south region. This cast doubt on the homogeneous standard deviations assumption. Note, however, that the west standard deviation is overly inflated because of the one extreme outlier.

Variable	region	N	Mean	Median	StDev
sites	midwest	12	13.50	10.50	13.53
	northeas	11	7.27	2.00	8.09
	south	15	11.67	12.00	7.05
	west	13	8.69	4.00	14.58

Minimum	Maximum	Q1	Q3
0.00	43.00	1.50	22.25
0.00	19.00	1.00	18.00
2.00	28.00	5.00	16.00
0.00	54.00	1.00	8.50

  d. The normality assumption is questionable only in the west region. This is caused by the one extreme outlier.
  e. The ordinary F-test can be applied to these data. We should be aware, however, of the one outlier in the west region.
  f. The F-test ($F = 0.75$, p-value $= 0.525$) is insignificant. There is no significant difference between the mean number of sites in the four regions.
  g. The F-test ($F = 2.36$, p-value $= 0.083$) is mildly significant. After removal of the outlier in the west there is mildly significant difference between the number of sites in the four regions.

10.37 a. When the underlying distributions are approximately normal and homogeneous.
  b. When the underlying distributions are similar in shape.
  c. When an F-test shows a significant difference between population means.

10.39 The Kruskal–Wallis test

10.41 a. There are five treatment levels.
  b.

```
Source         SS   df     MS      F
Treatment     428    4    107   2.903
Error        1032   28  36.857
Total        1460   32
.025 < p-value < .05
```

  c. There is a significant difference between the treatment levels.

10.43

Source	SS	df	MS	F	p-value
Treatment	756.5	3	252.167	42.45	0.000
Error	261.3	44	5.94		
Total	1017.8	47			

$$W = 4.70\sqrt{(5.94/12)} = 3.31$$

```
μ₄  μ₂  μ₃  μ₁
--------------

--------------
```
μ_4 μ_2 μ_3 μ_1

μ_4 is significantly greater than μ_1.

10.45 a. All three distributions are symmetric long-tailed distributions.

b. The ordinary F-test is not robust against long-tailed distributions. Because the distributions are long tailed and similarly shaped we should use the Kruskal–Wallis test.

10.47 a. There is probably not a significant difference between the median ages for adjacent phases but examining the medians for phase 1 and phase 4 we see a difference that is significant. There is no unusual behavior in the boxplots that would suggest that the assumptions for the ordinary F-test are not satisfied.

b. Based on the p-value $= .000$ ($< .005$) there is highly significant evidence that the null hypothesis should be rejected. Phases 4 and 3 are nonoverlapping, phases 3 and 1 are nonoverlapping, but phases 3 and 2 appear to overlap. It is difficult to assess whether or not phases 2 and 1 overlap.

10.49 Recall that there was one extreme outlier in the Texas data. This one outlier distorts both the mean and standard deviation of the sample. The mean of 1452 is not a realistic measure of the center of the distribution. The standard deviation of 6270 is not a realistic measure of the variability of the distribution. Moreover, it causes the MSE to be unusually large which in turn causes the F-statistic to be unusually small (MSE is in the denominator). Consequently, the mean for Texas is inflated but the F-test does not detect it because of the unusually large MSE. It is clear that we have violated two basic assumptions for the F-test. First, the distribution is not normally distributed and second the standard deviations are not

homogeneous. It is for this reason that we recommended removing the outlier and analyzing with the Kruskal–Wallis test.

10.51 a. $H_0: \mu_1 = \mu_2 = \mu_3$ versus H_a: at least 2 μ_i's differ

b. The standard deviations are homogeneous. We need normal probability plots to check the normality assumption, but with so few observations it would be difficult to detect departures from normality.

d. Based on the F-test there is a significant difference between the mean dissolving times for the different levels of impurity. In other words, the level of impurity has a significant effect on the solubility of the aspirin tablet.

10.53 a. The distributions look reasonably symmetric without unusually long tails.

b. The normal probability plots show no significant departures from normality.

c. The assumptions for the F-test are satisfied and thus it should be used to compare the centers of the distributions.

10.55 a. Based on the normal probability plots and the appearance of the side-by-side boxplots we classify the four distributions as skewed right.

b. Because the normality assumption is in doubt and the fact that the distributions are similarly skewed we should use the Kruskal–Wallis test to compare the centers of the distributions.

10.57 Because the difference is insignificant there is no need to conduct a multiple comparison test.

10.59 Based on the p-value the difference between the relief times for the three treatments for arthritis is insignificant.

Index

A

acceptance sampling, 304

additive law, 144
  for mutually exclusive events, 141

alpha (α), level of significance, 303–304, 306

alternative hypothesis (H_a)
  decision making and, 306–308
  described, 300–301
  interpreting results and, 304–305

American Journal of Epidemiology, 300

America's Smallest School: The Family (Educational Testing Service), 78

ANOVA. *See* one-way analysis of variance (ANOVA)

association, 354

assumptions
  for linear regression model, 457–458
  for one-way analysis of variance, 519–521
  for *t*-procedures, 368, 370

B

b_0 and b_1, 102–105

Baldus, David, 122

bar graph
  for categorical data, 11–12, 13
  computer tips for, 16
  of time series data, 14

before-after experiment, 397–398

bell curve, 28

Bernoulli, Jacob, 154

Bernoulli population
  binomial distribution and, 154–156
  characteristics of, 209–210
  comparing two, 357–361
  confidence interval estimation for, 248–258
  number of successes in, 211
  population proportion, estimating, 238–239
  population proportion testing of, 310–316

Bernoulli trials, 155, 159

between-sample variability, 516, 517

bias, 193

binomial data, simulating, 211, 253

binomial distribution
  confidence interval based on, 276–277
  normal approximation of, 214–215
  population median testing and, 330–331

binomial distribution, probability of, 153–167
  Bernoulli population, 154–156
  determining, 158–161
  exercises for, 164–167
  experiments and random variables, 156–158
  mean and standard deviation of, 161–163
  random variable and, 153–154

binomial experiments, 156–158

binomial random variable, 211–215

bivariate data, 119–122. *See also* variables, relationships between

bivariate outliers, 79–80

boxplot, 50–58, 381

Buchanan, Pat, 77

Bush, George W., 77, 190

C

categorical data
  chi-square goodness-of-fit test, 420–428
  chi-square test of homogeneity, 439–445
  chi-square test of independence, 428–439
  exercises for, 445–454
  in general, 11–13

categorical variables
  described, 4–6
  vs. numerical variables, 82–83
  relationships among, 118–127

causation, 354

cell count, expected, 429, 431

census, 192

center
  of distribution, 52–53
  of long-tailed distribution, 240–242
  measures of, 34–38
  of parent distribution, 229–231
  of skewed distribution, 242

Central Limit Theorem
  Bernoulli population and, 238
  confidence interval and, 249, 259, 260
  independent samples and, 407
  sampling distribution of proportion p, 210–211, 213–214
  sampling distribution of $\bar{x}$, 202–207

Chart Wizard, 122

chi-square (χ^2), 421–425, 429–432

chi-square goodness-of-fit test, 420–428

chi-square test of homogeneity, 439–445

chi-square test of independence, 428–439

class limits, 25

cluster sample, 195

coefficient of determination
  for evaluating residuals, 486, 487
  regression line and, 114–115, 475

comparative experiment, 352

comparison bar graph, 121

complement law, 142–143

confidence interval
  testing with, 314–316
  validity, precision of, 254–256

confidence interval estimation
  for Bernoulli population, 248–258
  for comparing population proportions, 357–359, 361
  exercises for, 284–297
  for $\mu_1-\mu_2$, 365–366
  for matched-pairs *t*-test, 395
  pooled *t*-procedures for, 376–377
  for population mean, 258–275
  for population median, 275–283

confidence interval estimation, Bernoulli population
  exercises for, 256–258
  in general, 248–251
  interpreting, 252–254
  upper critical values for z^*, 251–252

confidence interval estimation, population mean
  exercises for, 268–275
  in general, 258–259
  for μ based on $\bar{x}$ when σ is known, 259–260
  for μ when σ is unknown, 265
  sample size, determining, 260–261
  students' *t*-distribution, 262–265
  t-interval, 262, 265–268

confidence interval estimation, population median
  exercises for, 279–283
  large sample, 277–279
  small sample, 275–277
confidence level, 248
confounding variable, 352
contingency table, 119–120
continuous random variable, 168–169
continuous variable, 4, 6
control group, 352
convenience samples, 192
correlation
  cause and, 94–96
  exercises for, 97–100
  numerical versus numerical variables, 89–90
  outliers, effect of, 94
  r, interpretation of, 91–94
correlation coefficient, 76, 89–94
count variable, 4, 5
criteria for rejection, 306–308
critical value, 251–252
cross-classification, 119
cross-tabulated data, 428, 433
cycles, 15

D

data. *See also* univariate data
  analysis, 224
  collection, 192–193
  display, 11–22
  transformation, 80–82
data set
  bivariate, 4, 119–122
  multivariate, 3, 4
  univariate, 4
decision making, 306–308
degrees of freedom
  in chi-square goodness-of-fit test, 422–423
  in chi-square test of independence, 430, 432
  described, 41
  of F-distribution, 518
  of pooled t-procedures, 376, 378
  regression analysis and, 462
  for t-procedures, 369, 376, 377
  for two-sample t-procedures, 370–371
density curve, 27–29
descriptive statistics
  boxplot, 50–52
  center and spread, 52–54
  exercises for, 58–63
  shape, 50
  side-by-side boxplots, 54–58

deviation, 40–41, 161–163. *See also* standard deviation
difference between
  two means, 363–375
  two proportions, 357–363
discrete data, displaying, 13–14
discrete random variable, 168
discrete variable, 4, 5
displaying data
  categorical, 11–13
  discrete, 13–14
  exercises for, 16–22
  numerical, 22–33
  time series, 15–16
distribution. *See also* population; sampling; sampling distribution
  binomial probability distribution, 153–167
  center of, 34–38, 239–242
  of chi-square, 422–423
  F-distribution, 517–519
  long-tailed, 28, 29, 38, 240–242, 383–385
  measures of variability, 40
  normal probability distribution, 167–182
  parent, describing, 224–236
  of proportion p, 209–217
  quartiles and, 39
  shapes, 27–29
  skewed, 28, 225–227, 229, 242, 381–383
  standard deviation and, 40–44
  students' t, 262–265
  symmetric, 225–227, 229, 239–240, 376–377, 383–385
  of variable, 6–7
  of $\bar{x}$, sampling, 196–209
distribution, describing, 50–63
  boxplot, 50–52
  center and spread, 52–53
  exercises for, 58–63
  shape, 50
  side-by-side boxplots, 54–58
distributions, comparing
  choosing procedure for, 406–410
  exercises for, 410–418
  in general, 351–356
  two population centers, matched samples, 394–406
  two population centers, variances equal, 375–393
  two population means, 363–375
  two population proportions, 357–363
dotplot, 6–7
Durbin-Watson test, 492

E

EDA (exploratory data analysis), 6, 224
Educational Testing Service, 78
empirical rule, 42–43, 169–170
EPA mileage charts, 299
error, 307–308, 457–458. *See also* margin of error
estimation
  of Bernoulli population, 248–258
  of population mean, 258–275
  of population median, 275–283
  of population parameters, 236–248
estimator
  confidence interval and, 248
  of population parameters, 237
  of population variance, 242–243
  variability of, 238
event
  defined, 137
  independent, 145–148
  mutually exclusive, 141–144, 147
  probability of, 139–141
Excel, Microsoft, 16, 122
expected cell count, 429, 431
expected number of outcomes, 422
experiment
  binomial, 156–158
  defined, 136
  described, 192–193
exploratory data analysis (EDA), 6, 224
extrapolation, 105–106

F

F-distribution, 517–519
F-statistic, 516, 517
F-test, 519–521, 541–545
Financial World, 92
Fisher, Sir Ronald
  F-distribution and, 518
  Iris Setosa data, 53
fit of a line assessment, 111–118
five-number summary diagram, 39
five-stem-and-leaf plot, 23–24
frequency curve, histogram, 125–126
frequency table, 11, 25–26, 306–308

G

galaxies, 227–229
Gallup, George, 125–126
general additive law, 144
goodness-of-fit test, chi-square, 420–428
Gore, Al, 77, 190
Gosset, W.S., 262, 263
grand mean, 514

graphs
  bar graph, 11–12, 13, 14, 16
  boxplot, 50–58, 381
  for categorical data, 11–13
  of confidence interval, 253
  for discrete data, 13–14
  dotplot, 6–7
  histograms, 25–27, 29
  for numerical data, 22–29
  pie chart, 12–13, 16
  scatterplots, 77–89
  stem-and-leaf plots, 22–24
  for time series data, 15–16
grouped frequency table, 25

H

Hall, Monty, 136
Harvard Medical School, 351
Hidalgo y Costilla, Miguel, 2
histograms, 25–27, 29
home cost, 223–224
homogeneity
  chi-square test of, 439–445
  testing for, 380
homogeneous standard deviations, 513
hypothesis testing
  for analysis of variance, 515
  for β_1, 469–471
  decision making for, 306–308
  exercises for, 308–310, 341–349
  for μ_1-μ_2, 366–369
  null and alternative hypotheses, 299–301
  of population mean, 318–328
  of population median, 328–341
  of population proportion, 310–318, 359–361
  results, interpretation of, 304–306
  test statistic, 301–304

I

independence, chi-square test of, 428–439
independence, regression model, 491–493
independent events, 145–148
independent samples, 407–409
individual outcome, 138–139
inference
  example of, 223–224
  of linear regression model, 468–485
  parent distribution, 224–236
  population parameters, 236–248
  procedures, 361
influential observations
  with outliers, 94

of regression analysis, 476
  regression line and, 114
interquartile range (IQR), 52–53

K

Kruskal–Wallis test
  description, examples of, 541–545
  exercises for, 545–553

L

large sample, 277–279
large-sample test
  comparing proportions, 359–361
  of population median, 331–334
  of population proportion, 315–316
least squares regression
  computer tip for, 106
  example of, 460–461
  exercises for, 107–111
  finding b_0 and b_1, 103–105
  method of, 102–103
  numerical variables and, 101
  outliers and, 114, 488–491
  for prediction and extrapolation, 105–106
left-tailed test, 319
level of significance (α), 303–304, 306
linear regression model
  checking adequacy of, 485–501
  in general, 457–468
  inference about, 468–485
linearity, residual plot, 486–488
linearly related, 89
location, measures of, 38–39
long-tailed distribution
  defined, 28
  example of, 38
  illustrated, 29
  population median and, 240–242
  Wilcoxon rank-sum test of, 383–385
Los Angeles Times, 190
lottery sampling, 193
lurking variable, 94–95

M

margin of error
  of confidence interval, 254–255, 358
  defined, 248
  finding, 251
  sample size and, 260–261
marginal percents, 119
marginal totals, 440
matched-pairs design, 352–353
matched-pairs experiment, 399–401

matched-pairs t-test, 394–399, 410
matched samples procedures, 409–410
Mayo Clinic study, 351
mean. *See* population mean (μ)
mean deviation, 161–163
mean, sample
  sampling distribution of, 202–207
  trimmed, 35, 36–38
mean square for error (MSE), 517
mean square for treatments (MST), 517
mean squares, 517, 518
measurement population, 224–225
measurement variable, 4
measures of center, 34–38
measures of location, 38–39
measures of variability, 40–44
median. *See also* population median
  calculating, 35
  confidence interval for, 277–278
  example of, 36–37
  Kruskal–Wallis test for, 541–542
method of least squares, 102–103
Microsoft Excel, 16, 122
Minitab
  graphs and charts in, 16
  histograms, 26
  one-way analysis of variance with, 523–524
  random data in, 147, 204
  regression analysis with, 462
  stem-and-leaf plots, 23, 25
  t-interval and, 267
mode, 28
Morabia, Alfredo, 300
MSE (mean square for error), 517
MST (mean square for treatments), 517
multimodal distribution, 28, 227–229
multinomial experiment, 421–425
  multiple comparison procedure, Tukey's, 523, 524–527
multiplication law for independent events, 145–147
multivariate data set, 3, 4
mutually exclusive events, 141–144, 147
μ_y, 472–475

N

National Oceanic and Atmospheric Administration (NOAA), 82
negative relationship, 79
nonnormal population
  Kruskal–Wallis test for, 543
  sampling distribution of $\bar{x}$ from, 200–201
  t-interval and, 265
normal curve, 169–171

normal distribution
  checking normality, 176
  confidence interval and, 259–260
  exercises for, 180–182
  finding percentile, 174–176
  illustrated, 28
  normal curve, 169–171
  normal probability plot, 177–180
  population mean of, 239–240
  probability density function and,
    168–169
  z-scores and, 172–174
normal population, 197–200
normal probability plot, 177–180
normal random variable, 169–171
normality, checking for, 176, 488
null hypothesis (H_0)
  chi-square test of homogeneity and,
    439–440
  for comparing population
    proportions, 359–360
  decision making and, 306–308
  defined, 299–300
  interpreting results and, 304–305
  for Kruskal–Wallis test, 543, 545
  matched-pairs t-test and, 397–398
  population mean testing and, 366–367
  population median testing and,
    329–330
  population proportion testing and,
    310–316
  test statistic and, 301–304
  for Wilcoxon rank-sum test, 382, 383
number of successes, 211
numerical data, displaying, 22–33
  with density curve, 27–29
  exercise for, 29–33
  with histograms, 25–27
  outliers, 24–25
  with stem-and-leaf plot, 22–24
numerical variables
  vs. categorical variables, 82–83
  defined, 4
  example of, 5
  least squares regression and, 101
  vs. numerical variables, 78–79, 89–90

O

objectives, 191
observational study, 354–355
Old Faithful geyser, 234–235
one-sided alternative, 300
one-way analysis of variance
    (ANOVA), 513–540
  assumptions, checking, 519–521
  between-sample variability, 517

confidence intervals, 521–527
  design of, 513–516
  exercises for, 528–540
  F-distribution, 517–519
  Kruskal–Wallis test, 543–544
  within-sample variability, 517
outcomes, 138–139, 422
outliers
  bivariate, 79–80
  checking for, 488–491
  displaying, 24–25
  effect on correlation, 94
  regression line and, 113–114
  when describing a distribution,
    50–53

P

p
  sampling distribution of, 209–217
  as test statistic, 310–316
p-value, 302–304, 318–320
paired difference test, 394–399
parameters
  choosing, 229–230
  defined, 34
  population, estimating, 236–248
parent distribution
  analysis of variance and, 513
  center and variability of, 229–231
  exercises for, 231–236
  in general, 224–225
  population median testing and,
    328–329
  shape of, 225–229
  test statistic and, 301, 302
Pauling, Linus, 351
Pearson, Karl, 421
percent histogram, 26–27
percentile, finding, 174–176
percents
  of categorical variables, 120–122
  from relative frequencies, 11
physical mixing, 193
pi (π)
  Bernoulli population and, 238–239
  binomial probabilities and, 160–163
  comparing distributions and,
    357–361
  confidence interval for, 249–251
  population proportion testing and,
    310–316
  sample proportion of, 254–255
  sample size for estimating, 254–256
  sampling distribution and, 209–215
pie chart, 12–13, 16
poll, 190, 192. *See also* sampling

pooled standard deviation, 521, 523
pooled t-procedures, 375–380, 407
pooled t- test, 519
population. *See also* Bernoulli
    population
  chi-square test of homogeneity and,
    439–445
  identifying, 191
  sampling distribution of $\bar{x}$
    measurement, 197–201
  simple random sample from, 193
population centers, matched samples,
    comparing, 394–406
  exercises for, 401–406
  matched-pairs t-test, 394–399
  Wilcoxon signed-rank test for,
    399–401
population centers, variances equal,
    comparing, 375–393
  exercises for, 386–393
  pooled t-procedures, 375–380
  Wilcoxon rank-sum test, 381–386
population distribution, 27
population mean (μ)
  confidence interval estimation for,
    258–275
  estimators for, 237–238
  for hypothesis testing, 299–302
  as measurement of center, 34–35
  population parameters, estimating,
    239–240
  sampling distribution of $\bar{x}$ and, 198,
    199
population mean (μ), testing, 318–328
  exercises for, 324–328
  t-test, 318–320
  z-test versus t-test, 320–323
population means (μ), comparing two
  confidence interval for, 365–366
  exercises for, 371–375
  in general, 363–364
  hypothesis testing for , 366–369
  pooled t-procedures for, 375–380
  two-sample t-procedures, 364–365,
    370–371
population median (θ)
  confidence interval estimation for,
    275–283
  as measurement of center, 34–35
  population parameters, estimating,
    240–242
  Wilcoxon rank-sum test and,
    381–382
population median (θ), testing, 328–341
  exercises for, 334–349
  large-sample test, 331–334
  small-sample test, 329–331

population of measurements, 5–6
population of units, 3
population parameters, estimating, 236–248
    of Bernoulli population, 238–239
    choosing estimator, 237
    exercise for, 243–248
    population mean, 239–240
    population median, 240–242
    population variance, 242–243
    standard deviation, 243
    unbiased estimator, 237
    variability of estimators, 238
population proportion, testing, 310–318
    with confidence intervals, 314–316
    exercises for, 316–318
    general H_0, 312–313
    of successes, 310–312
    two-tailed test, 313–314
population proportions, comparing two, 357–363
population regression line, 457, 460–461
population variance, 242–243
positive relationship, 79
precision, 254
prediction, 105–106
predictor variable, 101
probability
    binomial distribution, 153–167
    concepts of, 136–153
    of confidence interval, 252
    example of, 136
    exercises for, 182–188
    normal distribution, 167–182
    p-value, 302–303
probability concepts, 136–153
    of event, 139–141
    exercises for, 148–153
    of independent events, 145–148
    of individual outcome, 138–139
    of mutually exclusive events, 141–144
    terminology of, 136–138
probability density function, 168–169
probability distribution, 153
procedure, choosing, 406–410
proportional sampling, 194–195
pth percentile, 175

Q

qualitative variable, 4
quantitative variable, 4
quartiles, 38–39
quasi experiment, 354–355

R

radiocarbon dating, 54
random data
    binomial, 157
    computer tip for, 147
    normal, 169–171, 176
Random Data command, 204, 211
random numbers, 193
random variable, 153–154, 156–158
randomization, 352–353
randomized block design, 353
rank-transform, 381, 383
rank-transform test, 399
ranking data, 385
ranks, 542
regression analysis
    checking model adequacy, 485–501
    estimation and prediction, 460–468
    example of, 456
    exercises for, 501–510
    linear regression model, 457–468
    linear regression model, inference about, 468–485
regression coefficient, 469–472
regression line. *See also* least squares regression
    example of, 474–475
    finding, 103–105
    fit of, 111–118
    prediction and extrapolation and, 105–106
regression output, 474–476
regression variables, 101–102
rejection region, 306–308
relationship
    linear, 76
    positive and negative, 79
relative frequency, 11, 138
relative standing, 43–44
representative sample, 192–193
research hypothesis. *See* alternative hypothesis
residual plots, 111–113, 486–493
residuals
    fitted line and, 103
    in regression analysis, 461–462, 476
    regression line and, 111–113
    standardized, 485–486
resistant line, 490
resistant statistic, 37
response variable, 101
results, interpretation of, 304–306
right-tailed test, 302, 311
Ronald H. Brown (ship), 82

S

sample
    cluster, 195
    defined, 3
    representative, 192–193
    simple random, 193
    stratified and systematic, 194
    term usage, 5–6
sample mean
    sampling distribution of, 202–207
    trimmed, 35, 36–38
sample median, 35
sample size
    determining, 254, 260–261
    effect on distribution of p, 212
    procedure choice and, 406–410
sample space, 136
sampling
    acceptance sampling, 304
    example of, 190
    exercises for, 217–221
    principles of, 191–196
sampling distribution
    of difference between sample means, 364
    of p, 311
    of p_1–p_2, 357
    of the sample mean, 202–207
    of a sample statistic, 197
    of test statistic, 301–302
sampling distribution of proportion p
    binomial random variable and, 211–215
    Central Limit Theorem and, 210–211
    exercises for, 215–217
    in general, 209–210
sampling distribution of $\bar{x}$, 196–209
    Central Limit Theorem, 202–207
    exercises for, 207–209
    from nonnormal population, 200–201
    from normal population, 197–200
sampling variability, 198, 209–210
Savant, Marilyn vos, 136
scatterplots, 77–89
    bivariate outliers and, 79–80
    categorical versus numerical variables, 82–83
    exercises for, 84–89
    numerical versus numerical variables, 78–79
    positive and negative relationships, 79
    transforming data and, 80–82
segment bar graph, 120
shape, distribution, 50, 225–229
short-tailed distribution, 28, 29

side-by-side boxplots, 35–43, 54–58, 512–513

sign test, 329–331

signed ranks, 401

significance, level of, 303–304, 306

simple random sample, 193

Simpson's Paradox, 56–58

skewed distribution
  center and variability of, 229
  center of, estimating, 242
  classifying, 225–227
  defined, 28
  Wilcoxon rank-sum test of, 381–383

small sample, 275–277

small-sample test, 329–331

SSE (sum of squares for error), 103, 462, 517

SST (sum of squares for treatments), 517

stamp collector, 1

standard deviation
  of binomial distribution, 161–163
  checking constant, 488
  confidence interval and, 250, 259–260, 265, 266
  described, 40–43
  estimating, 261–262, 461–462
  estimator variability and, 238
  of F test, 519–520
  to measure variability, 229, 230
  of normal distribution, 171–172
  of population, 303
  population parameters, estimating, 243
  of sampling distribution of p, 211, 212, 213
  of sampling distribution variable, 198, 199
  testing population mean and, 318–319

standard error of a statistic, 250

standard normal distribution, 172–173

standardized residuals, 485–486

standardized test statistic, 303

statistic elements, 2–10
  concepts of, 3
  exercises for, 7–10
  exploratory data analysis, 6
  population of measurements, 5–6
  variable classification, 4–5
  variable distribution, 6–7
  variables and data sets, 3–4

statistical design, 192

statistical inference, 191

statistical models, 457

statistics
  data analysis of, 224
  defined, 3, 34
  as estimator of parameter, 237
  standard error of, 250

statistics, summarizing data with, 33–49
  exercises for, 44–49
  five-number summary diagram, 39
  measures of center, 34–38
  measures of variability, 40
  quartiles, 38–39
  relative standing, 43–44
  standard deviation, 40–43

stem-and-leaf plots
  computer tip for, 29
  displaying numerical data with, 22–24
  of outliers, 25

straight line, 102–103

stratified sample, 194

Studentized residuals, 486

students' t-distribution, 262–265

success, 211, 249–251

sum of squares for error (SSE), 103, 462, 517

sum of squares for treatments (SST), 517

sum of squares, standard deviation, 40

summary measures, 34–44
  of center, 34–38
  quartiles, 39
  standard deviation, 40–44
  of variability, 40

sunspot number, Wolfer, 15–16

survey sampling, 194–195

surveys, 192

symmetric distribution
  center and variability of, 229
  classifying, 225–227
  defined, 28
  pooled t- procedures for, 376–377
  population mean of, 239–240
  Wilcoxon rank-sum test of, 383–385

symmetric long-tailed distribution, 240–242

systematic sample, 194

T

t-interval, 262, 265–268

t-procedures, pooled, 375–380, 407

t-procedures, two-sample
  for μ_1-μ_2, 368–369
  population means, comparing two, 364–365
  robustness of, 370–371

t-test
  pooled, 519
  for population mean, 318–320
  versus z-test, 320–323

tables
  for categorical data, 11–13
  contingency, 119–120
  for discrete data, 13–14
  exercises for, 16–22
  frequency, 11
  for time series data, 15–16

television, 78–79

test of significance, 304

test statistic
  level of significance, 303–304
  p-value, 302–303
  sampling distribution of, 301–302
  z-test, 303

testing. *See* hypothesis testing; variance, analysis of

time series data, 14, 15–16

trends, 15, 79

trimmed mean, 35, 36–38

Tukey's multiple comparison procedure, 523, 524–527

two-sample t-test, 369–370, 407, 409

two-sample Wilcoxon rank-sum test, 381

two-sided alternative, 300

two-tailed test, 313–314

Type I error, 307–308, 512

Type II error, 307–308

U

unbiased estimator, 237, 238

underlying distribution. *See* parent distribution

unimodal distribution, 28, 225–227

univariate data
  displaying with table and graph, 11–22
  distribution, describing, 50–63
  essential statistic elements, 2–10
  exercises for, 63–75
  numerical data, displaying, 22–33
  summarizing data with statistics, 33–49

University of Arizona Cancer Center, 351

upper critical values for z^*, 251–252

U.S. space shuttle *Challenger*, 35–43, 95–96, 225–227, 231

V

validity, 254
value, 79
variability
  between-sample, 516, 517
  of estimators, 238
  of parent distribution, 229–231
  standard deviation, 40–44
variable classification, 4–5
variable distribution, 6–7
variables
  categorical, 4–6, 82–83, 118–127
  chi-square test of homogeneity and,
    439–445
  choosing for sampling, 191–192
  confounding, 352
  continuous, 4, 6, 168–169
  defined, 3–4
  discrete, 4, 5, 168
  lurking, 94–95
  numerical, 4, 5, 78–79, 82–83,
    89–90, 101

population of measurements, 5–6
predictor and response, 101
variables, relationships between
  categorical variables, 118–127
  correlation, 89–100
  exercises for, 127–134
  fit of a line, assessing, 111–118
  least squares regression, 100–111
  scatterplots, 77–89
variance, analysis of
  exercises for, 553–565
  in general, 512–513
  Kruskal–Wallis test, 541–553
  one-way analysis of variance, 513–540
variance, population, 242–243,
  375–380
volunteer sample, 192

W

Wilcoxon rank-sum test, 381–386,
  407–408
Wilcoxon signed-rank test, 399–401, 410

Winkler, Walton van, 2
within-sample variability, 516, 517
Wolf, Rudolf, 15
Wolfer sunspot number, 15–16
Woodworth, George, 122

X

$\bar{x}$
  confidence interval and, 259–260
  population mean and, 239–240
  population median and, 240–242
  sampling distribution of, 196–209
  t-interval and, 262
  as test statistic, 301

Z

z^*, 251–252, 254
z-interval, 259–260
z-scores, 44, 172–174
z-test, 303, 320–323